Lecture Notes in Computer Science

Lecture Notes in Artificial Intelligence   **14176**

Founding Editor

Jörg Siekmann

The series Lecture Notes in Artificial Intelligence (LNAI) was established in 1988 as a topical subseries of LNCS devoted to artificial intelligence.

The series publishes state-of-the-art research results at a high level. As with the LNCS mother series, the mission of the series is to serve the international R & D community by providing an invaluable service, mainly focused on the publication of conference and workshop proceedings and postproceedings.

Xiaochun Yang · Heru Suhartanto ·
Guoren Wang · Bin Wang · Jing Jiang · Bing Li ·
Huaijie Zhu · Ningning Cui

Editors

Advanced Data Mining and Applications

19th International Conference, ADMA 2023
Shenyang, China, August 21–23, 2023
Proceedings, Part I

 Springer

Editors
Xiaochun Yang
Northeastern University
Shenyang, China

Heru Suhartanto
The University of Indonesia
Depok, Indonesia

Guoren Wang
Beijing Institute of Technology
Beijing, China

Bin Wang
Northeastern University
Shenyang, China

Jing Jiang
University of Technology Sydney
Sydney, NSW, Australia

Bing Li
Agency for Science, Technology
and Research (A*STAR)
Singapore, Singapore

Huaijie Zhu
Sun Yat-sen University
Guangzhou, China

Ningning Cui
Anhui University
Hefei, China

ISSN 0302-9743 ISSN 1611-3349 (electronic)
Lecture Notes in Artificial Intelligence
ISBN 978-3-031-46660-1 ISBN 978-3-031-46661-8 (eBook)
https://doi.org/10.1007/978-3-031-46661-8

LNCS Sublibrary: SL7 – Artificial Intelligence

This Springer imprint is published by the registered company Springer Nature Switzerland AG
The registered company address is: Gewerbestrasse 11, 6330 Cham, Switzerland

Paper in this product is recyclable.

Preface

The 19th International Conference on Advanced Data Mining and Applications (ADMA 2023) was held in Shenyang, China, during August 21–23, 2023. Researchers and practitioners from around the world came together at this leading international forum to share innovative ideas, original research findings, case study results, and experienced insights into advanced data mining and its applications. With the ever-growing importance of appropriate methods in these data-rich times, ADMA has become a flagship conference in this field. ADMA 2023 received a total of 503 submissions from 22 countries across five continents. After a rigorous double-blind review process involving 318 reviewers, 216 regular papers were accepted to be published in the proceedings, 123 were selected to be delivered as oral presentations at the conference, 85 were selected as poster presentations, and 8 were selected as industry papers. This corresponds to a full oral paper acceptance rate of 24.4%. The Program Committee (PC), composed of international experts in relevant fields, did a thorough and professional job of reviewing the papers submitted to ADMA 2023, and each paper was reviewed by an average of 2.97 PC members. With the growing importance of data in this digital age, papers accepted at ADMA 2023 covered a wide range of research topics in the field of data mining, including pattern mining, graph mining, classification, clustering and recommendation, multi-objective, optimization, augmentation, and database, data mining theory, image, multimedia and time series data mining, text mining, web and IoT applications, finance and healthcare. It is worth mentioning that ADMA 2023 was organized as a physical-only event, allowing for in-person gatherings and networking. We thank the PC members for completing the review process and providing valuable comments within tight schedules. The high-quality program would not have been possible without the expertise and dedication of our PC members. Moreover, we would like to take this valuable opportunity to thank all authors who submitted technical papers and contributed to the tradition of excellence at ADMA. We firmly believe that many colleagues will find the papers in these proceedings exciting and beneficial for advancing their research. We would like to thank Microsoft for providing the CMT system, which is free to use for conference organization, Springer for their long-term support, the host institution, Northeastern University, for their hospitality and support, Niu Translation and Shuangzhi Bo for their sponsorship. We are grateful for the guidance of the steering committee members, Osmar R. Zaiane, Chengqi Zhang, Michael Sheng, Guodong Long, Xue Li, Jianxin Li, and Weitong Chen. With their leadership and support, the conference ran smoothly. We also would like to acknowledge the support of the other members of the organizing committee. All of them helped to make ADMA 2023 a success. We appreciate the local arrangements, registration and finance management from the local arrangement chairs, registration management chairs and finance chairs Kui Di, Baoyan Song, Junchang Xin, Donghong Han, Guoqiang Ma, Yuanguo Bi, and Baiyou Qiao, the time and effort of the proceedings chairs, Bing Li, Huaijie Zhu, and Ningning Cui, the effort in advertising the conference by the publicity chairs and social network and social media coordination chairs, Xin Wang, Yongxin

Tong, Lina Wang, and Sen Wang, and the effort of managing the Tutorial sessions by the tutorial chairs, Zheng Zhang and Shuihua Wang, We would like to give very special thanks to the web chair, industry chairs, and PhD school chairs Faming Li, Chi Man Pun, Sen Wang, Linlin Ding, M. Emre Celebi, and Zheng Zhang, for creating a successful and memorable event. We also thank sponsorship chair Hua Shao for his sponsorship. Finally, we would like to thank all the other co-chairs who have contributed to the conference.

August 2023

Xiaochun Yang
Bin Wang
Jing Jiang

Organization

Chair of the Steering Committee

Xue Li   University of Queensland, Australia

Steering Committee

Osmar R. Zaiane   University of Alberta, Canada
Chengqi Zhang   Sydney University of Technology, Australia
Michael Sheng   Macquarie University, Australia
Guodong Long   Sydney University of Technology, Australia
Xue Li   University of Queensland, Australia
Jianxin Li   Deakin University, Australia
Weitong Chen   Adelaide University, Australia

Honor Chairs

Xingwei Wang   Northeastern University, China
Xuemin Lin   Shanghai Jiao Tong University, China
Ge Yu   Northeastern University, China

General Chairs

Xiaochun Yang   Northeastern University, China
Heru Suhartanto   The University of Indonesia, Indonesia
Guoren Wang   Beijing Institute of Technology, China

Program Chairs

Bin Wang   Northeastern University, China
Jing Jiang   University of Technology Sydney, Australia

Local Arrangement Chairs

Kui Di	Northeastern University, China
Baoyan Song	Liaoning University, China
Junchang Xin	Northeastern University, China

Registration Management Chairs

Donghong Han	Northeastern University, China
Guoqiang Ma	Northeastern University, China
Yuanguo Bi	Northeastern University, China

Finance Chair

Baiyou Qiao	Northeastern University, China

Sponsorship Chair

Hua Shao	Shenyang Huaruibo Information Technology Co., Ltd., China

Publicity Chairs

Xin Wang	Tianjin University, China
Yongxin Tong	Beihang University, China
Lina Wang	Wuhan University, China

Social Network and Social Media Coordination Chair

Sen Wang	University of Queensland, Australia

Proceeding Chairs

Bing Li	Agency for Science, Technology and Research (A*STAR), Singapore
Huaijie Zhu	Sun Yat-sen University, China
Ningning Cui	Anhui University, China

Tutorial Chairs

Zheng Zhang   Harbin Institute of Technology, Shenzhen, China
Shuihua Wang   University of Leicester, UK

Web Chair

Faming Li   Northeastern University, China

Industry Chairs

Chi Man Pun   University of Macau, China
Sen Wang   University of Queensland, Australia
Linlin Ding   Liaoning University, China

PhD School Chairs

M. Emre Celebi   University of Central Arkansas, USA
Zheng Zhang   Harbin Institute of Technology, Shenzhen, China

Program Committee

Meta Reviewers

Bohan Li   Nanjing University of Aeronautics and
                                    Astronautics, China
Can Wang   Griffith University, Australia
Chaokun Wang   Tsinghua University, China
Cheqing Jin   East China Normal University, China
Guodong Long   University of Technology Sydney, Australia
Hongzhi Wang   Harbin Institute of Technology, China
Huaijie Zhu   Sun Yat-sen University, China
Jianxin Li   Deakin University, Australia
Jun Gao   Peking University, China
Lianhua Chi   La Trobe University, Australia
Lin Yue   University of Newcastle, Australia
Tao Shen   University of Technology Sydney, Australia

Wei Emma Zhang	University of Adelaide, Australia
Weitong Chen	Adelaide University, Australia
Xiang Lian	Kent State University, USA
Xiaoling Wang	East China Normal University, China
Xueping Peng	University of Technology Sydney, Australia
Xuyun Zhang	Macquarie University, Australia
Yanjun Zhang	Deakin University, Australia
Zheng Zhang	Harbin Institute of Technology, Shenzhen, China

Reviewers

Abdulwahab Aljubairy	Macquarie University, Australia
Adita Kulkarni	SUNY Brockport, USA
Ahoud Alhazmi	Macquarie University, Australia
Akshay Peshave	GE Research, USA
Alex Delis	Univ. of Athens, Greece
Alexander Zhou	Hong Kong University of Science and Technology, China
Baoling Ning	Heilongjiang University, China
Bin Zhao	Nanjing Normal University, China
Bing Li	Institute of High Performance Computing, A*STAR, Singapore
Bo Tang	Southern University of Science and Technology, China
Carson Leung	University of Manitoba, Canada
Changdong Wang	Sun Yat-sen University, China
Chao Zhang	Tsinghua University, China
Chaokun Wang	Tsinghua University, China
Chaoran Huang	University of New South Wales, Australia
Chen Wang	Chongqing University, China
Chengcheng Yang	East China Normal University, China
Chenhao Zhang	University of Queensland, Australia
Cheqing Jin	East China Normal University, China
Chuan Ma	Zhejiang Lab, China
Chuan Xiao	Osaka University and Nagoya University, Japan
Chuanyu Zong	Shenyang Aerospace University, China
Congbo Ma	University of Adelaide, Australia
Dan He	University of Queensland, Australia
David Broneske	German Centre for Higher Education Research and Science Studies, Germany

Dechang Pi	Nanjing University of Aeronautics and Astronautics, China
Derong Shen	Northeastern University, China
Dima Alhadidi	University of New Brunswick, Canada
Dimitris Kotzinos	ETIS, France
Dong Huang	South China Agricultural University, China
Dong Li	Liaoning University, China
Dong Wen	University of New South Wales, Australia
Dongxiang Zhang	Zhejiang University, China
Dongyuan Tian	Jilin University, China
Dunlu Peng	University of Shanghai for Science and Technology, China
Eiji Uchino	Yamaguchi University, Japan
Ellouze Mourad	University of Sfax, Tunisia
Elsa Negre	LAMSADE, Paris-Dauphine University, France
Faming Li	Northeastern University, China
Farid Nouioua	Université Mohamed El Bachir El Ibrahimi de Bordj Bou Arréridj, Algeria
Genoveva Vargas-Solar	CNRS, France
Gong Cheng	Nanjing University, China
Guanfeng Liu	Macquarie University, Australia
Guangquan Lu	Guangxi Normal University, China
Guangyan Huang	Deakin University, Australia
Guannan Dong	University of Macau, China
Guillaume Guerard	ESILV, France
Guodong Long	University of Technology Sydney, Australia
Haïfa Nakouri	ISG Tunis, Tunisia
Hailong Liu	Northwestern Polytechnical University, China
Haojie Zhuang	University of Adelaide, Australia
Haoran Yang	University of Technology Sydney, Australia
Haoyang Luo	Harbin Institute of Technology (Shenzhen), China
Hongzhi Wang	Harbin Institute of Technology, China
Huaijie Zhu	Sun Yat-sen University, China
Hui Yin	Deakin University, Australia
Indika Priyantha Kumara Dewage	Tilburg University, The Netherlands
Ioannis Konstantinou	University of Thessaly, Greece
Jagat Challa	BITS Pilani, India
Jerry Chun-Wei Lin	Western Norway University of Applied Sciences, Norway
Jiabao Han	NUDT, China
Jiajie Xu	Soochow University, China
Jiali Mao	East China Normal University, China

Jianbin Qin	Shenzhen University, China
Jianhua Lu	Southeast University, China
Jianqiu Xu	Nanjing University of Aeronautics and Astronautics, China
Jianxin Li	Deakin University, Australia
Jianxing Yu	Sun Yat-sen University, China
Jiaxin Jiang	National University of Singapore, Singapore
Jiazun Chen	Peking University, China
Jie Shao	University of Electronic Science and Technology of China, China
Jie Wang	Indiana University, USA
Jilian Zhang	Jinan University, China, China
Jingang Yu	Shenyang Institute of Computing Technology, Chinese Academy of Sciences
Jing Du	University of New South Wales, Australia
Jules-Raymond Tapamo	University of KwaZulu-Natal, South Africa
Jun Gao	Peking University, China
Junchang Xin	Northeastern University, China
Junhu Wang	Griffith University, Australia
Junshuai Song	Peking University, China
Kai Wang	Shanghai Jiao Tong University, China
Ke Deng	RMIT University, Australia
Kun Han	University of Queensland, Australia
Kun Yue	Yunnan University, China
Ladjel Bellatreche	ISAE-ENSMA, France
Lei Duan	Sichuan University, China
Lei Guo	Shandong Normal University, China
Lei Li	Hong Kong University of Science and Technology (Guangzhou), China
Li Li	Southwest University, China
Lin Guo	Changchun University of Science and Technology, China
Lin Mu	Anhui University, China
Linlin Ding	Liaoning University, China
Lizhen Cui	Shandong University, China
Long Yuan	Nanjing University of Science and Technology, China
Lu Chen	Swinburne University of Technology, Australia
Lu Jiang	Northeast Normal University, China
Lukui Shi	Hebei University of Technology, China
Maneet Singh	IIT Ropar, India
Manqing Dong	Macquarie University, Australia

Mariusz Bajger	Flinders University, Australia
Markus Endres	University of Applied Sciences Munich, Germany
Mehmet Ali Kaygusuz	Middle East Technical University, Turkey
Meng-Fen Chiang	University of Auckland, New Zealand
Ming Zhong	Wuhan University, China
Minghe Yu	Northeastern University, China
Mingzhe Zhang	University of Queensland, Australia
Mirco Nanni	CNR-ISTI Pisa, Italy
Misuk Kim	Sejong University, South Korea
Mo Li	Liaoning University, China
Mohammad Alipour Vaezi	Virginia Tech, USA
Mourad Nouioua	Mohamed El Bachir El Ibrahimi University, Bordj Bou Arreridj, Algeria
Munazza Zaib	Macquarie University, Australia
Nabil Neggaz	Université des Sciences et de la Technologie d'Oran Mohamed Boudiaf, Algeria
Nicolas Travers	Léonard de Vinci Pôle Universitaire, Research Center, France
Ningning Cui	Anhui University, China
Paul Grant	Charles Sturt University, Australia
Peiquan Jin	University of Science and Technology of China, China
Peng Cheng	East China Normal University, China
Peng Peng	Hunan University, China
Peng Wang	Fudan University, China
Pengpeng Zhao	Soochow University, China
Philippe Fournier-Viger	Shenzhen University, China
Ping Lu	Beihang University, China
Pinghui Wang	Xi'an Jiaotong University, China
Qiang Yin	Shanghai Jiao Tong University, China
Qing Liao	Harbin Institute of Technology (Shenzhen), China
Qing Liu	Data61, CSIRO, Australia
Qing Xie	Wuhan University of Technology, China
Quan Chen	Guangdong University of Technology, China
Quan Z. Sheng	Macquarie University, Australia
Quoc Viet Hung Nguyen	Griffith University, Australia
Rania Boukhriss	University of Sfax, Tunisia
Riccardo Cantini	University of Calabria, Italy
Rogério Luís Costa	Polytechnic of Leiria, Portugal
Rong Zhu	Alibaba Group, China
Ronghua Li	Beijing Institute of Technology, China
Rui Zhou	Swinburne University of Technology, Australia

Rui Zhu	Shenyang Aerospace University, China
Sadeq Darrab	Otto von Guericke University Magdeburg, Germany
Saiful Islam	Griffith University, Australia
Sayan Unankard	Maejo University, Thailand
Senzhang Wang	Central South University, China
Shan Xue	University of Wollongong, Australia
Shaofei Shen	University of Queensland, Australia
Shi Feng	Northeastern University, China
Shiting Wen	Zhejiang University, China
Shiyu Yang	Guangzhou University, China
Shouhong Wan	University of Science and Technology of China, China
Shuhao Zhang	Singapore University of Technology and Design, Singapore
Shuiqiao Yang	UNSW, Australia
Shuyuan Li	Beihang University, China
Silvestro Roberto Poccia	University of Turin, Italy
Sonia Djebali	Léonard de Vinci Pôle Universitaire, Research Center, France
Suman Banerjee	IIT Jammu, India
Tao Qiu	Shenyang Aerospace University, China
Tao Zhao	National University of Defense Technology, China
Tarique Anwar	University of York, UK
Thanh Tam Nguyen	Griffith University, Australia
Theodoros Chondrogiannis	University of Konstanz, Germany
Tianrui Li	Southwest Jiaotong University, China
Tianyi Chen	Peking University, China
Tieke He	Nanjing University, China
Tiexin Wang	Nanjing University of Aeronautics and Astronautics, China
Tiezheng Nie	Northeastern University, China
Uno Fang	Deakin University, Australia
Wei Chen	University of Auckland, New Zealand
Wei Deng	Southwestern University of Finance and Economics, China
Wei Hu	Nanjing University, China
Wei Li	Harbin Engineering University, China
Wei Liu	University of Macau, Sun Yat-sen University, China
Wei Shen	Nankai University, China
Wei Song	Wuhan University, China

Weijia Zhang	University of Newcastle, Australia
Weiwei Ni	Southeast University, China
Weixiong Rao	Tongji University, China
Wen Zhang	Wuhan University, China
Wentao Li	Hong Kong University of Science and Technology (Guangzhou), China
Wenyun Li	Harbin Institute of Technology (Shenzhen), China
Xi Guo	University of Science and Technology Beijing, China
Xiang Lian	Kent State University, USA
Xiangguo Sun	Chinese University of Hong Kong, China
Xiangmin Zhou	RMIT University, Australia
Xiangyu Song	Swinburne University of Technology, Australia
Xianmin Liu	Harbin Institute of Technology, China
Xianzhi Wang	University of Technology Sydney, Australia
Xiao Pan	Shijiazhuang Tiedao University, China
Xiaocong Chen	University of New South Wales, Australia
Xiaofeng Gao	Shanghai Jiaotong University, China
Xiaoguo Li	Singapore Management University, Singapore
Xiaohui (Daniel) Tao	University of Southern Queensland, Australia
Xiaoling Wang	East China Normal University, China
Xiaowang Zhang	Tianjin University, China
Xiaoyang Wang	University of New South Wales, Australia
Xiaojun Xie	Nanjing Agricultural University, China
Xin Cao	University of New South Wales, Australia
Xin Wang	Southwest Petroleum University, China
Xinqiang Xie	Neusoft, China
Xiuhua Li	Chongqing University, China
Xiujuan Xu	Dalian University of Technology, China
Xu Yuan	Harbin Institute of Technology, Shenzhen, China
Xu Zhou	Hunan University, China
Xupeng Miao	Carnegie Mellon University, USA
Xuyun Zhang	Macquarie University, Australia
Yajun Yang	Tianjin University, China
Yanda Wang	Nanjing University of Aeronautics and Astronautics, China
Yanfeng Zhang	Northeastern University, China
Yang Cao	Hokkaido University, China
Yang-Sae Moon	Kangwon National University, South Korea
Yanhui Gu	Nanjing Normal University, China
Yanjun Shu	Harbin Institute of Technology, China
Yanlong Wen	Nankai University, China

Yanmei Hu	Chengdu University of Technology, China
Yao Liu	University of New South Wales, Australia
Yawen Zhao	University of Queensland, Australia
Ye Zhu	Deakin University, Australia
Yexuan Shi	Beihang University, China
Yicong Li	University of Technology Sydney, Australia
Yijia Zhang	Jilin University, China
Ying Zhang	Nankai University, China
Yingjian Li	Harbin Institute of Technology, Shenzhen, China
Yingxia Shao	BUPT, China
Yishu Liu	Harbin Institute of Technology, Shenzhen, China
Yishu Wang	Northeastern University, China
Yixiang Fang	Chinese University of Hong Kong, Shenzhen, China
Yixuan Qiu	The University of Queensland, Australia
Yong Zhang	Tsinghua University, China
Yongchao Liu	Ant Group, China
Yongpan Sheng	Southwest University, China
Yongqing Zhang	Chengdu University of Information Technology, China
Youwen Zhu	Nanjing University of Aeronautics and Astronautics, China
Yu Gu	Northeastern University, China
Yu Liu	Huazhong University of Science and Technology, China
Yu Yang	Hong Kong Polytechnic University, China
Yuanbo Xu	Jilin University, China
Yucheng Zhou	University of Technology Sydney, Australia
Yue Tan	University of Technology Sydney, Australia
Yunjun Gao	Zhejiang University, China
Yunzhang Huo	Hong Kong Polytechnic University, China
Yurong Cheng	Beijing Institute of Technology, China
Yutong Han	Dalian Minzu University, China
Yutong Qu	University of Adelaide, Australia
Yuwei Peng	Wuhan University, China
Yuxiang Zeng	Hong Kong University of Science and Technology, China
Zesheng Ye	University of New South Wales, Sydney, Australia
Zhang Anzhen	Shenyang Aerospace University, China
Zhaojing Luo	National University of Singapore, Singapore
Zhaonian Zou	Harbin Institute of Technology, China
Zheng Liu	Nanjing University of Posts and Telecommunications, China

Zhengyi Yang   University of New South Wales, Australia
Zhenying He   Fudan University, China
Zhihui Wang   Fudan University, China
Zhiwei Zhang   Beijing Institute of Technology, China
Zhixin Li   Guangxi Normal University, China
Zhongnan Zhang   Xiamen University, China
Ziyang Liu   Tsinghua University, China

Contents – Part I

Time Series

An Adaptive Data-Driven Imputation Model for Incomplete Event Series 3
Jiadong Chen, Hengyu Ye, Xiaofeng Gao, Fan Wu, Linghe Kong,
and Guihai Chen

From Time Series to Multi-modality: Classifying Multivariate Time Series
via Both 1D and 2D Representations 19
Chao Yang, Xianzhi Wang, Lina Yao, Guodong Long, and Guandong Xu

Exploring the Effectiveness of Positional Embedding
on Transformer-Based Architectures for Multivariate Time Series
Classification 34
Chao Yang, Yakun Chen, Zihao Li, and Xianzhi Wang

Modeling of Repeated Measures for Time-to-event Prediction 48
Jianfei Zhang, Lifei Chen, and Shengrui Wang

A Method for Identifying the Timeliness of Manufacturing Data Based
on Weighted Timeliness Graph 64
Zehua Liu, Xuefeng Ding, Yuming Jiang, and Dasha Hu

STAD: Multivariate Time Series Anomaly Detection Based
on Spatio-Temporal Relationship 73
Keyu Chen, Guoping Zhao, Zhenfeng Yao, and Zhihong Zhang

Recommendation I

Refined Node Type Graph Convolutional Network for Recommendation 91
Wei He, Guohao Sun, Jinhu Lu, Xiu Fang, Guanfeng Liu, and Jian Yang

Multi-level Noise Filtering and Preference Propagation Enhanced
Knowledge Graph Recommendation 107
Ge Zhao, Shuaishuai Zu, Li Li, and Zhisheng Yang

Enhancing Knowledge-Aware Recommendation with Contrastive Learning 123
Xinyue Zhang and Hui Gao

Knowledge-Rich Influence Propagation Recommendation Algorithm
Based on Graph Attention Networks 138
  Yuping Yang, Guifei Jiang, and Yuzhi Zhang

A Novel Variational Autoencoder with Multi-position Latent Self-attention
and Actor-Critic for Recommendation 155
  Jiamei Feng, Mengchi Liu, Song Hong, and Shihao Song

Fair Re-Ranking Recommendation Based on Debiased Multi-graph
Representations 168
  Fangyu Han, Shumei Wang, Jiayu Zhao, Renhui Wu, Xiaobin Rui,
  and Zhixiao Wang

Information Extraction

FastNER: Speeding up Inferences for Named Entity Recognition Tasks 185
  Yuming Zhang, Xiangxiang Gao, Wei Zhu, and Xiaoling Wang

CPMFA: A Character Pair-Based Method for Chinese Nested Named
Entity Recognition 200
  Xiayan Ji, Lina Chen, Fangyao Shen, Hongjie Guo, and Hong Gao

STMC-GCN: A Span Tagging Multi-channel Graph Convolutional
Network for Aspect Sentiment Triplet Extraction 213
  Chao Yang, Jiajie Xing, and Xianguo Zhang

Exploring the Design Space of Unsupervised Blocking with Pre-trained
Language Models in Entity Resolution 228
  Chenchen Sun, Yuyuan Jin, Yang Xu, Derong Shen, Tiezheng Nie,
  and Xite Wang

Joint Modeling of Local and Global Semantics for Contrastive Entity
Disambiguation 245
  Yuhua Ke, Shaojie Xue, Ziqi Chen, and Rui Meng

KFEA: Fine-Grained Review Analysis Using BERT with Attention:
A Categorical and Rating-Based Approach 260
  Liting Huang, Yongyue Yang, Xingli Tang, Hui Zhou, and Chunyang Ye

Emotional Analysis

Discovery of Emotion Implicit Causes in Products Based on Commonsense
Reasoning 277
*Qiutong Guo, Jianxing Yu, Yufeng Zhang, Haowei Jiang, Wei Liu,
and Jian Yin*

Multi-modal Multi-emotion Emotional Support Conversation 293
*Guangya Liu, Xiao Dong, Meng-xiang Wang, Jianxing Yu,
Mengjiao Gan, Wei Liu, and Jian Yin*

Exploiting Pseudo Future Contexts for Emotion Recognition
in Conversations 309
Yinyi Wei, Shuaipeng Liu, Hailei Yan, Wei Ye, Tong Mo, and Guanglu Wan

Generating Enlightened Suggestions Based on Mental State Evolution
for Emotional Support Conversation 324
*Mengjiao Gan, Jianxing Yu, Xiao Dong, Shuang Qiu, Wei Liu,
and Jian Yin*

Deep One-Class Fine-Tuning for Imbalanced Short Text Classification
in Transfer Learning 339
Saugata Bose, Guoxin Su, and Li Liu

EmoKnow: Emotion- and Knowledge-Oriented Model for COVID-19
Fake News Detection 352
*Yuchen Zhang, Xing Su, Jia Wu, Jian Yang, Hao Fan,
and Xiaochuan Zheng*

Popular Songs: The Sentiment Surrounding the Conversation 368
Julian Stefanzick and Xin Zhao

Market Sentiment Analysis Based on Social Media and Trading Volume
for Asset Price Movement Prediction 383
*Jiahao Li, Yuyun Gong, Qinghua Zhao, Yufan Xie, Simon Fong,
and Jerome Yen*

Data Mining

Efficient Mining of High Utility Co-location Patterns Based on a Query
Strategy 401
Vanha Tran, Lizhen Wang, Jinpeng Zhang, and Thanhcong Do

Point-Level Label-Free Segmentation Framework for 3D Point Cloud
Semantic Mining 417
    Anan Du, Shuchao Pang, and Mehmet Orgun

CD-BNN: Causal Discovery with Bayesian Neural Network 431
    Huaxu Han, Shuliang Wang, Hanning Yuan, and Sijie Ruan

A Preference-Based Indicator Selection Hyper-Heuristic for Optimization
Problems 447
    Adeem Ali Anwar, Irfan Younas, Guanfeng Liu, and Xuyun Zhang

An Elastic Scalable Grouping for Stateful Operators in Stream Computing
Systems 463
    Si Lei, Dawei Sun, and Atul Sajjanhar

Incremental Natural Gradient Boosting for Probabilistic Regression 479
    Weiwen Wu, Hui Zhang, Chunming Yang, Bo Li, and Xujian Zhao

Discovering Skyline Periodic Itemset Patterns in Transaction Sequences 494
    Guisheng Chen and Zhanshan Li

Double-Optimized CS-BP Anomaly Prediction for Control Operation Data 509
    Ming Wan, Xueqing Liu, and Yang Li

Bridging the Interpretability Gap in Coupled Neural Dynamical Models 524
    Mingrong Xiang, Wei Luo, Jingyu Hou, and Wenjing Tao

Multidimensional Adaptive kNN over Tracking Outliers (Makoto) 535
    Jessy Colonval and Fabrice Bouquet

Traffic

MANet: An End-To-End Multiple Attention Network for Extracting
Roads Around EHV Transmission Lines from High-Resolution Remote
Sensing Images 553
    Yaru Ren, Xiangyu Bai, Yu Han, and Xiaoyu Hu

Deep Reinforcement Learning for Solving the Trip Planning Query 569
    Changlin Zhao, Ying Zhao, Jiajia Li, Na Guo, Rui Zhu, and Tao Qiu

MDCN: Multi-scale Dilated Convolutional Enhanced Residual Network
for Traffic Sign Detection 584
    Yan Ke, Wanghao Mo, Zhe Li, Ruyi Cao, and Wendong Zhang

Identifying Critical Congested Roads Based on Traffic Flow-Aware Road
Network Embedding   598
 Jing Zhao, Peng Cheng, Qixiang Ge, Xun Zhu, Lei Chen, Xi Guo,
 Jinshan Sun, and Yangfang Yang

A Cross-Region-based Framework for Supporting Car-Sharing   614
 Rui Zhu, Xuexin Zhang, Xin Wang, Jiajia Li, Anzhen Zhang,
 and Chuanyu Zong

Attention-Based Spatial-Temporal Graph Convolutional Recurrent
Networks for Traffic Forecasting   630
 Haiyang Liu, Chunjiang Zhu, Detian Zhang, and Qing Li

Transformer Based Driving Behavior Safety Prediction for New Energy
Vehicles   646
 Hao Lin and Junjie Yao

Graph Convolution Recurrent Denoising Diffusion Model for Multivariate
Probabilistic Temporal Forecasting   661
 Ruikun Li, Xuliang Li, Shiying Gao, S. T. Boris Choy, and Junbin Gao

A Bottom-Up Sampling Strategy for Reconstructing Geospatial Data
from Ultra Sparse Inputs   677
 Marco Landt-Hayen, Yannick Wölker, Willi Rath, and Martin Claus

Recommendation II

Feature Representation Enhancing by Context Sensitive Information
in CTR Prediction   695
 Haibo Liu, Yafang Guo, Liang Wang, and Xin Song

ProtoMix: Learnable Data Augmentation on Few-Shot Features
with Vector Quantization in CTR Prediction   709
 Haijun Zhao, Ronghai Xu, Chang-Dong Wang, and Ying Jiang

When Alignment Makes a Difference: A Content-Based Variational
Model for Cold-Start CTR Prediction   724
 Jianyu Ren and Ruoqian Zhang

Dual-Granularity Contrastive Learning for Session-Based
Recommendation   740
 Zihan Wang, Gang Wu, and Haotong Wang

Efficient Graph Collaborative Filtering with Multi-layer Output-Enhanced
Contrastive Learning   755
    *Keke Li, Shaoqing Wang, Shun Zheng, Xia Wu, Yao Zhang,
    and Fuzhen Sun*

Influence Maximization with Tag Revisited: Exploiting
the Bi-submodularity of the Tag-Based Influence Function   772
    Atharva Tekawade and Suman Banerjee

Multi-Interest Aware Graph Convolution Network for Social
Recommendation   787
    Zhengyi Guo, Yanmin Zhu, Zhaobo Wang, and Mengyuan Jing

Enhancing Multimedia Recommendation Through Item-Item Semantic
Denoising and Global Preference Awareness   802
    Yanlong Zhang, Shangfei Zheng, Qian Zhou, Wei Chen, and Lei Zhao

Resident-Based Store Recommendation Model for Community
Commercial Planning   818
    Kaiwen Wu, Yanhu Li, and Xiaofeng He

Author Index   831

Time Series

An Adaptive Data-Driven Imputation Model for Incomplete Event Series

Jiadong Chen, Hengyu Ye, Xiaofeng Gao$^{(\boxtimes)}$, Fan Wu, Linghe Kong, and Guihai Chen

MoE Key Lab of Artificial Intelligence, Department of Computer Science and Engineering, Shanghai Jiao Tong University, Shanghai, China
{chenjiadong998,cs_22_yhy,linghe.kong}@sjtu.edu.cn,
{gao-xf,fwu,gchen}@cs.sjtu.edu.cn

Abstract. Event sequences play as a general fine-grained representation for temporal asynchronous event streams. However, in practice, event sequences are often fragmentary and incomplete with censored intervals or missing data, making it hard for downstream prediction and decision-making tasks. In this work, we propose a fresh extension on the definition of the temporal point process, which conventionally characterizes chronological prediction based on historical events, and introduce *inverse point process* that characterizes counter-chronological attribution based on future events. These two point process models allow us to impute missing events for one partially observed sequence with conditional intensities in two symmetric directions. We further design a peer imitation learning algorithm that lets two models cooperatively learn from each other, leveraging imputed sequences given by the counterpart as the supervised signal. The training process consists of iterative learning of two models and facilitates them to achieve a consensus. We conduct extensive experiments on both synthetic and real-world datasets, which demonstrate that our model can recover incomplete event sequences very close to the ground-truth, with averagely 49.40% improvement compared with related competitors measured by normalized optimal transport distance.

Keywords: Event Sequence · Temporal Point Process · Missing Data · Sequence Imputation

1 Introduction

Event sequences record temporal events by a sequence of tuples (m_i, t_i), where m_i denotes discrete event markers (i.e., types) and t_i denotes continuous timestamps. Event sequences play as a general fine-grained representation for event streams that are ubiquitous in many real-life applications, e.g., 1) medical records for patients in hospital [5]; 2) user's activity and behaviors in social networks [16]; 3) individual's visiting points of interests in a large city [18], etc. One common way for modeling event sequences is to treat arrival of each event as a random variable and formulate an event sequence as a stochastic point process of which a (conditional) intensity function is introduced to characterize the conditional distribution of next event given historical events. For point process, prior works [27] attempt to model intensity functions in statistical aspects, while some recent works [22] leverage deep neural networks to learn a more expressive neural intensity model. All of these methods assume that input event sequences are complete based on which one can learn a robust representation via the point process model.

However, in most practical situations, the collected event sequences are usually fragmentary and partially observed due to common missing-data mechanisms. One typical example could be medical records which chronologically mark down symptoms, diagnoses, and medications for patients. Data from one or few hospitals cannot guarantee completeness since patients usually come over to different hospitals for medical assistance. Therefore, it is greatly in demand to build a theoretically sound and practically effective approach to recover partially observed event sequences. To solve the above-mentioned problem, one may confront three non-trivial challenges. The first question is *how to build a model that possesses enough expressiveness to estimate missing data with any possible timestamp and marker for any given observed sequence (Q1)*. Second, there is no ground-truth information that can be used as supervision for learning, so a following question arises: *how to design a self-supervised approach for model learning on event sequence imputation (Q2)* ? Third, even if one manages to define or construct some artificial signals as supervision, it is hard to characterize the discrepancy between an estimated sequence and a target one, which induces another question: *how to properly define optimization objective (Q3)*?

In this paper, we propose a novel Peer Imitation Learning point process model for Event Sequence imputation (PILES) that efficiently solve the above three challenging questions. To answer Q1, we propose an acceptance-rejection strategy that can adaptively sample missing events based on a neural intensity function parametrized by bidirectional Long-Short Term Meomory (LSTM) network. The neural intensity model guarantees enough representation capacity for event sequences with any latent process, while the acceptance-rejection strategy can incrementally sample missing events between any two observed events in an input sequence. To answer Q2, we extend the definition of temporal point process, which conventionally characterizes chronological prediction for next events based on historical ones, and pioneeringly propose inverse point process model that characterizes counter-chronological attribution for previous events based on future ones. The two point process models allow us to tackle one partially observed sequence in a bidirectional view where we sample missing events via conditional intensities in two directions. Then the sampling results given by one model can naturally be used as supervision for another model, through which we can achieve self-supervised learning.

To answer Q3, i.e., measuring distance between imputed event sequences given by two models, we propose a novel peer imitation learning algorithm that enforces consistency between one model's output (as model policy) and the counterpart's (as expert policy). Similar to GAIL [9], we introduce discriminators and adversarial training to learn better consistency measurement, which equivalently minimizes the Jensen-Shannon divergence between two model distributions. Extensive experiment results on four synthetic datasets and two real-world datasets demonstrate that PILES is capable of giving much closer reconstructed event sequences to ground-truth event sequences compared with unsupervised competitors, with an averaged improvement of 49.40% measured by normalized optimal transport distance. Also, ablation studies show the necessity of proposed units for superior performance.

Our contributions are summarized as follows.

i) **New Aspect:** We propose an inverse point process model that characterizes conditional intensities for previous events in a direction of attribution, which can cooperate with conventional (forward) point process model to deal with sequence imputation in unsupervised situation. To our knowledge, this is the first work on fully unsupervised imputation for asynchronous event sequences.

ii) **Methodology:** We propose a novel peer imitation learning approach to synergize two point process models via pushing them to achieve a consensus, which enables self-supervised learning in unsupervised case.

iii) **Experiments:** We conduct experiments on six datasets with different latent structures and demonstrate convincing superiority of new model and its generality in three missing-data situations. All source code and data will be made publicly available.

2   Related Works

In this section, we briefly discuss related works and highlight their differences from our work. For event sequence inference, the work [6] proposes an importance sampling-based algorithm to analyze marked event sequences under the condition of continuous time Bayesian Networks [15], and develops filtering and smoothing algorithms to deal with incomplete data. Similarly, [17] designs an inference framework based on Markov Chain Monte Carlo (MCMC), and particularly employs a jumping strategy for missing data in sequences. However, these two approaches heavily rely on specific parametric models with strong prior assumptions, suffering from limitations for generalization to event streams with different latent structures. Besides, [12] considers a more general neural intensity function, integrating Z-transform [1] for learning model parameters. Such a method could only handle uni-dimensional events and would fail when it comes to high-dimensional event markers.

For event sequence imputation, [19] and [21] focus on specific application scenarios. The former targets intrapolated estimation for blood glucose concentration given specific time in an observed sequence of patient's medical records, while the latter aims to predict interactions between two users in social networks given a timestamp in a sequence of user activity. In these methods, notably, the arrival time of target events is assumed to be known in advance, which plays as informative features and substantially reduces the problem difficulty. Moreover, [20] puts forward an MCMC-based model to recover incomplete event sequences driven by Hawkes process [8]. However, the method assumes time intervals that contain missing events to locate specific positions for event imputation. Also, the Hawkes process used in this model introduces strong inductive bias and suffers from limitations when it comes to more general point processes. Recently, [14] develops the Neural Hawkes Particle Smoothing method for imputing missing events, and however, it requires a large number of ground-truth complete sequences for training an intensity model that provides sufficient information for latent process in input event sequences as important prior information. Unfortunately, in most practical scenarios, one has no access to such complete event sequences.

3   Problem Formulation

An event sequence consists of a series of tuples $e_i = (m_i, t_i)$, where m_i stands for the type of an event (a.k.a. event marker) and t_i denotes continuous event timestamp. We use E to represent an event sequence composed of N marker-time tuples.

$$E = \{e_i\}_{i=1}^N, \quad \forall\, 1 \leqslant i \leqslant N: \quad m_i \in \mathcal{M}, \ t_i \in [0, T].$$

where $\mathcal{M}$ is a marker set of size M.

Temporal Point Process. One common way to model event sequences is via point process [3] which treats the arrival of each event as a random variable given the history of previous events. To characterize the probability for when the next event would happen, a (conditional) intensity function is defined as $\lambda(t|\mathcal{H}_t) := \frac{\mathbb{P}(N(t+dt)-N(t)=1|\mathcal{H}_t)}{dt}$, where $\mathcal{H}_t$ and $N(t)$ denote the history of previous events and number of events before time t, respectively. The intensity function induces conditional density distribution for arrival time of the next event, $P(t|\mathcal{H}_t) = \lambda(t|\mathcal{H}_t) \exp(-\int_{t_n}^t \lambda(\tau|\mathcal{H}_t)d\tau)$. The marker of a new event usually obeys a certain categorical distribution $P(m|\mathcal{H}_t)$. There are various ways to specify the intensity function as parametric forms. For instance, the simplest case is Poisson process which has a constant intensity over time. Also, many previous works leverage various forms to parameterize the intensity function by assuming different inductive bias for event sequences, such as inhomogeneous Poisson process, Hawkes process, self-correcting process, etc. Recent works propose to use deep neural networks to model the intensities [25]. Beyond these expressive networks, [25] proposes to train a neural point process via generative adversarial networks, and [13, 22] adopt reinforcement learning or imitation learning to further improve learning on asynchronous sequences.

Real-life event sequences are often partially discovered. Given an observed sequence $E = \{(m_i, t_i)\}$, we define U_i as a subsequence of missing events in time interval of two adjacent observed events $[t_i, t_{i+1}]$ and each missing event is denoted by $\varepsilon_{i,j} = (\mu_{i,j}, \tau_{i,j})$, where $\mu_{i,j}$ and $\tau_{i,j}$ denote the marker and time of the j-th unobserved event in U_i respectively. $U_i = \{\varepsilon_{i,j}\}_{j=1}^{n_i}, \ \forall\, 1 \leqslant j \leqslant n_i: \ \mu_{i,j} \in \mathcal{M}, \ \tau_{i,j} \in [t_i, t_{i+1}]$. Note that U_i can be an empty set ($n_i = 0$) or with arbitrary length n_i. Here we define $t_0 = 0$ and $t_{N+1} = T$. The ground-truth complete event sequence $\overline{E}$ is the union of observed sequence E and unobserved event sequences U_i, $\overline{E} = E \cup U_1 \cup U_2 \cup \cdots \cup U_N$.

Given a collection of partially observed event sequences $\{E_k\}_{k=1}^K$, our target is to build a model that can impute all the missing events (including timestamps of markers) in E_k, as an estimation for $U_{k,i}$, $i = 1, \cdots, N_k$, $k = 1, \cdots, K$. Note that the ground-truth complete sequences $\{\overline{E}_k\}$ are not available as input, and one can only leverage $\{E_k\}$ for model learning and inference. Besides, other prior information about missing data is unknown, such as total number of missing events or missing-data mechanisms (interval-censored or missing-at-random). We call this as extremely or fully unsupervised situation.

4 Proposed Method

We propose PILES, a Peer Imitation Learning point process model for Event Sequence imputation, to solve the problem in fully unsupervised setting. An illustration of our proposed framework is depicted in Fig. 1.

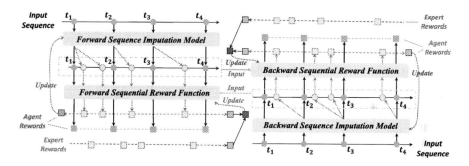

Fig. 1. An illustration of our peer imitation learning method for unsupervised event sequence imputation. Given input event sequence, missing events are imputed in both forward and backward directions.

4.1 Forward Imputation Model

Given a partially observed event sequence $E = \{e_1, e_2, \cdots, e_N\}$, our target is to estimate missing events between each pair of adjacent observed events e_i and e_{i+1}. Note that missing events may exist at any position in $[0, T]$, so we need an imputation model with enough expressiveness to accommodate all possible results. To this end, our forward imputation model chronologically samples new event markers and timestamps in an iterative manner from $t = 0$ to T, using newly imputed events to update historical sequence which gives conditional intensities for subsequent imputation.

Suppose we finish imputing events before observed event e_i and currently focus on imputing missing events between e_i and e_{i+1} and denote the latest imputed event as $\hat{\varepsilon}_{i,j}$, and then the current status of reconstructed sequence can be represented as $\hat{E}[i,j] = E \cup \hat{U}_1 \cup \hat{U}_2 \cup \cdots \cup \hat{U}_{i-1} \cup \{\hat{\varepsilon}_{i,1}, \cdots, \hat{\varepsilon}_{i,j}\}$, where $\hat{U}_p = \{\hat{\varepsilon}_{p,q}\}$ denotes set of imputed events in $[t_p, t_{p+1}]$ for $p = 1, \cdots, i-1$, and $\hat{\varepsilon}_{p,q} = (\hat{\mu}_{p,q}, \hat{\tau}_{p,q}) \in \hat{U}_p$ denotes the q-th imputed event between observed events e_p and e_{p+1}. Then the problem boils down to how to sample the next imputed event $\hat{\varepsilon} = (\hat{\mu}, \hat{\tau})$. We next introduce embeddings for events and neural intensity model that play as building blocks for our forward imputation model.

Event Embeddings. We embed each marker-time pair (m,t) into low-dimensional vectors via learnable embedding matrices. For event marker $m \in \mathcal{M}$, we first generate a one-hot representation $\mathbf{v} \in \{0,1\}^M$, and then multiply it with an embedding matrix $\mathbf{W}_M \in \mathbb{R}^{d \times M}$ to obtain an embedding vector $\mathbf{m} = \mathbf{W}_M \mathbf{v}$. To encode continuous timestamp $t \in \mathbb{R}$, we adopt a linear transformation $\mathbf{t} = \mathbf{w}_T t + \mathbf{b}_T$ to get its representation. Here $\mathbf{W}_M \in \mathbb{R}^{d \times M}$, $\mathbf{b}_T \in \mathbb{R}^d$, and $\mathbf{w}_T \in \mathbb{R}^d$ are all trainable parameters. The final representation vector $Emb(e)$ for event $e = (m,t)$ is the combination of marker and time representations with a integrating parameter β:

$$Emb(e) = \beta \cdot \mathbf{W}_M \mathbf{v} + (1 - \beta) \cdot (\mathbf{w}_T t + \mathbf{b}_T). \tag{1}$$

Neural Intensity Model. We further model intensity function over an event sequence using bidirectional LSTMs to concurrently utilize both history and future information. Such Bi-LSTM based encoding technique is widely used in event sequence modeling, such as recent work [14]. Given the current status of imputed sequence in Sect. 4.1 we define history subsequence $H[i,j]$ and future subsequence $F[i,j]$:

$$\begin{aligned} H[i,j] &= \{e_1, e_2, \cdots, e_i\} \cup \hat{U}_1 \cup \cdots \cup \hat{U}_{i-1} \cup \{\hat{\varepsilon}_{i,1}, \cdots, \hat{\varepsilon}_{i,j}\} \\ F[i,j] &= \{e_N, e_{N-1}, \cdots, e_{i+1}\} \end{aligned} \tag{2}$$

we encode $H[i,j]$ and $F[i,j]$ using two LSTMs $g_h(\cdot)$ and $g_f(\cdot)$ respectively to get the hidden states $\mathbf{h}_{i,j}, \mathbf{z}_{i,j} \in \mathbb{R}^H$,

$$\mathbf{h}_{i,j} = g_h\left(Emb(H[i,j])\right), \quad \mathbf{z}_{i,j} = g_f\left(Emb(F[i,j])\right), \tag{3}$$

where $Emb(H[i,j])$ and $Emb(F[i,j])$ denote the matrices generated by mapping each event in the sequence to an embedding vector using Eq. (1). From Eq. (2), we can see that the historical events, including both observed events and imputed events, are encoded to get the hidden state $\mathbf{h}_{i,j}$. In contrast, the future events, which only include observed events, are encoded to get $\mathbf{z}_{i,j}$. Then final hidden state $\mathbf{x}_{i,j} \in \mathbb{R}^{2H}$ can be set as the concatenation of $\mathbf{h}_{i,j}$ and $\mathbf{z}_{i,j}$: $\mathbf{x}_{i,j} = [\mathbf{h}_{i,j}, \mathbf{z}_{i,j}]$. Similar to [4], we define latent intensities $\boldsymbol{\lambda}_{i,j} \in \mathbb{R}^M$ over M marker classes $\boldsymbol{\lambda}_{i,j} = \mathbf{W}_h \mathbf{x}_{i,j} + \mathbf{b}_h$. where $\mathbf{W}_h \in \mathbb{R}^{(2H) \times M}$ and $\mathbf{b}_h \in \mathbb{R}^M$ are trainable parameters. Here each scalar element in $\boldsymbol{\lambda}_{i,j}$ represents conditional intensity for a certain class of marker.

Sampling Events for Imputation. We proceed to present how to sample next imputed events $\hat{\varepsilon} = (\hat{\mu}, \hat{\tau})$ given the conditional intensities. In general, we first sample marker $\hat{\mu}$ and obtain the corresponding intensity value from $\lambda_{i,j}$ to estimate timestamp $\hat{\tau}$. Given the hidden state vector $\mathbf{x}_{i,j}$, the next marker $\hat{\mu} \in \mathcal{M}$ can be sampled from a softmax distribution

$$\hat{\mu} \sim P\left(m \mid H[i,j], F[i,j]\right) = \frac{\exp\left((\mathbf{W}_p\mathbf{x}_{i,j} + \mathbf{b}_p)_{[m]}\right)}{\sum_{m'=1}^{M} \exp\left((\mathbf{W}_p\mathbf{x}_{i,j} + \mathbf{b}_p)_{[m']}\right)}. \tag{4}$$

where $\mathbf{a}_{[k]}$ denotes the k-th element of vector $\mathbf{a}$. Here $\mathbf{W}_p \in \mathbb{R}^{(2H) \times M}$ and $\mathbf{b}_h \in \mathbb{R}^M$. Then the corresponding intensity λ can be indexed from $\lambda_{i,j}$. Given the latest imputed timestamp $\hat{\tau}_{i,j}$, we can sample the time interval of next imputed event $\hat{\delta t}$ from a continuous distribution $\mathcal{T}(\delta t; \lambda)$ parametrized by $\lambda = \lambda_{i,j[\hat{\mu}]}$. Similar to [22], $\mathcal{T}$ can be chosen as Exponential distribution and the time interval can be sampled from a conditional density:

$$\hat{\delta t} \sim P(\delta t \mid H[i,j], F[i,j]) = \mathcal{T}(\delta t; \lambda). \tag{5}$$

Then we have timestamp for next event as $\hat{\tau} = \hat{\tau}_{i,j} + \hat{\delta t}$. Here one can also consider estimation for $\hat{\delta t}$ using approximated expectation [4] or deterministic mapping [24]. Once we obtain a sampled event $\hat{\varepsilon} = (\hat{\mu}, \hat{\tau})$ as candidate, we need to check its validity by taking the following acceptance-rejection method to judge if it can be properly added to the current status of reconstructed sequence. If $\hat{\tau} < t_{i+1}$, i.e., the timestamp of the sampled event is before the next observed event, we keep it in the sequence, set $\hat{\varepsilon}_{i,j+1} = \hat{\varepsilon}$, and update the history and future subsequences for next imputation step:

$$\begin{aligned} H[i, j+1] &= H[i,j] \cup \{\hat{\varepsilon}_{i,j+1}\}, \\ F[i, j+1] &= F[i,j]. \end{aligned} \tag{6}$$

If the timestamp $\hat{\tau}$ exceeds the time of next observed event t_{i+1}, we discard $\hat{\varepsilon} = (\hat{\mu}, \hat{\tau})$ and continue to impute missing events between observed events e_{i+1} and e_{i+2} by updating the history and future subsequences:

$$\begin{aligned} H[i+1, j] &= H[i,j] \cup \{e_{i+1}\}, \\ F[i+1, j] &= F[i,j] \setminus \{e_{i+1}\}. \end{aligned} \tag{7}$$

Sequence Reconstruction. When we go to the next step with updated history and future subsequences given by Eq. (6) or Eq. (7), we can encode $H[i, j+1]$, $F[i, j+1]$ (resp. $H[i+1, j]$, $F[i+1, j]$) by Eq. (3) and continue to impute the next missing event. Note that here we do not need to encode the whole sequence again, since for historical events, we can simply update the hidden state $\mathbf{h}_{i,j+1}$ (resp. $\mathbf{h}_{i+1,j}$) using the newly included event $\hat{\varepsilon}_{i,j+1}$ (resp. e_{i+1}). For future events, we can inversely encode the whole sequence $\{e_N, \cdots, e_2, e_1\}$ before the imputation process, and store and reuse the results during the imputation process. Therefore, our imputation model still maintains a linear time complexity w.r.t. the length of sequence. The whole process is executed iteratively until sampled timestamp reaches T, and we use $\hat{E}^F$ to denote the final imputed event sequence in forward direction: $\hat{E}^F = E \cup \hat{U}_1^F \cup \hat{U}_2^F \cup \cdots \cup \hat{U}_N^F$, where $\hat{U}_i^F$ denotes a subsequence of imputed events between e_i and e_{i+1}, i.e., $\hat{U}_i^F = \{(\hat{\mu}_{i,j}, \hat{\tau}_{i,j})\}$, $\forall\, 1 \leqslant j \leqslant \hat{n}_i$. Here we use a superscript F to highlight its forward direction for imputation to distinguish from the backward version in Sect. 4.2.

4.2   Inverse Point Process and Backward Model

While we obtain reconstructed event sequences, it is hard for model optimization since we have no ground-truth data as supervision. To solve the obstacle, we take a different view from traditional perspective. In temporal point process model, it conventionally characterize conditional intensities for future events given a history of events. In essence, it considers *prediction* in a chronological order. Why cannot we take an inverse perspective and tackle conditional intensities for historical events given future ones, which focuses on *attribution* in a counter-chronological order.

To this end, we introduce the concept of *Inverse Point Process*, which could be identified as a complementary latent model for traditional point process. We define conditional intensity function for inverse temporal point process as follows.

Definition 1. (Inverse Temporal Point Process): *Given future events $F(t)$ after time t, conditional intensity $\lambda(t|F(t))$ is defined as the probability density of a possible hidden event existing in $[t - \Delta t, t]$, where $N'(t)$ represents the number of future events after time t.*

$$\lambda\left(t|F(t)\right) := \lim_{\Delta t \to 0} \frac{\mathbb{P}(N'(t) - N'(t - \Delta t) = 1\,|\,F(t))}{\Delta t}. \qquad (8)$$

One can realize that the definitions of inverse point process and traditional point process are symmetric in nature. Both of them are based on countings and probability density of event occurrence within a unit period of time. The only difference lies in that the original definition uses countings of historical events, while inverse point process model considers future ones. Therefore, it is natural to consider a (symmetric) backward imputation model that can sample imputed events in input sequences from the future to the beginning.

Backward Imputation Model. Given a partially observed event sequence $E = \{e_1, e_2, \cdots, e_N\}$, similar to Sect. 4.1, we suppose the current imputation status represented by $\hat{E}[i, j] = E \cup \hat{U}_N \cup \hat{U}_{N-1} \cup \cdots \cup \hat{U}_{i+1} \cup \{\hat{\varepsilon}_{i,-j}, \cdots, \hat{\varepsilon}_{i,-1}\}$ and the latest imputed event is $\hat{\varepsilon}_{i,-j}$. We employ another two LSTMs $g'_f(\cdot)$ and $g'_h(\cdot)$ to encode history and future subsequences defined as

$$H'[i, -j] = \{e_1, e_2, \cdots, e_i\}$$
$$F'[i, -j] = \{e_N, \cdots, e_{i+1}\} \cup \hat{U}_{N-1} \cup \cdots \cup \hat{U}_{i+1} \cup \{\hat{\varepsilon}_{i,-j}, \cdots, \hat{\varepsilon}_{i,-1}\},$$

based on which we can obtain hidden state vector $\mathbf{x}'_{i,j}$ and intensity vector $\boldsymbol{\lambda}'_{i,j}$. After that, we sample a newly imputed event marker $\hat{\mu}$ and time interval $\delta\hat{t}$ using softmax distribution and exponential distribution as in Eq. (4) and Eq. (5) respectively, and the timestamp of next imputed event is calculated by $\hat{\tau} = \hat{\tau}_{i,-j} - \delta\hat{t}$. Then we conduct a similar acceptance-rejection strategy for the newly sampled $\hat{\varepsilon} = (\hat{\mu}, \hat{\tau})$. If $\hat{\tau} > t_i$, we set $\hat{\varepsilon}_{i,-j-1} = \hat{\varepsilon}$ and update history and future subsequences by

$$
\begin{aligned}
H'[i, -j-1] &= H'[i, -j], \\
F'[i, -j-1] &= F'[i, -j] \cup \{\hat{\varepsilon}_{i,-j-1}\}.
\end{aligned} \tag{9}
$$

Otherwise, we have

$$
\begin{aligned}
H'[i-1, -j] &= H'[i, -j] \setminus \{e_i\}, \\
F'[i-1, -j] &= F'[i, -j].
\end{aligned} \tag{10}
$$

Finally, the model outputs a reconstructed sequence $\hat{E}^B$ where we use the superscript B to highlight its backward direction. We denote trainable parameters in backward imputation model as θ_B.

Given imputed sequences $\hat{E}^F$ and $\hat{E}^B$ by forward and backward models respectively, we can build a self-supervised learning approach by enforcing consistency between the results of two models. One straightforward way is to consider Maximum Likelihood Estimation (MLE) that uses output of the counterpart as observed 'ground-truth' labels, which equivalently minimizes the Kullback-Leibler (KL) divergence [11] between two model distributions. The objective for forward imputation model can be $\max_{\theta_F} \left(\log P_{\theta_F} \left(\hat{E}^B \right) \right)$, where

$$
P_{\theta_F}(\hat{E}^B) = \prod_{\varepsilon=(\mu,\tau)\in\hat{E}^B} P_{\theta_F}(\mu) \cdot P_{\theta_F}(\delta t). \tag{11}
$$

Here $P_{\theta_F}(\mu)$ is given by Eq. (4) and $P_{\theta_F}(\delta t)$ (where δt denoting time interval between event ε and the next imputed event) is given by Eq. (5). Similarly, the objective for backward imputation model can be $\max_{\theta_B} \left(\log P_{\theta_B} \left(\hat{E}^F \right) \right)$, where

$$
P_{\theta_B}(\hat{E}^F) = \prod_{\varepsilon=(\mu,\tau)\in\hat{E}^F} P_{\theta_B}(\mu) + P_{\theta_B}(\delta t) \tag{12}
$$

and $P_{\theta_B}(\mu)$ as well as $P_{\theta_B}(\delta t)$ are given by the backward model.

4.3   Peer Imitation Learning

The MLE-based self-supervised learning enables two models to learn from each other in unsupervised situation. However, one limitation is that MLE loss enforces hard match of two models and assigns each imputed event in the sequence with equal importance. Some prior works focusing on learning point process models for complete event sequences propose to use reinforcement learning [13] and, especially, imitation learning [22,24] for optimization by 1) treating the prediction as model policy and ground-truth data as expert policy and 2) guiding the model to generate policy close to the expert. The imitation learning approach brings about some tolerance on local mismatch of two sequences and would adaptively allocate different importance to event pairs via a learnable discriminator that is jointly optimized to provide an adequate measurement.

In our case, we have no ground-truth data as expert policy. Like the MLE method, we can use the output of one model as 'ground-truth' data for target of the other model's training. The intuition is from real-world scenarios when students take classes: 1) students can directly learn from what teachers taught in class (ground-truth expert policy), which corresponds to the case in [22,24] with complete sequences in training set, and 2) students can also learn from the peers through discussion and knowledge sharing when teachers are absent. Inspired by this, we propose a novel *peer imitation learning* approach to improve the MLE-based optimization. We let one model's output (as model policy) be close to the counterpart's (as expert policy) through a measurement given by a neural reward function, which is with the same spirit of the discriminator in GAIL [9]. In this way, we free the model optimization from distance-based objectives via joint learning of sequence imputation model and reward function in an adversarial manner.

The objective for forward model can be written as

$$\min_{\theta_F} \max_{w_F} \mathbb{E}_{\hat{E}^B \sim P_{\theta_B}} \left(\log D_{w_F}(\hat{E}^B) \right) + \mathbb{E}_{\hat{E}^F \sim P_{\theta_F}} \left(1 - \log D_{w_F}(\hat{E}^F) \right), \quad (13)$$

where P_{θ_F} and P_{θ_B} are given in Eq. (11) and (12) and we introduce a discriminator $D_{w_F}(\cdot)$ (with parameters w_F) that aims to maximize reward for backward model's imputation and minimize reward for forward model's imputation. As shown in [7], Eq. (13) is equivalent to minimizing Jensen-Shannon divergence between two model distributions P_{θ_F} and P_{θ_B}. Similarly, for backward model, the objective is

$$\min_{\theta_B} \max_{w_B} \mathbb{E}_{\hat{E}^F \sim P_{\theta_F}} \left(\log D_{w_B}(\hat{E}^F) \right) + \mathbb{E}_{\hat{E}^B \sim P_{\theta_B}} \left(1 - \log D_{w_B}(\hat{E}^B) \right), \quad (14)$$

where $D_{w_B}(\cdot)$ is a discriminator with parameters w_B.

Neural Reward Functions. We further present the specification of two discriminators. We take $D_{w_F}(\cdot)$ for illustration and $D_{w_B}(\cdot)$ can be specified in a similar way. Assume $\hat{E}$ as an imputed sequence with $\hat{n}$ imputed events in total. $D_{w_F}(\hat{E})$ can be a sequence-to-sequence model, mapping $\hat{E}$ to a sequence of reward values for each imputed event: $[r_1, r_2, \cdots, r_{\hat{n}}] = d_F(\hat{E})$, where $d_F(\cdot)$ is a Bi-LSTM.

Policy Gradients. To optimize Eq. (13), we REINFORCE algorithm and compute policy gradient for θ_F,

$$\nabla_{\theta_F} \mathbb{E}_{\hat{E}^F \sim P_{\theta_F}} \left(\log D_{w_F}(\hat{E}^F) \right) \approx \frac{1}{C} \sum_{c=1}^{C} \sum_{\varepsilon_i \in \hat{E}_c^F} \gamma^i r_i \cdot \nabla_{\theta_F} \log P_{\theta_F}(\varepsilon_i), \quad (15)$$

where we sample C imputed sequences $\{\hat{E}_c^F\}$ from the forward imputation model P_{θ_F}, $\gamma \in (0, 1]$ is a discount factor, $\varepsilon_i = (\mu_i, \tau_i)$ denotes the i-th imputed event in $\hat{E}_c^F$, and $P_{\theta_F}(\varepsilon_i) = P_{\theta_F}(\mu_i) \cdot P_{\theta_F}(\tau_i)$ (given by Eq. (4) and (5)).

Also, the policy gradient for the discriminator w_F can be computed by

$$
\nabla_{w_F} \mathbb{E}_{\hat{E}^B \sim P_{\theta_B}} \left(\log D_{w_F}(\hat{E}^B) \right) + \mathbb{E}_{\hat{E}^F \sim P_{\theta_F}} \left(1 - \log D_{w_F}(\hat{E}^F) \right)
$$

$$
\approx \frac{1}{C} \sum_{c=1}^{C} \sum_{\varepsilon_i \in \hat{E}_c^B} \gamma_i r_i \cdot \nabla_{w_F} \log P_{\theta_B}(\varepsilon_i) + \frac{1}{C} \sum_{c'=1}^{C} \sum_{\varepsilon_{i'} \in \hat{E}_{c'}^B} \gamma_{i'} r_{i'} \cdot \nabla_{w_F} \left(1 - \log P_{\theta_F}(\varepsilon_{i'}) \right).
$$

$$(16)$$

The policy gradients for θ_B and w_B in Eq. (14) can be computed in similar ways. The training process is conducted iteratively. In each iteration of the algorithm, the parameters of forward and backward model are updated alternately, during which both of the modules (generator and discriminator) are adversarially optimized. In specific, we optimize over θ_F and w_F in N_s steps and then we turn to θ_B and w_B using N_s steps of updates.

5   Experiments

5.1   Experimental Setup

Datasets. For synthetic data, we adopt typical point processes for data generation: *Hawkes Process* [8], *Self-Correcting Process* [10], *Inhomogeneous Poisson Process*, and neural point process *RMTPP* [4]. For real-world datasets, we adopt *NYC-Taxi* [23] and *Elevator* [2] as widely used in previous works. We adopt the following hyper-parameter settings: learning rate $lr = 10^{-4}$, embedding dimension $d = 10$, batch size 128, integrating parameter in the event embeddings $\beta = 0.3$, discount factor $\gamma = 0.95$, update steps $N_s = 3$. All these hyperparameters are optimized by grid search.

Missing-Data Mechanisms. We adopt three common missing-data mechanisms to generate incomplete event sequences.

1. Random Missing (RM). We consider a fixed missing probability p and each event in sequences has the same probability to be masked out. We adopt $p = 0.2$ in our experiments.
2. Time Intervals (TI). Another typical real-life missing-data mechanism could be the unavailability of data between a certain time interval. Correspondingly, given time upper bound T of event sequences, we cut off all events in $[\alpha_1 T, \alpha_2 T]$. We consider $\alpha_1 = 0.4$ and $\alpha_2 = 0.6$ in our experiments.
3. Marker Types (MT). In some cases, one may exactly know the arrival time of events but event markers (i.e., types) are unobserved. In our experiments, we randomly choose one marker class in each dataset and drop all the events with such class of markers to generated incomplete sequences.

Baselines. It is hard to provide a fair comparison between our unsupervised model and other weakly supervised or semi supervised models, such as [14,19–21], as we discussed in Sect. 2. Therefore, for a convincing comparison, we turn to migrate state-of-the-art neural intensity model and consider four different variants namely *Uni-RMTPP(F)*, *Uni-RMTPP(B)*, *Bi-RMTPP(F)*, and *Bi-RMTPP(B)* from Recurrent Marked Temporal Point Process (RMTPP) [4] as competitors for unsupervised sequence imputation. In addition, we also simplify our imitation learning as directly using MLE for optimization, which constructs two variants of our models **PILES-MLE(F)** and **PILES-MLE(B)** for ablation studies. Moreover, to investigate the gap between our imputation results in unsupervised learning to those of supervised learning, we apply ground-truth complete event sequences for training Bi-RMTPP(F), called **Oracle-SUP**.

Table 1. The comparison results evaluated by **Normalized OTD**

Methods	Hawkes			Self-Correcting			Inhomogeneous			Recurrent			Elevator			NYC Taxi		
	RM	TI	MT	RM	TI	MT	RM	TI	MT	RM	TI	MT	RM	TI	MT	RM	TI	MT
Oracle-SUP	10.58	11.59	18.66	3.239	4.214	7.980	0.011	0.085	0.017	0.172	0.193	0.491	0.368	0.623	0.056	2.172	2.107	5.989
PILES(F)	**15.24**	**13.30**	33.01	**4.508**	**6.276**	16.30	0.039	**0.098**	**0.068**	**0.177**	**0.204**	**0.586**	**0.505**	0.908	0.197	**2.229**	3.016	**9.091**
PILES(B)	24.67	19.92	42.29	9.132	13.21	**14.55**	0.047	0.111	0.101	0.526	0.285	0.701	0.648	0.912	0.118	2.917	3.011	13.47
PILES-MLE(F)	23.26	21.12	**32.60**	12.31	10.33	16.15	**0.036**	0.102	0.074	0.487	0.324	1.116	0.582	0.816	0.322	3.453	4.126	10.41
PILES-MLE(B)	28.43	26.57	49.37	15.43	17.29	25.37	0.055	0.124	0.148	0.619	0.721	1.134	0.665	0.694	0.136	3.314	3.164	16.32
Bi-RMTPP(F)	34.37*	34.52*	50.09	19.23	20.63	37.34	0.105*	0.101*	0.111	1.147*	1.141*	3.864 *	0.863	**0.797***	0.110	11.32*	10.37	35.65*
Bi-RMTPP(B)	39.76	38.55	50.23	20.58	22.57	39.86	0.147	0.176	0.196	1.837	1.546	4.932	0.419	0.817	0.086	11.41	10.16*	38.36
Uni-RMTPP(F)	36.02	35.20	48.08*	18.31	18.67	33.04	0.108	0.110	0.106*	1.911	1.443	5.532	0.765	0.816	0.133	12.64	15.23	58.38
Uni-RMTPP(B)	36.52	40.11	50.28	17.33*	14.07*	30.00*	0.187	0.193	0.178	2.204	1.703	5.921	0.674*	0.811	**0.091***	13.21	16.32	60.43
Improvement	55.6%	61.5%	31.3%	74.0%	55.4%	51.5%	62.9%	3.0%	35.8%	84.6%	82.1%	84.8%	25.1%	0.0%	0.0%	80.3%	70.4%	74.5%

5.2 Evaluation Metrics

Evaluating event sequence imputation involves the difficulty for a well-defined distance metric between two asynchronous event sequences. It is intractable to align events in two sequences with asynchronous timestamps and different numbers of events. Recently, [14] proposes to use Optimal Transport Distance (OTD) for describing the event sequence similarity, which is computed by dynamic programming with pre-defined insertion and deletion costs. The fundamental intuition for OTD is to describe the distance between sequences by how much it cost to "modify" one sequence to become identical to another (similar to minimum edit distance in NLP [26]), where smaller OTD indicates better imputation performance. In specific, given a complete event sequence $E = \{e_1, e_2, \cdots, e_P\}$, and an imputed event sequence $\hat{E} = \{\hat{e}_1, \hat{e}_2, \cdots, \hat{e}_Q\}$, where $e_i = (m_i, t_i)$ and $\hat{e}_j = (\hat{\mu}_j, \hat{\tau}_j)$, we define an alignment set $\mathcal{S}$ for pairing events from two sequences: $\mathcal{S} = \{(i, j) \mid m_i = \hat{\mu}_j, i \in [1, P], j \in [1, Q]\}$, where we can discover that in our settings, only events with the same types of markers are allowed to be aligned.

We can also discover that given two event sequences, the alignment set $\mathcal{S}$ is not unique, since different events (say e_1, e_2 here) in one sequence could have the same type of markers so that given an event $\hat{e}$ in another sequence with identical marker, we can actually align $\hat{e}$ with either e_1 or e_2. Given two event sequences

E and $\hat{E}$, the distance D is defined as $OTD(E, \hat{E}) = \min_{\mathcal{S}} \sum_{(i,j) \in \mathcal{S}} |t_i - \hat{\tau}_j| + \sum_{i \notin \mathcal{S}} C_{insert} + \sum_{j \notin \mathcal{S}} C_{delete}$ Besides, we propose a normalized optimal transport distance (normalized OTD) as follows: $nOTD(E, \hat{E}) = \min_{\mathcal{S}} \sum_{(i,j) \in \mathcal{S}} \frac{|t_i - \hat{\tau}_j|}{T} + \sum_{j \notin \mathcal{S}} \frac{|T - \hat{\tau}_j|}{T} + \sum_{i \notin \mathcal{S}} \frac{|t_i|}{T}$.

5.3   Results and Analysis

Comparison on OTD. Figure 2(a) and Fig. 2(b) show OTD results on NYC Taxi and Elevator datasets respectively. Here we set the insertion and deletion cost C from 2^{-5} to 2^4 and compare the performances of our model and several baselines. On both datasets, the OTD values increase approximately linearly as C grows in logarithmic space, and notably, PILES constantly outperforms others with a minimal gap to supervised method Oracle-SUP. Among comparative methods, bi-directional model generally performs better than the uni-directional counterpart due to better utility of event data in sequences. There are significant improvements achieved by PILES over two RMTPP models: 76.2% on NYC Taxi dataset and 42.4% on Elevator dataset.

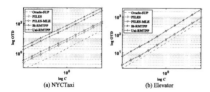

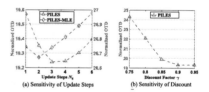

Fig. 2. Optimal Transport Distance (OTD) with delete and insert cost $C \in [-5, 2^4]$ in logarithmic axes. Each data point is average of forward and backward reconstructions.

Fig. 3. Sensitivity of update steps N_s and discount factor γ evaluated on NYC-Taxi in an RM setting. We fix $\gamma = 0.95$, and set $N_s = 3$ to evaluate the sensitivity of γ, where the learning rate is fixed at 10^{-4}.

Comparison on Normalized OTD. Table 2 presents results of normalized OTD on four synthetic datasets and two real-world datasets. As we can see, PILES generally provides the best performances in the majority of cases with different missing-data mechanisms on six datasets. The results demonstrate the superiority of proposed model. Notably, compared with RMTPP models, PILES manages to achieve 80.3% improvement on NYC-Taxi dataset with the setting of random missing (RM), which shows that the peer imitation learning approach contributes to better self-supervised learning by enforcing consistency between forward and backward imputation rather than training the model directly over imcomplete sequences. Comparing different missing-data mechanisms, we can see that missing marker type (MT) appears to be much more difficult than other two missing-data mechanisms and there exist obvious gaps between unsupervised methods and Oracle-SUP in MP settings. Moreover, in terms of different data generating processes, we can see that when using neural intensity model to generate data, our model PILES gives very close results to Oracle-SUP with much smaller gaps compared to other data generating processes. This phenomenon indicates that our method has capability to capture complex latent process in input sequences and also, the inductive bias shared by both the model and data generation affects the performance a lot.

Ablation Study. One more straight-forward approach for mutual learning is to directly use outputs from another model as supervised training data, based on which one can use MLE for optimization. Here we compare our PILES with peer imitation learning with the model using MLE as ablation study. As shown in Fig. 2 where we report results for PILES and PILES-MLE on six datasets, we can find that although slightly, PILES outperforms PILES-MLE on both datasets. Also in Table 1, the results show that PILES turn out to be superior than PILES-MLE in the majority of cases, which indicates that our proposed peer imitation learning is effective for event sequence imputation.

Parameter Sensitivity. We also analyze the sensitivity of our model on two hyper-parameters, the number of update steps N_s and discount factor γ, in Fig. 3. As shown in Fig. 3(a), PILES performs the best when N_s equals to 3. In fact, N_s controls how sufficiently two models' results are matched in each step of iterative training. When N_s is too small, two models are insufficiently learned from each other; when N_s is too large, they would 'over-fit' the estimation from the counterpart and get stuck by some local optima. In Fig. 3(b), we plot normalized OTD of PILES v.s. different discount factor γ's. We can see that the model performance declines faster when γ decreases. One possible reason is that when using relatively small γ, the rewards from previously imputed events would be weakened and the model focuses more on a few imputed events in a sequence, ignoring global information.

6   Conclusion

In this paper, we focus on reconstructing partially observed event sequences in a fully unsupervised setting. We propose a novel peer imitation learning point process framework which entails two-fold contributions on methodological level. First, we extend the definition of conventional temporal point process and propose inverse point process model. Second, we design a new peer imitation learning approach that lets two point process models in different directions cooperatively learn from each other to achieve a consensus on imputation results. Comprehensive experimental results on six datasets with multifarious missing-data mechanisms show the superiority of our method.

Acknowledgement. This work was supported by the National Key R&D Program of China [2020YFB1707900]; the National Natural Science Foundation of China [62272302, 62172276], and Shanghai Municipal Science and Technology Major Project [2021SHZDZX0102].

References

1. Antoniou, A.: Digital Signal Processing. McGraw-Hill, New York (2016)
2. Crites, R.H., Barto, A.G.: Improving elevator performance using reinforcement learning. In: NeurIPS, pp. 1017–1023 (1995)
3. Daley, D.J., Vere-Jones, D.: An Introduction to the Theory of Point Processes: Volume II: General Theory and Structure. Springer, Heidelberg (2007). https://doi.org/10.1007/978-0-387-49835-5
4. Du, N., Dai, H., Trivedi, R., Upadhyay, U., Gomez-Rodriguez, M., Song, L.: Recurrent marked temporal point processes: embedding event history to vector. In: SIGKDD, pp. 1555–1564 (2016)
5. Enguehard, J., Busbridge, D., Bozson, A., Woodcock, C., Hammerla, N.Y.: Neural temporal point processes for modelling electronic health records (2020)
6. Fan, Y., Xu, J., Shelton, C.R.: Importance sampling for continuous time bayesian networks. J. Mach. Learn. Res. **11**(Aug), 2115–2140 (2010)
7. Goodfellow, I.J., et al.: Generative adversarial nets. In: NeurIPS, pp. 2672–2680 (2014)
8. Hawkes, A.G.: Spectra of some self-exciting and mutually exciting point processes. Biometrika **58**(1), 83–90 (1971)
9. Ho, J., Ermon, S.: Generative adversarial imitation learning. In: NeurIPS, pp. 4565–4573 (2016)
10. Isham, V., Westcott, M.: A self-correcting point process. Stochastic Process. Appl. **8**(3), 335–347 (1979)
11. Kullback, S., Leibler, R.A.: On information and sufficiency. Ann. Math. Stat. **22**(1), 79–86 (1951)
12. Lee, Y., Vo, T.V., Lim, K.W., Soh, H.: Z-transforms and its inference on partially observable point processes. In: IJCAI, pp. 2369–2375 (2018)
13. Li, S., Xiao, S., Zhu, S., Du, N., Xie, Y., Song, L.: Learning temporal point processes via reinforcement learning. In: NeurIPS, pp. 10781–10791 (2018)
14. Mei, H., Qin, G., Eisner, J.: Imputing missing events in continuous-time event streams. In: ICML, pp. 4475–4485 (2019)
15. Nodelman, U., Shelton, C.R., Koller, D.: Continuous time bayesian networks. arXiv preprint arXiv:1301.0591 (2012)
16. Pan, Z., Huang, Z., Lian, D., Chen, E.: A variational point process model for social event sequences. In: AAAI, pp. 173–180 (2020)
17. Rao, V., Teh, Y.W.: MCMC for continuous-time discrete-state systems. In: NeurIPS, pp. 701–709 (2012)
18. Reinhart, A.: A review of self-exciting spatio-temporal point processes and their applications. Stat. Sci. **33**(3), 299–318 (2018)
19. Schaubel, D.E., Cai, J.: Multiple imputation methods for recurrent event data with missing event category. Can. J. Stat. **34**(4), 677–692 (2006)
20. Shelton, C.R., Qin, Z., Shetty, C.: Hawkes process inference with missing data. In: AAAI, pp. 6425–6432 (2015)

21. Stomakhin, A., Short, M.B., Bertozzi, A.L.: Reconstruction of missing data in social networks based on temporal patterns of interactions. Inverse Prob. **27**(11), 115013 (2011)
22. Upadhyay, U., De, A., Rodriguez, M.G.: Deep reinforcement learning of marked temporal point processes. In: NeurIPS, pp. 3172–3182 (2018)
23. Whong, C.: Foiling nyc's taxi trip data (2014)
24. Wu, Q., Zhang, Z., Gao, X., Yan, J., Chen, G.: Learning latent process from high-dimensional event sequences via efficient sampling. In: NeurIPS, pp. 3842–3851 (2019)
25. Xiao, S., Yan, J., Yang, X., Zha, H., Chu, S.M.: Modeling the intensity function of point process via recurrent neural networks. In: AAAI, pp. 1597–1603 (2017)
26. Zhao, Y., Jiang, H., Wang, X.: Minimum edit distance-based text matching algorithm. In: NLPKE, pp. 1–4 (2010)
27. Zhou, K., Zha, H., Song, L.: Learning triggering kernels for multi-dimensional hawkes processes. In: ICML, pp. 1301–1309 (2013)

From Time Series to Multi-modality: Classifying Multivariate Time Series via Both 1D and 2D Representations

Chao Yang[1(✉)], Xianzhi Wang[1], Lina Yao[2,5], Guodong Long[3],
and Guandong Xu[4]

[1] School of Computer Science, University of Technology Sydney, Ultimo, Australia
chao.yang@student.uts.edu.au
[2] School of Computer Science and Engineering, UNSW, Kensington, Australia
[3] Australian AI Institute, University of Technology Sydney, Ultimo, Australia
[4] Data Science Institute, University of Technology Sydney, Ultimo, Australia
[5] CSIRO Data61, Canberra, Australia

Abstract. Multivariate time series classification is crucial for various applications such as activity recognition, disease diagnosis, and brain-computer interfaces. Deep learning methods have recently achieved promising performance thanks to their powerful representation learning capacity. However, existing deep learning-based classifiers rely solely on temporal information while disregarding clues from the frequency perspective. In this regard, we propose a novel method for classifying multivariate time series leveraging both temporal and frequency information. We first apply Short-Time Fourier Transform (STFT) to transform time series into spectrograms, which contain a 2D representation of frequency components and their temporal positions. In particular, for each variable, we generate spectrograms with varying frequencies and temporal resolutions under different window sizes. The transformation essentially adds a new modality to 1D time series and converts the multivariate time series classification into a multi-modality data classification task, making it possible to bring powerful backbones from computer vision fields to solve the time series classification problem. We then construct a dual-stream network based on the ResNet architecture that takes in both 1D and 2D representations for accurate multivariate time series classification. Our extensive experiments on 30 public datasets show our method outperforms multiple competitive state-of-the-art baselines.

Keywords: Multivariate time series classification · multimodal learning · deep learning

1 Introduction

Multivariate time series is a type of data that exists across multiple domains and have broad applications in human activity recognition [40], heart disease

X. Yang et al. (Eds.): ADMA 2023, LNAI 14176, pp. 19–33, 2023.
https://doi.org/10.1007/978-3-031-46661-8_2

diagnosis [21], and brain-computer interfaces [5]. A typical multivariate time series contains a sequence of data points at regular time intervals, where values of multiple variables or measurements from multiple sensors exist at the same time points. Compared with traditional univariate time series classification, multivariate time series classification is inherently more challenging due to the temporal variations and correlations among multiple variables. It has thus attracted the increasing attention of researchers as a sheer amount of time series data are collected by sensors in the era of Industry 4.0 [28].

While traditional methods for multivariate time series classification have been based on statistical or machine learning methods, deep learning-based methods, represented by Long Short-Term Memory (LSTM) [14], Inception-time [9], and Time Series Transformer (TST) [42] have gained prevalence recently thanks to their outstanding capability to extract effective features and learn representation in complex scenarios. Until now, all the existing approaches have been focusing on the temporal information of multivariate time series data while disregarding the underlying frequency information, which is proven invaluable in many domains like signal processing [23]. Intuitively, real-world multivariate time series data often exhibit periodicity that is challenging to detect and model from a purely temporal perspective. This highlights the necessity of incorporating frequency information into the classifier to model and classify multivariate time series data accurately. All the above inspires us to develop a novel approach that can leverage temporal and frequency information comprehensively for more accurate multivariate time series classification.

Existing methods that extract frequency information from time series data generally aim for time series forecasting, represented by ETSformer [35] and COST [34]. These methods are commonly based on Fourier Transform [2], which decomposes time series into a set of sine functions representing different frequencies, with the amplitude of each sine function indicating the intensities of the frequency components. Fourier Transform, however, can only observe time series' global frequency components without their temporal positions, resulting in insufficient frequency information that limits the accuracy of multivariate time series classification. Therefore, it calls for new approaches that can incorporate more comprehensive frequency information to improve classification performance.

In light of the above, we aim to classify multivariate time series sequences by leveraging both temporal and frequency information. Specifically, we adopt the Short-Time Fourier Transform (STFT) [11] to address the limitations of the Fourier Transform. STFT divides time series into overlapping segments, applies a Fourier transform to each segment, and finally concatenates the resulting 2D frequency domain representations to provide more comprehensive information that covers both the frequency components and their temporal positions. In particular, we use three different window sizes to generate spectrograms for each variable; these spectrograms carry multi-resolution frequency information that reflects multi-scale temporal patterns of time series which is crucial for modeling time series [3]. Through the above transformation, we create a new data modality and transform the time series classification task into a multi-modality classification task. This further allows us to bring computer vision backbones into time series classification, which have shown effectiveness in exploiting 2D representations [36]. We further construct a dual-stream architecture based on ResNet [13], a widely used computer vision method, to leverage the power of both 2D representations (with frequency information) and 1D representations (with temporal information) of time series. The combination of 2D and 1D representations enables us to classify time series effectively, demonstrated by our proposed method consistently outperforming state-of-the-art baselines on 30 public multivariate time series datasets.

In a nutshell, we make the following contributions in this paper:

- We employ Short-Time Fourier Transform (STFT) with varying window sizes to generate 2D representations containing frequency components and corresponding temporal positions at multiple resolutions.
- We propose a dual-stream architecture based on ResNet to leverage both temporal and frequency information. A fully-connected layer with softmax function takes the fusion of the output feature maps from two streams to map them to a probability distribution of classes.
- We conducted extensive experiments on 30 public datasets and demonstrated the superior performance of our method to state-of-the-art baselines. We offer further insights by investigating convolutional backbone selection sensitivity, impact as a plugin, and ablation studies.

2   Related Work

2.1   Traditional Machine Learning Methods

Statistical and traditional machine learning methods have been extensively employed for multivariate time series classification. Distance-based approaches, such as k-Nearest Neighbors (KNN) [32] combined with Dynamic Time Warping (DTW) [32], as well as feature-based methods, including Support Vector Machine (SVM) [43], TS-CHIEF [26], HIVE-COTE [19], and ROCKET [7], have been used. However, these methods typically rely on manually-crafted features and face difficulties in capturing complex relationships efficiently from high-dimensional data [1].

2.2   Deep Learning Methods

Deep learning methods are prevalent for multivariate time series classification due to the ability to capture high-dimensional non-linear relationships [17]. Convolutional Neural Networks (CNNs) [20, 46] are used to capture local temporal variations, while variants include Inception-time [9], Attentional Gated Res2Net [39], and OS-CNN [30]. As CNN lacks the ability to capture long-range dependencies, Recurrent Neural Networks (RNNs) [27] that can memorize the temporal patterns are used to classify multivariate time series, while variants include Long Short-Term Memory (LSTM) [14] and Gated Recurrent Unit (GRU) [4]. Ensemble models of RNN and CNN such as LSTM-FCN [15] and

CNN-RNN Cascade model [38] incorporate both of them to exploit the CNN's ability to harness local temporal information and RNN's ability to leverage long-range dependencies for multivariate time series classification. Transformer [33] is a recently proposed method in natural language processing [8] that realizes parallel computation and multi-scale temporal information utilization, making it competitive on various tasks. Variants that are designed for multivariate time series classification include Time Series Transformer (TST) [42], Gated Transformer [22], and AutoTransformers [24]. However, all the existing methods for multivariate time series classification only focus on temporal information, ignoring the time series's inherent frequency information, which limits the capacity to classify various time series sequences.

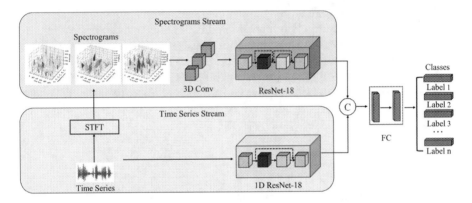

Fig. 1. The architecture of the proposed method. We use the time series with one variable to illustrate our method for simplicity. We employ Short-Time Fourier Transform with three different window sizes to generate a set of 2D spectrograms with multiple resolutions. We construct a dual-stream architecture based on ResNet to leverage both 1D representations and 2D representations. In the spectrogram stream, we use a 3D convolutional layer to fuse the spectrogram information from the resolution perspective and feed the output to ResNet-18. In the time series stream, we follow the architecture of ResNet-18 while replacing the 2D convolutional kernels using 1D convolutional kernels to adapt the shape of time series data. Finally, the output feature maps from two streams are concatenated (C in this Figure means concatenation) and fed into a fully-connected layer to map the output to the probability distribution of the classes.

3   Methodology

The proposed method is based on a dual-stream architecture consisting of a spectrogram stream and a time series stream, as illustrated in Fig. 1. We first implement the STFT using three different window sizes to generate a set of 2D spectrograms with varying temporal and frequency resolutions. Following this, a 3D convolutional layer is utilized to fuse the resolution-wise information of the spectrograms, while the output is fed into a ResNet-18 network to leverage 2D representations. Concurrently, the time series data is fed into a 1D ResNet-18 network that leverages 1D representations using 1D convolutional kernels. The output feature maps of both streams are concatenated, and a fully-connected layer with softmax function is applied to map the output to the probability distribution of the classes. We elaborate on each component of the proposed method in the following sections.

3.1   Short-Time Fourier Transform

Real-world time series data are typically sampled from continuous data streams at specific sampling rates. In signal processing, the Discrete Fourier Transform (DFT) is commonly used to extract frequency components from time series data, which can be described as follows:

$$X(k) = \sum_{t=0}^{T-1} x(t)e^{-i2\pi kt/T} \tag{1}$$

where x_t is the time series sequence, and $t \in (0, T-1)$, T is the length of the time series. $X(k)$ is the frequency component obtained after DFT, while k is the index. However, the DFT lacks temporal position information of the frequency components, resulting in insufficient frequency information. To address this limitation, the Short-Time Fourier Transform (STFT) is performed, which involves using a sliding window to divide a time series sequence into short time intervals and performing the Fourier Transform on each interval to obtain the frequency components and their temporal positions. The STFT can be described as:

$$X(j, \omega) = \sum_{t=0}^{L-1} x(t)w(j-t)e^{-i\omega t} \tag{2}$$

where $x(t)$ represents the input time series in the time domain, $w(j-t)$ represents truncating the time series $x(t)$ with a window function in time to obtain the short time interval $x(t)w(j-t)$, L is the length of the window, j represents the center position of the current window, and ω represents the frequency of interest. In this case, STFT provides more sufficient frequency information including the frequency components and their temporal positions compared with the Fourier Transform. We applied the Fourier Transform and STFT to a sequence sampled from the **Handwriting** dataset to illustrate the differences between the spectrograms obtained through the Fourier Transform and STFT, and the results are shown in Fig. 2.

The STFT requires a balance between frequency and temporal resolutions, which presents a challenge in selecting an optimal window size. A larger window size provides more precise frequency information but results in poorer temporal resolution, while a smaller window size provides better temporal resolution but less precise frequency information. We follow a traditional signal processing approach [12] to address this issue, where the window size is chosen based on

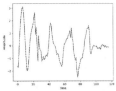
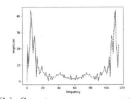

(a) Time Domain   (b) Spectrogram generated   (c) Spectrogram generated
                    by Fourier Transform   by STFT

Fig. 2. The time domain of a time series sampled from the Handwriting dataset and the spectrograms generated by the Fourier Transform and Short-Time Fourier Transform (STFT). The spectrogram generated by the STFT provides more comprehensive frequency information, including both the frequency components and their temporal positions, in contrast to the spectrogram generated by the Fourier Transform

the time series's bandwidth. To select an appropriate window size, we calculate the maximum bandwidth among all variables, which can be described as:

$$\text{Bandwidth }_n = \lceil f_{\max}^n - f_{\min}^n \rceil$$
$$\text{Bandwidth } = \max(\text{ Bandwidth }_0, \text{ Bandwidth }_1, \ldots, \text{ Bandwidth }_N)$$

(3)

where f_{max}^n and f_{min}^n are the maximum frequency and the minimum frequency present in the time series's nth variable, respectively, and N is the variable number. We then use three window sizes: the two, three, and four times the time series's frequency bandwidth, respectively, with an overlap of 50%, to generate three spectrograms with multi-level resolutions. For a time series sequence $x \in \mathbb{R}^{N \times T}$, where N is the variable number and T is the sequence length, the corresponding spectrogram generated by STFT is $s \in \mathbb{R}^{N \times 3 \times H \times W}$ where 3 means three different window sizes that we use, and H and W are the spectrogram's height and width. In this way, we extract the frequency components and their temporal positions from the time series and create a new data modality by converting the 1D time series sequence into a set of 2D representations, enabling us to borrow the powerful backbones from the computer vision field for leveraging 2D representation.

3.2   ResNet-18

We propose a dual-stream architecture based on ResNet [13] to leverage both the 2D representations in the spectrogram stream and the 1D representations in the time series stream for representation learning. ResNet is a popular deep neural network architecture that addresses the issue of vanishing gradients, which arises when the gradients become too small to effectively update the weights during backpropagation, particularly in very deep networks. This property has made it a competitive backbone for various computer vision tasks, motivating us to adopt it in our approach. In the spectrogram stream, the sets of 2D representations generated by the STFT are first fed into a 3D convolutional layer for resolution-wise information fusion. This layer down-samples the input spectrograms from the resolution perspective and generates a single 2D representation for each variable. The calculation process can be described as:

$$y = W * x + b \tag{4}$$

where $x \in \mathbb{R}^{N \times 3 \times H \times W}$ and $y \in \mathbb{R}^{N \times H \times W}$, and W and b are the convolutional kernel and bias term, respectively.

The resulting 2D representation and the original 1D time series are then fed into two separate neural networks, namely ResNet-18 and 1D ResNet-18, respectively. The architecture of ResNet-18, illustrated in Fig. 3, comprises six residual blocks, each consisting of two convolutional layers with a kernel size 3×3. The output feature maps are fed into an average pooling layer for downsampling from the spatial perspective, generating the latent vector of the input feature maps. In the 1D ResNet-18, we follow the same architecture as ResNet-18 but replace the 2D convolutional kernels with 1D convolutional kernels to accommodate the shape of the 1D representations. This enables us to process the time series data while retaining the advantages of ResNet-18's architecture. We then concatenate the output of the two streams and feed them into the fully-connected layer with a softmax function to map them to the probability distribution of the classes.

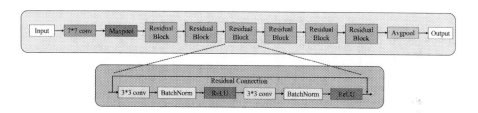

Fig. 3. The architecture of ResNet-18.

4 Experiments

4.1 Datasets

We evaluated our method using the UEA Time Series Classification Repository [6], which contains 30 public multivariate time series datasets. These datasets concern different domains and reflect diverse data characteristics in terms of sequence lengths and variable numbers, etc. All datasets had been preprocessed and split into training and test sets. The detailed statistics of each dataset are summarized in Table 1.

We further normalized them to zero mean and unit standard deviation and applied zero paddings to ensure that each dataset contains sequences of the same lengths.

Table 1. Statistics of the 30 UEA datasets used in experimentation.

Dataset	Train Cases	Test Cases	Dimensions	Length	Classes
ArticularyWordRecognition	275	300	9	144	25
AtrialFibrillation	15	15	2	640	3
BasicMotions	40	40	6	100	4
CharacterTrajectories	1,422	1,436	3	182	20
Cricket	108	72	6	1,197	12
DuckDuckGeese	60	40	1345	270	5
EigenWorms	128	131	6	17,984	5
Epilepsy	137	138	3	206	4
EthanolConcentration	261	263	3	1,751	4
ERing	30	30	4	65	6
FaceDetection	5,890	3,524	144	62	2
FingerMovements	316	100	28	50	2
HandMovementDirection	320	147	10	400	4
Handwriting	150	850	3	152	26
Heartbeat	204	205	61	405	2
JapaneseVowels	270	370	12	29	9
Libras	180	180	2	45	15
LSST	2,459	2,466	6	36	14
InsectWingbeat	30,000	20,000	200	78	10
MotorImagery	278	100	64	3,000	2
NATOPS	180	180	24	51	6
PenDigits	7,494	3,498	2	8	10
PEMS-SF	267	173	963	144	7
Phoneme	3,315	3,353	11	217	39
RacketSports	151	152	6	30	4
SelfRegulationSCP1	268	293	6	896	2
SelfRegulationSCP2	200	180	7	1,152	2
SpokenArabicDigits	6,599	2199	13	93	10
StandWalkJump	12	15	4	2,500	3
UWaveGestureLibrary	120	320	3	315	8

4.2 Baselines

We consider several popular machine learning methods and recently proposed deep learning models as baselines. The selected competitive methods include ROCKET [7], Time Series Transformer (TST) [42], ShapeNet [18], Dynamic Time Warping (DTW), TS2Vec [41], MLSTM-FCN [15], OS-CNN [30], TapNet [45], Temporal Neighborhood Coding (TNC) [31], and WEASEL+ MUSE [25].

4.3 Model Configuration and Evaluation Metric

We trained our model for 500 training epochs using Adam [16] optimizer. The learning rate is initialized to 0.001; it scales down with a coefficient of 0.1 every

50 epochs after the first 100 epochs. We repeated the training and test processes five times and took the average of multiple runs as the final results to mitigate the impact of randomized parameter initialization. We used dropout to avoid possible overfitting. Training and testing are done on a single Nvidia GTX 3080 Ti.

We use *accuracy*, which is currently used by all baseline methods, as the metric for comparison. We additionally use macro *precision*, *recall*, and *F1-Score* in our parameter and ablation studies to gain further insights into our model's performance.

Table 2. Accuracy of different models on 30 benchmark datasets. The best performance values are bolded, and the second-best performance values are underlined.

Dataset	Ours	WEASEL+MUSE	TST	ROCKET	DTW	TS2Vec	MLSTM-FCN	OS-CNN	TapNet	TNC	ShapeNet
ArticularyWordRecognition	**0.996**	0.990	0.977	**0.996**	0.987	0.987	0.973	0.988	0.987	0.973	0.987
AtrialFibrillation	**0.524**	0.333	0.067	0.249	0.200	0.200	0.267	0.233	0.333	0.133	0.400
BasicMotions	**1.000**	**1.000**	0.975	0.990	0.975	0.975	0.950	**1.000**	**1.000**	0.975	**1.000**
CharacterTrajectories	0.997	0.990	0.975	0.967	0.989	0.995	0.985	0.998	0.997	0.967	0.980
Cricket	**1.000**	**1.000**	**1.000**	**1.000**	**1.000**	0.972	0.917	0.993	0.958	0.958	0.986
DuckDuckGeese	**0.767**	0.575	0.562	0.461	0.492	0.680	0.675	0.540	0.575	0.460	0.725
EigenWorms	**0.897**	0.890	0.748	0.863	0.618	0.847	0.504	0.414	0.489	0.840	0.878
Epilepsy	**1.000**	**1.000**	0.949	0.991	0.964	0.964	0.761	0.980	0.971	0.957	0.987
Ering	0.875	0.133	**0.964**	0.447	0.133	0.874	0.133	0.881	0.133	0.852	0.133
EthanolConcentration	**0.476**	0.430	0.326	0.452	0.323	0.308	0.373	0.240	0.323	0.297	0.312
FaceDetection	**0.683**	0.545	0.681	0.647	0.529	0.501	0.545	0.575	0.556	0.536	0.602
FingerMovements	**0.601**	0.490	0.560	0.553	0.530	0.480	0.580	0.568	0.530	0.470	0.580
HandMovementDirection	0.443	0.365	0.243	**0.446**	0.231	0.338	0.365	0.443	0.378	0.324	0.338
HandWriting	**0.672**	0.605	0.359	0.567	0.286	0.515	0.286	0.668	0.357	0.249	0.451
HeartBeat	**0.863**	0.727	0.776	0.717	0.717	0.515	0.663	0.489	0.751	0.746	0.756
JapaneseVowels	0.967	0.984	**0.994**	0.962	0.949	0.984	0.976	0.991	0.965	0.978	0.984
Libras	**0.981**	0.973	0.656	0.906	0.870	0.867	0.856	0.950	0.850	0.817	0.856
LSST	0.782	**0.878**	0.408	0.632	0.551	0.537	0.373	0.413	0.568	0.595	0.590
MotorImagery	**0.632**	0.590	0.500	0.531	0.500	0.510	0.510	0.535	0.590	0.500	0.610
NATOPS	0.941	0.500	0.850	0.885	0.883	0.928	0.889	**0.968**	0.939	0.911	0.883
PEMS-SF	**0.932**	0.870	0.919	0.751	0.711	0.682	0.699	0.760	0.751	0.699	0.751
PenDigits	0.991	0.968	0.560	**0.996**	0.977	0.989	0.978	0.985	0.980	0.979	0.977
Phoneme	0.287	0.190	0.085	0.284	0.151	0.233	0.110	**0.299**	0.175	0.207	0.298
RacketSports	**0.934**	0.190	0.809	0.928	0.803	0.855	0.803	0.877	0.868	0.776	0.882
SelfRegulationSCP1	**0.961**	0.934	0.925	0.908	0.775	0.812	0.874	0.835	0.652	0.799	0.782
SelfRegulationSCP2	**0.738**	0.710	0.589	0.533	0.539	0.578	0.472	0.532	0.550	0.550	0.578
SpokenArabicDigits	0.994	0.460	0.993	0.712	0.963	0.988	0.990	**0.997**	0.983	0.934	0.975
StandWalkJump	**0.659**	0.333	0.267	0.456	0.200	0.467	0.067	0.383	0.400	0.400	0.533
UWaveGestureLibrary	**0.951**	0.916	0.903	0.944	0.903	0.906	0.891	0.927	0.894	0.759	0.906
InsectWingBeat	**0.697**	0.163	0.105	0.168	0.105	0.466	0.167	0.667	0.208	0.469	0.250
Average Accuracy	**0.808**	0.658	0.658	0.698	0.628	0.698	0.621	0.704	0.657	0.670	0.699
Average Rank	**1.57**	4.93	6.63	4.93	7.83	6.17	7.77	4.77	6.07	8.00	4.83

4.4   Comparison Results

The performance comparison results (shown in Table 2) reveal that our model has demonstrated superior performance to all the baseline methods across a wide range of experimental datasets. Specifically, our model achieved the best results on 21 datasets, the second-best performance on six datasets, and the third-best on two datasets out of 30 experimental datasets. It demonstrated superior performance compared to all baselines, achieving a 14.7% increase in average classification accuracy compared to the second-best method, OS-CNN, and a 15.6% increase compared to the third-best method, ShapeNet. Furthermore, our model achieved an average rank of 1.57, outperforming the second-best method, OS-CNN, which had an average rank of 4.77. Figure 4 shows the result of the Wilcoxon signed-rank test (with a confidence level of 95%) on the baseline methods' performance, consistently showing that our method achieved the highest classification performance among all the compared methods.

Traditional machine learning methods, including WEASEL+MUSE, DTW, and ROCKET, are limited in handling such large datasets, reflected in their inferior performance on datasets including InsectWingBeat and FaceDetection, which contain 50,000 and 9,114 samples, respectively. Furthermore, existing deep learning models often ignore the inherent frequency information in time series data, which can be crucial for accurately classifying time series with significant differences in the frequency domain rather than the time domain.

We attribute this improvement to two key factors. First, our method's dual-stream architecture effectively captures both temporal and frequency information, enhancing its ability to discriminate time series sequences between different classes. Second, by utilizing the Short-Time Fourier Transform (STFT), our method leverages the frequency components and their temporal locations of the time series to provide more comprehensive frequency information compared to the Fourier Transform. Our results from the Wilcoxon signed-rank test, conducted with a confidence level of 95%, further confirm that our method achieved the best classification performance among all compared methods.

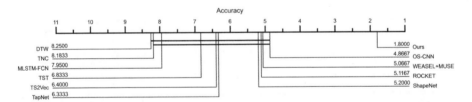

Fig. 4. Critical Difference (CD) diagram of the selected baselines and our method with a confidence level of 95%.

4.5   Convolutional Backbone Selection Sensitivity

We replaced the ResNet with other popular computer vision backbones including ResNeXt [37], Res2Net [10], ResNeSt [44], and Inception [29] to explore the impact of the backbone selection on the performance. We conducted experiments on three datasets including DuckDuckGeese, HeartBeat, and HandWriting. The results can be found in Table 3. The tested backbones have more complex architectures and parameters compared to ResNet, leading to better performance

Table 3. The Training and test results of different backbones from the computer vision field. The best performance values are bolded.

Dataset	Models	Training				Test			
		Accuracy	Recall	Precision	F1-Score	Accuracy	Recall	Precision	F1-Score
DuckDuckGeese	ResNet	0.862	0.813	0.764	0.788	**0.767**	**0.741**	0.736	**0.738**
	ResNeXt	0.866	0.809	**0.866**	0.837	0.673	0.645	0.639	0.642
	Res2Net	0.894	0.815	0.827	0.821	0.639	0.613	0.632	0.622
	ResNeSt	**0.907**	**0.915**	0.855	**0.884**	0.692	0.704	**0.761**	0.731
	Inception	0.859	0.897	0.811	0.852	0.734	0.729	0.736	0.732
HeartBeat	ResNet	0.906	0.891	0.882	0.886	0.863	0.772	0.795	0.783
	ResNeXt	0.916	0.902	0.914	0.908	0.741	0.726	0.719	0.722
	Res2Net	**0.931**	**0.919**	0.922	0.920	0.714	0.678	0.669	0.673
	ResNeSt	0.928	0.917	**0.927**	**0.922**	0.665	0.640	0.608	0.624
	Inception	0.909	0.874	0.907	0.890	**0.782**	**0.738**	**0.806**	**0.771**
HandWriting	ResNet	0.735	0.702	0.744	0.722	**0.672**	**0.654**	**0.661**	**0.657**
	ResNeXt	0.849	0.865	0.872	0.868	0.533	0.542	0.591	0.565
	Res2Net	0.856	0.802	0.874	0.836	0.592	0.607	0.586	0.596
	ResNeSt	**0.857**	**0.886**	**0.883**	**0.884**	0.557	0.573	0.605	0.589
	Inception	0.764	0.753	0.773	0.763	0.597	0.612	0.596	0.604

Table 4. The experimental results when using our frequency stream with 2D representations as a plugin. W/o means that the method does not contain the spectrogram stream and vice versa. The best performance values are bolded.

Dataset	Method	Train				Test			
		Accuracy	Recall	Precision	F1-Score	Accuracy	Recall	Precision	F1-Score
EigenWorms	MLSTM-FCN (w/o)	0.587	0.574	0.624	0.598	0.504	0.519	0.479	0.498
	MLSTM-FCN (w)	**0.721**	**0.714**	**0.677**	**0.695**	**0.629**	**0.624**	**0.595**	**0.609**
	TST (w/o)	0.839	0.832	0.816	0.824	0.748	0.791	0.778	0.784
	TST (w)	**0.882**	**0.893**	**0.885**	**0.889**	**0.826**	**0.828**	**0.819**	**0.823**
RacketSports	MLSTM-FCN (w/o)	0.828	0.779	**0.833**	0.805	0.803	0.709	0.702	0.705
	MLSTM-FCN (w)	**0.843**	**0.811**	0.805	**0.808**	**0.814**	**0.727**	**0.751**	**0.739**
	TST (w/o)	0.854	0.819	0.822	0.820	0.809	0.712	0.705	0.708
	TST (w)	**0.894**	**0.833**	**0.882**	**0.857**	**0.824**	**0.762**	**0.793**	**0.777**

during the training phase but overfitting on the test sets. We believe that with the increase of the dataset scale, implementing more complicated backbones may enhance the classifier's classification capacity. As most of the datasets we use contain limited samples in the training set (fewer than 1000), we selected ResNet as the optimal solution based on our evaluation of the performance metrics.

4.6 Impact of Our Spectrogram Stream as a Plugin

We incorporate the spectrogram stream as a plugin into the existing architectures including TST [42] and MLSTM-FCN [15] to evaluate the effectiveness of the 2D representations with frequency information in improving the performance of the existing methods. The outcomes of our investigation, as presented in Table 4, indicate a significant improvement in the average classification accuracy and the F1-Score of both methods during both the training and testing phases. Specifically, we observed an increase of 8.6% and 7.3% in the average classification accuracy and F1-Score, respectively, during the training phase, and an increase of 9.6% and 10.4% in the average classification accuracy and F1-Score, respectively, during the test phase. These findings suggest that the utilization of 2D representations with frequency information can enhance the performance of existing methods.

Table 5. Ablation test for our method. Fourier Transform means we use Fourier Transform instead of STFT to extract frequency information. Single window size means we only use one window size (three times the bandwidth) to generate the spectrogram. Time Series and Spectrogram Stream only mean using information from one stream separately instead of both to classify time series. The best performance values are bolded.

Dataset	Model	Accuracy	Precision	Recall	F1-Score
DuckDuckGeese	Fourier Transform	0.675	0.669	0.688	0.678
	Single Window Size	0.689	0.707	0.725	0.716
	Time Series Stream Only	0.632	0.619	0.661	0.639
	Spectrogram Stream Only	0.718	0.711	0.724	0.717
	Ours	**0.767**	**0.741**	**0.736**	**0.738**
FaceDetection	Fourier Transform	0.575	0.602	0.552	0.576
	Single Window Size	0.627	0.585	0.673	0.626
	Time Series Stream Only	0.630	0.615	0.622	0.618
	Spectrogram Stream Only	0.647	0.651	0.642	0.646
	Ours	**0.681**	**0.622**	**0.716**	**0.666**
PEMS-SF	Fourier Transform	0.751	0.643	0.637	0.640
	Single Window Size	0.794	0.718	0.698	0.708
	Time Series Stream Only	0.819	0.822	0.803	0.812
	Spectrogram Stream Only	0.874	0.856	0.877	0.866
	Ours	**0.932**	**0.957**	**0.889**	**0.922**

4.7 Ablation Study

We conducted ablation studies on three datasets, including DuckDuckGeese, FaceDetection, and PEMS-SF, to investigate the effectiveness of individual components of our proposed method. We compared the performance of the method with the use of Fourier Transform instead of STFT to extract frequency information. Besides, for STFT, we use a single window size (three times the bandwidth) for spectrogram generation instead of three window sizes. Additionally, we tried to use information from one single stream (either the time series or spectrogram stream) individually to classify time series instead of both. The experimental results are summarized in Table 5.

Our analysis reveals that each component improves the classifier's performance. Notably, STFT demonstrates a more significant impact on the classification accuracy of the model on two of the datasets. This finding implies that the utilization of 2D representations with frequency information, provided by STFT, is crucial for enhancing the classification capacity of the model.

5   Conclusion and Future Work

This study proposes a novel dual-stream architecture for accurately classifying multivariate time series sequences. The method leverages the inherent frequency information in the time series data by implementing STFT to obtain the frequency components and their temporal positions. We construct a dual-stream architecture based on ResNet, which can leverage both 1D and 2D representations effectively to classify multivariate time series sequences. We evaluate the proposed model on diverse datasets containing sequences of various lengths and variable numbers. The experimental results show that our method outperforms several baseline and state-of-the-art methods by a significant margin. We also conduct a thorough investigation of the effect of different components and settings on the model's performance. Our future work includes exploring the interpretability of our proposed method through visualization technologies for convolutional neural networks from the computer vision field. Additionally, we plan to extend our work to more time series-related tasks, such as time series imputation, forecasting, and abnormal detection.

References

1. Bengio, Y., LeCun, Y., et al.: Scaling learning algorithms towards AI. Large-Scale Kernel Mach. **34**(5), 1–41 (2007)
2. Bracewell, R.N., Bracewell, R.N.: The Fourier Transform and its Applications, vol. 31999. McGraw-Hill, New York (1986)
3. Chen, Z., Ma, Q., Lin, Z.: Time-aware multi-scale RNNs for time series modeling. In: IJCAI, pp. 2285–2291 (2021)
4. Cho, K., et al.: Learning phrase representations using RNN encoder-decoder for statistical machine translation. arXiv preprint arXiv:1406.1078 (2014)
5. Coyle, D., Prasad, G., McGinnity, T.M.: A time-series prediction approach for feature extraction in a brain-computer interface. IEEE Trans. Neural Syst. Rehabil. Eng. **13**(4), 461–467 (2005)
6. Dau, H.A., et al.: Hexagon-ML: the UCR time series classification archive (2018)
7. Dempster, A., Petitjean, F., Webb, G.I.: Rocket: exceptionally fast and accurate time series classification using random convolutional kernels. Data Min. Knowl. Disc. **34**(5), 1454–1495 (2020)
8. Devlin, J., Chang, M.W., Lee, K., Toutanova, K.: Bert: pre-training of deep bidirectional transformers for language understanding. arXiv preprint arXiv:1810.04805 (2018)
9. Fawaz, H.I., et al.: Inceptiontime: finding alexnet for time series classification. Data Min. Knowl. Disc. **34**(6), 1936–1962 (2020)

10. Gao, S., Cheng, M.M., Zhao, K., Zhang, X.Y., Yang, M.H., Torr, P.H.: Res2net: a new multi-scale backbone architecture. IEEE Trans. Pattern Anal. Mach. Intell. **43**, 652–662 (2019)
11. Griffin, D., Lim, J.: Signal estimation from modified short-time Fourier transform. IEEE Trans. Acoust. Speech Signal Process. **32**(2), 236–243 (1984)
12. Harris, F.J.: On the use of windows for harmonic analysis with the discrete Fourier transform. Proc. IEEE **66**(1), 51–83 (1978)
13. He, K., Zhang, X., Ren, S., Sun, J.: Deep residual learning for image recognition. In: Proceedings of the IEEE Conference on Computer Vision and Pattern Recognition, pp. 770–778 (2016)
14. Hochreiter, S., Schmidhuber, J.: Long short-term memory. Neural Comput. **9**(8), 1735–1780 (1997)
15. Karim, F., Majumdar, S., Darabi, H., Harford, S.: Multivariate LSTM-FCNS for time series classification. Neural Netw. **116**, 237–245 (2019)
16. Kingma, D.P., Ba, J.: Adam: a method for stochastic optimization. arXiv preprint arXiv:1412.6980 (2014)
17. LeCun, Y., Bengio, Y., Hinton, G.: Deep learning. Nature **521**(7553), 436–444 (2015)
18. Li, G., Choi, B., Xu, J., Bhowmick, S.S., Chun, K.P., Wong, G.L.H.: Shapenet: a shapelet-neural network approach for multivariate time series classification. In: Proceedings of the AAAI Conference on Artificial Intelligence, vol. 35, pp. 8375–8383 (2021)
19. Lines, J., Taylor, S., Bagnall, A.: Time series classification with hive-cote: the hierarchical vote collective of transformation-based ensembles. ACM Trans. Knowl. Disc. Data **12**(5), 1–35 (2018)
20. Liu, C.L., Hsaio, W.H., Tu, Y.C.: Time series classification with multivariate convolutional neural network. IEEE Trans. Ind. Electron. **66**(6), 4788–4797 (2018)
21. Liu, M., Kim, Y.: Classification of heart diseases based on ECG signals using long short-term memory. In: 2018 40th Annual International Conference of the IEEE Engineering in Medicine and Biology Society (EMBC), pp. 2707–2710. IEEE (2018)
22. Liu, M., et al.: Gated transformer networks for multivariate time series classification. arXiv preprint arXiv:2103.14438 (2021)
23. Purwins, H., Li, B., Virtanen, T., Schlüter, J., Chang, S.Y., Sainath, T.: Deep learning for audio signal processing. IEEE J. Sel. Topics Signal Process. **13**(2), 206–219 (2019)
24. Ren, Y., Li, L., Yang, X., Zhou, J.: Autotransformer: automatic transformer architecture design for time series classification. In: Advances in Knowledge Discovery and Data Mining: 26th Pacific-Asia Conference, PAKDD 2022, Chengdu, China, 16–19 May 2022, Proceedings, Part I, pp. 143–155. Springer, Heidelberg (2022). https://doi.org/10.1007/978-3-031-05933-9_12
25. Schäfer, P., Leser, U.: Multivariate time series classification with weasel+ muse. arXiv preprint arXiv:1711.11343 (2017)
26. Shifaz, A., Pelletier, C., Petitjean, F., Webb, G.I.: TS-chief: a scalable and accurate forest algorithm for time series classification. Data Min. Knowl. Disc. **34**(3), 742–775 (2020)
27. Smirnov, D., Nguifo, E.M.: Time series classification with recurrent neural networks. Adv. Anal. Learn. Temp. Data **8** (2018)
28. Spiegel, S., Gaebler, J., Lommatzsch, A., De Luca, E., Albayrak, S.: Pattern recognition and classification for multivariate time series. In: Proceedings of the Fifth International Workshop on Knowledge Discovery from Sensor Data, pp. 34–42 (2011)

29. Szegedy, C., Vanhoucke, V., Ioffe, S., Shlens, J., Wojna, Z.: Rethinking the inception architecture for computer vision. In: Proceedings of the IEEE Conference on Computer Vision and Pattern Recognition, pp. 2818–2826 (2016)
30. Tang, W., Long, G., Liu, L., Zhou, T., Blumenstein, M., Jiang, J.: Omni-scale cnns: a simple and effective kernel size configuration for time series classification. In: International Conference on Learning Representations (2021)
31. Tonekaboni, S., Eytan, D., Goldenberg, A.: Unsupervised representation learning for time series with temporal neighborhood coding. In: International Conference on Learning Representations (2020)
32. Tran, T.M., Le, X.M.T., Nguyen, H.T., Huynh, V.N.: A novel non-parametric method for time series classification based on k-nearest neighbors and dynamic time warping barycenter averaging. Eng. Appl. Artif. Intell. **78**, 173–185 (2019)
33. Vaswani, A., et al.: Attention is all you need. In: Advances in Neural Information Processing Systems, pp. 5998–6008 (2017)
34. Woo, G., Liu, C., Sahoo, D., Kumar, A., Hoi, S.: Cost: contrastive learning of disentangled seasonal-trend representations for time series forecasting. arXiv preprint arXiv:2202.01575 (2022)
35. Woo, G., Liu, C., Sahoo, D., Kumar, A., Hoi, S.: Etsformer: exponential smoothing transformers for time-series forecasting. arXiv preprint arXiv:2202.01381 (2022)
36. Wu, H., Hu, T., Liu, Y., Zhou, H., Wang, J., Long, M.: Timesnet: temporal 2d-variation modeling for general time series analysis. arXiv preprint arXiv:2210.02186 (2022)
37. Xie, S., Girshick, R., Dollár, P., Tu, Z., He, K.: Aggregated residual transformations for deep neural networks. In: Proceedings of the IEEE Conference on Computer Vision and Pattern Recognition, pp. 1492–1500 (2017)
38. Yang, C., Jiang, W., Guo, Z.: Time series data classification based on dual path CNN-RNN cascade network. IEEE Access **7**, 155304–155312 (2019)
39. Yang, C., Wang, X., Yao, L., Long, G., Jiang, J., Xu, G.: Attentional gated res2net for multivariate time series classification. Neural Process. Lett. **55**, 1–25 (2022)
40. Yang, J., Nguyen, M.N., San, P.P., Li, X., Krishnaswamy, S.: Deep convolutional neural networks on multichannel time series for human activity recognition. In: IJCAI, Buenos Aires, Argentina, vol. 15, pp. 3995–4001 (2015)
41. Yue, Z., et al.: Ts2vec: towards universal representation of time series. arXiv preprint arXiv:2106.10466 (2021)
42. Zerveas, G., Jayaraman, S., Patel, D., Bhamidipaty, A., Eickhoff, C.: A transformer-based framework for multivariate time series representation learning. In: Proceedings of the 27th ACM SIGKDD Conference on Knowledge Discovery & Data Mining, pp. 2114–2124 (2021)
43. Zhang, D., Zuo, W., Zhang, D., Zhang, H.: Time series classification using support vector machine with gaussian elastic metric kernel. In: 2010 20th International Conference on Pattern Recognition, pp. 29–32. IEEE (2010)
44. Zhang, H., et al.: Resnest: split-attention networks. In: Proceedings of the IEEE/CVF Conference on Computer Vision and Pattern Recognition, pp. 2736–2746 (2022)
45. Zhang, X., Gao, Y., Lin, J., Lu, C.T.: Tapnet: multivariate time series classification with attentional prototypical network. In: Proceedings of the AAAI Conference on Artificial Intelligence, vol. 34, pp. 6845–6852 (2020)
46. Zhao, B., Lu, H., Chen, S., Liu, J., Wu, D.: Convolutional neural networks for time series classification. J. Syst. Eng. Electron. **28**(1), 162–169 (2017)

Exploring the Effectiveness of Positional Embedding on Transformer-Based Architectures for Multivariate Time Series Classification

Chao Yang[✉], Yakun Chen, Zihao Li, and Xianzhi Wang

School of Computer Science, University of Technology Sydney, Ultimo, Australia
chao.yang@student.uts.edu.au

Abstract. Positional embedding is an effective means of injecting position information into sequential data to make the vanilla Transformer position-sensitive. Current Transformer-based models routinely use positional embedding for their position-sensitive modules while no efforts are paid to evaluating its effectiveness in specific problems. In this paper, we explore the impact of positional embedding on the vanilla Transformer and six Transformer-based variants. Since multivariate time series classification requires distinguishing the differences between time series sequences with different labels, it risks causing performance degradation to inject the same content-irrelevant position token into all sequences. Our experiments on 30 public multivariate time series classification datasets show positional embedding positively impacts the vanilla Transformer's performance yet negatively impacts Transformer-based variants. Our findings reveal the varying effectiveness of positional embedding on different model architectures, highlighting the significance of using positional embedding cautiously in Transformer-based models.

Keywords: Positional embedding · Multivariate time series classification · Deep learning

1 Introduction

Multivariate time series classification plays a critical role in various fields, such as gesture recognition [29], disease diagnosis [15], and brain-computer interfaces [6]. Recent years have witnessed Transformer-based methods making remarkable breakthroughs in numerous disciplines, such as natural language processing [19,24], computer vision [1,9], and visual-audio speech recognition [20,22]. This success has inspired an increasing application of Transformer-based architectures [4,16,30] to multivariate time series classification. Besides, Transformer's ability to perform parallel computation and leverage long-range dependencies in sequential data make it especially suitable for modeling time series data [26].

Since Transformer is position-insensitive, positional embedding was introduced to allow the model to learn the relative position of tokens. Positional embedding generally injects position information into sequence data [23]. It takes

X. Yang et al. (Eds.): ADMA 2023, LNAI 14176, pp. 34–47, 2023.
https://doi.org/10.1007/978-3-031-46661-8_3

the form of sinusoidal functions of different frequencies, with each embedding dimension corresponding to a sinusoid whose wavelengths form a geometric progression. To date, positional embedding has been a routine for Transformer-based models that deal with multivariate time series classification problems.

Despite the widespread use, there have been debates [27,31] around the necessity of positional embedding, and a comprehensive investigation of positional embedding's effectiveness on various Transformer-based models is still to be developed. Firstly, Transformer-based models [14,33] that contain position-sensitive modules (e.g., convolutional and recurrent layers) can automatically learn the position information, making positional embedding redundant to some extent. This point is supported by studies in other fields [18,28] suggesting positional embedding may be unnecessary and replaced with position-sensitive layers. Secondly, positional embedding has inherent limitations that may potentially impair the classifier's performance. Since positional embedding is hand-crafted, it may bring inductive bias that may adversely impact the model's performance in some cases. While positional embedding injects the same position tokens into time series of different classes, it poses additional challenges to the classifier in figuring out the differences between sequences with different class labels.

In this paper, we explore the impact of positional embedding on various Transformer-based models to facilitate researchers and practitioners in making informed decisions on whether to incorporate positional embedding in their models for multivariate time series classification. In a nutshell, we make the following contributions in this paper:

- We comprehensively review existing Transformer-based models that contain position-sensitive layers and summarize them into six types of Transformer-based variants.
- We conduct extensive experiments on 30 multivariate time series classification datasets and evaluate the impact of positional embedding on the vanilla Transformer and Transformer-based variants.
- Our results show that positional embedding positively impacts the performance of the vanilla Transformer while negatively influencing the performance of the Transformer-based variants in multivariate time series classification.

2   Background

2.1   Positional Embedding

Positional embedding was first proposed for Transformer in [23], which uses fixed sine and cosine functions of different frequencies to represent the position information, as described below:

$$PE_{(pos,2i)} = \sin\left(pos/10000^{2i/d_{\text{model}}}\right)$$
$$PE_{(pos,2i+1)} = \cos\left(pos/10000^{2i/d_{\text{model}}}\right)$$

(1)

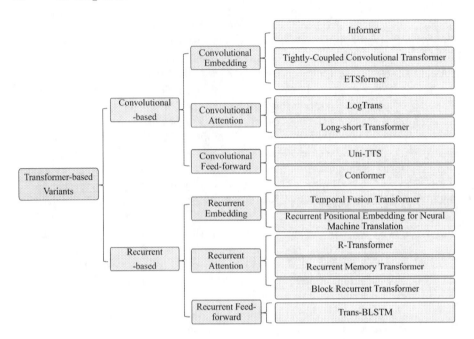

Fig. 1. A summary of Transformer-based variants for modeling sequential data.

where *pos* and *i* are the position and the dimension indices, respectively. d_{model} is the dimensionality of the input time series, where each dimension of thepositional embedding corresponds to a sinusoid. For any fixed offset k, $PE_{\text{pos}+k}$ is represented as a linear function of PE_{pos}; this enables the model to learn the relative positions easily. The positional embedding is then added to the time series as the input to Transformer.

Considering hand-crafted positional embedding is generally less expressive and adaptive [26], Time Series Transformer (TST) [32] enhances the vanilla Transformer [23] by implementing learnable positional embedding. Specifically, TST shares the same architecture as the vanilla Transformer, which stacks several basic blocks, each consisting of scaled dot-product multi-head attention and a feed-forward network (FFN) to leverage temporal data. But it differs in initializing the positional embedding using fixed values and then updating the embedding jointly with other model parameters through the training procedure.

2.2   Transformer-Based Variants

Current studies often incorporate convolutional or recurrent layers in the vanilla Transformer architecture in dealing with sequence-related tasks, including time series analysis. Figure 1 summarize these Transformer-based variants into six categories of representative methods, as detailed below.

– **Convolutional Embedding**: Methods in this category, namely Informer [33], Tightly-Coupled Convolutional Transformer (TCCT) [21], and ETSformer [27], implement a convolutional layer to obtain convolutional embeddings, which map the raw input sequences to a latent space before feeding them to the transformer block.

– **Convolutional Attention**: Instead of calculating the point-wise attention, LogTrans [13] and Long-short Transformer [34] use the convolutional layer to calculate the attention matrix (including queries, keys, and values) of segments to leverage the local temporal information.
– **Convolutional Feed-forward**: Uni-TTS [17] and Conformer [8] implement a convolutional layer after the multi-head attention as the feed-forward layer (or part of the feed-forward layer) to capture local temporal correlations.
– **Recurrent Embedding**: Temporal Fusion Transformer (TFT) [14] and the work in [3] use a recurrent layer to encode content-based order dependencies into the input sequence.
– **Recurrent Attention**: Recurrent Memory Transformer [2], Block Recurrent Transformer [11], and R-Transformer [25] use a recurrent neural net to calculate the attention matrix, which harnesses the temporal information more effectively when compared with the point-wise attention.
– **Recurrent Feed-forward**: Instead of point-wise feed-forward, TRANS-BLSTM [10] uses a recurrent layer after multi-head attention to harness non-linear temporal dependencies.

3 Problem Definition

A multivariate time series sequence can be described as: $X = \{x_1, x_2, \ldots x_T\}$, where $x_i \in \mathbb{R}^N$, $i \in \{1, 2, \cdots, T\}$, T is the maximum length of the sequence, and N is the number of variables. A dataset contains multiple (sequence, label) pairs and is denoted by $D = \{(X_1, y_1), (X_2, y_2), \ldots, (X_n, y_n)\}$, where each y_k denotes a label, $k \in \{1, 2, \cdots, n\}$. The objective of multivariate time series classification is to train a classifier to map the input sequences to probability distributions over the classes for the dataset D.

4 Methodology

We call TST [32] the **basic model**. To avoid having to compare all the related studies exhaustively, we design six Transformer-based variants based on the six types of techniques that are incorporated in the related studies (shown in Fig. 1), respectively. We further identify three configurable components of a Transformer architecture (shown in Fig. 2) as the input embedding layer (which projects the input time series into the latent space), the projection layer (which calculates the attention matrix), and the feed-forward layer (which leverages non-linear relationships). For each variant, we try different techniques (e.g., a Linear layer, a Convolutional layer, or a Gated Recurrent Unit) in each layer/component, as detailed in Table 1.

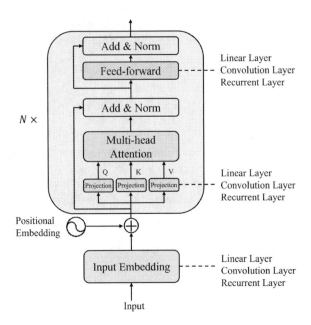

Fig. 2. A general architecture of transformer-based variants for modeling sequential data. The corresponding relations between configurable components and the respective candidate techniques are indicated by dashed lines.

4.1 Basic Model

The **basic model** adopts linear layers in all three components. In this case, for each sample $\mathbf{x_t} \in \mathbb{R}^M$: $\mathbf{X} \in \mathbb{R}^{M \times T} = [\mathbf{x}_1, \mathbf{x}_2, \dots, \mathbf{x_T}]$, where T is the sequence length and M is the variable number. The input embedding can be described as:

$$U_t = W^x x_t + b^x \tag{2}$$

where $t = 0, 1, \dots, T$ is the time stamp index, $W^x \in \mathbb{R}^{M \times d_k}$ and $b^x \in \mathbb{R}^{d_k}$ are learnable parameters. The projection layer can be described as:

$$\begin{aligned}
Q &= W^Q U_t + b^Q \\
K &= W^K U_t + b^K \\
V &= W^V U_t + b^V
\end{aligned} \tag{3}$$

where $W^Q \in \mathbb{R}^{d_k \times d_k}$, $W^K \in \mathbb{R}^{d_k \times d_k}$, $W^V \in \mathbb{R}^{d_k \times d_k}$, $b^Q \in \mathbb{R}^{d_k}$, $b^K \in \mathbb{R}^{d_k}$, and $b^V \in \mathbb{R}^{d_k}$ are are learnable parameters. We use standard scaled Dot-Product attention proposed in the vanilla Transformer [23] for self-attention calculation:

$$\text{Attention}(Q, K, V) = \text{softmax}\left(\frac{QK^T}{\sqrt{d_k}}\right) V. \tag{4}$$

The feed-forward layer can be described as:

$$FFN(x) = \text{ReLU}\left(W_1 x + b_1\right) W_2 + b_2 \tag{5}$$

Table 1. Configurations for the basic model and six variants. *ConvEmbedding* means Convolutional Embedding Variant; *RecEmbedding* means Recurrent Embedding Variant; the same naming rule applies to other models.

Model	Input Embedding	Projection	Feed-forward
Basic Model	Linear	Linear	Linear
ConvEmbedding	Convolutional Layer	Linear	Linear
ConvAttention	Linear	Convolutional Layer	Linear
ConvFFD	Linear	Linear	Convolutional Layer
RecEmbedding	Gated Recurrent Unit	Linear	Linear
RecAttention	Linear	Gated Recurrent Unit	Linear
RecFFD	Linear	Linear	Gated Recurrent Unit

where $W_1 \in \mathbb{R}^{d_k \times d_k}$, $W_2 \in \mathbb{R}^{d_k \times d_k}$, $b_1 \in \mathbb{R}^{d_k}$, and $b_2 \in \mathbb{R}^{d_k}$ are all leanable parameters.

4.2 Convolutional-Based Variants

We refer to the architectures that employ convolutional layers in any of the three components (input embedding layer, projection layer, or feed-forward layer) as convolutional-based variants. Here, we utilize a one-dimensional convolutional layer with a kernel size of 3. We also set the padding to 1 to preserve the lengths of representations. In the following, we illustrate our convolutional-based variants one by one.

Convolutional Embedding Variant replaces the linear layer with the convolution layer in the input embedding layer, which is formulated below:

$$U_t = W^x * x_t + b^x \tag{6}$$

where $*$ is the convolutional operation, $W^x \in \mathbb{R}^{M \times d_k \times P}$ and $b^x \in \mathbb{R}^M$ are learnable parameters, and P is the kernel size.

Convolutional Attention Variant replaces the linear layer with the convolution layer in the projection layer, which is formulated below:

$$\begin{aligned}
Q &= W^Q * U_t + b^Q \\
K &= W^K * U_t + b^K \\
V &= W^V * U_t + b^V
\end{aligned} \tag{7}$$

where $W^Q \in \mathbb{R}^{d_k \times d_k \times P}$, $W^K \in \mathbb{R}^{d_k \times d_k \times P}$, $W^V \in \mathbb{R}^{d_k \times d_k \times P}$, $b^Q \in \mathbb{R}^{d_k}$, $b^K \in \mathbb{R}^{d_k}$, and $b^V \in \mathbb{R}^{d_k}$ are learnable parameters.

Convolutional Feed-forward Variant formulated the linear layer with the convolution layer in the feed-forward layer, which is described below:

$$FFN(x) = \text{ReLU}\left(W_1 * x + b_1\right) * W_2 + b_2 \tag{8}$$

where $W_1 \in \mathbb{R}^{d_k \times d_k \times P}$, $W_2 \in \mathbb{R}^{d_k \times d_k \times P}$, $b_1 \in \mathbb{R}^{d_k}$, and $b_2 \in \mathbb{R}^{d_k}$ are the leanable parameters.

4.3   Recurrent-Based Variants

We name the architectures that use recurrent layers in any of the three components (input embedding layer, projection layer, or feed-forward layer) as recurrent-based variants. Here, we use Gate Recurrent Unit (GRU) [5] as the recurrent layer. In the following, we illustrate our recurrent-based Variants one by one.

Recurrent Embedding Variant replaces the linear layer with the GRU in the input embedding layer, which is formulated below:

$$
\begin{aligned}
r_t &= \sigma\left(W_{ir}^x x_t + b_{ir}^x + W_{hr}^x U_{(t-1)} + b_{hr}^x\right) \\
z_t &= \sigma\left(W_{iz}^x x_t + b_{iz}^x + W_{hz}^x h_{(t-1)} + b_{hz}^x\right) \\
n_t &= \tanh\left(W_{in}^x x_t + b_{in}^x + r_t \circ \left(W_{hn}^x h_{(t-1)} + b_{hn}^x\right)\right) \\
h_t &= (1 - z_t) \circ n_t + z_t \circ h_{(t-1)} \\
U_t &= Concat(h_1, h_2, ..., h_T)
\end{aligned} \tag{9}
$$

where $W_{ir}^x \in \mathbb{R}^{M \times d_k}$, $W_{iz}^x \in \mathbb{R}^{M \times d_k}$, $W_{in}^x \in \mathbb{R}^{M \times d_k}$, $W_{hr}^x \in \mathbb{R}^{d_k \times d_k}$, $W_{hz}^x \in \mathbb{R}^{d_k \times d_k}$, $W_{hn}^x \in \mathbb{R}^{d_k \times d_k}$, $b_{ir}^x \in \mathbb{R}^{d_k}$, $b_{hr}^x \in \mathbb{R}^{d_k}$, $b_{iz}^x \in \mathbb{R}^{d_k}$, $b_{hz}^x \in \mathbb{R}^{d_k}$, $b_{in}^x \in \mathbb{R}^{d_k}$, and $b_{hn}^x \in \mathbb{R}^{d_k}$ are learnable parameters, $\circ$ is the Hadamard product.

Recurrent Attention Variant replaces the linear layer with the GRU in the projection layer. Since the calculation processes of all the matrices are similar, for simplicity, we only present the calculation process of the query matrix Q in the projection layer below:

$$
\begin{aligned}
r_t &= \sigma\left(W_{ir}^Q U_t + b_{ir}^Q + W_{hr}^Q U_{(t-1)} + b_{hr}^Q\right) \\
z_t &= \sigma\left(W_{iz}^Q U_t + b_{iz}^Q + W_{hz}^Q h_{(t-1)} + b_{hz}^Q\right) \\
n_t &= \tanh\left(W_{in}^Q U_t + b_{in}^Q + r_t \circ \left(W_{hn}^Q h_{(t-1)} + b_{hn}^Q\right)\right) \\
h_t &= (1 - z_t) \circ n_t + z_t \circ h_{(t-1)} \\
Q &= Concat(h_1, h_2, ..., h_T)
\end{aligned} \tag{10}
$$

where $W_{ir}^Q \in \mathbb{R}^{d_k \times d_k}$, $W_{iz}^Q \in \mathbb{R}^{d_k \times d_k}$, $W_{in}^Q \in \mathbb{R}^{d_k \times d_k}$, $W_{hr}^Q \in \mathbb{R}^{d_k \times d_k}$, $W_{hz}^Q \in \mathbb{R}^{d_k \times d_k}$, $W_{hn}^Q \in \mathbb{R}^{d_k \times d_k}$, $b_{ir}^Q \in \mathbb{R}^{d_k}$, $b_{hr}^Q \in \mathbb{R}^{d_k}$, $b_{iz}^Q \in \mathbb{R}^{d_k}$, $b_{hz}^Q \in \mathbb{R}^{d_k}$, $b_{in}^Q \in \mathbb{R}^{d_k}$, and $b_{hn}^Q \in \mathbb{R}^{d_k}$ are learnable parameters.

Recurrent Feed-forward Variant replaces the linear layer with the GRU in the feed-forward layer, which is formulated below:

$$
\begin{aligned}
r_t &= \sigma\left(W_{ir} U_t + b_{ir} + W_{hr} U_{(t-1)} + b_{hr}\right) \\
z_t &= \sigma\left(W_{iz} U_t + b_{iz} + W_{hz} h_{(t-1)} + b_{hz}\right) \\
n_t &= \tanh\left(W_{in} U_t + b_{in} + r_t \circ \left(W_{hn} h_{(t-1)} + b_{hn}\right)\right) \\
h_t &= (1 - z_t) \circ n_t + z_t \circ h_{(t-1)} \\
O &= Concat(h_1, h_2, ..., h_T)
\end{aligned} \tag{11}
$$

where $W_{ir} \in \mathbb{R}^{d_k \times d_k}$, $W_{iz} \in \mathbb{R}^{d_k \times d_k}$, $W_{in} \in \mathbb{R}^{d_k \times d_k}$, $W_{hr} \in \mathbb{R}^{d_k \times d_k}$, $W_{hz} \in \mathbb{R}^{d_k \times d_k}$, $W_{hn} \in \mathbb{R}^{d_k \times d_k}$, $b_{ir} \in \mathbb{R}^{d_k}$, $b_{hr} \in \mathbb{R}^{d_k}$, $b_{iz} \in \mathbb{R}^{d_k}$, $b_{hz} \in \mathbb{R}^{d_k}$, $b_{in} \in \mathbb{R}^{d_k}$, and $b_{hn} \in \mathbb{R}^{d_k}$ are learnable parameters, and O is the final output of the feed-forward layer.

5   Experiments

We empirically evaluate the impact of positional embedding on the performance of the basic model and transformer-based variants (illustrated in Sect. 4) for multivariate time series classification. We report our experimental configurations and discuss the results in the following subsections.

5.1   Datasets

We selected 30 public multivariate time series datasets from the UEA Time Series Classification Repository [7]. All datasets were pre-split into training and test sets[1]. We normalized all datasets to zero mean and unit standard deviation and applied zero padding to ensure all the sequences in each dataset bear the same length.

5.2   Model Configuration and Evaluation Metrics

We trained the basic model and six variants for 500 epochs using the Adam optimizer [12] on all the datasets with and without the learnable positional embedding. Besides, we applied an adaptive learning rate, which was reduced by a factor of 10 after every 100 epochs, and employed dropout regularization to prevent overfitting. Table 2 summarizes our model configurations for each dataset.

We evaluate the models using two metrics: *accuracy* and macro *F1-Score*. To mitigate the effect of randomized parameter initialization, we repeated the training and test procedures five times and took the average as the final results.

5.3   Results and Analysis

Table 3 and Table 4 show the methods' performance on the 30 datasets. The results show positional embedding positively impacts the basic model—with positional embedding, the basic model's performance improves by 17.5% and 14.3% in accuracy and macro F1-Score, respectively. This reveals the significance of enabling the basic model to leverage the position information (e.g., via positional embedding) in solving the multivariate time series classification problem.

In contrast, positional embedding negatively impacts the performance of the Transformer-based variants. Without positional embedding, convolutional

---

[1] Details about datasets and train-test split can be found at http://www. timeseriesclassification.com/dataset.php.

Table 2. Model configurations.

Dataset	Learning Rate	#Layer	Batch Size	Dropout	#Attention head
ArticularyWordRecognition	0.01	3	32	0.01	2
AtrialFibrillation	0.01	2	16	0.01	2
BasicMotions	0.00001	2	16	0.01	2
CharacterTrajectories	0.01	2	16	0.01	2
Cricket	0.01	2	16	0.01	2
DuckDuckGeese	0.001	4	8	0.3	5
EigenWorms	0.01	1	1	0.01	2
Epilepsy	0.00001	4	16	0.01	2
EthanolConcentration	0.001	2	16	0.01	4
ERing	0.00001	2	16	0.01	2
FaceDetection	0.00001	2	16	0.01	2
FingerMovements	0.001	2	16	0.01	2
HandMovementDirection	0.01	2	16	0.1	2
Handwriting	0.01	5	16	0.01	2
Heartbeat	0.00001	2	16	0.01	2
JapaneseVowels	0.01	3	16	0.3	2
Libras	0.01	5	16	0.01	2
LSST	0.01	2	16	0.01	2
MotorImagery	0.00001	2	16	0.01	2
NATOPS	0.00001	3	16	0.01	2
PenDigits	0.001	2	16	0.01	2
PEMS-SF	0.00001	2	16	0.01	2
Phoneme	0.00001	3	16	0.01	2
RacketSports	0.00001	2	16	0.1	4
SelfRegulationSCP1	0.00001	3	16	0.1	2
SelfRegulationSCP2	0.00001	2	16	0.01	2
SpokenArabicDigits	0.00001	3	16	0.1	2
StandWalkJump	0.01	3	16	0.01	2
UWaveGestureLibrary	0.01	2	16	0.01	2

embedding (i.e., ConvEmbedding in Table 1) and recurrent embedding (i.e., RecEmbedding in Table 1) models outperformed all other variants, achieving the best accuracy of 56.21% and 56.17%, respectively, and the best macro F1-Scores of 0.528 and 0.5375, respectively. These two models differ from all other models in that their input embedding layers encode the position information when projecting the raw data to a latent space, making the position information accessible by subsequent layers for feature extraction and resulting in superior performance. Incorporating positional embedding decreased the average accuracy of the variants by 12.7% (convolutional embedding), 9.1% (convolutional attention), 18.6% (convolutional feed-forward), 22.1% (recurrent embedding),

Table 3. Accuracy of different models on 30 benchmark datasets.

	ArticularyWordRecognition	AtrialFibrillation	BasicMotions	CharacterTrajectories	Cricket	DuckDuckGeese
BasicModel (w/ PE)	0.4788	0.4524	1.0000	0.5140	0.9643	0.7667
BasicModel (w/o PE)	0.1280	0.2949	1.0000	0.4684	0.9395	0.5847
ConvEmbedding (w/ PE)	0.5125	0.4333	1.0000	0.4560	0.7468	0.6922
ConvEmbedding (w/o PE)	0.6774	0.7037	1.0000	0.6207	0.8016	0.7145
ConvAttention (w/ PE)	0.5413	0.5238	1.0000	0.2120	0.6515	0.6257
ConvAttention (w/o PE)	0.5091	0.4000	1.0000	0.1975	0.7433	0.3902
ConvFFD (w/ PE)	0.5232	0.4524	1.0000	0.2834	0.8775	0.5858
ConvFFD (w/o PE)	0.6483	0.5238	1.0000	0.3000	0.9623	0.6472
RecEmbedding (w/ PE)	0.5580	0.4524	1.0000	0.4227	0.8442	0.4800
RecEmbedding (w/o PE)	0.7321	0.6111	1.0000	0.5478	0.9339	0.6608
RecAttention (w/ PE)	0.7312	0.6444	1.0000	0.3842	0.6938	0.3120
RecAttention (w/o PE)	0.7548	0.6768	1.0000	0.7450	0.8371	0.6444
RecFFD (w/ PE)	0.6052	0.4167	1.0000	0.3405	0.6660	0.6065
RecFFD (w/o PE)	0.6890	0.5500	1.0000	0.5001	0.7978	0.6419

	EigenWorms	Epilepsy	JapaneseVowels	Libras	LSST	MotorImagery
BasicModel (w/ PE)	0.4447	0.8153	0.9616	0.0113	0.1716	0.5801
BasicModel (w/o PE)	0.4186	0.7753	0.9346	0.1527	0.1770	0.4332
ConvEmbedding (w/ PE)	0.3556	0.8156	0.9526	0.0317	0.1719	0.4264
ConvEmbedding (w/o PE)	0.4843	0.8725	0.9633	0.0760	0.1086	0.7525
ConvAttention (w/ PE)	0.3792	0.7377	0.7418	0.0878	0.1365	0.5000
ConvAttention (w/o PE)	0.3844	0.5564	0.7491	0.1352	0.1845	0.6420
ConvFFD (w/ PE)	0.4716	0.8041	0.9688	0.0298	0.0987	0.4391
ConvFFD (w/o PE)	0.4872	0.8954	0.9721	0.0490	0.1269	0.4585
RecEmbedding (w/ PE)	0.6724	0.8037	0.9566	0.0691	0.0461	0.5525
RecEmbedding (w/o PE)	0.7053	0.8538	0.9669	0.1070	0.0943	0.6326
RecAttention (w/ PE)	0.4171	0.6734	0.9668	0.1023	0.1006	0.6302
RecAttention (w/o PE)	0.6420	0.8163	0.9608	0.1897	0.0904	0.7551
RecFFD (w/ PE)	0.4199	0.7711	0.9702	0.1014	0.1370	0.4688
RecFFD (w/o PE)	0.6545	0.8353	0.9709	0.1159	0.1664	0.4719

	NATOPS	PenDigits	PEMS-SF	Phoneme	EthanolConcentration	ERing
BasicModel (w/ PE)	0.2834	0.1639	0.6204	0.0164	0.3880	0.6659
BasicModel (w/o PE)	0.1198	0.1203	0.8091	0.0110	0.0627	0.6065
ConvEmbedding (w/ PE)	0.2249	0.6416	0.8076	0.0126	0.1086	0.6982
ConvEmbedding (w/o PE)	0.3074	0.6452	0.8770	0.0291	0.1979	0.7695
ConvAttention (w/ PE)	0.2111	0.0784	0.8717	0.0170	0.0632	0.6660
ConvAttention (w/o PE)	0.2699	0.5021	0.8953	0.0211	0.1254	0.6913
ConvFFD (w/ PE)	0.2339	0.2036	0.8155	0.0142	0.0627	0.4966
ConvFFD (w/o PE)	0.2796	0.3388	0.8816	0.0142	0.1340	0.7357
RecEmbedding (w/ PE)	0.2951	0.3561	0.6314	0.0059	0.0618	0.4318
RecEmbedding (w/o PE)	0.4215	0.4066	0.8277	0.0106	0.4755	0.8008
RecAttention (w/ PE)	0.2638	0.1491	0.5946	0.0111	0.0985	0.6291
RecAttention (w/o PE)	0.4958	0.1711	0.8348	0.0243	0.1236	0.6488
RecFFD (w/ PE)	0.2797	0.1003	0.5953	0.0074	0.0618	0.7222
RecFFD (w/o PE)	0.4231	0.2755	0.8333	0.0123	0.1254	0.7735

	FaceDetection	FingerMovements	HandMovementDirection	Handwriting	Heartbeat	RacketSports
BasicModel (w/ PE)	0.6285	0.6075	0.2691	0.0239	0.7528	0.0879
BasicModel (w/o PE)	0.5487	0.3925	0.1830	0.0460	0.8627	0.0546
ConvEmbedding (w/ PE)	0.6461	0.5405	0.3032	0.0266	0.8663	0.4439
ConvEmbedding (w/o PE)	0.5589	0.5861	0.3224	0.0307	0.7506	0.4530
ConvAttention (w/ PE)	0.6762	0.7082	0.2908	0.0154	0.7238	0.2111
ConvAttention (w/o PE)	0.6816	0.7552	0.3442	0.0664	0.8663	0.5574
ConvFFD (w/ PE)	0.6457	0.6725	0.2858	0.0364	0.7258	0.1691
ConvFFD (w/o PE)	0.6347	0.7242	0.2950	0.4540	0.7272	0.3109
RecEmbedding (w/ PE)	0.6793	0.5187	0.4222	0.0400	0.3610	0.1716
RecEmbedding (w/o PE)	0.5500	0.5233	0.3779	0.0570	0.7096	0.2121
RecAttention (w/ PE)	0.6797	0.5028	0.2083	0.0318	0.3610	0.1970
RecAttention (w/o PE)	0.5425	0.6731	0.2559	0.0513	0.7528	0.3110
RecFFD (w/ PE)	0.6648	0.2550	0.4335	0.0189	0.7238	0.1253
RecFFD (w/o PE)	0.5527	0.4167	0.4454	0.0352	0.7435	0.2003

	SelfRegulationSCP1	SelfRegulationSCP2	SpokenArabicDigits	StandWalkJump	UWaveGestureLibrary	Average
BasicModel (w/ PE)	0.8494	0.5259	0.1111	0.5694	0.5284	0.4915
BasicModel (w/o PE)	0.8176	0.5125	0.1111	0.2778	0.2865	0.4183
ConvEmbedding (w/ PE)	0.8720	0.5847	0.4433	0.2778	0.3656	0.4986
ConvEmbedding (w/o PE)	0.8846	0.5694	0.5988	0.5500	0.3938	0.5621
ConvAttention (w/ PE)	0.8188	0.5207	0.4681	0.4524	0.4763	0.4623
ConvAttention (w/o PE)	0.8592	0.5000	0.6620	0.5500	0.3818	0.5042
ConvFFD (w/ PE)	0.8375	0.5250	0.5591	0.1944	0.3472	0.4607
ConvFFD (w/o PE)	0.8979	0.6080*	0.6418	0.5500	0.5496	0.5465
RecEmbedding (w/ PE)	0.7935	0.2500	0.5833	0.4778	0.4031	0.4600
RecEmbedding (w/o PE)	0.8472	0.2500	0.7232	0.8182	0.4338	0.5617
RecAttention (w/ PE)	0.8494	0.5000	0.5112	0.5500	0.4113	0.4553
RecAttention (w/o PE)	0.8746	0.5710	0.5420	0.7167	0.3374	0.5531
RecFFD (w/ PE)	0.8658	0.4486	0.4491	0.4667	0.3620	0.4512
RecFFD (w/o PE)	0.8843	0.5752	0.4555	0.4615	0.5370	0.5222

Table 4. Macro F1-Score of different models on 30 benchmark datasets.

	ArticularyWordRecognition	AtrialFibrillation	BasicMotions	CharacterTrajectories	Cricket	DuckDuckGeese
BasicModel (w/ PE)	0.5395	0.3523	1.0000	0.5783	0.9581	0.7177
BasicModel (w/o PE)	0.0988	0.3297	1.0000	0.5687	0.9156	0.4478
ConvEmbedding (w/ PE)	0.6170	0.4139	1.0000	0.5554	0.7208	0.5365
ConvEmbedding (w/o PE)	0.7226	0.6035	1.0000	0.6734	0.8285	0.6379
ConvAttention (w/ PE)	0.5094	0.4000	1.0000	0.1879	0.6382	0.3959
ConvAttention (w/o PE)	0.5821	0.5157	1.0000	0.3141	0.7758	0.4760
ConvFFD (w/ PE)	0.5921	0.3529	1.0000	0.3892	0.8719	0.5365
ConvFFD (w/o PE)	0.7152	0.3333	1.0000	0.4318	0.9582	0.6360
RecEmbedding (w/ PE)	0.6272	0.3591	1.0000	0.4510	0.8194	0.4748
RecEmbedding (w/o PE)	0.7605	0.4577	1.0000	0.6335	0.9025	0.5304
RecAttention (w/ PE)	0.7599	0.4603	1.0000	0.4473	0.7108	0.3354
RecAttention (w/o PE)	0.7767	0.4553	1.0000	0.7868	0.8600	0.5338
RecFFD (w/ PE)	0.6615	0.4142	1.0000	0.4351	0.7121	0.5338
RecFFD (w/o PE)	0.7314	0.5161	1.0000	0.5712	0.8176	0.6114

	EigenWorms	Epilepsy	JapaneseVowels	Libras	LSST	MotorImagery
BasicModel (w/ PE)	0.4020	0.8169	0.9670	0.0321	0.2009	0.4325
BasicModel (w/o PE)	0.3912	0.7667	0.9439	0.1307	0.1297	0.5799
ConvEmbedding (w/ PE)	0.2471	0.7568	0.9514	0.0148	0.1504	0.4265
ConvEmbedding (w/o PE)	0.2849	0.8345	0.9600	0.0403	0.1220	0.4500
ConvAttention (w/ PE)	0.2460	0.5912	0.7222	0.0581	0.0555	0.4120
ConvAttention (w/o PE)	0.3592	0.7100	0.6509	0.0158	0.1317	0.6393
ConvFFD (w/ PE)	0.3695	0.7667	0.9656	0.0568	0.0585	0.4390
ConvFFD (w/o PE)	0.4923	0.8941	0.9744	0.0744	0.0829	0.4623
RecEmbedding (w/ PE)	0.5984	0.7479	0.9566	0.1339	0.0752	0.5498
RecEmbedding (w/o PE)	0.6606	0.8557	0.9679	0.1508	0.0681	0.6291
RecAttention (w/ PE)	0.3231	0.5496	0.9686	0.0690	0.0950	0.4630
RecAttention (w/o PE)	0.5197	0.8072	0.9673	0.2292	0.0830	0.5526
RecFFD (w/ PE)	0.3738	0.7462	0.9721	0.1448	0.1312	0.4533
RecFFD (w/o PE)	0.6519	0.8410	0.9741	0.1512	0.0724	0.4547

	NATOPS	PenDigits	PEMS-SF	Phoneme	EthanolConcentration	ERing
BasicModel (w/ PE)	0.3127	0.0869	0.7693	0.0236	0.2152	0.6486
BasicModel (w/o PE)	0.2291	0.0747	0.6615	0.0191	0.1432	0.5273
ConvEmbedding (w/ PE)	0.3257	0.6173	0.7872	0.0234	0.1739	0.6498
ConvEmbedding (w/o PE)	0.3974	0.6262	0.8655	0.0247	0.2096	0.7679
ConvAttention (w/ PE)	0.2333	0.1291	0.8543	0.0099	0.0777	0.5992
ConvAttention (w/o PE)	0.3031	0.4512	0.8900	0.0259	0.2224	0.6600
ConvFFD (w/ PE)	0.2121	0.2591	0.7932	0.0249	0.1432	0.5523
ConvFFD (w/o PE)	0.3564	0.3129	0.8763	0.0207	0.1581	0.7092
RecEmbedding (w/ PE)	0.3635	0.4070	0.6814	0.0142	0.1419	0.4622
RecEmbedding (w/o PE)	0.3715	0.4176	0.7929	0.0149	0.2600	0.7814
RecAttention (w/ PE)	0.1894	0.2101	0.6375	0.0205	0.1483	0.5415
RecAttention (w/o PE)	0.3423	0.2118	0.8277	0.0234	0.2217	0.6511
RecFFD (w/ PE)	0.2344	0.1380	0.6328	0.0165	0.1419	0.6853
RecFFD (w/o PE)	0.4271	0.3123	0.7916	0.0238	0.2224	0.7615

	FaceDetection	FingerMovements	HandMovementDirection	Handwriting	Heartbeat	RacketSports
BasicModel (w/ PE)	0.6270	0.5580	0.2603	0.0441	0.6119	0.1502
BasicModel (w/o PE)	0.5483	0.4029	0.1593	0.0680	0.6167	0.1304
ConvEmbedding (w/ PE)	0.6454	0.4931	0.2703	0.0549	0.6167	0.2044
ConvEmbedding (w/o PE)	0.5563	0.5310	0.2960	0.0598	0.6119	0.3170
ConvAttention (w/ PE)	0.6814	0.5710	0.3052	0.0369	0.7351	0.2400
ConvAttention (w/o PE)	0.6762	0.5181	0.1970	0.0622	0.5950	0.2582
ConvFFD (w/ PE)	0.6396	0.5612	0.3102	0.0454	0.6144	0.2732
ConvFFD (w/o PE)	0.5544	0.5923	0.2349	0.0631	0.6093	0.2814
RecEmbedding (w/ PE)	0.6793	0.5154	0.2097	0.0713	0.4561	0.1677
RecEmbedding (w/o PE)	0.5499	0.5211	0.3298	0.0795	0.5702	0.2394
RecAttention (w/ PE)	0.6793	0.4997	0.1763	0.0594	0.4561	0.2551
RecAttention (w/o PE)	0.5419	0.5457	0.2305	0.0653	0.6091	0.2499
RecFFD (w/ PE)	0.6642	0.4031	0.3617	0.0465	0.5950	0.1940
RecFFD (w/o PE)	0.5523	0.4174	0.3231	0.0609	0.5986	0.3186

	SelfRegulationSCP1	SelfRegulationSCP2	SpokenArabicDigits	StandWalkJump	UWaveGestureLibrary	Average
BasicModel (w/ PE)	0.8167	0.4669	0.2222	0.4260	0.5309	0.4748
BasicModel (w/o PE)	0.8120	0.5040	0.2222	0.2947	0.3282	0.4153
ConvEmbedding (w/ PE)	0.8632	0.5131	0.5001	0.2871	0.3785	0.4757
ConvEmbedding (w/o PE)	0.8839	0.5448	0.6328	0.4611	0.3677	0.5280
ConvAttention (w/ PE)	0.2548	0.4346	0.3021	0.2353	0.3867	0.3898
ConvAttention (w/o PE)	0.8119	0.4434	0.4141	0.2559	0.4812	0.4633
ConvFFD (w/ PE)	0.8249	0.5217	0.6121	0.2508	0.3023	0.4600
ConvFFD (w/o PE)	0.8976	0.5740	0.6497	0.4496	0.5897	0.5167
RecEmbedding (w/ PE)	0.7074	0.4000	0.5586	0.4680	0.4077	0.4657
RecEmbedding (w/o PE)	0.8132	0.4000	0.6771	0.7370	0.4143	0.5375
RecAttention (w/ PE)	0.8167	0.4511	0.4883	0.4128	0.3663	0.4341
RecAttention (w/o PE)	0.8736	0.5425	0.5469	0.5220	0.3617	0.5151
RecFFD (w/ PE)	0.8524	0.4133	0.5315	0.2353	0.3765	0.4517
RecFFD (w/o PE)	0.8839	0.5533	0.5451	0.4163	0.5811	0.5235

21.5% (recurrent attention), and 15.7% (recurrent feed-forward), respectively. Results for the macro F1-Score show similar trends.

Since the convolutional and recurrent layers can inherently capture the position information from sequential data, it is natural to consider positional embedding redundant for Transformer-based variants. Besides, positional embedding risks introducing inductive bias and contaminating the original data. Specifically, positional embedding injects the same information into sequences of different classes, bringing new challenges to the classifiers; this may also contribute to performance degradation.

Further reflecting on the results, we suggest that positional embedding may not be necessary for Transformer-based variants that already contain position-sensitive modules. In particular, for time series classification tasks, while the classifier focuses on the differences between time series sequences across different classes, positional embedding is content-irrelevant, adding the same position information to all sequences regardless of their class labels. As position-sensitive modules generally consider content information when encoding the position information, redundant content-irrelevant positional embedding may lead the model towards capturing spurious correlations that potentially hinder the classifier's performance.

6   Conclusion

Existing Transformer-based architectures generally contain position-sensitive layers while routinely incorporating positional embedding without comprehensively evaluating its effectiveness on multivariate time series classification. In this paper, we investigate the impact of positional embedding on the vanilla Transformer architecture and six types of Transformer-based variants in multivariate time series classification. Our experimental results on 30 public time series datasets show that positional embedding lifts the performance of the vanilla Transformer while adversely impacting the performance of Transformer-based variants on classification tasks. Our findings refute the necessity of incorporating positional embedding in Transformer-based architectures that already contain position-sensitive layers, such as convolutional or recurrent layers. We also advocate applying position-sensitive layers directly on the input for any Transformer-based architecture that considers using position-sensitive layers to gain better results in multivariate time series classification.

References

1. Arnab, A., Dehghani, M., Heigold, G., Sun, C., Lučić, M., Schmid, C.: Vivit: a video vision transformer. In: Proceedings of the IEEE/CVF International Conference on Computer Vision, pp. 6836–6846 (2021)
2. Bulatov, A., Kuratov, Y., Burtsev, M.S.: Recurrent memory transformer. arXiv preprint arXiv:2207.06881 (2022)

3. Chen, K., Wang, R., Utiyama, M., Sumita, E.: Recurrent positional embedding for neural machine translation. In: Proceedings of the 2019 Conference on Empirical Methods in Natural Language Processing and the 9th International Joint Conference on Natural Language Processing (EMNLP-IJCNLP), pp. 1361–1367 (2019)

4. Chen, Z., Chen, D., Zhang, X., Yuan, Z., Cheng, X.: Learning graph structures with transformer for multivariate time series anomaly detection in IoT. IEEE Internet Things J. **9**, 9179–9189 (2021)

5. Cho, K., et al.: Learning phrase representations using rnn encoder-decoder for statistical machine translation. arXiv preprint arXiv:1406.1078 (2014)

6. Coyle, D., Prasad, G., McGinnity, T.M.: A time-series prediction approach for feature extraction in a brain-computer interface. IEEE Trans. Neural Syst. Rehabil. Eng. **13**(4), 461–467 (2005)

7. Dau, H.A., et al.: Hexagon-ML: the UCR time series classification archive (2018)

8. Gulati, A., et al.: Conformer: convolution-augmented transformer for speech recognition. arXiv preprint arXiv:2005.08100 (2020)

9. Han, K., Wang, Y., Chen, H., Chen, X., Guo, J., Liu, Z., Tang, Y., Xiao, A., Xu, C., Xu, Y., et al.: A survey on vision transformer. IEEE Trans. Pattern Anal. Mach. Intell. **45**(1), 87–110 (2022)

10. Huang, Z., Xu, P., Liang, D., Mishra, A., Xiang, B.: Trans-blstm: transformer with bidirectional LSTM for language understanding. arXiv preprint arXiv:2003.07000 (2020)

11. Hutchins, D., Schlag, I., Wu, Y., Dyer, E., Neyshabur, B.: Block-recurrent transformers. arXiv preprint arXiv:2203.07852 (2022)

12. Kingma, D.P., Ba, J.: Adam: a method for stochastic optimization. arXiv preprint arXiv:1412.6980 (2014)

13. Li, S., et al.: Enhancing the locality and breaking the memory bottleneck of transformer on time series forecasting. Adv. Neural Inf. Process. Syst. **32** (2019)

14. Lim, B., Arık, S.Ö., Loeff, N., Pfister, T.: Temporal fusion transformers for interpretable multi-horizon time series forecasting. Int. J. Forecast. **37**(4), 1748–1764 (2021)

15. Liu, M., Kim, Y.: Classification of heart diseases based on ecg signals using long short-term memory. In: 2018 40th Annual International Conference of the IEEE Engineering in Medicine and Biology Society (EMBC), pp. 2707–2710. IEEE (2018)

16. Liu, M., et al.: Gated transformer networks for multivariate time series classification. arXiv preprint arXiv:2103.14438 (2021)

17. Liu, Y., et al.: Delightfultts: the microsoft speech synthesis system for blizzard challenge 2021. arXiv preprint arXiv:2110.12612 (2021)

18. Pan, Z., Cai, J., Zhuang, B.: Fast vision transformers with hilo attention. arXiv preprint arXiv:2205.13213 (2022)

19. Raganato, A., Tiedemann, J.: An analysis of encoder representations in transformer-based machine translation. In: Proceedings of the 2018 EMNLP Workshop BlackboxNLP: Analyzing and Interpreting Neural Networks for NLP. The Association for Computational Linguistics (2018)

20. Serdyuk, D., Braga, O., Siohan, O.: Transformer-based video front-ends for audio-visual speech recognition, p. 15. arXiv preprint arXiv:2201.10439 (2022)

21. Shen, L., Wang, Y.: TCCT: tightly-coupled convolutional transformer on time series forecasting. Neurocomputing **480**, 131–145 (2022)

22. Song, Q., Sun, B., Li, S.: Multimodal sparse transformer network for audio-visual speech recognition. IEEE Trans. Neural Netw. Learn. Syst. (2022)

23. Vaswani, A., et al.: Attention is all you need. Adv. Neural Inf. Process. Syst. **30** (2017)

24. Wang, Q., et al.: Learning deep transformer models for machine translation. arXiv preprint arXiv:1906.01787 (2019)
25. Wang, Z., Ma, Y., Liu, Z., Tang, J.: R-transformer: recurrent neural network enhanced transformer. arXiv preprint arXiv:1907.05572 (2019)
26. Wen, Q., et al.: Transformers in time series: a survey. arXiv preprint arXiv:2202.07125 (2022)
27. Woo, G., Liu, C., Sahoo, D., Kumar, A., Hoi, S.: Etsformer: exponential smoothing transformers for time-series forecasting. arXiv e-prints arXiv:2202.01381 (2022)
28. Wu, H., et al.: CVT: introducing convolutions to vision transformers. In: Proceedings of the IEEE/CVF International Conference on Computer Vision, pp. 22–31 (2021)
29. Yang, C., Jiang, W., Guo, Z.: Time series data classification based on dual path CNN-RNN cascade network. IEEE Access **7**, 155304–155312 (2019)
30. Yuan, Y., Lin, L.: Self-supervised pretraining of transformers for satellite image time series classification. IEEE J. Sel. Topics Appl. Earth Obs. Remote Sens. **14**, 474–487 (2020)
31. Zeng, A., Chen, M., Zhang, L., Xu, Q.: Are transformers effective for time series forecasting? arXiv preprint arXiv:2205.13504 (2022)
32. Zerveas, G., Jayaraman, S., Patel, D., Bhamidipaty, A., Eickhoff, C.: A transformer-based framework for multivariate time series representation learning. In: Proceedings of the 27th ACM SIGKDD Conference on Knowledge Discovery & Data Mining, pp. 2114–2124 (2021)
33. Zhou, H., et al.: Informer: beyond efficient transformer for long sequence time-series forecasting. In: Proceedings of the AAAI Conference on Artificial Intelligence, vol. 35, pp. 11106–11115 (2021)
34. Zhu, C., et al.: Long-short transformer: efficient transformers for language and vision. Adv. Neural. Inf. Process. Syst. **34**, 17723–17736 (2021)

Modeling of Repeated Measures for Time-to-event Prediction

Jianfei Zhang[1], Lifei Chen[2], and Shengrui Wang[1,2(✉)]

[1] Université de Sherbrooke, Sherbrooke, Québec, Canada
`jianfei.zhang@usherbrooke.ca`
[2] Fujian Normal University, Fuzhou, Fujian, China

Abstract. Predicting the time for an event to occur while simultaneously exploring the coexisting effects of various risk factors has captivated considerable research interest. However, the profusion of repeated measurements involving a diverse array of risk factors has outpaced the capabilities of current methods for analyzing time-to-event data. In this paper, we propose a novel approach that entails the conversion of the time-to-event analysis conundrum into a sequence of discrete survival learning and prediction tasks, each approached autonomously. Our innovative strategy for modeling repeated measures facilitates the quantification of measurement impacts on projected outcomes at distinct junctures. When extrapolating the trajectory of health status over time, our method harnesses both censored and uncensored data to refine logistic regression parameters. Through a series of comparative experiments and meticulous ablation studies conducted on two real-life health datasets, we underscore the intrinsic practical promise of our method. Notably, our approach showcases its efficacy in prognosticating the temporal aspects of breast cancer patient mortality and the onset of disabilities among the elderly.

1 Introduction

Time-to-event data materializes when attention zeroes in on the passage of time (measured in years, months, weeks, or days) from the inception of a follow-up study until the occurrence of a specific event. In medical research, this event of significance is typically an adverse incident encompassing facets like injury, the inception or resurgence of an ailment, (re)hospitalization, and even demise. The pursuit of time-to-event prediction seeks to tackle inquiries such as *"How much time remains before the event takes place?"* - a question that has held enduring practical import. Foreseen time-to-event insights empower clinicians to promptly address patients' queries about potential outcomes, while decision-makers glean information about the probable timing of disease-related rehospitalizations. To illustrate, envision prognosticating outcomes for individuals diagnosed with breast cancer. One could discern that a 60-year-old patient boasts a 70% chance of surviving one year and a 40% probability of reaching the three-year milestone. These prognostic revelations can exert a pivotal influence on

X. Yang et al. (Eds.): ADMA 2023, LNAI 14176, pp. 48–63, 2023.
https://doi.org/10.1007/978-3-031-46661-8_4

treatment choices, lifestyle adaptations, and, on occasion, considerations concerning end-of-life care.

Within time-to-event data, the outcome encompasses not only the occurrence or non-occurrence of an event but also the specific timing of that event. The events sometimes cannot be observed throughout the follow-up period due to various censorships [10]. Hence, predicting time-to-event accurately is challenging and even impossible. Standard regression models fall short in their capacity to incorporate both the event occurrence and the timing aspects as outcomes within the same model. In the recent past, a significant multitude of methods has been devised to confront this quandary, facilitating predictions that effectively consider both dimensions concurrently, e.g., statistical models [16] and machine-learning-based methods [25]. To make the problem simple, we can answer the question *"When and how probably will be free of failure?"*

Investigating the simultaneous impacts of risk factors on the outcome holds significant relevance for both researchers and clinicians, and has been a focal point of extensive research endeavors. However, the measurement of risk factors often occurs over time or through real-time devices during the follow-up period, particularly in the context of extended-term follow-ups. Consequently, this results in an abundance of recurrent measurements, leading to fluctuations in the corresponding risk factors over time. Consider breast cancer patients who are monitored over a span of time, undergoing monthly assessments like white blood cell counts. Similarly, in studies focusing on cognitive changes related to aging, researchers gather data on participants' cognitive capabilities along with details about their date of death or diagnosis of dementia. These instances involve time-dependent risk factors like mental health, injuries caused by falls, and smoking. It is a great challenge for current time-to-event prediction methods to take care of these complex repeated measures, although they can use a single measure; of course, this will ignore the effect caused by historical measures.

Analyzing repeated measures and inference needs special techniques. In this context, approaches such as time-varying analysis [30] can be employed, where a single measurement evolves over time and is subsequently substituted by successive measurements to explore short-term relationships. Alternatively, measure-averaged analysis [29] can be employed, wherein a participant's initial measurement is solely represented by risk factor values at a specific point in time or a single estimate derived from the mean of measurements over a defined period. It's noteworthy that within the literature [21,28], the impact of repeated measurements on time-to-event outcomes has been demonstrated. Consider an observational study focusing on the effects of a drug on specific health indicators. In this scenario, a patient's present health status could influence their future drug exposure or dosage. Consequently, establishing a model capable of automatically accommodating repeated measurements is of paramount importance.

In this paper, we undertake a transformation of the time-to-event prediction task, reconfiguring it into a sequence of challenges involving survival probability estimates (SSP) at distinct time junctures. For each of these estimation challenges, we methodically gauge the influence of recurrent measurements on the survival probability. The recently devised comprehensive data-driven learning approach leverages all available data points to compute the cumulative event risk, thereby optimizing data utilization. To ascertain the parameters, we engage in the optimization of an objective function, which involves the regularization of model parameters and the facilitation of their gradual transitions. Empirical assessments are conducted using real-world time-to-event data featuring repeated measurements. The new technique exhibits superior performance when compared to other cutting-edge survival models. We summarize the contributions as follows:

- Our method is characterized by its simplicity, both computationally and conceptually. It entails a series of logistic regression fits, and the operations can be readily comprehended within the framework of regression modeling.
- Empirical evidence attests to the effectiveness of our approach. We applied it to predict survival rates in breast cancer patients and a diverse Canadian cohort, encompassing individuals without disabilities.

2   Related Work

The essence of this paper revolves around the realm of time-to-event data coupled with the inclusion of repeated measurements. Consequently, our focus in this review shall encompass endeavors related to time-to-event prediction models and the analysis of time-varying data. Broadly delineated, methods for analyzing time-to-event data can be categorized into two primary groups: statistical methods and machine-learning-based approaches.

2.1   Statistical Methods

Statistical methods concentrate on the intricacies of event time distributions and parameter estimation. These methods can be categorized into parametric, nonparametric, and semi-parametric approaches. Nonparametric methods empirically estimate the survival probability through Nelson-Aalen estimator [1] and Kaplan-Meier estimator [11]. While they yield unbiased descriptive insights, they lack the capability to evaluate the impact of multiple risk factors on outcomes. In contrast, parametric methods posit that the time-to-event adheres to a specific distribution, such as exponential, Weibull, (log-)logistic, etc. For example, the accelerated failure time model [26] operates under the premise that the underlying time-to-event distribution has been accurately specified. To circumvent this assumption, the semi-parametric Cox model [2] has gained broader traction in time-to-event data analysis due to its user-friendliness, established efficacy, and interpretability of results [28]. This model assumes a specific influence of risk factors on events, an assumption that is frequently violated in practical scenarios.

2.2   Machine-Learning-Based Approaches

As the utilization of time-to-event data, particularly those featuring time-varying risk factors, continues to rise, machine-learning-based methods have increasingly supplanted traditional statistical models within the realm of survival analysis. This transition is particularly evident in the context of time-to-event studies, where machine learning finds synergy with survival settings. Feed-forward neural networks [5], recurrent neural networks [8,27], and deep neural networks [12,15] have all been employed for time-to-event analysis. Additionally, multi-task learning has emerged as a common strategy for time-to-event prediction. For instance, in [18], a pioneering approach involved the creation of a multi-task logistic regression model to learn patient-specific survival distributions. Gaussian processes have also made their mark in survival analysis tasks. For instance, [13] introduced scalable variational inference algorithms for a Gaussian-process-based survival model, accounting for uncertainty in hazard function modeling. Furthermore, [6] proposed a semi-parametric Bayesian model for survival analysis employing a Gaussian process. This model modulates the hazard function through the multiplication of a parametric baseline hazard and a nonparametric component.

2.3   Time-Varying Models

Recognizing the necessity to accommodate the evolving impact of factors on survival outcomes as time progresses, especially in the context of extended follow-up periods [7], there has been a notable surge of interest in the pursuit of learning time-varying coefficients rather than adhering to fixed ones. These models, characterized by varying coefficients, allow for the exploration of dynamic patterns and naturally extend the scope of classical parametric models. Their growing popularity in data analysis [4] is attributed to their robust interpretability. To estimate the time-varying coefficients within the Cox model framework, diverse strategies have emerged. [23] maximized a kernel-weighted partial likelihood, while [22] employed local empirical partial likelihood smoothing. Another approach, as seen in [19], leveraged time-varying coefficients to elucidate the evolving effects of risk factors on breast cancer failure. However, the practical applicability of the proportionality assumption may be limited, particularly when the effects of risk factors undergo temporal shifts. To address this, [27] introduced a survival recurrent neural network, which yielded superior predictions compared to other state-of-the-art survival models. This advancement underscores the efficacy of modern machine learning techniques in capturing the intricate dynamics of time-varying factors affecting survival outcomes.

3   Problem Statement

We identify each individual using two variables, that is, the response 'censor' $C \in \{0, 1\}$ indicating whether or not the individual time is censored and the response 'stamp' $S \in \mathbb{R}^+$ showing the time when the event happened or the individual is lost to follow-up. If $C = 0$, we have the time of event occurrence, say T, which is uncensored. The event occurred right at the last recorded time S, we thus have $T = S$. When T is censored, $C = 1$. For this situation, S underestimates the true but unknown T, i.e., $T > S$. Table 1 gives an example: three old Canadians aged over 65. We have $T = 7$ months for the 77-year-old Québecois. $C = 1$ indicates that T is censored due to dropout or early end of follow-up. Here, every individual can be presented by D factors that are time-invariant (e.g., 'province' and 'age') or time-varying (e.g., 'depression', 'sleep', and 'fall'). Generally, these measures can be written as $\mathbf{X} = (\mathbf{x}_1, \ldots, \mathbf{x}_V) \in \mathbb{R}^{D \times V}$, where, at the vth time of the V different times $\tau_1 < \cdots < \tau_V$, we have $\mathbf{x}_v = (x_{v1}, \ldots, x_{vD})^\top$. The time-invariant factor d is $x_{1d} = x_{2d} = \cdots = x_{Vd}$. We aim to identify a mapping function $f_{\mathbf{W}}$ ($\mathbf{W}$ is the vector of parameters to learn) for probability prediction at $t_1 < t_2 \cdots < t_K$.

Table 1. An example: time-to-event data and repeated measures for three aging Canadians living in Québec, Ontario, and Alberta, respectively.

Risk Factors						Response	
time-invariant		··	time-varying (Repeated Measures)			Censor	Stamp
Province	Age	··	Depression	·· Fall	Smoking	(C)	(S)
Québec	77	··	mild	·· no	sometimes	0	7
			mild	·· yes	sometimes		
			··	·· ··	··		
			no	·· no	seldom		
Ontario	69	··	severe	·· no	often	1	11
			moderate	·· no	seldom		
			··	·· ··	··		
			moderate	·· no	sometimes		
Alberta	86	··	no	·· yes	never	1	18
			no	·· no	never		
			··	·· ··	··		
			mild	·· no	never		

4 Our Approach

4.1 Health Status

The initial step involves encoding the response variables (C and S) into health statuses represented by Y. The value at a specific time point t can be expressed as $y[t] = (-\mathbb{1}(C = 1))^{\mathbb{1}(S<t)}$. This function assumes the value of 1 if the event

occurred on or before time t, which can be represented as $S \geq t$. Otherwise, it takes the value of 0 when $C = 0$, or -1 if $C = 1$. Here, $\mathbb{1}(judgment)$ denotes an indicator function that equals 1 when the *judgment* is true, and 0 otherwise.

Example: The health statuses for the three individuals are illustrated in Table 2. Each health status $y[t_k]$ denotes whether a specific event (such as disability) has occurred by time t_k: it takes on a value of 1 if the event has occurred, 0 if it hasn't, and -1 if the information is unknown. Once $y[t_k]$ transitions to "0", it remains so and does not revert back to "1". Consequently, there exist $K+1$ valid sequences following the pattern $(1, 1, \ldots, 0, 0, \ldots)$, which encompasses sequences consisting entirely of "1" s and sequences entirely of "0"s. For instance, let's consider the case of the 77-year-old Québecois: his health status remains "1" until $S = 7$, at which point it transitions to "0". In contrast, for the other two, the exact onset times are censored. Consequently, their health statuses remain "1" until the corresponding time stamp, and subsequently shift to "-1" beyond that point.

Table 2. An example of health status for three Canadian seniors aged over 60.

Health Status										Response	
$y[1]$	$\cdot\cdot$ $y[7]$	$y[8]$	$\cdot\cdot$	$y[11]$	$y[12]$	$\cdot\cdot$	$y[18]$	$y[19]$	$\cdot\cdot$	C	S
1	1 1	0	0	0	0	0	0	0	0	0	7
1	1 1	1	1	1	-1	-1	-1	-1	-1	1	11
1	1 1	1	1	1	1	1	1	-1	-1	1	18

4.2   Training

For an individual indexed as $i \in \mathcal{G}_0 = \{i | \forall i : C_i = 0\}$, possessing known health statuses denoted as $Y = (y[t_1], \ldots, y[t_K])$ at time instances $t_1 < \ldots < t_K$, along with associated measurements $\mathbf{X}$ $(\tau_V \leq t_K)$, we estimate the probability of observing Y through the utilization of generalized logistic regression.

$$\Pr\left(Y \mid \mathbf{X}; \mathbf{W}\right)_0 = \frac{\exp\left(\mathbf{W} * \mathbf{X} \cdot \boldsymbol{\Delta} \cdot \mathbb{1}(\mathbf{y} \leq 0)\right)}{\exp\left(\mathbf{W} * \mathbf{X} \cdot \boldsymbol{\Delta} * \mathbf{A}\right) \cdot \mathbf{1}} \tag{1}$$

$$\mathbf{W} = (\mathbf{w}_1, \ldots, \mathbf{w}_V) \in \mathbb{R}^{D \times V}$$

$$\mathbf{w}_v = (w_{v1}, \ldots, w_{vD})^\top \in \mathbb{R}^{D \times 1}, \forall v = 1, 2, \ldots, V$$

$$\mathbf{W} * \mathbf{X} = (\mathbf{w}_1 \cdot \mathbf{x}_1, \ldots, \mathbf{w}_V \cdot \mathbf{x}_V) \in \mathbb{R}^{1 \times V}.$$

The regression coefficient $\mathbf{W}$ serves to quantify the influence of factors and their recurrent measurements on the probability of an individual's event-free status. In this context, the coefficients $\mathbf{w}_v$ signify the contribution of measurements denoted by D at the specific time τ_v. The summation of transformed measurements over V time instances is computed through a column-wise Hadamard product [20]. To assess the impact of the D factors and their V repeated measurements on the manifestation of Y, we introduce the concept of the time-decay ratio. This ratio is shaped by the passage of time and can be expressed as follows:

$$\boldsymbol{\Delta} = \mathbf{exp}(\delta(k, v)) \in \mathbb{R}^{K \times V} \tag{2}$$

$$\delta(k, v) = -(t_k - \tau_v) \times \mathbb{1}(t_k \geq \tau_v). \tag{3}$$

The symbol $\mathbf{exp}$ denotes the element-wise exponential operation applied to a matrix. The decay ratio, represented as $\exp(\sigma(k, v))$, captures the exponential of the time difference between t_k and τ_v, reflecting how the impact of a measurement at time τ_v on the probability at time t_k diminishes over time. As time passes, this impact reduces gradually. For instance, the effect of a fall-caused injury on the onset of disability would attenuate as time progresses. To delve into event-free probabilities, we employ the lower triangular identity matrix denoted as $\mathbf{A} = (\boldsymbol{\alpha}_1, \ldots, \boldsymbol{\alpha}_K) \in \mathbb{R}^{K \times K}$, where $\alpha_{ij} = 1$ if $i \geq j$ and 0 otherwise. This matrix aids us in investigating probabilities of remaining event-free. For individuals belonging to $\mathcal{G}_1 = \{i | \text{for all } i : C_i = 1\}$, who possess an unknown event time, their health statuses remain consistent prior to the given time stamp. Consequently,

$$\Pr\left(Y \mid \mathbf{X}; \mathbf{W}\right)_1 = \frac{\exp(\mathbf{W} * \mathbf{X} \cdot \boldsymbol{\Delta} * \mathbf{A} \cdot \mathbb{1}(\mathbf{y} \leq 0)) \cdot \mathbf{1}}{\exp\left(\mathbf{W} * \mathbf{X} \cdot \boldsymbol{\Delta} * \mathbf{A}\right) \cdot \mathbf{1}}. \tag{4}$$

The numerator represents the accumulation of the risks associated with the occurrence of target responses. To learn the matrix $\mathbf{W}$, we undertake minimization of the negative logarithm of the likelihood across all individuals through an expectation-maximization process [3]. This involves suitable initialization and can be outlined as follows:

$$\min_{\mathbf{W}} \; P(\mathbf{W}) - \sum_{i \in \mathcal{G}_0} \log(\Pr(Y_i | \mathbf{X}_i; \mathbf{W})_0) - \sum_{i \in \mathcal{G}_1} \log(\Pr(Y_i | \mathbf{X}_i; \mathbf{W})_1), \tag{5}$$

The incorporation of the elastic-net penalty $P(\mathbf{W}) = \lambda_1 |\mathbf{W}|_1 + \lambda_2 |\mathbf{W}|_2^2$ into the loss function enhances its strong convexity, resulting in a unique minimum. To establish suitable values for the hyperparameters λ_1 and λ_2, we rely on an independent validation set. The learning objective function exhibits strict convexity within the feasible domain $\mathbf{w}_k \in \mathbb{R}^D, \forall k$, ensuring a unique globally optimal solution.

4.3   Prediction

Assuming the availability of known measurements $\mathbf{X}'$ for a new individual at time instances $\tau_1 < \cdots < \tau_{V'}$ before time t_k, and utilizing the learned parameters $\widehat{\mathbf{W}} \in \mathbb{R}^{D \times V}$, the prediction of the event-free probability at time t_k effectively amounts to predicting the health statuses $Y'[t_k] = (1, 1, \ldots, 1) \in \mathbb{R}^{1 \times k}$. This probability, denoted as $\Pr\left(Y'[t_k] \mid \mathbf{X'}; \widehat{\mathbf{W}}\right)$, is amenable to estimation through the regression model depicted in Eq. 1. (In the subsequent discussion, the notation involving a "'" signifies that it is redefined for the test set.)

4.4   Estimate of Time-to-Event

Upon obtaining the predicted survival probabilities for different time points from our model, the subsequent task entails estimating the time-to-event for the new individual. To initiate this process, we establish the relative error at time t using the following definition: $E(t) = \sum_{k=1}^{K}(|\log t - \log t_k|)\Pr\left(Y'[t_k] \mid \mathbf{X'}; \widehat{\mathbf{W}}\right)$, thereby seeking the time at which E is the lowest, i.e., the time-to-event is $\widehat{T} = \operatorname{argmin}_{t \in \{t_1, \ldots, t_K\}} E(t)$.

5   Experiments

5.1   Data

The dataset comprises the Breast Cancer dataset sourced from the prognostic data repository provided by the PCoE of NASA Ames, in addition to the Aging Canada dataset extracted from the Canadian Community Health Survey (CCHS) statistical surveys. In the context of the Breast Cancer dataset, the times-to-event were computed by subtracting the date of diagnosis from the date of last contact, which corresponds to the study cutoff. As for the Aging Canada dataset, our focus was on respondents surveyed between 2009 and 2010. This dataset centers on the health of individuals aged 45 and above, probing various factors influencing healthy aging. Over 2,000 valid interviews covering the population residing in all ten provinces were included. Regarding the factors under consideration, we engaged in pairwise correlation analysis between independent factors via a correlation matrix. If the Pearson correlation coefficient [14] exceeded 0.75 for two factors, we made the decision to retain the more pertinent one and discard the other. This process yielded 16 factors for the Breast Cancer dataset and 24 for the Aging Canada dataset. For the categorization of factors into numerical representations (with the exception of the two existing numerical factors: age and BMI), we employed Softmax normalization. For a comprehensive overview of the data statistics, please refer to Table 3.

Table 3. Statistics of the two lifetime datasets, such as data size (number of individuals) N, dimensionality (number of risk factors) V, censoring rate (number of individuals with censored time-to-event) C, missing-value percentage M, and event of interest

Data	N	V	C	M	Event
Breast Cancer	2,165	12	23.3%	12.7%	death
Aging Canada	4,470	24	42.6%	27.5%	disability

5.2   Baselines

During the experimental phase, we executed a comparative analysis involving our method and nine distinct time-to-event prediction models. Cox model with elastic-net regularization (Cox-EN) [31]: This model optimizes regression coefficients through the incorporation of elastic-net regularization. Cox model with neural network (CoxNN) [5]: CoxNN departs from the traditional linear exponent of the Cox hazard by introducing a nonlinear artificial neural network output. Accelerated failure time (AFT) model [26]: AFT assumes a Weibull survival distribution to model survival times. Random survival forest (RSF) [9]: RSF aggregates tree-based Nelson-Aalen estimators to derive estimates for the conditional cumulative failure hazard. Multi-task logistic regression (MTLR) [18]: MTLR models the survival distribution by employing multi-task logistic regression in a dependent manner, with the regularization parameter determined via an additional 10-fold cross-validation (10CV). Accumulative-hazard-based joint likelihood (AHJ) model [28]: AHJ captures the connection between survival probability and time-varying factors of longitudinal data in a succinct yet potent manner. DeepSurv [12]: DeepSurv is a deep learning extension of the Cox proportional hazards model. DeepHit [15]: DeepHit employs a deep neural network to directly learn the distribution of survival times. It avoids presumptions about the underlying stochastic process, allowing for the possibility of evolving relationships between risk factors and risks over time. Survival neural network (SNN) [27]: SNN computes binary classification scores at fixed time intervals to estimate survival outcomes. For all models, the hyperparameters were optimized using a validation dataset. The chosen settings are those that yield optimal outcomes as per the respective work's findings.

5.3   Performance Evaluation Metrics

To assess the predictive efficacy, we employed three distinct evaluation metrics: the area under the ROC curve (AUC), the concordance index (C-index), and the Brier score (BS), which we adapted to align with our specific context.

AUC. AUC measures the predictive ability of the model at a particular time. It qualifies the ability of a model to address questions such as "Whether individual i would be likely to die one year after index discharge?" It can be defined as

$$\text{AUC} = \frac{1}{|\mathcal{G}_0'| \times |\mathcal{G}_1'|} \sum_{i \in \mathcal{G}_0'} \sum_{j \in \mathcal{G}_1'} \mathbb{1}\left(\Pr(Y_i'[\mathcal{T}^*]|\mathbf{X}_i) < \Pr(Y_j'[\mathcal{T}^*]|\mathbf{X}_j)\right).$$

C-Index. C-index is a generalization of the concept of AUC. It measures how accurately a model can answer the questions such as "Which one of the two patients i and j is more likely to die or be rehospitalized?" We define C-index as

$$\text{C-index} = \frac{1}{|\mathcal{P}|} \sum_{i,j \in \mathcal{P}} \mathbb{1}\left(\Pr(Y_i'[T_i]|\mathbf{X}_i) \leq \Pr(Y_j'[S_j]|\mathbf{X}_j)\right),$$

where $\mathcal{P} = \{(i,j)|\forall i, j : T_i \le S_j\}$ is the number of comparable pairs of patients. (Refer to [28] for more details regarding this metric.) Similar to AUC, C-index takes values from 0.5 (completely random) to 1.0 (perfect prediction).

BS. Precisely, the Brier score (BS) operates as a mean squared error for time-to-event predictions, serving as an indicator of the caliber of survival probability predictions – in essence, quantifying prediction accuracy. The Brier score allows for an assessment of the model's capability to address queries such as "How precise is the prognosis regarding the recurrence of the disease for patient i?" This measure is computed as an overall error gauge spanning all the time points, enabling a comprehensive evaluation of predictive accuracy, as follows:

$$\text{BS} = \frac{1}{|\mathcal{G}_0' \cup \mathcal{G}_1'|} \sum_{i \in \mathcal{G}_0' \cup \mathcal{G}_1'} \left(C_i - \Pr(Y_i'[S_i]|\mathbf{X}_i)\right)^2,$$

where ϵ_i represents the event indicator for individual i, taking the value of 1 when an event occurs and 0 otherwise. Notably, the Brier score (BS) is constrained to fall within the interval $[0, 1]$. A smaller BS value corresponds to a heightened precision in prognostication, indicating greater accuracy in the predictions.

RE for Survival Time. In order to assess the accuracy of the predicted onset time, we introduce the concept of relative error (RE), which quantifies the disparity between the estimated time $\widehat{T}$ and the actual ground truth T for all individuals in the test set who have experienced an event:

$$\text{RE} = \frac{1}{|\mathcal{G}_0'|} \sum_{\forall i \in \mathcal{G}_0'} \min\left\{|(\widehat{T}_i - T_i)/T_i|, 1\right\}.$$

Here, a smaller RE means a more accurate estimate of time-to-event.

5.4   Results and Discussion

Comparison of Time-to-Event Prediction. Table 4 presents the 10 cross-validation results on the two datasets. Our model outperforms all the other models but the relative error of the predicted time-to-event on the Aging Canada test set. SSP achieves not only accurate predictions of survival probability but accurate estimates of time-to-event, where the RE of the survival estimates for cancer patients is the lowest and for aging people is the second lowest (only inferior to AHJ). In most cases, SNN is the second best, e.g., it achieves the second-best C-index, BS, and RE for the time-to-event estimate on the Breast Cancer dataset, and the second-best AUC and BS on the Aging Canada dataset. Compared to SNN, we can see that SSP achieves an over 2% AUC improvement on Breast Cancer data and 1% on Aging Canada data. Meantime, it yields accurate results with an over 2% C-index improvement, while achieving a lower BS and more accurate time-to-event prediction. The comparison between SSP and MTLR demonstrates that our approach designed for repeated measures performs more effectively. ML-based models perform better than the statistic models, such as Cox-EN and AFT, revealing the strong processing capabilities and applicability of ML-based models to deal with complex time-to-event data. The performance of Cox-EN, CoxNN, and AHJ, is not as good as SSN, MTLR, and SSP; the most possible reason is that these three models assume that Cox's proportional risk meets. Although AHJ considers the change in the risk of the event, it does not perform better than SSP. This is mainly because of its proportional hazard assumptions (N.b.: AHJ is a typical Cox-based model). The performance of the parametric model AFT on two data sets is the worst since the Weibull distribution assumption does not fit most of the time distribution well. The models (e.g., SSP, SNN, and AHJ) that are specifically designed for handling time-varying risk factors perform better, in comparison with those non-time-varying models.

Table 4. Comparison of the 10 cross-validation (10CV) results, and RE of the estimated survival times, on the test data, in the form of the mean (standard deviation). The best results are in **bold** and the second-best performances are underlined.

Dataset	Model	AUC	C-index	BS	RE
Breast Cancer	Cox-EN	.676(.047)	.695(.027)	.283(.033)	.473(.068)
	CoxNN	.693(.034)	.673(.028)	.313(.025)	.542(.073)
	AFT	.670(.016)	.651(.056)	.260(.038)	.450(.062)
	RSF	.687(.021)	.682(.034)	.274(.036)	.410(.094)
	MTLR	<u>735</u>(.041)	.701(.029)	.264(.020)	.392(.085)
	AHJ	.712(.024)	.682(.017)	.302(.012)	.325(.073)
	DeepSurv	.673(.031)	.669(.028)	.372(.025)	.583(.059)
	DeepHit	.715(.014)	.677(.035)	.278(.029)	.339(.061)
	SNN	.724(.035)	<u>.739</u>(.030)	<u>.196</u>(.021)	<u>.311</u>(.053)
	SSP	**.757**(.022)	**.763**(.029)	**.192**(.019)	**.308**(.058)
Aging Canada	Cox-EN	.671(.023)	.682(.020)	.372(.023)	.420(.053)
	CoxNN	.674(.049)	.707(.013)	.341(.023)	.413(.084)
	AFT	.632(.027)	.665(.022)	.282(.019)	.439(.082)
	RSF	.634(.033)	.643(.044)	.302(.033)	.475(.075)
	MTLR	.742(.027)	.733(.019)	.298(.043)	.346(.082)
	AHJ	.729(.025)	<u>.738</u>(.023)	.251(.015)	**.334**(.064)
	DeepSurv	.703(.031)	.695(.027)	.299(.022)	384(.079)
	DeepHit	.698(.032)	.725(.031)	.291(.033)	.429(.058)
	SNN	<u>.772</u>(.018)	.723(.024)	<u>.234</u>(.041)	.369(.064)
	SSP	**.781**(.019)	**.743**(.018)	**.223**(.028)	<u>.341</u>(.073)

Predicted Survival Probability. Figure 1 illustrates the average predicted survival probabilities for all individuals across a 12-month period. Observing the graph, it becomes evident that the survival probabilities generated by our model (SSP) markedly diverge from the predictions of the other models. Notably, our model consistently yields lower probabilities at each time point (with one time point corresponding to each month). The survival curves produced by SNN, AHJ, and MTLR appear closely aligned, with their probabilities demonstrating significant discrepancies from the other six models (Cox-EN, CoxNN, AFT, RSF, DeepSurv, and DeepHit), which yield comparatively higher probabilities. While these curves don't conclusively indicate the superior model, the predictions provided by SSP, SNN, AHJ, and MTLR effectively facilitate the differentiation between individuals at high risk and those at low risk. Regarding the Aging Canada dataset, most models generate closely aligned monthly average survival probabilities. Notably, both SSP and SNN yield notably lower survival probabilities (dipping below 0.7) for the last three months of the follow-up period.

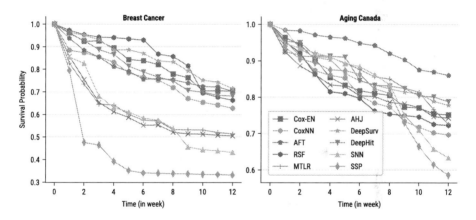

Fig. 1. Comparison of the average of predicted survival probability

Case Study. Figure 2 showcases the predicted survival probabilities for two distinct patients: one categorized as high risk and the other as low risk. (Note: In this context, high-risk patients are those who succumbed during the follow-up period, while low-risk patients remained unaffected.) Here are the particulars of the patients: 1) The high-risk is a 58-year-old female who was discharged from the hospital on July 10, 2010, following a 13-day hospital stay. Tragically, she passed away due to breast cancer 185 d after her discharge. Notably, there were no instances of rehospitalization prior to her demise; 2) The low-risk is a 75-year-old male discharged from the hospital on April 2, 2012, after 23 d of stay. He was still alive by the end of the 1-year follow-up period and had never been rehospitalized before his death. Only the SNN, MTLR, and SSP models are able to distinctly differentiate between these two patients. Notably, the SSP model manifests the most pronounced distinction between the two cases. Notably, the SSP model is capable of projecting an exceedingly low survival probability for the high-risk patient at the 185-day mark. This functionality holds immense value as it can enable timely warnings and provision of advice regarding early interventions for high-risk patients.

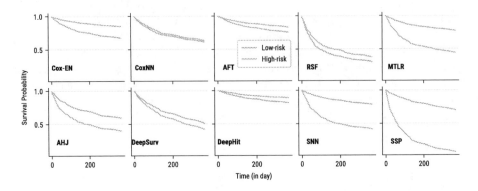

Fig. 2. Comparison of the predicted survival probability for two patients

Ablation Study. Given that SSP amalgamates several crucial components, including censoring likelihood (see Eq. 4), time-sensitive risk considerations (as depicted in Eq. 2), and regularization within the learning objective, we undertake exhaustive ablation studies to dissect the contributions of distinct components. Four distinct variants are considered for analysis: 1) SSP-censor disregards the censoring likelihood and employs only Eq. 1 to determine model parameters; 2) SSP-decay omits the time-dependent decay in cumulative risk computation, with $\eta(u, \tau)$ set to 1 in Eq. 2; 3) SSP-static neglects repeated measures, utilizing Cox proportional hazard instead of Eq. 2; 4) SSP-reg: In this variant, regularization is excluded, rendering $P(\mathbf{W}) = 0$ in the learning objective function as presented in Eq. 5. The results (see Fig. 3) consistently underscore the supremacy of SSP relative to all the variants. Notably, the improvement of SSP over SSP-censor underscores the enhancement in predictive capacity gained by estimating the censoring likelihood. The superiority of SSP in contrast to SSP-decay underscores the effectiveness of the time-decay setting. Furthermore, the comparison between SSP and SSP-static accentuates the significance of incorporating all repeated measures. In comparison with SSP-reg, SSP exhibits higher AUC and C-index values, illustrating the potential efficacy of regularization.

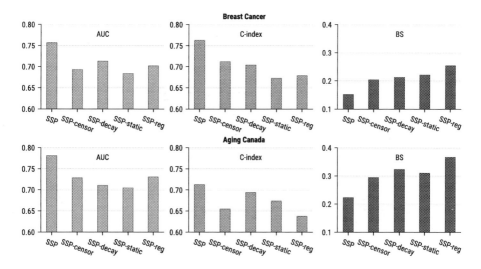

Fig. 3. Comparison of variants' performance

6   Conclusion

This paper proposes a new repeated measure modeling method that quantifies the impact of measures on health status and time-to-event. The new method optimizes the parameters on full data and the regularization approach. The model learning is performed to transform survival prediction into multiple logistic regression learning tasks at different time points. This approach eliminates the need for any assumptions regarding the distribution of unknown data. The outcomes of comparative experiments conducted on real-world data substantiate that the proposed method surpasses state-of-the-art models. It demonstrates superior performance and efficacy in predicting both survival probability and time-to-event outcomes.

Acknowledgments. This work was supported by the Natural Sciences and Engineering Research Council of Canada (NSERC) under Discovery Grant RGPIN-2020-07110 and Discovery Accelerator Supplements Grant RGPAS-2020-00089, and the National Natural Science Foundation of China (NSFC) under Grant No. U1805263.

References

1. Aalen, O.: Nonparametric estimation of partial transition probabilities in multiple decrement models. Ann. Stat., 534–545 (1978)
2. Cox, D.R.: Regression models and life tables. J. R. Stat. Soc. Ser. B Stat. Methodol. **34**, 187–220 (1972)
3. Dempster, A.P., Laird, N.M., Rubin, D.B.: Maximum likelihood from incomplete data via the EM algorithm. J. R. Stat. Soc. Ser. B Stat. Methodol., 1–38 (1977)
4. Fan, J., Zhang, W.: Statistical methods with varying coefficient models. Stat. Interface **1**(1), 179 (2008)

5. Faraggi, D., Simon, R.: A neural network model for survival data. Stat. Med. **14**(1), 73–82 (1995)
6. Fernández, T., Rivera, N., Teh, Y.W.: Gaussian processes for survival analysis. In: NeurIPS, pp. 5021–5029 (2016)
7. Fisher, L.D., Lin, D.Y.: Time-dependent covariates in the Cox proportional-hazards regression model. Annu. Rev. Public Health **20**(1), 145–157 (1999)
8. Giunchiglia, E., Nemchenko, A., van der Schaar, M.: RNN-SURV: a deep recurrent model for survival analysis. In: ICANN, pp. 23–32 (2018)
9. Ishwaran, H., Kogalur, U.B., Blackstone, E.H., Lauer, M.S.: Random survival forests. Ann. Appl. Stat., 841–860 (2008)
10. Jenkins, S.P.: Survival analysis unpublished Manuscript, Institute for Social and Economic Research, Chapter 3, University of Essex, Colchester, UK (2005)
11. Kaplan, E.L., Meier, P.: Nonparametric estimation from incomplete observations. J. Amer. Statist. Assoc. **53**(282), 457–481 (1958)
12. Katzman, J., Shaham, U., Bates, J., Cloninger, A., Jiang, T., Kluger, Y.: DeepSurv: personalized treatment recommender system using a Cox proportional hazards deep neural network. BMC Med. Res. Methodol., 18–24 (2018)
13. Kim, M., Pavlovic, V.: Variational inference for gaussian process models for survival analysis. In: UAI, pp. 435–445 (2018)
14. Kirch, W. (ed.): Pearson's Correlation Coefficient, pp. 1090–1091. Springer, Netherland(2008). https://doi.org/10.1007/978-1-4020-5614-7_2569
15. Lee, C., Zame, W.R., Yoon, J., van der Schaar, M.: Deephit: a deep learning approach to survival analysis with competing risks. In: AAAI, pp. 2314–2321 (2018)
16. Lee, E.T., Wang, J.W.: Statistical Methods for Survival Data Analysis. John Wiley & Sons, 4th edn. (2013)
17. Li, Y., Wang, L., Wang, J., Ye, J., Reddy, C.K.: Transfer learning for survival analysis via efficient l2,1-norm regularized Cox regression. In: ICDM, pp. 231–240 (2017)
18. Lin, H.c., Baracos, V., Greiner, R., Chun-nam, J.Y.: Learning patient-specific cancer survival distributions as a sequence of dependent regressors. In: NeurIPS, pp. 1845–1853 (2011)
19. Liu, M., Lu, W., Shore, R.E., Zeleniuch-Jacquotte, A.: Cox regression model with time-varying coefficients in nested case - control studies. Biostatistics **11**(4), 693–706 (2010)
20. Liu, S., Trenkler, G.: Hadamard, khatri-rao, kronecker and other matrix products. Int. J. Inf. Syst. Sci. **4**(1), 160–177 (2008)
21. Moghaddass, R., Rudin, C.: The latent state hazard model, with application to wind turbine reliability. Ann. Appl. Stat. **9**(4), 1823–1863 (2014)
22. Sun, Y., Sundaram, R., Zhao, Y.: Empirical likelihood inference for the Cox model with time-dependent coefficients via local partial likelihood. Scand. J. Stat. **36**(3), 444–462 (2009)
23. Tian, L., Zucker, D., Wei, L.: On the Cox model with time-varying regression coefficients. J. Amer. Statist. Assoc. **100**(469), 172–183 (2005)
24. Vinzamuri, B., Li, Y., Reddy, C.K.: Active learning based survival regression for censored data. In: CIKM, pp. 241–250 (2014)
25. Wang, P., Li, Y., Reddy, C.K.: Machine learning for survival analysis: A survey. ACM Comput. Surv. **51**(6), 1–36 (2019)
26. Wei, L.J.: The accelerated failure time model: a useful alternative to the Cox regression model in survival analysis. Stat. Med. **11**(14–15), 1871–1879 (1992)
27. Zhang, J., Wang, S., Chen, L., Guo, G., Vanasse, A.: Time-dependent survival neural network for remaining useful life prediction. In: PAKDD, pp. 441–452 (2019)

28. Zhang, J., Wang, S., Courteau, J., Chen, L., Bach, A., Vanasse, A.: Predicting COPD failure by modeling hazard in longitudinal clinical data. In: ICDM, pp. 639–648 (2016)
29. Zhang, Z., Reinikainen, J., Adeleke, K.A., Pieterse, M.E., Groothuis-Oudshoorn, C.G.: Time-varying covariates and coefficients in Cox regression models. Annals Trans, Med. **6**(7), 121 (2018)
30. Zhou, M.: Understanding the Cox regression models with time-change covariates. Am. Stat. **55**(2), 153–155 (2001)
31. Zou, H., Hastie, T.: Regularization and variable selection via the elastic net. J. R. Stat. Soc. Ser. B Stat. Methodol. **67**(2), 301–320 (2005)

A Method for Identifying the Timeliness of Manufacturing Data Based on Weighted Timeliness Graph

Zehua Liu[1,2], Xuefeng Ding[1,2], Yuming Jiang[1,2], and Dasha Hu[1,2(✉)]

[1] College of Computer Science, Sichuan University, Chengdu 610065, China
hudasha@scu.edu.cn
[2] Big Data Analysis and Fusion Application Technology Engineering Laboratory of Sichuan Province, Chengdu 610065, China

Abstract. Timeliness is one of the important indicators of data quality. In industrial production processes, a large amount of dependent data is generated, often resulting in unclear timestamps. Therefore, this article combines the conclusion dependency graph into a process dependency graph to determine the identification order of the timeliness of each process data; By constructing a weighted timeliness graph (WTG) and path single flux, a data timeliness identification method that does not completely rely on timestamps is proposed. Finally, a time-effectiveness identification method based on weighted time-effectiveness graph was discussed through an example and 9 dependency rules, and the effectiveness of the method was verified through a set of experiments.

Keywords: Manufacturing Big Data · Quality Appraisal · Dependency Rules · Data Timeliness

1 Introduction

A large number of sensors will collect various types of data in industrial production processes. Before using data, it is necessary to evaluate its quality, and the timeliness of data is an important indicator of data quality [1]. Therefore, a time dependent method for identifying timeliness is proposed. This method can effectively identify the timeliness of data when the timestamp is incomplete. The prerequisite for assessing the timeliness of data is that the data is correct and can meet work needs. Compared to the correctness of data, the timeliness of data does not necessarily need to be tested in the real world [2]. Therefore, the measurement of data timeliness should be an estimate, not a validation statement under certainty, to determine the probability of data validity [3]. For a large amount of data, it is reasonable to quantify timeliness through this estimation when the validity of the data is unclear [4].

Reference [5] proposes a probability base metric for calculating the timeliness of Wiki articles related to timeliness events. Reference [6] defines a recurrent timeliness rules (RTR) to evaluate the timeliness of periodic data generated during the manufacturing process. Reference [7] define a measure as a function that depends on the age of the attribute value at the time the currency is evaluated and the sensitivity parameters that make the measure suitable for the application environment. Reference [8] assumes that timestamps are available and mentions using the current time variable to represent the latest time. Reference [9] developed an uncertain constraint database scheme based on the constraint database scheme, and abstracted the important example of the complexity of query identification in uncertain time constraint databases. Reference [10] was used to investigate the work on time constraint satisfaction problem (TCSP) by finding estimates that satisfy a set of time variables with time constraints. Reference [11] first studied the use of rules $\forall t_1, \cdots, t_j : R(t_1[EID] = t_j[EID] \land \varphi \rightarrow t_u \prec_A t_v), j \in [1, k]$ to help identify the timeliness of data when there is no clear timestamp in the database. Document [12] proposed a dynamic functional dependency relationship, which stipulates that a copy function can require some attributes that cannot be independently changed to be copied together. It is necessary to judge the time series relationship describing different attribute values of the same entity based on a small amount of time series rules obtained from domain knowledge, so as to identify which values are outdated. The disadvantage is that the current method cannot identify whether a value is outdated or invalid at a given time point.

In summary, the current research on data timeliness is not suitable for direct application to the identification of industrial data timeliness. There are two main reasons: firstly, it is unable to effectively provide the timeliness of the current data without clear timestamps. Secondly, it is impossible to quantify the timeliness of data at a given time point. Therefore, this article proposes a method for evaluating the timeliness of manufacturing data based on weighted timeliness maps, which is used to identify the timeliness of industrial data through time dependency relationships without a clear timestamp.

2   Data Timeliness Identification Method

This article presents the dependency relationships between various conclusions by constructing a conclusion dependency sequence diagram. Then, merge the conclusion dependency graph into a process dependency sequence graph to determine the calculation order of the timeliness of each process data. Finally, the timeliness of data in a certain process is determined through a weighted timeliness graph.

2.1   Limited Timeliness Dependency Rules

Time dependent rules refer to the identification of the timeliness of production data through the dependency relationships between processes.

Definition 1. (Limited timeliness dependency rule): In Rule r, there is a dependency relationship between each process, and the data generated in Process A will be limited by its related processes. The degree of limitation is quantified by the strength of the dependency. The rules that have limited dependencies between these process data are defined as limited time dependency rules, and their dependencies can be represented as:

$$r : \forall t_i, t_j(\psi_B \rightarrow t_i \prec_A t_j, \beta) \tag{1}$$

Among them, $t_i \prec_A t_j$ represents the generation of date t_i before the generation of date t_j, β It is the dependency strength of the rule, ψ Represents the temporal relationship or constraint conditions in process B. The dependency strength of rule represents the likelihood of the rule represented by r. Because in actual industrial production processes, each process or employee is relatively independent, probability can be used to represent the relationship between different process data.

Let $lhs(r)$ represents the left part of the rule, and $rhs(r)$ represents the right part of the rule, that is, in formula (1), $lhs(r) = \psi$, $rhs(r) = t_i \prec_A t_j$. For the right part of the rule, $rhs(r) = tml(lhs(r)) \times tml(r)$, where the dependency strength calculation of the left part $lhs(r) = \psi$ follows the following rule:(a). If ψ is the condition $t_i[A]opt_j[A]$ or $t_k[A]opa$ is determined, then if ψ is true, then $tml(\psi) = 1$ is satisfied, otherwise $tml(\psi) = 0$, where $op \in \{=, \neq, <, >, \leq, \geq\}$ and a are constants; (b). If ψ is $t_i \prec_A t_j$ or $t_k \prec_A \tau$ and ψ is not the right part of any rule r, then the value of $tml(\psi)$ is obtained statistically; (c). If ψ is the right part of other rules $r\prime$, the value of $tml(\psi)$ is obtained by the calculation method of the right part of rules $r\prime$ dependence strength; (d). If $\psi = \psi_1 \wedge \psi_2$, then ψ represents the conjunction of ψ_1 and ψ_2, and the dependence strength $tml(\psi) = min\{tml(\psi_1), tml(\psi_2)\}$ after the conjunction.

2.2 Process Dependency Sequence Graph

To determine the calculation order of the timeliness of each process data, it is necessary to construct a process dependency graph. To construct a process dependency graph, it is first necessary to construct a conclusion dependency graph based on dependency rules, and then merge them. The specific steps for constructing a dependency sequence diagram are as follows. Firstly, based on the type of conclusion, determine whether the conclusion needs to be included in the dependency sequence diagram. The conclusion of $t_i[A]opt_j[A]$ or $t_k[A]opa$ is a deterministic conclusion that satisfies the condition of 1, otherwise it is 0 and does not need to be added to the conclusion dependency diagram; The conclusions of $t_i \prec_A t_j$ and $t_i \prec_A \tau$ are non-deterministic and need to be added to the conclusion dependency sequence diagram. Then, when the conclusion is determined to be non-deterministic, two types of nodes, $(\prec, A, *, *)$ and $(\prec, A, \tau)$, are constructed for each process A to represent the conclusions of type $t_i \prec_A t_j$ and type $t_i \prec_A \tau$. Finally, scan the rule set $\Sigma(r)$, is there a rule r_k that makes $\psi_1 = lhs(r_k)$, $\psi_2 = rhs(r_k)$? If so, add directed edges from ψ_1 to ψ_2 to the dependency order graph.

The overall principle for merging conclusion dependency graphs into process dependency graphs is: (a). Merge conclusion nodes containing multiple identical processes in the diagram into one process node; (b). The directed edge no longer represents the dependency relationship between conclusions, but rather represents the dependency relationship between various processes.

2.3 Weighted Timeliness Graph

To evaluate the timeliness of all data items in process A, it is necessary to determine the values of $tml(t_i \prec_A t_j)$ and $tml(t_i \prec_A \tau)$ separately. Therefore, weighted timelines graph (WTG) has been defined.

Definition 2. (weighted timeliness graph, WTG): Let D represent a set of data, Σ Represents a collection of time dependent rules, T representative Σ All sets containing time in the time rule, where A represents a certain process, θ Is the effective time threshold of D, and the weighted time graph of process A is WTG_A, which is defined as:

1. WTG_A contains data items t and time τ in two types of conclusions $t_i \prec_A t_j$ and $t_i \prec_A \tau$, and both take a certain time τ as the initial node of the WTG_A graph, where τ includes the original time in the dataset and the threshold we set θ;
2. aggregate Σ Each rule r in is defined as the initial node of the WTG_A graph if $\tau \in T \cup \{\theta\}$ causes $\tau \subset rhs(r)$;
3. aggregate Σ For each rule r in, if $t_1, t_2 \in D$ causes $rhs(r) = t_1 \prec_A t_2$, $tml(lhs(r)) > 0$ or $tml(r) > 0$, update the weight of the directed edge from t_2 to t_1;
4. The weight of the directed edge t_1, t_2 in the WTG_A graph is denoted as $weight(t_1, t_2) = tml(r) \times tml(lhs(r))$.

The implementation details of constructing a weighted timeliness graph are shown in Algorithm 1.

Algorithm 1: Weighted timeliness graph construction algorithm

---

Input: Dataset D, Rule Set Σ, Process A, designated time θ

Output: Weighted timeliness graph

```
01. V = new ArrayList();E = new ArrayList();visited(r) = 0// Initialization.
02. for τ in T:                    // Scan time set, whether there is a specified time.
03.   if τ_k = θ:
04.     createNode(τ_k)
05.   for r in Σ:
06.     if visited(r) = 0:
07.       if rhs(r_x) = t ≺_A τ_k: // Scan Rule Collection Σ. Is it a conclusion containing time τ.
08.         for t in D:
09.           if tml(lhs(r_x(t_k))) ≠ 0:
10.             V.add(createNode(t_k)); ω_y = weight(τ_k, t_k); E.add(createEdge(τ_k, t_k, ω))
11.           visited(r_x) = 1 //Set the access location of rule r to 1.
12.       if rhs(r_x) = t_i ≺_A t_j:
13.         for t_i, t_j in D:
14.           if tml(lhs(r_x(t_i, t_j))) ≠ 0:
15.             V.add(createNode(t_i)); V.add(createNode(t_j))
16.             ω_y = weight(t_j, t_i); E.add(createEdge(t_j, t_i, ω_y))
17.           visited(r_x) = 1
18.     else:    // If the rule r has already been accessed.
19.       α,β = ExtractVertex(rhs(r_x)); ω' = weight(α, β)
20.       ω = Max{ω',ω_y}; E.updateEdge(β, α, ω)//
```

2.4 Path Single Flux of Weighted Timeliness Graph

The timeliness evaluation of data item $t_i[A]$ requires determining the possibility of conclusion $t_i \prec_A \tau$ being established, that is, the value of $tml(t_i \prec_A \tau)$. The possibility of conclusion $t_i \prec_A \tau$ can be obtained by immediate inference of a rule r_k or there is a directed path $Path(\tau, t_i)$ from τ to t_i in the weighted time effect graph.

Definition 3. (Path single flux of weighted timeliness graph): The single flux of path $Path(\tau, t_i)$ in the WTG_A graph is denoted as $sFlux(Path(\tau, t_i))$, which is defined as the weight of the maximum directed edge that can flow in a single path from τ to t_i.

The weighted time effect graph path single flux can be mainly divided into two meanings:

1. (a). In the weighted efficiency graph of process A, when there is only one path from τ to t_i in path $Path(\tau, t_i)$, we take the directed edge with the highest weight in the path as the flux of this path. Namely, $Flux(Path(\tau, t_i)) = min\{weight(t_i \prec_A v_1), weight(v_1 \prec_A v_2), \cdots, weight(v_k \prec_A \tau)\}$.
2. (b). In the weighted efficiency chart of process A, when there are multiple paths from τ to t_i in path $Path(\tau, t_i)$, we take the maximum flux value of the multiple paths as the single flux of path $Path(\tau, t_i)$ in the graph WTG_A. Namely, $sFlux(Path) = Max\{Flux(Path_1), Flux(Path_2), \cdots, Flux(Path_k)\}$.

3 Example Discussion and Experimental Verification

In order to further explain the defined method for assessing timeliness based on weighted timeliness graph, this article selects an example for discussion. Finally, the effectiveness of the method was verified by testing its recall and accuracy on real datasets.

Table 1. Process data example.

Number	PID	Load test	Hot test	High Voltage test	Vibration test	Warranty
t_1	SC18091	2000 W	280°C	8.08×10^5 pa	20000 Hz	2022
t_2	SC18092	2000 W	200°C	6.06×10^5 pa	10000 Hz	2021
t_3	SC18093	2000 W	160°C	6.05×10^5 pa	8000 Hz	2020

As shown in Table 1, an example D of a process flow with three tuples is given. The process flow example D includes four process data examples of load testing, high pressure test, high temperature test and vibration test, as well as the identification code PID and Warranty of each product.

In addition, its corresponding set of restricted failure dependency rules has been defined $\Sigma(r)$, $\Sigma(r)$ contains 5 dependency rules, denoted as r_1 to r_5, as shown in Table 2.

There are 5 rules in Table 2 that represent the dependency relationships between different process data in Table 1. It can be seen that in practice, there may be situations where multiple dependency rules derive the same conclusion. Due to the stronger dependency strength of rules, they often have stronger persuasiveness. Therefore, the

Table 2. Formulation rule representation of dependency intensity uncertainty.

Number	Formulaic representation
r_1	$\forall t_i, t_j(t_i[Loadtest] < t_j[Loadtest] \bigwedge t_i[Warranty] < t_j[Warranty] \rightarrow t_i \prec_{Loadtest} t_j, 0.9)$
r_2	$\forall t_i, t_j(t_i \prec_{Loadtest} t_j \rightarrow t_i \prec_{Highvoltagetest} t_j, 0.8)$
r_3	$\forall t(t[Loadtest] = 2000W \bigwedge t[Highvoltagetest] = 280°C \bigwedge t[BID] = SC18091 \rightarrow t \prec_{Loadtest} 2023, 1)$
r_4	$\forall t(t \prec_{Loadtest} 2023 \rightarrow t \prec_{Highvoltagetest} 2023, 0.8)$
r_5	$\forall t_i, t_j(tml(t_j \prec_{Loadtest} 2023) = min\{tml(t_j \prec_{Loadtest} t_i), tml(t_i \prec_{Loadtest} 2023)\})$

value with higher dependency strength is chosen as the final dependency strength of this conclusion. The formula is:

$$rhs(r) \Rightarrow tml(Q) = max\{tml(rhs(r_1)), tml(rhs(r_2)), \cdots, tml(rhs(r_k))\}, i \in [1, k] \quad (2)$$

Among them, $tml(rhs(r_i)) = tml(r_i) \times tml(lhs(r_i))$.

On the basis of the original r_2 and r_4, three rules r_6 : $\forall t_i, t_j(t_i \prec_{Vibrationtest} t_j \rightarrow t_i \prec_{Loadtest} t_j, 0.8)$, r_7 : $\forall t_i, t_j(t_i \prec_{Vibrationtest} t_j \rightarrow t_i \prec_{Highvoltagetest} t_j, 0.8)$, r_8 : $\forall t_i, t_j(t_i \prec_{Loadtest} 2023 \rightarrow t_i \prec_{Vibrationtest} t_j)$ have been added. Since the left part of rules r_1 and r_3 belong to deterministic conclusions, they do not need to appear in the dependency order diagram. Based on these five rules, a conclusion dependency order diagram was constructed, as shown in Fig. 1.

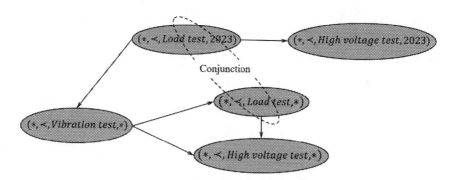

Fig. 1. Dependency sequence diagram.

In the conclusion dependency sequence diagram shown in Fig. 1, $t_i \prec_{Loadtest} t_j$ and $t_j \prec_{Loadtest} 2023$ can be combined to obtain a new conclusion $t_i \prec_{Loadtest} 2023$. At this point, the three conclusions $(*, \prec, Loadtest, \tau)$, $(*, \prec, Loadtest, *)$, and $(*, \prec, Vibrationtest, *)$ interact with each other (there is a hidden loop), so the unique identification order for the timeliness of the conclusion cannot be determined. At this point, it is necessary to merge the conclusion dependency sequence diagram, as shown in Fig. 2.

As shown in Fig. 2, the calculation order for the timeliness of each process data can be determined based on the merged process dependency sequence diagram of the conclusion dependency sequence diagram as follows: *Vibrationtest, Loadtest, Highvoltagetest*.

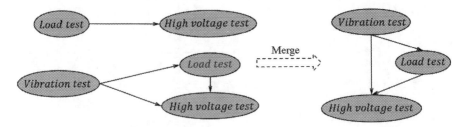

Fig. 2. Merging conclusion dependency sequence graphs.

In order to make the case more convincing, a new rule $r_9 = \forall t(t[Vibrationtest]$ $= 800\,\text{Hz} \rightarrow t \prec_{Loadtest} 2023, 0.8)$ is added to Σ based on rule sets r_1 and r_3. By using these three rules, the weighted efficiency graph $WTG_{Loadtest}$ of process $Loadtest$ at time 2023 can be obtained, as shown in Fig. 3 (a). The timeliness relationship between $t_1, t_2,$ and t_3 data items in the WTG diagram of the $Loadtest$ process at time $\tau = 2023$ in Fig. 3 (a) is shown in Fig. 3(a). In Fig. 3(b), there is $weight(t_1, t_2) = tml(t_1 \prec_{Highvoltagetest} t_2) \times tml(r_2) = 0.9 \times 0.8 = 0.72$.

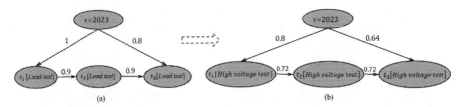

Fig. 3. Weighted timeliness graph of process $Loadtest$ and process $Highvoltagetest$ at 2023.

Using $t_3 \prec_{Loadtest} 2023$ in Fig. 3 as an example to illustrate how to calculate the single flux of a path, if there are two paths from $\tau = 2023$ to t_3. Therefore, $Flux(Path_1(2023, t_3)) = \min\{1, 0.9, 0.9\} = 0.9$, $Flux(Path_2(2023, t_3)) = \min\{0.8\} = 0.8$ and $tml(t_3[Loadtest]) = sFlux(Path(2023, t_3)) = Max\{0.9, 0.8\} = 0.9$.

In order to verify the effectiveness of the time effective identification method based on the weighted timeliness graph, this paper conducts simulation experiments on real datasets from the industrial big data innovation platform(https://www.industrial-bigdata. com/Data). The experimental results are shown in Fig. 4.

As shown in Fig. 4, it can be seen that the recall and precision both increase with the increase of the number of rules. However, when the number of rules reaches 4, the recall rate and accuracy remain stable at a certain value and do not change.

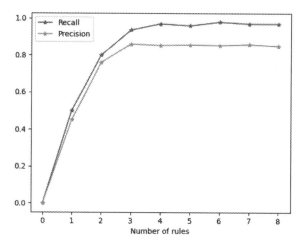

Fig. 4. The trend of recall and precision with the number of rules.

4   Conclusion

The quality appraisal method of manufacturing big data based on timeliness-dependent rules effectively evaluates the quality of manufacturing big data from the perspective of data timeliness. This model can effectively identify the timeliness of manufacturing big data through the dependency relationship between production process data without a clear timestamp. The quantification of rule dependency intensity was achieved by limiting the time dependent rules. The calculation order of the timeliness of each process data was determined through the process order dependency graph. The timeliness of the data was evaluated through a weighted timeliness graph (WTG) and its path single flux. Finally, the identification method was discussed through three process data instances and nine dependency rules, and its effectiveness was verified on real datasets (The recall rate can reach around 0.97 and the precision rate can reach around 0.82).

In future research on timeliness identification, the weight values of edges in the weighted timeliness graph can be set to dynamically change to adapt to different industrial application scenarios.

Acknowledgement. This work was supported in part by the National Key R&D Program of China under Grant No. 2020YFB1707900 and 2020YFB1711800; the National Natural Science Foundation of China under Grant No. 62262074, U2268204 and 62172061; the Science and Technology Project of Sichuan Province under Grant No. 2022YFG0159, 2022YFG0155, 2022YFG0157.

References

1. Li, M., Li, J., Cheng, S., Sun, Y.: Uncertain rule based method for determining data currency. IEICE Transactions on Information and Syst. E101.D (10), 2447–2457(2018)
2. Batini, C., Scannapieco, M.: Data and Information Quality: Dimensions, Principles and Techniques. Springer Publishing Company, Incorporated (2016)

3. Even, A., Shankaranarayanan, G., Berger, P.D.: Evaluating a model for cost-effective data quality management in a real-world CRM setting. Decis. Support. Syst. **50**(1), 152–163 (2010)
4. Firmani, D., Mecella, M., Scannapieco, M., Batini, C.: On the meaningfulness of "big data quality" (invited paper). Data Science Eng. **1**(1), 6–20 (2015)
5. Klier, M., Moestue, L., Obermeier, A.A., Widmann, T.: Event-driven assessment of currency of wiki articles: a novel probability-based metric. In: International Conference on Interaction Sciences (2021)
6. Liu, Z., Ding, X., Tang, J., Jiang, Y., Hu, D.: Anomaly monitoring of process based on recurrent timeliness rules (AMP-RTR). Applied Sciences **12**(24), 12917 (2022)
7. Ballou, D., Wang, R., Pazer, H., Tayi, G.K.: Modeling information manufacturing systems to determine information product quality. Manage. Sci. **44**(4), 462–484 (1998)
8. Dyreson, C.E., Jensen, C.S., Snodgrass, R.T.: Now in temporal databases. In: Encyclopedia of Database Systems. Springer, New York (2018)
9. Koubarakis, M.: The complexity of query evaluation in indefinite temporal constraint databases. Theoret. Comput. Sci. **171**(1), 25–60 (1997)
10. Bodirsky, M., Kára, J.J.J.A.: The complexity of temporal constraint satisfaction problems. Association for Computing Machinery **57**(9), 1–41 (2010)
11. Fan, W., Geerts, F., Wijsen, J.: Determining the currency of data. ACM Trans. Database Systems (TODS) **37**(4), 25–29 (2012)
12. Vianu, V.J.J.A.: Dynamic functional dependencies and database aging. Association for Computing Machinery **34**(1), 28–59 (1987)

STAD: Multivariate Time Series Anomaly Detection Based on Spatio-Temporal Relationship

Keyu Chen[1], Guoping Zhao[2(✉)], Zhenfeng Yao[2], and Zhihong Zhang[1]

[1] School of Computer and Artificial Intelligence, Zhengzhou University, Zhengzhou, China

[2] Esunny Information Technology, Zhengzhou Commodity Exchange, Zhengzhou, China

zhaoguoping@esunny.cc

Abstract. Anomaly detection for multivariate time series is a very complex problem that requires models not only to accurately identify anomalies, but also to provide explanations for the detected anomalies. However, the majority of existing models focus solely on the temporal relationships of multivariate time series, while ignoring the spatial relationships among them, which leads to the decrease of detection accuracy and the defects of anomaly interpretation. To address these limitations, we propose a novel model, named spatio-temporal relationship anomaly detection (STAD). This model employs a novel graph structure learning strategy to discover spatial features among multivariate time series. Specifically, Graph Attention Networks (GAT) and graph structure are used to integrate each time series with its neighboring series. The temporal features of multivariate time series are jointly modeled by using Transformers. Furthermore, we incorporate an anomaly amplification strategy to enhance the detection of anomalies. Experimental results on four public datasets demonstrate the superiority of our proposed model in terms of anomaly detection and interpretation.

Keywords: Anomaly Detection · Multivariate Time Series · GAT · Transformers · Spatio-Temporal Relationship · Graph Structure Learning

1 Introduction

Anomaly detection for multivariate time series has emerged as a prominent research topic in recent times. In the areas of production and IT systems, time series data can directly reflect the working status and operating conditions of the system, which is an important basis for anomaly detection. In the past, domain experts usually utilize their expertise to establish thresholds for each indicator based on empirical observations. However, with the unprecedented explosion in data complexity and scale due to rapid technological advancements, traditional techniques have become insufficient to effectively address the challenges posed by anomaly detection. To tackle this problem, a lot of unsupervised methods based on classical machine learning have been developed over the previous years, including density estimation-based methods [6] and distance-based methods [3,14]. Nevertheless, these approaches fail to capture the intricate and high-dimensional relationships that exist among time series.

Recently, Methods based on deep learning have contributed to the enhancement of anomaly detection for multivariate time series. For example, AutoEncoders (AE) [5], VAE [12], GAN [17], and Transformers [23] are recent popular anomaly detection methods that employ sequence reconstruction to encode time series data. In addition, Long Short-Term Memory (LSTM) networks [9] and Recurrent Neural Networks (RNN) [22] have also displayed promising results for detecting anomalies in multivariate time series. However, most of these methods fail to consider the association between various time series, moreover, they do not offer a clear explanation of which time series are correlated with each other, thus impeding the interpretation of detected anomalies. A complex set of multivariate time series are often intrinsically linked to each other.

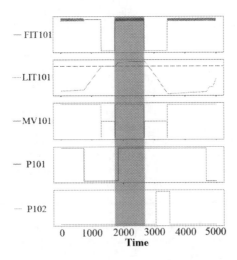

Fig. 1. Multivariate time series segments from the SWaT dataset, with anomalies shaded in red. (Color figure online)

In Fig. 1, the time series are obtained from five sensors of the same process at the SWaT water treatment testbed [16]. The red shaded region corresponds to an anomaly, indicating that the LIT101 value has exceeded the threshold. In addition, the readings of FIT101, MV101, and P101 all changed during this period, P102 changed after the anomaly has ended. Based on the fault log, we know that LIT101 serves as the level transmitter responsible for measuring the water level of the tank, and the anomalous segment corresponds to the overflow of the tank.

The fundamental reason of this anomaly is because of the premature opening of MV101 (inlet valve) in the same process. Given that the state of MV101 is limited to only two possibilities (open and closed) and irregular, identifying abnormal for it through the temporal features is a challenging task. Consequently, integrating spatial features become crucial to detect and explain the anomaly. Several methods have employed graph neural networks for anomaly detection because of its remarkable capability to leverage spatial structural information, such as MTAD-GAT [26] and GDN [8]. However, MTAD-GAT assumes a complete graph structure for the spatial characteristics of multivariate time series, which may not accurately reflect their asymmetric correlations in real-world scenarios. GDN [8] is limited to a single time point and fails to catch the detailed associations between a time point and a whole sequence. GTA [7] combines graph structures for spatial feature learning and Transformers for temporal modeling. However, it utilizes Gumbel-Softmax, which is insufficient in accurately representing the spatial relationships among multivariate time series.

This paper presents a method for anomaly detection by leveraging the spatio-temporal relationships among multivariate time series. The proposed approach leverages the joint optimization of Graph Attention Networks (GAT) and Transformers for unsupervised anomaly detection. In order to explore complex temporal and spatial dependencies among diverse time series, a novel graph structure learning strategy is proposed, which considers multivariate time series as separate nodes and learns attention weights of each node to obtain a bidirectional graph structure. The proposed method employs GAT and graph structure to integrate information of nodes with their neighbors, while the temporal features of time series are modeled utilizing Transformers. The utilization of Transformers in the proposed approach is motivated by their capability to capture long-term dependencies, compute global dependencies, and enable efficient parallel computation. To further enhance the detection performance, an anomaly amplification strategy based on local and global differences is also introduced. In summary, this paper makes the following major contributions:

- We propose a new method for learning the graph structure in multivariate time series.
- We propose an novel method for multivariate time series anomaly detection, which efficiently captures spatio-temporal information using GAT and Transformers.
- Extensive experiments are conducted on four popular datasets to demonstrate the effectiveness of our proposed method. And, ablation studies are conducted to understand the impact of each component in our architecture.

2   Related Work

2.1   Traditional Anomaly Detection for Multivariate Time Series

Traditional methods for time series anomaly detection typically contain distance-based methods and clustering-based methods. LOF (Local Outlier Factor) [6] is a density-based method, which determines the degree of anomaly by comparing the local density between each data point and its surrounding neighboring data points. KNN [3] is a distance-based outlier detection method, which detects anomalies by calculating the distances between each data point and its K nearest neighbors. IsolationForest [14] uses a tree structure to decompose data and quantifies the distances between nodes to identify outliers. Traditional unsupervised methods for anomaly detection are limited in their ability to identify anomalies, as they do not take into account the spatio-temporal relationships inherent in the data.

2.2   Deep Learning Anomaly Detection for Multivariate Time Series

Prediction-Based Models: LSTMNDT [10] leverages LSTM [9] network to predict time series collected from spacecraft, but it ignores the spatial correlations. MTAD-GAT [26] employs two GAT layers to model the spatio-temporal relationships simultaneously, but MTAD-GAT assumes that the spatial characteristics of multivariate time series are a complete graph, in most cases, time series are typically associated in an asymmetric manner. GDN [8] uses node embedding for graph structure learning, encodes spatial information using GAT. However, GDN is limited to a single time point and cannot catch the detailed associations between a time point and a whole sequence.

Reconstruction-Based Models: LSTM-VAE [19] utilizes a LSTM network and a variational autoencoder (VAE) [12] for the reconstruction of time series. DAGMM [27] combines a deep autoencoder with Gaussian Mixture Model. But the Gaussian Mixture Model is not suitable for complex distributed datasets. OmniAnomaly [22] employs a new stochastic RNN based on the LSTM-VAE model for anomaly detection. GANS [13,17,20] uses generators for reconstruction. Anomaly transformer [25] leverages a prior-association and series-association and compares them to better identify anomalies. USAD [4] uses a deep autoencoder trained with adversarial training to learn and detect anomalies in new data. However, all the methods mentioned above only take into account either temporal or spatial associations, without learning both associations, and they may lack sufficient ability to accurately localize anomalies. GTA [7] combines graph structures for spatial feature learning and Transformers for temporal modeling. However, it utilizes Gumbel-Softmax, which is insufficient in accurately representing the spatial relationships between multivariate time series.

3   Method

In this section, we give the details of the proposed spatio-temporal relationship anomaly detection (STAD) for multivariate time series. At first, we present the problem statement and the overall architecture of STAD. Next, we will elaborate the particulars of the graph structure learning, the GAT-based spatial model, Transformers and anomaly amplification modules.

3.1   Problem Statement

In our study, time series is represented by $\{\mathbf{X}_1, \mathbf{X}_2, \cdots, \mathbf{X}_d\}$. For the time series i, $\mathbf{X}_i = [x_{1i}, x_{2i}, \cdots, x_{Ni}]$, where $\mathbf{X}_i \in \mathbb{R}^N$ denotes the observed value of time series i. N is the length of $\mathbf{X}$ and d is the number of multivariate time series. Our goal is to model multivariate time series data in order to identify any anomalous behaviors.

3.2   Overview

The overall architecture of the model in this paper is shown in Fig. 2. It consists of three main components:

(1) Graph Structure Learning: Learn a graph structure that represents spatial relationships between multivariate time series.
(2) GAT-based spatial model: Fusing time series with spatial features using GAT and graph structure.
(3) Transformers based on anomaly amplification strategy: Transformers are used to reconstruct the spatio-temporal relationships of each time series. Anomaly amplification strategy is used to amplify anomalies.

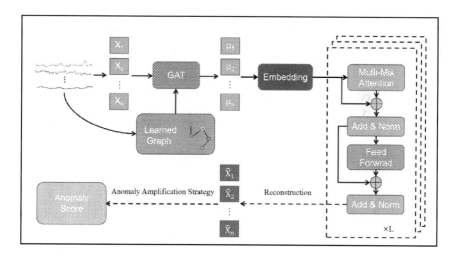

Fig. 2. An overview of the proposed STAD method.

3.3   Graph Structure Learning

For our model, the primary task is to reconstruct spatial and temporal relationships for multivariate time series. For spatial modeling, we utilize a learnable graph to represent the relationships between multivariate time series. We consider each time series as a node, and the relationships between the time series are represented as edges in the graph. An adjacency matrix $A \in \mathbb{R}^{d \times d}$ is used to express this graph, where A_{ij} denotes that there are edges between node i to node j. Our proposed framework has flexibility and can automatically learn the relationships of the graph without prior knowledge about the graph structure. In order to obtain the hidden dependencies between nodes, we designed a framework that, unlike the GDN [8], does not use the node embedding learning graph structure. We learn a weight matrix that assigns a weight score to each node based on its own features and similarity to other nodes, and then use top k to filter the most relevant sets for the graph structure:

$$e_{ij} = \text{LeakyReLU} \left(\mathbf{w}^{\mathrm{T}} \cdot (\mathbf{X}_i \oplus \mathbf{X}_j) \right) \tag{1}$$

$$\rho_{ij} = \frac{\exp(e_{ij})}{\sum_{k=1}^{d} \exp(e_{ik})} \tag{2}$$

$$A_{ij} = 1 \left\{ j \in \text{TopK} \left(\{ \rho_{ik} : k \in C_i \} \right) \right\} \tag{3}$$

where $\oplus$ stands for stitching two nodes together. $\mathbf{X}_i \in \mathbb{R}^N$ is the feature vector of node i, $\mathbf{w} \in \mathbb{R}^{2N}$ is a learnable parameter vector, LeakyReLU is a nonlinear activation function, $\rho_{ij} \in \mathbb{R}^{d \times d}$ is the weight score between source node i and target node j. Next, we define a GAT-based spatial model that utilizes the learned adjacency matrix A to model the spatial features of multivariate time series.

3.4 GAT-Based Spatial Model

We use GAT and graph structure learning to fuse the information of the nodes with their neighbors. For the input multivariate time series $\mathbf{X} \in \mathbb{R}^{N \times d}$, we compute the aggregated representation μ_i of node i as follows:

$$\mu_i = \text{ReLU} \left(\alpha_{i,i} \mathbf{W} \mathbf{X}_i + \sum_{j \in N(i)} \alpha_{i,j} \mathbf{W} \mathbf{X}_j \right) \tag{4}$$

where, $\mathbf{X}_i \in \mathbb{R}^N$ is the input feature of node i, $N(i) = \{ j | A_{ij} > 0 \}$ represents the neighborhood set of node i and its values are obtained from matrix A, $\mathbf{W} \in \mathbb{R}^{N \times N}$ is the trainable weight matrix with a linear transformation for each node. Unlike GDN [8], we connect the node features to the weight score ρ so that not only the local spatial dependencies but also the global spatial dependencies in the graph can be captured. The attention coefficient $\alpha_{i,j}$ is computed using the following calculation method:

$$Concat_i = \rho_i \oplus \mathbf{W} \mathbf{X}_i \tag{5}$$

$$\pi_{i,j} = \text{LeakyReLU}\Big(\mathbf{a}^{\text{T}}\big(Concat_i \oplus Concat_j\big)\Big) \tag{6}$$

$$\alpha_{i,j} = \frac{\exp\left(\pi_{i,j}\right)}{\sum_{k \in N(i) \cup \{i\}} \exp\left(\pi_{i,k}\right)} \tag{7}$$

where $\oplus$ denotes concatenation, $Concat_i$ concatenates the weight scores ρ_i and $\mathbf{WX}_i$, the vector $\mathbf{a}$ represents the learnable coefficients of the attention mechanism. LeakyReLU is used to calculate the attention coefficients and we employ the Softmax function to normalize the computed coefficients. Next, we use Transformers to model temporal features.

3.5   Transformers Based on Amplifying Anomalies Strategy

We supply $\mu \in \mathbb{R}^{N \times d}$ to the Transformers for reconstruction by alternately stacking Multi-Mix Attention and feedforward layers. This structure better captures the details and patterns present in time series data. Among them, the overall equation of layer l is as follows:

$$\mathbf{Z}^l = \text{Add\&Norm}\Big(\text{Multi-Mix Attention}\big(\mu^{l-1}\big) + \mu^{l-1}\Big) \tag{8}$$

$$\mu^l = \text{Add\&Norm}\Big(\text{Feed-Forward}\big(\mathbf{Z}^l\big) + \mathbf{Z}^l\Big) \tag{9}$$

where $\mu^l \in \mathbb{R}^{N \times d_{model}}$, $l \in \{1, 2, \cdots L\}$ represents the output of layer l, featuring d_{model} channels. Initial input $\mu^0 = \text{Embedding}(\mu)$. $\mathbf{Z}^l \in \mathbb{R}^{N \times d_{model}}$ is the hidden representation of layer l.

Multi-mix Attention: Inspired by Anomaly Transformer [25], we propose the Multi-Mix Attention with local associations and global associations to amplify anomalies. Local associations are derived from a learnable Gauss function. The Gauss function can focus on adjacent layers and amplify local associations. To prevent the weights from decaying too rapidly or overfitting, we design the scale parameter σ as a learnable parameter, which allows the function to better adapt to different patterns of time series. In addition, we use Transformers' self-attentive scores as the global associations, which can adaptively find the most effective global distributions. The Multi-Mix Attention of layer l is as follows:

$$\mathbf{Q}, \mathbf{K}, \mathbf{V}, \sigma = \mu^{l-1}\mathbf{M}_{\mathbf{Q}}^l, \mu^{l-1}\mathbf{M}_{\mathbf{K}}^l, \mu^{l-1}\mathbf{M}_{\mathbf{V}}^l, \mu^{l-1}\mathbf{M}_{\sigma}^l \tag{10}$$

$$\text{Local-Association :} \mathbf{G}^l = \text{Rescale}\left(\left[\frac{1}{\sqrt{2\pi}\sigma_i}\exp\left(-\frac{|j-i|^2}{2\sigma_i^2}\right)\right]_{i,j \in \{1,\cdots,N\}}\right) \tag{11}$$

$$\text{Global-Association :} \mathbf{S}^l = \text{Softmax}\left(\frac{\mathbf{Q}\mathbf{K}^T}{\sqrt{d_{model}}}\right) \tag{12}$$

$$\text{Reconstruction}: \widehat{\mathbf{Z}}^l = \mathbf{S}^l \mathbf{V} \tag{13}$$

where $\mathbf{Q}, \mathbf{K}, \mathbf{V} \in \mathbb{R}^{N \times d_{model}}$, $\sigma \in \mathbb{R}^{N \times 1}$ denote query, key, self-attentive value and learning scale respectively. $\mathbf{M}_{\mathbf{Q}}^l, \mathbf{M}_{\mathbf{K}}^l, \mathbf{M}_{\mathbf{V}}^l \in \mathbb{R}^{d_{model} \times d_{model}}$, $\mathbf{M}_{\sigma}^l \in \mathbb{R}^{d_{model} \times 1}$ denote the parameter matrices of the $\mathbf{Q}, \mathbf{K}, \mathbf{V}$ and σ in the l-th layer respectively. We use Gaussian kernels to calculate the association weights between each two points, and then convert these weights into a discrete distribution through row-wise normalization with Rescale to obtain $\mathbf{G}^l \in \mathbb{R}^{N \times N}$. $\mathbf{S}^l \in \mathbb{R}^{N \times N}$ is the attention map of Transformers. We found that it contains abundant information and can be utilized as a global learning association. $\widehat{\mathbf{Z}}^l \in \mathbb{R}^{N \times d_{model}}$ is the hidden representation after the Multi-Mix Attention in the l-th layer.

We use KL divergence to represent the difference between local and global associations [18]. By averaging multiple layers of association differences, more information can be fused, and the combined association differences is:

$$\text{Dis}\,(\mathbf{G}, \mathbf{S}) = \left[\frac{1}{L} \sum_{l=1}^{L} \left(\text{KL}(\mathbf{G}_{i,:}^l \,\|\, \mathbf{S}_{i,:}^l) + \text{KL}(\mathbf{S}_{i,:}^l \,\|\, \mathbf{G}_{i,:}^l) \right) \right]_{i=1,\cdots,N} \tag{14}$$

where, $\text{KL}(\cdot \,\|\, \cdot)$ corresponds to the Kullback-Leibler divergence between the associations of $\mathbf{G}^l$ and $\mathbf{S}^l$ for each row. $\text{Dis}\,(\mathbf{G}, \mathbf{S}) \in \mathbb{R}^{N \times 1}$ is the degree of deviation of input time series with local-association $\mathbf{G}$ and global-association $\mathbf{S}$. Since the Gaussian function has local single-peakedness, so that the Gaussian distribution will show fluctuations on both normal and anomalous data, while normal data tends to exhibit smoother performance with the global association, which indicates that the Dis value of the abnormal points will be smaller than the Dis value of the normal points, so Dis has good anomaly differentiation.

3.6 Joint Optimization

Finally, we optimize the spatio-temporal model. We employ additional losses to amplify the Dis, which can further amplify the difference. The loss functions are:

$$\mathbf{L}_1 = \|\mu - \mathbf{X}\|_F^2 \tag{15}$$

$$\mathbf{L}_2 = \left\| \widehat{\mathbf{X}} - \mu \right\|_F^2 \tag{16}$$

$$\mathbf{L}_{total} = \beta \times \mathbf{L}_1 + (1 - \beta) \times \mathbf{L}_2 - \lambda \times \|\text{Dis}\,(\mathbf{G}, \mathbf{S})\|_1 \tag{17}$$

where $\widehat{\mathbf{X}}$ represents the reconstruction of μ through the use of Transforms. $\|\cdot\|_F$, $\|\cdot\|_K$ represents the Frobenius and k-norms, β denotes a balance parameter that lies within the interval $[0, 1]$, λ represents the weighting of the loss terms. When $\lambda > 0$, the optimization is to amplify Dis.

Note that excessively amplifying differences can compromise the accuracy of Gaussian kernel [18], rendering the local-association devoid of meaningful interpretation. To avoid this, Anomaly Transforms proposes a minimax strategy [25].

In the minimization phase, the local association $\mathbf{G}$ is optimized to approximate the sequence association $\mathbf{S}$ learned from the original sequence. For the maximization stage, we optimize the global association to increase the difference. The loss functions of the two stages are as follows:

$$\text{MinimizePhase}:\mathbf{L}_{total} = \beta \times \mathbf{L}_1 + (1 - \beta) \times \mathbf{L}_2 + \lambda \times \|\text{Dis}(\mathbf{G}, \mathbf{S}_{detach})\|_1 \quad (18)$$

$$\text{MaximizePhase}:\mathbf{L}_{total} = \beta \times \mathbf{L}_1 + (1 - \beta) \times \mathbf{L}_2 - \lambda \times \|\text{Dis}(\mathbf{G}_{detach}, \mathbf{S})\|_1 \quad (19)$$

where detach refers to the discontinuation of backpropagating the gradient and $\lambda > 0$. During the minimization phase, the backpropagation of the gradient of $\mathbf{S}$ is halted, enabling $\mathbf{G}$ to approximate $\mathbf{S}$. Conversely, during the maximization phase, the gradient backpropagation of $\mathbf{G}$ is stopped while $\mathbf{S}$ is optimized to amplify anomalies.

Anomaly Score: By combining association differences with joint optimization, we obtain the anomaly score:

$$Score = \text{Softmax}\Big(-\text{Dis}(\mathbf{G}, \mathbf{S})\Big) \odot \Big(\beta \times \mathbf{L}_1 + (1 - \beta) \times \mathbf{L}_2\Big) \quad (20)$$

where $\odot$ is the element multiplication method. This design allows the reconstruction error and anomaly amplification strategies to synergistically improve the detection performance.

4 Experiments

4.1 Datasets

To evaluate our method, we carry out detailed experiments on four datasets. The characteristics of these datasets are summarized in Table 1.

- Secure Water Treatment Testbed (SWaT): The SWaT dataset is derived from genuine industrial control system data obtained from a water treatment plant [16]. It contains 51 sensors.
- Water Distribution Testbed (WADI): This is an extension of the SWaT system, but has a larger and more complex data scale compared to the SWaT dataset [2].
- Server Machine Dataset (SMD) [22]: SMD consisting of 38-dimensional data collected over a 5-week period from a major Internet corporation. Only a subset of the dataset is used for evaluation due to Service Changes, which affected some machines in the dataset. The subset consists of 7 entities (machines) that did not undergo any service changes.
- Pooled Server Metrics (PSM) [1]: The PSM dataset is provided by eBay, reflects the status of servers, 25 dimensions in total.

Table 1. Details of the datasets.

Dataset	SWaT	WADI	SMD	PSM
Training size	396000	838857	144546	105984
Validation size	99000	209714	36135	26497
Testing size	449919	172801	180682	87841
Number of Sensors	51	123	38	25
Anomaly rate(%)	12.13	5.99	8.80	27.76

4.2 Baseline and Evaluation Metrics

We compared our STAD with several baseline approaches, including traditional methods: Isolation Forest [14], and deep-learning-based models: USAD [4], GDN [8], OmniAnomaly [22], LSTM-VAE [19], DAGMM [27], and Anomaly transformer [25]. We use Precision, Recall, and F1 scores to evaluate the performance of our method, which are widely used in anomaly detection.

4.3 Implementation Details

Adhering to the established protocol in Anomaly Transformer [25], we use a non-overlapping sliding window approach to obtain subsets. The fixed size of the sliding window is uniformly set to 100. We utilized grid search to obtain the anomaly threshold and hyperparameters that result in the highest F1 score. The top-K values for SWaT, PSM, WADI, and SMD are 10, 5, 30, and 15 respectively. The Transformer model consists of 3 layers, we set the number of heads to 8 and the d_{model} dimension to 512. The value of λ is set to 3, β to 0.5 and we employ the Adam optimizer [11] with the learning rate of 10^{-4}. Training process employs an early stopping strategy and batch size is set to 32. All experiments were conducted using a single NVIDIA Titan RTX 12GB GPU in PyTorch. To ensure that any timestamps during an anomaly event can be detected, we utilized a widely adopted point adjustment strategy [21,22,24]. In order to maintain fairness, the same point adjustment strategy was implemented across all baseline experiments.

4.4 Result Analysis

In many real-world anomaly detections, failure to detect anomalies can result in severe consequences. Therefore, detecting all genuine attacks or anomalies is more crucial than achieving high accuracy. As shown in Table 2, our proposed STAD outperforms other methods in terms of F1 performance. It is noteworthy that while most methods perform well on datasets such as SWaT, PSM, and SMD, as their anomalies are more easily detectable, our model still outperforms them in F1 score. When dealing with more complex MTS datasets like WADI, most existing methods yield poor results, while our model shows a significant

improvement compared to others. We also observe that: (1) Compared to traditional unsupervised methods, deep learning-based techniques generally demonstrate superior detection performance; (2) Compared to models that solely learn
a single relationship, the concurrent acquisition of temporal and spatial relationships significantly amplifies the anomaly detection efficacy.

Table 2. Experimental results on four public datasets.(%)

Method	SWaT			WADI			SMD			PSM		
	P	R	F1	P	R	F1	P	R	F1	P	R	F1
iForest [14]	49.29	44.95	47.02	62.41	61.55	61.98	59.45	85.64	68.31	76.09	92.45	83.48
LSTM-VAE [19]	76.00	89.50	82.20	46.32	32.20	37.99	87.36	79.63	83.84	73.62	89.92	80.96
DAGMM [27]	89.92	57.84	70.40	22.28	19.76	20.94	69.13	87.25	76.67	93.49	70.03	80.08
OmniAnomaly [22]	81.42	84.30	82.83	26.52	97.99	41.74	96.79	94.37	96.20	88.39	74.46	80.83
GDN [8]	99.35	68.12	81.17	97.35	40.11	57.17	67.83	95.78	77.01	54.92	99.92	70.88
USAD [4]	98.70	74.02	84.60	64.51	32.20	42.96	93.46	95.65	90.24	56.44	92.69	70.15
Anomaly-Transformer [25]	91.55	96.73	94.07	79.70	93.83	85.91	95.86	94.71	95.15	96.91	98.90	97.89
Ours	93.97	99.84	**96.46**	85.57	97.98	**91.34**	95.94	96.64	**96.43**	98.45	98.42	**98.32**

4.5 Ablation Experiments

To investigate the efficacy of each constituent of our methodology, we conducted
ablation experiments to observe how the model performance varies on the four
datasets. Firstly, we investigated the significance of using GAT to model spatial dependency relationships. We directly applied the raw data as input to the
Transformers. Secondly, we used a static graph to replace the learned graph to
prove the effectiveness of our proposed graph structure learning. Finally, to validate the necessity of Multi-Mix Attention, we removed it and only use spatial
relations to reconstruct.

Table 3. Experimental results of STAD and its variants.(%)

Method	SWaT			WADI			SMD			PSM		
	P	R	F1	P	R	F1	P	R	F1	P	R	F1
Without-GAT	90.31	93.73	93.76	70.13	91.67	85.37	95.31	94.27	93.47	94.51	96.75	96.29
Without-Graph Learning	90.15	93.20	91.09	80.61	84.34	78.32	93.82	91.97	92.43	92.49	93.87	96.42
Without-Multi-Mix Attention	95.94	66.45	78.52	81.95	47.45	60.01	68.49	99.65	80.13	58.65	99.46	73.79
Ours	93.97	99.84	**96.46**	85.57	97.98	**91.34**	95.94	96.64	**96.43**	98.45	98.42	**98.32**

The summarized results are presented in the Table 3. Furthermore, the following observations are provided based on the results: (1) The difference between
the models that do not learn graph structure and our proposed model highlights
the significance of spatial features in addressing anomaly detection for multivariate time series data. (2) Our structure learning is more effective than using

a static graph as the graph structure. (3) The transformer architecture with the Multi-Mix Attention performs remarkable performance in handling time series data. Overall, It is evident that each component of our model is effective and indispensable, thereby endowing the framework with powerful capabilities for detecting anomalies in multivariate time series.

4.6   Interpretability

We visualize the anomaly amplification strategy section, as seen in Fig. 3, for real-world datasets, our model can correctly detect anomalies. For the SWaT dataset, our approach has shown the ability to detect anomalies at an early stage, indicating its potential for practical applications such as providing early warning for faults.

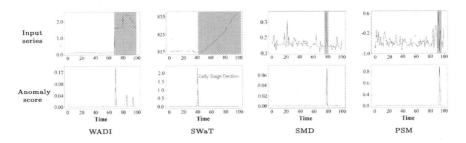

Fig. 3. Visualization of model learning in a real-world dataset. Anomalies are marked by red shading. (Color figure online)

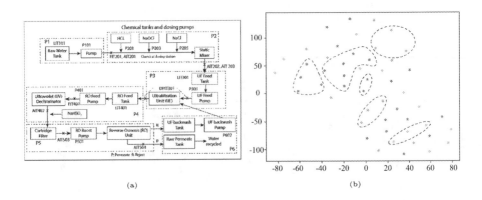

Fig. 4. Visualization of graph structure learning.

Additionally, the visualization of the learned graph structure further demonstrates the effectiveness of our proposed model. Figure 4(a) is the process diagram of the secure water treatment testbed [16]. It can be observed that the SWaT system is mainly divided into 6 processes, and sensors in the same process stage are more likely to be interdependent. Figure 4(b) displays the t-SNE [15] plot of the sensor embeddings learned by our model on the SWaT dataset, where most nodes belonging to the same process cluster together. This demonstrates the effectiveness of our graph structure learning.

4.7   Case Analysis

We use the example in Fig. 1 to illustrate why our model helps with anomaly interpretation. From the previous anomaly analysis, we know that the anomaly is manifested as water tank overflow, but the root anomaly is caused by the early opening of MV101. It is hard to find anomalies of MV101 with its irregular switch status. However, through Fig. 5(a), we can see that our model successfully detected the anomaly in MV101.

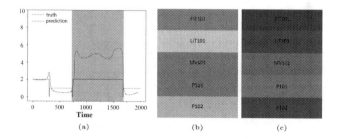

Fig. 5. Case study showing the attack in SWaT.

In addition, other sensors are expected to be correlated with MV101 when the system is functioning normally. Figure 5 presents the weight scores between the other sensors of the same process and MV101. As depicted in Fig. 5(b), our model effectively learns the features associated with MV101 under normal conditions. When anomalies occur (corresponding to the red section in Fig. 1), the sensors weight scores are visualized in Fig. 5(c). It is evident that the sensor under attack (MV101) is more closely associated (darker in color) with other sensors in the same subprocess. This is reasonable, as when an anomaly occurs, the sensors associated with the anomaly are more strongly affected.

5   Conclusion

This paper proposes a novel approach for multivariate time series anomaly detection by leveraging spatio-temporal relationships. The proposed approach utilizes a graph attention network (GAT) and a graph structure learning strategy to capture spatial associations among multivariate time series. Additionally, Transformers are used to model temporal relationships within the time series. An anomaly amplification strategy is also employed to enhance the anomaly scores. Experimental results demonstrate that the proposed method outperforms existing approaches in identifying anomalies and is effective in explaining anomalies. Future work may involve incorporating online training techniques to better handle complex real-world scenarios.

Acknowledgements. This paper was supported by the National Natural Science Foundation of China(Grant No.U22B2051).

References

1. Abdulaal, A., Liu, Z., Lancewicki, T.: Practical approach to asynchronous multivariate time series anomaly detection and localization. In: Proceedings of the 27th ACM SIGKDD Conference on Knowledge Discovery & Data Mining, pp. 2485–2494 (2021)
2. Ahmed, C.M., Palleti, V.R., Mathur, A.P.: Wadi: a water distribution testbed for research in the design of secure cyber physical systems. In: Proceedings of the 3rd International Workshop on Cyber-physical Systems for Smart Water Networks, pp. 25–28 (2017)
3. Angiulli, F., Pizzuti, C.: Fast outlier detection in high dimensional spaces. In: Elomaa, T., Mannila, H., Toivonen, H. (eds.) PKDD 2002. LNCS, vol. 2431, pp. 15–27. Springer, Heidelberg (2002). https://doi.org/10.1007/3-540-45681-3_2
4. Audibert, J., Michiardi, P., Guyard, F., Marti, S., Zuluaga, M.A.: Usad: unsupervised anomaly detection on multivariate time series. In: Proceedings of the 26th ACM SIGKDD International Conference on Knowledge Discovery & Data Mining, pp. 3395–3404 (2020)
5. Baldi, P.: Autoencoders, unsupervised learning, and deep architectures. In: Proceedings of ICML Workshop on Unsupervised and Transfer Learning, pp. 37–49. JMLR Workshop and Conference Proceedings (2012)
6. Breunig, M.M., Kriegel, H.P., Ng, R.T., Sander, J.: Lof: identifying density-based local outliers. In: Proceedings of the 2000 ACM SIGMOD International Conference on Management of Data, pp. 93–104 (2000)
7. Chen, Z., Chen, D., Zhang, X., Yuan, Z., Cheng, X.: Learning graph structures with transformer for multivariate time-series anomaly detection in iot. IEEE Internet Things J. **9**(12), 9179–9189 (2021)
8. Deng, A., Hooi, B.: Graph neural network-based anomaly detection in multivariate time series. In: Proceedings of the AAAI Conference on Artificial Intelligence, vol. 35, pp. 4027–4035 (2021)
9. Hochreiter, S., Schmidhuber, J.: Long short-term memory. Neural Comput. **9**(8), 1735–1780 (1997)
10. Hundman, K., Constantinou, V., Laporte, C., Colwell, I., Soderstrom, T.: Detecting spacecraft anomalies using lstms and nonparametric dynamic thresholding. In: Proceedings of the 24th ACM SIGKDD International Conference on Knowledge Discovery & Data Mining, pp. 387–395 (2018)
11. Kingma, D.P., Ba, J.: Adam: a method for stochastic optimization. iclr. 2015. arXiv preprint arXiv:1412.6980 9 (2015)
12. Kingma, D.P., Welling, M., et al.: An introduction to variational autoencoders. Foundat. Trends® Mach. Learn. **12**(4), 307–392 (2019)

13. Li, D., Chen, D., Jin, B., Shi, L., Goh, J., Ng, S.-K.: MAD-GAN: multivariate anomaly detection for time series data with generative adversarial networks. In: Tetko, I.V., Kůrková, V., Karpov, P., Theis, F. (eds.) ICANN 2019. LNCS, vol. 11730, pp. 703–716. Springer, Cham (2019). https://doi.org/10.1007/978-3-030-30490-4_56

14. Liu, F.T., Ting, K.M., Zhou, Z.H.: Isolation forest. In: 2008 Eighth IEEE International Conference on Data Mining, pp. 413–422. IEEE (2008)

15. Van der Maaten, L., Hinton, G.: Visualizing data using t-sne. J. Mach. Learn. Res. **9**(11) (2008)

16. Mathur, A.P., Tippenhauer, N.O.: Swat: a water treatment testbed for research and training on ics security. In: 2016 International Workshop on Cyber-physical Systems for Smart Water Networks (CySWater), pp. 31–36. IEEE (2016)

17. Mirza, M., Osindero, S.: Conditional generative adversarial nets. arXiv preprint arXiv:1411.1784 (2014)

18. Neal, R.M.: Pattern recognition and machine learning. Technometrics **49**(3), 366 (2007)

19. Park, D., Hoshi, Y., Kemp, C.C.: A multimodal anomaly detector for robot-assisted feeding using an lstm-based variational autoencoder. IEEE Robot. Autom. Lett. **3**(3), 1544–1551 (2018)

20. Schlegl, T., Seeböck, P., Waldstein, S.M., Langs, G., Schmidt-Erfurth, U.: f-anogan: fast unsupervised anomaly detection with generative adversarial networks. Med. Image Anal. **54**, 30–44 (2019)

21. Shen, L., Li, Z., Kwok, J.: Timeseries anomaly detection using temporal hierarchical one-class network. Adv. Neural. Inf. Process. Syst. **33**, 13016–13026 (2020)

22. Su, Y., Zhao, Y., Niu, C., Liu, R., Sun, W., Pei, D.: Robust anomaly detection for multivariate time series through stochastic recurrent neural network. In: Proceedings of the 25th ACM SIGKDD International Conference on Knowledge Discovery & Data Mining, pp. 2828–2837 (2019)

23. Vaswani, A., et al.: Attention is all you need. In: Advances in Neural Information Processing Systems 30 (2017)

24. Xu, H., et al.: Unsupervised anomaly detection via variational auto-encoder for seasonal kpis in web applications. In: Proceedings of the 2018 World Wide Web Conference, pp. 187–196 (2018)

25. Xu, J., Wu, H., Wang, J., Long, M.: Anomaly transformer: time series anomaly detection with association discrepancy. arXiv preprint arXiv:2110.02642 (2021)

26. Zhao, H., et al.: Multivariate time-series anomaly detection via graph attention network. In: 2020 IEEE International Conference on Data Mining (ICDM), pp. 841–850. IEEE (2020)

27. Zong, B., et al.: Deep autoencoding gaussian mixture model for unsupervised anomaly detection. In: International Conference On Learning Representations (2018)

Recommendation I

Refined Node Type Graph Convolutional Network for Recommendation

Wei He[1], Guohao Sun[1], Jinhu Lu[1], Xiu Fang[1(✉)], Guanfeng Liu[2], and Jian Yang[2]

[1] Donghua University, Shanghai, China
xiu.fang@dhu.edu.cn
[2] Macquarie University, Sydney, Australia

Abstract. Recently, because of the remarkable performance in alleviating the data sparseness problem in recommender systems, Graph Convolutional (Neural) Networks (GCNs) have drawn wide attention as an effective recommendation approach. By modeling the user-item interaction graph, GCN iteratively aggregates neighboring nodes into embeddings of different depths according to the importance of each node. However, the existing GCN-based methods face the common issues that, they do not consider the node information and graph structure during aggregating nodes, such that they cannot assign reasonable weights to the neighboring nodes. Additionally, they ignore the differences in node types in the user-item interaction graph and thus, cannot explore the complex relationship between users and items, resulting in a suboptimal result. To solve these problems, a novel GCN-based framework called **RNT-GCN** is proposed in this paper. RNT-GCN integrates the structure of the graph and node information to assign reasonable importance to different nodes. In addition, RNT-GCN refines the node types, such that the heterogeneous properties of the user-item interaction graph can be better preserved, and the collaborative information of users and items can be effectively extracted. Extensive experiments prove the **RNT-GCN** achieved significant performance compared to **SOTA** methods.

Keywords: Recommender system · Graph convolutional network · User-item interaction graph

1 Introduction

Personalized recommender systems play a vital role in the era of big data and can extract pivotal information from a mass of data and effectively alleviate information overload [4]. Recommender systems face the common issue that the data in the real world is extremely sparse and noisy [7,22], in other words, numerous items interact with only a few users. Therefore, it is hard to extract

---

W. He and G. Sun—Contributed equally to this research.

© The Author(s), under exclusive license to Springer Nature Switzerland AG 2023
X. Yang et al. (Eds.): ADMA 2023, LNAI 14176, pp. 91–106, 2023.
https://doi.org/10.1007/978-3-031-46661-8_7

enough information to predict user preferences accurately, and the performance of recommendations is limited in real-world applications.

Recently, Graph Convolutional (Neural) Networks (GCNs) with their powerful representation learning capabilities [11] has attracted wide attention in many areas, such as Natural Language Processing (NLP) [15], etc. By modeling the historical interaction data as a graph (e.g., the user-item interaction graph), GCN-based methods successfully extracted the high-order collaboration signals between nodes and effectively alleviated the data sparsity problem [7,13,22].

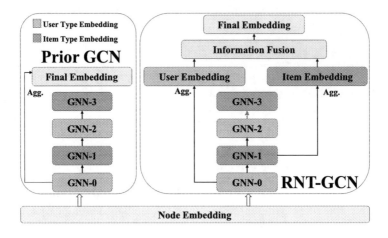

Fig. 1. The difference between the prior GCN-based methods and RNT-GCN.

Although remarkable progress has been made with GCN-based methods, there are still some significant disadvantages. When aggregating the information of neighbor nodes, GCN-based methods cannot assign reasonable importance to different neighbor nodes because they only consider the importance of neighbor nodes based on the structure of the graph or node information. To be specific, the methods of using the structure of the graph aggregate neighbor node according to the degree of the node, the popular nodes (with larger degrees) have a higher possibility to be recommended. The methods of using the node information aggregate neighbor node according to the embedded similarity between the central node and the neighbor node, the neighbor nodes with low similarity probably contain important information and will be assigned a low weight. These disadvantages will result in the neglect of important information and thus, obtaining poor recommendations. In addition, the existing methods directly process the user-item interaction graph and do not consider the intrinsic differences between user and item nodes, resulting in low-quality embeddings. For example, as shown in Fig. 1, the prior GCN-based methods do not distinguish the user and item nodes, they indiscriminately aggregate embeddings at different depths. Therefore, due to the lack of distinguishing the node type, the existing methods cannot explore the complex relationship between users and items, which results in a suboptimal result. These disadvantages limit the expression power of GCN, which influences the recommendation performance of the GCN-based methods.

To solve the above-mentioned problems, we propose a novel **R**efined **N**ode **T**ype **G**raph **C**onvolutional **N**etwork for Recommendation, called **RNT-GCN**. The main contributions of this paper can be summarized as follows:

– In RNT-GCN, we propose an attention-refinement network (ARN) to assign the appropriate aggregation importance for different neighbor nodes. ARN learns the relationship between neighbor nodes and central nodes at the graph level and node level, such that the important information of local neighbor nodes can be reasonably extracted.
– In RNT-GCN, we propose a type-aware embedding aggregation strategy. By this strategy, the nodes of each layer are aggregated according to their type, such that the heterogeneous properties of the user-item interaction graph can be better preserved.
– In RNT-GCN, we propose a new information fusion layer that captures user and item connections during prediction. This layer considers the connections between different attribute nodes on different sides during prediction and effectively extracts the collaborative information of users and items.
– Extensive experiments on four public datasets demonstrate that RNT-GCN outperforms the SOTA models. The code is published on GitHub[1].

2   Related Work

Some scholars [3,7,13,14,17,20,22] realized that the historical interaction between users and items can be modeled as a user-item interaction graph. It can provide plentiful information about the graph structure and nodes and effectively improve the performance of the recommender system. Early works related to graphs mainly explored the information of the graph through random walks such as randwalk [6] etc. However, the performance of these methods is heavily influenced by the quality of interactions generated by the randomwalk [13]. As a consequence, models become difficult to train.

In recent years, many GCN-based methods have been proposed to explore graph-structured data. For example, by performing embedding propagating on the graph, NGCF [22] explicitly explores the higher-order connections between nodes. To exploit the heterogeneous nature of the user-item bipartite graph, NIA-GCN [17] parallel independently learns the representation of the user and item from different neighborhood depths and calculates the inner product between the first and second layer embeddings, resulting in only the node information of the specific range of receptivity fields can be captured. These GCN-based methods face many issues, such as complexity, bulkiness, and over-smoothing [2,13], which result in the shallow depth of the convolutional layer and then make these methods hard to extract the information of the high-order node. To solve these problems, some works try to simplify GCN [3,7,13,14]. For example, in order to

---

[1] https://github.com/WeiHeCnSH/RNTGCN-TF-master.

better learn the attributes of neighbors, JK-NET [23] aggregates the embedding of each layer in the jump concatenate and adaptively selects the range of aggregated neighborhood information. Similar to JK-Net, LR-GCCF [3] uses residuals to concatenate the embeddings of the layers and removes nonlinearities. Meanwhile, LightGCN [7] realized that nodes encoded with ID contain only a small amount of semantic information, and feature transformation is also redundant for GCN. Hence, GCN is further simplified. Later, based on LightGCN, IMP-GCN [13] and UltraGCN [14] are proposed. Compared with LightGCN, IMP-GCN [13] thinks that the indiscriminate use of higher-order neighbor nodes allows harmful information to be propagated in the graph. To solve this problem, they propose to divide the subgraph according to the user's interest and perform high-order convolution operations in the subgraph. UltraGCN [14] thinks that the messaging mechanism in GCN greatly slows down the convergence of the model, so they further simplify LightGCN by directly approximating the limit of infinite-layer graph convolutions via a constraint loss. In addition, GDCF [24] think existing approaches model interactions within the euclidean space be inadequate to fully uncover the intent factors between user and item. Meanwhile, SGNR [18] separates the user-item interaction graph into two weighted homogeneous graphs to resolve the heterogeneous information aggregation problem, but doesn't establish the distinction and relationship between different types of embeddings of the users and items.

Although these graph-based approaches have achieved good performance, they still face the problems proposed in the Introduction section.

3   Problem Statement

User-item Interaction Matrix: An user-item interaction matrix is denoted as $A \in \mathbb{R}^{M \times N}$, where M indicates the number of users and N indicates the number of items. If there is an interaction between user u and item i, then $a_{u,i} = 1$, otherwise $a_{u,i}$ will be 0.

Bipartite Graph: A bipartite graph $\mathcal{G}(\mathcal{U}, \mathcal{I}, \mathcal{E})$ is constructed based on the user-item interaction matrix, where $\mathcal{U}$ indicates user set, $\mathcal{I}$ indicates item set, and $\mathcal{E}$ indicates edge set if user u interacte with item i. Each node in the graph can be represented using a vector of dimension d. $e_u^{(0)} \in \mathbb{R}^d$ and $e_i^{(0)} \in \mathbb{R}^d$ represents the initialized embedding of the user u and the item i, respectively.

Target Problem: Our goal is to train a model to predict whether an interaction will occur between the user u and the item i in the bipartite graph $\mathcal{G}(\mathcal{U}, \mathcal{I}, \mathcal{E})$.

4   Methodology

In this section, we introduce RNT-GCN in detail. As shown in Fig. 2. The proposed RNT-GCN consists of three parts: the attention refinement network, the type-aware embedding aggregation, and the information fusion layer. The user (left) and the item (right) first perform multi-layer convolution to compute the representation of each layer, then aggregate the information of each layer according to the node type, and finally fuse the embedding information of both node types to form the final node representation to make predictions.

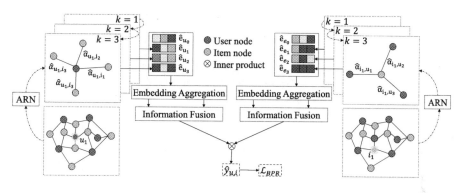

Fig. 2. The overall framework of the proposed RNT-GCN model.

4.1   Attention Refinement Network

To retain important information and weaken the influence of unimportant nodes, by using the attention refinement network (ARN) in Fig. 3.a, we combine the structure of the graph and the information of the nodes to assign reasonable aggregate importance to neighboring node from the graph level and the node level, respectively.

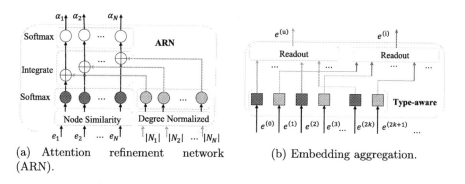

(a) Attention   refinement   network (ARN).

(b) Embedding aggregation.

Fig. 3. Illustration of proposed attention refinement network (left) and embeddings aggregation (right).

Importance of Graph-Level. Inspired by the famous GCN-based methods [3, 7,22], we use symmetric normalization terms to express the connection strength between different neighbors i and the center u at the graph level. In this way, we combine the information of the degree of the central node and the neighboring node. The importance of the graph level can be represented as follows:

$$\hat{\alpha}_{u,i}^{(g)} = \frac{|N_u \cap N_i|}{\sqrt{|N_u|}\sqrt{|N_i|}} = \frac{1}{\sqrt{|N_u|}\sqrt{|N_i|}}. \tag{1}$$

where N_u and N_i indicate the number of the first-order neighbors of users and items, respectively. The intersection between the neighbors of users and items can be represented as the interaction generated by users and items.

Importance of Node-Level. To better aggregate node information, we further consider the relationship of nodes represented in the embedding space to assign importance to different neighbor nodes. To be specific, in the embedding space, the distance between two nodes can reflect the correlations between the nodes to some extent. The nodes with closer distances indicate more similarities, so more information should be aggregated from these nodes [19,21]. We assign the importance of node level according to the distance between nodes. The importance of node level can be represented as follows:

$$\alpha_{u,i}^{(n)} = sim_n \left(e_u^{(0)}, e_i^{(0)} \right). \tag{2}$$

For simplicity, we use cosine similarity to measure the strength of the connection between nodes. In the future, we will research more node-level importance calculation methods. The importance of node level is represented as follows:

$$\alpha_{u,i}^{(n)} = \cos \left(e_u^{(0)}, e_i^{(0)} \right) = \frac{|e_u^{(0)} \odot e_i^{(0)}|}{|e_u^{(0)}||e_i^{(0)}|}. \tag{3}$$

where $\odot$ refers to the element-wise product. We normalize all node cosine similarities as node-level importance using the softmax function. The normalized node level importance can be represented as follows:

$$\hat{\alpha}_{u,i}^{(n)} = softmax \left(\alpha_{u,i}^{(n)} \right) = \frac{\exp \left(\alpha_{u,i}^{(n)} \right)}{\sum_{j \in N_u} \exp \left(\alpha_{u,j}^{(n)} \right)}. \tag{4}$$

The node-level importance of the item can also be obtained in the same way.

Importance of Integration. To integrate the importance of two different neighbor nodes, we designed a dual attention integration strategy to combine the structure of the graph and the relationship between nodes as the importance of different neighboring nodes. It can be represented as follows:

$$\alpha_{u,i} = \lambda \hat{\alpha}_{u,i}^{(n)} + (1 - \lambda) \alpha_{u,i}^{(g)}. \tag{5}$$

where λ is an adjustable super-parameter for balancing the weight of the importance of the two different aspects. The importance of different neighbor nodes during aggregated can be represented as follows:

$$\hat{\alpha}_{u,i} = softmax\,(\alpha_{u,i}/\tau) = \frac{\exp\,(\alpha_{u,i}/\tau)}{\sum_{j \in N_u} \exp\,(\alpha_{u,j}/\tau)}. \tag{6}$$

where τ indicates the temperature of the normalization function, which can make the training process smoother. The normalized importance of nodes can avoid the scale increase of embedding in the convolution.

Same as [7], we also abandon feature transformation and nonlinear activation. Meanwhile, we have also removed the self-loop, such that there is only one type of neighbor node per layer. Therefore, the graph convolution operation in RNT-GCN can be represented as follows:

$$e_u^{(l+1)} = \sum_{i \in N_u} \hat{\alpha}_{u,i} e_i^{(l)}, \; e_i^{(l+1)} = \sum_{u \in N_i} \hat{\alpha}_{u,i} e_u^{(l)}. \tag{7}$$

where $e_u^{(l)}$ and $e_i^{(l)}$ denote the embedding representation of user node u and item node i at l layer, respectively.

4.2   Type-Aware Embedding Aggregation

To capture neighbor node embedding information at different depths, we perform L convolution operations to obtain the embedding of the user (item). Because of the heterogeneous property of the user-item interaction graph, the embedding of different layers contains different semantic information. As shown in Fig. 1, the embedding of user types is obtained from the even layer (i.e., 0, 2,...), and item type is obtained from the odd layer (i.e., 1, 3,..). On the item side, similar findings can be derived by analyzing the convolution process of the items. To preserve the embedding information of different types of nodes, we designed a node-type-aware information aggregation strategy, as shown in Fig. 3.b. Based on the node types in each layer, we divide the embedding of the convolution layers into two types, i.e., the embedding of user type $e^{(u)}$ and embedding of item type $e^{(i)}$. There, we aggregate embeddings of different depths according to the node type to generate embeddings of different types of nodes. On the user side, the odd layer is the embedding of item types $e_u^{(i)}$, and the even layer is the embedding of user types $e_u^{(u)}$. The user and item type embeddings for user nodes can be represented as follows:

$$e_u^{(u)} = \sum_{k=0,1,2,\ldots} a_k e_u^{(2k)}, \; e_u^{(i)} = \sum_{k=0,1,2,\ldots} b_k e_u^{(2k+1)}. \tag{8}$$

On the item side, the odd layer is the embedding of user types $e_i^{(u)}$, and the even layer is the embedding of item types $e_i^{(i)}$. Similarly, the user type embeddings and item type embeddings for item nodes can be represented as follows:

$$e_i^{(i)} = \sum_{k=0,1,2,\ldots} c_k e_i^{(2k)}, \; e_i^{(u)} = \sum_{k=0,1,2,\ldots} d_k e_i^{(2k+1)}. \tag{9}$$

where a_k, b_k, c_k, d_k indicate the aggregation weights of k layer embeddings of the same type of nodes on the user and item sides, respectively. Here, we average the embedding of each layer.

4.3   Information Fusion

The existing embedding aggregation methods [1,3,7,22] do not distinguish node types. Therefore, the user-item relationship will appear in the prediction process, which will affect the recommendation performance. To solve this problem, we design two novel information fusion methods, i.e., sequence concatenate and reverse sequence concatenate, considering two different convolutional processes, to explore the impact of different combinatorial relationships of embedding types on the results.

Sequence Concatenate. The embedding of two node types is concatenated according to the order of embedding generated by the convolution process to generate the final representation of nodes. The final representation of the node can be represented as follows:

$$e_u = e_u^{(u)} \| e_u^{(i)}, \ e_i = e_i^{(i)} \| e_i^{(u)}. \tag{10}$$

where $\|$ denotes the concatenate operation. The informational fusion strategy of sequence concatenate is simply denoted as RNT-sc.

Reverse Sequence Concatenate. Consequently, we adjust the order of concatenating embeddings so that the same types of embedding correspondence are used during prediction. The final representation of the node can be represented as follows:

$$e_u = e_u^{(u)} \| e_u^{(i)}, \ e_i = e_i^{(u)} \| e_i^{(i)}. \tag{11}$$

The informational fusion strategy of reverse-sequence concatenate is simply denoted as RNT-rsc.

4.4   Model Prediction

Similar to the existing GCN-based methods [1,3,7,13,17,22], we use the inner product of the final representation of the user and the item as the preference prediction of the user u for the item i. It can be represented as follows:

$$\hat{r}_{u,i} = e_u^T e_i. \tag{12}$$

We expand the scoring function to simply investigate the impact of two different information-fusion strategies on the results. The scoring prediction for sequential concatenate expands can be represented as follows:

$$\hat{r}_{u,i} = e_u^{(u)} \odot e_i^{(i)} + e_u^{(u)} \odot e_i^{(i)}. \tag{13}$$

The prediction score for the reverse concatenate expands can be represented as follows:

$$\hat{r}_{u,i} = e_u^{(u)} \odot e_i^{(u)} + e_u^{(i)} \odot e_i^{(i)}. \tag{14}$$

Sequential concatenate uses embeddings of user types and item types to perform inner product operations and force them to be close to each other in different semantic spaces. Inverse concatenate performs inner product operations in homogeneous embeddings, such that the intrinsic properties of different types of embeddings are preserved.

4.5 Model Training

The Bayesian Personalized Ranking (BPR) [16] loss is applied as the loss function of RNT-GCN. The loss function can be represented as follows:

$$\mathcal{L}_{BPR} = \frac{1}{|\mathcal{O}|} \sum_{(u,i,j)\in\mathcal{O}} \ln \sigma \left(\hat{r}_{u,i} - \hat{r}_{u,j} \right) + \gamma \left\| \Theta \right\|_2^2. \tag{15}$$

where $\mathcal{O}$ indicates the training set, i indicates the observed positive samples, and j is a negative sample randomly sampled from items with which the user has not interacted. The γ indicates the L_2 regularization weights and the Θ indicates the embedding of the user with the item nodes and the trainable parameters in the model. The σ indicates the Sigmoid function, and the Adam algorithm [10] is used to optimize the model parameters.

5 Experimental

5.1 Experimental Setup

In this section, we conduct extensive experiments to compare the performances of our proposed RNT-GCN models. The experiments will answer the following several questions:

- Q1: How is the performance of RNT-GCN compared with the SOTA graph-based methods?
- Q2: What is the impact of different information fusion strategies on the performance of RNT-GCN?
- Q3: What is the performance impact of interaction groups with different sparsities levels?
- Q4: How effect of different parameters (number of layers L, Balance Weight W, Node Dropout Ratio P) on RNT-GCN performance?
- Q5: How do the ARN, type-aware embedding aggregation, and the information fusion layer contribute to the performance of RNT-GCN?
- Q6: Whether RNT-GCN can improve the embedding quality of nodes?

DataSet. In order to objectively evaluate our proposed method, we conducted experiments on four publicly available datasets, **Kindle Store** (14,356 users, 15,885 items, 357,477 interactions), **Gowalla** (29,858 users, 40,981 items, 1,027,370 interactions), **ml-1m** (6,022 users, 3,043 items, 895,699 interactions), and **Book-Cross** (6,768, 13,683 items, 374,563 interactions)[2]. The preprocessed datasets of the first three are provided by **IMP-GCN** [13], **LightGCN** [7], and **UltraGCN** [14], respectively. For the Book-Cross dataset, we treat all ratings as interaction records and remove the users and items with fewer than 10 interaction records. For each dataset, 80 percent of each user's interaction records is randomly selected as the training set, and the remaining interactions are used as the test set. The items with observed interactions with the user are used as positive samples for the user, and items with unobserved interactions are randomly sampled for users as negative samples that form training pairs with each positive sample.

Benchmarks and Evaluation Metrics. To reveal the value of our proposed method, we compared it with several **SOTA** benchmark methods, covering **MF**-based (BPR-MF [12]), **MLP**-based (NeuMF [8]), **GCN**-based (GCMC [1], NGCF [17,22], LR-GCCF [3], LightGCN [7], IMP-GCN [13]). The all-ranking strategy used to calculate each metric and use all items that do not interact with the user as a negative sample of users. We calculate the average result of all users as the prediction result of the model and present the results of Recall@N and NDCG@N [9], where N is set to 10, 20, respectively.

Implementation Details. All embedding sizes of the model were fixed to 64, and the embeddings of the model were initialized using Xavier [5]. The default learning rates and mini-batches are 1e-3 and 1024, respectively. The temperature τ is set to 0.1. For other parameters, we use a grid search to find the optimal values. The depth of convolutional L was within $\{1, \ldots, 6\}$, the balance weights W were set to $\{0.1, \ldots, 0.9\}$, L_2 normalization coefficient was searched within $\{1e-6, \ldots, 1e-2\}$, and the node dropout rating P was also introduced within $\{0.0, \ldots, 0.9\}$ for searching within. For all benchmarks, the other parameters are set as in the original paper, and we do our best to adjust the parameters to obtain the optimal values. When the results of Recall@20 on the test set are lower than the optimal value for 50 consecutive rounds, the training is stopped.

5.2   Performance Comparison with the SOTA Methods (Q1 & Q2)

The detailed results of all methods are shown in Table 1. Compared with the SOTA methods, the main observations are as follows:
(1) RNT-GCN consistently exceeds all benchmarks in all datasets. Compared to the strongest benchmark IMP-GCN, RNT-GCN outperforms by an average

---

[2] http://www2.informatik.uni-freiburg.de/~cziegler/BX/.

Table 1. Overall Performance Comparison With SOTA Models.

DatasSet	Kindle Store				Gowalla				ml-1m				Book-Cross			
Metric	Recall		NDCG		Recall		NDCG		Recall		NDCG		Recall		NDCG	
Method	10	20	10	20	10	20	10	20	10	20	10	20	10	20	10	20
BPR-MF	5.50	4.16	8.62	5.31	9.72	10.62	13.73	11.71	15.09	21.03	24.17	22.87	3.73	4.53	5.72	5.03
NeuMF	3.92	2.96	6.22	3.79	7.71	7.52	11.94	8.91	14.84	21.15	23.74	22.93	2.17	2.88	3.50	3.20
GCMC	5.22	3.93	8.35	5.08	9.53	10.30	14.18	11.65	16.17	22.62	26.43	24.72	3.32	3.95	5.74	4.65
NGCF	6.29	4.68	9.88	6.00	10.68	11.27	15.55	12.71	16.64	22.18	26.54	24.47	5.46	6.15	8.47	7.02
NIA-GCN	5.39	3.95	8.60	5.13	10.47	11.09	15.39	12.54	15.56	21.52	24.81	23.46	4.01	2.86	6.55	3.89
LR-GCCF	6.34	4.75	10.00	6.11	12.12	13.26	17.22	14.66	17.08	23.23	26.73	25.19	5.62	6.41	8.70	7.26
LightGCN	6.51	4.86	10.23	6.24	12.79	14.00	18.11	15.46	17.21	23.43	26.95	25.41	5.76	6.52	9.12	7.45
IMP-GCN	7.04	5.31	10.80	6.69	13.06	14.24	18.65	15.82	17.37	23.56	27.11	25.56	6.33	7.19	9.77	8.09
RNT-sc	6.64	4.99	10.11	6.29	12.27	13.47	17.51	14.93	17.25	23.39	27.07	25.45	5.90	6.70	9.13	7.57
RNT-rsc	7.96	6.00	11.94	7.48	13.19	14.21	18.78	15.80	17.66	24.19	27.48	26.13	6.79	7.73	10.15	8.62
Imp.(%)	13.07	12.99	10.56	11.81	1.00	-	0.70	-	1.67	2.67	1.36	2.23	7.27	7.51	3.89	6.55

of 13% on the Kindle Stores dataset in Recall@10 and NDCG@10. Substantial improvements have also been made on other datasets. Experimental results confirm that our proposed method can reasonably aggregate neighbor nodes and effectively extract heterogeneous information from the user-item interaction graph, preserve the information of different types of nodes, and effectively improve the quality of user and item representations.

(2) RNT-GCN yields more improvement in smaller positions (i.e., the top 10 ranks) than in larger positions (i.e., the top 20 ranks), indicating that RNT-GCN tends to rank the relevant items higher, which means the proposed model can batter adapt the real-world recommendation scenario.

(3) RNT-GCN shows advantages on sparse datasets like Kindle Store and Book-Cross. This indicates that, even in the sparse dataset, RNT-GCN can still extract more useful information than methods.

(4) RNT-rsc performs better than RNT-sc. The reason is that RNT-rsc adjusts the concatenated order of different types of embedding such that the predicted ones are matched with the same type of embedding.

5.3 Parameter Sensitivity Analysis (Q3)

In this section, we analyze the effects of different parameters on the RNT-GCN. Due to the trend is the same on other datasets and page limitations, only the results on the Kindle Store and the ml-1m dataset are reported.

Balance Weights. We adjusted the balance weights to analyze the impact of two different levels of node importance on the result. The results are reported in Fig. 4. The ARN that fuses two kinds of information can obtain better results on the Kindle Store dataset. This is because ARN can more reasonably measure the importance of different neighbor nodes and aggregate more important information. The influence of balance weights on the Kindle store is greater than that of the ml-1m dataset. We speculate that because ml-1m is a dense dataset, and the weight values assigned to different neighboring nodes are very close to each other regardless of degree normalization strategy or node-level attention with

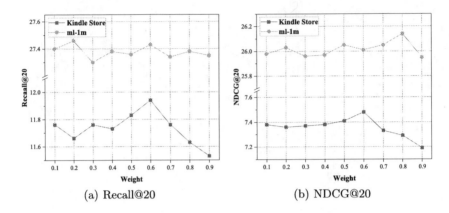

Fig. 4. Effect of Balance Weights.

normalization, the influence of fusing two different types of significance on the final results is slight. Thus, RNT-GCN results fluctuate moderately on ml-1m.

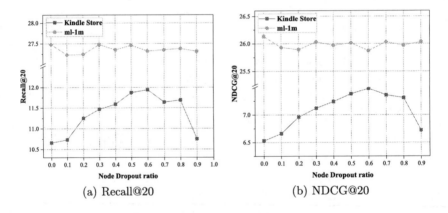

Fig. 5. Effect of Node dropout Ratio

Node Dropout Ratio. We adjust the node dropout ratio P to analyze the effect. The results are shown in Fig. 5. A proper node drop ratio for the Kindle Store dataset can improve performance because it effectively avoids overfitting and excessive node drop rates that lead to poor performance due to missing important information. The impact of varying node dropout ratings on the results is less significant for the ml-1m dataset, as it is denser than the Kindle store and can extract more comprehensive information from the interaction graph. The effectiveness of node dropout in improving recommended performance is varied and should be adjusted based on the dataset.

Number of Layers. We vary the depth of the model to investigate the effect of convolution depth on RNT-GCN. The experimental results are summarized in Table 2. When increasing the number of layers, the performance of RNT-GCN increases rapidly and comes to the peak at the 4th layer on the Kindle dataset and the 3rd layer on ml-1m dataset. With layers increasing, the performance of RNT-GCN decrease. The reason is that, more layers can obtain neighboring nodes with a larger range of perceptual fields, such that more node neighbor information can be obtained. However, when the number of convolutional layers exceeded 4 on the Kindle dataset and 3 on the ml-1m dataset, increasing the number of layers will lead to a slight decrease in performance. We hypothesize that with increasing depth, the node embeddings converge and become indistinguishable from each other.

Table 2. Effect of Number of Layers.

DataSet	Kindle Store				ml-1m			
Metric	Recall@10	Recall@20	NDCG@10	NDCG@20	Recall@10	Recall@20	NDCG@10	NDCG@20
1	7.30	11.01	5.47	6.84	16.57	26.03	22.76	24.73
2	7.60	11.53	5.76	7.22	17.20	26.72	23.57	25.45
3	7.61	11.45	5.77	7.34	**17.66**	**27.48**	**24.19**	**26.13**
4	**7.96**	**11.94**	**6.00**	**7.48**	17.56	27.20	24.08	25.98
5	7.74	11.61	5.88	7.31	16.87	26.49	23.39	25.30
6	7.64	11.51	5.76	7.19	16.47	25.64	23.00	24.71

5.4 Data Sparsity Effect (Q4)

In this section, we investigate the efficacy of RNT-GCN in mitigating the issue of data sparsity. We divide the test set of the Kindle Store into 4 groups based on the number of interactions per user and display the Recall@20 and NDCG@20 results based on different sparsities, the detailed results are shown in Fig. 6. Due to the page limitation, we only report the results in the Kindle Store (the similar trend in other datasets). The results show that the Recall and NDCG of RNT-GCN achieve more improvement in sparse user groups than in dense user groups (i.e., the performance in the user groups with fewer than 17 and 33 times interactions gains more improvement than in the user groups with fewer than 71 and 664 times interactions). This indicates RNT-GCN is beneficial for learning the inactive users and items and can effectively alleviate data sparsity.

5.5 Ablation Experiments (Q5)

To validate the effect of each part of the model, we perform ablation experiments. The results are reported in Table 3. "W/o n-i" and "W/o g-i" refer to "without graph-level importance" and "without node-level importance", respectively. The "Cat" and "Avg" represent the pooling of embedding of each layer by concatenating and averaging, respectively. We obtained the following observations: All parts

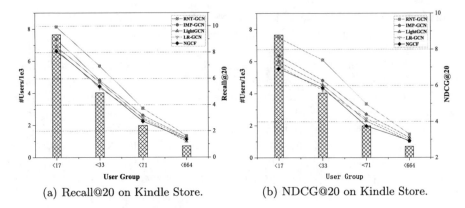

(a) Recall@20 on Kindle Store.   (b) NDCG@20 on Kindle Store.

Fig. 6. The comparison of user group sparsity distributions on the Kindle Store with partial baselines. The background histograms reflect the number of users for each group, while the lines depict the results.

Table 3. Ablation experiments result on each proposed component.

DataSet	Kindle Store				ml-1m			
Architecture	Recall@10	Recall@20	NDCG@10	NDCG@20	Recall@10	Recall@20	NDCG@10	NDCG@20
W/o n-i	7.72	11.75	5.89	7.37	17.43	27.34	23.87	25.89
W/o g-i	7.79	11.86	5.96	7.46	**17.68**	27.35	24.15	26.07
Cat	6.47	10.05	4.84	6.16	17.13	26.82	23.23	25.17
Avg	6.93	10.64	5.21	6.57	17.50	27.08	23.86	25.69
All	**7.96**	**11.94**	**6.00**	**7.48**	17.66	**27.48**	**24.19**	**26.13**

of RNT-GCN prove to be workable and effective. Compared with using only the importance of the graph level or the node level, ARN that incorporates both can assign more reasonable importance to different nodes and learn higher-quality embedding. It can contribute to the RNT-GCN and improve the recommendation performance. Compared to the summation-based and concatenate-based strategies, RNT-GCN has achieved significant improvement. This shows that type-aware aggregate strategies and the information fuse layer can better capture the specific information of different nodes and effectively extract the high-order collaborative signals at different types of nodes.

5.6   Embedding Visualization (Q6)

To show a more intuitive understanding and comparison of RNT-GCN, we randomly select 10 users and visualize the embeddings of the user and item using t-SNE. The visualization is shown in Fig. 7. We can find that, compared with MF, items in RNT-GCN are more closely related to each other and have fewer items clustered. The results show that RNT-GCN improves the distribution of nodes in the embedding space and can obtain higher-quality embeddings.

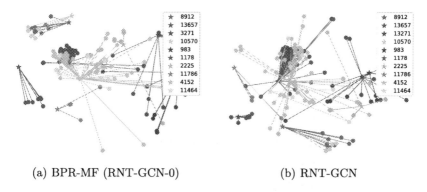

(a) BPR-MF (RNT-GCN-0)   (b) RNT-GCN

Fig. 7. Visualization of the learned t-SNE transformed representations derived from MF and RNT-GCN. Each star represents a user from the Kindle Store dataset, while the points with the same color denote the relevant items.

6   Conclusion

In this paper, we propose a novel recommender system framework, RNT-GCN. In RNT-GCN, we first develop an attention refinement network that can combine information at both the graph-level and node-level to reasonably assign importance to nodes. Then, to preserve the specific properties of different types of nodes, we propose a type-aware aggregation strategy that aggregates embeddings of different depths by node type. Finally, to capture the relationship between user and item, we develop an information fusion layer that fuses two different types of embedding. Extensive experiments on four real datasets confirm the effectiveness of our method, and ablation experiments confirm that each part makes an important contribution.

Acknowledgement. This work was supported by Shanghai Science and Technology Commission (No. 22YF1401100), Fundamental Research Funds for the Central Universities (No. 22D111210, 22D111207), and National Science Fund for Young Scholars (No. 62202095).

References

1. van den Berg, R., Kipf, T.N., Welling, M.: Graph Convolutional Matrix Completion (2018)
2. Chen, H., et al.: Graph neural transport networks with non-local attentions for recommender systems. In: WWW, pp. 1955–1964 (2022)
3. Chen, L., et al.: Revisiting graph based collaborative filtering: a linear residual graph convolutional network approach. In: AAAI, pp. 27–34 (2020)
4. Covington, P., Adams, J., Sargin., E.: Deep neural networks for YouTube recommendations. In: RecSys, pp. 191–198 (2016)
5. Glorot, X., Bengio, Y.: Understanding the difficulty of training deep feedforward neural networks. In: AISTATS, pp. 249–256 (2010)

6. Gori, M., Pucci, A.: ItemRank: a random-walk based scoring algorithm for recommender engines". In: IJCAI, pp. 2766–2771 (2007)

7. He, X., et al.: LightGCN: simplifying and powering graph convolution network for recommendation. In: SIGIR, pp. 639–648 (2020)

8. He, X., et al.: Neural Collaborative Filtering. In: WWW, pp. 173–182 (2017)

9. He, X., et al.: TriRank: review-aware explainable recommendation by modeling aspects. In: CIKM, pp. 1661–1670 (2015)

10. Kingma, D.P., Ba, J.: Adam: a method for stochastic optimization. In: ICLR (2015)

11. Kipf, T.N., Welling, M.: Semi-supervised classification with graph convolutional networks. In: ICLR (2017)

12. Koren, Y., Bell, R.M., Volinsky, C.: Matrix factorization techniques for recommender systems. Computer **8**, 30–37 (2009)

13. Liu, F., et al.: Interest-aware message-passing GCN for recommendation. In: WWW, pp. 1296–1305 (2021)

14. Mao, K., et al.: UltraGCN: ultra simplification of graph convolutional networks for recommendation. In: CIKM, pp. 1253–1262 (2021)

15. Mehta, N., Pacheco, M.L., Goldwasser, D.: Tackling fake news detection by continually improving social context representations using graph neural networks. In: ACL, pp. 1363–1380 (2022)

16. Rendle, S., et al.: BPR: bayesian personalized ranking from implicit feedback. In: UAI, pp. 452–461 (2009)

17. Sun, J., et al.: Neighbor interaction aware graph convolution networks for recommendation. In: SIGIR, pp. 1289–1298 (2020)

18. Sun, J., et al.: Separated graph neural networks for recommendation systems. In: IEEE TII, pp. 382–393 (2023)

19. Velickovic, P., et al.: Graph attention networks. In: ICLR (2018)

20. Wang, X., et al.: Disentangled graph collaborative filtering. In: SIGIR, pp. 1001–1010 (2020)

21. Wang, X., et al.: KGAT: knowledge graph attention network for recommendation. In: KDD, pp. 950–958 (2019)

22. Wang, X., et al.: Neural graph collaborative filtering. In: SIGIR, pp. 165–174 (2019)

23. Xu, K., et al.: Representation learning on graphs with jumping knowledge networks. In: ICML, pp. 5449–5458 (2018)

24. Zhang, Y., et al.: Geometric disentangled collaborative filtering. In: SIGIR 2022, Madrid, Spain, pp. 80–90 (2022)

Multi-level Noise Filtering and Preference Propagation Enhanced Knowledge Graph Recommendation

Ge Zhao, Shuaishuai Zu, Li Li[✉], and Zhisheng Yang

School of Computer and Information Science, Southwest University, Chongqing, China
lily@swu.edu.cn

Abstract. Knowledge Graph (KG) can provide semantic information about items, which can be used to mitigate the sparsity problem in recommendation systems. In recent years, the trend in knowledge-aware recommendation methods has been to leverage Graph Neural Networks (GNNs) to aggregate node information in KG. However, many of these methods focus on mining the item knowledge association on KG, but ignore the potential item auxiliary information in user's history interaction outside KG. Furthermore, these methods equally aggregate all neighbor entities of the item on KG, which will inevitably introduce irrelevant entity-interaction behaviors. To address these issues, we propose a novel model, called *Multi-level Noise Filtering and Preference Propagation Enhanced Recommendation* (MNFP). Technically, we employ self-attention mechanisms to model the user's interaction sequence to mine the item's auxiliary information. Then, we design a twin-tower preference propagation mechanism that iteratively expands item auxiliary information on KG. Additionally, we propose a multi-level noise filtering mechanism. By learning the relationship consistency between the item and its neighbor entities, the model can guide the item to selectively link highly related neighbors in preference propagation, thus reducing the introduction of noise. We evaluate MNFP on three real-world datasets: MovieLens-1M, Last.FM and Book-Crossing. Results show that MNFP significantly outperforms state-of-the-art methods on *AUC* and *F1*.

Keywords: Recommendation systems · Knowledge graph · Preference propagation

1 Introduction

Recommendation systems have proven to be effective in addressing the issue of information overload by aiding users in selecting suitable products. Collaborative Filtering (CF) recommendation [4,8] is a popular and successful technique used in recommendation systems, which models the similarity between users or items. However, limited user-item interaction data can lead to data sparsity, as well as cold-start related issues, thereby limiting the efficacy of the CF algorithm.

© The Author(s), under exclusive license to Springer Nature Switzerland AG 2023
X. Yang et al. (Eds.): ADMA 2023, LNAI 14176, pp. 107–122, 2023.
https://doi.org/10.1007/978-3-031-46661-8_8

Supplementing the system with additional information is a common solution [23]. Such supplementary information may include images, texts, knowledge graphs, etc. The knowledge graph [9] is composed of nodes and edges, representing a heterogeneous graph that encompasses extensive semantic information. The structured knowledge it provides has been proven valuable in mitigating the cold-start problem in recommender systems.

One challenge of incorporating KG as auxiliary information into recommendation systems is how to effectively utilize the semantic information in KG. For example, KGCN [17] extends non-spectral GCN methods to knowledge graphs by aggregating neighborhood information. KGIN [20] models user-item interactions at a fine-grained intent level and aggregates remote relation paths using GNN. In addition, there are some methods [16,17] that use GNN [3,7,13] to enhance recommendations by aggregating information about entities in KG.

However, the aforementioned approaches suffer from two limitations. (1) KG has a vast number of interactions that are independent of recommendations. Existing techniques tend to aggregate all of these neighboring entities equally during propagation. This kind of operation will inevitably introduce too many irrelevant entities and interactions. (2) The current methods solely rely on mapping a series of user history click items to the KG while overlooking the item auxiliary information that can be extracted from the user's interaction data.

This paper proposes the *multi-level noise filtering and preference propagation enhanced recommendation* (MNFP) model, which seeks to address the limitations discussed above through two key components: (1) a multi-level noise filtering mechanism, and (2) a twin-tower preference propagation. Firstly, our model employs a relation-aware method to learn the relationship consistency between items and their neighboring entities in KG. This relationship consistency is utilized to retain the top-k related neighbor entities for subsequent processing. Additionally, we extract relationship information within the item-entity interaction, assigning greater importance to relations with higher relationship consistency. Our model also filters irrelevant interactions in high-order propagation based on relationship consistency and important relation. Secondly, we incorporate a self-attention mechanism to model the potential auxiliary information present in a user's interaction sequence. Twin-tower preference propagation is then employed to explicitly encode this auxiliary information, thereby enabling a more natural linkage between item knowledge and semantic information in KG, and consequently leading to improved item representation.

We evaluate MNFP performance on three real-world datasets: MovieLens-1M, Last.FM and Book-Crossing. Experimental results show that our proposed method MNFP significantly outperforms state-of-the-art algorithms. Our contributions are summarized as follows:

(1) We propose an end-to-end model that effectively captures potential item auxiliary information from user interaction data. Furthermore, we have explicitly encoded this information within the KG through the use of a twin-tower preference propagation structure, thus enhancing item representation.

(2) We design a multi-level noise filtering mechanism that is finely embedded within the twin-tower preference propagation. The main purpose of the mechanism is to filter out irrelevant entities and interactions.

(3) We conduct experiments on three real-world datasets with KGs of different sizes to show the robustness and superiority of MNFP.

2   Related Work

According to current research, the knowledge-aware recommendation can be primarily classified into three methods: embedding-based methods, path-based methods, and propagation-based methods.

The embedded-based methods [1,15] utilize knowledge graph embedding techniques to obtain the representations of entities and relationships within KG and then integrate the learned representation into the recommendation module as auxiliary information. On the other hand, the path-based methods [5,11] establish remote connections between entities in KG by constructing meta paths or meta graphs. Propagation-based methods [14,17,19,20] utilize GNNs to aggregate multi-hop neighbors of item nodes, thereby mining higher-order semantic information from KG. Among them, KGCN [17] captures the neighbor structure of each entity through neighborhood aggregation operations. KGAT [19] builds a user-item-entity (UIKG) by mixing the user-item graph and the knowledge graph, then using an attention mechanism to selectively aggregate neighbor nodes. CG-KGR [2] uses collaborative signals combined with KG polymerization for richer representation.

3   Preliminaries

In this section, we introduce the structured definition of the user-item interaction data and knowledge graph and then formulate our research task.

Interaction Data. In a typical recommendation scenario, according to the definition of the recommendation model, the sets of users and items are denoted as $\mathcal{U} = \{u_1, u_2, \ldots\}$ and $\mathcal{V} = \{v_1, v_2, \ldots\}$, where user-item interaction matrix $\mathcal{Y} \in |\mathcal{U}| \times |\mathcal{V}|$ is defined according to users' implicit feedback. In interaction matrix $\mathcal{Y}$, $y_{u,v} = 1$ indicates that the user u has been exposed to the item v by clicking, browsing, or purchasing, etc.

Knowledge Graph (KG). In order to introduce knowledge graph $\mathcal{G}$ to assist recommendation, we formally define the knowledge graph, which is comprised of entities-relations triples $\{(h, r, t) \mid h, t \in \mathcal{E}, r \in \mathcal{R}\}$. Here h, r, t denotes the head, relation, and tail in the knowledge triple, and $\mathcal{E}$ and $\mathcal{R}$ are the set of entities and relations in the knowledge graph, respectively. We employ a set $E = \{(v, e) \mid v \in \mathcal{V}, e \in \mathcal{E}\}$, where (v, e) indicates that item v can be linked to entity e in the knowledge graph $\mathcal{G}$ to clarify the mapping and alignment between items and entities.

Task Description. Given interaction matrix $\mathcal{Y}$ and knowledge graph $\mathcal{G}$, our task of knowledge-aware recommendation is to learn a function $\hat{y}_{uv} = \mathcal{F}(u, v; \Theta, \mathcal{Y}, \mathcal{G})$ that can predict how likely a user would adopt an item, where $\hat{y}_{uv}$ denotes the probability that user u has interacted in item v, and Θ denotes the model parameters of the function $\mathcal{F}$.

4   Method

In this section, we will detail our proposed MNFP model. Its overall model framework is shown in Fig. 1. The model framework consists of four main parts: (1) Seed entity set construction. (2) Multi-level noise filtering. (3) Twin-tower preference propagation. (4) Model prediction.

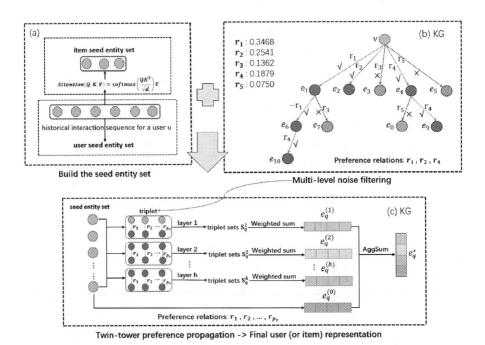

Fig. 1. Illustration of the proposed MNFP model for recommendation. Best viewed in color. Among Fig. 1, Fig. 1(b) is a toy example, where the decimal represents the relationship consistency between the item and its neighbor entity. In Fig. 1(c), all entities have the same filtering mechanism as in Fig. 1(b).

4.1   Multi-level Noise Filtering Based on Relation-Aware

Unlike the previous knowledge-aware based methods [12,14,19,20] that link all neighbor entities of the item in KG. Our purpose is to fix the center of gravity

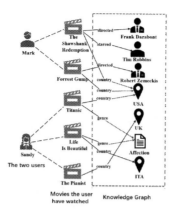

Fig. 2. Illustration of knowledge graph enhanced movie recommender systems. Movies watched by two of the users link different entities by relation on KG.

of the link of the item, screen out the neighbors that are highly relevant to the item, and reduce the introduction of irrelevant interactions. For example, as shown in Fig. 2, both movies watched by *Mark* were released in the USA while *Sandy*'s three watched movies were from different countries, with two of them from the same genre. Consequently, it is clear that *Mark* prefers movies according to their country of origin while *Sandy* is more interested in movie genres. Hence, the entity based on the director relation, *Robert Zemeckis*, is not that relevant to *Mark*. This leads to the problem that when the item has many neighbor entities, however, there are few highly correlated neighbors.

First-Order Noise Filtering. By aligning an item with an entity on KG, it can be connected to various neighboring entities through different relationships. Hence, we propose a noise filtering mechanism that utilizes the relation-aware method to derive the relationship consistency between the items and entities from their relationship perspective. Precisely, the relationship consistency between the item and the i-th neighboring entity is defined as:

$$\beta(v,t_i) = \text{softmax}(\sigma\,(W_2(\sigma\,(W_1(\sigma\,(W_0\,((e_v \odot e_{t_i})\,\|e_{r_i}) + b_0) + b_1)) + b_2))) \quad (1)$$

where $e_v \in R^d$, $e_{r_i} \in R^d$, and $e_{t_i} \in R^d$ are embeddings of the item, relation, and the item's i-th neighbor entity, respectively. $\odot$ is the element-wise product, $\|$ is the concatenation operation. In addition, W_0, W_1, W_2 and b are trainable weight matrices and biases, the relationship consistency $\beta(v,t_i)$ reflects the degree of correlation between the item v and its i-th neighbor entity under relation r_i.

We further update $\beta(v,t_i)$ with truncated probability p_τ, aiming to mitigate the low-value effect.

$$\beta'(v,t_i) = \max\,(\beta\,(v,t_i)\,,p_\tau) \quad (2)$$

Considering the updated relationship consistency, we select only the top p_v most relevant neighboring entities for each current entity. Here, p_v is a hyperparameter set by the experiment. This means that each entity can have at most

p_v highly related neighboring entities. This process is referred to as first-order noise filtering.

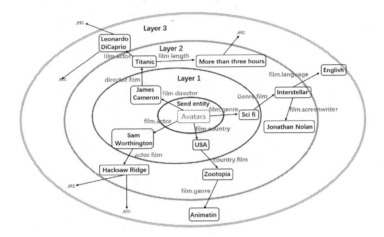

Fig. 3. An example of high order interaction for *Avatars* on KG, the ellipse represents the triple set of *Avatars* in different propagation layers.

High-Order Noise Filtering. Propagation-based methods [16,17,20] mine information about multi-hop nodes in KG through aggregators without distinguishing which higher-order interactions are useless.

In Fig. 3. Taking the movie *Avatars* watched by the user as an example, it has a second-order neighbor entity *Zootopia*, and a third-order triple [*Zootopia*, film.genre, *Animation*] composed of *Zootopia*. This triple indicates that the movie type of *Zootopia* is *Animation*, which is completely different from *Avatars*. So *Zootopia* and [*Zootopia*, film.genre, *Animation*] may be useless to the user as a kind of high-order interactive information.

To solve the above problems, we propose the concept of high-order noise filtering based on first-order noise filtering. In social recommendation, users' immediate friends are considered to have the closest relationship with those users. Drawing inspiration from this idea, we extract relational information from the interactions between the item and its highly related first-order neighbor entities. We refer to this relational information as the preference relation. Finally, we get the preference relation set of each item defined as follows:

$$R_v = \left\{ r_{vt_i} \mid v \in V \ and \ t_i \in \mathcal{E}_v^1 \right\}, \quad i = 1, 2, \cdots, p_v \tag{3}$$

where r_{vt_i} is the relation in the interaction between item v and its i-th highly neighbor entity t_i, $\mathcal{E}_v^1$ is the entity set consisting of item's neighbor entities.

In the context of higher-order propagation, items are only aggregated with higher-order neighbor entities that are connected through their preference relationships. For instance, as shown in Fig. 1(b), r_1, r_2, and r_4 represent the preference relationships of item v. Thus, only the multi-hop nodes e_6, e_9, and e_{10} are aggregated by the item v. This is because the relationships in the triple of these nodes belong to the preference relationship set of item v. This process is referred to as high-order noise filtering, which helps the model fix the center of gravity of high-order propagation and reduce the introduction of noise.

4.2   Twin-Tower Preference Propagation

In recent years, the twin-tower structure [6,18,21]. has become popular in recommendation systems, utilizing two towers to process relevant features of both users and items, respectively, and finally learning embedding vectors for users and items. In preference propagation, the user's interacted items represent their preferences, which are then iteratively propagated as the user's seed entity set on KG to extend their interests. To enable item representations to incorporate both the auxiliary information and the semantic properties of the item, we propose an innovative approach that combines the joint twin-tower structure with preference propagation. Specifically, we utilize the item auxiliary information to construct seed entity sets for each item to propagate on KG.

Initial Seed Entity Set. Before preference propagation, both users and items are required to construct their seed entity set on KG. The user's seed entity set is formed by the items with which they have interacted. These items are then mapped to KG to form the seed entity set. Firstly, the historical interaction item sequence of user u can be defined in the following way:

$$T_u = \{v_u \mid (u \in \mathcal{U} \ and \ v \in V) \ and \ v \in \{v \mid y_{uv} = 1\}\} \tag{4}$$

We perform a mapping and alignment operation between the item sequence and entities in the knowledge graph, to obtain the user's seed entity set defined as follows:

$$\mathcal{E}_u^0 = \{e \mid (v, e) \in E \ and \ v \in \{v \mid y_{uv} = 1\}\} \tag{5}$$

where (v, e) indicates that item v can be aligned with an entity e in the KG.

We have observed that T_u preserves not only the user's preferences but also auxiliary information related to the items. To illustrate this, let's consider a sample historical sequence of items: basketball shoes, shoe cleaners, socks, fruit, and mobile phone case. Shoe cleaners and socks are auxiliary items of basketball shoes as they share a similar application scenario and can provide complementary information. Thus, including shoe cleaners and socks as auxiliary information for basketball shoes can reflect some degree the preferences of the users and the characteristics of the items.

The self-attention mechanism is employed to model T_u for mining auxiliary information as follows:

$$A\left(Q, K, V\right) = \sum_j \frac{k\left(\mathbf{q}_i, \mathbf{k}_j\right)}{\sum_l k\left(\mathbf{q}_i, k_l\right)} \mathbf{v}_j = \mathbf{v}_j P_{(\mathbf{k}_j | \mathbf{q}_i)} \tag{6}$$

where $k\left(\mathbf{q}_i, \mathbf{k}_j\right) = \left(\mathbf{q}_i \mathbf{k}_j^\top / \sqrt{d}\right)$, q_i, k_i, v_i denote the i-th row in Q, K, V, respectively. $P_{(\mathbf{k}_j | \mathbf{q}_i)} = k\left(\mathbf{q}_i, \mathbf{k}_j\right) / \sum_l k\left(\mathbf{q}_i k_l\right)$ denotes the similarity between item i and item j. Then, we select the top-k most similar item for each item as its auxiliary information.

By mapping the item's auxiliary information and the item itself to the KG as a seed entity set for the item v.

$$\mathcal{E}_v^0 = \{e \mid ((v, v_s), e) \in E\} \tag{7}$$

where v_s is the auxiliary information of v.

Preference Propagation. We can explicitly encode user preference and item auxiliary information on KG by utilizing preference propagation alongside a multi-level noise filtering mechanism.

In preference propagation, the neighbor entity set in the different layers and the k-th layer triple sets for user u and item v under the multi-level noise filtering mechanism is defined as follows:

$$\mathcal{E}_q^k = \left\{t \mid (h, r, t) \in \mathcal{G} \ and \ (h \in \mathcal{E}_q^{k-1}, r \in R_v)\right\}, \quad k = 1, 2, \ldots, h \tag{8}$$

$$\mathcal{S}_q^k = \left\{(h, r, t) \mid (h, r, t) \in \mathcal{G} \ and \ (h \in \mathcal{E}_q^{k-1}, r \in R_v)\right\}, \quad k = 1, 2, \ldots, h \tag{9}$$

where k represents the layer number of propagation, $\mathcal{S}_q^k$ shows the k-th layer neighbor entity set of the seed entity set. q in $\mathcal{S}_q^k$ is a unified placeholder for user u or item v, R_v is the preference relation set of the item in the seed entity sets of the user or the item. $\mathcal{S}_q^k$ consists of triples that contain only the preference relation. Please note that when $k = 1$, the entity contained in $\mathcal{S}_q^1$ is the highly related neighbor entity of the seed entity.

One step further, giving (h, r, t) the i-th triple of the k-th layer triple set, we can get the k-th layer's embedding of the user u or the item v as follows:

$$e_q^{(k)} = \sum_{i=1}^{Num(\mathcal{S}_q^k)} \alpha\left(e_h^i, e_r^i\right) e_t^i \tag{10}$$

$$\alpha\left(e_h^i, e_r^i\right) = \text{softmax}(\sigma(W_2\sigma(W_1\sigma(W_0(e_h^i \| e_r^i) + b_0) + b_1) + b_2)) \tag{11}$$

where $e_h^i \in R^d$ and $e_r^i \in R^d$ is the embedding of the head entity and relation in the i-th triple, $\alpha\left(e_h^i, e_r^i\right)$ is the attentive weight, and can reveal which triple sets the model should focus on.

For the embedding of the seed entity set, we regard it as $e_q^{(0)} \in R^d$. $e_q^{(0)}$ represents the user's original preference information and the item's auxiliary information, which defines as follows:

$$e_q^{(0)} = \text{mean}\left(\sum_{e \in \mathcal{E}_q^0} e\right) \qquad (12)$$

where $e \in R^d$ is the embedding of the seed entity.

Finally, we aggregate the multiple representations of user u and item v as follows:

$$I_u = \left\{e_u^{(0)}, e_u^{(1)}, \ldots, e_u^{(h)}\right\}, I_v = \left\{e_v, e_v^{(0)}, e_v^{(1)}, \ldots, e_v^{(h)}\right\} \qquad (13)$$

where $e_v \in R^d$ is ID embedding of the item. To be specific, we successfully encode user interaction, item auxiliary, and semantic information from the KG into Iu and Iv using the multi-level noise filtering mechanism.

4.3 Model Prediction

We obtain the representation I_u and I_v of user u and item v after h layers. Then, we sum them up to form the final representations.

$$\mathbf{e}_u^* = e_u^{(0)} + \cdots + e_u^{(h)}, \quad \mathbf{e}_v^* = e_v + v_{up}^{(h)} + e_v^{(0)} + \cdots + e_v^{(h)} \qquad (14)$$

Finally, we use the representation of user and item to calculate the click probability $\hat{y}_{uv}$ as:

$$\hat{y}_{uv} = \mathbf{e}_u^{*\top} \mathbf{e}_v^* \qquad (15)$$

4.4 Model Optimization

In order to enhance the performance of the MNFP model, we have implemented negative sampling during the training phase. To elaborate, non-interactive items are randomly chosen as negative samples, and the chosen number matches that of positive samples. The loss function used for this process is represented as follows:

$$\mathcal{L} = \sum_{u \in U} \left(\sum_{v:y_{uv}=1}^{(u,v) \in P^+} \mathcal{J}\left(y_{uv}, \hat{y}_{uv}\right) - \sum_{v:y_{uv}!=1}^{(u,v) \in p^-} \mathcal{J}\left(y_{uv}, \hat{y}_{uv}\right) \right) + \lambda \|\Theta\|_2^2 \qquad (16)$$

where $\mathcal{J}$ is the cross-entropy loss, P^+ is a positive sampling distribution, including the set of items that the user has interacted with, P^- is a negative sampling distribution, $\Theta = \{\mathbf{E}, \mathbf{W}, \mathbf{W}_i, \forall i \in \{0, 1, 2 \cdots \}\}$ is the model parameter set, $\mathbf{E}$ is the embedding for all entities and relations, λ controls the L_2 regularization term.

5   Experiments

5.1   Dataset Description

In the evaluation, we utilized three real-world datasets: MovieLens-1M, Last.FM, Book-Crossing, which are publicly available [14].

- MovieLens-1M[1]: it is a widely used benchmark dataset in movie recommendations, including the user's 1–5 points' grading for movies on the MovieLens website.
- Last-FM[2]: it is a music listening dataset collected from Last.fm online music systems, which include artists, albums, tracks, etc.
- Book-Crossing[3]: it is a dataset about book scoring, and the books are scored from 0 to10 points, which is collected from the book-crossing community.

As for the sub-KG construction, we follow the work in RippleNet, KGAT [14,19] to construct sub-KG from KG, where use Microsoft Satori[4] to construct subgraph for MovieLens-1M datasets, use Freebase to construct subgraph for Book-Crossing, and Last.FM datasets. Also for each constructed sub-KG, its confidence level is set to be greater than 0.9. Table 1 summarizes the detailed statistics of the three datasets: MovieLens-1M, Last.FM, Book-Crossing.

Table 1. Statistics and hyper-parameter settings for the three datasets. (p_τ: truncated probability, p_v: quantity threshold value of retained neighbor entities, user_size: user triple set size, item_size: item triple set size, $L2_{weight}$: regularization term.)

	MovieLens-1M	Last.FM	Book-Crossing
#users	6,036	1,872	17,860
#items	2,445	3,846	14,967
#interactions	753,772	42,346	139,746
#entities	182,011	9,366	77,903
#relations	12	60	25
#triples	1,241,996	15,518	151,500
#p_τ	0.1800	0.1500	0.1500
#p_v	3	3	4
#user_size	80	62	44
#item_size	64	64	64
#$L2_{weight}$	1e-6	5e-5	1e-4

---

[1] https://grouplens.org/datasets/movielens/1m/.
[2] https://grouplens.org/datasets/hetrec-2011/.
[3] http://www2.informatik.uni-freiburg.de/~cziegler/BX/.
[4] https://searchengineland.com/library/bing/bing-satori.

5.2   Evaluation Metrics

For each dataset, we split the ratio of training, evaluation, and test sets into 6:2:2. We evaluate our method in CTR scenarios: In click-through rate (CTR) prediction, we apply the trained model to each interaction in the test set. Given the predicted interaction probability, we use AUC and $F1$ to evaluate CTR predictions.

5.3   Baselines Models

we compare the proposed MNFP with the following baseline models:

- LibFM [10] is a feature-based factorization model, which is widely used in feature combinations.
- CKE [23] introduces structural information, text data, image data, and other knowledge base information to improve the quality of recommended systems.
- PER [22] builds meta-paths between items to model user and item representations.
- RippleNet [14] propagates user preferences on a knowledge graph and explores users' potential interests.
- KGAT [19] explicitly models higher-order connectivity in KG in an end-to-end manner by introducing KG and building collaborative knowledge graphs.
- KGCN [17] extends non-spectral GCN methods to knowledge graphs by selectively aggregating neighborhood information.
- KGNN-LS [16] converts the knowledge graph to a user-specific weighted graph and aggregates neighborhood information with different weights.
- KGIN [20] is a state-of-the-art propagation-based model that models user intent and further refines the representation by integrating relational information of multi-hop paths through relational path-aware aggregation.
- CG-KGR [2] is a state-of-the-art propagation-based model that merges collaborative information with KG information and customizes the knowledge extraction from external KG.

Table 2. The result of AUC and $F1$ in CTR prediction.

Model	MovieLens-1M		Book-Crossing		Last.FM	
	AUC	F1	AUC	F1	AUC	F1
LibFM	0.8923 (−4.44%)	0.7964 (−7.51%)	0.6851 (−11.49%)	0.6183 (−8.33%)	0.7786 (−8.00%)	0.7102 (−8.80%)
CKE	0.9065 (−2.95%)	0.8024 (−6.82%)	0.6759 (−12.67%)	0.6235 (−7.56%)	0.7471 (−11.72%)	0.6740 (−13.45%)
PER	0.7124 (−23.71%)	0.6670 (−22.54%)	0.6048 (−21.86%)	0.5726(−15.11%)	0.6414 (−24.21%)	0.6033 (−22.52%)
RippleNet	0.9190 (−1.61%)	0.8422 (−2.19%)	0.7211 (−6.83%)	0.6472(−4.05%)	0.7762 (−8.28%)	0.7025 (−9.79%)
KGCN	0.9090 (−2.68%)	0.8366 (−2.85%)	0.6841 (−11.61%)	0.6313(−6.40%)	0.8027 (−5.15%)	0.7086 (−9.00%)
KGNN−LS	0.9140 (−2.15%)	0.8410 (−2.33%)	0.6762 (−12.64%)	0.6314(−6.39%)	0.8052 (−4.86%)	0.7224 (−7.23%)
KGAT	0.9140 (−2.15%)	0.8440 (−2.01%)	0.7314 (−5.50%)	0.6544(−2.98%)	0.8293 (−2.01%)	0.7424 (−4.66%)
KGIN	0.9191 (−1.60%)	0.8441 (−1.99%)	0.7273 (−6.03%)	0.6614 (−1.94%)	0.8431 (−0.38%)	0.7596 (−2.45%)
CG-KGR	0.9110 (−2.47%)	0.8359 (−2.93%)	0.7498 (−3.13%)	0.6689(−0.83%)	0.8336 (−1.50%)	0.7433 (−4.54%)
MNFP	**0.9341**	**0.8613**	**0.7740**	**0.6745**	**0.8463**	**0.7787**

5.4 Parameter Settings

We implement our MNFP model in PyTorch. We choose Adam as the optimizer, set the batch size to 2048, and use the Xavier method to initialize the embedding parameters. In addition, the learning rate is set as 2×10^{-3}, the embedding size s is searched in $\{16, 32, 64, 128, 256\}$, the layer number of propagation k is chosen in $\{1, 2, 3, 4, 5\}$. In twin-tower preference propagation, users and items get their own triple sets, we search for triple set sizes for users and items in $\{26, 44, 62, 80, 98\}$ and $\{8, 16, 32, 64, 128\}$, respectively. Other hyper-parameter settings are provided in Table 1.

5.5 Performance Comparison

We report the empirical results of all methods on the three datasets in Table 2. Based on these performance comparisons, we have the following observations:

- Our model MNFP achieves the best results on all three datasets. Specifically, it achieves significant improvements over the strongest baselines w.r.t. *AUC* by 3.13% and 1.60% in Book, Movie, and *F1* by 2.45% in Music respectively.
- It can be seen from Table 2 that CKE and LibFM have their advantages and disadvantages on different datasets. It shows that KG is not used, which limits the performance of the CF-based algorithm. CKE simply integrates the KG embedding method into the model and does not make full use of the semantic information in KG, which limits its performance. The performance of the path-based method PER is even much worse than that of CKE and LibFM, indicating the limitations of manually defining meta-paths on KG.
- Propagation-based methods outperform the above methods on all three datasets, illustrating the effectiveness of modeling neighbor entities and higher-order connectivity. Among them, Ripplenet does not make full use of item information, which is why its performance is lower than several other propagation-based methods. KGAT, KGIN, CG-KGR, and others can establish remote connections through aggregation mechanisms and learn the global graph information to enhance the recommendation effect.
- In the study, MNFP utilizes the twin-tower preference propagation mechanism to encode the item auxiliary information and designs a multi-level filtering mechanism to filter irrelevant entities. As a result, MNFP can minimize noise effects and enhance the embedding representation of users and items.

5.6 Ablation Studies

Impact of Multi-level Noise Filtering and Twin-Tower Preference Propagation. We investigate the impact of our model's multi-level noise filtering mechanism and twin-tower preference propagation on its overall performance by comparing MNFP against its three variants. (1) We remove the multi-level noise filtering mechanism and irrelevant interactions are not filtered in the twin-tower preference propagation, which is called MNFP$_{w/o\ F}$. (2) We disable the

Table 3. Impact of multi-level noise filtering and twin-tower preference propagation.

	Movie		Book		Music	
	AUC	F1	AUC	F1	AUC	F1
MNFP$_{w/o\ F}$	0.9221	0.8489	0.7423	0.6401	0.8203	0.7465
MNFP$_{w/o\ Fh}$	0.9285	0.8547	0.7612	0.6545	0.8342	0.7568
MNFP$_{w/o\ P}$	0.9241	0.8508	0.7234	0.6325	0.8219	0.7488
MNFP	**0.9341**	**0.8613**	**0.7740**	**0.6745**	**0.8463**	**0.7787**

high-order noise filtering mechanism, leaving only the first-order noise filtering, which is called MNFP$_{w/o\ Fh}$. (3) We forgo building the item seed entity set by mining item auxiliary information, that is, only letting the user seed entity set propagate on KG, which is called MNFP$_{w/o\ P}$.

- The results of three variants and MNFP are reported in Table 3, from which we have the following observations: (1) Disabling the multi-level noise filtering mechanism significantly degrades model performance, this is because all the items in the model do not selectively link to their highly related neighbor entities, thus introducing a lot of noise. (2) MNFP$_{w/o\ Fh}$ performance is between MNFP$_{w/o\ F}$ and MNFP, which shows that there are a lot of irrelevant interactions in the process of high order propagation and this irrelevant information will seriously affect the model performance. (3) Compared with the complete model of MNFP in Table 3, the results for MNFP$_{w/o\ P}$ drop significantly, which illustrates the importance of encoding items and their auxiliary information through twin-tower preference propagation on KG. Specifically, in MNFP$_{w/o\ P}$, the twin-tower propagation degenerates into the propagation of user seed entity sets, items do not effectively combine semantic information and knowledge association between different entities in KG.

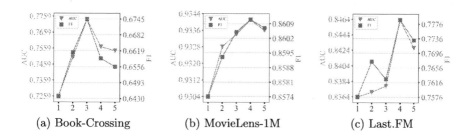

(a) Book-Crossing   (b) MovieLens-1M   (c) Last.FM

Fig. 4. Result of MNFP at different layer number of propagation k.

5.7 Sensitivity Analysis

Impact of the Layer Number of Propagation k. We change the layer number of propagation k to study the effect of different numbers of layers on the performance of MNFP. The experimental results are shown in Fig. 4. It can be seen that when the layer on music and movie is 4, and the layer on the book is 3, MNFP achieves the best performance. In addition, by increasing the layer on the three datasets, the performance of MNFP will be better, which indicates the importance of modeling high-order interaction information of entities on KG. However, when the layer is too large, the high-order propagation will introduce too much noise. Therefore, maintaining a reasonable layer can improve the recommendation performance.

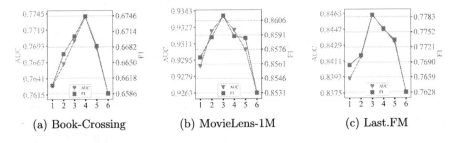

(a) Book-Crossing   (b) MovieLens-1M   (c) Last.FM

Fig. 5. Result of MNFP at different number of reserved neighbor entities p_v.

Impact of the Quantity Threshold Value p_v **of Retained Neighbor Entities.** To study the influence of p_v on model performance, we vary p_v in $\{1, 2, 3, 4, 5, 6\}$, the results shown in Fig. 5. As can be seen from the figure, when p_v is set to 3 or 4, it can bring the best performance to the model, the reason for this is that during the propagation, the item retains the highly related neighbor entities and filters out the irrelevant interactions. Meanwhile p_v is too small, and the item retains very few associated neighbor entities, that is, MNFP cannot adequately associate the semantic information on KG.

Impact of the Dimension of Embedding s. When the model embedding size s is different, the performance of MNFP is shown in Table 4. It can be seen that by increasing the embedding size, the model performance will also improve, because the larger the embedding vector dimension, the more useful entity and item information it contains, but too large an embedding size can cause overfitting.

Table 4. Impact of different embedding size s.

	Movie		Book		Music	
	AUC	F1	AUC	F1	AUC	F1
$s = 16$	0.9307	0.8560	0.7343	0.6361	0.8283	0.7479
$s = 32$	0.9313	0.8569	0.7497	0.6521	0.8238	0.7424
$s = 64$	**0.9341**	**0.8613**	0.7637	0.6653	0.8378	0.7628
$s = 128$	0.9302	0.8585	**0.7740**	**0.6745**	**0.8463**	**0.7787**
$s = 256$	0.9297	0.8532	0.7612	0.6637	0.8413	0.7728

6   Conclusion and Future Work

In this paper, we propose MNFP, a novel propagation-based knowledge-aware recommendation model. MNFP emphasizes the importance of explicitly encoding the item auxiliary information in the historical interaction item sequence of the user. In particular, auxiliary information and user preference information is propagated recursively on KG through a kind of twin-tower preference propagation mechanism, enhancing the latent representation of users and items. In addition, MNFP uses a multi-level noise filtering mechanism to eliminate unrelated entities and interactions and reduce the introduction of noise during propagation. Extensive experiments demonstrate the superiority of MNFP on three datasets.

In future work, we intend to incorporate timestamps of user-item interactions to add time-series modeling to further explore the process of changing user preferences and interests.

Acknowledgements. This work was supported by NSFC (grant No.61877051).

References

1. Cao, Y., Wang, X., He, X., Hu, Z., Chua, T.S.: Unifying knowledge graph learning and recommendation: towards a better understanding of user preferences. In: WWW, pp. 151–161 (2019)
2. Chen, Y., Yang, Y., Wang, Y., Bai, J., Song, X., King, I.: Attentive knowledge-aware graph convolutional networks with collaborative guidance for personalized recommendation. In: 2022 IEEE 38th International Conference on Data Engineering (ICDE), pp. 299–311. IEEE (2022)
3. Dwivedi, V.P., Joshi, C.K., Laurent, T., Bengio, Y., Bresson, X.: Benchmarking graph neural networks. arXiv preprint arXiv:2003.00982 (2020)
4. He, X., Liao, L., Zhang, H., Nie, L., Hu, X., Chua, T.S.: Neural collaborative filtering. In: WWW, pp. 173–182 (2017)
5. Hu, B., Shi, C., Zhao, W.X., Yu, P.S.: Leveraging meta-path based context for top-n recommendation with a neural co-attention model. In: KDD, pp. 1531–1540 (2018)

6. Huang, P.S., He, X., Gao, J., Deng, L., Acero, A., Heck, L.: Learning deep structured semantic models for web search using clickthrough data. In: CIKM, pp. 2333–2338 (2013)
7. Kipf, T.N., Welling, M.: Semi-supervised classification with graph convolutional networks. arXiv preprint arXiv:1609.02907 (2016)
8. Lian, J., Zhou, X., Zhang, F., Chen, Z., Xie, X., Sun, G.: xDeepFM: combining explicit and implicit feature interactions for recommender systems. In: KDD, pp. 1754–1763 (2018)
9. Paulheim, H.: Knowledge graph refinement: a survey of approaches and evaluation methods. Semant. Web **8**(3), 489–508 (2017)
10. Rendle, S.: Factorization machines with LIBFM. ACM Trans. Intell. Syst. Technol. (TIST) **3**(3), 1–22 (2012)
11. Shi, C., Hu, B., Zhao, W.X., Philip, S.Y.: Heterogeneous information network embedding for recommendation. IEEE Trans. Knowl. Data Eng. **31**(2), 357–370 (2018)
12. Tang, X., Wang, T., Yang, H., Song, H.: Akupm: attention-enhanced knowledge-aware user preference model for recommendation. In: KDD, pp. 1891–1899 (2019)
13. Veličković, P., Cucurull, G., Casanova, A., Romero, A., Lio, P., Bengio, Y.: Graph attention networks. arXiv preprint arXiv:1710.10903 (2017)
14. Wang, H., et al.: Ripplenet: propagating user preferences on the knowledge graph for recommender systems. In: CIKM, pp. 417–426 (2018)
15. Wang, H., Zhang, F., Xie, X., Guo, M.: DKN: deep knowledge-aware network for news recommendation. In: WWW, pp. 1835–1844 (2018)
16. Wang, H., et al.: Knowledge-aware graph neural networks with label smoothness regularization for recommender systems. In: KDD, pp. 968–977 (2019)
17. Wang, H., Zhao, M., Xie, X., Li, W., Guo, M.: Knowledge graph convolutional networks for recommender systems. In: WWW, pp. 3307–3313 (2019)
18. Wang, J., Zhu, J., He, X.: Cross-batch negative sampling for training two-tower recommenders. In: SIGIR, pp. 1632–1636 (2021)
19. Wang, X., He, X., Cao, Y., Liu, M., Chua, T.S.: KGAT: knowledge graph attention network for recommendation. In: KDD, pp. 950–958 (2019)
20. Wang, X., et al.: Learning intents behind interactions with knowledge graph for recommendation. In: WWW, pp. 878–887 (2021)
21. Yang, J., et al.: Mixed negative sampling for learning two-tower neural networks in recommendations. In: WWW, pp. 441–447 (2020)
22. Yu, X., et al.: Personalized entity recommendation: a heterogeneous information network approach. In: WSDM, pp. 283–292 (2014)
23. Zhang, F., Yuan, N.J., Lian, D., Xie, X., Ma, W.Y.: Collaborative knowledge base embedding for recommender systems. In: KDD, pp. 353–362 (2016)

Enhancing Knowledge-Aware Recommendation with Contrastive Learning

Xinyue Zhang and Hui Gao[✉]

School of Computer Science and Engineering, University of Electronic Science and Technology of China, Chengdu, China
kepa0107@std.uestc.edu.cn, huigao@uestc.edu.cn

Abstract. Knowledge graph serves as a side information, bringing diversity and interpretability to the recommendation. A well-developed recommender system can efficiently capture user and item characteristics, accurately reflecting user preferences. However, supervised signals with graph structure are extraordinarily sparse, and the collaborative and knowledge graphs contain irrelevant edges, exacerbating noise propagation and reducing the robustness of recommendations. To address the above issues, we propose a model for enhancing **K**nowledge-aware **R**ecommendation with **C**ontrastive **L**earning (**KRCL**), including two contrastive learning tasks and three functional modules. Specifically, we construct two views, using TransR and TATEC to optimize knowledge representations from distance and semantic aspects, respectively. After the item-side knowledge is augmented, we remove unreliable interaction edges from collaborative graph to reduce noise propagation. We then perform contrastive learning on the output node representations of different views through graph propagation. To further tap the latent interest of users, we consider users/items that exhibit similar representations as semantic neighbors, treating them as positive pairs in contrastive learning. The structural and semantic contrastive tasks are eventually integrated in a multi-task learning manner to jointly boost the recommendation performance. To validate the effectiveness of our method, we conduct extensive experiments on three benchmark datasets. Experimental results demonstrate that our KRCL significantly outperforms previous state-of-the-art baselines.

Keywords: Recommender Systems · Knowledge Graph · Contrastive Learning

1 Introduction

Nowadays, recommender systems have gained significant importance in mining potential interests of users by analyzing their historical behavior. As a fundamental algorithm, the traditional Collaborative Filtering (CF) method [10,18]

X. Yang et al. (Eds.): ADMA 2023, LNAI 14176, pp. 123–137, 2023.
https://doi.org/10.1007/978-3-031-46661-8_9

captures user preferences by modeling the similarity of the user-item interaction matrix, but suffers from the data sparsity and noise problems.

A widely adopted solution is to utilize auxiliary data sources, such as the Knowledge Graph (KG) [12,30], which contains rich factual attributes and connections to learn high-quality representations. Earlier works [1,34] generate embeddings by mapping KG entities and relations into a low-dimensional space, this approach preserves the structural information of KG but ignores the semantic associations of user preferences. Several studies [2,22,30] propose path-based approaches to learn the similarity of path connections. However, constructing meta-path requires strong domain knowledge and heavily depends on manually designed path patterns. In contrast, Graph Neural Networks (GNN) [11,21,25] have shown promising prospects in recommendation, which aggregate embeddings of multi-hop neighborhoods through graph propagation, leading to a substantial improvement in recommendation performance.

Despite the remarkable success of GNN-based models, they still have some limitations: (1) Sparse supervised signals: Most recommender systems rely sorely on sparse supervised signals [27], which are insufficient to generate high-quality node representations due to the sparsity of observed target behavior in real-world scenarios. In severe cases, it can cause the degradation problem [7], i.e., node embeddings degenerate to no discrimination. (2) Long-tail effect: Interactions are typically power-law distributed [6], in which case user embeddings tend to update towards popular items [4] while ignoring the personalization of potential users in low-degree (long-tail) items. (3) Noisy data interference: KG contains many uncorrelated connections between items and attributes. In addition, users' feedback in collaborative graph [3] is mostly implicit (e.g., clicking, browsing) and may not accurately reflect their true preferences. In fact, GNN-based models are more susceptible to noise than conventional methods, as the neighborhood aggregation scheme amplifies the impact of noisy interactions on node representations. Therefore, it is crucial to improve the robustness of GNN-based models against potential noise.

Recently, some researchers [14,24,28] have successfully incorporated contrastive learning into recommendation. SGL [27] generates multiple views by randomly dropping edges to maximize the agreement among views of the same node, while minimizing that of the different nodes. Despite the effectiveness, this approach has a certain probability of discarding crucial edges, amplifying the effect of noise propagation and resulting in semantic bias. Moreover, such structure-based contrastive learning approaches fail to leverage higher-order relationships among nodes and tend to overlook collaborative and semantic signals.

To handle these challenges, we propose a **K**nowledge-aware **R**ecommendation framework with **C**ontrastive **L**earning (**KRCL**), designing two different contrastive learning tasks and a joint training paradigm to optimize our model. Our framework includes a relation-aware aggregation mechanism that complements KG information to enhance item-side knowledge. To alleviate noise propagation, we apply a graph denoising technique that identifies highly reliable interactions and denoise low-scoring interaction edges. Furthermore, we exploit higher-order semantic connections to develop the prototypical contrastive objective. To conclude, we summarize our contributions as:

- We propose KRCL, an innovative model that incorporates rich information in KG, introducing self-supervised signals and graph denoising technique to alleviate the data sparsity and noise propagation issues in recommendation.
- We construct two views to perform contrastive learning in both structural and semantic aspects, which effectively enhances the node representations in knowledge-aware recommendation.
- Experimental results and further analysis illustrate that our KRCL achieves significant improvements over previous state-of-the-art baselines.

2   Related Work

2.1   Graph-Based Collaborative Filtering

GNN-based methods have shown excellent performance in recommendation scenarios, primarily by leveraging the neighborhood aggregation mechanism to learn node representations. NGCF [25] exploits higher-order connectivity in collaborative graph to propagate embeddings, whereas DGCF [21] decouples user and item representations based on users' intent. LightGCN [11] removes the nonlinear activation and feature transformation modules from NGCF [25], allowing for a simple and effective recommendation process. More recently, Knowledge Graph (KG) provides rich knowledge for item-side information and enhances collaborative representations. KGAT [23] applies an attention mechanism and incorporates KG as a collaborative knowledge graph for neighborhood propagation. CKAN [26] employs heterogeneous propagation and knowledge graph propagation strategies to propagate the collaborative and knowledge-aware signals. KGCL [33] designs a knowledge-guided contrastive learning method for both KG and collaborative graph, giving additional weight to unbiased interaction edges.

2.2   Contrastive Learning

Contrastive learning has achieved great success in the field of computer vision [5] and has subsequently been adapted to natural language processing [9] and recommender systems [31]. The primary goal of contrastive learning is to maximize the consistency among positive pairs while minimizing the similarity of negative samples. SGL [27] augments graph data with node dropout, edge dropout, and random walk operations to achieve self-supervised tasks. MCCLK [36] constructs three different views to explore cross-view collaboration in a self-supervised manner. HCCF [29] compares the global hypergraph-enhanced cross-view with the GCN-encoded local graph to enhance the quality of node representations.

3   Preliminaries

Collaborative Graph. The collaborative graph includes interactions between users and items. Traditional collaborative filtering methods recommend items based on the historical behavior of users (e.g., view, click and purchase). Given user and item sets $\mathcal{U} = \{u_1, u_2, \ldots, u_M\}$, $\mathcal{V} = \{v_1, v_2, \ldots, v_N\}$, the interaction data can be defined as $Y \in \mathcal{R}^{M \times N}$, where $y_{uv} = 1$ indicates that user u interacts with item v, otherwise $y_{uv} = 0$.

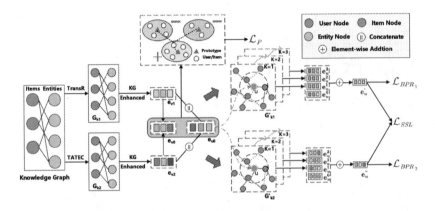

Fig. 1. The overall architecture of our proposed KRCL.

Knowledge Graph. Knowledge Graph (KG) contains abundant real-world facts, such as associative attributes and external auxiliary knowledge. Let $\mathcal{E}$ be the entity set and $\mathcal{R}$ be the relation set, KG is defined as $\mathcal{G} = \{(h, r, t) \mid h, t \in \mathcal{E}, r \in \mathcal{R}\}$, where (h, r, t) refers to an edge from head entity h to the tail entity t with relation r. To avoid renaming of head and tail entities, we create alignment of items and entities $\mathcal{A} = \{(v, e) \mid v \in \mathcal{V}, e \in \mathcal{E}\}$. In this way, the knowledge graph facilitates richer item representations and complements the single interaction data.

Task Formulation. Based on the above definitions, our knowledge-aware recommendation task can be formulated as:

Input. Interaction data Y in collaborative graph and the knowledge graph $\mathcal{G}$.

Output. A function that predicts the interaction probability y_{uv} between a user u and an item v.

4   Methodology

Figure 1 illustrates the overall architecture of KRCL. Our model mainly consists of four parts: the knowledge enhancement module, the graph denoising module, the semantic enhancement module and the joint training module. We design two types of contrastive learning tasks to capture the structural and semantic features of user-item interactions.

4.1   Knowledge Enhancement Module

Relation-Aware Item Information Aggregation. As a complement to the item-side information, KG contains abundant relational edges. Nodes with different neighbors vary in importance. For instance, when recommending movies for users, the relation "Category" holds more value than the relation "Time". Inspired by the Graph Attention Network [19], we propose a relation-aware method to aggregate neighborhood nodes:

$$w_{i,r,e} = \frac{\alpha\left(v_i, v_r, v_e\right)}{\sum_{e' \in N_i} \alpha\left(v_i, v_r, v_{e'}\right)} \tag{1}$$

$$\alpha\left(v_i, v_r, v_e\right) = \exp\left[\sigma\left(v_r^T\left(W\left(v_e\|v_i\right) + b\right)\right)\right] \tag{2}$$

$$v_i^{l+1} = \sigma\left(v_i^l + \sum_{e \in N_i} w_{i,r,e} v_e\right) \tag{3}$$

where $w_{i,r,e}$ denotes the weight of item i on the neighbor node $e \in N_i$ under relation r, $\sigma(\cdot)$ denotes the LeakyReLU [32] function, v_i^{l+1} denotes the aggregated representation of item i from layer l to layer $l+1$, v_r and v_e denote the embedding of relations and entities, respectively.

Knowledge Representation Enhancement. We generate knowledge representations in two ways: the translational distance model TransR [13] and the semantic matching model TATEC [8]. TransR assumes that similar entities and relationships are close in the common semantic space while TATEC satisfactorily combines the two-way (e.g., (h, r), (r, t)) and three-way (e.g., (h, r, t)) patterns to predict links. Specifically, TransR emphasizes the rationality of KG triples, whereas TATEC concentrates on extracting the underlying semantic relations, which collectively enhance the node representations of KG:

$$f_1(h, r, t) = -\|M_r v_h + v_r - M_r v_t\|_2^2 \tag{4}$$

$$f_2(h, r, t) = v_h^T M_r v_t + v_h^T v_r + v_t^T v_r + v_h^T D v_t \tag{5}$$

where v_h, v_r, v_t represents a triple embeddings in KG, $M_r \in \mathbf{R}^{d \times d}$ is a projection matrix from the entity space to the r-relation space, and D is a diagonal matrix shared across all relations. The initialization yields two different graph representations $\mathcal{G}_{k1}$ and $\mathcal{G}_{k2}$. And the loss functions to optimize node representations can be denoted as:

$$\mathcal{L}_{TransR} = \sum_{(h,r,t,t') \in \mathcal{G}_{k1}} -ln\sigma(f_1(h, r, t') - f_1(h, r, t)) \tag{6}$$

$$\mathcal{L}_{TATEC} = \sum_{(h,r,t,t') \in \mathcal{G}_{k2}} -ln\sigma(f_2(h, r, t') - f_2(h, r, t)) \tag{7}$$

where $\sigma(\cdot)$ is the Sigmoid function, (h, r, t') is a negative triplet by randomly replacing the tail t of a valid triplet (h, r, t). By modeling entities and relations in

KG triples, knowledge can be directly injected into item embeddings, facilitating the knowledge representations from different aspects.

4.2 Graph Denoising Module

Collaborative graph suffers from a certain degree of noise (e.g., hand slip and surrogate shopping), which causes deviations between users' behavior lists and their actual preferences. Recently, SGL [27] has introduced auxiliary signals for representation learning by randomly dropping interaction edges. However, this approach has a certain probability of discarding key edges, leading to unsatisfactory results. According to the homogeneity theory [15], nodes that exhibit similar structures or features are more likely to interact with each other compared to the ordinary nodes. Therefore, we propose to evaluate the sampling score based on the structural similarity between users and items.

Given the user-item interaction data $Y \in \mathcal{R}^{M \times N}$, we project user and item representations onto a common hidden space and extract the one-hop neighbors as their structural features:

$$X_U = Y E_I, \quad X_I = Y^\top E_U \tag{8}$$

where E_U and E_I are the initial embedding matrices, X_U and X_I represent the structural feature matrices of users and items, respectively. We employ the cosine similarity and perform min-max normalization to calculate the interactive sampling score:

$$s_{u,i} = \frac{X_u^T X_i}{\|X_u\|\|X_i\|} \tag{9}$$

$$r_{u,i} = \frac{s_{u,i} - s^{min}}{s^{max} - s^{min}} \tag{10}$$

where $r_{u,i}$ denotes the interactive sampling score, X_u and X_i denote the u-th row of X_U and the i-th row of X_I, respectively. Edges with low sampling scores are more likely to be noise, so we define $p_{u,i}$ as the denoising probability to remove noisy interactions:

$$p_{u,i} = \Gamma\left(r_{u,i} > \varepsilon\right) \cdot r_{u,i} \tag{11}$$

where $\Gamma(\cdot)$ is a binary indicator that returns 1 when the sampling score $r_{u,i}$ exceeds the predefined threshold value ε. Otherwise, $p_{u,i}$ returns 0 and the edge is dropped. During the process, $\mathcal{G}_{k1}$ and $\mathcal{G}_{k2}$ are denoised to the corrupted graphs $\mathcal{G}'_{k1}$ and $\mathcal{G}'_{k2}$. In general, the graph denoising technique prunes the noisy interactions and thus avoids noisy message propagation throughout the GNN.

4.3 Semantic Enhancement Module

Some nodes may not be directly reachable on the graph, but have similar semantic features. We identify high-level semantic neighbors by capturing latent prototypes of users and items. Figure 1 illustrates that similar users/items tend to fall in adjacent embedding spaces, with prototypes serving as the centre of semantic clusters. To this end, we employ the K-means clustering algorithm to divide users and items into K clusters. During the training process, nodes minimize the distance from their prototype centres and maximize the difference from other prototypes. The prototypical objective is optimized based on InfoNCE [16]:

$$\mathcal{L}_P^U = \sum_{u \in \mathcal{U}} - \log \frac{\exp(v_u \cdot c_i / \tau)}{\sum_{c_t \in C} \exp(v_u \cdot c_t / \tau)} \tag{12}$$

$$\mathcal{L}_P^I = \sum_{i \in \mathcal{I}} - \log \frac{\exp(v_i \cdot c_j / \tau)}{\sum_{c_t \in C} \exp(v_i \cdot c_t / \tau)} \tag{13}$$

$$\mathcal{L}_P = \alpha \mathcal{L}_P^U + \beta \mathcal{L}_P^I \tag{14}$$

where v_u is the initial embedding of users and v_i is the mean vector of the initial item representations in $\mathcal{G}_{k1}$ and $\mathcal{G}_{k2}$. c_u and c_i are the prototypes of user u and item i, respectively. Incorporating semantic neighbors as comparison pairs increases the interpretability of recommendations and further enhances user representations through graph convolution operations.

4.4 Joint Training

According to the Eq. (3), the output item representation after relation-aware augmentation is used as the initial item representation x_i^0 of the two corrupted graphs $\mathcal{G}'_{k1}$ and $\mathcal{G}'_{k2}$. For simplicity and efficiency, we utilize LightGCN [11] as backbone to perform message propagation in collaborative graph:

$$\mathbf{x}_u^{(l+1)} = \sum_{i \in N_u} \frac{\mathbf{x}_i^{(l)}}{\sqrt{|N_u||N_i|}} \ , \ \mathbf{x}_i^{(l+1)} = \sum_{u \in N_i} \frac{\mathbf{x}_u^{(l)}}{\sqrt{|N_i||N_u|}} \tag{15}$$

where $\mathbf{x}_u^{(l+1)}$ denotes the user embeddings of the $(l+1)$-th layer that aggregates information from neighboring items N_u, and this is equally applicable to items with users as neighbors. After graph propagation, KRCL performs contrastive learning on the output node representations, maximizing the consistency of positive pairs and minimizing that of the negative pairs:

$$\mathcal{L}_{ssl} = \sum_{x \in \mathcal{U} \cup \mathcal{V}} - \log \frac{\exp(v_x^1 \cdot v_x^2 / \tau)}{\sum_{t \in \mathcal{U} \cup \mathcal{V}} \exp(v_x^1 \cdot v_t^2) / \tau)} \tag{16}$$

During the training phase, KRCL employs the Bayesian Personalized Ranking (BPR) loss as the primary task loss to optimize parameters:

$$\mathcal{L}_{bpr} = \sum_{u \in \mathcal{U}} \sum_{i \in N_u} \sum_{i' \notin N_u} - \log \sigma(\hat{y}_{u,i} - \hat{y}_{u,i'}) \tag{17}$$

where N_u represents the neighbor items of user u and i' represents a random sample of non-interactive items of user u. We propose a multi-task learning strategy that jointly trains the structural and semantic contrastive tasks and the primary BPR loss:

$$\mathcal{L} = \mathcal{L}_{bpr} + \lambda_1\mathcal{L}_{cl} + \lambda_2\mathcal{L}_P + \lambda_3||\Omega||^2 \tag{18}$$

where Ω denotes the set of model parameters, λ_1, λ_2 and λ_3 represent hyperparameters that control the weight of self-supervised signals and regularization terms, respectively.

Table 1. Statistics of the datasets

Graph	Stats	Yelp 2018	Amazon-Book	MovieLens-1M
User-Item Interaction	#users	45919	70679	6035
	#items	45538	24915	2347
	#Interactions	1183610	846434	376743
	#Density	5.7×10^{-4}	4.8×10^{-4}	2.7×10^{-2}
Knowledge Graph	#Entities	47472	29714	181603
	#Relations	42	39	12
	#Triplets	869603	686516	41083

5   Experiments

5.1   Experimental Settings

We first present the experimental datasets and various state-of-the-art baselines. Afterwards, We discuss the variable factors and demonstrate the effectiveness and robustness of KRCL.

Datasets. Table 1 summarizes three benchmark datasets with varying degrees of sparsity and the associated knowledge graphs.

Yelp 2018[1]. Collected by famous American business review website, it covers numerous areas such as restaurants and shopping malls, and is widely used for business venue recommendation.

Amazon-Book[2]. It is Amazon's most categorized dataset, containing user rating data and book metadata (e.g. category, price, brand).

MovieLens-1M[3]. It includes user ratings, movie attributes and tag features, and is the most commonly used recommendation dataset in the movie domain.

Similar to RippleNet [20], we use the Microsoft Satori4[4] to construct knowledge datasets for MovieLens-1M. By contrast, Yelp 2018 and Amazon-Book are

---

[1] https://www.yelp.com/dataset.
[2] https://jmcauley.ucsd.edu/data/amazon/.
[3] https://grouplens.org/datasets/movielens/1m/.
[4] https://searchengineland.com/library/bing/bing-satori.

relatively sparse, and we follow the similar settings in [23] to map items to Freebase entities [35]. To ensure the quality of the constructed graph, we consider entities within two hops of items as neighbors.

Compared Methods. we compare our proposed KRCL with various competitive baseline methods for performance evaluation:

- BPRMF [17]. It is a typical matrix factorization model that learns latent node representations through BPR-optimized ranking loss pairs.
- NFM [10] It is a factorization-based approach that combines multiple-layer perceptron and Deep Neural Network (DNN) to model the nonlinear user-item interactions.
- NGCF [25]. It is a collaborative graph neural network that exploits higher-order connectivity to aggregate user and item embeddings.
- LightGCN [11]. It simplifies the design of GCN by removing the feature transformation and non-linear activation modules, which improves the generalization ability and makes training more efficient.
- SGL [27]. It introduces self-supervised signals into the graph collaborative filtering framework to perform contrastive learning on the structural features.
- NCL [14]. It discards augmented view construction for self-supervised learning and improves recommendation performance by learning representative embeddings for structural and semantic neighbors.
- KGAT [23]. It achieves high-order relation modeling based on GNNs by iteratively integrating neighbors on the collaborative knowledge graph through an attention mechanism.
- KGIN [24]. It disentangles user-item interactions with fine-grained intent granularity, combines user intent with KG triplets and explores the relational path-aware aggregation.
- CKAN [26]. It employs knowledge-aware attention mechanism and enriches node representations through the combination of collaborative information and knowledge.
- KGCL [33]. Under the joint self-supervised learning paradigm, it creates different KG structural views and selects items with high consistency scores to guide the contrastive learning of the collaborative graph.

5.2   Hyperparameters

All experiments are implemented with PyTorch on a single NVIDIA RTX 3090 GPU. The embedding size is fixed to 64 for all models. We apply the Adam optimizer with a learning rate of $5e^{-4}$ and $1e^{-3}$. The L2 regularization with decaying weights is explored between $\{10^{-5}, 10^{-4}, 10^{-3}\}$. The range of semantic neighbors is in $\{100, 200, 500, 1000, 1500\}$. We explore the number of propagation layers in $\{1, 2, 3, 4\}$ and vary α, β, λ_1 and λ_2 in $\{10^{-7}, 10^{-6}, 10^{-4}, 10^{-2}, 10^{-1}, 1\}$. We employ the early stopping strategy to prevent overfitting. All our experimental results are evaluated in terms of Recall@20 and NDCG@20.

5.3   Results

As shown in Table 2, our proposed KRCL yields the best performance on three benchmark datasets. From the results, we have some observations:

Compared to traditional methods such as MF and NFM, GNN-based models achieve superior performance as GNNs can iteratively propagate higher-order connectivity information to enhance node representations. NGCF directly applies Graph Convolutional Network (GCN) to capture higher-order relationships, while LightGCN is lighter and more efficient, demonstrating the effectiveness and necessity of simplifying the GCN structure.

SGL and NCL design an auxiliary self-supervised task to the collaborative filtering, achieving remarkable improvements over traditional methods. Furthermore, knowledge-aware approaches such as KGIN, KGCL show relatively superior performance compared to the models that only use GNNs, indicating that the introduction of KG is essential to overcome data sparsity and enhance the diversity of recommendations.

KRCL consistently achieves the best results on datasets with various densities and recommendation scenarios, which demonstrates the excellent generalizability of our method. In general, the improved performance of KRCL can be attributed to three aspects: (1) Knowledge graph serves as an auxiliary information that enhances item-side knowledge and increases the diversity and interpretability of the recommendation. (2) The target denoising strategy on noisy interactions to mitigate the negative impact of noises in the collaborative graph. (3) The introduction of self-supervised signals in both structural and semantic aspects to further enhance node representations.

Table 2. Performance comparison of different recommendation models

Model	Yelp 2018		Amazon-Book		MovieLens-1M	
	Recall	NDCG	Recall	NDCG	Recall	NDCG
MF	0.0484	0.0397	0.0976	0.0726	0.2427	0.2537
NFM	0.0435	0.0376	0.0883	0.0657	0.2518	0.2400
NGCF	0.0579	0.0477	0.0988	0.0580	0.2698	0.2566
LightGCN	0.0648	0.0473	0.1208	0.0734	0.2793	0.2538
SGL	0.0668	0.0483	0.1434	0.0763	0.2846	0.2684
NCL	0.0735	0.0492	0.1492	0.0787	0.2935	0.2755
KGAT	0.0675	0.0432	0.1380	0.0726	0.2786	0.2625
CKAN	0.0689	0.0441	0.1365	0.0708	0.2811	0.2746
KGIN	0.0712	0.0462	0.1436	0.0748	0.2958	0.2836
KGCL	0.0748	0.0486	0.1486	0.0783	0.3060	0.2910
KRCL	**0.0794**	**0.0517**	**0.1558**	**0.0827**	**0.3195**	**0.2970**

5.4   Further Analysis of KRCL

Ablation Study. We set up three variants to explore the impact of two contrastive tasks on KRCL performance, where "w/o S" and "w/o P" denote the removal of contrastive learning on output nodes and semantic neighbors, respectively, and "w/o $both$" signifies the removal of the two self-supervised signals.

Figure 2 shows that both contrastive learning tasks contribute to the model performance. The effect of "w/o S" is significant, especially on the MovieLens-1M dataset, indicating the effectiveness of graph denoising and structural contrastive learning. In contrast, "w/o P" has a relatively limited impact on KRCL, mainly due to the lack of fine-grained semantic neighbors and insufficient data connectivity, but it still has a beneficial effect on KRCL. To sum up, the structural and prototypical contrastive tasks complement each other and jointly enhance the performance of KRCL.

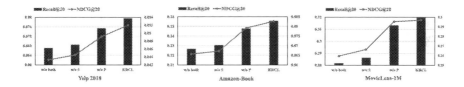

Fig. 2. Ablation study on three datasets

Table 3. Performance under different aggregation layers

	Yelp 2018		Amazon-Book		Movielens-1M	
	Recall	NDCG	Recall	NDCG	Recall	NDCG
KRCL - 1	0.0700	0.0456	0.1293	0.0681	0.2957	0.2757
KRCL - 2	0.0762	0.0499	0.1506	0.0800	0.3136	0.2934
KRCL - 3	0.07921	0.0517	**0.1558**	**0.0827**	**0.3195**	**0.2970**
KRCL - 4	**0.07949**	**0.0520**	0.1544	0.0816	0.2162	0.1914

Performance Under Different Semantic Neighbors. Figure 3 shows the impact of semantic neighbors on the prototypical contrastive objective. Due to the sparsity of the Amazon-Book and Yelp 2018 datasets, the introduction of semantic neighbors results in an overall upward trend, which plays a more significant role in accurate division and efficient learning during clustering. Conversely, the results of MovieLens-1M dataset are volatile and sometimes cause negative results. Specifically, when $k = 50$ or 100, the results are dissatisfactory in contrast to $k = 0$ (the "w/o P" in Sect. 5.4). This can be attributed to the insufficient sparsity of the MovieLens-1M dataset, when the value of k is small, the semantic neighbors are over-detailed, which may introduce additional noise. Furthermore, we also observe that the performance of KRCL degrades on all datasets with large values of k. As k grows, the semantic neighbors are roughly divided and are virtually useless. This conclusion aligns with the common knowledge that only a portion of users are similar in real life.

Performance Under Different Aggregation Layers. We vary the number of aggregation layers in $\{1, 2, 3, 4\}$ to investigate the impact on model performance. As presented in Table 3, increasing the depth of KRCL can significantly improve the model performance, as the stacked layers incorporate higher-order neighbor information into node representations, which is critical for modeling user preferences. KRCL achieves the best result with $k = 4$ in Yelp 2018 while $k = 3$ in Amazon-Book and MovieLens-1M, respectively. When $k = 4$, the performance of MovieLens-1M suffers a dramatic decline due to the over-smoothing problem caused by stacking too many layers. As such, we emphasize the importance of properly controlling the number of aggregation layers to optimize the utilization of neighbor information.

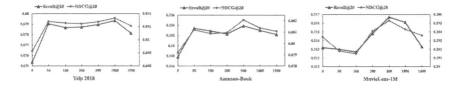

Fig. 3. Performance under different semantic neighbors

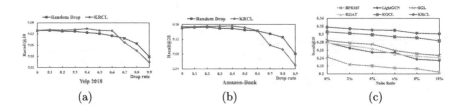

Fig. 4. Parameter settings for different ratios of noise

Discussion on Graph Denoising. Although the introduction of KG boosts the recommendation performance, it inevitably introduces some noise. Therefore, we explore the optimal sampling strategy by adjusting the denoising threshold ε. Figure 4(a) and 4(b) illustrate the difference between the methods of random drop and drop based on the sampling score on Yelp 2018 and Amazon-Book datasets, from which our method shows superiority at low dropout ratios, reaching an optimal value at 0.4. When the ratio exceeds 0.6, our performance drops sharply, even worse than that of the random dropout method. This indicates that noisy interactions only exist in a small part of the graph and that moderate denoising is essential to boost the model performance.

To validate the effect of noisy data, we randomly replace 0%–10% of the noisy edges in the training set, while leaving the test set unchanged. We conduct experiments on the MovieLens-1M dataset. From Fig. 4(c), we can observe that the performance of all models degrades significantly with the addition of noise. The current most stably performing model, KGCL, decreases by an average of 7.39% after the introduction of 10% noise, while our model only reduces the performance by 5.30%, indicating that our proposed KRCL not only consistently outperforms various baseline models, but also suffers the least variation in noise. This experiment validates the necessity of the graph denoising module and further illustrates the effectiveness and robustness of KRCL.

6   Conclusion

In this work, we propose KRCL, a novel model that exploits contrastive learning to boost the performance of knowledge-aware recommendation. KRCL incorporates relation-aware knowledge into the collaborative graph and applies the graph denoising technique to alleviate the data sparsity and noise propagation problems in recommendation. We design two contrastive learning tasks and a joint training paradigm to learn node representations in different aspects. Extensive experiments on three benchmark datasets verify that KRCL is particularly effective and robust. And we will continue to explore multimodal knowledge and time-series information to extract user preferences at a more fine-grained level.

Acknowledgements. Our work was supported by Sichuan Science and Technology Program (No. 2023YFG0021, No. 2022YFG0038 and No. 2021YFG0018), and by Xinjiang Science and Technology Program (No. 2022D01B185).

References

1. Ai, Q., Azizi, V., Chen, X., Zhang, Y.: Learning heterogeneous knowledge base embeddings for explainable recommendation. Algorithms **11**(9), 137 (2018)
2. Chen, H., Li, Y., Sun, X., Xu, G., Yin, H.: Temporal meta-path guided explainable recommendation. In: Proceedings of the 14th ACM International Conference on Web Search and Data Mining, pp. 1056–1064 (2021)
3. Chen, H., Wang, L., Lin, Y., Yeh, C.C.M., Wang, F., Yang, H.: Structured graph convolutional networks with stochastic masks for recommender systems. In: Proceedings of the 44th International ACM SIGIR Conference on Research and Development in Information Retrieval, pp. 614–623 (2021)
4. Chen, J., Dong, H., Wang, X., Feng, F., Wang, M., He, X.: Bias and debias in recommender system: a survey and future directions. ACM Trans. Inf. Syst. **41**(3), 1–39 (2023)
5. Chen, T., Kornblith, S., Norouzi, M., Hinton, G.: A simple framework for contrastive learning of visual representations. In: International Conference on Machine Learning, pp. 1597–1607. PMLR (2020)

6. Clauset, A., Shalizi, C.R., Newman, M.E.J.: Power-law distributions in empirical data. SIAM Rev. **51**(4), 661–703 (2009)
7. Gao, J., He, D., Tan, X., Qin, T., Wang, L., Liu, T.-Y.: Representation degeneration problem in training natural language generation models. arXiv preprint arXiv:1907.12009 (2019)
8. García-Durán, A., Bordes, A., Usunier, N.: Effective blending of two and three-way interactions for modeling multi-relational data. In: Calders, T., Esposito, F., Hüllermeier, E., Meo, R. (eds.) ECML PKDD 2014. LNCS (LNAI), vol. 8724, pp. 434–449. Springer, Heidelberg (2014). https://doi.org/10.1007/978-3-662-44848-9_28
9. Giorgi, J., Nitski, O., Wang, B., Bader, G.: Declutr: deep contrastive learning for unsupervised textual representations. arXiv preprint arXiv:2006.03659 (2020)
10. He, X., Chua, T.-S.: Neural factorization machines for sparse predictive analytics. In: Proceedings of the 40th International ACM SIGIR Conference on Research and Development in Information Retrieval, pp. 355–364 (2017)
11. He, X., Deng, K., Wang, X., Li, Y., Zhang, Y., Wang, M.: Lightgcn: simplifying and powering graph convolution network for recommendation. In: Proceedings of the 43rd International ACM SIGIR Conference on Research and Development in Information Retrieval, pp. 639–648 (2020)
12. Huang, C., et al.: Knowledge-aware coupled graph neural network for social recommendation. In: Proceedings of the AAAI Conference on Artificial Intelligence, vol. 35, pp. 4115–4122 (2021)
13. Lin, Y., Liu, Z., Sun, M., Liu, Y., Zhu, X.: Learning entity and relation embeddings for knowledge graph completion. In: Proceedings of the AAAI Conference on Artificial Intelligence, vol. 29 (2015)
14. Lin, Z., Tian, C., Hou, Y., Zhao, W.X.: Improving graph collaborative filtering with neighborhood-enriched contrastive learning. In: Proceedings of the ACM Web Conference 2022, pp. 2320–2329 (2022)
15. McPherson, M., Smith-Lovin, L., Cook, J.M.: Birds of a feather: homophily in social networks. Annu. Rev. Sociol. **27**(1), 415–444 (2001)
16. van den Oord, A., Li, Y., Vinyals, O.: Representation learning with contrastive predictive coding. arXiv preprint arXiv:1807.03748 (2018)
17. Rendle, S., Freudenthaler, C., Gantner, Z., Schmidt-Thieme, L.: BPR: Bayesian personalized ranking from implicit feedback. arXiv preprint arXiv:1205.2618 (2012)
18. Sarwar, B., Karypis, G., Konstan, J., Riedl, J.: Item-based collaborative filtering recommendation algorithms. In: Proceedings of the 10th International Conference on World Wide Web, pp. 285–295 (2001)
19. Veličković, P., Cucurull, G., Casanova, A., Romero, A., Lio, P., Bengio, Y.: Graph attention networks. arXiv preprint arXiv:1710.10903 (2017)
20. Wang, H., et al.: Ripplenet: propagating user preferences on the knowledge graph for recommender systems. In: Proceedings of the 27th ACM International Conference on Information and Knowledge Management, pp. 417–426 (2018)
21. Wang, X., Jin, H., Zhang, A., He, X., Xu, T., Chua, T.-S.: Disentangled graph collaborative filtering. In: Proceedings of the 43rd International ACM SIGIR Conference on Research and Development in Information Retrieval, pp. 1001–1010 (2020)
22. Wang, X., Wang, D., Canran, X., He, X., Cao, Y., Chua, T.-S.: Explainable reasoning over knowledge graphs for recommendation. In: Proceedings of the AAAI Conference on Artificial Intelligence, vol. 33, pp. 5329–5336 (2019)

23. Wang, X., He, X., Cao, Y., Liu, M., Chua, T.-S.: KGAT: knowledge graph attention network for recommendation. In: Proceedings of the 25th ACM SIGKDD International Conference on Knowledge Discovery & Data Mining, pp. 950–958 (2019)
24. Wang, X., et al.: Learning intents behind interactions with knowledge graph for recommendation. In: Proceedings of the Web Conference 2021, pp. 878–887 (2021)
25. Wang, X., He, X., Wang, M., Feng, F., Chua, T.-S.: Neural graph collaborative filtering. In: Proceedings of the 42nd International ACM SIGIR Conference on Research and Development in Information Retrieval, pp. 165–174 (2019)
26. Wang, Z., Lin, G., Tan, H., Chen, Q., Liu, X.: CKAN: collaborative knowledge-aware attentive network for recommender systems. In: Proceedings of the 43rd International ACM SIGIR Conference on Research and Development in Information Retrieval, pp. 219–228 (2020)
27. Wu, J., et al.: Self-supervised graph learning for recommendation. In: Proceedings of the 44th International ACM SIGIR Conference on Research and Development in Information Retrieval, pp. 726–735 (2021)
28. Wu, Y., et al.: Multi-view multi-behavior contrastive learning in recommendation. In: Bhattacharya, A., et al. (eds.) DASFAA 2022. LNCS, vol. 13246, pp. 166–182. Springer, Cham (2022). https://doi.org/10.1007/978-3-031-00126-0_11
29. Xia, L., Huang, C., Xu, Y., Zhao, J., Yin, D., Huang, J.: Hypergraph contrastive collaborative filtering. In: Proceedings of the 45th International ACM SIGIR Conference on Research and Development in Information Retrieval, pp. 70–79 (2022)
30. Xian, Y., Fu, Z., Muthukrishnan, S., De Melo, G., Zhang, Y.: Reinforcement knowledge graph reasoning for explainable recommendation. In: Proceedings of the 42nd International ACM SIGIR Conference on Research and Development in Information Retrieval, pp. 285–294 (2019)
31. Xie, X., et al.: Contrastive learning for sequential recommendation. In: 2022 IEEE 38th International Conference on Data Engineering (ICDE), pp. 1259–1273. IEEE (2022)
32. Xu, J., Li, Z., Du, B., Zhang, M., Liu, J.: Reluplex made more practical: leaky ReLU. In: 2020 IEEE Symposium on Computers and communications (ISCC), pp. 1–7. IEEE (2020)
33. Yang, Y., Huang, C., Xia, L., Li, C.: Knowledge graph contrastive learning for recommendation. In: Proceedings of the 45th International ACM SIGIR Conference on Research and Development in Information Retrieval, pp. 1434–1443 (2022)
34. Zhang, F., Yuan, N.J., Lian, D., Xie, X., Ma, W.-Y.: Collaborative knowledge base embedding for recommender systems. In: Proceedings of the 22nd ACM SIGKDD International Conference on Knowledge Discovery and Data Mining, pp. 353–362 (2016)
35. Zhuang, C., Ma, Q.: Dual graph convolutional networks for graph-based semi-supervised classification. In: Proceedings of the 2018 World Wide Web Conference, pp. 499–508 (2018)
36. Zou, D., et al.: Multi-level cross-view contrastive learning for knowledge-aware recommender system. In: Proceedings of the 45th International ACM SIGIR Conference on Research and Development in Information Retrieval, pp. 1358–1368 (2022)

Knowledge-Rich Influence Propagation Recommendation Algorithm Based on Graph Attention Networks

Yuping Yang[1], Guifei Jiang[1,2,3]([✉]), and Yuzhi Zhang[1,2,3]

[1] Nankai University, Tianjin, China
yypSandra@mail.nankai.edu.cn {g.jiang,zyz}@nankai.edu.cn
[2] Haihe Laboratory of Information Technology Application Innovation (HL-IT), Tianjin, China
[3] Tianjin Key Laboratory of Operating System, Tianjin, China

Abstract. One of the biggest challenges in Recommendation Algorithms (RA) is how to obtain user and item embeddings from sparse interaction history. To take this challenge, most graph neural network based RAs explicitly incorporate high-order collaborative filtering signals on the user-item bipartite graph with either multi-layer semantics on the Knowledge Graph (KG) or multi-level neighbors on the social network. However, none of them fully integrate these three types of graph-structured data, which decreases embeddings' precision. Based on this consideration, this paper integrates the three types of data by proposing a knowledge-rich influence propagation RA based on the graph attention mechanism. Specifically, in the semantic propagation, we categorize user preferences into deep interest obtained by multiple graph attention message propagations on related KG parts, and shallow interest generated from the interaction history. Moreover, the influence weight between items is determined by the number of co-interactions and the semantic similarity. These two factors as well as social relations together decide the influence weight between users. With these influence weights, final user and item embeddings are calculated through multi-layer message propagation. The experimental results show that the proposed recommendation algorithm outperforms several compelling baselines on six scaled-down real-world datasets. This work has confirmed the effectiveness of combining these three types of data to increase RAs' coverage and accuracy.

Keywords: Recommendation Algorithm · Knowledge Graph · Social Network · Graph Attention Network · User Preference

1 Introduction

As one of the most popular recommendation Algorithms (RA), Collaborative Filtering (CF) is able to detect inter-user or inter-item similarities to make recommendations for users accordingly. Since user-item interaction history is represented as a bipartite graph, Graph Neural Networks (GNN) are utilized to predict the missing user-item interactions [3]. However, user preference inference based only on the sparse user-item interaction history yields insufficient and coarse-grained results. The widely used approach of solving the sparsity problem is to combine side information collected from the system. Side information includes attributes in the form of text or images, social relationships among users and users' comments on items and other contents [10]. In particular, as shown in Fig. 1, GNNs can be used to fully merge two kinds of graph-structured side information: triplets between items and attributes on the Knowledge Graph (KG) and user-user connections on the social network with the bipartite graph [4].

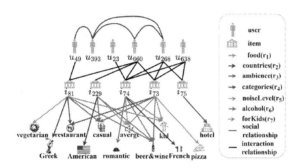

Fig. 1. An example with three kinds of graph-structured data in Yelp2018

Specifically, GNN-based collaborative filtering RAs usually combine the bipartite graph with either the KG or the social network. The former incorporates rich semantics to enhance the accuracy and interpretability of recommendations, and the latter leverages high-order social effects to modify user interests [17,21]. Yet, neither of them is fully satisfactory as they mainly rely on limited data. On the one hand, the KG-based RAs fail to account for user attribute-level interests despite that users choose items for specific reasons. On the other hand, they neglect to explicitly model high-order CF signals reflecting directly inter-user and inter-item influences [19]. In addition, they do not appreciate the importance of social diffusion. Without mixing high-order CF signals expressing user interest similarities with social influences, the social network-based RAs also fail to take advantage of available item attributes to infer user preferences.

To address these issues, this paper proposes a **K**nowledge-Rich **I**nfluence **P**ropagation **Rec**ommendation (**KIPRec**) that seamlessly fuses the KG, the social network, and the user-item bipartite graph. To obtain reasonably precise user and item embeddings, multi-layer Graph ATtention networks (GAT) are employed to aggregate neighbor messages. There are two main components in KIPRec. The first part is to explore semantic information, where user preferences are divided into shallow-level interests derived from interacted item sequences and deep-level interests obtained by aggregating semantics from their KG subgraphs. The second part is to propagate influence information. User-user connections in the social network are combined with user-user co-interactions from the bipartite graph to organize an influence graph and refine user interests. In particular, the main contributions of this paper are summarized as follows:

- A comprehensive RA is proposed by integrating three types of side information with graph structures to alleviate the sparsity problem in RAs.
- The validity of the proposed RA is fully and extensively investigated in two commonly used recommendation scenarios: Click-Through-Rate (CTR) prediction and Top-K recommendation. Specifically, for the Recall@20 metric, our RA improves by 1.92% and 5.64% on the Last-FM and Yelp2018 datasets, respectively, compared to the state-of-the-art (SOTA) KG-based RA, and it improves by 10.05% and 11.54% over the SOTA social network-based RA, respectively.
- Ablation experiments are conducted to analyze the functions of KIPRec's two main components.

The rest of this paper is organized as follows: Sect. 2 reviews the related work on GNN-based collaborative filtering RAs based on the KG or the social network. Section 3 presents the main problem addressed in this paper. The structure of proposed model is detailedly explained in Sect. 4. Extensive comparisons among KIPRec and its baselines as well as other studies are provided in Sect. 5. Finally, we conclude this paper with a discussion of the future directions in Sect. 6.

2 Related Work

By explicitly modeling high-order CF signals on the user-item bipartite graph via multiple forms of message passing, the current graph convolutional neural network-based methods such as LightGCN [5], GAT-based methods, and graph sampling aggregate network-based methods have achieved significant improvements compared with the traditional matrix decomposition methods [12] and most deep learning-based methods [26]. However, these GNN-based approaches fail to satisfactorily perform the task of acquiring useful information on heterogeneous graphs with multiple node or edge types. We will discuss the most related work in terms of the following aspects.

GNN-based RAs on the KG are divided into the following three categories: embedding-based methods, path-based methods, and propagation-based methods [6]. Items in the bipartite graph are mapped to entities in the KG, hence embedding-based approaches like MKR [14] combine the KG embedding algorithms and the recommendation module in a certain sequence. However, it might be difficult to accurately correlate the embeddings learned by these two modules. Path-based approaches, such as KPRN [18], merge the bipartite graph and the KG into a unified graph according to the correlations between items and entities. Paths between users and items are extracted from the graph to generate embeddings. Yet, path-based approaches require expert knowledge to decrease space and time consumption for processing huge volumes of paths.

Differently, propagation-based approaches apply GNNs directly to the KG, deriving node semantic embeddings by message propagation. Typical procedures for this sort of algorithms are outlined below. RippleNet [13] initializes item embeddings and refines user embeddings based on related multi-layer triplets. Starting from the user-interacted items, these triplets are obtained by aligning head and tail entities in triplets on the KG. KGCN [15] first randomly initializes user embeddings but constructs item embeddings for the target user by aggregating the neighbor messages from the items' receptive fields with GAT. By correlating intents with relations on the KG, KGIN [17] provides an explicit interpretation of intentions and a finer granularity of user and item embeddings. Besides exploring semantics in the KG, CKAN [19] additionally employs CF signals on the item side. CKAN first refreshes entity embeddings on the KG using a multi-layer GAT, then considers embeddings of the user-interacted items on the KG as the user embedding, and finally pools embeddings of all items on the KG that share at least one users on the bipartite graph into the item embedding. In this paper, the semantic propagation component adopts a KGCN-like technique to construct semantic propagation trees and refine embeddings, which provides a reasonable explanation of users' preferences and items' features. CKAN does not involve the interest similarities between users with co-interactions.

GNN-based RAs on the social network aim to rationally leverage the bipartite graph and the social network with diverse topologies to strengthen user interests. GATs are employed to evaluate the impacts of social connections and CF signals on each user's interest, as some users are easily swayed by their peers while others tend to adhere to their own preferences [1]. GraphRec [2] only considers users' first-order CF signals and social neighbors, and Diffnet++ [21] takes into account users' high-order contexts. Furthermore, some approaches like DISGCN [9] are proposed to disentangle the correspondence between the social connection and interest similarity into multiple dimensions, and some self-supervised approaches such as SEPT [24] are designed to eliminate fragile and noisy social relationships.

In addition, GSIRec [8] is equipped with a knowledge assistance module as the bridge between knowledge-aware recommendation and social-aware recommendation. However, this multi-task method may encounters the problem of negative transfer when knowledge gained in one task lowers performance in another [27]. Social-RippleNet [7] additionally utilizes the KG to adjust item embeddings and applies the social network to modify user preferences. Yet, users' profiles would be more transparent if the KG were used to extend the tags that define users.

3   Problem Formulation

In this paper, the user-item bipartite graph is denoted as a binary-valued symmetric matrix $Y \in \mathbb{R}^{m \times n}$ as usual, where $y_{ui} = y_{iu} = 1$ indicates there is an observed interaction between a user $u \in U$ and an item $i \in I$. The KG $\mathcal{G} = \{(h, \imath, t) \mid h, t \in \mathcal{E}, \imath \in \mathcal{R}\}$ is a directed graph made up of the entity set

$\mathcal{E}$ and relation set $\mathcal{R}$, where $(\hbar, \iota, t)$ refers to a triplet containing a head entity $\hbar$, a tail entity t and a relation ι connecting them. Most items in the bipartite graph have corresponding entities in the KG, i.e., $I \subseteq \mathcal{E}$. The social network is an undirected graph represented by $S \in \mathbb{R}^{m \times m}$, with $s_{uu'} = s_{u'u} = 1$ indicating the existence of a following or followed relationship for users $u, u' \in U$. Given a bipartite graph Y, a KG $\mathcal{G}$ and a social network S, our model optimizes parameters in Θ to learn embeddings of user u_p and item i_q, which in turn predict the likelihood $\hat{y}_{pq}$ of the interaction between them via the function $\mathcal{F}$: $\hat{y}_{pq} = \mathcal{F}(u_p, i_q \mid Y, \mathcal{G}, S, \Theta)$. We will specify Θ in Sect. 4.4.

4   Methodology

This paper focuses on implicit binary-valued user behaviors without considering the effect of time. As mentioned before, the goal is to predict the probability of an interaction between a user-item pair (p, q). To this end, Fig. 2 and Fig. 3 illustrate the overall structure of the proposed RA. It consists of four components: an initialization layer, a semantic propagation layer, an influence propagation layer, and a prediction layer. We will elaborate them in the following.

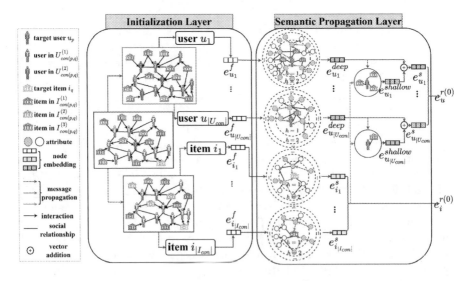

Fig. 2. The initialization layer and semantic propagation layer in KIPRec

4.1   The Initialization Layer

To begin with, we represent all nodes and edges on all graphs with trainable embeddings of a fixed length d. The notations $e_{u_p}^f$, $e_{i_q}^f$, $e_{\mathfrak{e}_c}^f$, $e_{r_d}^f \in \mathbb{R}^d$ respectively denote the free embedding of user u_p, item i_q, entity $\mathfrak{e}_c$, relation r_d. The influence propagation layer's initial embeddings are derived from the semantic propagation layer's outputs. Before proceeding to the semantic propagation layer, we need to calculate the context information for the user-item pair (p, q) in the influence propagation layer, i.e., all the users and items related to (p, q).

The inter-user and inter-item influence matrices $Y_{inf} \in \mathbb{R}^{m \times m}$ and $A_{inf} \in \mathbb{R}^{n \times n}$ are calculated by multiplying the user-item interaction matrix $Y \in \mathbb{R}^{m \times n}$ with the item-user interaction matrix $A \in \mathbb{R}^{n \times m}$. Values in Y_{inf} and A_{inf} reflect numbers of common interactions. (p, q)'s user-level context information is represented as $U_{con(p,q)} = U^1_{con(p,q)} \cup U^2_{con(p,q)} \cup U^3_{con(p,q)}$. In detail, $U^1_{con(p,q)}$ contains the users who have shared items with u_p. $U^2_{con(p,q)}$ represents the users that might share similar interests with u_p in the absence of co-interactions. And $U^3_{con(p,q)}$ denotes users who have social connections with u_p. $U^t_{con(p,q)} = \cup_{k=1}^{K} U^{t(k)}_{con(p,q)}$, where $U^0_{con(p,q)} = \{u_p\}$, $k \in \{1, 2, ..., K\}$ and $t \in \{1, 2, 3\}$. Each user group contains K parts, since the influence propagation procedure is repeated K times.

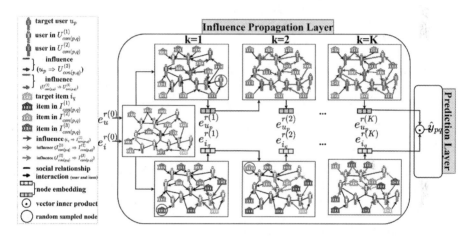

Fig. 3. The influence propagation layer and the interaction prediction layer in KIPRec

Specifically, during the k-th influence propagation, $U^{1(k)}_{con(p,q)}$ consists of the first-order neighbors of all users in $U^{1(k-1)}_{con(p,q)}$, with column subscripts of the non-zero elements in Y_{inf}. Similarly, $U^{3(k)}_{con(p,q)}$ is formed by all users in the first-order neighbors of $U^{3(k-1)}_{con(p,q)}$ under the direction of the social matrix S. Regarding $U^{2(k)}_{con(p,q)}$, user interaction histories are frequently restricted to a narrow area, resulting in users with similar interests rarely having shared items, so we randomly choose a L fraction from $U - U^{1(k)}_{con(p,q)}$ to generate $U^{2(k)}_{con(p,q)}$.

The item-level context information of (p, q) lies in $I_{con(p,q)} = I^1_{con(p,q)} \cup I^2_{con(p,q)} \cup I^3_{con(p,q)}$. Alternatively, we use a similar method as before to obtain

$I^1_{con(p,q)}$ and $I^2_{con(p,q)}$. During the k-th influence propagation, $I^{3(k)}_{con(p,q)}$ is formed by first-order neighbors of all users in $U^{(k-1)}_{con(p,q)}$ under the direction of the interaction matrix Y. $I^3_{con(p,q)} = \cup^K_{k=1} I^{3(k)}_{con(p,q)}$, where $k \in \{1, 2, ..., K\}$. Figure 2 displays how to obtain the third-order context information for the target user and item.

4.2   The Semantic Propagation Layer

This section shows how to generate semantic propagation trees and update their roots' embeddings. For all users from the user contexts and items from the item contexts of the target user-item pair (p, q) described in Sect. 4.1, we establish their semantic propagation trees according to the user-item bipartite graph Y and the KG $\mathcal{G}$. Initial embeddings of these users and items are then updated through extended GAT operations on the trees.

Constructing Semantic Propagation Trees. $T^h_{q'}$ and $T^h_{p'}$ respectively represent triplet sets related to an item $i_{q'}$ in $I_{con(p,q)}$ and a user $u_{p'}$ in $U_{con(p,q)}$ during the h-th semantic propagation. From corresponding entity of $i_{q'}$ on the KG as root, related subgraph of $i_{q'}$ is extracted by aligning the tail of one triplet with the head of another for H times as described in Eq. (1), where the symbol $*$ indicates an arbitrary entity or relationship, and $(\hbar, \imath, t)$ is a triplet on the KG $\mathcal{G}$. We limit the number of entities per layer $| T^h_{q'} |$ as w_1 to save memory space.

$$\begin{cases} T^h_{q'} = \{(\hbar, \imath, t) \mid (\hbar, \imath, t) \in \mathcal{G} \text{ and } \hbar = i_{q'} \}, & h = 1 \\ T^h_{q'} = \left\{ (\hbar, \imath, t) \mid (\hbar, \imath, t) \in \mathcal{G} \text{ and } (*, *, \hbar) \in T^{h-1}_{q'} \right\}, & h = 2, ..., H \end{cases} \quad (1)$$

As described in Eq. (2), for each user $u_{p'}$ in $U_{con(p,q)}$, we sample w_2 attributes from the first-order entity neighbors of $u_{p'}$'s interacted items, i.e., $|T^1_{p'}| = w_2$, and connect $u_{p'}$ with these entities. $y_{p'\hbar} = 1$ means there is an interaction between $u_{p'}$ and the item corresponding to entity $e_\hbar$. Afterwards, the semantic propagation trees are built for users in the same way, i.e., $| T^h_{p'} | = w_1$ when $h \geq 2$. Figure 2 illustrates four semantic propagation trees with $H, w_1 = 2$ and $w_2 = 4$.

$$\begin{cases} T^h_{p'} = \{(u_{p'}, \imath, t) \mid (\hbar, \imath, t) \in \mathcal{G} \text{ and } y_{p'\hbar} = 1 \}, & h = 1 \\ T^h_{p'} = \left\{ (\hbar, \imath, t) \mid (\hbar, \imath, t) \in \mathcal{G} \text{ and } (*, *, \hbar) \in T^{h-1}_{p'} \right\}, & h = 2, ..., H \end{cases} \quad (2)$$

Implementing Semantic Propagation Operations. We represent an arbitrary user or item with the uniform symbol o. The semantic propagation procedure is executed $(H - v)$ times for each node in N^v_o. N^v_o includes all nodes on the v-th layer of the semantic tree with node o as the root, where $v \in \{0, 1, ..., H\}$. Assumed that the l-th message propagation is to be performed ($l \in \{1, ..., H-v\}$) and the g-th node on the v-th layer is $e_g \in N^v_o$, related triplets of e_g are contained in $T^{v+1}_{o(g)} = \{(\hbar, \imath, t) \mid (\hbar, \imath, t) \in T^{v+1} \text{ and } \hbar = e_g\}$. We first map the tail entity e_t's embedding derived from the $(l-1)$-th message propagation onto the same level with relation $r_\imath$ via the Hadamard product operation $\odot$. This not only decouples e_t's features, but also preserves semantics along message propagation paths, as described in KGIN [17]. Further, the contribution of the t-th ($t \in \{1, ..., |T^{v+1}_{o(g)}|\}$) triplet to the head entity $\mathfrak{e}_g$, i.e., α^l_t, is valued in Eq. (3) as the similarity between the root o and $\mathfrak{e}_t$ by concatenating the two embeddings labeled with $||$ and then obtaining a numerical value via a two-layer perceptron.

$$\alpha^l_t = W_2\left(LeakyReLU\left(W_1(e^{l-1}_o \;||\; (e_{r_t} \odot e^{l-1}_{\mathfrak{e}_t})) + b_1\right)\right) + b_2 \tag{3}$$

Moreover, $\widetilde{\alpha}^l_t$ is the result of softmax normalization on α^l_t. The embedding of $\mathfrak{e}_g$ after the l-th message propagation results from aggregated relevant features and the embedding of $\mathfrak{e}_g$ after the $(l-1)$-th propagation, shown in Eq. (4).

$$e^l_{\mathfrak{e}_g} = e^{l-1}_{\mathfrak{e}_g} + \sum_{t=1}^{|T^{v+1}_{o(g)}|} \widetilde{\alpha}^l_t(e_{r_t} \odot e^{l-1}_{\mathfrak{e}_t}), \quad where \quad \widetilde{\alpha}^l_t = \frac{e^{\alpha^l_t}}{\sum_{t'=1}^{|T^{v+1}_{o(g)}|} e^{\alpha^l_{t'}}} \tag{4}$$

The symbol n denotes a user, an item or an entity. $\imath$ refers to a relation. In general, the initial embeddings of n and $\imath$ in the semantic propagation layer come from the initialization layer, i.e., $e^0_n = e^f_n$ and $e_{r_\imath} = e^f_{r_\imath}$. After the propagation procedure is performed H times, semantics related to the root nodes ($u_{p'} \in U_{con(p,q)}$ and $i_{q'} \in I_{con(p,q)}$) on the semantic propagation trees, is distilled into their embeddings $e^H_{u_{p'}}$ and $e^H_{i_{q'}}$. Primary characteristics of $i_{q'}$ are indicated by $e^s_{i_{q'}}$ (equal to $e^H_{i_{q'}}$). However, as in Eqs. (5) and (6), interests of $u_{p'}$ are also expressed by user-item interactions $\mathcal{Y}_{u_{p'}}$ except for deep-level user interests marked by $e^{deep}_{u_{p'}}$ (equal to $e^H_{u_{p'}}$) [20]. The contribution $\widetilde{\beta}_{p'v}$ of $u_{p'}$'s interactive item i_v to shallow interest embedding $e^{shallow}_{u_{p'}}$ is generated through an extended GAT operation and normalization. Eventually, deep interest $e^{deep}_{u_{p'}}$ and shallow interest $e^{shallow}_{u_{p'}}$ of $u_{p'}$ both contribute to the preliminary embedding $e^s_{u_{p'}}$.

$$\beta_{p'v} = W_4\left(LeakyReLU\left(W_3(e^{deep}_{u_{p'}} \;||\; e^s_{i_v}) + b_3\right)\right) + b_4 \tag{5}$$

$$e^{shallow}_{u_{p'}} = \sum_{v \in \mathcal{Y}_{u_{p'}}} \widetilde{\beta}_{p'v} e^s_{i_v}, \quad where \quad \widetilde{\beta}_{p'v} = \frac{e^{\beta_{p'v}}}{\sum_{v' \in \mathcal{Y}_{u_{p'}}} e^{\beta_{p'v'}}} \tag{6}$$

4.3   The Influence Propagation Layer

For further adjustment of user and item embeddings, the influence propagation is required to be performed over K iterations by absorbing relevant information from their K-nearest neighbors. The inital embedding of any user u in context $U_{con(p,q)}$ of u_p, indicated by $e^{r(0)}_u$, is equal to e^s_u. Similarly, each item i in context $I^1_{con(p,q)} \cup I^2_{con(p,q)}$ of i_q has the condition $e^{r(0)}_i = e^s_i$. K-nearest neighbors of u_p and i_q are respectively included in $U_{con(p,q)}$ and $I^1_{con(p,q)} \cup I^2_{con(p,q)}$.

During the k-th message propagation, item i_p needs to gain influence from its first-order neighbors in $I^{(1)}_{con(p,q)} = I^{1(1)}_{con(p,q)} \cup I^{2(1)}_{con(p,q)}$, where $k \in \{1, 2, ..., K\}$.

$I_{con(p,q)}^{1(1)}$ denotes items with which i_p shares users, while $I_{con(p,q)}^{2(1)}$ indicates items with similar semantics but no shared user. These items that may affect form $I_{con(p,q)}^{(1)}$. For an item $i_q \in I_{con(p,q)}^{(1)}$, we measure the semantic similarity between i_q and i_q via the inner product of $e_{i_q}^{r(k-1)}$ and $e_{i_q}^{r(k-1)}$, then amplify it by multiplying with the number of their co-interactions denoted as a_{qq} from A_{inf} in Eq. (7), and finally get the influence degree $\gamma_{qq}^{(k)}$. As shown in Eq. (8), the embedding of i_q after the k-th influence propagation is the total of the $(k-1)$-th outcome and the weighted sum of its first-order neighbors' embeddings during this propagation.

$$\gamma_{qq}^{(k)} = a_{qq}(e_{i_q}^{r(k-1)} \cdot e_{i_q}^{r(k-1)}) \tag{7}$$

$$e_{i_q}^{r(k)} = e_{i_q}^{r(k-1)} + \sum_{i_q \in I_{con(p,q)}^{(1)}} \widetilde{\gamma}_{qq}^{(k)} e_{i_q}^{r(k-1)}, \quad where \quad \widetilde{\gamma}_{qq}^{(k)} = \frac{e^{\gamma_{qq}^{(k)}}}{\sum_{i_{q'} \in I_{con(p,q)}^{(1)}} e^{\gamma_{qq'}^{(k)}}} \tag{8}$$

During the k-th propagation, the embedding of u_p is affected by users with social relationships, in addition to users with co-interactions or similar semantics, i.e., $U_{con(p,q)}^{(1)} = U_{con(p,q)}^{1(1)} \cup U_{con(p,q)}^{2(1)} \cup U_{con(p,q)}^{3(1)}$. Therefore in Eq. (9), the semantic similarity between u_p and $u_p \in U_{con(p,q)}^{(1)}$ is amplified by the number of co-interactions $\mathfrak{n}_{pp}$ and the social influence $\mathfrak{s}_{pp}$, where we set $\mathfrak{s}_{pp}$ to λ times the most co-interactions as described in Eq. (10) since friends tend to orient user interests more. $\mathfrak{s}_{pp} > 0$ implies that there is a social relationship between u_p and u_p. As described in Eq. (11), the embedding of u_p after the k-th propagation combines each neighbor's embedding based on an weight $\widetilde{\delta}_{pp}^{(k)}$ and its own embedding on the $(k-1)$-th propagation together.

$$\mathfrak{s}_{pp} = \begin{cases} 0, & if \ \mathfrak{s}_{pp} = 0 \\ \max_{u_{p'} \in U_{con(p,q)}^{1(1)} \cup U_{con(p,q)}^{2(1)}} \lambda \mathfrak{n}_{pp'}(\lambda \in \mathbb{N} \ and \ 1 \leq \lambda \leq 10), & if \ \mathfrak{s}_{pp} > 0 \end{cases} \tag{9}$$

$$\delta_{pp}^{(k)} = (\mathfrak{n}_{pp} + \mathfrak{s}_{pp})(e_{u_p}^{r(k-1)} \cdot e_{u_p}^{r(k-1)}) \tag{10}$$

$$e_{u_p}^{r(k)} = e_{u_p}^{r(k-1)} + \sum_{u_p \in U_{con(p,q)}^{(1)}} \widetilde{\delta}_{pp}^{(k)} e_{u_p}^{r(k-1)}, \quad where \quad \widetilde{\delta}_{pp}^{(k)} = \frac{e^{\delta_{pp}^{(k)}}}{\sum_{u_{p'} \in U_{con(p,q)}^{(1)}} e^{\delta_{pp'}^{(k)}}} \tag{11}$$

4.4   The Interaction Prediction Layer

The results after K iterations of influence propagation represented by $e_{u_p}^{r(K)}$ and $e_{i_q}^{r(K)}$ are supposed as the final embeddings of u_p and i_q. They fully incorporate the semantics of the KG, social influences amongst users, and the numbers of co-interactions. The inner product of $e_{u_p}^{r(K)}$ and $e_{i_q}^{r(K)}$ predicts whether there is an interaction between them. As shown in Eq. (12), this paper applies the gradient descent to the BPR loss function to optimize parameters in Θ, including embeddings of all users, items, relationships as well as entities denoted as Θ_1 and the perceptrons' factors labeled as Θ_2, i.e., $\Theta_1 = \{e_{u_p}^f, e_{i_q}^f, e_{\mathfrak{e}_c}^f, e_{r_d}^f \mid u_p \in U, i_q \in I, \mathfrak{e}_c \in \mathcal{E}, r_d \in \mathcal{R}\}$ and $\Theta_2 = \{[W_i, b_i]_{i=1,2,3,4}\}$. We sample one item i_j without observed interaction for each observed user-item pair (u_p, i_q) in the training set, contributing to the collection $\mathcal{O}$. Moreover, the L2 regularization technique is served to avoid overfitting, where η is the related hyper-parameter.

$$\mathcal{L} = \min_{\Theta} \frac{1}{|\mathcal{O}|} \sum_{(u_p, i_q, i_j) \in \mathcal{O}} -\ln Sigmoid(\hat{y}_{pq} - \hat{y}_{pj}) + \eta|| \Theta ||_2^2 \qquad (12)$$

5 Experiments

In this section, we conduct experiments on six datasets including KGs or social networks to evaluate our proposed method and address the three questions: **RQ1**: Can KIPRec outperform the SOTA KG-based RAs and social network-based ones on real-world datasets ? **RQ2**: What exactly is the function of each component of KIPRec ? **RQ3**: How do user-user influence, item-to-item influence and multi-layer user interest affect KIPRec based on a real-world example ?

5.1 Experimental Settings

Dataset Description. Six real datasets are applied: Last-FM (music), Yelp2018 (business), Movie-Lens20M (film), Book-Crossing (book), Epinions (e-commerce) and Flickr (image). Last-FM, Movie-Lens20M and Book-Crossing are collected from CKAN [19], which contains the interaction history and the KG. Epinions and Flickr are obtained from GraphRec [2] and Diffnet++ [21] separately. KGAT [16] provides the bipartite graph and the KG of Yelp2018. We further extract its social network from the official dataset. We finally divide all users and items into four groups according to the number of their observed interactions, and narrow down all the datasets by randomly picking about 10% from each group. The statistics of all datasets are summarized in Table 1. All the datasets are randomly divided into training (60%), validation (20%) and testing (20%).

Evaluation Metrics. We investigate the performance of the proposed model with the Area Under Curve (AUC) metric in the CTR prediction, and the Recall (Recall@20) as well as Normalized Discounted Cumulative Gain (NDCG@20) metrics in top-K recommendation. From items other than those in the training set, the model is required to recommend 20 items to each user.

Table 1. Statistics of the datasets used in the experiments (# means the number, ✕ means the dataset doesn't involve this part, the inter-density means the density of each user-item interaction matrix, the triplet-avg means the average triplets per item and the social-density means the density of each social matrix.)

Content		Last-FM	Yelp2018	Movie-Lens20M	Book-Crossing	Epinions	Flickr
Bipartite Graph	# users	1,892	3,884	5,734	5,931	5,183	5,791
	# items	3,846	4,553	1,695	7,483	5,589	8,212
	# interactions	21,173	13,059	28,194	17,603	9,457	32,432
	inter-density	0.291%	0.074%	0.290%	0.040%	0.033%	0.068%
Knowledge Graph	# entities	9,366	95,514	87,310	70,419	✕	✕
	# relations	60	42	32	25	✕	✕
	# triplets	15,518	1,071,279	50,188	133,892	✕	✕
	triplet-avg	4	235	30	18	✕	✕
Social Network	# links	12,532	3,418	✕	✕	44,613	116,570
	social-density	0.350%	0.023%	✕	✕	0.166%	0.348%

Baselines. Seven GNN-based RAs on the KG are chosen, including embedding-based approaches (CKE, MKR, KGAT) and propagation-based methods (RippleNet, KGCN, CKAN and KGIN), and Diffnet and Diffnet++ in the second category are selected. In addition, BPRMF is not dependent on either the KG or the social network. CKE [25] is an embedding-based approach in the co-learning category that integrates the loss functions in the KG embedding and recommendation task into a whole, while MKR [14] is a multi-task learning strategy with the cross&compress units. KGAT [16] proposes a novel GAT-based approach inspired by transR to maintain the adjacency matrix between entities. RippleNet [13] emphasizes description of user interests, on the contrary, KGCN [15] focuses on the exploration of items' semantics. KGIN [17] explains recommendation actions with intents. In contrast to the KG-based RA mentioned above, CKAN [19] also takes into account items' first-order CF signals. Diffnet [22] conveys high-order impacts between users on the social network to enhance user embeddings. Diffnet++ [21] additionally weights the contributions of the user's high-order CF signals and high-order social neighbors.

Parameter Settings. Our KIPRec model is implemented with the Pytorch deep learning framework. To be fair, the embedding size of all nodes and edges is 64, the mini-batch size is 128, and the maximum number of training epochs is 250 for all methods across all datasets. We initialize all parameters with the Xavier initializer and optimize them with the Adam optimizer. We tune the hyper-parameters within given ranges via a grid search. The training phrase is terminated if Recall@20 on the training set does not improve in the subsequent 10 epochs. Each result in this section is the average of five repeated experiments with different random seeds under the best hyper-parameter setup.

5.2   RQ1: Performance Evaluation

We compare KIPRec with its baselines with Recall@20 and NDCG@20 as shown in Table 2 and Table 3, and AUC in Fig. 4. The main experimental results are summarized as follows.

Table 2. Overall comparison of 9 KG-based methods in the top-K recommendation (The %Improv. symbol indicates the gap between the bolded best performance and the underlined suboptimal result from KIPRec's perspective.)

Model	Last-FM		Yelp2018		Movie-Lens20M		Book-Crossing	
	Recall	NDCG	Recall	NDCG	Recall	NDCG	Recall	NDCG
BPRMF	0.2492	0.1446	0.0546	0.0266	0.4077	0.2086	0.0467	0.0285
CKE	0.2557	0.1444	0.0602	0.0296	0.4137	0.2108	0.0491	0.0323
MKR	0.2663	0.1469	0.0923	0.0461	0.4305	0.2139	0.0752	0.0534
RippleNet	0.2483	0.1387	0.0667	0.0282	0.4242	0.1981	0.0513	0.0366
KGCN	0.2864	0.1514	0.1370	0.0566	0.4965	0.2259	0.0853	0.0396
KGAT	0.2903	0.1702	0.1588	0.0681	0.4180	0.2098	0.0571	0.0324
CKAN	0.2449	0.1368	0.1486	0.0693	0.4670	0.2669	0.0779	0.0483
KGIN	<u>0.3342</u>	<u>0.1922</u>	<u>0.2004</u>	<u>0.0875</u>	**0.5720**	**0.3018**	**0.1476**	**0.0863**
KIPRec	**0.3406**	**0.1946**	**0.2117**	**0.0900**	<u>0.5718</u>	<u>0.2877</u>	<u>0.1472</u>	<u>0.0846</u>
%Improv.	1.92%	1.25%	5.64%	2.86%	-0.03%	-4.67%	-0.27%	-1.97%

Compared with the KG-based KGIN, Recall@20 and NDCG@20 evaluated on KIPRec are respectively improved by 1.92% and 1.25% on Last-FM, and by 5.64% and 2.86% on Yelp2018. In comparison with the social network-based Diffnet++, the two metrics on KIPRec increases by 10.05% and 12.81% on Last-FM, and by 11.54% and 7.91% on Yelp2018. Side information is reasonably utilized by KIPRec to generate more accurate user and item embeddings. In particular, the gap between KIPRec and Diffnet++ is more noticeable than that between KIPRec and KGIN, proving that the KG provides richer materials than the social network. It is feasible in KIPRec to obtain the preliminary embeddings from the KG. Moreover, BPRMF is the least effective, because it makes absolutely no use of any side information other than the bipartite graph.

On the other hand, on Movie-Lens20M and Book-Crossing including only the KG, KIPRec decreases by 0.03% and 0.27% respectively in terms of Recall@20, and by 4.67% and 1.97% relative to NDCG@20 compared with KGIN. This is because KIPRec incorporates the intention dimension associated with relationships on the KG. Yet, KIPRec achieves the similar goal, slightly less effective, by developing the multi-layer user interests structure. Figure 4 illustrates that CKAN or KIPRec is always the optimal or suboptimal model in the CTR prediction on Last-FM, Yelp2018 and Book-Crossing, since CF signals on the bipartite graph modify embeddings from the KG.

Table 3. Overall comparison of 4 social based methods in the top-K recommendation

Model	Last-FM		Yelp2018		Epinions		Flickr	
	Recall	NDCG	Recall	NDCG	Recall	NDCG	Recall	NDCG
BPRMF	0.2492	0.1446	0.0546	0.0266	0.0614	0.0234	0.0109	0.0047
Diffnet	0.2730	0.1504	0.1770	0.0772	0.1310	0.0609	0.0261	0.0129
Diffnet++	0.3095	0.1725	0.1898	0.0834	0.1448	0.0651	0.0306	0.0148
KIPRec	**0.3406**	**0.1946**	**0.2117**	**0.0900**	**0.1471**	0.0602	**0.0351**	**0.0156**
%Improv.	10.05%	12.81%	11.54%	7.91%	1.59%	-7.53 %	17%	6.85%

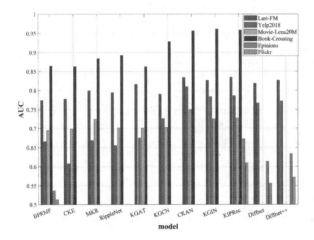

Fig. 4. The result of AUC in the CTR prediction scenario

Diffnet++ and KIPRec surpass Diffnet by integrating high-order CF signals and social influence. Recall@20 of KIPRec on Epinions and Flickr increases by 1.59% and 17%, and NDCG@20 is improved by -7.53% and 6.85% compared with Diffnet++. Inter-user and inter-item influences directly enable more accurate embeddings with less noise in KIPRec. In addition, KIPRec promotes embedding learning by broadening influence ranges through random sampling.

5.3   RQ2: Ablation Experiments

Table 4. Comparisons among variants of KIPRec

Category	Last-FM	Yelp2018	Movie-Lens20M	Book-Crossing	Setting
$\text{KIPRec}_{w/o\ DI}$	0.3189	0.2053	0.5607	0.1274	$e^s_{u_{p'}} = e^{shallow}_{u_{p'}}$
$\text{KIPRec}_{w/o\ SI}$	0.3197	0.2061	0.5683	0.1290	$e^s_{u_{p'}} = e^{deep}_{u_{p'}}$
$\text{KIPRec}_{w/o\ Semantic}$	0.3098	0.1938	-	-	Eq. (8):$e^{r(0)}_{t_q} = e^f_{t_q}$, Eq. (11):$e^{r(0)}_{u_p} = e^f_{u_p}$
$\text{KIPRec}_{w/o\ Sample}$	0.3203	0.1516	0.5432	0.1074	$L = 0, I^2_{con(p,q)} = \varnothing, U^2_{con(p,q)} = \varnothing$
$\text{KIPRec}_{w/o\ Social}$	0.3301	0.2040	**0.5718**	**0.1472**	Eq. (10):$s_{pp} = 0$
$\text{KIPRec}_{w/o\ Propagation}$	0.3018	0.1878	0.5325	0.1161	Sect. 4.3: $e^{r(K)}_{u_p} = e^{r(0)}_{u_p} = e^s_{u_p}, e^{r(K)}_{t_q} = e^{r(0)}_{t_q} = e^s_{t_q}$
KIPRec	**0.3406**	**0.2117**	**0.5718**	**0.1472**	-

We conduct two ablation experiments to estimate two layer's effects in KIPRec ($\text{KIPRec}_{\text{w/o Semantic}}$ and $\text{KIPRec}_{\text{w/o Propagation}}$). Their performances have significantly been decreased compared to KIPRec. As shown in Table 4, their performances indicates that both are indispensable. Further, both $\text{KIPRec}_{\text{w/o DI}}$ and $\text{KIPRec}_{\text{w/o SI}}$ outperform the $\text{KIPRec}_{\text{w/o Semantic}}$ on Last-FM and Yelp2018, but they are less effective than KIPRec, which reveals that capturing multi-layer user interests is essential. The effectiveness improvement in $\text{KIPRec}_{\text{w/o Sample}}$ or $\text{KIPRec}_{\text{w/o Social}}$ is significant over $\text{KIPRec}_{\text{w/o Propagation}}$, since RAs cannot adequately capture user interests from semantics without rich interaction history.

In addition, this paper explores the impact of different embedding dimensions on information capture ability through the experiment results in Table 5. It can be seen that KIPRec's recommendation effect performs best when $d = 64$ on three datasets. The optimal performances on the other three datasets are not particularly better than that of 64 dimensional setting. It is worth mentioning that larger dimensions on the large-scale dataset (Yelp2018) lead to GPU's usage exceeding the limit.

Table 5. Embedding dimension (d) settings' influence in KIPRec

Dataset	16	32	64	128	256	512
Last-FM	0.3287	0.3324	0.3406	0.3418	0.3436	**0.3439**
Yelp2018	0.1428	0.2094	0.2117	**0.2151**	-	-
Movie-Lens20M	0.5585	0.5627	**0.5718**	0.5586	0.5677	0.5713
Book-Crossing	0.0954	0.1130	**0.1472**	0.1425	0.1346	0.1447
Epinions	0.1471	0.1442	**0.1471**	0.1471	0.1399	0.1471
Flickr	0.0325	0.0334	0.0351	0.0345	**0.0356**	0.0339

5.4   RQ3: Case Study

To illustrate the effect of KIPRec's semantic propagation layer at the attribute level, we select users and items associated with u_{268} and i_{73}, as shown in Fig. 5.

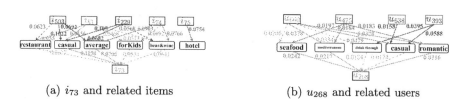

(a) i_{73} and related items   (b) u_{268} and related users

Fig. 5. A real-word example from Yelp2018 related to the semantic propagation layer (The solid line connects a user or item with the attributes that it values the most. Number on the line indicates how much the importance the attribute is placed.)

From the perspectives of whether users or items attach the same level of importance to each attribute and whether their most valued attributes are the same, we find that the semantic similarities to i_{73} and u_{475} in descending order are $i_{229} > i_{503} > i_{81} > i_{74} > i_{75}$ and $u_{475} > u_{660} > u_{393} > u_{638}$, which is consistent with the statistics in Table 6 and Table 7.

Table 6. Relations between i_{73} and its related nodes in influence propagation layer

Neighbors	i_{503}	i_{81}	i_{229}	i_{74}	i_{75}
Semantic similarity	0.9082	0.8393	0.9755	0.6429	0.5286
Co-interactions (training)	0	$1\{u_{660}\}$	$1\{u_{660}\}$	$3\{u_{660}, u_{268}, u_{638}\}$	$2\{u_{268}, u_{638}\}$
Overall similarity	0.0001	0.0018	0.1359	0.0493	0.0252

KIPRec aggregates messages from neighbors based on the overall similarity among users or items. The overall similarity is mainly determined by the semantic similarity, followed by co-interactions and social relationships. Thus, compared to i_{74}, i_{229} has a higher semantic similarity to i_{73} and thus a higher overall similarity; compared to i_{75}, i_{81} has a lower number of interactions in common with i_{73}, and thus a lower overall similarity.

Table 7. Relations between u_{268} and its related nodes in influence propagation layer

Neighbors	u_{475}	u_{393}	u_{638}	u_{660}
Semantic similarity	0.7986	0.5822	0.5690	0.7732
Co-interactions (training)	0	0	$3\{i_{73}, i_{74}, i_{75}\}$	$2\{i_{73}, i_{74}\}$
Social relationships	0	1	0	1
Overall similarity	3.20×10^{-35}	5.41×10^{-12}	1.74×10^{-32}	1.14×10^{-6}
Co-interactions (testing)	$1\{i_{484}\}$	$3\{i_{225}, i_{265}, i_{652}\}$	$1\{i_{235}\}$	$1\{i_{229}\}$

Because social influence plays a greater role in user representations, users that have social relationships with u_{268} are more similar in general to those have no social relationship. Furthermore, i_{73} and u_{268} respectively absorb user information from i_{503} and u_{475}, which are only semantically similar. The number of users' co-interactions in the testing set is essentially consistent with their overall similarity. The exception to this is that u_{660} and u_{268} have the highest overall similarity yet with one common interaction. This is because there are a few interaction records in the dataset for u_{660}.

6 Conclusion

This paper has proposed a RA based on GATs that naturally combines the knowledge graph, the bipartite graph and the social network to enrich user and item embeddings. A multi-layer user interest structure is designed at the attribute and item levels to achieve user embeddings with rich semantics. Besides the similarity from the semantic propagation layer, the inter-user and inter-item influence weights in the influence propagation layer are further acquired based on the bipartite graph and the social network. Multiple graph attention message propagation are performed in the two layers to obtain more precise user and item embeddings. Extensive experiments and comparative analysis on two datasets demonstrate the advantage of the proposed model over the SOTA baselines.

The future work is manifold. It would be interesting to study how to reduce GAT operations' time and space costs in the semantic propagation layer on large-scale graphs using sample strategies [11,23]. It is also worth investigating self-supervised methods to generate more accurate recommendations by identifying relationships among the refined item embeddings from the user-item bipartite graph, the item-item co-interaction graph, and the KG.

Acknowledgments. We are grateful to the reviewers of this paper for their constructive and insightful comments. The research reported in this paper was partially supported by the National Key R&D Program of China (NO. 2021YFB0300104), and the ANR project AGAPE ANR-18-CE23-0013.

References

1. Chen, T., Guo, J., et al.: Graph representation learning for popularity prediction problem: a survey. Discrete Math., Algorithms Appl. **14**(7), 2230003 (2022)
2. Fan, W., Ma, Y., et al.: A graph neural network framework for social recommendations. TKDE **34**(5), 2033–2047 (2020)
3. Gao, C., Wang, X., et al.: Graph neural networks for recommender system. In: WSDM, pp. 1623–1625 (2022)
4. Guo, Q., Zhuang, F., et al.: A survey on knowledge graph-based recommender systems. TKDE **34**(8), 3549–3568 (2020)
5. He, X., Deng, K., et al.: LightGCN: simplifying and powering graph convolution network for recommendation. In: SIGIR, pp. 639–648 (2020)
6. Huang, C.: Recent advances in heterogeneous relation learning for recommendation. In: IJCAI, pp. 4442–4449 (2021)
7. Jiang, W., Sun, Y.: Social-RippleNet: jointly modeling of ripple net and social information for recommendation. Appl. Intell. **53**(3), 3472–3487 (2022). https://doi.org/10.1007/s10489-022-03620-2
8. Li, A., Yang, B.: GSIRec: Learning with graph side information for recommendation. World Wide Web **24**(5), 1411–1437 (2021). https://doi.org/10.1007/s11280-021-00910-6
9. Li, N., Gao, C., et al.: Disentangled modeling of social homophily and influence for social recommendation. TKDE **35**(6), 5738–5751 (2022)
10. Liu, T., Wang, Z., et al.: Recommender systems with heterogeneous side information. In: WWW, pp. 3027–3033 (2019)
11. Liu, X., Yan, M., et al.: Sampling methods for efficient training of graph convolutional networks: a survey. IEEE/CAA J. Automatica Sinica **9**(2), 205–234 (2022)
12. Rendle, S., Freudenthaler, C., et al.: BPR: bayesian personalized ranking from implicit feedback. In: UAI, pp. 452–461 (2009)

13. Wang, H., Zhang, F., et al.: RippleNet: propagating user preferences on the knowledge graph for recommender systems. In: CIKM, pp. 417–426 (2018)
14. Wang, H., Zhang, F., et al.: Multi-task feature learning for knowledge graph enhanced recommendation. In: WWW, pp. 2000–2010 (2019)
15. Wang, H., Zhao, M., et al.: Knowledge graph convolutional networks for recommender systems. In: WWW, pp. 3307–3313 (2019)
16. Wang, X., He, X., et al.: KGAT: knowledge graph attention network for recommendation. In: SIGKDD, pp. 950–958 (2019)
17. Wang, X., Huang, T., et al.: Learning intents behind interactions with knowledge graph for recommendation. In: WWW, pp. 878–887 (2021)
18. Wang, X., Wang, D., et al.: Explainable reasoning over knowledge graphs for recommendation. In: AAAI, pp. 5329–5336 (2019)
19. Wang, Z., Lin, G., et al.: CKAN: collaborative knowledge-aware attentive network for recommender systems. In: SIGIR, pp. 219–228 (2020)
20. Wang, Z., Wang, Z., et al.: Exploring multi-dimension user-item interactions with attentional knowledge graph neural networks for recommendation. IEEE Trans. Big Data, TBD **9**(1), 212–226 (2023)
21. Wu, L., Li, J., et al.: DiffNet++: a neural influence and interest diffusion network for social recommendation. TKDE **34**(10), 4753–4766 (2020)
22. Wu, L., Sun, P., et al.: A neural influence diffusion model for social recommendation. In: SIGIR, pp. 235–244 (2019)
23. Xu, X., Feng, W., et al.: Sampling methods for efficient training of graph convolutional networks: a survey. In: ICLR (2020)
24. Yu, J., Yin, H., et al.: Socially-aware self-supervised tri-training for recommendation. In: SIGKDD, pp. 2084–2092 (2021)
25. Zhang, F., Yuan, N.J., et al.: Collaborative knowledge base embedding for recommender systems. In: SIGKDD, pp. 353–362 (2016)
26. Zhang, S., Yao, L., et al.: Deep learning based recommender system: a survey and new perspectives. ACM Comput. Surv. **52**(1), 1–38 (2019)
27. Zhu, F., Wang, Y., et al.: Cross-domain recommendation: challenges, progress, and prospects. In: IJCAI, pp. 4721–4728 (2021)

A Novel Variational Autoencoder with Multi-position Latent Self-attention and Actor-Critic for Recommendation

Jiamei Feng, Mengchi Liu$^{(\boxtimes)}$, Song Hong , and Shihao Song

School of Computer Science, South China Normal University, Guangzhou 510631,
China
liumengchi@scnu.edu.cn

Abstract. Variational Autoencoder (VAE) has been extended as a representative nonlinear latent method for collaborative filtering recommendation. As a high-dimensional representation of data, latent vectors play a vital role in the transmission of important information in a VAE model. However, VAE-based models suffer from a common limitation that the transmission ability of the latent vectors' important information is limited, resulting in lower quality of global information representation. To address this, we present a novel VAE model with multi-position latent self-attention and reinforcement learning' actor-critic algorithm. We first build a multi-position latent self-attention model, which can learn richer and more complex latent vectors and strengthens the transmission of important information at different positions. At the same time, we use reinforcement learning to enhance the interactive learning process of collaborative filtering recommendation training. Specifically, our model is stable and can be easy applied in the recommendation. We observed significant improvements over the previous state-of-the-art baselines on three social media datasets, where the largest improvement can reach 26.10%.

Keywords: Recommendation · Variational autoencoder ·
Actor-Critic · Latent attention

1 Introduction

Collaborative filtering [2] models are the most commonly used and widely applied methods for recommendation. After decades of development, recommendation techniques have shifted from the latent linear models to deep nonlinear models for modeling latent features and feature interactions among sparse features. Variational AutoEncoder (VAE) [11] has been extended to Mult-VAE [12], a representative nonlinear method for collaborative filtering, and has received widespread attention in the recommender system community in recent years. VAE-based methods [1,3,13,14,16] are significantly outperform the classical latent variable models.

© The Author(s), under exclusive license to Springer Nature Switzerland AG 2023
X. Yang et al. (Eds.): ADMA 2023, LNAI 14176, pp. 155–167, 2023.
https://doi.org/10.1007/978-3-031-46661-8_11

However, it remains a question whether these models can effectively learn representations of user preferences. Latent vectors, which serve as a high-dimensional representation of user-item scoring data, play a vital role in transmitting important information. Attention mechanism can solve the transmission of important information, but it is usually applied to sequence recommendation [21], which results depend on sequence with poorly parallelism. With the development of study, there are better applications in the non-sequenced data. Here the latent vector transmission of VAE model is independent of sequence. And as a deep learning model, VAE also lacks interactivity, which is its inherent question. Reinforcement learning is a good way to face this question, but only reinforcement learning is difficult to use in VAE directly.

In order to address the above problem, we propose a novel VAE-based model for collaborative filtering recommendation, which we call the VAE with multi-position latent self-attention and actor-critic or acMLAVAE for short. As the name suggests, our model consider multi-position latent self-attention, where each position is independent. Considering that different positions have different important information nodes, this paper designs a latent self-attention mechanism, which combines feedforward network and residual network to make it suitable for latent vectors at different positions. On the one hand, it can adaptively assign larger weight values to the nodes containing important information in the latent vector, and conversely, assign smaller weight values. On the other hand, the improved VAE latent function transfer does not consider sequence, and mining information through the similarity between vectors can achieve the attention of global information. Especially we add reinforcement learning of actor-critic [13] on the training process. This is a good way to solve the problem of poor interactivity of VAE. The main contributions of this work are as follows:

1. A novel model of VAE for collaborative filtering using multi-position and latent self-attention is proposed, which allows the model to learn richer and more complex latent vectors and strengthens the transmission of important information.
2. We use VAE with multi-position latent self-attention as the actor and a new neural network as the critic to enhance interaction in the reinforcement learning process. The model realizes the common advantages of deep learning and reinforcement learning.
3. Our model acMLAVAE is comprehensively evaluated on numerical experiments and shows that it consistently outperforms the eight existing baselines across all metrics. Especially, we demonstrate good stability of our model and trains latent self-attention's weights automatically.

2   Related Works

Actor-Critic. Reinforcement learning encompasses a large category of machine learning. Actor-critic is one main class of reinforcement learning that is based on estimating only the values of states. In actor-critic, the actor implements a policy and the critic attempts to estimate the value of each state under this policy [15]. With the development of actor-critic, recent papers [9,19] have added attention to actor-critic algorithm to improve reinforcement learning effects. But only reinforcement learning improvement cannot be directly applied to collaborative filtering recommendation. Some of the combination of actor-critic and VAE are more applied to the automatic driving collar [5,6], so we need to explore the reinforcement field of collaborative filtering.

Attention. The attention mechanism has a relatively wide and mature application in the field of graph attention networks [18] and translation [17]. In text processing, the traditional attention is generally used in the encoder-decoder of the sequence model Long Short Term Memory (LSTM) [7]. Such as DA-LSTM-VAE [22] is a model, which attention is based sequence LSTM that the weight calculation needs target. However, those models of processing the sequence depend on the order between words and cannot be calculated in parallel. The transformer [17] greatly promotes the development of self-attention mechanism. Therefore, it provides us with a good idea to solve VAE's important latent information transmission problem.

3 Preliminaries

Under massive sparse data, VAE modeling has been investigated for collaborative filtering to achieve good performance. We employ the notation $u \in 1, \ldots, U$ to denote users and $i \in 1, \ldots, I$ to denote items. x_u is a bag-of-words vector from user u in matrix X.

3.1 Variational Autoencoder for Collaborative Filtering

VAE is a typical encoder-decoder architecture and have powerful generation capability. The model of VAE is shown in Fig. 1. VAE uses the variational reasoning framework to learn the approximate posterior of latent variables.

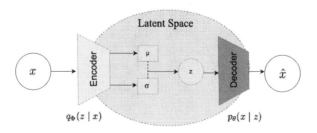

Fig. 1. VAE of Encoder and Decoder.

It is clear to see that VAE consists of two parts, the encoder and decoder. In encoder part, maps the input data x to a latent vector z in the latent space, and outputs the mean μ and variance σ^2 of the latent vector.

$$z = \mu + \epsilon \odot \sqrt{exp(\log(\sigma^2))} = q_\Phi(z \mid x) \tag{1}$$

Here, the Φ-parameteried neural network encoder is expressed as $q_\Phi(z \mid x)$. ϵ is a vector sampled from the standard normal distribution $N(0,1)$, $\odot$ denotes element-wise multiplication, and $\log(\sigma^2)$ represents the logarithm of the variance.

The decoder part generates the reconstructed data $\hat{x}$. Here, θ represents the parameters of the decoder. The θ-parameteried neural network decoder is expressed as $p_\theta(x \mid z)$.

$$\hat{x} = Decoder(z; \theta) = p_\theta(x \mid z) \tag{2}$$

The loss function of VAE consists of two parts, the reconstruction error and the KL divergence.

$$L_{VAE} = L_{reconstruction}(x, \hat{x}) - KL(q(z|x)||p(z)) \tag{3}$$

Here, $L_{reconstruction}$ represents the reconstruction error [10], KL represents the KL divergence, $q(z|x)$ represents the posterior distribution, and $p(z)$ represents the prior distribution.

$$L_{reconstruction} = E(q_\Phi(z|x)\left[\log p_\theta(x \mid z)\right] \tag{4}$$

In Eq. 5, k represents the dimension of the latent vector z, that is, the length of the latent vector.

$$KL(q(z|x)||p(z)) = -\frac{1}{2}\sum_k (1 + \log((\sigma_k)^2) - (\mu_k)^2 - (\sigma_k)^2) \tag{5}$$

A traditional VAE model typically utilizes Gaussian log-likelihood [4] for latent space collaborative filtering. Specifically, for each user u, the corresponding latent representation $z_u = 0$ follows a standard Gaussian prior distribution, i.e., $z_u \sim N(0, I)$. We cannot get distribution from z_u directly. In order to solve this problem, we use $q_\Phi(z_u \mid x_u)$ instead of $p_\theta(x_u \mid z_u)$ to realize the process of variational inference. VAE is trained using the backpropagation algorithm to minimize the loss function L_{VAE} to update the Φ neural network of the encoder and θ neural network of the decoder.

$$L_{VAE}(x_u; \Phi, \theta) = E(q_\Phi(z_u|x_u)\left[\log p_\theta(x_u \mid z_u)\right] - \\ KL(q_\Phi(z_u \mid x_u))||p_\theta(z_u)). \tag{6}$$

3.2 Self-attention

The transformer [17] greatly promotes the development of self-attention. It consists self-attention and multi-head self-attention. Multi-head self-attention is

only a distributed and parallel expression of self-attention, which does not change
the internal structure of attention, but improves the training efficiency. So this
paper uses multi-head self-attention showing as self-attention, because parallel
does not affect the final result. A self-attention function can be described as
mapping a query and a set of key-value pairs to an output, where the query,
keys, values, and output are all vectors. From the definition of attention [17], we
know that the queries, keys and values are packed together into matrices Q, K
and V. Here, d is the dimension of queries and keys. M is the masks. We know
the matrix of outputs as:

$$Attention(Q, K, V) = softmax(\frac{QK^\top}{\sqrt{d}} + M)V \qquad (7)$$

From Eq. 8, we know that the calculation of self-attention is not dependent on
the sequence of words, but on the similarity of words to information mining.

4   Proposed Model

4.1   The Framework

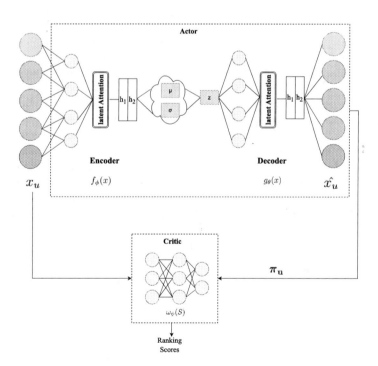

Fig. 2. The Overall Framework of acMLAVAE.

VAE is a deep learning model that can represent latent vectors friendly, but lacks interactivity. The reinforcement learning algorithm actor-critic learns the optimal strategy through interaction. Through the joint learning of the two models, the deep relationship representation of the latent vector is obtained to complete the final recommendation task. We set Φ-parameteried neural network encoder as $g_{\Phi}(x)$ and θ-parameteried neural network decoder as $f_{\theta}(x)$ as actor part. we propose multi-position latent self-attention that is used in the actor process. The location of latent attention in Fig. 2 corresponds to the position of the multi-position latent self-attention in the encoder and decoder. Critic process uses Ψ-parameteried neural network function ω, which takes in the prediction π to compare with the ground-truth x. The critic process uses Ψ-parameteried neural network function ω: $\{\pi; x\} \rightarrow y \in [0, 1]$, and outputs a scalar y to rate the prediction quality. The overall framework is shown in Fig. 2.

4.2   Variational Autoencoder with Multi-position Latent Self-attention (MLAVAE)

A novel model that we extend a VAE-based model [12] by adding multi-position latent self-attention in latent space, which is denoted as MLAVAE. The traditional VAE models only result in activation of a few latent vectors which will become more serious as the depth of the network deepens. As a high-dimensional representation of user-item scoring data, latent vectors play an important role in reconstructing data. For the problem that the inactive latent vector affects the quality of the reconstructed data, this paper designs a multi-position latent self-attention in Fig. 2's encode part that conforms to the latent vector, assigns a larger weight value to the nodes containing important information in the latent vector, enhances the role of important information in the reconstructed data, automatically gives a smaller weight value to the unimportant information, weakens its impact on the reconstructed data, and reduces noise. The general idea of our multi-position latent self-attention is to first utilize self-attention to generate an attention-weight matrix and then obtain output by applying element wise multiplication between the attention-weight matrix and the original input. Multi-position latent self-attention can quantify the importance of different information through the weight of latent vector in the process of data learning, and the weight is automatically adjusted during the learning process by random initialization without manual modification of parameters. This dynamic adjustment of the learning process makes it possible to select important information from different datasets, finally improving the accuracy of recommendation. In the following, we shall explain this process step by step.

Step 2 to step 7 is the process of VAE with multi-position latent self-attention in Algorithm 1. In step 2, when the data enters the encoder, it becomes a vector in the latent space, where h takes the latent vector. In order to strengthen the influence of key latent vectors on the collaborative filtering recommendation, and taking into account the different key information carried by latent vectors at different positions, we propose a multi-position method, which can be encoder, decoder, or encoder+decoder. In step 3, we assign (h, h, h) to (K, Q, V) of Eq. 8 in encoder. Here, M is the masks of 10 percent. First, we do not need to adjust the parameters manually. Second, the weights of MLAVAE are adjusted adaptively, and the nodes containing important information in the latent vectors are given larger weights. We get multi-position latent self-attention denoted as MLA.

$$MLA = softmax(\frac{hh^\top}{\sqrt{d}} + M)h \tag{8}$$

Step 4 is feedforward network that W_1, W_2, b_1 and b_2 are randomly initialized matrices to ensure weight independent learning. Step 5 is residual network to avoid gradient disappearance and network degradation, so as to reduce the training difficulty of deep neural network. Here $f_u(\cdot)$ is full connection layer, which weight is static. In step 6, the latent vector is passed to the decoder. Then, we add the above multi-position latent self-attention to Eq. 8, and get Eq. 9.

$$L_{MLAVAE}(x_u; \Phi, \theta) = E_{(q_\Phi(z_u|x_u)}[\log p_\theta(x_u \mid z_u, MLA)] - \\ KL(q_\Phi(z_u \mid x_u))\|p_\theta(z_u, MLA)). \tag{9}$$

Finally, we use L_{MLAVAE} to update the Φ and θ neural network.

4.3   The Design of MLAVAE with Actor-Critic(acMLAVAE)

Algorithm 1. acMLAVAE

Initialization: Randomly initialize actor network parameters Φ, θ and critic parameters Ψ, $f_u(\cdot)$ is full connection layer, time steps T.
Input: State interaction matrix' vector x_u into actor network, W_1, W_2, b_1 and b_2 are randomly initialized matrices, probability distribution prediction $\pi_\mathbf{u}$ is obtained as action.

1: **for** $t = 1$ to T **do**
2:     $h \leftarrow encoder : f_u(x_u)$;
3:     $h_0 \leftarrow Attention(h, h, h)$;
4:     $h_1 \leftarrow (W_1 h_0 + b_1)W_2 + b_2$;
5:     $h_2 \leftarrow h_1 + f_u(h_1)$;
6:     $decoder \leftarrow h_2$;
7:     Update Φ, θ with $\frac{\partial L_{MLAVAE}}{\partial \Phi}$ $and \frac{\partial L_{MLAVAE}}{\partial \theta}$ in Eq. 9;
8:     $\pi_\mathbf{u} \leftarrow decoder$;
9:     Decompose $\pi_\mathbf{u}$ into S;
10:     Compute $L_{critic} = \| \omega_\psi(S) - y \|^2$;
11:     Update ψ with $\frac{\partial L_{critic}}{\partial \psi}$.
12: **end for**

Finally the overall framework learning process is provided in Algorithm 1. The process of step 2 to step 7 is the actor training process, and step 8 to step 11 is the critic training process. Critic part decomposes probability $\pi_\mathbf{u}$ into S, and computes the mean square error $\| \omega_\psi(S) - y \|^2$ to get L_{critic}. In the mass,

Φ-parameteried neural network in encoder and θ-parameteried neural network in decoder obtain more useful features, and Ψ-parameteried neural network in critic minimizes the mean square error to complete the recommendation. Specifically, the above three neural network architectures are as follows. Encoder: linear $\rightarrow$ tanh $\rightarrow$ latent self-attention $\rightarrow$ feedforward network $\rightarrow$ residual network $\rightarrow$ Linear. Decoder: linear $\rightarrow$ tanh $\rightarrow$ latent self-attention $\rightarrow$ residual network $\rightarrow$ feedforward network $\rightarrow$ residual network $\rightarrow$ softmax. Criric: batch normalization $\rightarrow$ relu $\rightarrow$ sigmoid. Tanh, relu and sigmoid are activation function. The sigmoid function is used to rescale the result to $[0, 1]$, which is more suitable for latent variables.

5   Experiments

5.1   Experiment Platform and Datasets

We develop the proposed algorithm acMLAVAE with tensorflow architecture in a CentOS 7 system (PowerLeader server with an Intel CPU and 1T memory). Batch size is 500. The dimension of the word embedding is 600, and the number of attention heads is 8.

We conducted our empirical evaluations on three real world social media and publicly available datasets from Mult-VAE [12]: ML-20M, Netflix and MSD. The ML-20M dataset is a user-movie rating dataset. Each user left their rating score on a scale of 5, which has been binarized for our study. The Netflix dataset comprises of user-movie ratings and was originally introduced for the Netflix Prize competition. The rating scale is the same as the ML-20M. The MSD dataset used in this study contains user-song play count information. Users who interacted with less than 5/20 movies/songs were removed, and the data were converted to binary as in the case of implicit feedback.

5.2   Evaluation Indicators

Two standard metrics are utilized for evaluating the quality of recommendation, including normalized discounted cumulative gain (NDCG@R) and Recall@R. Recall@R and NDCG@R are the universal evaluation metrics based on ranking, where R is a parameter. Recall@R measures the fraction of the positive items in the test data. The R in Recall@R keeps the first R items equally important. A higher NDCG@R represents the positive items in the test data are ranked higher in the ranking list. The R in NDCG@R supports multi-level similarity, which makes retrieval more reasonable when designing multi-table data. $m(r)$ is the item of rank r. H_0 is hold-out items that user u clicks on interaction. Recall@R for a given user u can be defined as Eq. 10.

$$\text{Recall@R}\,(u, m(r)) = \frac{\sum\limits_{r=1}^{R} \delta\,[m(r) \in H_0]}{\min\,(R, |H_0|)}. \tag{10}$$

Discounted cumulative gain (DCG@R) for user u is defined as Eq. 11. NDCG@R linearly normalizes DCG@R to $[0, 1]$.

$$\text{DCG@R}\,(u, m(r)) = \sum_{r=1}^{R} \frac{2^{\delta[m(r) \in H_0]} - 1}{\log(r + 1)}. \tag{11}$$

5.3 Experimental Results and Analysis

Overall Performance of AcMLAVAE. The results in Table 1 clearly show that the performance of acMLAVAE on each dataset surpasses other baselines.

Table 1. acMLAVAE compare with eight kinds of models on three datasets. Bold indictes the overall best.

Datasets	ML-20M			Netflix			MSD		
Models	Recall@20	Recall@50	NDCG@100	Recall@20	Recall@50	NDCG@100	Recall@20	Recall@50	NDCG@100
acMLAVAE	**0.4448**	**0.5729**	**0.5018**	**0.4098**	**0.4797**	**0.4667**	**0.3017**	**0.3872**	**0.3909**
RaCT	0.3960	0.5368	0.4274	0.3578	0.4490	0.3906	0.2587	0.3560	0.3190
SE-VAE	0.4178	0.5555	0.4471	0.3771	0.4631	0.4090	NULL	NULL	NULL
RecVAE	0.4113	0.5502	0.4399	0.3597	0.4481	0.3930	0.2711	0.3719	0.3195
MacridVAE	0.3964	0.5290	0.4249	0.3458	0.4347	0.3798	NULL	NULL	NULL
Mult-VAE	0.3912	0.5323	0.4228	0.3509	0.4428	0.3857	0.2600	0.3560	0.3100
Mult-DAE	0.3870	0.5240	0.4190	0.3440	0.4380	0.3800	0.2660	0.3630	0.3130
CDAE	0.391	0.523	0.418	0.343	0.428	0.376	0.188	0.283	0.237
WMF	0.360	0.498	0.386	0.316	0.404	0.351	0.211	0.312	0.257

It also proves that the addition of multi-position latent self-attention with actor-critic can improve the performance of VAE generally, because multi-position latent self-attention with actor-critic integrates multiple layers to achieve latent vector to transmit important information. Mult-VAE [12] is currently the most advanced collaborative filtering model based on encoder and decoder network. The experiment evaluated the model in this paper from three indicators: Recall@20, Recall@50 and NDCG@100, and the indicators increased by [13.70%, 7.63%, 18.68%] on ML-20M, [16.79%, 8.33%, 21.00%] on Netflix, and [16.04%, 8.76%, 26.10%] on MSD. Especially in NDCG@100 's performance is excellent in three datasets. RaCT [13], MacridVAE [14], SE-VAE [3] and RecVAE [16] are all improved, and those four models are based on Mult-VAE. Macrid-VAE [14] adds auxiliary information based on Mult-VAE. SE-VAE [3] focuses on sampling based on Mult-VAE. RecVAE [16] changes the prior distribution based on Mult-VAE. Compared with Mult-VAE, although their Recall@20, Recall@50 and NDCG@100 values have been improved, but not fully utilized the latent vectors containing important information. RaCT [13] adds actor-critic based on Mult-VAE. Although reinforcement learning can improve interaction, it ignores the internal connection of the model. Compared with WMF [8] and CDAE [20] models, the evaluation indicators on the ML-20M, Netflix and MSD datasets have significantly improved.

Improvement on Different R. The other baselines take Mult-VAE [12] as the baseline, so we mainly focus on the comparison with Mult-VAE. Table 2 shows under different numbers of R (1, 3, 5, 20, 50, 100, 200) can be improved significantly, which indicates acMLAVAE's stability of neural network training in recommendation.

Table 2. Improvement on different R.

model	acMLAVAE	Mult-VAE
Recall@50	0.5729	0.5323
Recall@20	0.4448	0.3912
NDCG@200	0.5362	0.4606
NGCG@100	0.5018	0.4228
NDCG@5	0.4449	0.3129
NDCG@3	0.4901	0.3271
NDCG@1	0.6131	0.3602

In collaborative filtering, we are interested in predicting the top-N items to the user in a sense that top-N items are more carefully observed in recommender systems. Recall@R is the proportion of relevant (clicked) items predicted in the top R items. This metric becomes useful considering that the online users focus on top-N recommended items on first page or upper area without scrolling down. However, Recall@R indicates the proportion of items brought in the first R position that were actually in the target subset, and does not consider the order. NDCG overcomes this issue by using the monotonically increasing discount. It emphasizes the importance of higher ranking than lower ones. Therefore, our model improves steadily under different R recommened, especially the NDCG@1 up to 61.31%.

Influence of Multi-position. As shown in Fig. 2, we known latent attention can be in the encoder or in the decoder or in the encoder and decoder (short as double) at the same time. Remove all other impression factors, and here only verify the influence of attention position. Table 3 shows the influence of multi-position.

On ML-20M, the effect of latent attention is the best in encoder at most cases, and the situation of using latent attention in both encoder and decoder is the second best. On Netflix, latent attention in encoder is the best, and in decoder is the second best. On MSD, second best appears in decoder and double. In datasets with different fields, sizes and sparsity, the effect of latent attention is different in different locations. So each location reflects different important information about latent vectors. On the whole, the effect of latent attention is best when the position in encoder.

Ablation Studies. In order to figure out the contribution of different components to the performance of our acMALVAE, we conduct some ablation studies.

Table 3. The different positions in encoder, decoder and double. Bold indictes the overall best, and underline indicates second best.

Dataset	ML-20M			Netflix			MSD		
Position	Decoder	Encoder	Double	Decoder	Encoder	Double	Decoder	Encoder	Double
NGCG@100	0.4243	**0.4301**	<u>0.4255</u>	<u>0.3781</u>	**0.3831**	0.3667	<u>0.3065</u>	**0.3117**	0.3057
Recall@50	0.5376	<u>0.5391</u>	**0.5405**	<u>0.4381</u>	**0.4405**	0.4258	<u>0.3570</u>	**0.3578**	0.3560
Recall@20	0.3962	**0.4021**	<u>0.3992</u>	<u>0.3448</u>	**0.3495**	0.3326	<u>0.2588</u>	**0.2615**	0.2581
NDCG@200	0.4610	**0.4672**	<u>0.4625</u>	<u>0.4254</u>	**0.4301**	0.4141	0.3247	**0.3396**	<u>0.3336</u>
NDCG@5	0.3092	**0.3239**	<u>0.3110</u>	<u>0.3199</u>	**0.3326**	0.3066	0.2248	**0.2369**	<u>0.2250</u>
NDCG@3	0.3187	**0.3375**	<u>0.3226</u>	<u>0.3298</u>	**0.3457**	0.3168	<u>0.2440</u>	**0.2586**	0.2438
NDCG@1	0.3489	**0.3763**	<u>0.3552</u>	<u>0.3522</u>	**0.3761**	0.3417	0.2764	**0.2985**	<u>0.2782</u>

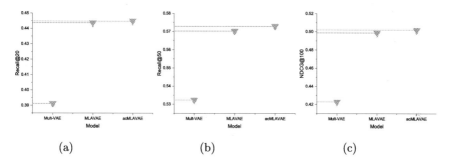

(a)   (b)   (c)

Fig. 3. Ablation studies on ML-20M dataset. (a), (b) and (c) correspond to Recall@20, Recall@50 and NDCG@100 respectively.

Notice that we only present the results on Recall@20, Recall@50 and NDCG@100 because the trends on other metrics are similar to theirs. Specifically, we remove the target representation enhancement model. Figure 3 shows ablation studies on ML-20M. acMLAVAE is a novel model proposed by this paper. When acMLAVAE does not use actor-critic, it is MLAVAE. This reflects the interaction of the combination of intensive learning and deep learning. When MLAVAE does not use multi-position latent self-attention, it is Mult-VAE. Which demonstrates that multi-position latent self-attention learns richer and complex latent vectors and strengthens the transmission of important information. From the comparison of the Recall@20, Recall@50 and NDCG@100, we can know that the addition of multi-position latent self-attention and actor-critic is meaningful. We only show the results on ML-20M because the results on Netflix and MSD are similar to Fig. 3. Therefore, the validation of three datasets can further illustrate two points. First, our model has good stability. Second, our model has good performance in collaborative filtering recommendation.

6   Conclusion

The acMLAVAE model can increase interaction, and enhance the training process to learn richer and complex latent vectors and strengthens the transmission of important information in collaborative filtering recommendation. Compared to other baselines, acMLAVAE shows superior performance and exhibits strong stability. The research in this paper is more inclined to strengthen the importance of the hidden vector itself. In future work, more complex data forms and data relationships will be incorporated, such as knowledge graph, graph attention networks to explore the impact on the recommendation effect.

Acknowledgements. This work was partly supported by the Guangzhou Key Laboratory of Big Data and Intelligent Education (No.201905010009) and National Natural Science Foundation of China (No.61672389).

References

1. Askari, B., Szlichta, J., Salehi-Abari, A.: Variational autoencoders for top-k recommendation with implicit feedback. In: Proceedings of the 44th International ACM SIGIR Conference on Research and Development in Information Retrieval, pp. 2061–2065. ACM (2021)
2. Chen, J., Lian, D., Jin, B., Huang, X., Zheng, K., Chen, E.: Fast variational autoencoder with inverted multi-index for collaborative filtering. In: WWW '22: The ACM Web Conference 2022, Virtual Event, Lyon, France, April 25–29, 2022, pp. 1944–1954. ACM (2022)
3. Cho, Y., Oh, M.: Stochastic-expert variational autoencoder for collaborative filtering. In: WWW '22: The ACM Web Conference 2022, Virtual Event, Lyon, France, April 25–29, 2022, pp. 2482–2490. ACM (2022)
4. Gopalan, P., Hofman, J.M., Blei, D.M.: Scalable recommendation with hierarchical poisson factorization. In: UAI, pp. 326–335 (2015)
5. Gupta, A., Khwaja, A.S., Anpalagan, A., Guan, L.: Safe driving of autonomous vehicles through state representation learning. In: 17th International Wireless Communications and Mobile Computing, IWCMC 2021, Harbin City, China, June 28 - July 2, 2021, pp. 260–265. IEEE (2021)
6. Gupta, A., Khwaja, A.S., Anpalagan, A., Guan, L., Venkatesh, B.: Policy-gradient and actor-critic based state representation learning for safe driving of autonomous vehicles. Sensors **20**(21), 5991 (2020)
7. Hochreiter, S., Schmidhuber, J.: Long short-term memory. Neural Comput. **9**(8), 1735–1780 (1997)
8. Hu, Y., Koren, Y., Volinsky, C.: Collaborative filtering for implicit feedback datasets. In: Proceedings of 8th IEEE International Conference on Data Mining, pp. 263–272. IEEE (2008)
9. Iqbal, S., Sha, F.: Actor-attention-critic for multi-agent reinforcement learning. arXiv (2018)
10. Karamanolakis, G., Cherian, K.R., Narayan, A.R., Yuan, J., Tang, D., Jebara, T.: Item recommendation with variational autoencoders and heterogeneous priors. In: Proceedings of the 3rd Workshop on Deep Learning for Recommender Systems (DLRS), pp. 10–14 (2018)
11. Kingma, D.P., Welling, M.: Auto-encoding variational bayes. In: 2nd International Conference on Learning Representations, ICLR 2014, Banff, AB, Canada, April 14–16, 2014, Conference Track Proceedings (2014)

12. Liang, D., Krishnan, R.G., Hoffman, M.D., Jebara, T.: Variational autoencoders for collaborative filtering. In: Proceedings of the 2018 World Wide Web Conference (WWW), pp. 689–698 (2018)
13. Lobel, S., Li, C., Gao, J., Carin, L.: Towards amortized ranking-critical training for collaborative filtering. In: International Conference on Learning Representation(ICLR) (2020)
14. Ma, J., Zhou, C., Cui, P., Yang, H., Zhu, W.: Learning disentangled representations for recommendation. In: Proceedings of the Advances in Neural Information Processing Systems (NeurIPS), pp. 5712–5723. Vancouver, BC, Canada (2019)
15. Mustapha, S.M., Lachiver, G.: A modified actor-critic reinforcement learning algorithm. In: Conference on Electrical and Computer Engineering (2000)
16. Shenbin, I., Alekseev, A., Tutubalina, E., Malykh, V., Nikolenko, S.I.: Recvae: A new variational autoencoder for top-n recommendations with implicit feedback. In: Proceedings of the Thirteenth ACM International Conference on Web Search and Data Mining (WSDM), pp. 528–536. ACM, Houston, TX, USA (2020)
17. Vaswani, A., et al.: Attention is all you need. In: Advances in Neural Information Processing Systems 30: Annual Conference on Neural Information Processing Systems 2017, December 4–9, 2017, Long Beach, CA, USA, pp. 5998–6008 (2017)
18. Velickovic, P., Cucurull, G., Casanova, A., Romero, A., Liò, P., Bengio, Y.: Graph attention networks. In: 6th International Conference on Learning Representations, ICLR 2018, Vancouver, BC, Canada, April 30 - May 3, 2018, Conference Track Proceedings. OpenReview.net (2018)
19. Yang, N., Lu, Q., Xu, K., Ding, B., Gao, Z.: Multi-actor-attention-critic reinforcement learning for central place foraging swarms. In: 2021 International Joint Conference on Neural Networks (IJCNN), pp. 1–6 (2021)
20. Yang, S.H., Long, B., Smola, A.J., Zha, H., Zheng, Z.: Collaborative competitive filtering: learning recommender using context of user choice. In: Proceedings of the 34th International ACM SIGIR Conference on Research and Development in Information Retrieval, pp. 295–304 (2011)
21. Zhao, J., Zhao, P., Zhao, L., Liu, Y., Sheng, V.S., Zhou, X.: Variational self-attention network for sequential recommendation. In: 2021 IEEE 37th International Conference on Data Engineering (ICDE), pp. 1559–1570. IEEE (2021)
22. Zhao, Y., Zhang, X., Shang, Z., Cao, Z.: Da-LSTM-VAE: Dual-stage attention-based LSTM-VAE for KPI anomaly detection. Entropy **24**(11), 1613 (2022)

Fair Re-Ranking Recommendation Based on Debiased Multi-graph Representations

Fangyu Han[1], Shumei Wang[2], Jiayu Zhao[1], Renhui Wu[1], Xiaobin Rui[1],
and Zhixiao Wang[1(✉)]

[1] China University of Mining and Technology, Xuzhou, China
{hanfy,zhaojy,wurh,ruixiaobin,zhxwang}@cumt.edu.cn
[2] Jiangsu Normal University, Xuzhou, China
plum_xz@jsnu.edu.cn

Abstract. The successful application of graph neural networks in recommendation scenarios causes serious exposure of sensitive information of users. Research shows that social bias such as sexism and ageism are prevalent in recommendations, and the use of multi-graph information even makes it worse. Existing fair recommendation algorithms only concentrate on users' sensitive attributes in user-item graph, failing to fully remove those attributes from multiple graphs. In addition, merely hiding sensitive information is not enough, there is still a gap in recommendation utility for different user groups. In this work, we propose a novel fair re-ranking recommendation model based on debiased multi-graph representations, which contains three functional layers. Multi-graph embedding layer iteratively propagates and aggregates both topological and interactive information on multiple graphs. Attribute hiding layer uses generative adversarial networks to hide user sensitive information and thus debias users' representations. Fair ranking layer adopts a re-ranking strategy with our proposed unfairness metric to further optimize the final recommendation list. Extensive experiments on real-world datasets demonstrate the performance of our proposed model in both recommendation utility and fairness, outperforming state-of-the-art models.

Keywords: Recommendation · Graph embedding · Fairness · Debias

1 Introduction

With the advancement of information technology, people can easily access a large number of online services. However, this also leads to a serious problem, information overload, that the number of items greatly exceeds the user's affordability. It is difficult for people to filter out what they really need from the mass information. In order to mitigate the impact of information overload, recommendation system (RS) came into being, which can capture users' potential interests and customize personalized services according to their historical interactions.

Recently, due to its outstanding advantages in handling non-European structural data, graph neural network (GNN) is widely used in RS. GNN-based models learn the embeddings of users and items by propagating and aggregating user information over graph structures, exploiting the rich multi-hop neighborhood information to achieve better recommendation results. NGCF [1] and Light-GCN [2] capture high-order connectivity of user-item interactions through message passing (i.e. neighborhood aggregation). Furthermore, homogeneous graphs also contain meaningful topological information for recommendation system. Therefore, GHCF [3] further enhances node representation by directly modeling multiple types of relationships for prediction.

In fact, the prevalence of social bias in data collection can cause GNN-based models to give unfair predictions during automated decision-making, reducing the trustworthiness of those methods [13]. Similarly, substantial evidence suggests that it suffers from this problem in the application of recommendation scenarios. For example, a GNN-based algorithm on book recommendation is found biased towards suggesting books with male authors [4]. This implies that GNN-based methods could amplify social bias, discriminate against disadvantaged groups and undermine cultural diversity. Therefore, how to make recommendation systems perceptively capture the real interests of users without social bias, so as to make the recommendation results independent of sensitive attributes such as user's gender and age, is the main research direction of fair recommendation systems nowadays.

However, existing graph-based fair recommendation algorithms only hide sensitive information in the user-item bipartite graph [5], failing to adequately remove sensitive attributes hidden in other graphs, and thus fail to protect user privacy deeply enough to achieve the desired fair recommendation. Although existing recommendation algorithms that fuse multiple graphs will improve recommendation performance, they also inevitably pose potential problems while tapping into deeper user interests. The rich topological information contained in the user-user homogeneous graph may aggregate users with similar interest preferences while simultaneously aggregating users with the same sensitive attributes [6]. Since the aggregation of user preferences contributes to the aggregation of user attributes, fusing user-user homogeneous graphs in RS will lead to the exposure of users' sensitive attributes from message aggregation of their neighbors. This can cause greater disparity in recommendation utility between two groups as well as more unfair recommendation results.

In addition, some existing research in fair recommendation algorithms only concentrate on learning debiased embeddings [7]. In fact, there is still a distance between hiding sensitive user attributes and achieving fairness in the ultimate recommendation results. We find that even if the user's sensitive information is not used explicitly in the model, two groups of users classified by the same sensitive attribute still suffer from the recommendation utility unfairness. It is probably due to the bias in data collection caused by inevitable social bias in the real world, making the advantaged users achieve a better recommendation utility than the disadvantaged users. Meanwhile, we note that current re-ranking algorithms generally use their own proposed fairness metrics when selecting the optimization objective function, which lacks reasonable explanation. Therefore, a more reasonable fairness metric should be proposed to quantify the unfairness of the recommendation utility.

In this paper we propose a fairness-aware representation learning and re-ranking model in the context of multi-graph recommendation. This fairness-aware model can not only fully hide the sensitive attributes exposed in both the interaction information from user-item heterogeneous graphs and the topological information from user-user homogeneous graphs, but also narrow the recommendation utility gap between distinct groups under the same sensitive attributes, and eventually achieve fairness on the user side in the ranking recommendation task.

The main contributions of this work are as follows:

- Our proposed model can simultaneously hide the sensitive attributes exposed in multiple graphs to learn debiased representations for RS. With adversarial learning technology, we design a set of discriminators for sensitive attributes to remove users' sensitive information from both heterogeneous graphs and homogeneous graphs. To the best of our knowledge, this is the first work to tackle unfairness problem in the context of multi-graph recommendation.
- We introduce the concept of Gini coefficient from the field of economics into fair ranking strategy to propose a new metric quantifying the unfairness of the recommendation utility. It provides a visual and effective way of reflecting and detecting disparities between groups, and also offers support for warning and preventing serious polarisation between the advantaged and the disadvantaged subsequently.
- We perform extensive experiments on two real-life datasets, and the results reveal that our model can mitigate unfairness against disadvantaged groups in RS while maintaining or even slightly improving recommendation utility.

2   Related Work

2.1   Recommendation Systems

Collaborative filtering [9] is one of the most successful methodologies to solve recommendation problems (including the top-N recommendation task in this paper). With the development of artificial intelligence and machine learning, collaborative filtering recommendation models can be divided into two types: shallow models and deep learning models.

The shallow model is specifically given a user-item rating matrix R that projects both users and items in the same low-dimensional potential space. Each user's predicted preference for an item can then be measured by the similarity of the user's potential vector to the item's potential vector. Bayesian personalize ranking (BPR) [26] is an implicit feedback handling technique based on latent factors and SVD++ [10] proposes to combine implicit and explicit feedback from users to model their latent interests.

Algorithms based on deep learning use a combination of fully-connected layers, convolution layers, inner products, and subnets to capture complex similarity relationships. At the same time, GNNs come to be used in recommendation algorithms due to their powerful structured data learning representation capabilities [11]. NGCF [1] propagates user and item embeddings by using multiple GCN layers to capture higher-order connectivity. Multi-GCCF [12] integrates the proximal information by explicitly building and processing user-user and item-item graphs.

2.2   Fair Recommendation Systems

Like humans, algorithms are vulnerable to social bias that cause the system to make unfair decisions [13]. Here, fairness means that the system is free from any prejudice or preference towards the inherent features and attributes of individuals or groups. To achieve fairness, researchers have made a lot of efforts to solve the bias of different tasks in machine learning.

As an important application of machine learning, the need for fairness in recommendation systems has also received great attention from both industry and academia. Many researchers have proposed various ways to mitigate the effects of bias or unfairness. Though there is no consensus on fairness definitions since the fairness demands can be different under different scenarios. Currently, Group Fairness (e.g., demographic parity [14] and equalized odds [15]), Individual Fairness (e.g., similarity-based fairness [16], ranking-based fairness [17]), and Hybrid Fairness are commonly used metrics for evaluating fairness in RS. Generally, we consider a model to be fair if no sensitive attributes (e.g., gender, race, age, etc.) are used explicitly in the modeling process of the RS and if the recommendation results are isolated from the protected attributes. Current work on methods for implementing fairness-aware recommender systems can be divided into four main categories, that is, re-ranking [18], regularization [19], causal methods [20], and adversarial learning [21].

3   Methodology

In this section, we describe the architecture of our fair **Re**-ranking **Re**commendation based on debiased multi-graph representations (ReRec) model.

As Fig. 1 shows, the proposed model ReRec consists of three main layers. First, the multi-graph embedding layer initializes user embeddings and item embeddings, and iteratively integrates the interaction information between heterogeneous nodes and the topological information between homogeneous nodes, in order to refine the user embeddings and item embeddings. Second, the attribute hiding layer adopts a generative adversarial network (GAN). Each sensitive attribute is associated with one discriminator to hide the corresponding sensitive information through adversarial training techniques, generating dabiased representations for RS. Finally, the fair ranking layer re-ranks the users' recalled list to narrow the recommendation utility gap between different groups under the fairness constraints and generate a fairer top-N recommendation list.

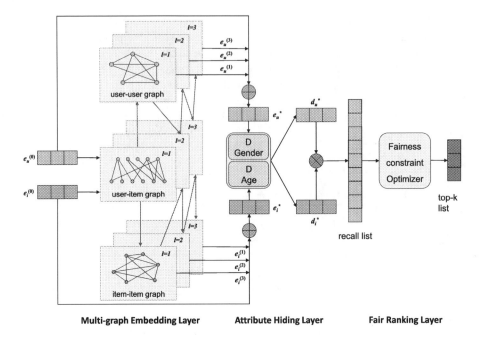

Fig. 1. The overall structure of our proposed ReRec model.

3.1 Multi-graph Embedding Layer

Assume that there are n users and m items, respectively. Following the mainstream recommendation model [1], ReRec transforms these user ids and item ids into embedding vectors $\mathbf{e}_u$ and $\mathbf{e}_i$, and d denotes the size of the embedding. Let $\mathbf{e}_u^{(0)}$ denote the ID embedding of user u and $\mathbf{e}_i^{(0)}$ denote the ID embedding of item i. The initial user embedding and item embedding in this layer will be optimized in an end-to-end fashion.

The multi-graph embedding layer makes full use of the message passing architecture of graph convolutional neural networks to capture collaborative information from different graph structures and refine the initial embeddings of users and items.

Heterogeneous Graphs. This part uses forward propagation to capture collaborative information between heterogeneous nodes and to minimize the loss of fine-grained feature information. The basic idea of GCN is to learn node representations by smoothing features on the graph, and it updates node representations in the lth iteration by normalizing and aggregating the representations of its neighbors in the $(l-1)^{th}$ iteration. Here, we adopt the propagation rules of LightGCN [2], where $\mathbf{h}_u^{(l)}$ and $\mathbf{h}_i^{(l)}$ respectively represent the representations of user u and item i in the lth iteration, as follows.

$$\mathbf{h}_u^{(l)} = \sum_{i \in \mathcal{N}_u} \frac{1}{\sqrt{|\mathcal{N}_u|}\sqrt{|\mathcal{N}_i|}} \mathbf{S}_i^{(l-1)}; \mathbf{S}_u^{(0)} = \mathbf{e}_u^{(0)}$$

$$\mathbf{h}_i^{(l)} = \sum_{u \in \mathcal{N}_i} \frac{1}{\sqrt{|\mathcal{N}_i|}\sqrt{|\mathcal{N}_u|}} \mathbf{S}_u^{(l-1)}; \mathbf{S}_i^{(0)} = \mathbf{e}_i^{(0)} \tag{1}$$

Interaction message propagation in heterogeneous graphs starts with user embedding $\mathbf{S}_u^{(l-1)}$ and item embedding $\mathbf{S}_i^{(l-1)}$ as input, which comes from the propagation of messages on homogeneous graphs in the $(l-1)^{th}$ iteration. It then uses the user-item graph to intersect features and integrate user-item interactions into the embedding function. Finally, the neighborhoods are aggregated to obtain fused embeddings with crossing messages.

Homogeneous Graphs. User-user and item-item relationships are the basic homogeneous information used to complement recommendations, which reveal behavioural similarities between users. Therefore, ReRec uses the user-item bipartite graph to construct the user-user graph and the item-item graph during the propagation of topological information in the homogeneous graph, and then iteratively learns embeddings from the graph.

We then follow the idea of graph convolutional networks through back propagation operations to explore potential information in homogeneous graphs. User (item) embeddings are generated by aggregating neighboring features through a one-hop graph convolution layer and aggregators.

$$\mathbf{S}_u^{(l)} = \sum_{v \in N_s'(u)} \frac{1}{\sqrt{|N_s'(u)|\,|N_s'(i)|}} \mathbf{h}_u^{(l)}$$

$$\mathbf{S}_i^{(l)} = \sum_{j \in N_s'(i)} \frac{1}{\sqrt{|N_s'(i)|\,|N_s'(u)|}} \mathbf{h}_i^{(l)} \tag{2}$$

where $N_s'(u)$ denotes the one-hop neighborhood in the user-user graph and $N_s'(i)$ denotes the one-hop neighborhood in the item-item graph. From this we obtain $\mathbf{S}_u^{(l)}$ and $\mathbf{S}_i^{(l)}$, which denote the homogeneous graph embedding of user and item in the lth iteration, respectively.

Multi-graph Fusion. Finally, we combine user (item) embeddings from heterogeneous and homogeneous graphs by introducing the element-wise sum:

$$\mathbf{e}_u^{(l)} = \mathbf{S}_u^{(l)} + \mathbf{h}_u^{(l)}; \mathbf{e}_i^{(l)} = \mathbf{S}_i^{(l)} + \mathbf{h}_i^{(l)} \tag{3}$$

ReRec uses two essential steps, interaction information propagation in heterogeneous graphs and topological information propagation in homogeneous graphs, to alternately propagate messages on multiple graphs and thus iteratively refine user and item embeddings.

ReRec employs holistic connection, using embeddings that evolve through all iterations (including the initial embeddings) as the final representation.

$$\mathbf{e}_u^* = \frac{1}{L+1} \sum_{l=0}^{L} \mathbf{e}_u^{(l)}; \mathbf{e}_i^* = \frac{1}{L+1} \sum_{l=0}^{L} \mathbf{e}_i^{(l)} \tag{4}$$

3.2 Attribute Hiding Layer

The attribute hiding layer needs to satisfy two goals: accurate for users' personalize preferences while debiased for the sensitive attributes. On one hand, the embeddings should represent the user's preferences in order to facilitate recommendation accuracy. On the other hand, these embeddings should be debiased and hide any information related to each user's sensitive attributes.

In this paper, we employ the adversarial training technique to achieve attribute hiding. Specifically, given the multi-graph fused user embedding $\mathbf{e}_u^*$ and item embedding $\mathbf{e}_i^*$ as inputs, each discriminator is designed for one corresponding sensitive attribute x (e.g., gender and age) to form an adversary network $d(\cdot)$, which outputs the hidden user embedding $\mathbf{d}_u^*$ and item embedding $\mathbf{d}_i^*$. Where the k-th discriminator tries to predict the value of the k-th sensitive attribute, i.e. each sub-discriminator $\mathcal{D}^k$ serves as a classifier to guess the corresponding attribute. Thus, the information aggregation network $f(\cdot)$ and the adversary network $d(\cdot)$ play the following two-player mini-max game with the following value function $V(\mathcal{F}, \mathcal{D})$.

$$\arg \min_f \ \arg \max_d V(\mathcal{F}, \mathcal{D}) = V_{\text{Rec}} - V_{\text{Adv}} \tag{5}$$

where V_{Rec} is the log likelihood of the rating distribution in implicit feedback and V_{Adv} is the log likelihood of the predicted attribute distribution.

For the user-item interaction distribution in implicit feedback, we assume it follows a Gaussian distribution, with the mean of the Gaussian distribution is the predicted rating of the interaction, as shown in Eq.(6). Thus, we model the value function of V_{Rec} as

$$V_{Rec} = -\sum_{u=1}^{M} \sum_{v=1}^{V} (y_{uv} - \hat{y}_{uv})^2 \tag{6}$$

Given the sensitive attribute vector x_i, a natural idea is to design the value function of the node embeddings after multi-graph fusion in the discriminator as:

$$V_{\text{Adv}} = \mathbb{E}_{(u,v,r,x_u)} \sum_{k=1}^{K} x_{uk} \ln \mathcal{D}^k (\mathbf{e}_u^*) \tag{7}$$

We adopt the inner product of the user embedding $\mathbf{d}_u^*$ and the item embedding $\mathbf{d}_i^*$ through the adversary network in this layer as the user's preference score for the item.

$$\hat{S}_{ui} = \mathbf{d}_u^{*T} \cdot \mathbf{d}_i^* \tag{8}$$

3.3   Fair Ranking Layer

In this layer, we propose a ranking algorithm that generates fairness-aware recommendation lists based on a fairness constraint method, which can effectively mitigate the unfairness of recommendation utility. The fair ranking strategy is divided into two steps: recall and re-ranking.

Firstly, given the user-item ratings after passing the adversary network in the previous layer, for each user i there will be a top-N recommendation list as the recall result. We define binary matrix $\mathbf{W}_{ij}$ to represent whether item j is recommended to user i after re-ranking. Specifically, if item j is recommended to user i, then we have $\mathbf{W}_{ij} = 1$, otherwise $\mathbf{W}_{ij} = 0$. We use $\mathbf{W}_i = [\mathbf{W}_{i1}, \mathbf{W}_{i2}, \cdots, \mathbf{W}_{iN}]^{\mathrm{T}}$ to denote the new top-K recommendation list for user i, where $\sum_{j=1}^{N} \mathbf{W}_{ij} = K, K \leq N$. The notation $\mathcal{M}$ is a metric that allows us to evaluate the recommendation utility, so we use $\mathcal{M}(\mathbf{W}_i)$ to denote the recommendation utility for user i.

We classify users into a disadvantaged group D and an advantaged group A based on sensitive attributes such as gender, age, etc. To minimize the utility gap, we invoke Gini coefficient [8], which takes the uniform degree of users into account and is typically used to measure pairwise differences. A well-balanced distribution has a Gini coefficient equal to zero, which is the ideal situation (the lower Gini coefficient, the better). Gini coefficient provides a visual and effective way of reflecting and detecting disparities between groups and also offers theoretical support for warning and preventing serious polarisation between the advantaged and the disadvantaged subsequently.

Definition 1. Recommendation Utility Unfairness:

$$\mathrm{RUU}(\mathcal{M}) = \frac{\left| \sum_{i \in D} \mathcal{M}(\mathbf{W}_i) - \sum_{i \in A} \mathcal{M}(\mathbf{W}_i) \right|}{2n \sum_i \mathcal{M}(\mathbf{W}_i)} \tag{9}$$

We then design the optimization objective function to accurately select K items from the top-N list recalled by each user, so as to maximize benefits while satisfying fairness constraints. We formulate the objective function with the fairness constraints as follows.

$$\max_{\mathbf{W}_{ij}} \sum_{i=1}^{n} \sum_{j=1}^{N} \mathbf{W}_{ij} \hat{S}_{i,j}$$

$$\text{s.t.} \quad \mathrm{RUU}(\mathcal{M}) = \frac{\left| \sum_{i \in D} \mathcal{M}(\mathbf{W}_i) - \sum_{i \in A} \mathcal{M}(\mathbf{W}_i) \right|}{2n \sum_i \mathcal{M}(\mathbf{W}_i)} \leq \varepsilon \tag{10}$$

$$\sum_{j=1}^{N} \mathbf{W}_{ij} = K, \ \mathbf{W}_{ij} \in \{0,1\}$$

where ε represents the strictness of the fairness requirement. We consider the fairest case in the following experiment by setting it to 0. After obtaining the set of recommended items under the fairness constraints, we rank these items according to their original preference scores $\hat{S}_{ij}$ to construct the final recommendation list.

4   Experiments

In this section, we first describe the experimental setup, then evaluate the recommendation utility and fairness of ReRec on two widely used real-world datasets.
    Specifically, we aim to answer the following three research questions:

- RQ1: Compared with state-of-the-art biased and debiased recommendation models and frameworks, how does ReRec perform?
- RQ2: What is the effect of different components in ReRec?
- RQ3: How does the hyper-parameter setting of the ReRec model affect the results of the experiment?

4.1   Experimental Settings

Datasets. We validated our experiments on two publicly available and widely used datasets, MovieLens-1M and Lastfm-360K, each containing user-item interactions and demographic information of users. Both datasets are summarized in Table 1.

Table 1. Statistics of the Datasets

Dataset	Users	Items	Interactions	Density
MovieLens-1M	6040	3416	999611	4.8448%
Lastfm-360K	359347	292589	17559530	0.0167%

Evaluation Metrics. Due to the overly time-consumption of iterating through all user-item interactions to generate a full rank, we use a negative sampling strategy to consider implicit feedback.

Recommendation Utility Metrics. As in mainstream methods, two widely used metrics are adopted to evaluate the recommendation utility: F_1 and NDCG (normalized discounted cumulative gain) in our experiments. Specifically, F_1 balances recall and precision to comprehensively assess the classification quality of the model. NDCG considers the hit positions of the items and gives a higher score if the hit items are in the top positions. Similar to the mainstream model, we show the results The larger value of the utility of the recommendation denotes better recommendation results.

Recommendation Utility Unfairness Metrics. In terms of evaluating the fairness of the recommendation results, we divided all users into two categories, the advantaged group and the disadvantaged group, according to their sensitive attributes (e.g. gender and age). For the gender attribute, we follow the previous papers [4] that consider the male group to be the advantaged group and the female group to be the disadvantaged group. For the age attribute, we consider those over the 60 s to be the elderly [22], i.e. the disadvantaged group, according to the definition of the World Health Organization, and the rest to be the advantaged group. We measure the utility of recommendations on the advantaged and disadvantaged groups separately and the Recommendation Utility Unfairness (RUU) is calculated as Eq.(9). The smaller value of RUU denotes fairer recommendation results.

Baseline Algorithms. We have implemented our ReRec model in PyTorch. To demonstrate the effectiveness of our proposed ReRec, we evaluate the performance of the ReRec method in terms of both recommendation utility and fairness. First, we compare ReRec with recommendation methods without fairness-aware, including the traditional shallow model PMF [23] and the deep model GCN [24]. In addition, we compare several other fairness-aware recommendation methods, such as: DeBayes [4], FairGo [5], and Adv-MultVAE [7].

Parameter Settings. All models in experiments were optimized by the Adam optimizer [25], with leaklyReLU used as the activation function between the layers. For all relevant models, the embedding size for users and items is fixed to 64, the initial learning rate is set to 0.001 to ensure fairness. We set $K = 10$ for both recommendation utility and unfairness metrics by default. For the baseline methods, we use the default hyper-parameters except for dimensions. The best models are selected based on the performance on the validation set within 300 epochs. In the re-ranking component, we apply Bayesian Personalized Ranking (BPR) loss for all the baseline models.

4.2 Performance Comparison (RQ1)

Tables 2 and 3 report the results of the overall performance comparison. We obtain the following observations:

1) Our proposed ReRec achieves the best performance on all metrics for all datasets. In terms of accuracy, ReRec obtains higher F_1 and NDCG values than other baseline methods, indicating that ReRec can capture users' interest preferences more accurately. In terms of fairness, ReRec achieves the fairest results by minimizing the gap in recommendation utility between the advantage and disadvantage groups for both sensitive attributes, gender and age.

2) In recommendation methods without considering fairness, we compare the results of the shallow model PMF and the deep learning model GCN. GCN clearly has better recommendation utility, and fairer recommendation behaviour . Disadvantaged groups tend to be less active, resulting in greater data sparsity,

Table 2. Performance on MovieLens-1M.

		PMF		GCN		DeBayes		FairGo		Adv-MultVAE		ReRec	
		Utility	Unfairness	Utility	Unfairness	Utility	Unfairness	Utility	Unfairness	Utility	Unfairness	Utility	Unfairness
Gender	F_1	7.36	3.60	14.44	2.07	15.33	0.79	15.39	0.78	15.50	0.75	**15.85**	**0.00**
	NDCG	31.20	2.84	69.45	2.08	68.09	1.18	75.10	1.11	76.28	1.10	**79.46**	**0.76**
Age	F_1	7.36	8.83	14.44	0.24	15.33	0.26	15.39	0.46	15.50	0.48	**15.85**	**0.76**
	NDCG	31.20	5.58	69.45	2.54	68.09	1.09	75.10	0.69	76.28	0.73	**79.46**	**0.45**

Our best results are highlighted in bold. The results are reported in percentage (%). Higher Utility values show better recommendation quality, while lower Unfairness values indicate better fairness.

Table 3. Performance on Lastfm-360K

		PMF		GCN		DeBayes		FairGo		Adv-MultVAE		ReRec	
		Utility	Unfairness	Utility	Unfairness	Utility	Unfairness	Utility	Unfairness	Utility	Unfairness	Utility	Unfairness
Gender	F_1	6.15	11.95	12.63	7.25	13.51	2.08	13.61	1.07	13.60	1.14	**14.66**	**0.00**
	NDCG	20.91	9.38	47.83	4.41	45.27	1.94	51.52	1.92	52.11	1.93	**53.56**	**1.23**
Age	F_1	6.15	25.61	12.63	6.30	13.51	4.73	13.65	4.25	13.60	4.52	**14.72**	**0.04**
	NDCG	20.91	35.11	47.83	16.05	45.27	10.42	51.54	10.53	52.11	9.31	**53.62**	**6.11**

so it is difficult to recommend accurately. In contrast, GCN directly models the graph structure for embedding learning, alleviating the sparsity issue and leading to more stable recommendation results.

3) Among three fairness-aware recommendation methods, Adv-MultVAE slightly outperforms the other two methods in terms of accuracy, probably because it is an adaptation of the MultVAE architecture. Experiments show that MultVAE performs consistently excellent, outperforming baseline methods such as basic GCN. Adv-MultVAE is a state-of-the-art recommendation model known for addressing the challenges of personalized and diverse recommendations. In terms of fairness, Adv-MultVAE and FairGo have similar performance and both outperform the DeBayes method, possibly because they both incorporate the core component of adversarial training, playing mini-max games between the generator and discriminator to achieve fairness, which demonstrates the effectiveness of adversarial training.

4.3   Ablation Analysis (RQ2)

To assess and verify the effectiveness of the components (i.e. multi-graph embedding layer, attribute hiding layer and fair ranking layer) of our proposed ReRec model, we derive three different models (namely ReRec-M, ReRec-A, ReRec-R) and conduct an ablation analysis on MovieLens-1M and Lastfm-360K.

Tables 4 and 5 illustrates the performance of models with different component combinations. The output embedding size is 64 for all ablation experiments.

In ReRec-M, the topological relationships between user-user and item-item are learned implicitly through the same message-passing layer as the user-item relationship. ReRec-M can only explicitly encode the interaction information in the user-item bipartite graph and fails to capture the relative importance

Table 4. Ablation Analys on MovieLens-1M

		ReRec-M		ReRec-A		ReRec-R		ReRec	
		Utility	Unfairness	Utility	Unfairness	Utility	Unfairness	Utility	Unfairness
Gender	F_1	15.39	0.49	15.74	0.13	15.80	0.76	**15.85**	**0.00**
	NDCG	72.27	1.05	74.14	1.12	79.10	1.17	**79.46**	**0.76**
Age	F_1	15.39	0.46	15.74	0.57	15.80	0.98	**15.85**	**0.06**
	NDCG	72.27	0.72	74.14	1.71	79.10	2.63	**79.46**	**0.45**

Table 5. Ablation Analys on Lastfm-360K

		ReRec-M		ReRec-A		ReRec-R		ReRec	
		Utility	Unfairness	Utility	Unfairness	Utility	Unfairness	Utility	Unfairness
Gender	F_1	12.63	1.15	13.62	1.61	13.65	3.04	**14.66**	**0.00**
	NDCG	47.83	2.07	50.80	2.79	53.54	3.95	**53.56**	**1.23**
Age	F_1	12.63	3.13	13.65	1.27	13.64	4.25	**14.72**	**0.04**
	NDCG	47.83	8.73	50.82	9.44	53.54	10.14	**53.62**	**6.11**

of the user-user and item-item relationships, resulting in poor recommendation performance.

In contrast, ReRec-A suffers a smaller loss in recommendation accuracy but a larger loss in recommendation fairness. ReRec-A follows the common multigraph fusion recommendation methods without fairness-awareness. Instead of projecting users and items into a new filter space, it pays no attention to masking the sensitive information of users. Here, users' sensitive attributes can be inferred rather accurately from users' embeddings, so users' sensitive information can be leaked from graph representation learning, which affects the utility of recommendations provided by the system to different groups.

ReRec-R uses the inner product of user embeddings and item embeddings as user preference scores for items after passing through the attribute hiding layer and directly ranks the top K items selected for recommendation. This variant has no redistribution of resources, resulting in huge disparities in the utility of recommendations received by different user groups, and is therefore the least fair one among the three variants.

It is worth noting that ReRec-R removes the fair ranking layer. However, its fairness score is still higher than baseline systems. One possible reason for this is that multi-graph embedding explores more sensitive information about users while deeply digging into their behavioral patterns and interest preferences. The rich user attributes extracted by the multi-graph embedding mechanism will work together with the generative adversary network in the next layer to maximize the removal of users' sensitive attributes and retain other information useful for RS.

4.4   Hyper-parameter Study (RQ3)

In this subsection, hyper-parametric experiments are carried out to investigate the effect of the recommended number of items K. We take the usual values of K as 1, 3, 5, 10, 20, and 30, and the corresponding results are presented in Fig. 2 . Here we only show the results on the MovieLens-1M dataset, the results on the Lastfm-360K dataset are similar.

We make the following observations: in terms of accuracy, the value of NDCG decreases with the increase of K. The value of F_1 tends to vary insignificantly and is generally close to optimal at $K = 5$. In terms of fairness, the fairest results are universally obtained at $K = 10$.

One possible reason for this is that if K is too small it increases coincidence and makes the recommendation results less stable, while if K is too large it introduces noise and aggravate the unfairness between different groups. Therefore, the balance between accuracy and fairness can be obtained when K is close to 10. It is worth noting that when the value of K is very small (less than 5), the optimizer in the fairness ranking layer tends to take longer to find the optimal solution, so if efficiency is a consideration, the value of K should be slightly larger.

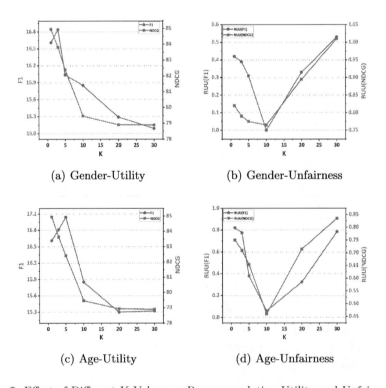

(a) Gender-Utility   (b) Gender-Unfairness

(c) Age-Utility   (d) Age-Unfairness

Fig. 2. Effect of Different K Values on Recommendation Utility and Unfairness.

5   Conclusion and Future Work

This work addresses the challenge of mitigating societal bias in the context of multi-graph based recommendation. To this end, we propose ReRec, a fair re-ranking recommendation method based on debiased multi-graph representations. Specifically, the method can not only deeply protect sensitive user attributes on multiple graphs, but also learn the debiased embedding representations of users and items, and finally adopt a fair re-ranking strategy on the recall list to alleviate the recommendation utility unfairness between different groups and improve the overall user satisfaction. We conduct experiments on two datasets (MovieLens-1M and Lastfm-360K) and evaluate the recommendation utility unfairness on user groups relevant to sensitive attributes (e.g., gender and age), validating the utility and fairness of our proposed model.

In the future, we want to explore the challenge of the explainability of fair recommendation systems. For example, using external information such as social information or knowledge graph information to explain the reasonableness of recommendation results in order to create a more user-friendly and trustworthy recommendation system.

References

1. Wang, X., He, X., Wang, M., Feng, F., Chua, T.S.: Neural graph collaborative filtering. In: Proceedings of the 42nd International ACM SIGIR Conference on Research and Development in Information Retrieval, pp. 165–174 (2019)
2. He, X., Deng, K., Wang, X., Li, Y., Zhang, Y., Wang, M.: LightGCN: simplifying and powering graph convolution network for recommendation. In: Proceedings of the 43rd International ACM SIGIR Conference on Research and Development in Information Retrieval, pp. 639–648 (2020)
3. Chen, C., et al.: Graph heterogeneous multi-relational recommendation. In: Proceedings of the AAAI Conference on Artificial Intelligence, vol. 35, no. 5, pp. 3958–3966 (2021)
4. Buyl, M., De Bie, T.: DeBayes: a Bayesian method for debiasing network embeddings. In: International Conference on Machine Learning, pp. 1220–1229. PMLR (2020)
5. Wu, L., Chen, L., Shao, P., Hong, R., Wang, X., Wang, M.: Learning fair representations for recommendation: a graph-based perspective. In: Proceedings of the Web Conference, vol. 2021, pp. 2198–2208 (2021)
6. Anagnostopoulos, A., Kumar, R., Mahdian, M.: Influence and correlation in social networks. In: Proceedings of the 14th ACM SIGKDD International Conference on Knowledge Discovery and Data Mining, pp. 7–15 (2008)
7. Ganhör, C., Penz, D., Rekabsaz, N., Lesota, O., Schedl, M.: Unlearning protected user attributes in recommendations with adversarial training. In: Proceedings of the 45th International ACM SIGIR Conference on Research and Development in Information Retrieval, pp. 2142–2147 (2022)
8. Gini, C.: Measurement of inequality of incomes. Econ. J. **31**(121), 124–125 (1921)
9. Deshpande, M., Karypis, G.: Item-based top-n recommendation algorithms. ACM Trans. Inf. Syst. (TOIS) **22**(1), 143–177 (2004)

10. Koren, Y.: Factorization meets the neighborhood: a multifaceted collaborative filtering model. In: Proceedings of the 14th ACM SIGKDD International Conference on Knowledge Discovery and Data Mining, pp. 426–434 (2008)

11. Wu, S., Sun, F., Zhang, W., Xie, X., Cui, B.: Graph neural networks in recommender systems: a survey. ACM Comput. Surv. **55**(5), 1–37 (2022)

12. Sun, J., et al.: Multi-graph convolution collaborative filtering. In: IEEE International Conference on Data Mining (ICDM), vol. 2019, pp. 1306–1311. IEEE (2019)

13. Mehrabi, N., Morstatter, F., Saxena, N., Lerman, K., Galstyan, A.: A survey on bias and fairness in machine learning. ACM Comput. Surv. (CSUR) **54**(6), 1–35 (2021)

14. Kim, J.S., Chen, J., Talwalkar, A.: FACT: a diagnostic for group fairness trade-offs. In: International Conference on Machine Learning, pp. 5264–5274. PMLR (2020)

15. Zafar, M.B., Valera, I., Gomez Rodriguez, M., Gummadi, K.P.: Fairness beyond disparate treatment and disparate impact: learning classification without disparate mistreatment. In: Proceedings of the 26th International Conference on World Wide Web, pp. 1171–1180 (2017)

16. Dong, Y., Kang, J., Tong, H., Li, J.: Individual fairness for graph neural networks: a ranking based approach. In: Proceedings of the 27th ACM SIGKDD Conference on Knowledge Discovery and Data Mining, pp. 300–310 (2021)

17. Dong, Y., Kang, J., Tong, H., Li, J.: Individual fairness for graph neural networks: a ranking based approach. In: Proceedings of the 27th ACM SIGKDD Conference on Knowledge Discovery and Data Mining, pp. 300–310 (2021)

18. Rahman, T.A., Surma, B., Backes, M., Zhang, Y.: Fairwalk: towards fair graph embedding. In: International Joint Conference on Artificial Intelligence (2019)

19. Yao, S., Huang, B.: Beyond parity: fairness objectives for collaborative filtering. In: Advances in Neural Information Processing Systems 30 (2017)

20. Kusner,M. J., Loftus, J., Russell, C., Silva, R.: Counterfactual fairness. In: Advances in Neural Information Processing Systems 30 (2017)

21. Wu, Y., Zhang, L. , Wu, X.: On discrimination discovery and removal in ranked data using causal graph. In: Proceedings of the 24th ACM SIGKDD International Conference on Knowledge Discovery and Data Mining, pp. 2536–2544 (2018)

22. Organization, W.H., et al.: Tackling Abuse of Older People: Five Priorities for the United Nations Decade of Healthy Ageing (2021–2030). World Health Organization (2022)

23. Mnih, A., Salakhutdinov, R.R.: Probabilistic matrix factorization. In: Advances in Neural Information Processing Systems 20 (2007)

24. Kipf, T. N., Welling, M.: Semi-supervised classification with graph convolutional networks. In: International Conference on Learning Representations (2017)

25. Kingma, D.P., Ba, J.: Adam: a method for stochastic optimization. CoRR, vol. abs/1412.6980 (2014)

26. Rendle, S., Freudenthaler, C., Gantner, Z., Schmidt-Thieme, L.: BPR: Bayesian personalized ranking from implicit feedback. In: Proceedings of the Twenty-Fifth Conference on Uncertainty in Artificial Intelligence, pp. 452–461 (2009)

Information Extraction

FastNER: Speeding up Inferences for Named Entity Recognition Tasks

Yuming Zhang[1], Xiangxiang Gao[2], Wei Zhu[3], and Xiaoling Wang[3(✉)]

[1] College of Computer Science and Software Engineering, Shenzhen University,
Shenzhen, China
[2] Shanghai Jiaotong University, Shanghai, China
[3] East China Normal University, Shanghai, China
xlwang@cs.ecnu.edu.cn

Abstract. BERT and its variants are the most performing models for named entity recognition (NER), a fundamental information extraction task. We must apply inference speedup methods for BERT-based NER models to be deployed in the industrial setting. Early exiting allows the model to use only the shallow layers to process easy samples, thus reducing the average latency. In this work, we introduce FastNER, a novel framework for early exiting with a BERT biaffine NER model, which supports both flat NER tasks and nested NER tasks. First, we introduce a convolutional bypass module to provide suitable features for the current layer's biaffine prediction head. This way, an intermediate layer can focus more on delivering high-quality semantic representations for the next layer. Second, we introduce a series of early exiting mechanisms for BERT biaffine model, which is the first in the literature. We conduct extensive experiments on 6 benchmark NER datasets, 3 of which are nested NER tasks. The experiments show that: (a) Our proposed convolutional bypass method can significantly improve the overall performances of the multi-exit BERT biaffine NER model. (b) our proposed early exiting mechanisms can effectively speed up the inference of BERT biaffine model. Comprehensive ablation studies are conducted and demonstrate the validity of our design for our FastNER framework.

Keywords: Early Exiting · Pre-trained language models · Inference speed-up

1 Introduction

Since the rise of BERT [3], the pre-trained language models (PLMs) are the state-of-the-art (SOTA) models for natural language processing (NLP) [13,31]. Many PLMs are developed by the academia and industry, such as GPT [15], XLNet [25], and ALBERT [10], and so forth. These BERT-style models achieved

Y. Zhang, X. Gao and W. Zhu—Equal contributions.

X. Yang et al. (Eds.): ADMA 2023, LNAI 14176, pp. 185–199, 2023.
https://doi.org/10.1007/978-3-031-46661-8_13

considerable improvements in many NLP tasks by self-supervised pre-training and transfer learning on labeled tasks, such as classification, text pair matching, named entity recognition (NER), etc. Despite their outstanding performances, their industrial usage is still limited by the high latency during inference.

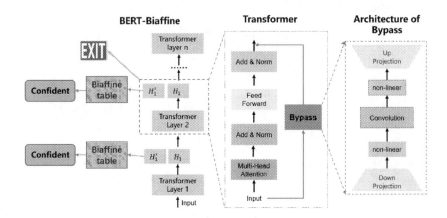

Fig. 1. The overall architecture of our FastNER framework.

NER and other sequence labeling tasks play a central role in many application scenarios, such as question answering, document search, document-level information extraction, etc. However, these applications require low latency. For example, an online search engine needs to respond to the user's query in less than 100 milo-seconds. Thus, a NER model should be accurate and efficient. In addition, at certain time intervals, consumer query traffic is very concentrated. For example, during dinner hours, food search engines will be used much often than usual. Thus, it is important for deployed NER models to flexibly adjust their latency.

Literature has focused on making PLMs' inference more efficient via adaptive inference [14,23,28,30]. The idea of adaptive inference is to process simple queries with lower layers of BERT and more difficult queries with deeper layers, thus speeding up inference on average without loss of accuracy. The speedup ratio can be flexibly controlled with certain hyper-parameters without re-deploying the model services. Early exiting is the representative adaptive inference methods [1]. As depicted in Fig. 1, it implements adaptive inference by installing an early exit, i.e., an intermediate prediction head, at each layer of PLMs (multi-exit PLMs) and early exiting "easy" samples to speed up inference. All the exits are jointly optimized at the training stage with BERT's parameters. At the inference stage, certain early exiting strategies are designed to decide which layer to exit [4,8,18,23,27,28]. In this mode, different samples can exit at different depths.

For our framework to be generally applicable, we mainly adopt the biaffine model [26] for NER. The biaffine model converts the NER task into a 2-dimensional table filling task, thus providing a solution to both the flat and nested NER problem. [26] shows that the biaffine model can achieve state-of-the-art (SOTA) performances on both nested NER tasks and flat NER tasks.

In this work, we propose a framework for the early exiting of BERT biaffine NER models, inspired by BADGE [29]. First, we add a convolutional bypass to the current transformer layer to provide different representations for the current layer's biaffine exit and the next transformer layer of the BERT backbone. In this way, the BERT backbone will not be distracted from different tasks, thus improving the cross-layer average performance of the multi-exit biaffine model. Second, we extend the commonly used early exiting mechanisms in sentence classification tasks, entropy-based early exiting and max-probability based early exiting, to the biaffine NER model, we can perform adaptive inferences for NER tasks. Intuitively, the decision of early exit is made when the intermediate biaffine exit is confident in its predictions.

Extensive experiments are conducted on the six benchmark NER tasks. Three of the tasks are nested NER tasks, ACE2004[1], ACE2005[2], GENIA [9]. We also experiment on three flat NER tasks, CONLL2003 [19], OntoNotes 4.0 Chinese[3] and the Chinese MSRA task [11]. We show that: (1) our FastNER training method consistently performs better than the baseline multi-exit model training methods. (2) We show that we can achieve 2–3x speedup with limited performance losses. In addition, we show that with the better multi-exit model trained with our FastNER, better efficiency-performance tradeoffs can be made. Ablation studies validate the architectural design of our FastNER methods.

The rest of the paper is organized as follows. First, we introduce the preliminaries for the Biaffine NER model and early exiting. Second, we elaborate on our FastNER method. Third, we conduct experiments on 6 NER tasks and conduct a series of ablations studies. Finally, we conclude with possible future works.

2 Preliminaries

This section introduces the background for PLMs and early exiting. Throughout this work, we consider a NER task with samples $\{(x, y), x \in \mathcal{X}, y \in \mathcal{Y}, i = 1, 2, ..., N\}$, e.g., sentences and their NER span information, and the number of entity categories is K (including the non-entity type label). The input sequence length after BERT's subword tokenization is L.

2.1 PLM Models

We use BERT as the backbone model. BERT is a Transformer [20] model pre-trained in a self-supervised manner on a large corpus. In the ablation studies, we also use ALBERT [10] as backbones. ALBERT is more lightweight than BERT since it shares parameters across different layers, and the embedding matrix is factorized. The number of layers of our PLM backbone is denoted as M, and the hidden dimension is d.

---

[1] https://catalog.ldc.upenn.edu/LDC2005T09.

[2] https://catalog.ldc.upenn.edu/LDC2006T06.

[3] https://catalog.ldc.upenn.edu/LDC2011T03.

2.2   The Biaffine Model for NER

The BERT-Biaffine model [26] transforms the NER task into a two-dimensional table filling task. It asks the model to identify whether the slot in the table with coordinate (s, e) corresponds to an entity with category k, that is, whether a pair of tokens (x_s, x_e) in the input sequence $x = (x_1, x_2, , , , , x_L)$ is the start and end tokens for an entity with category k. Formally, after BERT encoding, the contextualized embedding of tokens s and e are h_s and h_e ($h_s, h_e \in \mathcal{R}^d$). Then h_s and h_e will go through two multi-layer perceptrons with Tanh activation function (denoted as MLP-start and MLP-end),

$$h_s = \text{Tanh}(h_s W_1^{start}) W_2^{start}, \tag{1}$$

$$h_e = \text{Tanh}(h_e W_1^{end}) W_2^{end}. \tag{2}$$

MLP-start and MLP-end transform the BERT's representations to adapt to the table-filling NER task. Then in a biaffine layer f, the score of span (s, e) is calculated by

$$f(s, e) = h_s^T U h_e + W(h_s \oplus h_e) + b. \tag{3}$$

Since we need to calculate the scores for K entity categories, U is a $d \times K \times d$ tensor, and W is a $2d \times K$ tensor. $f(s, e) \in \mathcal{R}^K$ is the scores (or logits). A softmax operation will transform $f(s, e)$ into a probability distribution $p(s, e)$, which represents how likely the span (s, e) is a category k entity.

The learning objective of the biaffine model is to assign a correct category (including the non-entity) to each valid span. Hence it is a multi-class classification problem at each slot of the two-dimensional table and can be optimized with cross-entropy loss:

$$\mathcal{L} = -\sum_{s=1}^{L}\sum_{e=s}^{L}\sum_{k=1}^{K} \mathcal{I}(y(s, e) = k) \log p_k(s, e), \tag{4}$$

where $y(s, e)$ is the ground-truth label of span (s, e), $p_k(s, e)$ is the predicted probability mass of (s, e) having label k, and $\mathcal{I}(\cdot)$ is the indicator function. After fine-tuning the BERT biaffine model, the inference procedure of the BERT biaffine model follows [26], which involves determining the final named entity spans. Since there may be conflicting spans, [26] rank the spans via their probability masses, and the span with a higher probability mass will be kept when it conflicts with other predicted spans.

2.3   Early Exiting

As depicted in Fig. 1, early exiting architectures, or multi-exit architectures, are networks with exits[4] at each transformer layer. Since the previous literature usually considers sentence-level classification tasks, the exits are sentence classifiers. However, since we are dealing with sequence labeling tasks formulated as two-dimensional table filling, with M exits, M separate biaffine modules $f^{(m)}$ are

---

[4] Some literature (e.g., DeeBERT [23]) also refers to exits as off-ramps.

installed right after each layer of BERT ($m = 1, 2, ..., M$), and the scores for span (s, e) at layer m is given by:

$$f^{(m)}(s, e) = h_s^T U^{(m)} h_e + W^{(m)}(h_s \oplus h_e) + b^{(m)}.$$ (5)

And the loss function at each layer becomes

$$\mathcal{L}^{(m)} = -\sum_{s=1}^{L}\sum_{e=s}^{L}\sum_{k=1}^{K} \mathcal{I}(y(s, e) = k) \log p_k^{(m)}(s, e),$$ (6)

where $p^{(m)}(s, e) = \text{Softmax}(f^{(m)}(s, e))$ is the predicted probability distribution at exit m.

Training. The most commonly used training method for early exiting architectures is joint training (JT). All the exits are jointly optimized at the training stage with a summed loss function. Following [6] and [28], the overall loss function is:

$$\mathcal{L}^{WA} = \frac{\sum_{m=1}^{M} m * \mathcal{L}^{(m)}}{\sum_{m=1}^{M} m}.$$ (7)

Note that the weight m corresponds to the relative inference cost of exit m.

Two other commonly used training methods are two-stage training [14,23] (2ST) and alternating training [24] (ALT). 2ST first fine-tunes the PLM backbone and the last exit till convergence in the first stage and then fine-tunes the intermediate exits in the second stage. ALT trains the backbone and the last exit at the even optimization steps, and the intermediate exits at the odd optimization steps.

Inference. At inference, the multi-exit PLMs can operate in two different modes, depending on whether or not the computational budget to classify an example is known.

Static Early Exiting. We can directly appoint a fixed exit m^* of PLM, $f^{(m^*)}$, to predict all the queries.

Dynamic Early Exiting. Under this mode, upon receiving a query input x, the model starts to predict on the exits $f^{(1)}$, $f^{(2)}$, ..., in turn. It will continue until it receives a signal to stop early at an exit $m^* < M$, or arrives at the last exit M. At this point, it will output a predictions by combining the current and previous predictions in a certain way. Different samples might exit at different layers under this early exit setting.

Speedup Ratio. Following PABEE [28], we mainly report the speedup ratio as the efficiency metric. For each test sample x_i, the inference time cost is t_i under early exiting, and is T_i without early exiting, then the average speedup ratio on the test set is calculated by

$$\text{Speedup} = 1 - \frac{\sum_{i=1}^{N} t_i}{\sum_{i=1}^{N} T_i}.$$ (8)

3   FastNER

3.1   A Lite Biaffine Module

Note that the original BERT biaffine NER model does not consider the early exiting scenarios. Each biaffine module (Eq. 5) introduces 5–7 million parameters. If we add this biaffine module at each layer, the added parameters will amount to 60 million or above. Introducing too many randomly initialized parameters would result in low efficiency, difficulty in optimization, and overfitting for shallow layers. Thus, we propose a modified version of biaffine module called the lite biaffine module.

In the lite biaffine module, MLP-start and MLP-end are substituted by a simple linear projection layer that project h_s and h_e from dimension d to $d_1 = d/4$,[5] and the down-projected $h-s$ and h_e are fed into Eq. 5 for logit calculations. This way, the parameters in a biaffine layer will be reduced to less than 0.5 million. In the experiments, we will show that our lite biaffine module performs better than the original one in the early exiting scenarios.

3.2   Motivation

Similar to the analysis in [29], training a multi-exit BERT biaffine model requires training multiple prediction heads of different depths simultaneously. Thus, under this setting, an intermediate layer has to fulfill two tasks at once: (a) providing semantic representations to the next layer and (b) providing proper token features to the biaffine module of the current layer. One may wonder whether these two tasks conflict with each other and result in poor optimizations. [12] investigate this problem in the sentence classification tasks and find that each layer's optimizations are often in conflict and can cause gradient instability. They provide a solution called gradient equilibrium, which is to adjust the gradients from each exit. However, in our experiments, we will show that this method does not provide significant improvements.

Another solution is to use different sub-networks for these two tasks, following the literature on sparse multi-task learning [17]. However, this method provides two different representations with two different, forward passes with different sub-networks, significantly slowing down the inference speed. Thus, this approach does not meet our purpose.

To summarize, we need a new method that can provide two different representations, one for the next layer and the other for the current layer's prediction, within a single forward pass.

3.3   Bypass Architecture

We now present the core of our FastNER framework: the convolutional bypasses (depicted in Fig. 1). The notation follows [29]. Denote the hidden states of the

---

[5] The reason why we set $d_1 = d/4$ is that smaller d_1 would result in significant performance drops, according to our initial experiments.

input from the last layer as H_{m-1}. H_{m-1} will go through the transformer layer's self-attention (MHSA) and positional feed-forward (FFN) modules, to become $H_{m,0}$, and then a LayerNorm operation to output H_m.

We want the efficient bypass B_m to adjust H_m to fit the task better. B_m is simple in architecture (On the right side of Fig. 1). After receiving the input H_{m-1}, B_m reduce its dimension from d to r, and obtain $H_{m,B}^{(1)}$ (where $r << d$, and r is called the bottleneck dimension) via a down-projection $W_{down} \in \mathbb{R}^{d \times r}$. $H_{m,B}^{(1)}$ will go through a non-linear activation function g_1, a convolutional layer with kernel size 3 (denoted as $conv$) and another activation function g_2, and become $H_{m,B}^{(2)}$. $H_{m,B}^{(2)}$ will then be up-projected to $H_{m,B}^{(3)}$ to recover the dimension, by an up-projection matrix $W_{up} \in \mathbb{R}^{r \times d}$. The literature usually refer to r as the bottleneck dimension. Formally, B_m can be expressed as:

$$H_{m,B}^{(3)} \leftarrow g_2(\text{conv}(g_1(H_{m-1}W_{down})))W_{up}. \tag{9}$$

Finally, the current layer will output two representations, H_m, which is the original hidden states, and H_m', which is modified by the bypass by:

$$H_m' = \text{LayerNorm}(H_{m,B}^{(3)} + H_{m,0}),$$
$$H_m = \text{LayerNorm}(H_{m,0}). \tag{10}$$

H_m is passed to the next transformer layers as input, and H_m^{by} will be the hidden states received by the intermediate biaffine exit. The bottleneck dimension r is very small, like 16, so that the extra parameters or flops introduced by the bypasses are less than 1% of the compared to the BERT backbone.

3.4 Early Exiting for Biaffine NER Model

Although the literature comprehensively studied the early exiting for sentence classification tasks, the early exiting mechanism of entity-level tasks like NER has been neglected. Based on the literature on early exiting on sentence classification tasks, this work proposes two plausible early exiting mechanisms for the biaffine NER model.

Entropy-Based Early Exiting (Entropy). This method is a directly extension of [18] and [23] from the sentence classification tasks to NER. We denote the table of distributions predicted by the biaffine exit m as $\mathcal{T}^{(m)} = \{p^{(m)}(s,e)|s,e \in 1,...,L\}$, which is a $L \times L \times K$ tensor. On each slot $p^{(m)}(s,e)$ of the biaffine table, we can calculate its entropy $\text{Ent}^{(m)}(s,e)$ via

$$\text{Ent}^{(m)}(s,e) = \frac{-1}{\log(K)} \sum_{k=1}^{K} p_k^{(m)}(s,e) \log(p_k^{(m)}(s,e)). \tag{11}$$

Intuitively, if the biaffine exit is confident with its prediction, the average entropy $\text{AvgEnt}^{(m)}$, calculated by

$$\text{AvgEnt}^{(m)} = \frac{\sum_{s=1}^{L} \sum_{e=1}^{L} \text{Ent}^{(m)}(s,e)}{L * L}, \tag{12}$$

will be smaller. Thus AvgEnt$^{(m)}$ can be treated as the early exiting criterion. A threshold τ_e is predefined. If at layer m, AvgEnt$^{(m)}$ is smaller than τ_e, the model will exit. Otherwise, the model will continue its forward pass.

Maximum-Probability-Based Early Exiting (Maxprob). This method is a direct extension of [16] from the sentence classification tasks to NER. Intuitively, if the biaffine exit is confident with its prediction, the table of predicted distributions $\mathcal{T}^{(m)}$ will concentrate their probability masses on single specific labels. Denote the maximum probability mass at slot (s, e) as $MP^{(m)}(s, e)$, then the average maximum probability is given by

$$\text{AvgMP}^{(m)} = \frac{\sum_{s=1}^{L} \sum_{e=1}^{L} MP^{(m)}(s, e)}{L * L}. \tag{13}$$

A threshold τ_{mp} is predefined. If at layer m, AvgMP$^{(m)}$ is larger than τ_{mp}, the model will exit. Otherwise, the model will continue its forward pass.

4   Experiments

4.1   Datasets

We evaluate our FastNER on both nested and flat NER tasks. For the nested NER task, we use the ACE2004 task[6], ACE2005 task[7], and GENIA task [9]. For the flat NER task, we evaluate our method on the CONLL 2003 task [19] (CONLL03), the OntoNotes 4.0 corpus[8] (Onto-4), and the Chinese MSRA NER (MSRA) task [11].

4.2   Baselines

For multi-exiting BERT biaffine fine-tuning, we compare our FastNER framework with the following baselines:

Two-Stage Training (2ST). This method is adapted by [23] and [14]. It first fine-tunes the backbone and the last exit till convergence. Then all the intermediate layers' exits (except the last layer's exit) will be trained on top of the frozen backbone.

Joint Training (JT). This method trains the BERT backbone and all the biaffine exits jointly. Literature has different variations for JT. PABEE [28] and RightTool [16] adopts increasing loss weights for higher layers. We will denote their version of joint training as JT-PABEE. BranchyNet [18] adopts different and gradually increasing learning rates for different exits during training. We will denote their version as JT-BranchyNet.

---

[6] https://catalog.ldc.upenn.edu/LDC2005T09.

[7] https://catalog.ldc.upenn.edu/LDC2006T06.

[8] https://catalog.ldc.upenn.edu/LDC2011T03.

Alternating Training. BERxit proposes to combine the training of JT-PABEE and 2ST, that is, conduct back-propagation via the loss from the last exit at the odd optimization steps and conduct back-propagation via the average loss from all the intermediate exits.

Gradient Equilibrium (GradEquil). [12] proposes to adjust the gradient norm of each intermediate layer during optimization, so that the training process will be more stable.

Sparse Multi-task (Sparse-MT). [17] As analyzed in Sect. 3.2, this method is not suitable for model inference speedup methods since it requires multiple forward passes to generate different representations suitable for different tasks. We include this method as a sanity check and show that our FastNER method performs better even with a single forward pass.

Early Exiting Mechanisms. To show that our FastNER method can effectively improve the model's early exiting performances, we will run dynamic early exiting with different early exiting mechanisms as described in Sect. 3.4: (a) Entropy-based method (entropy); (b) maximum probability-based method (maxprob). Early exiting will be run on different backbones to show that our FastNER framework can improve the efficiency-performance tradeoffs.

4.3 Experimental Settings

Training. English NER tasks use the open-sourced Google BERT [3][9] as the backbone, and the Chinese tasks adopt the BERT-www-ext released from [2][10] as the backbone model. We also use ALBERT-base and ALBERT base Chinese by [10] as the backbone models for ablation studies. We add a lite biaffine NER layer or an original biaffine layer after each intermediate layer of the pre-trained models as the intermediate classifiers. The convolutional bypasses' activation function is set to be GELU [5]. We fine-tune models for at most 25 epochs; early stopping with patience eight is performed, and the best checkpoint is selected based on the dev set performances. We perform grid search over batch sizes of 16, 32, 128, learning rates of 1e−5, 2e−5, 3e−5, 5e−5 with an Adam optimizer, and the convolutional bypasses' bottleneck dimension of 8, 16, 32. We implement FastNER and all the baselines on the base of Hugging Face's Transformers [22]. Experiments are conducted on four Nvidia V100 16 GB GPUs.

Inference. Following prior work on input-adaptive inference [8,18], inference is on a per-instance basis, i.e., the batch size for inference is set to 1. This is a common scenario in the industry where individual requests from different users [16] come at different time points. We report the median performance over five runs with different random seeds.

---

[9] https://huggingface.co/bert-base-uncased.

[10] https://github.com/ymcui/Chinese-BERT-wwm.

4.4 Evaluation Metrics

Entity-level F1 is the most widely used metric for NER tasks [7]. For multi-exit PLMs, each exit has a performance score. Thus, to properly evaluate multi-exit PLMs, we propose the following three derived metrics: (a) F1-avg, which denotes the cross-layer average F1 score; (c) F1-best, which is the best F1 score among all the layers. We use F1-avg as our primary metric for experimental result reporting and checkpoint selection during training.

Table 1. Experimental results of models with BERT backbone on the six benchmark NER datasets.

	ACE2004		ACE2005		GENIA		CONLL2003		MSRA		OntoNotes4.0	
BERT biaffine model												
	F1		F1		F1		F1		F1		F1	
BERT + original Biaffine	83.5		83.9		77.4		90.7		95.3		81.7	
BERT + Lite Biaffine (ours)	83.6		83.8		77.6		91.0		95.2		82.2	
Multi-exit BERT biaffine model												
	F1-avg	F1-best	F1-avg	F1-best	F1-avg	F1-best	F1-avg	F1-best	F1-avg	F1-best	F1-avg	F1-best
2ST	69.6	83.6	73.0	83.8	70.2	77.9	82.8	91.1	88.1	94.3	71.3	75.9
JT-PABEE	77.5	82.5	78.7	82.8	70.3	77.8	86.6	90.3	89.5	94.5	69.3	79.8
JT-BranchyNet	77.6	82.8	78.6	82.9	70.5	77.4	86.9	90.5	89.1	94.0	70.8	80.6
GradEquil	77.8	83.2	78.4	83.8	70.7	77.9	87.1	90.7	89.5	94.3	71.2	81.1
ALT	78.1	83.3	78.4	83.3	70.5	77.8	86.8	90.6	88.7	93.7	71.9	81.8
Sparse-MT	77.9	83.0	78.1	82.9	70.8	78.2	86.3	90.2	88.4	93.6	71.6	80.5
FastNER + original biaffine	78.5	84.1	79.4	83.9	71.1	78.1	87.7	91.3	90.5	94.7	73.2	82.1
FastNER	**78.7**	84.0	**79.7**	84.1	**71.5**	78.8	**87.8**	91.5	**91.1**	95.1	**73.8**	82.3

4.5 Overall Comparison

We compare our FastNER with the previous SOTA training methods of multi-exit BERT-biaffine models. Table 1 reports the performance on the six benchmark NER datasets when using BERT as the backbone model. The upper half of the table reports the performances of the last transformer layer's biaffine exit or the 6-th layer's exit. With fewer randomly initialized parameters, our lite biaffine layer can outperform comparably with the original biaffine layer.

The cross-layer average and best performances are reported in the lower half of Table 1. The following takeaways can be made:

– Our FastNER method consistently outperforms the existing multi-exit BERT biaffine model training methods in terms of F1-avg by a clear margin. Note that as modifications to the joint training methods, GradEquil and Sparse-MT perform comparably to JT-BranchyNet and JT-PABEE under our settings. Although ALT [24] and 2ST [23] perform well in sentence-level tasks like the GLUE benchmarks [21], it does not perform very well when training the BERT biaffine NER models.
– With the help of the bypasses, our FastNER method improves the average performances across all the intermediate layers and improves the F1-best scores compared with JT-PABEE or JT-BranchyNet by a large margin. This result is consistent with our motivation: introducing the convolutional bypasses can

help the intermediate transformer layer to concentrate on providing hidden representations. In contrast, the bypasses can provide a modified version of the current layer's hidden states that are more suitable for the current layer's biaffine exit. In this way, both the F1-best and F1-avg scores can improve. As a direct result, the best layer's score under our FastNER framework is comparable to or performs better than vanilla fine-tuning.

- To show that our FastNER does not achieve such performance improvements merely by adding more parameters, we also run FastNER with the original biaffine module (the FastNER + original biaffine setting in Table 1). With much more additional parameters, FastNER + original biaffine still underperforms the FastNER setting in terms of F1-avg. We think this is because the original biaffine modules are too parameter cumbersome for the shallow layers to learn.

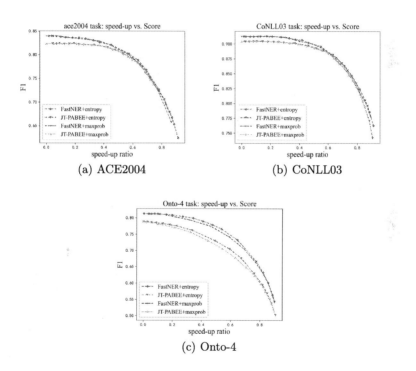

Fig. 2. The speedup-score curves with different dynamic early exiting methods, on the ACE, CoNLL03 and Onto-4 datasets. The multi-exit BERT biaffine models are trained with FastNER or JT-PABEE.

4.6   Dynamic Early Exiting Performances

With the improved overall performances on each layer, intuitively, the model's early exiting performances can be improved. We run early exiting with different

Table 2. Experimental results for the ablation study of whether to pass H_i' to the next layer. The F1-avg (cross-layer average F1) scores are reported. The performance differences between FastNER and FastNER-reduced are significant for both tasks.

	ACE2004		Onto-4	
	F1-avg	F1-best	F1-avg	F1-best
FastNER	78.7	84.0	73.8	82.3
FastNER-reduced	76.6	80.7	71.3	75.9

confidence thresholds on multi-exit BERT-biaffine models trained by FastNER or JT-PABEE. The early exiting mechanisms are the entropy-based method and the maxprob-based method. The resulting speedup-performance curves are plotted in Fig. 2, where the x-axis represents the speedup ratio, and the y-axis is the F1 score achieved on the test set.

From Fig. 2, we can see that with our FastNER training and early exiting methods, a BERT biaffine model can achieve 2x–3x speedups with limited performance degradations. The apparent gaps between our FastNER model and JT-PABEE shown in Fig. 2 can also be observed in Fig. 2, proving that improving the averaging performances across layers in better efficiency-performance tradeoffs during early exiting. In addition, the entropy-based and maxprob-based methods can perform comparably with each other for the early exiting of BERT biaffine models.

4.7   Ablation on Whether to Pass H_i' to the Next Layer

The core idea of FastNER is to provide different intermediate hidden states for different purposes via our convolutional bypasses. As a sanity check and to demonstrate that our design is necessary, we now consider the following setting: reducing our design of bypasses by passing H_i' to the next layer and using it for prediction. We will denote this setting as FastNER-reduced. We use FastNER-reduced for training on ACE2004 and Onto-4 datasets, and the results are reported in Table 2.

From Table 2, we can see that FastNER-reduced asks H_i' to complete two tasks at once, resulting in a significant drop in overall performances. Note that the performance difference between FastNER and FastNER-reduced is significant on ACE2004 and Onto-4. The results show that our design of providing different representations for different purposes via convolutional passes is the key to improving the overall performances of multi-exit BERT biaffine models.

4.8   Comparisons of Different Bottleneck Dimensions r

The bottleneck dimension r is 16 for the main experiments. To investigate the effects of bottleneck dimensions on the model performances, We conducted ablation experiments on the ACE2004 and Onto-4 tasks. Table 3 reports the F1-avg

Table 3. Experimental results for comparing different bottleneck dimensions. The F1-avg (cross-layer average F1) scores are reported.

	ACE2004		Onto-4	
	F1-avg	F1-best	F1-avg	F1-best
r = 64	78.5	83.3	73.7	80.8
r = 32	78.7	83.9	**73.8**	82.3
r = 16	**78.8**	84.0	73.6	82.4
r = 8	78.7	83.8	73.3	81.9
r = 4	78.3	83.5	72.7	81.8

scores under different values of r. From Table 3, we can see that smaller bottleneck dimensions do not result in significant performance drops. An intriguing observation is that larger bottleneck dimensions do not provide performance improvements, demonstrating that the superior performances of FastNER do not come from introducing additional model parameters.

5 Conclusions

In this work, we first design the FastNER-bypass framework, consisting of convolutional bypasses, to enhance the overall performances of multi-exit BERT biaffine NER models. Second, the existing literature does not investigate the problem of early exiting for the NER tasks. Thus, we transfer the early exiting methods for sentence-level tasks to the biaffine NER model and propose two early exiting mechanisms: the entropy-based method and the maxprob-based method. Experiments are conducted on six benchmark NER datasets. The experimental results show that: (a) our FastNER framework can effectively improve the overall performances of multi-exit BERT biaffine models, thus providing stronger backbones for dynamic early exiting; (b) the early exiting mechanisms we designed for the BERT biaffine NER models can achieve 2–3 times inference speedups with quite limited performances drops.

Acknowledgement. This work was supported by NSFC grants (No. 61972155 and 62136002), National Key R&D Program of China (No. 2021YFC3340700), Shanghai Trusted Industry Internet Software Collaborative Innovation Center.

References

1. Bolukbasi, T., Wang, J., Dekel, O., Saligrama, V.: Adaptive neural networks for efficient inference. In: ICML (2017)
2. Cui, Y., Che, W., Liu, T., Qin, B., Wang, S., Hu, G.: Revisiting pre-trained models for Chinese natural language processing. arXiv:abs/2004.13922 (2020)

3. Devlin, J., Chang, M.W., Lee, K., Toutanova, K.: BERT: pre-training of deep bidirectional transformers for language understanding. arXiv preprint arXiv:1810.04805 (2018)

4. Gao, X., Zhu, W., Gao, J., Yin, C.: F-PABEE: flexible-patience-based early exiting for single-label and multi-label text classification tasks. In: ICASSP 2023–2023 IEEE International Conference on Acoustics, Speech and Signal Processing (ICASSP), pp. 1–5. IEEE (2023)

5. Hendrycks, D., Gimpel, K.: Bridging nonlinearities and stochastic regularizers with gaussian error linear units. arXiv:abs/1606.08415 (2016)

6. Huang, G., Chen, D., Li, T., Wu, F., Maaten, L.V.D., Weinberger, K.Q.: Multi-scale dense convolutional networks for efficient prediction. arXiv:abs/1703.09844 (2017)

7. Huang, Z., Xu, W., Yu, K.: Bidirectional LSTM-CRF models for sequence tagging. arXiv:abs/1508.01991 (2015)

8. Kaya, Y., Hong, S., Dumitras, T.: Shallow-deep networks: understanding and mitigating network overthinking. In: ICML (2019)

9. Kim, J.D., Ohta, T., Tateisi, Y., Tsujii, J.: GENIA corpus-a semantically annotated corpus for bio-textmining. Bioinformatics (Oxford, England) **19**(Suppl. 1), i180-2 (2003). https://doi.org/10.1093/bioinformatics/btg1023

10. Lan, Z., Chen, M., Goodman, S., Gimpel, K., Sharma, P., Soricut, R.: ALBERT: a lite BERT for self-supervised learning of language representations. arXiv:abs/1909.11942 (2020)

11. Levow, G.A.: The third international Chinese language processing bakeoff: word segmentation and named entity recognition. In: Proceedings of the Fifth SIGHAN Workshop on Chinese Language Processing, pp. 108–117. Association for Computational Linguistics, Sydney, Australia, July 2006. https://aclanthology.org/W06-0115

12. Li, H., Zhang, H., Qi, X., Yang, R., Huang, G.: Improved techniques for training adaptive deep networks. In: 2019 IEEE/CVF International Conference on Computer Vision (ICCV), pp. 1891–1900 (2019)

13. Li, X., et al.: Pingan smart health and SJTU at COIN - shared task: utilizing pre-trained language models and common-sense knowledge in machine reading tasks. In: Proceedings of the First Workshop on Commonsense Inference in Natural Language Processing, pp. 93–98. Association for Computational Linguistics, Hong Kong, China, November 2019. https://doi.org/10.18653/v1/D19-6011, https://aclanthology.org/D19-6011

14. Liu, W., Zhou, P., Zhao, Z., Wang, Z., Deng, H., Ju, Q.: FastBERT: a self-distilling BERT with adaptive inference time. arXiv:abs/2004.02178 (2020)

15. Radford, A., Wu, J., Child, R., Luan, D., Amodei, D., Sutskever, I., et al.: Language models are unsupervised multitask learners. OpenAI Blog **1**(8), 9 (2019)

16. Schwartz, R., Stanovsky, G., Swayamdipta, S., Dodge, J., Smith, N.A.: The right tool for the job: matching model and instance complexities. In: Proceedings of the 58th Annual Meeting of the Association for Computational Linguistics, pp. 6640–6651. Association for Computational Linguistics, Online, July 2020. https://doi.org/10.18653/v1/2020.acl-main.593, https://aclanthology.org/2020.acl-main.593

17. Sun, T., et al.: Learning sparse sharing architectures for multiple tasks. In: AAAI (2020)

18. Teerapittayanon, S., McDanel, B., Kung, H.T.: BranchyNet: fast inference via early exiting from deep neural networks. In: 2016 23rd International Conference on Pattern Recognition (ICPR), pp. 2464–2469 (2016)

19. Tjong Kim Sang, E.F., De Meulder, F.: Introduction to the CoNLL-2003 shared task: language-independent named entity recognition. In: Proceedings of the Seventh Conference on Natural Language Learning at HLT-NAACL 2003, pp. 142–147 (2003). https://aclanthology.org/W03-0419

20. Vaswani, A., et al.: Attention is all you need. arXiv:abs/1706.03762 (2017)

21. Wang, A., Singh, A., Michael, J., Hill, F., Levy, O., Bowman, S.R.: GLUE: a multi-task benchmark and analysis platform for natural language understanding. arXiv preprint arXiv:1804.07461 (2018)

22. Wolf, T., et al.: Transformers: state-of-the-art natural language processing. In: Proceedings of the 2020 Conference on Empirical Methods in Natural Language Processing: System Demonstrations, pp. 38–45. Association for Computational Linguistics, Online, October 2020. www.aclweb.org/anthology/2020.emnlp-demos.6

23. Xin, J., Tang, R., Lee, J., Yu, Y., Lin, J.: DeeBERT: dynamic early exiting for accelerating BERT inference. arXiv preprint arXiv:2004.12993 (2020)

24. Xin, J., Tang, R., Yu, Y., Lin, J.: BERxiT: early exiting for BERT with better fine-tuning and extension to regression. In: Proceedings of the 16th Conference of the European Chapter of the Association for Computational Linguistics: Main Volume, pp. 91–104 (2021)

25. Yang, Z., Dai, Z., Yang, Y., Carbonell, J., Salakhutdinov, R., Le, Q.V.: XLNet: generalized autoregressive pretraining for language understanding. In: NeurIPS (2019)

26. Yu, J., Bohnet, B., Poesio, M.: Named entity recognition as dependency parsing. In: Proceedings of the 58th Annual Meeting of the Association for Computational Linguistics, pp. 6470–6476. Association for Computational Linguistics, Online, July 2020. https://doi.org/10.18653/v1/2020.acl-main.577, https://aclanthology.org/2020.acl-main.577

27. Zhang, Z., Zhu, W., Zhang, J., Wang, P., Jin, R., Chung, T.S.: PCEE-BERT: accelerating BERT inference via patient and confident early exiting. In: Findings of the Association for Computational Linguistics: NAACL 2022, pp. 327–338. Association for Computational Linguistics, Seattle, United States, July 2022. https://doi.org/10.18653/v1/2022.findings-naacl.25, https://aclanthology.org/2022.findings-naacl.25

28. Zhou, W., Xu, C., Ge, T., McAuley, J., Xu, K., Wei, F.: BERT loses patience: fast and robust inference with early exit. arXiv:abs/2006.04152 (2020)

29. Zhu, W., Wang, P., Ni, Y., Xie, G., Wang, X.: BADGE: speeding up BERT inference after deployment via block-wise bypasses and divergence-based early exiting. In: ACL Industry (2023)

30. Zhu, W., Wang, X., Ni, Y., Xie, G.: GAML-BERT: improving BERT early exiting by gradient aligned mutual learning. In: Proceedings of the 2021 Conference on Empirical Methods in Natural Language Processing, pp. 3033–3044. Association for Computational Linguistics, Online and Punta Cana, Dominican Republic, November 2021. https://aclanthology.org/2021.emnlp-main.242

31. Zhu, W., et al.: PANLP at MEDIQA 2019: pre-trained language models, transfer learning and knowledge distillation. In: Proceedings of the 18th BioNLP Workshop and Shared Task, pp. 380–388. Association for Computational Linguistics, Florence, Italy, August 2019. https://doi.org/10.18653/v1/W19-5040, https://aclanthology.org/W19-5040

CPMFA: A Character Pair-Based Method for Chinese Nested Named Entity Recognition

Xiayan Ji[1], Lina Chen[2(✉)], Fangyao Shen[2], Hongjie Guo[2], and Hong Gao[2]

[1] College of Physics and Electronic Information Engineering,
Zhejiang Normal University, Jinhua, China
[2] School of Computer Science and Technology, Zhejiang Normal University,
Jinhua, China
chenlina@zjnu.cn

Abstract. Chinese Nested Named Entity Recognition (CNNER) faces several challenges due to the language diversity phenomena, the complexity of the language, and the imbalanced distribution of entity types in Chinese text. To address these challenges in CNNER, we propose a new method called CPMFA (Character Pair-based method with Multi-feature representation and Attention mechanism). The CPMFA method predicts the predefined relations of character pairs in a sentence, and identifies nested named entities based on these relations. First, our method utilizes the pre-trained language model LERT (Linguistically-motivated Bidirectional Encoder Representation from Transformer), and Bidirectional Long Short-Term Memory (BiLSTM) to generate comprehensive and precise character representations. Second, our method uses multi-feature representation to capture complex semantic information within the text, and employs the Pyramid Squeeze Attention (PSA) module to emphasize key features. Finally, to overcome the challenge of the imbalanced distribution of entity types, PolyLoss function is integrated into our model training process. Results of experiments show that the proposed CPMFA method achieves an F1 score of 83.79%. Compared to other mainstream span-based methods, the proposed CPMFA method has excellent performance in CNNER.

Keywords: Chinese character pair · Chinese nested named entity recognition · Multi-feature representation · Attention mechanism · Pre-trained language model

1 Introduction

Nested named entities are entities that have overlapping structures. The majority of existing named entity recognition models have difficulty in accurately identifying such complex nested named entities, and cannot capture specific and detailed entity information in the text. Therefore, recognizing nested named entities has always been a highly challenging task.

X. Yang et al. (Eds.): ADMA 2023, LNAI 14176, pp. 200–212, 2023.
https://doi.org/10.1007/978-3-031-46661-8_14

In recent years, researchers have increasingly focused on the application of deep learning models in Chinese Nest Named Entity Recognition (CNNER). However, the literature in this field remains limited. Zhang et al. [1] proposed a novel boundary-aware layered neural model (BLNM) with segmentation attention, which captures the potential word information and enhance Chinese character representation, but is ineffective when dealing with a combination of Chinese and English text. Yu et al. [2] introduced a layered regional exhaustive model (LREM), which utilizes a neural network to explore exhaustive combinations of sentences; however, it requires an improved understanding of Chinese semantic language and does not fully utilize critical information within the text. Li et al. [3] developed a multi-layer joint learning model that uses a self-attention mechanism to effectively aggregate entity information features and identify nested entities layer by layer. However, the method faces challenges in handling imbalanced entity classes.

Previous studies on nested named entity recognition (NNER) primarily focus on English texts. There are notable differences between Chinese and English. Models that perform well in English NNER often encounter challenges when applied to Chinese texts, resulting in unsatisfactory outcomes. Existing researches has identified the following difficulties in CNNER:

1. Language diversity phenomena in Chinese text: with the constant integration and evolution of language and culture, the language diversity phenomena in Chinese technical materials and reference documents continue to increase. The mixture of Chinese, English, numbers, symbols, and other linguistic expressions presents a significant challenge for CNNER.
2. Complexity of the Chinese language: in Chinese text, the presence of multiple layers and high frequency of nested named entities, along with polysemous phenomena, results in a significantly challenging task of CNNER.
3. Entity type imbalance in Chinese text: for practical applications, entity type numbers distribution in Chinese text often follows a long-tail distribution where only a few entity types occupy the majority of data, significantly impeding model recognition performance.

To address the previously outlined challenges, this study proposes a Character Pair-based method with Multi-feature representation and an Attention mechanism (CPMFA). The proposed method makes the following contributions:

1. To overcome the challenge of language diversity phenomena in Chinese text, we introduce the linguistically-motivated pre-trained language model called LERT (Linguistically-motivated Bidirectional Encoder Representation from Transformer), as well as Bidirectional Long Short-Term Memory (BiLSTM), to vectorize the text. This approach improves the quality of character representation in Chinese text.
2. To address the complexity of Chinese language, this study utilizes multi-feature representation to incorporate comprehensive information from the text, as well as adopts the Pyramid Squeeze Attention (PSA) module to prioritize key features.

3. To deal with the long-tail distribution problem of entity class numbers in Chinese text, PolyLoss loss function is employed to improve the model's recognition performance.

This paper provides an overview of related work in Sect. 2, introduces the CPMFA method in Sect. 3, evaluates its performance on a Chinese nested named entity dataset with a detailed performance analysis in Sect. 4, and draws conclusions in Sect. 5.

2   Related Work

NNER utilizes two primary methods: sequence labeling-based and span-based.

Using a sequence labeling-based method, a label sequence with the highest probability is generated, which infers the boundaries and types of named entities more efficiently. Huang et al. [4] first utilized the BiLSTM-CRF model for named entity recognition (NER), which enables the capture of contextual information and dependencies between labels. To handle nested entities, Strakova et al. [5] combined multiple labels to create new ones. However, because characters in nested entities can have multiple labels, decoding named entities using the sequence labeling-based approach is more complex.

Using a span-based method, named entities are identified by categorizing the subsequences of the text sequence. Li et al. [6] detected entity fragments by exploring every possible text span and applying relationship classification to discover possible relationships among sets of entity fragments, thereby achieving recognition of nested entities. Xia et al. [7] proposed the Multi-Grained Named Entity Recognition (MGNER) model, which comprises a detection system and a classifier. The model aims to identify and categorize all potential fragments of entities. Li et al. [8] accomplished entity boundary determination and entity type recognition by predicting word relationships. However, span-based methods focus primarily on contextual information and do not fully explore the underlying information of the text.

Basic NNER methods have limitations in mining deep textual information. Therefore, some scholars have suggested incorporating attention mechanisms in NNER models to improve their performance and effectiveness [9]. Cui et al. [10] proposed a Multi-Head Adjacent Attention-based Pyramid Layered model to capture the dependency relationships between adjacent characters in the input text. Similarly, Rodríguez et al. [11] proposed an attention mechanism based specifically on the use of elements of the noun syntactic type to capture syntactic information in the text. These models show that attention mechanisms can be used to extract deep textual information.

The pre-trained language models, represented by Bidirectional Encoder Representation from Transformers (BERT) [12], have shown remarkable performance in many NER tasks. To address the issue of polysemy in Chinese, Li et al. [13] suggested a syntactic dependency guided BERT-BiLSTM-GAM-CRF model. Yu et al. [14] incorporated BERT into a previously utilized NER model to extract entities from mineral literature.

Xu et al. [15] proposed an approach that utilized BERT and a supervised multi-head self-attention mechanism to capture lexical correlations. This approach combined both attention mechanism and pre-trained language model, achieving excellent performance in the task of nested named entity recognition. In this study, the LERT pre-trained language model and Pyramid Squeeze Attention (PSA) module were employed to further enhance performance of the named entity recognition model. In real-world scenarios, the uneven distribution of entity quantities is a common issue. Various types of entities can exhibit distinct frequencies and distributions within the text. Such type imbalance can significantly impair the performance of NNER models.

The loss function measures the discrepancy between predicted and true labels during training. Focal Loss [16], an adaptive loss function utilized in the field of image recognition, aims to address type imbalance by reducing the impact of more frequently appearing samples while increasing the weight of the less frequent ones. Leng et al. [17] introduced PolyLoss, a linear combination of polynomial functions that can be customized to match the unique characteristics of the dataset. This method improves upon traditional loss functions by accounting for the nuances of the data, resulting in more accurate predictions. Although there has been extensive research conducted in the field of NNER, there are currently no literature references that examine the effectiveness of implementing PolyLoss within this area. We present the first study to use PolyLoss in NNER and demonstrate its ability in enhancing performance.

3   CPMFA Method

Our inspiration for defining the task of CNNER as the prediction of relations between characters comes from Li et al. [8]. Therefore, we propose a Character Pair-based method with Multi-feature Representation and Attention mechanism (CPMFA).

CPMFA model predicts the relations of character pairs in a sentence, based on three predefined relations: None, Next Neighboring Character (NNC), and Tail-Head Character-* (THC-*, where * represents the entity type). None indicates no relation between two characters. Next Neighboring Character (NNC) indicates whether two characters are adjacent within an entity. Tail-Head Character-* (THC-*) identifies the entity boundary and entity type. THC-* denotes the tail and head boundaries, while * represents the entity type.

3.1   CPMFA Model

Figure 1 depicts the architecture of the CPMFA model, consisting of three discrete components: Encoder Layer, Feature Extraction Layer, and Decoder Layer.

Encoder Layer. The Encoder Layer incorporates LERT and BiLSTM to produce superior character representations. LERT, a pre-trained language model, is applied to attain a comprehensive representation of textual information.

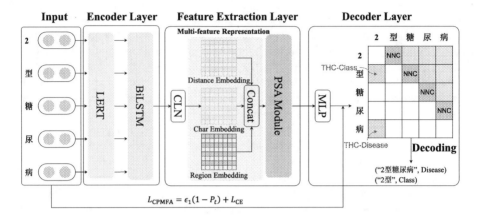

Fig. 1. The architecture of the CPMFA model.

For an input sentence $C = [c_1, c_2, ..., c_n]$, LERT produces vectorized text $X = [x_1, x_2, ..., x_n]$, as demonstrated in Eq. (1). To further enhance the model's understanding of textual context, the BiLSTM is employed to produce the result $H = [h_1, h_2, ..., h_n] \in R^{N \times d_h}$, as mentioned in Eq. (2). Here, d_h is the dimensional aspects of character representation.

$$X = LERT(C) \tag{1}$$

$$H = BiLSTM(X) \tag{2}$$

Feature Extraction Layer. The Feature Extraction Layer is used to extract pertinent features of character pair relations, thereby enabling accurate prediction of named entities. The layer comprises three components: a conditional layer normalization (CLN), multi-feature representation, and a PSA module.

Conditional Layer Normalization. The CLN generates grid representations between two characters, which are fundamental for extracting pertinent features related to character pairs, as illustrated in Eq. (3). This matrix $V \in \mathbb{R}^{N \times N \times d_h}$ is the result of the CLN, where V_{ij} stands for the representation of the character pair (c_i, c_j). Since both NNC and THC-* relations are directional, V_{ij}, which represents the character pair (c_i, c_j), can be considered as a combination of the representations of c_i and c_j, denoted by h_i and h_j respectively. Here, h_i is the condition for producing the gain parameter $\gamma_{ij} = W_\alpha h_i + b_\alpha$ as well as bias $\lambda_{ij} = W_\beta h_i + b_\beta$ of layer normalization. As mentioned in Eq. (4), μ and σ represent the mean and standard deviation of the elements present in h_j.

$$V_{ij} = CLN(h_i, h_j) = \gamma_{ij} \odot \left(\frac{h_j - \mu}{\sigma}\right) + \lambda_{ij} \tag{3}$$

$$\mu = \frac{1}{d_h} \sum_{k=1}^{d_h} h_{jk}, \sigma = \sqrt{\frac{1}{d_h} \sum_{k=1}^{d_h} (h_{jk} - \mu)^2} \tag{4}$$

Multi-feature Representation. Constructing multi-feature representations facilitates the integration of features from various perspectives and improves the accuracy of predicting character pair relations. Referring to Eq. (5), the multi-feature representation $E \in \mathbb{R}^{N \times N \times d_e}$ is obtained through concatenating distance embedding $E_D \in \mathbb{R}^{N \times N \times d_d}$, region embedding $E_R \in \mathbb{R}^{N \times N \times d_r}$, and character embedding $V \in \mathbb{R}^{N \times N \times d_h}$. The distance embedding shows the relative position of characters; the region embedding displays up-down triangle area information on the grid; and character embedding conveys semantic information.

$$E = Concat([E_D, E_R, V]) \tag{5}$$

PSA module. The PSA module can efficiently focus on key features in multi-feature representations and process character-to-character interaction information. The PSA module consists of two modules: the Squeeze and Concat (SPC) module and the SEWeight module [18].

As shown in Fig. 2, the SPC module is composed of multiple parallel branches that operate independently. Each branch takes E' as input and contains a number of channels d_e. E' is obtained by permuting the dimensions of the input multi-feature representation E. Grouped convolutions and convolution kernels of various sizes are utilized to compress the channels and capture spatial information across different scales. The resulting feature maps from the SPC module are represented by $F \in \mathbb{R}^{d_e \times N \times N}$, as depicted in Eq. (6).

$$F = SPC(E') \tag{6}$$

The SEWeight module utilizes input F, as shown in Eq. (7), to generate an attention vector Z_i for that branch.

$$Z_i = SEWeight(F_i), \quad i \in 1, 2, 3, 4 \tag{7}$$

Equation (8) shows the concatenation of attention weights from each branch, which creates the multi-scale channel weight. Multiplying the multi-scale fusion feature map with the multi-scale fusion channel weight in a channel-wise operation creates an adaptive channel weight for the feature map. The "Concat" denotes the concatenate operator, while $\odot$ denotes the element-wise product operator.

$$M_F = Concat([W_1, W_2, \ldots, W_4]) \odot F \tag{8}$$

Decoder Layer. The Decoder Layer comprises two main parts: predicting character pair relations and decoding identified named entities.

Predict. Our proposed model is designed to predict character pair relations by calculating the relation's probability of belonging to a specific class. The feature extraction layer and dimensional transposition produce the feature grid representation of character pairs, denoted as M_F'. The relation score y_{ij}' of the

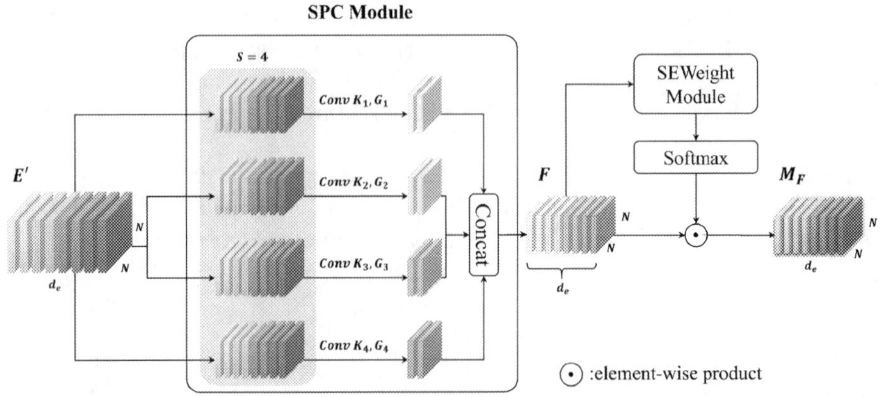

Fig. 2. The PSA module's structure diagram and a detailed description of the SPC module at S=4. K refers to the convolution kernel's size, G represents the group size, and "Concat" represents the feature concatenation in the channel dimension.

character pair (c_i, c_j) is computed by the Multilayer Perceptron (MLP), as shown in Eq. (9). To evaluate the probability $\hat{y}_{ij}$ of character pair (c_i, c_j) belonging to specific classes, the Softmax function is employed, as indicated in Eq. (10).

$$y'_{ij} = MLP(M'_{F_{ij}}) \tag{9}$$

$$\hat{y}_{ij} = Softmax(y'_{ij}) \tag{10}$$

Decode. We extract named entities by decoding the relations of character pairs. Relations of character pairs establish a directed graph, including nodes for characters and edges for relations. The model identifies specific pathways connecting distinct characters to one another, with each pathway mapped to a unique entity.

3.2 Training

Loss Function. We integrate the PolyLoss framework into our CPMFA model to resolve the long-tail entity type distribution issue in Chinese text. Tuning the polynomial coefficients in the PolyLoss-based loss function can optimize the model's performance for different datasets and tasks.

Our objective during training is to minimize L_{CPMFA}, as shown in Eq. (11). Here, L_{CE}, P_t, and ϵ represent the cross-entropy loss function, the probability of the model's true class label, and an adjustable hyperparameter, respectively.

In Eq. (12) and Eq. (13), the symbols used include N, representing the number of characters present in the sentence. Additionally, y_{ij} denotes a binary vector used to represent the actual relation of the character pair (c_i, c_j). The predicted probability vector, in contrast, is denoted as $\hat{y}_{ij}$. Lastly, r signifies the r-th relation contained in a predefined relation set, $\mathcal{R}$.

$$L_{CPMFA} = L_{CE} + \epsilon(1 - P_t) \tag{11}$$

$$L_{CE} = \sum_{i=1}^{N} \sum_{j=1}^{N} \sum_{r=1}^{|\mathcal{R}|} y_{ij} log \hat{y}_{ij} \qquad (12)$$

$$P_t = \sum_{i=1}^{N} \sum_{j=1}^{N} \sum_{r=1}^{|\mathcal{R}|} y_{ij} \hat{y}_{ij} \qquad (13)$$

4   Experiments and Results

4.1   Dataset

We utilized DiaKG, the authoritative dataset for the Chinese diabetes domain, and followed the division protocol established by Chang et al. [19] to extract the train, validation, and test sets.

Table 1 displays the statistical information for different granularities of the DiaKG dataset. The dataset contains a proportion of nested entities, up to 22% of it. Improving the precision of nested named entity recognition has the potential to enhance the overall model's performance.

Table 1. The statistical information on the granularity of the DiaKG dataset.

Granularity	Statistics	Train	Dev	Test	Total
Sentence	Total	4906	1636	1636	8178
	Sentences with nested named entities	3550 (72.36%)	1205 (73.66%)	1164 (71.15%)	6255 (76.49%)
	Avg. sentence length	151.34	153.68	150.63	151.68
Entity	Total	65774	22417	21496	109687
	nested named entities	**11101(16.88%)**	**3771(16.82%)**	**3516(16.36%)**	**24155(22.02%)**
	Avg. entity length	4.37	4.37	4.38	4.37
	Max number of nested layers	3	3	2	3

Figure 3 presents the frequency of annotations for the 18 entity types in the DiaKG dataset. The figure shows that the number of entity types are diverse and follow a long-tailed distribution, indicating data imbalance.

4.2   Evaluation Metrics

In this study, we follow the exact matching pattern to evaluate the performance of NNER. That is to say, a predicted entity is considered as correctly identified only when its predicted boundaries and types exactly match the annotated results in the dataset. Currently, in CNNER tasks, precision (P), recall (R), and F1 score (F1) are commonly used to evaluate the performance [20]. Precision measures the model's ability to correctly predict entities, while Recall measures the model's ability to identify all entities. The F1 score is the harmonic mean of precision and recall.

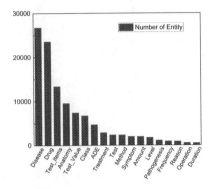

Fig. 3. Distribution of entities and number of entities in DiaKG dataset.

Precision, recall, and F1 score are calculated based on the number of true positives (TP), false positives (FP), and false negatives (FN), as shown in Eq. (14), Eq. (15) and Eq. (16).

$$P = \frac{TP}{TP + FP} \tag{14}$$

$$R = \frac{TP}{TP + FN} \tag{15}$$

$$F1 = \frac{2 \times P \times R}{P + R} \tag{16}$$

4.3   Experimental Setting

The development language used in this study is Python and the deep learning framework used is Pytorch, which was trained on an NVIDIA 3090 Ti graphics card. Regarding the model hyper-parameters, we set the dimension of the LERT embedding to 768, the dimension of the LSTM hidden layer to 478, the batch size to 8, and the initial learning rate to 0.001. To prevent the model from overfitting, early stopping criteria and a dropout rate of 0.5 were employed in this study.

4.4   Results

Comparison of Loss Function. This study evaluates the performance of three loss functions: L_{CPMFA}, cross-entropy (L_{CE}), and focal (L_{FL}). The hyperparameter ϵ of L_{CPMFA} can be adjusted. Figure 4(a) and Fig. 4(c) demonstrate that models trained with L_{CPMFA} perform better in F1 score and recall than those using L_{CE} and L_{FL} functions. Figure 4(b) illustrates that, when $\epsilon \in 1, 2, 4$, models trained with L_{CPMFA} have lower precision compared to models trained with L_{CE}. However, models trained with L_{CPMFA} have higher precision than models trained with L_{FL}. In conclusion, using the L_{CPMFA} loss function and customizing the hyperparameters based on dataset characteristics could improve the CPMFA model's ability to identify named entities.

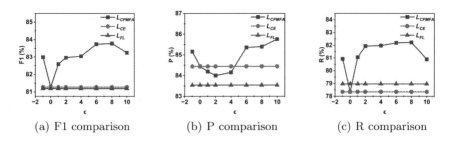

(a) F1 comparison   (b) P comparison   (c) R comparison

Fig. 4. Performances of our model trained with different loss functions.

Necessity of Multi-feature Representation. We conducted ablation experiments on the embeddings to verify the significance of multi-feature representation. The results are presented in Table 2. Using only character embedding for input features yielded the lowest F1 score. Integrating distance and region embedding resulted in a noteworthy improvement in the performance of the model. The integration of character and region embedding attained optimal precision, with a score of 85.41%. However, the multi-feature representation recorded only 0.01% lower precision, indicating a negligible difference between them. Our multi-feature representation, which integrated character, distance, and region embedding, achieved excellent results in both recall and F1 score. Therefore, the incorporation of multi-feature representation could effectively represent the relations of character pairs and enhance the model's recognition performance.

Table 2. Performances of our model with different feature embedding. The bold value indicates the optimal results.

	F1 (%)	P (%)	R (%)
Char Embedding	80.61	82.99	78.36
Char Embedding+Distance Embedding	82.90	84.11	81.73
Char Embedding+Region Embedding	83.05	**85.41**	80.81
Ours (Char Embedding+Distance Embedding+Region Embedding)	**83.79**	85.40	**82.23**

Comparison with Baselines. Table 3 provides a comparison between our proposed model and previous work on the DiaKG dataset. The span-based method outperformed the sequence labeling-based method. Our CPMFA model achieved the highest F1 score and recall at 83.79% and 82.23%, respectively. However, the model's precision was suboptimal, at 85.40%, which was 1.58% lower than that of the Efficient Global Pointer model. The analysis indicated that the pointer-based architecture of the Efficient Global Pointer model had a higher accuracy in identifying longer entities in input sequences. It successfully identified spans by learning patterns of head and tail word pairs, even if the intermediate words had changed. In contrast, our CPMFA model comprehensively recognized entities by classifying the relation between each character pair.

Table 3. Performances of our model and baseline models on DiaKG dataset. The bold value indicates the optimal results.

	Method	F1 (%)	P (%)	R (%)
Sequence labeling-based	Cascade-CRF [21]	62.58	59.52	65.97
Span-based	W2NER [8]	80.51	81.52	79.52
	Global Pointer [22]	69.35	73.36	65.76
	Efficient Global Pointer [23]	82.95	**86.98**	79.28
Ours	CPMFA	**83.79**	85.40	**82.23**

Ablation Experiments. We selected the DiaKG dataset to evaluate the effectiveness of our model's components. We conducted an analysis by removing one component at a time to observe the impact on performance. Table 4 shows the performance of the model's variations with "w/o" representing "without." The results signify that all model components are essential for optimal performance.

First, we replaced the LERT pre-trained language model with the BERT pre-trained language model in our CPMFA model, causing a 2.26% decrease in the F1 score. This indicates that pre-trained language model with rich linguistic features can significantly tackle the challenge of linguistic diversity phenomena in Chinese text.

Then, we removed the multi-feature representation from our model CPMFA. This resulted in a 3.18% decrease in the F1 score, indicating that the multi-feature representation module can perform deeper information mining as a complement to pre-trained language models.

Lastly, we evaluated the effectiveness of using the PSA module by removing it from the experiment, resulting in a 1.18% reduction in F1-score. The result indicates that the PSA module can help to focus on the most essential aspects of the input features.

Table 4. Ablation study on DiaKG dataset. The bold value indicates the optimal results.

	F1 (%)	P (%)	R (%)
CPMFA (ours)	**83.79**	**85.40**	**82.23**
w/o LERT	81.52	84.38	78.85
w/o multi-feature representation	80.61	82.99	78.36
w/o PSA	82.61	84.03	81.23

5   Conclusion

This paper proposes a CPMFA model for CNNER and evaluate its performance in medical nested text related to diabetes. First, our model utilizes a pre-trained language model LERT and BiLSTM to acquire high-quality character embeddings that effectively tackle the challenge of linguistic diversity phenomena in Chinese text. Second, the model integrates multi-feature representation and the PSA module to capture critical features for effective text mining. To address the issue of imbalanced entity types, the PolyLoss-based function is employed during training. Ablation Experiments validate the effectiveness of each model component. Notably, the CPMFA model outperforms existing NNER models in terms of F1 score, demonstrating its potential to enhance CNNER performance for Chinese medical text and offer a novel technical solution in other domains.

Acknowledgement. This study was supported by the Key Project of Regional Innovation and Development Joint Fund of National Natural Science Foundation of China (Grant No. U22A2025).

References

1. Rujia, Z., Lu, D., Peng, G., Bang, W.: Chinese nested named entity recognition algorithm based on segmentation attention and boundary-aware. Comput. Sci. **50**(01), 213–220 (2023)
2. Shiyuan, Y., Shuming, G., Ruiyang, H., Jianpeng, Z., Nan, H.: Layered regional exhaustive model for Chinese nested named entity recognition. Comput. Technol. Dev. **32**(09), 161–166+179 (2022)
3. Li, H., Xu, H., Qian, L., Zhou, G.: Multi-layer joint learning of Chinese nested named entity recognition based on self-attention mechanism. In: Natural Language Processing and Chinese Computing: 9th CCF International Conference, NLPCC 2020, Zhengzhou, China, 14–18 October 2020, Proceedings, Part II, pp. 144–155. Springer, Cham (2020). https://doi.org/10.1007/978-3-030-60457-8_12
4. Huang, Z., Xu, W., Yu, K.: Bidirectional LSTM-CRF models for sequence tagging. arXiv preprint arXiv:1508.01991 (2015)
5. Straková, J., Straka, M., Hajič, J.: Neural architectures for nested NER through linearization. arXiv preprint arXiv:1908.06926 (2019)
6. Li, F., Lin, Z., Zhang, M., Ji, D.: A span-based model for joint overlapped and discontinuous named entity recognition. In: Proceedings of the 59th Annual Meeting of the Association for Computational Linguistics and the 11th International Joint Conference on Natural Language Processing (Volume 1: Long Papers), pp. 4814–4828 (2021)
7. Xia, C., et al.: Multi-grained named entity recognition. In: 57th Annual Meeting of the Association for Computational Linguistics, ACL 2019, pp. 1430–1440. Association for Computational Linguistics (ACL) (2020)
8. Li, J., et al.: Unified named entity recognition as word-word relation classification. In: Proceedings of the AAAI Conference on Artificial Intelligence, vol. 36, pp. 10965–10973 (2022)

9. Islam, T., Zinat, S.M., Sukhi, S., Mridha, M.F.: A comprehensive study on attention-based NER. In: Khanna, A., Gupta, D., Bhattacharyya, S., Hassanien, A.E., Anand, S., Jaiswal, A. (eds.) International Conference on Innovative Computing and Communications. AISC, vol. 1388, pp. 665–681. Springer, Singapore (2022). https://doi.org/10.1007/978-981-16-2597-8_57

10. Cui, S., Joe, I.: A multi-head adjacent attention-based pyramid layered model for nested named entity recognition. Neural Comput. Appl. **35**(3), 2561–2574 (2023)

11. Rodríguez, A.J.C., Castro, D.C., García, S.H.: Noun-based attention mechanism for fine-grained named entity recognition. Expert Syst. Appl. **193**, 116406 (2022)

12. Devlin, J., Chang, M.W., Lee, K., Toutanova, K.: BERT: pre-training of deep bidirectional transformers for language understanding. arXiv preprint arXiv:1810.04805 (2018)

13. Li, D., Yan, L., Yang, J., Ma, Z.: Dependency syntax guided BERT-BiLSTM-GAM-CRF for Chinese NER. Expert Syst. Appl. **196**, 116682 (2022)

14. Yu, Y., et al.: Chinese mineral named entity recognition based on BERT model. Expert Syst. Appl. **206**, 117727 (2022)

15. Xu, Y., Huang, H., Feng, C., Hu, Y.: A supervised multi-head self-attention network for nested named entity recognition. In: Proceedings of the AAAI Conference on Artificial Intelligence, vol. 35, pp. 14185–14193 (2021)

16. Lin, T.Y., Goyal, P., Girshick, R., He, K., Dollár, P.: Focal loss for dense object detection. In: Proceedings of the IEEE International Conference on Computer Vision, pp. 2980–2988 (2017)

17. Leng, Z., et al.: PolyLoss: a polynomial expansion perspective of classification loss functions. arXiv preprint arXiv:2204.12511 (2022)

18. Zhang, H., Zu, K., Lu, J., Zou, Y., Meng, D.: EPSANet: an efficient pyramid squeeze attention block on convolutional neural network. In: Wang, L., Gall, J., Chin, T.J., Sato, I., Chellappa, R. (eds.) Proceedings of the Asian Conference on Computer Vision, pp. 1161–1177. Springer, Cham (2022). https://doi.org/10.1007/978-3-031-26313-2_33

19. Chang, D., et al.: DiaKG: an annotated diabetes dataset for medical knowledge graph construction. In: Qin, B., Jin, Z., Wang, H., Pan, J., Liu, Y., An, B. (eds.) Knowledge Graph and Semantic Computing: Knowledge Graph Empowers New Infrastructure Construction: 6th China Conference, CCKS 2021, Guangzhou, China, 4–7 November 2021, Proceedings, vol. 1466, pp. 308–314. Springer, Cham (2021). https://doi.org/10.1007/978-981-16-6471-7_26

20. Li, J., Sun, A., Han, J., Li, C.: A survey on deep learning for named entity recognition. IEEE Trans. Knowl. Data Eng. **34**(1), 50–70 (2020)

21. Wei, Z., Su, J., Wang, Y., Tian, Y., Chang, Y.: A novel cascade binary tagging framework for relational triple extraction. arXiv preprint arXiv:1909.03227 (2019)

22. Su, J., et al.: Global pointer: novel efficient span-based approach for named entity recognition. arXiv preprint arXiv:2208.03054 (2022)

23. Su, J.: Efficient globalpointer: fewer parameters, more effects, January 2022. https://spaces.ac.cn/archives/8877

STMC-GCN: A Span Tagging Multi-channel Graph Convolutional Network for Aspect Sentiment Triplet Extraction

Chao Yang, Jiajie Xing, and Xianguo Zhang[✉]

College of Computer Science, Inner Mongolia University, Hohhot 010021,
People's Republic of China
1031291901@qq.com, 563076103@qq.com, 2595083628@qq.com

Abstract. Aspect-Based Sentiment Triplet Extraction (ASTE) is a rapidly growing field in sentiment analysis. While most research has focused on processing the ASTE task either in a pipeline or end-to-end manner, both methods have their limitations. Pipeline methods may accumulate errors in practical applications, while sequence labeling methods in end-to-end approaches may overlook important feature information of the three elements themselves. Additionally, various features in sentences and emotional word markers have not been effectively explored in these methods. To address these limitations, we propose a novel solution called Span Tagging Multi-Channel Graph Convolutional Network (STMC-GCN) that explicitly combines multiple prominent features to extract span-level sentiment triplets, where each span may consist of multiple words and play different roles. Specifically, we designed a three-channel graph fusion model that converts sentences into multiple channels of graphs. These channels extract node text features, centrality features, and position features, which are then extracted through cross-channel convolution operations to obtain a common graph representation shared by different channels. To optimize downstream classification with better results, we use consistency and difference constraints to enhance common attributes and independence. Finally, we explore span-level information and constraints to generate more accurate aspect-based sentiment triplet extractions. Experimental results illustrate that STMC-GCN performs well on multiple datasets, proving the effectiveness and robustness of the model.

Keywords: Aspect-based sentiment triplet extraction · Span tagging · Three-channel graph fusion model · Cross-channel convolution

1 Introduction

Aspect-Based Sentiment Triplet Extraction (ASTE) [6,12] aims to extract opinion triplets from comments. ASTE is related to other tasks such as Aspect Term

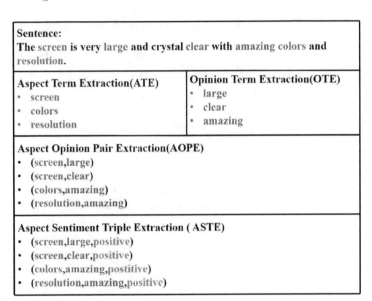

Sentence: The screen is very large and crystal clear with amazing colors and resolution.	
Aspect Term Extraction(ATE) • screen • colors • resolution	**Opinion Term Extraction(OTE)** • large • clear • amazing

Aspect Opinion Pair Extraction(AOPE)
- (screen,large)
- (screen,clear)
- (colors,amazing)
- (resolution,amazing)

Aspect Sentiment Triple Extraction (ASTE)
- (screen,large,positive)
- (screen,clear,positive)
- (colors,amazing,postitive)
- (resolution,amazing,positive)

Fig. 1. An example of ASTE. Blue for aspect items, brown for opinion items, green for emotional polarity. (Color figure online)

Extraction(ATE) [4,21], Opinion Term Extraction (OTE) [8,17], and Aspect-Opinion Pair Extraction (AOPE) [2]. For example, as shown in Fig. 1.

In prior research on the Aspect-Based Sentiment Analysis (ASTE) task, two mainstream approaches have been identified. Pipeline method extracts aspect, opinion terms & sentiment polarity independently in 1st approach [7,12]. Although this method has achieved great success, it has low computational efficiency, ignores the relationship and interaction between individual word pairs. The second approach is the end-to-end method [20,22,23]: this approach heavily relies on word interaction to predict the sentiment relationship of the target opinion pair. Furthermore, most end-to-end methods focus on word-level tagging schemes, which capture interactions at the word level but not at the span level, ignoring the semantic information of multi-word aspect/opinion terms. This treatment of multi-word terms as a group of independent words makes it difficult to ensure the consistency of sentiment when extracting triplets. Although previous work has achieved excellent results, ASTE still faces some challenges.

Recent research in ASTE has failed to compare with Pre-trained Language Models (PLMs) as necessary. However, recent progress has found that using PLMs can lead to remarkable performance due to the implicit structure captured by these models [10,11]. Although general PLMs have been applied in sentiment analysis, they have limited capabilities for different tasks.

To address the issues mentioned earlier, we propose a new approach called the Span Tagging Multi-Channel Graph Convolutional Network (STMC-GCN). Our approach is based on a sentiment-aware multi-level pre-training model called SentiWSP [5] as the encoding layer, and it design corresponding pre-training tasks at both the word-level and sentence-level to enhance the model's ability to capture sentiment knowledge in the text. Inspired by CNNs, our proposed approach combines sentence semantics, importance, position, and their combinations in the graph. Using an attention mechanism, we obtain representations based on word combinations and syntactic dependencies to enhance the semantic relationships between inputs. Finally, we input the representations into the span tagging method and formulate it as a multi-class classification problem for each span. By exploring span-level information and constraints, our approach effectively generates more accurate aspect-based sentiment triplets. This paper makes the following contributions:

- A novel multi-channel graph convolutional network is proposed to explicitly aggregate three channels to comprehensively study the aggregation of different features, using a weighted graph to aggregate node features of neighbors in the word-sentence graph, and utilizing a common convolutional channel to extract common features shared by three specific channels.
- Replacing traditional pre-training models with new sentiment pre-training models, this approach is more sample-efficient and beneficial for the ASTE task.
- A new span tagging method is proposed, which includes a span tagging scheme considering span role diversity to overcome the limitations of existing labeling schemes, and a greedy inference strategy considering span-level constraints to effectively generate more accurate triples.
- The proposed STMC-GCN method exhibits high predictive accuracy and state-of-the-art performance, as demonstrated by its excellent results on four classic datasets.

2   Approach

This section aims to provide a detailed introduction to the architecture of the STMC-GCN method.

2.1   Task Formulation

Enter a sentence $X = \{w_1, w_2, ..., w_n\}$ where w_i represents a word and n is the total number of words in the sentence, our task is to extract sentiment triplets T= $\{(a, o, s)_m\}_{m=1}^{|T|}$ from the sentence X, where a, o, and s represent the aspect term, opinion term, and sentiment polarity. The Sentence X has a total of $|T|$ triples.

2.2   Basic Module

Figure 2 illustrates that STMC-GCN is composed of three components: sentence encoding, multi-channel GCN module, and tagging decoding module. At the outset, the sentence is fed into the sentence encoding module, which employs a

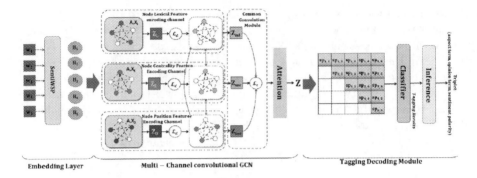

Fig. 2. The model's architecture comprises several critical components, including a sentiment-aware multi-level pre-training model called SentiWSP, a multi-channel graph convolutional network, and a novel span tagging method. These components work together to capture sentiment knowledge in text and generate aspect-based sentiment triplets with high accuracy.

sentiment pre-training model called SentiWSP, as the encoder to extract hidden contextual representations. This process results in an output sequence of hidden representations. In the multi-channel GCN module, we introduce three distinct channels includes semantic, centrality, and position encoding channels. These channels help in learning the linguistic, vital, and positional features of the sentence respectively. Simultaneously, we utilize a convolution module with consistency constraints to extract common properties in three feature spaces. In the tagging decoding module, we generate a more precise aspect-sentiment triplet extraction by exploring span-level information and constraint conditions. This module includes two parts: span tagging scheme and greedy inference strategy.

2.3   Embedding Layer

The SentiWSP model is a pre-training model designed to enhance the ability of models to capture emotional knowledge in text. It achieves this by incorporating pre-training tasks at both the word and sentence levels. Unlike the BERT model, which has a broader focus, the SentiWSP model emphasizes emotional-related tasks and language representation, allowing it to extract emotional triplets more effectively by learning emotional knowledge from large data volumes.

To extract hidden contextual representations, we utilize SentiWSP as the sentence encoder. The pre-trained SentiWSP model generates context word embeddings, which enhances the model's ability to understand emotional words and capture emotional information more precisely. The formula for the method is presented below:

$$H_i = \ SentiWSP\,(w_i) \tag{1}$$

where w_i is the i input of sentence S, and H_i stands for the contextual representation of the output of the pre-trained model. The symbol SentiWSP($\cdot$) represents all operations of the SentiWSP model.

2.4 Multi-channel Convolutional GCN

Our proposed STMC-GCN uses graph neural networks to model relationships between words and integrate multiple features. It consists of three graph channels that encode node text, centrality, and position features, and uses cross-channel convolution to extract a common graph representation. These representations are subsequently fused to produce the final output.

The input is fed to the convolutional graph modules, the semantic graph, centrality graph and position graph $\mathbb{G}_L = (\mathbb{A}, X_L)$, $\mathbb{G}_E = (\mathbb{A}, X_E)$, $\mathbb{G}_O = (\mathbb{A}, X_O)$. Each word node linked to related nodes, and the edge feature as relationship weight. Due to their unique topological structures and common features, we designed a shared convolutional module to extract their shared embeddings by considering the common features of the three spaces. Therefore, this module is used to learn the common embeddings Z_{ml}, Z_{me}, and Z_{mo}. To optimize downstream sentence classification, we use consistency and difference constraints to enhance the common properties and independence.

Node-Based Lexical Feature Encoding Channel. The lexical features are set to $X_L = H_L^{(0)}$, where $H_L^{(0)} \in R^{n \times d}$. To construct the graph of a sentence with n words, we use dependency parsing to create an adjacency matrix to represents the graph of the sentence.

$$Z_L^{k+1} = \sigma \left(\widetilde{D}^{-\frac{1}{2}} \widetilde{A} \widetilde{D}^{-\frac{1}{2}} Z_L^{(k-1)} W_L^{(k)} \right) \tag{2}$$

In this equation, the *ReLU* function is used as the activation function denoted by σ. The parameter matrix for the vocabulary channel is $W_L^{(k)} \in R^{d_{k-1} \times d_k}$, while $\widetilde{D}$ represents the degree matrix of each vertex in $\widetilde{A}$. To ensure the stability and accuracy of the model, we have normalized $\widetilde{A}$ and employed operation $\widetilde{D}^{-\frac{1}{2}} \widetilde{A} \widetilde{D}^{-\frac{1}{2}}$. The input representation is $Z_L^{(0)} = X_L$, where each row $Z_i \in R^d$ represents the d-dimensional feature of the i node. $\mathbb{Z}_L$ represents the embedding of the output in the vocabulary space. We also incorporate TF-IDF values as edge weights into $A_{i,j}$ to consider the edge weights in the semantic graph channel.

Node-Based Centrality Feature Encoding Channel. In graph theory, centrality features are used to measure the importance of nodes in a network and identify nodes that occupy central positions and have a high impact. In STMC-GCN, Centrality feature encoding channel learns word importance. The centrality feature $X_E = H_E^{(0)}$ and $H_E^{(0)} \in R^{n \times d}$ are initialized with four typical centrality features, and each feature embedding has a dimension of $R^{n \times (d/4)}$. We encode each discrete feature value of centrality and add it as additional information $H_E^{(0)}$ in the final representation. Thus, we define the centrality channel as follows:

$$Z_E^{k+1} = \sigma \left(\widetilde{D}^{-\frac{1}{2}} \widetilde{A} \widetilde{D}^{-\frac{1}{2}} Z_E^{(k-1)} W_E^{(k)} \right) \tag{3}$$

In this model, the parameter matrix $W_E{}^{(k)} \in R^{d_{k-1} \times d_k}$ of the centrality channel is used to calculate the centrality feature of nodes. The input representation matrix $Z_E{}^{(0)} = X_E$ is used as the model input. Additionally, we incorporate edge weights calculated by edge centrality and inject them into the weight matrix $A_{i,j}$. This approach enables us to more accurately measure the strength of relationships between nodes.

Node-Based Position Feature Encoding Channel. Work by [9] showed that creating three different positional encodings removes randomness, on top of which we compute positional correlations using separate channels. We learn the $X_O = H_O{}^{(0)}$ position feature to capture the order of tokens in the original text, and initialize the $H_O{}^{(0)} \in \mathbb{R}^{n \times d}$ feature. The positional channels are as follows:

$$Z_O{}^{k+1} = \sigma \left(\tilde{D}^{-\frac{1}{2}} \tilde{A} \tilde{D}^{-\frac{1}{2}} Z_o{}^{(k-1)} W_o{}^{(k)} \right) \tag{4}$$

In this model, the parameter matrix $W_O{}^{(k)} \in R^{d_{k-1} \times d_k}$ of the position channel is used to calculate position-relatedness. The input representation matrix $Z_o{}^{(0)} = X_o$ is used as the model input, and the final output embedding is represented as Z_O, which is learned through the position feature to capture the order of tokens. This channel can effectively capture the positional relationships between adjacent nodes. We expect that the embeddings of adjacent nodes will be closer to each other, while distant nodes will have smaller weights and will be further apart in the embedding.

Common Convolution Module. The shared convolutional module consists of a shared convolutional channel with shared parameters, which aims to extract the features common to the three specific channels. Specifically, the vocabulary features embedded in Z_{ml} in the vocabulary graph channel are transformed into the shared convolutional channel.

$$Z_{ml}{}^{k+1} = \sigma \left(\tilde{D}^{-\frac{1}{2}} \tilde{A} \tilde{D}^{-\frac{1}{2}} Z_{ml}{}^{(k-1)} W_m{}^{(k)} \right) \tag{5}$$

Here, $Z_{ml}{}^{k+1}$ represents the word feature embedding of the k-1 layer, while $Z_{ml}{}^{(0)} = X_L$. Similarly, The rest of the channels are similar:

$$Z_{me}{}^{k+1} = \sigma \left(\tilde{D}^{-\frac{1}{2}} \tilde{A} \tilde{D}^{-\frac{1}{2}} Z_{me}{}^{(k-1)} W_m{}^{(k)} \right) \tag{6}$$

$$Z_{mo}{}^{k+1} = \sigma \left(\tilde{D}^{-\frac{1}{2}} \tilde{A} \tilde{D}^{-\frac{1}{2}} Z_{mo}{}^{(k-1)} W_m{}^{(k)} \right) \tag{7}$$

Then, the final output of the shared embedding is:

$$Z_M = \left(Z_{ml} + Z_{me} + Z_{mo} \right) / 3 \tag{8}$$

Disparity Constraint and Consistency Constraint. We employed the differential constraint and consistency constraint from AM-GCN [18]. The differential constraint ensures the independence between Z_L and Z_{ml}, Z_E and Z_{me}, and Z_O and Z_{mo}, while the consistency constraint enforces consistency between the output of the shared convolutional channel and the three specific channels.

$$\mathcal{L}_d = HSIC(Z_L, Z_{ml}) + HSIC(Z_E, Z_{me}) + HSIC(Z_O, Z_{mo}) \quad (9)$$

In this study, we employed the Hilbert-Schmidt Independence Criterion (HSIC) [16], a kernel-based measure of statistical dependence between two variables, to enhance their distinguishability from each other.

To increase the commonality between the embedding matrices Z_{ml}, Z_{me}, and Z_{mo}, we employed a consistency constraint. Firstly, we normalized these matrices and transformed them into similarity matrices, resulting in three corresponding similarity matrices: S_L, S_E, and S_o. Specifically, we ensured that $S_L = Z_{ml} \cdot Z_{ml}^T, S_E = Z_{me} \cdot Z_{me}^T, S_O = Z_{mo} \cdot Z_{mo}^T$, and then utilized an L2-based normalization method to optimize the consistency constraint on these three similarity matrices:

$$\mathcal{L}_c = \| S_L - S_E \|_F^2 + \| S_E - S_O \|_F^2 + \| S_O - S_L \|_F^2 \quad (10)$$

Attention Mechanism. The model consists of three specific representations, Z_L, Z_E, and Z_O, as well as a shared representation, Z_M since node labels may be related to one or multiple representations. To integrate these representations, we utilized an attention mechanism to adaptively fuse them into the final sentence representation, Z. Specifically, the representations are weighted and summed using an attention mechanism to derive the final sentence representation, Z.

$$Z = a_L \cdot Z_L + a_E \cdot Z_E + a_O \cdot Z_O + a_M \cdot Z_M \quad (11)$$

$a_L, a_E, a_O, a_M \in R^{n \times 1}$ are the learnable attention values of n nodes, which respectively represent Z_L, Z_E, Z_O, and a shared representation Z_M.

Table 1. Descriptions of sub tags

Tags	Meaning
NEG	Word pairs express negative emotions.
NEU	Word pairs express neutral emotions.
POS	Word pairs express positive emotions.
A	an aspect term
O	an opinion term
N	nothing in the specific dimension

2.5   Tagging Decoding Module

Span Tagging. To account for the diversity of span roles, our labeling scheme defines each span $(SP_{i,j})$ in three role dimensions, namely, whether the $SP_{i,j}$ is a valid aspect term, a valid opinion term, and a valid sentiment fragment. We use $role^a_{i,j}$, $role^o_{i,j}$, and $role^s_{i,j}$ to represent these sub-labels, where $role^a_{i,j} \in \{N, A\}$, $role^o_{i,j} \in \{N, O\}$, $role^s_{i,j} \in \{N, NEG, NEU, POS\}$. Table 1 lists the meanings of these sub-labels.

To independently consider the three dimensions, we propose a three-dimensional span tagging method with the label set $\{N, A\} - \{N, O\} - \{N, NEG, NEU, POS\}$. The resulting labeling is represented as an upper triangular table T, where $T[i][j]$ denotes the label of span $SP_{i,j}$, as shown in Fig. 3. It should be noted that the label of the span "amazing color" is "$N - O - POS$" because this span represents both opinion terms and aspect terms for positive sentiment.

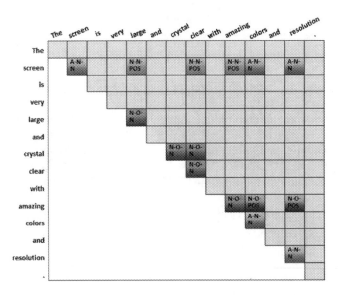

Fig. 3. Example of span tagging. Cells with gray background represent "N-N-N".

We form the span representation in the following way:

$$SP_{i,j} = H_i \oplus H_j \oplus Z \tag{12}$$

We use $SP = \{SP_{1,1}, SP_{1,2}, ..., SP_{i,j}, ..., SP_{n,n}\}$ to represent all possible enumerated spans in X, where i and j denote the start and end positions in X, respectively, and $\oplus$ indicates vector concatenation. Z denotes the sentence representation obtained earlier.

Therefore, $SP_{i,k}$ is a sentiment fragment, which means that $(SP_{i,k}, SP_{l,j})$ it is a valid (aspect, opinion) or (opinion, aspect) pair.

Greedy Inference. A greedy inference strategy is used to improve triplet accuracy in span tagging schemes. Under this strategy, "greedy" refers to retrieving the aspect/opinion terms with the maximum length. The specific implementation is as follows: First, all possible aspects, opinions, and sentiment segments are obtained from the tagged results. Next, each sentiment segment is analyzed while considering two cases: 1) aspect words precede opinion words, and 2) aspect words follow opinion words. By considering the order in which aspects and viewpoints appear, the accuracy of the extracted sentiment triplets is ensured through mutual span-level constraints, and the sentiment fragments representing aspect and viewpoint pairings are extracted and verified for correct sentiment polarity and valid aspects and viewpoints. Finally, an inference module generates sentiment triplets from the labeled results. Compared to previous decoding algorithms, this strategy guarantees the accuracy of the extracted sentiment triplets through constraints, thereby improving the efficiency and accuracy of span tagging schemes.

2.6   Training

Finally, the span representation is inputted to a fully-connected layer (i.e., classifier) with a softmax activation function, which outputs probabilities for character labels:

$$\hat{Y} = softmax\left(W_p SP_{i,j} + b_p\right) \tag{13}$$

where W_p and b_p are learnable weights and biases.

$$\mathcal{L}_y = -\sum_{i=1}^{n}\sum_{j=i}^{n} loss\left(\hat{Y}, y_{i,j}\right) \tag{14}$$

where $y_{i,j}$ represents the golden label of $sp_{i,j}$ and loss is the standard cross-entropy loss. The following is the overall objective function:

$$\mathcal{L} = \mathcal{L}_y + \alpha\mathcal{L}_d + \beta\mathcal{L}_c \tag{15}$$

where α and β are parameters of consistency and difference constraints.

3   Experiments

3.1   Dataset

The dataset was derived from SemEval Challenges [13–15] and was collected by [22], who improved the dataset provided by [12]. The datasets cover the fields of restaurants and laptops, dividing into training set, development set and test set. The specific statistical data can be found in Table 2. The labels A-O-N, N-O-S, A-N-S, and A-O-S indicate that a span may play different roles in a sentence.

Table 2. The statistics for the dataset are as follows: #S and #T represent the total number of sentences and triples, respectively. In this dimension, N has no specific meaning.

Dataset		#S	#T	Role distributions of spans						
				A-N-N	N-N-S	N-O-N	A-O-N	N-O-S	A-N-S	A-O-S
14res	Train	1266	2337	2041	2318	2071	1	14	7	1
	Dev	310	577	486	570	495	0	5	3	0
	Test	492	994	843	991	851	1	3	7	0
14lap	Train	906	1460	1281	1450	1281	1	4	1	0
	Dev	219	345	292	343	308	0	0	0	0
	Test	328	541	462	539	471	0	1	0	0
15res	Train	605	1013	860	1011	940	0	0	1	0
	Dev	148	249	214	248	235	0	0	0	0
	Test	322	485	435	485	463	0	0	0	0
16res	Train	857	1394	1201	1391	1304	1	0	1	1
	Dev	210	339	289	339	318	0	0	0	0
	Test	326	514	452	512	472	1	1	0	0

3.2 Experiment Settings

We utilized SentiWSP as our sentence encoder, with AdamW optimizer set as the default optimizer and the maximum sequence length set to 128. The number of hidden dimensions x is set to 768 and 300 for SentiWSP and GCN, respectively. The learning rate for SentiWSP fine-tuning is set to 2×10^{-5}. To prevent model overfitting, set dropout to 0.5. The model is trained for 100 epochs with a batch size of 16. In the center channel of STMC-GCN, feature scores are decomposed into 10 bins with evenly distributed widths. The best scores for parameters α and β are 0.1 and 0.01, respectively. The optimal parameters are selected according to the validation set F1 score.

3.3 Baselines

We compared our **STMC-GCN** with state-of-the-art baselines, which can be briefly divided into three categories. The first category is the pipeline method, including **CMLA+**, **RINANTE+**, **Li-unified-R**, and **Peng-two-stage**, which were proposed by [12]. The second category is end-to-end methods, including **OTE-MTL** [25], **JET-BERT** [22], and **GTS-BERT** [20]. The third category is MRC-based methods, including **BMRC** [3], which is a multi-round MRC-based model that is end-to-end during the training phase but works in a pipeline during the inference phase. **ASTE-RL** [24] employs a hierarchical reinforcement learning framework. **ES-ASTE** [19] uses an enhanced span-level framework. **EMC-GCN** [1] proposes to fully exploit the relationship between words and different linguistic features.

3.4   Overall Experiments Performance

The overall performance of our proposed framework is shown in Table 3. According to the results, our model outperforms all baseline results. Compared to the EMC-GCN framework, our framework achieves significant absolute F1-score improvements of 2.06%, 2.52%, 3.04%, and 2.53% respectively. Our F1 score are better than the state-of-the-art baseline. As shown in Fig. 4, all of our models significantly and consistently outperform the two state-of-the-art marker-based

Table 3. The experimental results on the test set are presented in the table below. The best F1 results are shown in bold. The letters "#" and "△" indicate that the results are retrieved from [22] and [3,25], respectively. The symbol "†" indicates that we reproduced the results using their own code.

Model	14res			14lap			15res			16res		
	P	R	F1	P	R	F1	P	R	F1	P	R	F1
CMLA+[#]	39.18	47.12	42.83	30.09	36.89	33.20	33.16	39.84	36.28	41.34	42.10	41.72
RINANTE+[#]	31.42	39.38	34.95	21.64	18.66	20.07	29.88	30.06	29.97	25.68	22.30	23.94
Li-unified-R[#]	41.04	67.35	51.00	40.58	44.12	42.27	44.48	51.39	47.65	37.12	54.51	44.45
Peng-two-stage[#]	43.24	63.66	51.46	37.38	50.19	42.84	47.61	57.24	51.93	46.96	64.24	54.46
OTE-MTL[△]	62.00	55.97	58.71	49.51	39.22	43.76	56.37	41.12	47.65	51.47	51.98	51.72
JET-BERT[#]	70.56	55.94	62.40	55.28	47.99	51.30	63.89	51.89	57.27	58.39	63.83	60.98
GTS-BERT[†]	68.09	68.77	68.43	59.41	51.94	55.30	59.28	57.89	58.58	66.48	67.43	66.95
BMRC[△]	75.61	61.77	67.91	70.55	48.98	57.81	67.19	53.40	60.39	61.08	65.75	63.30
ASTE-RL[†]	68.60	66.58	67.57	64.80	53.41	58.58	63.85	60.31	62.03	66.91	69.89	68.36
ES-ASTE[†]	70.01	65.23	67.54	66.12	51.76	58.17	60.26	63.27	61.72	65.07	63.76	64.41
EMC-GCN[†]	70.58	72.64	71.59	60.05	58.25	59.13	61.54	62.29	61.91	64.36	71.40	67.71
Ours	78.15	69.58	**73.65**	70.56	54.78	**61.65**	72.34	58.89	**64.95**	72.42	68.18	**70.24**

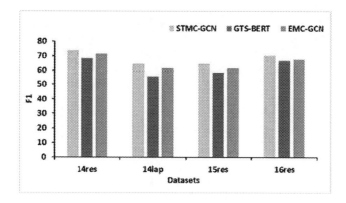

Fig. 4. The predicted results of different models for sentences.

baselines, GTS-BERT and EMC-GCN. Although our precision results are high, our recall results are slightly below optimal. One reason for this is that considering mutual span-level constraints makes the process of retrieval and pairing during inference more rigorous. Our graph model shows significant improvement over EMC-GCN with the same multi-channel graph convolutional network, demonstrating the effectiveness of STMC-GCN, which allows the model to learn more by modeling other valuable features and captures sentence relationships effectively.

4 Model Analysis

4.1 Ablation Experiments

Table 4. F1 scores of ablation study.

Model	14res	14lap	15res	16res
STMC-GCN	**73.65**	**61.65**	**64.95**	**70.24**
STMC-GCN w/o $\mathbb{G}_L$	71.99	59.67	62.74	67.87
STMC-GCN w/o $\mathbb{G}_E$	72.87	61.24	64.26	69.55
STMC-GCN w/o $\mathbb{G}_O$	73.01	60.88	63.95	68.76
STMC-GCN w/o Common Graph Channel	72.97	61.08	64.58	69.43

We performed ablation study on different modules in STMC-GCN. The results are presented in Table 4. According to the results, the semantic graph channel is the most important. In the absence of semantic features, the performance of STMC-GCN drops slightly on 14res and 14lap, which are 1.66% and 1.98%, respectively. However, it decreased by 2.21% and 2.37% on 15res and 16res, respectively. Since 15res and 16res contain less training data, language features can provide additional information when training data is insufficient. In addition, the common convolutional map channel plays a crucial role in extending the advantages of STMC-GCN due to the presence of "common" attributes in sentences.

4.2 Case Study

Figure 5 presents two case studies. In this example, aspect terms and opinion terms are highlighted in green and brown, respectively. The green line indicates that the aspect term and opinion term match and form a triplet with positive sentiment. In sentence (a), "priced" has two roles, namely as aspect term and opinion term. Obviously, STMC-GCN can handle the multiple relationships of word pairs very well. In sentence (b), both the first and second models incorrectly predicted sentiment polarity. Most likely because "wait" and "no" express only independent negative emotions. Considering only their independent representations, the above two models failed to correctly learn their semantic combinations. However, our STMC-GCNperforms better by treating them as whole spans with specific roles.

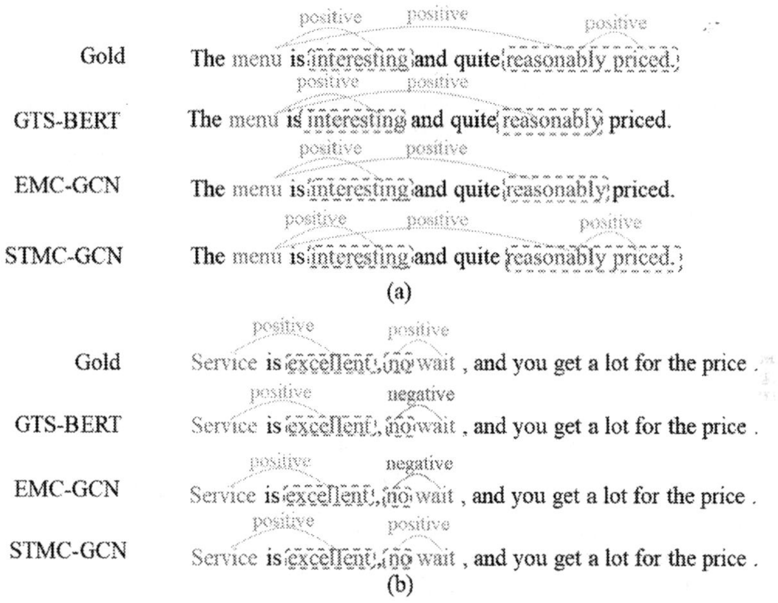

Fig. 5. The predicted results of different models for sentences.

5 Conclusions

In this paper, we present the STMC-GCN architecture for the ASTE task. By utilizing an emotional pre-training model, we enhance the recognition of emotional vocabulary. The multi-channel graph convolutional network explicitly combines sentence semantic features, importance features, and positional features, as well as their combinations in graphs, and fuses more refined features. Lastly, the extracted triplet module explores span-level information. We validate the effectiveness of the proposed STMC-GCN model on multiple datasets.

References

1. Chen, H., Zhai, Z., Feng, F., Li, R., Wang, X.: Enhanced multi-channel graph convolutional network for aspect sentiment triplet extraction. In: Proceedings of the 60th Annual Meeting of the Association for Computational Linguistics (Volume 1: Long Papers), pp. 2974–2985 (2022)

2. Chen, S., Liu, J., Wang, Y., Zhang, W., Chi, Z.: Synchronous double-channel recurrent network for aspect-opinion pair extraction. In: Proceedings of the 58th Annual Meeting of the Association for Computational Linguistics, pp. 6515–6524 (2020)

3. Chen, S., Wang, Y., Liu, J., Wang, Y.: Bidirectional machine reading comprehension for aspect sentiment triplet extraction. In: Proceedings of the AAAI Conference on Artificial Intelligence, vol. 35, pp. 12666–12674 (2021)

4. Fadel, A.S., Saleh, M.E., Abulnaja, O.A.: Arabic aspect extraction based on stacked contextualized embedding with deep learning. IEEE Access **10**, 30526–30535 (2022)

5. Fan, S., et al.: Sentiment-aware word and sentence level pre-training for sentiment analysis. arXiv preprint arXiv:2210.09803 (2022)

6. Hu, M., Liu, B.: Mining and summarizing customer reviews. In: Proceedings of the Tenth ACM SIGKDD International Conference on Knowledge Discovery and Data Mining, pp. 168–177 (2004)

7. Huang, L., et al.: First target and opinion then polarity: enhancing target-opinion correlation for aspect sentiment triplet extraction. arXiv preprint arXiv:2102.08549 (2021)

8. Jiang, J., Wang, A., Aizawa, A.: Attention-based relational graph convolutional network for target-oriented opinion words extraction. In: Proceedings of the 16th Conference of the European Chapter of the Association for Computational Linguistics: Main Volume, pp. 1986–1997 (2021)

9. Ke, G., He, D., Liu, T.Y.: Rethinking positional encoding in language pre-training. arXiv preprint arXiv:2006.15595 (2020)

10. Li, Z., Zou, Y., Zhang, C., Zhang, Q., Wei, Z.: Learning implicit sentiment in aspect-based sentiment analysis with supervised contrastive pre-training. arXiv preprint arXiv:2111.02194 (2021)

11. Liu, J., Zheng, S., Xu, G., Lin, M.: Cross-domain sentiment aware word embeddings for review sentiment analysis. Int. J. Mach. Learn. Cybern. **12**, 343–354 (2021)

12. Peng, H., Xu, L., Bing, L., Huang, F., Lu, W., Si, L.: Knowing what, how and why: a near complete solution for aspect-based sentiment analysis. In: Proceedings of the AAAI Conference on Artificial Intelligence, vol. 34, pp. 8600–8607 (2020)

13. Pontiki, M., Galanis, D., Pavlopoulos, J., Papageorgiou, H., Manandhar, S.: SemEval-2014 task 4: aspect based sentiment analysis. In: Proceedings of International Workshop on Semantic Evaluation (SemEval 2014) (2014)

14. Pontiki, M., Galanis, D., Papageorgiou, H., Manandhar, S., Androutsopoulos, I.: Semeval-2015 task 12: aspect based sentiment analysis. In: Proceedings of the 9th International Workshop on Semantic Evaluation (SemEval 2015), pp. 486–495 (2015)

15. Pontiki, M., et al.: Semeval-2016 task 5: aspect based sentiment analysis. In: ProWorkshop on Semantic Evaluation (SemEval-2016), pp. 19–30. Association for Computational Linguistics (2016)

16. Song, L., Smola, A., Gretton, A., Borgwardt, K.M., Bedo, J.: Supervised feature selection via dependence estimation. In: Proceedings of the 24th International Conference on Machine Learning, pp. 823–830 (2007)

17. Veyseh, A.P.B., Nouri, N., Dernoncourt, F., Dou, D., Nguyen, T.H.: Introducing syntactic structures into target opinion word extraction with deep learning. arXiv preprint arXiv:2010.13378 (2020)

18. Wang, X., Zhu, M., Bo, D., Cui, P., Shi, C., Pei, J.: AM-GCN: adaptive multi-channel graph convolutional networks. In: Proceedings of the 26th ACM SIGKDD International Conference on Knowledge Discovery & Data Mining, pp. 1243–1253 (2020)

19. Wang, Y., Chen, Z., Chen, S.: ES-ASTE: enhanced span-level framework for aspect sentiment triplet extraction. J. Intell. Inf. Syst. **60**, 593–612 (2023)
20. Wu, Z., Ying, C., Zhao, F., Fan, Z., Dai, X., Xia, R.: Grid tagging scheme for aspect-oriented fine-grained opinion extraction. arXiv preprint arXiv:2010.04640 (2020)
21. Xu, H., Liu, B., Shu, L., Yu, P.S.: Double embeddings and CNN-based sequence labeling for aspect extraction. arXiv preprint arXiv:1805.04601 (2018)
22. Xu, L., Li, H., Lu, W., Bing, L.: Position-aware tagging for aspect sentiment triplet extraction. arXiv preprint arXiv:2010.02609 (2020)
23. Yan, H., Dai, J., Qiu, X., Zhang, Z., et al.: A unified generative framework for aspect-based sentiment analysis. arXiv preprint arXiv:2106.04300 (2021)
24. Yu Bai Jian, S., Nayak, T., Majumder, N., Poria, S.: Aspect sentiment triplet extraction using reinforcement learning. In: Proceedings of the 30th ACM International Conference on Information & Knowledge Management, pp. 3603–3607 (2021)
25. Zhang, C., Li, Q., Song, D., Wang, B.: A multi-task learning framework for opinion triplet extraction. arXiv preprint arXiv:2010.01512 (2020)

Exploring the Design Space of Unsupervised Blocking with Pre-trained Language Models in Entity Resolution

Chenchen Sun[1(✉)], Yuyuan Jin[1], Yang Xu[1], Derong Shen[2], Tiezheng Nie[2],
and Xite Wang[3]

[1] School of Computer Science and Engineering, Tianjin University of Technology, Tianjin, China
suncc_db@163.com
[2] School of Computer Science and Engineering, Northeastern University, Shenyang, China
{shendr,nietiezheng}@mail.neu.edu.cn
[3] College of Information Science and Technology, Dalian Maritime University, Dalian, China
wangxite@dlmu.edu.cn

Abstract. Entity resolution (ER) finds records that refer to the same entities in the real world. Blocking is an important task in ER, filtering out unnecessary comparisons and speeding up ER. Blocking is usually an unsupervised task. In this paper, we develop an unsupervised blocking framework based on pre-trained language models (B-PLM). B-PLM exploits the powerful linguistic expressiveness of the pre-trained language models. A design space for B-PLM contains two steps. (1) The Record Embedding step generates record embeddings with pre-trained language models like BERT and Sentence-BERT. (2) The Block Generation step generates blocks with clustering algorithms and similarity search methods. We explore multiple combinations in above two dimensions of B-PLM. We evaluate B-PLM on six datasets (Structured + dirty, and Textual). The B-PLM is superior to previous deep learning methods in textual and dirty datasets. We perform sufficient experiments to compare and analyze different combinations of record embedding and block generation. Finally, we recommend some good combinations in B-PLM.

Keywords: entity resolution · unsupervised blocking · pre-trained language models · data integration · deep learning

1 Introduction

Entity resolution (ER) is a fundamental problem in data integration and data governance. Blocking [1] is a key issue in ER, reducing unnecessary comparisons and improving ER efficiency. Blocking methods place similar records in the same block and then perform intra-block comparisons. Traditional blocking methods are based on blocking keys, which require human selections for each dataset. How to reduce human involvement? That is, how to get high-quality blocks without block keys. Meanwhile, traditional blocking techniques are difficult to achieve good blocking results on textual and dirty datasets. Thus, how to perform blocking on text and dirty datasets is another issue.

© The Author(s), under exclusive license to Springer Nature Switzerland AG 2023
X. Yang et al. (Eds.): ADMA 2023, LNAI 14176, pp. 228–244, 2023.
https://doi.org/10.1007/978-3-031-46661-8_16

Many deep learning (DL) works make contributions to reducing human efforts. As DL continues to evolve, DL has been widely used in ER [2, 3]. These works apply similarities between record embeddings to calculate the likelihood of matches. For efficiency, some DL-based ER methods [2, 3] use hash functions for blocking. DL-based blocking approaches are rare [4–6]. Meanwhile, blocking is still seen as a step in ER, rather than as a separate problem. Most current DL-based blocking approaches [2–4] use labeled record pairs to train representation models for blocking. Other DL-based blocking approaches [5] use the generated auxiliary labels for training. However, blocking essentially filters irrelevant record pairs, similar to indexing in databases. Therefore, blocking shouldn't use labeled data. Moreover, as described in [6], in most blocking scenarios there is no labeled data available. To this end, this paper explores the design space of unsupervised blocking solutions.

We propose an unsupervised blocking framework based on pre-trained language models (B-PLM). B-PLM is a simple but strong baseline for deep unsupervised blocking. Pre-trained language models (PLMs), like BERT [7], Sentence-BERT [8], Doc2Vec [9], and fastText [10], can provide rich semantic information for blocking. The potential of PLMs in unsupervised blocking has not been systematically explored. In addition, the PLMs are outstanding in handling long text data. At the same time, PLMs can handle other types of datasets (e.g., structured and dirty). Therefore, based on PLMs, B-PLM consists of two major steps: record embedding and block generation. Record embedding obtains the semantic features of records based on PLMs and aggregation methods. PLMs have strong capabilities of semantic representation and discrimination, which are useful for record representation. For block generation, the essence is to group records according to estimated similarities. Clustering algorithms [11–13] are often used for grouping tasks; top-k query and range query in databases also have similar capabilities. Therefore, we provide clustering algorithms and similarity search methods for block generation. Our experimental results demonstrate effectiveness of B-PLM. We design sufficient experiments to investigate impacts of different record embedding methods and different block generation methods. Finally, we recommend some blocking solutions in our framework.

The main contributions are as follows.

- We define a solution space for an unsupervised blocking framework based on pre-trained language models (B-PLM), consisting of five record embedding methods and seven block generation options.
- We demonstrate B-PLM's effectiveness with experimental evaluations on six datasets. Experimental results show that optimal methods in B-PLM greatly outperform the state-of-the-art work DeepBlocker [5]. This result demonstrates B-PLM on a textual dataset.
- We systematically compare and analyze characteristics and effectiveness of different record embedding methods and different block generation methods in B-PLM. Furthermore, we further studied the different layers of BERT and Sentence-BERT. Finally, we compared parameter sensitivity of block generation methods.

Organization. Section 2 introduces the B-PLM framework in detail. Section 3 conducts extensive experiments to compare different combinations and analyze each component in B-PLM. Section 4 presents related work. Section 5 concludes the paper.

2   The Design Space of Unsupervised Blocking with Pre-Trained Language Models

2.1   Framework Overview

Given a set of records from a single data source or multiple data sources, blocking filters out unnecessary comparisons by dividing the records that are likely to match into the same block as much as possible. To this end, we proposed an unsupervised blocking framework based on pre-trained language models (B-PLM), which contains two core steps, as shown in Fig. 1.

Step 1: **Record Embedding**. This step generates record embeddings for later block generation. First, each record transforms into a token sequence. Second, token sequences input a PLM to get record embeddings. Record embeddings are either directly generated or produced with aggregations of token embeddings. We explored the current dominant PLMS. This step provides five options: BERT + Average, Sentence-BERT, Doc2Vec, fastText + Average, and fastText + SIF, which are introduced in Sect. 2.2.

Step 2: **Block Generation**. With record embeddings, similar records should be grouped into the same blocks. In Fig. 1, block generation methods are divided into two categories: one is based on clustering and the other is based on similarity search. We explore the current common clustering algorithms and similarity search methods. Clustering algorithms include AHC, BIRCH, DBSCAN, AP and k-means; and similarity search methods include top-k query and range query, which are specified in Sect. 2.3.

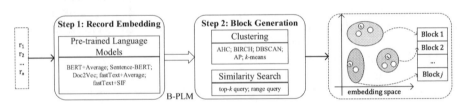

Fig. 1.   Unsupervised blocking framework with pre-trained language models.

2.2   Record Embedding

This step provides five options for record embedding, as Fig. 1 shows. We convert records into sequences for input into PLMs to obtain record embedding. Formally, a record $r = \{attr_i, val_i\}_{1 \le i \le k}$ converts to a sequence $S(r)$. This paper considers both the case with and without schema information. In the case with schema information, $S(r) = $ [CLS] [ATT] $attr_1$ [VAL] val_1 ...[ATT] $attr_k$ [VAL] val_k [SEP]. In the case without schema information, $S(r) = $ [CLS] val_1 ... val_k [SEP]. [ATT] and [VAL] indicate the beginning of the attribute name and attribute value, respectively.

BERT (Bidirectional Encoder Representations from Transformers) [7] adopts a multi-layer transformer encoder architecture, as Fig. 2 shows. BERT generates an embedding for each token, thus averaging all token embeddings in a sequence to get

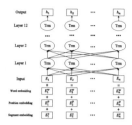

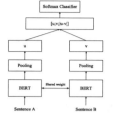

Fig. 2. Architecture of BERT.   **Fig. 3.** Architecture of Sentence-BERT.   **Fig. 4.** Architecture of Doc2Vec.

record embeddings. BERT tends to encode all records into a smaller space region, which results in most record pairs having a high similarity score.

SBERT (Sentence-BERT) [8] is a modification of BERT, as shown in Fig. 3. Sentence A and Sentence B are put into shared weight BERT models, respectively. Going by pooling gets the embedding u and embedding v corresponding to sentence A and sentence B. In contrast to BERT, SBERT enables semantically similar sentences to be similar in the embedding space as well.

Doc2Vec [9] learns embeddings from text fragments with variable lengths. As shown in Fig. 4, a paragraph vector is introduced in Doc2Vec. In B-PLM, Doc2Vec uses paragraph vectors as record embeddings.

FastText [10] is a character-level pre-trained word embedding model. FastText can generate embeddings for tokens outside the vocabulary and is robust to some spelling errors. B-PLM provides two fastText methods, fastText + Average and fastText + SIF. The average method does not consider the importance of each token in the record. In contrast, SIF [14] is a state-of-the-art solution for weighted averaging.

2.3   Block Generation

The block generation step contains two types of methods, clustering algorithms and similarity search methods.

Clustering Algorithms. There are five clustering algorithms: AHC, AP, DBSCAN, BIRCH and k-means.

AHC (Agglomerative Hierarchical Clustering) [11] is a bottom-up hierarchical clustering algorithm. In the beginning, each sample point is treated individually as a cluster. Clusters with high similarities are merged until all sample points form a cluster or a certain similarity threshold is reached.

BIRCH (Balanced Iterative Reducing and Clustering using Hierarchies) [12] is a hierarchical clustering algorithm. The main step in BIRCH is the creation of CF (clustering feature) tree. BIRCH does not work well if the clusters are not spherical, as it uses the concept of radius to control cluster boundaries.

DBSCAN (Density-Based Algorithm for Discovering Clusters in Large Spatial Databases with Noise) [11] is a density-based clustering algorithm that can discover clusters of arbitrary shape. It defines a cluster as the largest set of density connected points.

AP (Affinity Propagation) [13] is a graph-based clustering algorithm. The algorithm starts by treating all sample points as cluster centers, and clustering is run by passing messages between sample points. Automatically it identifies the number of clusters from sample points by maximizing the similarity sum of all sample points to their nearest cluster centers.

***k*-means** [11] is a partitioning-based clustering algorithm. First, k initial clustering centers are randomly selected. Next, each point is arranged into the cluster closest to that point. Subsequently, the cluster centers are recalculated for each cluster. It is repeated until the result is stable.

Similarity Search Methods. Similarity search methods include top-k query and range query.

Top-*k* query. For each record r, similarities between r and other records are first computed, and then records with the k highest similarities are selected to form a block with r. We use top-k for short in this paper.

Range Query. For each record r, similarities between r and other records are first computed, and then records with similarity scores above a given threshold are selected to form a block with r.

3   Comparative Experimental Evaluation and Analysis

3.1   Experiment Setup

Datasets. We evaluate B-PLM on six ER datasets, which are shown in Table 1. Most of the datasets in this paper are textual datasets, and previous blocking methods have struggled to achieve high-quality blocking on textual datasets. Cora is a common ER dataset. Notebook and Altosight are both from Blocking Contest in SIGMOD 2022[1]. Notebook is a dataset about laptops, with only one title attribute. Altosight is a dataset about electronic products, with attribute values that may be misplaced or missing. WDC_cameras is from the Web Data Commons project [15] and we use the cameras_small version. In WDC_cameras, we only use the title attribute. Both Abt-Buy and Amazon-Google2 are dual-source datasets about products.

We further analyze the characteristics of datasets. We obtain the entity redundancy figure for each dataset, which details the number of records corresponding to each entity, as shown in Fig. 5. In Cora, most entities have a large amount of redundancy. In Notebook, Altosight and WDC_cameras, few entities have more than 8 corresponding records. Almost all entities in Abt-Buy and Amazon-Google2 have only 1–3 corresponding records. Cora has the highest entity redundancy while Abt-Buy and Amazon-Google have the lowest.

Metrics. We use pair completeness (PC), reduction ratio (RR), $F\alpha$ and PC-RR curve as evaluation metrics. $PC = |G \cap C|/|G|$, $RR = 1 - (|C|/|D|)$, where G denotes all truly matched pairs, C denotes candidate pairs generated by blocking, and D denotes all record pairs

---

[1] http://sigmod2022contest.eastus.cloudapp.azure.com.

Table 1. Dataset details.

	Dataset	#records	#attributes	Type
single-source	Cora	1295	12	Structured + Dirty
	Notebook	1661	1	Textual
	Altosight	1993	5	Structured + Dirty
	WDC_cameras	1904	1	Textual
dual-source	Abt-Buy	1081 + 1092	3	Textual
	Amazon-Google2	1363 + 3226	2	Textual

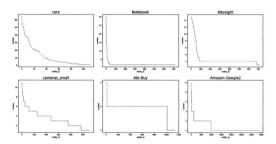

Fig. 5. Entity redundancy figures of datasets.

in a dataset. PC measures how many truly matched pairs are retained in blocks, while RR measures how many comparisons are reduced. We use $F\alpha$ (the harmonic mean of PC and RR) to measure the overall performance, $F\alpha = 2 \cdot PC \cdot RR/(PC + RR)$. In the PC-RR curve, the closer the curve is to the upper right, the better the performance of a blocking method is.

Baselines. DeepBlocker [5] is a state-of-the-art deep blocking work. DeepBlocker proposes eight blocking solutions. DeepBlocker first uses fastText to get the word embedding. The word embeddings are then aggregated into record embeddings and finally paired based on cosine similarity between vectors. We select the three best solutions: SIF, AE, and CTT as the baseline. AE uses full connection layers as encoder and decoder and the output of encoder as the record embedding. CTT uses a data generation procedure to automatically generate labeled record pairs. Then CTT learns record embeddings through classification tasks. In block generation, SIF, AE, and CTT all use top-k query.

Settings. Experiments are implemented with following Python libraries: Scikit-learn, transformer, sentence-transformer, fastText and gensim. All experiments are run on a server with an Inter Core I9-10900K CPU @ 3.70 GHz, 32 G RAM and an NVIDIA Quadro RTX 4000 GPU. Record embedding uses SBERT in default. All PLMs are used in this paper without fine-tuning. For BERT, we use the pre-trained version of "bert-base-uncased". For SBERT we use the pre-trained version of "all-mpnet-base-v2". For Doc2Vec we use the pre-trained version of "English Wikipedia DBOW". For fastText, we use the pre-trained version of "Wiki.En". Except for k-means the other block generation methods do not specify the number of clusters, which is more in line with real-world situations. The k of k-means is chosen from 10 to 50 and the step size is set to 2. Similarity search methods use cosine similarity.

3.2 Overall Performance

For each dataset, we report the best result from SIF, AE, and CTT as DeepBlocker. Overall comparison results are presented in Table 2 and Table 3. Overall B-PLM has advantages in textual and dirty datasets. The best performer in our framework is SBERT + top-k. SBERT + top-k has an average Fα of 93.65% on six datasets and a 5.55% improvement compared to the baseline. SBERT + range query has an average Fα of 92.85% on the six datasets, and a 4.75% improvement compared to the baseline. SBERT + AHC and SBERT + k-means also achieve better results than the baseline. In addition, except for Notebook, on the rest datasets, the best, second, and third solutions are all from our framework, demonstrating B-PLM's effectiveness.

The powerful semantic expressiveness of SBERT is reflected in the results. The results of DeepBlocker and SBERT + top-k, which both take the same block generation method, and we find that SBERT + top-k performs significantly better than DeepBlocker on most datasets.

For Cora, the clustering algorithms perform better overall. As illustrated in Fig. 5, many entities in Cora have high redundancy. In the remaining datasets, most entities have low redundancy (no more than five records). Clustering methods can generate blocks of different sizes according to the redundancy level of each entity, which is a self-adaption process. Therefore, clustering methods are more advantageous on datasets with various high entity redundancy.

The dual-source datasets Abt-Buy and Amazon-Google2 are more suitable for top-k query and range query. Because the dual-source dataset assumes that there are no duplicates within each data source. The executions of top-k query and range query naturally ensure that each record of data source A finds the most similar records in data source B to form a block. Top-k query and range query are not interfered with by similar records within data sources, making them easier to get high-quality blocks. Dual-source datasets are more difficult for clustering-based block generation methods. Because clustering-based methods run in the entire set of two data sources, but do not only conduct comparisons between two data sources.

3.3 Analysis of Record Embedding

Effectiveness of Record Embedding Methods. We compare record embeddings with different PLMs in blocking. The results are shown in Tables 4, 5, 6, 7, 8, 9 and 10. For both BERT and SBERT, we report the results at layer 12, and for *BERT and *SBERT, we report their results at optimal layers.

Overall, the best record embedding method is SBERT. The overall performance of SBERT is better than other PLMs for each block generation method. SBERT's strong linguistic power is demonstrated. The overall results of BERT are generally worse than those of SBERT, mainly because the focus of BERT in the pre-training step is to produce

Table 2. Overall Performance on six datasets (part 1 in 2). Bold, single underline and double underline represent the best, second and third, respectively.

Method	Dataset											
	Cora			Notebook			Altosight			WDC_cameras		
	PC(%)	RR(%)	Fα(%)	PC(%)	RR(%)	Fα(%)	PC(%)	RR(%)	Fα(%)	PC(%)	RR(%)	Fα(%)
DeepBlocker	88.08	91.73	90.38	96.80	90.68	93.64	68.76	79.49	73.74	83.65	91.83	87.55
SBERT + AHC	93.18	94.03	93.60	93.99	97.76	95.84	70.08	79.74	74.60	89.64	87.66	88.63
SBERT + BIRCH	93.47	96.08	94.76	89.52	97.84	93.49	70.77	82.57	76.21	72.97	91.93	81.36
SBERT + DBSCAN	92.10	96.94	94.46	97.30	48.54	64.77	73.63	55.17	63.08	81.22	81.28	81.25
SBERT + AP	93.51	96.77	95.11	88.88	87.69	88.28	64.57	83.05	72.65	81.22	88.12	84.53
SBERT + k-means	93.46	97.07	95.23	81.49	97.16	88.64	74.52	90.00	81.53	85.19	89.54	87.31
SBERT + top-k	93.18	94.80	93.98	95.42	88.72	91.95	86.95	88.00	87.47	92.13	91.83	91.98
SBERT + range query	91.74	95.44	93.55	94.95	97.58	96.25	79.49	82.36	80.90	91.36	90.79	91.07

Table 3. Overall Performance on six datasets (part 2 in 2).

Method	Dataset						
	Abt-Buy			Amazon-Google2			AVG of Six
	PC(%)	RR(%)	Fα(%)	PC(%)	RR(%)	Fα(%)	Fα(%)
DeepBlocker	95.04	95.24	95.14	82.86	94.17	88.16	88.10
SBERT + AHC	97.08	95.41	96.24	86.29	97.38	91.50	90.07
SBERT + BIRCH	91.44	93.55	92.48	67.10	95.88	78.95	86.21
SBERT + DBSCAN	91.83	84.34	87.92	86.46	69.71	77.19	78.11
SBERT + AP	88.23	93.87	90.96	61.95	96.35	75.42	84.49
SBERT + k-means	92.12	96.64	94.33	72.32	94.96	82.11	88.19
SBERT + top-k	99.22	97.25	98.23	98.46	98.14	98.30	93.65
SBERT + range query	98.64	96.77	97.70	97.00	98.27	97.63	92.85

better token embeddings rather than sentence embeddings. In contrast, SBERT learns embeddings of whole sentences directly, requiring that semantically similar sentences

Table 4. Results of AHC using different record embedding methods. Bold, single underline and double underline represent the best, second and third, respectively.

Embedding	Dataset						
	Cora Fα(%)	Notebook Fα(%)	Altosight Fα(%)	WDC_cameras Fα(%)	Abt-Buy Fα(%)	Amazon-Google2 Fα(%)	AVG Fα(%)
BERT	74.38	94.15	54.98	65.55	55.61	65.60	68.38
SBERT	93.60	95.84	**74.60**	**88.63**	**96.24**	**91.50**	90.07
Doc2Vec	90.99	94.86	65.54	80.90	66.40	64.60	77.22
fastText + Average	75.03	93.41	58.49	72.26	54.85	56.54	68.43
fastText + SIF	88.80	95.64	43.45	74.81	86.42	73.74	77.14
*BERT	87.56(l = 1)	94.71(l = 5)	63.82(l = 0)	78.45(l = 0)	69.92(l = 1)	66.62(l = 1)	76.85
*BERT + Schema	1.53(l = 12)	96.60(l = 6)	63.96(l = 1)	74.17(l = 0)	54.41(l = 2)	62.91(l = 11)	58.93
*SBERT	**96.17(l = 0)**	**96.65(l = 0)**	**74.60(l = 12)**	**88.63(l = 12)**	**96.24(l = 12)**	**91.50(l = 12)**	**90.63**
*SBERT + Schema	92.43(l = 12)	96.64(l = 0)	76.40(l = 12)	85.65(l = 12)	93.45(l = 12)	85.41(l = 12)	88.33

are also similar in the embedding space. The performance differences between BERT and *BERT are large. The reasons for this will be analyzed later.

Table 5. Results of BIRCH using different record embedding methods.

Embedding	Dataset						
	Cora Fα(%)	Notebook Fα(%)	Altosight Fα(%)	WDC_cameras Fα(%)	Abt-Buy Fα(%)	Amazon-Google2 Fα(%)	AVG Fα(%)
BERT	76.60	92.80	52.89	63.75	36.88	55.08	63.00
SBERT	94.76	93.49	**76.21**	**81.36**	**92.48**	78.95	86.21
Doc2Vec	26.39	49.95	24.95	6.80	0.19	0.51	18.13
fastText + Average	76.75	90.18	25.89	44.07	31.10	52.13	53.35
fastText + SIF	77.09	94.69	48.84	60.63	63.68	61.41	67.72
*BERT	88.15(l = 3)	94.70(l = 2)	64.65(l = 0)	64.48(l = 9)	50.60(l = 3)	57.19(l = 5)	69.96
*BERT + Schema	79.29(l = 2)	93.81(l = 4)	55.17(l = 8)	67.68(l = 12)	47.49(l = 4)	61.12(l = 11)	67.43
*SBERT	**96.39(l = 0)**	**94.82(l = 0)**	**76.21(l = 12)**	**81.36(l = 12)**	**92.48(l = 12)**	**78.95(l = 12)**	**86.70**
*SBERT + Schema	92.93(l = 12)	94.36(l = 5)	73.77(l = 12)	80.44(l = 12)	89.17(l = 12)	74.93(l = 12)	84.27

FastText + Average is less effective than fastText + SIF. This suggests that the core words in each record play dominant roles. Another aspect, Doc2Vec achieves good performances in AHC, top-k, and range query (top 3 or so). But it has poorer results in all other block generation methods, which are all Euclidean distance-based block generation

Table 6. Results of DBSCAN using different record embedding methods.

Embedding	Dataset						
	Cora Fα(%)	Notebook Fα(%)	Altosight Fα(%)	WDC_cameras Fα(%)	Abt-Buy Fα(%)	Amazon-Google2 Fα(%)	AVG Fα(%)
BERT	71.07	62.84	55.73	66.58	66.43	58.51	63.53
SBERT	**94.46**	64.77	**63.08**	**81.25**	**87.92**	**77.19**	78.11
Doc2Vec	48.12	17.07	11.75	1.66	0.91	6.71	14.37
fastText + Average	76.86	66.63	48.81	53.79	55.67	53.07	59.14
fastText + SIF	64.97	65.52	56.08	61.75	61.24	58.04	61.27
*BERT	87.15(l = 1)	**67.95(l = 8)**	59.61(l = 0)	71.46(l = 4)	68.63(l = 0)	59.30(l = 1)	69.02
*BERT + Schema	83.43(l = 3)	67.55(l = 2)	58.77(l = 3)	74.34(l = 2)	64.95(l = 4)	57.88(l = 3)	67.82
*SBERT	**94.46(l = 12)**	67.43(l = 8)	**63.08(l = 12)**	**81.25(l = 12)**	**87.92(l = 12)**	**77.19(l = 12)**	**78.56**
*SBERT + Schema	92.48(l = 12)	67.72(l = 0)	62.67(l = 12)	77.92(l = 3)	85.33(l = 12)	73.47(l = 12)	76.60

Table 7. Results of AP using different record embedding methods.

Embedding	Dataset						
	Cora Fα(%)	Notebook Fα(%)	Altosight Fα(%)	WDC_cameras Fα(%)	Abt-Buy Fα(%)	Amazon-Google2 Fα(%)	AVG Fα(%)
BERT	82.05	90.28	58.54	72.43	71.51	54.93	71.62
SBERT	**95.11**	88.28	72.65	**84.53**	**90.96**	**75.42**	84.49
Doc2Vec	53.43	68.73	44.26	54.20	4.93	11.01	39.43
fastText + Average	81.58	89.36	40.28	62.43	60.32	52.46	64.41
fastText + SIF	70.19	77.11	54.20	62.21	70.17	68.50	67.06
*BERT	89.96(l = 2)	92.10(l = 1)	69.97(l = 0)	74.30(l = 1)	71.51(l = 12)	54.93(l = 12)	75.46
*BERT + Schema	72.35(l = 12)	91.97(l = 3)	64.87(l = 1)	73.03(l = 12)	52.71(l = 11)	50.95(l = 11)	67.65
*SBERT	**95.11(l = 12)**	93.18(l = 0)	72.65(l = 12)	**84.53(l = 12)**	**90.96(l = 12)**	**75.42(l = 12)**	**85.31**
*SBERT + Schema	94.51(l = 12)	**93.34(l = 0)**	**75.88(l = 12)**	81.21(l = 12)	87.54(l = 12)	68.89(l = 12)	83.56

methods, so we infer that Doc2Vec is not suitable for block generation methods using Euclidean distance. All in all, SBERT shows its great advantage in record embedding, compared to other PLMs. In addition, this paper conducts experiments on BERT and SBERT which consider schema information. In most cases, schema information does not improve blocking results. Therefore, the schema information is not considered in this paper.

Table 8. Results of k-means using different record embedding methods.

Embedding	Dataset						
	Cora Fα(%)	Notebook Fα(%)	Altosight Fα(%)	WDC_cameras Fα(%)	Abt-Buy Fα(%)	Amazon-Google2 Fα(%)	AVG Fα(%)
BERT	82.80	92.62	60.52	69.64	53.26	<u>58.49</u>	69.56
SBERT	<u>95.23</u>	88.64	**81.53**	**87.31**	**94.33**	**82.11**	<u>88.19</u>
Doc2Vec	88.20	92.04	64.43	<u>79.18</u>	50.49	47.06	70.23
fastText + Average	80.60	90.43	51.64	73.37	<u>65.54</u>	54.49	69.35
fastText + SIF	82.01	89.93	48.44	66.68	<u>82.33</u>	<u>66.89</u>	72.71
*BERT	89.74(l = 1)	<u>94.57(l = 5)</u>	<u>69.42(l = 1)</u>	76.09(l = 0)	58.29(l = 1)	<u>58.49(l = 12)</u>	74.43
*BERT + Schema	79.74(l = 1)	94.22(l = 1)	57.56(l = 12)	74.99(l = 0)	32.78(l = 12)	40.35(l = 12)	63.27
*SBERT	**96.07(l = 0)**	**94.67(l = 4)**	**81.53(l = 12)**	**87.31(l = 12)**	**94.33(l = 12)**	**82.11(l = 12)**	**89.34**
*SBERT + Schema	<u>93.40(l = 12)</u>	94.52(l = 0)	<u>78.81(l = 12)</u>	84.43(l = 12)	<u>92.33(l = 12)</u>	<u>79.12(l = 12)</u>	87.10

Table 9. Results of top-k using different record embedding methods.

Embedding	Dataset						
	Cora Fα(%)	Notebook Fα(%)	Altosight Fα(%)	WDC_cameras Fα(%)	Abt-Buy Fα(%)	Amazon-Google2 Fα(%)	AVG Fα(%)
SBERT	<u>93.98</u>	91.95	**87.47**	**91.98**	**98.23**	**98.30**	<u>93.65</u>
Doc2Vec	91.32	91.21	<u>82.32</u>	89.64	<u>95.28</u>	<u>90.72</u>	90.08
fastText + Average	88.77	<u>92.89</u>	68.66	87.26	84.68	76.58	83.14
fastText + SIF	90.36	**93.29**	70.52	87.37	95.08	88.13	87.46
*BERT	92.17(l = 1)	92.48(l = 0)	82.29(l = 1)	<u>90.09(l = 0)</u>	90.68(l = 0)	80.57(l = 1)	88.05
*BERT + Schema	87.98(l = 1)	92.36(l = 0)	77.54(l = 1)	88.47(l = 0)	86.24(l = 12)	76.27(l = 12)	84.81
*SBERT	**94.98(l = 0)**	92.51(l = 0)	**87.47(l = 12)**	**91.98(l = 12)**	**98.23(l = 12)**	**98.30(l = 12)**	**93.91**
*SBERT + Schema	<u>94.10(l = 0)</u>	<u>92.95(l = 0)</u>	<u>87.39(l = 12)</u>	90.92(l = 12)	<u>98.21(l = 12)</u>	<u>95.12(l = 12)</u>	93.12

Layer Analysis of Sentence-BERT and BERT. We investigate effectiveness of each layer in SBERT and BERT for blocking on Cora, Altosight, and Amazon-Google2. We get output token embeddings from each layer of BERT and SBERT and apply an average operation to obtain record embeddings.

In Fig. 6, 7 and 8, for BERT, the highest Fα is often found in layers 0–3. Therefore, shallow layers are superior to other layers in most cases. A study [16] has shown that surface information features are in the shallow network, syntactic information features and semantic information features are in the middle and higher layer network. Surface

Table 10. Results of range query using different record embedding methods.

Embedding	Dataset						
	Cora Fα(%)	Notebook Fα(%)	Altosight Fα(%)	WDC_cameras Fα(%)	Abt-Buy Fα(%)	Amazon-Google2 Fα(%)	AVG Fα(%)
BERT	63.52	80.66	62.78	**74.54**	81.72	73.33	72.76
SBERT	93.55	96.25	80.90	**91.07**	**97.70**	**97.63**	92.85
Doc2Vec	89.13	95.46	78.33	87.66	92.37	85.86	88.14
fastText + Average	73.94	95.85	62.44	81.79	83.22	70.00	77.87
fastText + SIF	90.27	96.15	72.12	87.07	94.92	87.83	88.06
*BERT	87.62(l = 1)	96.50(l = 0)	75.61(l = 0)	85.91(l = 0)	88.58(l = 0)	74.82(l = 1)	84.84
*BERT + Schema	14.77(l = 12)	96.85(l = 3)	67.13(l = 12)	85.21(l = 0)	76.96(l = 2)	76.01(l = 12)	69.49
*SBERT	**93.64(l = 0)**	**97.43(l = 0)**	80.90(l = 12)	**91.07(l = 12)**	**97.70(l = 12)**	**97.63(l = 12)**	**93.06**
*SBERT + Schema	92.29(l = 12)	**97.43(l = 0)**	83.45(l = 12)	88.92(l = 0)	97.05(l = 12)	96.40(l = 12)	92.59

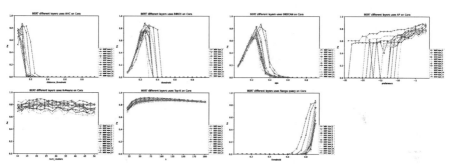

Fig. 6. Results of different BERT layers on Cora, layer = 0 indicates the input of BERT (token embedding + segment embedding + position embedding).

information is some intuitive information including sentence length etc. Syntactic information and semantic information include a great deal of grammar. In blocking, if two records are similar in surface information, then the two records should be placed in the same block. The syntactic and semantic details are more meaningful in matching of ER. For most experimental results, our intuition is correct. On Amazon-Google2, the optimal layers of BIRCH, AP, and k-means fall into other layers because Amazon-Google2's records are longer than other datasets. Longer records can result in common tokens being repeated in many records, and this is where syntax information is needed to help get better record embeddings. In addition, the shallower the layer is, the wider the focus of each token is [17]. Thus, each token in the shallow layers aggregates with each other, leading to drops in performance in the shallow layers.

In Fig. 9, 10 and 11, we observe that for SBERT, good blocking results can be seen at the 0-th layer and the 11-th layer, but the optimal layer is usually the 12-th layer. We

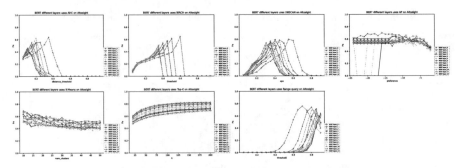

Fig. 7. Results of different BERT layers on Altosight.

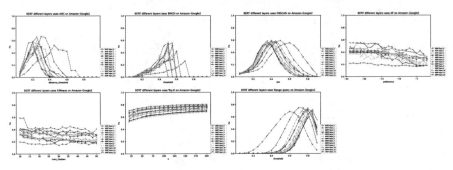

Fig. 8. Results of different BERT layers on Amazon-Google2.

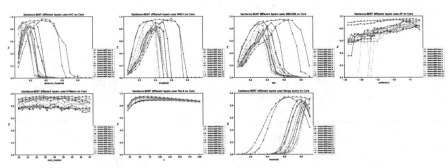

Fig. 9. Results of different SBERT layers on Cora, layer = 0 indicates the input of SBERT (token embedding + position embedding).

can draw a general conclusion: the optimal layer comes most from the last layer, and the first two layers generally outperform the middle layers. SBERT benefits from its whole-sentence learning, with the 12-th layer output showing a clear advantage on the blocking task. Of course, due to the nature of the blocking task itself, layers that focus on surface information can also work well.

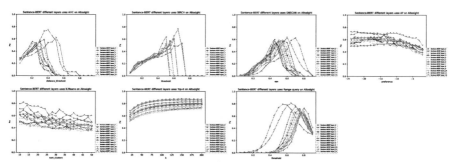

Fig. 10. Results of different SBERT layers on Altosight.

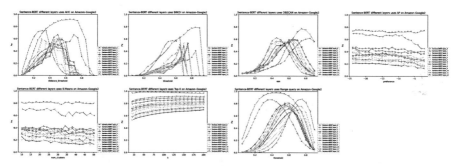

Fig. 11. Results of different SBERT layers on Amazon-Google2.

3.4 Analysis of Block Generation

Effectiveness of Block Generation Methods. In Table 2, we find that the best block generation methods in B-PLM are top-k and range query. The simple similarity search method achieves optimal performance. AHC performs better than other clustering algorithms. Firstly, AHC is a bottom-up clustering algorithm. Thus naturally more similar records are preferentially grouped into the same blocks. Also, AHC and k-means use cosine distances, whereas other clustering algorithms use Euclidean distances. In higher dimensional spaces, the Euclidean distance loses efficacy, whereas the cosine distance does not have this problem. In addition, k-means also achieves good results, especially on Cora, a more redundant dataset, and on Altosight, a dirty dataset, so we believe that k-means may be suitable for more difficult settings. BIRCH, DBSCAN, and AP perform poorly overall. To explore the characteristics of each block generation method more fully, we present the PC-RR curves for different block generation variants on six datasets, as shown in Fig. 12. The optimal block generation method is top-k, as the top-k method has a more slow decrease in RR as PC increases. The performance of range query is second only to that of top-k. For AHC, RR declines sharply if PC exceeds a certain value. In general, the top four block generation methods in terms of RC-RR curve are top-k, range query, AHC and k-means. Furthermore, we believe that DBSCAN is less suitable for blocking. Firstly, density measurement is inaccurate in high dimensional spaces where Euclidean distance's accuracy declines. Secondly, we cannot guarantee that cluster densities in a dataset are similar, but DBSCAN is only applicable to data with similar densities across different clusters. Finally, DBSCAN has several parameters that need to be adjusted, which decreases its usability for blocking.

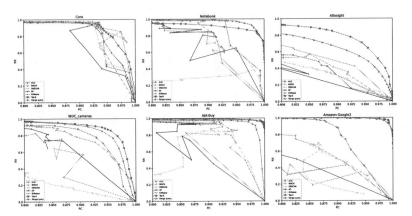

Fig. 12. PC-RR curves of different block generation methods on different datasets.

Parameter Sensitivity in Block Generation. We investigate parameters of different block generation methods. With SBERT, how parameter adjustments affect blocking qualities in different block generation methods are presented in Fig. 9, 10 and 11. AP, k-means and top-k are relatively insensitive to parameters, where blocking qualities on most layers change mildly with parameter adjustments. BIRCH is more sensitive to parameters. In general, top-k, k-means, and AP are robust to parameter changes, and their appropriate parameters are easy to catch.

4 Related Work

Non-deep learning blocking has been extensively studied [1]. There are six types of blocking techniques. 1. The Traditional Blocking method inserts all records with the same blocking key value (BKV) into the same block. 2. The basic idea of Sorted Neighborhood Blocking is to sort the database according to BKVs and then generate blocks by moving windows. 3. The basic idea of Q-Gram Based Blocking creation is to use q-grams for each BKV variation. 4. Suffix Array-Based Blocking is to insert the BKVs and their suffixes into a suffix array-based inverted index. 5. Canopy Clustering is based on the idea of using a computationally cheap clustering approach to create high-dimensional overlapping clusters. 6. String-Map-Based Blocking is based on a mapping of strings to objects in a multi-dimensional Euclidean space, where the distances between strings are preserved. Recently, DL-based blocking methods have emerged, which obtain record embedding by applying DL models, and then perform embedding-based block generation. DeepER [2] uses an RNN or LSTM to learn record embeddings for facilitating ER, and LSH is employed for blocking. BERT-ER [3] is based on BERT learn record embeddings and uses learnable hash functions for blocking. AutoBlock [4] uses fast-Text [10] to obtain token embeddings and learns how to combine token embeddings into record embeddings with labeled data. The above methods all use labeled data. Fabio Azzalini et al. [6] use fastText to obtain token embeddings and generate record embeddings by averaging or recurrent neural network (RNN). Then block generation is performed by LSH or clustering-based methods. DeepBlocker [5] utilizes fastText for token embeddings. To obtain record embeddings, DeepBlocker proposes four self-supervised tasks to accomplish the aggregation of token embeddings. Please refer to Sect. 3.1 for details of DeepBlocker. Most DL-based approaches still see blocking as part of ER, whereas B-PLM sees blocking as a separate task. B-PLM leverages the powerful linguistic expressiveness of PLM compared to the above methods, providing a wealth of external information for unsupervised blocking tasks.

5   Conclusion

This paper explores a design space of unsupervised blocking solutions based on PLMs. B-PLM is available as a baseline for unsupervised blocking. We compare seven representative solutions. The proposed solutions are experimentally proven to outperform the state-of-the-art DL-based blocking framework. The experimental results proved the advantages of B-PLM on textual and dirty datasets. Based on thorough experiments, we suggest recommended methods for record embedding (SBERT) and block generation (top-k). In the next step, we will consider some more advanced PLM models, and consider fine-tuning PLMs and dimension reduction.

Acknowledgments. This work is supported by the National Natural Science Foundation of China (Grant Nos. 62002262, 62172082, 62072086, 62072084).

References

1. Christen, P.: A survey of indexing techniques for scalable record linkage and deduplication. IEEE Trans. Knowl. Data Eng. **24**(9), 1537–1555 (2011)
2. Ebraheem, M., Thirumuruganathan, S., Joty, S., Ouzzani, M., Tang, N.: Distributed representations of tuples for entity resolution. Proc. VLDB Endowment **11**(11), 1454–1467 (2018)
3. Li, B., Miao, Y., Wang, Y., Sun, Y., Wang, W.: Improving the efficiency and effectiveness for BERT-based entity resolution. In: Proceedings of the AAAI Conference on Artificial Intelligence, vol. 35, no. 15, pp. 13226–13233 (2021)
4. Zhang, W., Wei, H., Sisman, B., Dong, X. L., Faloutsos, C., Page, D.: AutoBlock: a hands-off blocking framework for entity matching. WSDM, pp. 744–752 (2020)
5. Thirumuruganathan, S., Li, H., Tang, N., Ouzzani, M., Govind, Y., Paulsen, D., Fung, G., Doan, A.: Deep learning for blocking in entity matching: a design space exploration. Proc. VLDB Endow. **14**(11), 2459–2472 (2021)
6. Azzalini, F., Jin, S., Renzi, M., Tanca, L.: Blocking techniques for entity linkage: a semantics-based approach. Data Sci. Eng. **6**(1), 20–38 (2020)
7. Devlin, J., Chang, M.W., Lee, K., Toutanova, K.: Bert: pre-training of deep bidirectional transformers for language understanding. arXiv preprint arXiv:1810.04805 (2018)

8. Reimers, N., Gurevych, I.: Sentence-BERT: sentence embeddings using siamese BERT-networks. In: EMNLP-IJCNLP, pp. 3980–3990 (2019)
9. Le, Q., Mikolov, T.: Distributed representations of sentences and documents. In: ICML, pp. 1188–1196 (2014)
10. Bojanowski, P., Grave, E., Joulin, A., Mikolov, T.: Enriching word vectors with subword information. Trans. Assoc. Comput. Linguistics, **5**, 135–146 (2017)
11. Han, J., Pei, J., Tong, H.: Data Mining: Concepts and Techniques. Morgan Kaufmann (2022)
12. Zhang, T., Ramakrishnan, R., Livny, M.: BIRCH: an efficient data clustering method for very large databases. ACM SIGMOD Rec. **25**(2), 103–114 (1996). https://doi.org/10.1145/235968.233324
13. Frey, B.J., Dueck, D.: Clustering by passing messages between data points. Science **315**(5814), 972–976 (2007)
14. Arora, S., Liang, Y., Ma, T.: A simple but tough-to-beat baseline for sentence embeddings. In: ICLR (Poster) (2017)
15. Primpeli, A., Peeters, R., Bizer, C.: The WDC training dataset and gold standard for large-scale product matching. In: Companion Proceedings of the 2019 World Wide Web Conference, pp. 381–386 (2019)
16. Jawahar, G., Sagot, B., Seddah, D.: What does BERT learn about the structure of language?. ACL **1**, 3651–3657 (2019)
17. Clark, K., Khandelwal, U., Levy, O., Manning, C.D.: What does BERT look at? An analysis of bert's attention: BlackboxNLP@ACL, 276–286 (2019)

Joint Modeling of Local and Global Semantics for Contrastive Entity Disambiguation

Yuhua Ke, Shaojie Xue, Ziqi Chen, and Rui Meng[✉]

Guangdong Provincial Key Laboratory of Interdisciplinary Research and Application for Data Science, BNU-HKBU United International College, Zhuhai, China
ruimeng@uic.edu.cn

Abstract. Entity disambiguation (ED) is a critical natural language processing (NLP) task that involves identifying and linking entity mentions in the text to their corresponding real-world entities in reference knowledge graphs (KGs). Most existing efforts perform ED by firstly learning the representations of mention and candidate entities using a variety of features and subsequently assessing the compatibility between mention and candidate entities as well as the coherence between entities. Despite advancements in the field, the limited textual descriptions of mentions and entities still lead to semantic ambiguity, resulting in suboptimal performance for the entity disambiguation task. In this work, we propose a novel framework *LogicED*, which considers both *L*ocal and *g*lobal semant*i*cs for *c*ontrastive *E*ntity *D*isambiguation. Specifically, we design a local contextual module, which utilizes a candidate-aware self-attention (CASA) model and the contrastive learning strategy, to learn robust and discriminative contextual embeddings for both mentions and candidate entities. Furthermore, we propose a global semantic graph module that takes into account both the local mention-entity compatibility and the global entity-entity coherence to optimize the entity disambiguation from a global perspective. Extensive experiments on benchmark datasets demonstrate that our proposed framework surpasses the state-of-the-art baselines.

Keywords: Entity Disambiguation · Contrastive Learning · Local and Global Semantics

1 Introduction

Entity Disambiguation (ED) refers to the process of accurately associating a mention in a given text with the appropriate entity from a set of candidate entities obtained from reference knowledge graphs (KGs). ED is a fundamental and critical task in various natural language processing (NLP) applications, including information extraction [22], question answering [33], text generation [21], and semantic parsing [2]. The task is, however, challenging owing to the semantic ambiguity that arises from the limited textual description of mentions and

X. Yang et al. (Eds.): ADMA 2023, LNAI 14176, pp. 245–259, 2023.
https://doi.org/10.1007/978-3-031-46661-8_17

entities. For instance, as depicted in Fig. 1, the sentence "In the 2022 World Cup final, Argentina wins against France 4-2 through a penalty shoot-out." comprises three mentions, namely "2022 World Cup", "Argentina" and "France", each of which has corresponding candidate entities. For example, the mention "Argentina" may refer to different entities such as the "Argentina national football team" or the "Argentina Republic" within the KG.

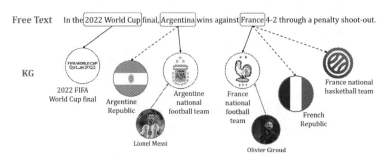

Fig. 1. An example of mentions in the free text and their candidate entities in the KG. The candidates are connected by dash lines, while the correct ones are in solid lines.

Local models typically disambiguate each mention independently, in which the semantic representations of both mention and candidate entities are firstly captured through a dual-encoder [13]. The disambiguation is then framed as a multi-class classification problem [4,25], where the correct referent entity is identified by assessing the compatibility between the mention and its candidate entities. Traditional approaches mainly utilize hand-crafted statistical features to obtain the representations of mention and candidate entities [14,24]. Subsequently, many approaches learn a domain specific word embedding using some embedding techniques, such as word2vec [8,25], CNN [18] and Bi-LSTM [6,23]. Recently, pretrained language models (PLMs) have shown remarkable effectiveness in natural language understanding, leading to their adoption in ED for learning contextual entity representations and achieving state-of-the-art performance [1,4,30]. While considering the local mention-entity semantics is crucial, relying solely on this information might not yield satisfactory disambiguation results. Therefore, the global entity-entity coherence among all document mentions has been proposed to enhance the disambiguation process from a global perspective [10,12,28,30].

Despite considerable advancements, existing approaches for ED task still face three key challenges. First, existing solutions predominantly treat the disambiguation of each mention as a multi-class classification problem over candidate entities. Such an approach disregards the contrastive degree between positive and negative samples and can be susceptible to overfitting when training samples are limited. Second, the semantic representations of mention and candidate entities are learned independently, which may overlook the valuable interactive and contextual information that is essential for facilitating the mutual comprehension between mention and entities. Moreover, while certain studies attempt to merge local mention-entity evidence with global entity-entity coherence during the process of ED, deriving global coherence either incurs a larger computational cost to derive document-level contextual embedding or necessitates external KG structural information, which is both challenging and time-consuming to obtain.

To address the aforementioned issues, we propose a novel framework, *LogicED*, which jointly models the Local and global semantics for contrastive Entity Disambiguation. Specifically, *LogicED* consists of two carefully designed modules, i.e., local contextual module and global semantic graph module, to consider both local contextual semantics between mention and candidate entities, as well as the global coherence semantics between coreferent entities for entity disambiguation. In the local contextual module, a pre-trained language model (PLM) based on the multi-layer transformer is first applied to obtain the contextual embeddings of both mentions and candidate entities. Furthermore, to enhance the mutual comprehension between mention and entities, we design a candidate-aware self-attention (CASA) model to capture the interactive and inter-contextual information of mention and entities. Next, in order to learn more robust and discriminative mention and entity representations, we formulate the ED problem as a supervised contrastive learning task with InfoNCE loss, in which the ground-truth entity of the mention is regarded as the positive instance and remaining candidate entities as negative instances. The learned representations of mention and entities are then utilized to construct the ED semantic graph, which integrates both the local mention-entity compatibility and global entity-entity coherence. The ED task considering both local and global semantics is then formulated as a dense subgraph detection problem. We prove that the problem is NP-hard and propose a Maximum Spanning Forest (MSF) greedy algorithm to derive the solution.

In summary, the main contributions of this paper are summarized as follows:

- We present *LogicED*, a novel framework for contrastive entity disambiguation by jointly modeling the local and global semantics.
- We design a local contextual module with entity-aware self-attention (CASA) and supervised contrastive learning mechanisms to derive robust and discriminative contextual representations of mention and candidate entities.
- We construct the ED semantic graph to unify local mention-entity compatibility and global entity-entity coherence and formulate the ED task as a dense subgraph detection problem. We prove the problem to be NP-hard and design a Maximum Spanning Forest (MSF) greedy algorithm to obtain the approximated disambiguation results.
- We conduct extensive experiments on several benchmark datasets. Experimental results verify that our proposed *LogicED* outperforms the state-of-the-art baselines.

2   Related Work

2.1   Entity Linking

Neural entity linking systems have been developed since 2015, with many models utilizing entity and mention/context word embeddings to represent their meanings. Advancements in entity embedding techniques have led to the success of neural methods in entity linking [31]. While static word embeddings like word2vec [17] and GloVe [19] are initially used to represent mentions and entities, recent works [4,23] have demonstrated the benefits of using pretrained language models like BERT [7] for contextual embeddings. Other methods are also involved to solve the entity linking problem. Botha et al. [3] proposes a dual-encoder architecture for multilingual entity linking (MEL), which uses two independent encoders for mentions and entities and maximizes the improvement of mention embeddings with their corresponding similarity between entity embeddings. This approach outperforms state-of-the-art results on a more limited task of cross-lingual linking. Meanwhile, Cao et al. [4] develops a system for retrieving entities by generating their unique names and proposes a context-conditioned GENRE model to retrieve entities one by one in an autoregressive manner. Building on this approach, Cao et al. [5] further improves it by proposing a sequence-to-sequence mGENRE system for the MEL problem, facilitating the connection between languages and entities.

2.2   Entity Disambiguation

The task of entity disambiguation aims at finding the matched mention-candidate pair, is primarily treated as the multi-label classification problem [23,25]. Many recent approaches utilize the dual-encoder paradigm, in which two separate encoders are trained to learn vector representations in a shared space for both mentions and entities [3,14,26]. With the availability of large pre-trained language models, such as BERT [7], the performance of the ED has been significantly improved. For instance, Cao et al. [4] and Barba et al. [1] have used BERT to classify each word in the document to the corresponding entity. Furthermore, Yamada et al. [29] proposes a masked entity prediction task, where a BERT-based model learns to predict randomly masked input entities.

A recently popular method for tackling ED problem combines the use of local and global models. In this approach, the local model serves as the encoder that captures textual information, while the global model leverages document-level coherence for disambiguation. Some works attempt to transform the ED task into an optimal subgraph problem by converting mentions and entities in a document into a *Mention-Entity* graph. They then apply adjusted Page Rank [20] or random walk [11,28] algorithms to iteratively calculate the weight of nodes. Similarly, Guo et al. [12] uses robust disambiguation to compute a dense subgraph that approximates the best joint mention-entity mapping.

In addition, some studies have investigated the impact of global semantics on mentions. Le and Titov [15] and Barba et al. [1] approach the ED problem as a relation prediction task and an extractive task, respectively. They use transformer-based pre-trained models to capture semantic features. Local neural attention is employed, and conditional random fields are used to model the global context for disambiguation [10]. Furthermore, Yang et al. [32] uses the structured gradient tree boosting algorithm for producing globally optimal entity assignments for mentions.

3 Methodology

3.1 Problem Formulation

Given a document D containing n mentions, denoted as $M = \{m_i | i = 1, 2, ..., n\}$, and their corresponding candidate entities $C = \{C_i | i = 1, 2, ..., n\}$ in a reference knowledge graph (KG), where $C_i = \{c_{ij} | j = 1, 2, .., |C_i|\}$ contains all candidate entities for mention m_i. The task of entity disambiguation (ED) is to correctly associate each mention m_i with the appropriate candidate entity in C_i.

3.2 Overview

We propose *LogicED*, a novel framework for contrastive entity disambiguation by jointly modeling the local and global semantics. As illustrated in Fig. 2, our framework consists of two carefully-designed modules:

- **Local Contextual Module.** This module utilizes the textual description of mention and candidate entities to derive their respective representations. The input sequences of both ends are firstly encoded by a pretrained language model (PLM). To enhance mutual comprehension, we propose a candidate-aware self-attention (CASA) model to further capture the interactive and inter-contextual information of both mentions and entities. Then, the supervised contrastive learning paradigm with InfoNCE loss is deployed to derive more discriminative mention and entity representations.
- **Global Semantic Graph Module.** This module utilizes the learned mention and entity representations from the Local Contextual Module to construct a semantic entity-disambiguation (ED) graph that integrates both local mention-entity compatibility and global entity-entity coherence. Subsequently, we formulate the ED task as a dense subgraph detection problem and propose a Maximum Spanning Forest (MSF) greedy algorithm to solve it. Finally, we derive the ED results based on the identified dense subgraph.

3.3 Local Contextual Module

In this module, we leverage the textual descriptions of mentions and candidate entities to learn robust and discriminative mention and entity contextual embeddings.

We firstly utilize a pretrained language model (PLM), i.e., RoBERTa [16], to encode the mention and its candidate entities, independently, to obtain the contextual representation.

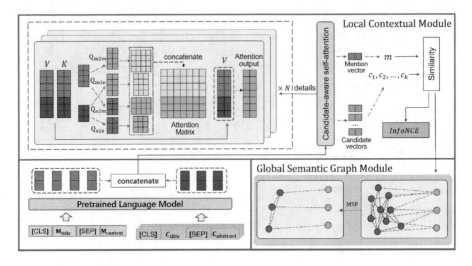

Fig. 2. The overall Architecture of *LogicED*.

The input sequences of a mention m and a candidate entity e are denoted as S_m and S_e, respectively. Specifically, S_m comprises the mention's title and its context, where the two components are concatenated using the separator token [#SEP]. Besides, the classifier token [#CLS] is added at the beginning of the concatenated sequence, denoted as "[#CLS] M_{title} [#SEP] $M_{context}$". Similarly, we concatenate the candidate entity title with its abstract to form S_e, denoted as "[#CLS] C_{title} [#SEP] $C_{abstract}$".

Candidate-Aware Self-attention (CASA). Although the PLM encoder can learn contextual embeddings of mentions and candidate entities, it may overlook the valuable interactive and contextual information between mention and entities as the encoding procedure is performed independently. Therefore, we propose to perform a candidate-aware self-attention (CASA) between mention and entities on top of the PLM encoder to enhance the mutual comprehension between mention and entities.

The self-attention mechanism is a fundamental component of the transformer model [27], allowing it to capture the relationships between input tokens by calculating attention scores between each pair of tokens. Given an input sequence of vectors $x_1, x_2, ..., x_n$, where $x_i \in \mathbb{R}^D$, the self-attention mechanism computes the weighted sum of the input sequence to obtain the output vector sequence $y_1, y_2, ..., y_n$, where $y_i \in \mathbb{R}^L$. Formally, the output vector y_i is computed as:

$$y_i = \sum_{j=1}^{n} \alpha_{ij} V x_j \tag{1}$$

$$e_{ij} = \frac{K x_j^T Q x_i}{\sqrt{L}} \tag{2}$$

$$\alpha_{ij} = softmax(e_{ij}) \tag{3}$$

Here, $\boldsymbol{Q} \in \mathbb{R}^{L \times D}, \boldsymbol{K} \in \mathbb{R}^{L \times D}, \boldsymbol{V} \in \mathbb{R}^{L \times D}$ are the query, key and value matrices, respectively.

In our ED system, we use a shared-parameters PLM model to embed two types of sequences separately. Denoting $\mathcal{H}_m = \{m_i | i = 1, 2, ..., n\}$, as the mention end embedded vectors and $\mathcal{H}_e = \{C_i | i = 1, 2, ..., n\}$ as the candidate end embedded vectors, where $C_k = \{C_{km} | m = 1, 2, .., |C_k|\}$. For each mention embedded vector m_k we concatenate it with its candidate embedded vector c_{kq} as $X = [m_k; c_{kq}] = \{x_i | i = 1, 2, ..., h\}$. Based on that, we propose an extra layer with enhanced self-attention to capture interactive knowledge between the embedded vectors of mentions and candidates. It has four types of matching cases for each pair of tokens within different sequences, and the attention score matrix can be represented as:

$$e_{ij} = \begin{cases} \boldsymbol{K}\boldsymbol{x}_j^T \boldsymbol{Q}_{m2m}\boldsymbol{x}_i, & \text{if both } \boldsymbol{x}_i \text{ and } \boldsymbol{x}_j \text{ are from } m_k; \\ \boldsymbol{K}\boldsymbol{x}_j^T \boldsymbol{Q}_{m2c}\boldsymbol{x}_i, & \text{if } \boldsymbol{x}_i \text{ is from } m_k \text{ and } \boldsymbol{x}_j \text{ is from } c_{kq}; \\ \boldsymbol{K}\boldsymbol{x}_j^T \boldsymbol{Q}_{c2m}\boldsymbol{x}_i, & \text{if } \boldsymbol{x}_i \text{ is from } c_{kq} \text{ and } \boldsymbol{x}_j \text{ is from } m_k; \\ \boldsymbol{K}\boldsymbol{x}_j^T \boldsymbol{Q}_{c2c}\boldsymbol{x}_i, & \text{if both } \boldsymbol{x}_i \text{ and } \boldsymbol{x}_j \text{ are from } c_{kq}. \end{cases} \tag{4}$$

Here, $\boldsymbol{Q}_{m2m}, \boldsymbol{Q}_{m2e}, \boldsymbol{Q}_{e2m}, \boldsymbol{Q}_{e2e} \in \mathbb{R}^{L \times D}$ are the query matrices.

Contrastive Learning. We approach entity disambiguation (ED) as a contrastive learning problem, which considers the contrastive degree between positive and negative samples to derive more discriminative mention and entity representations. To this end, we employ the InfoNCE loss function, a classic contrastive loss used for training textual semantic models. Given a query vector $q \in \mathbb{R}^D$ and a set of key vectors $\boldsymbol{K} = \{k_1, k_2, .., k_n\}$ containing one positive sample, denoted as k_+, and $n - 1$ negative samples, where $k_i \in \mathbb{R}^D$. Contrastive learning aims to increase the similarity between q and the positively labeled key k_+, while decreasing the similarity between q and the negatively labeled keys. In our system, we consider the mention as the query and its candidate entities as the keys, where the ground-truth entity of the mention is regarded as the positive sample and the remaining candidate entities as negative samples. Adopting the InfoNCE loss can force the model to learn to push the negative samples away while pulling the positive sample closer in the vector space. The InfoNCE loss is defined as follows:

$$\mathcal{L}_q = -log\left(\frac{exp(\frac{q \cdot k_+}{\tau})}{\sum_{i=1}^{n} exp(\frac{q \cdot k_i}{\tau})}\right) \tag{5}$$

where τ is the hyper-parameter temperature to control the shape of the output logits' distribution and allows us to adjust the model's capability to discriminate negative samples. Specifically, a larger value of τ will make the model focus more on hard negative samples.

3.4   Global Semantic Graph Module

Given a set of documents with the mention-candidate compatibility scores generated from the Local Contextual Module, our Global Semantic Module firstly construct an ED semantic graph for each document to integrate the local entity-mention compatibility and global entity-entity coherence. Then, we formulate ED as a dense subgraph detection problem that aims at finding the dense subgraph. Considering the NP-hard nature of the problem and the associated high computational cost, we propose a greedy Minimum Spanning Forest (MSF) algorithm to detect the dense subgraph and obtain the final disambiguation results.

ED Semantic Graph Construction. We refer to the ED semantic graph built in this section as $G = (V, E)$. Specifically, V denotes the set of vertices, and E denotes the set of edges among the vertices in G. Given a document, the ED semantic graph contains two different types of vertices in V, i.e., $V = \mathcal{M} \cap \mathcal{C}$, and two different types of edges, i.e., $E = \mathcal{E}_m \cap \mathcal{E}_e$.

Mention Vertices ($\mathcal{M}$). A mention vertex $m_i \in \mathcal{M}$ represents the ith mention of the document that needs to be disambiguated.

Candidate Vertices ($\mathcal{C}$). A candidate vertex $c_{ij} \in \mathcal{C}$ represents the jth candidate for the ith mention in the document. Since the number of candidates for each mention may differ, we sort the mention's candidates by their local score and restrict the maximum number of candidates in order to reduce computational complexity.

Mention-Candidate Edges ($\mathcal{E}_m$). Each edge between a mention vertex and a candidate vertex is weighted by calculating the dot product similarity between the mention end and the candidate end embedded vectors. The weight of the mention-candidate edges between the ith mention and its jth candidate is denoted by $\phi_m(m_i, c_{ij})$. The value of $\phi_m(m_i, c_{ij})$ reflects the likelihood from the local model that the candidate c_{ij} is the correct entity for the mention m_i.

$$\phi_m(m_i, c_{ij}) = \boldsymbol{v}_i \cdot \boldsymbol{v}_{ij} \tag{6}$$

where v_i is the embedded vector of ith mention and v_{ij} is the embedded vector of jth candidate of ith mention.

Candidate-candidate Edges ($\mathcal{E}_e$). The weight of a candidate-candidate edge between two candidate vertices $c_{ij}, c_{mn} \in \mathcal{C}$ is denoted as $\phi_e(c_{ij}, c_{mn})$. This value measures the semantic coherence between the two candidates entities by computing the cosine similarity between their hidden states generated by the local contextual module:

$$\phi_e(c_{ij}, c_{mn}) = \cos(\boldsymbol{v}_{ij}, \boldsymbol{v}_{mn}) = \frac{\boldsymbol{v}_{ij} \cdot \boldsymbol{v}_{mn}}{\|\boldsymbol{v}_{ij}\|\|\boldsymbol{v}_{mn}\|} \tag{7}$$

where v_{ij}, v_{mn} are the candidate end embedded vectors with respect to c_{ij} and c_{mn}. Note that, in order to reduce the computational complexity, we set a threshold to filter out edges whose weight is lower than the threshold.

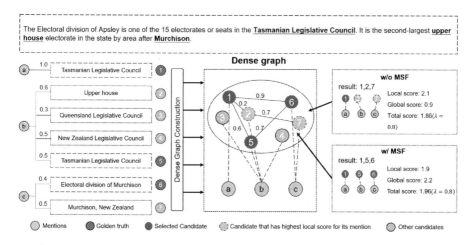

Fig. 3. An example to illustrate how MSF optimizes the ED process.

The Dense Subgraph Problem. Our goal is to detect a dense subgraph within an ED semantic graph that contains all mention vertices and exactly one mention-candidate edge for each mention, resulting in the disambiguation of all mentions in a document. However, the dense subgraph problem is NP-hard as they generalize the Steiner-tree problem. Hence, exact algorithms on large input graphs are infeasible. Therefore, we define the density of a subgraph as the summation of its edge weights to capture the global coherence and propose a greedy algorithm for detecting the dense subgraph within a given ED semantic graph.

Maximum Spanning Forest Global Model. Our objective is to find a dense subgraph that maximizes the density within an ED semantic graph. We propose a Minimum Spanning Forest (MSF) algorithm inspired by the traditional Maximum Spanning Tree (MST) algorithm to achieve the target. Starting with a full ED semantic graph, the MSF algorithm iteratively selects the mention-candidate pair that maximizes the density in the current step. The selected vertices become part of the subgraph and contribute to the next-step selection, similar to the MST algorithm. However, since some candidate vertices and candidate-candidate edges are removed to improve computational efficiency, the current subgraph in the ED semantic graph cannot be connected to other parts. In this scenario, we randomly select another mention to restart the MSF algorithm.

The MSF algorithm is summarized in Algorithm 1. Initially, the algorithm randomly selects a mentioned vertex and chooses its candidates with the highest local score (lines 2–3). The main loop then adds the mention-candidate pair into the subgraph iteratively (lines 5–16), either by selecting the pair with the maximum coherence score s_{ij} or by re-randomizing the mention-candidate pair as initialization.

Algorithm 1 MSF

Require: ED semantic graph G
1: $G_s \leftarrow$ empty Subgraph
2: $m_r \leftarrow$ randomly selected mention
3: $c_r \leftarrow m_r's$ candidate with a highest local score
4: adding $\{m_r, c_r\}$ pair to G_s
5: **while** not all mentions selected **do**
6:      **if** exists edges connect G_s to $G \setminus G_s$ **then**
7:          **for** $\{m, c\} \in G_s$ **do**
8:              computes s_{ij} for c using Equation 8
9:          **end for**
10:         $c_n \leftarrow \max (s_{ij})$
11:         $m_n \leftarrow$ mention of c_n
12:     **else**
13:         $m_n, c_n \leftarrow$ re-randomize
14:     **end if**
15:     add $\{m_n, c_n\}$ into G_s
16: **end while**

The coherence score of candidate j of mention i, denoted as s_{ij}, is calculated by:

$$s_{ij} = (1 - \lambda)\phi_m(m_i, c_{ij}) + \lambda * \underset{c \in G_s}{\text{Avg}} \; \phi_e(c, c_{ij}) \tag{8}$$

where λ is the hyper-parameter, c is the selected candidates in the current semantic graph G_s, ϕ_m, and ϕ_e are functions that compute the mention-candidate and candidate-candidate scores, respectively. The first term in s_{ij} calculates the mention-candidate score between m_i and c_{ij}. The second term in s_{ij} is the average candidate-candidate score between c_{ij} and all the other selected candidate vertices in the subgraph. Figure 3 illustrates the benefits of employing MSF algorithm. Without MSF, the system cannot correctly disambiguate mentions "upper house" and "Murchison". Incorporating MSF enables the system to rectify this mistake by considering the candidate-candidate coherence.

In order to validate the effectiveness of our algorithm, we conduct an analysis of its time complexity. Assume a document contains N mentions and each mention has at most k candidates. The most time-consuming part of the algorithm is the while loop from lines 5–15, which is executed N times. Assuming there are l selected mention-candidate pairs in the subgraph, then for each candidate vertex in G_s, the algorithm traverses to all $(k - l)$ candidate vertices in $G \setminus G_s$, which takes $l * (k - l)$ and reaches the peak $k^2/4$, when $c = k/2$ for a fully connected graph. Therefore, the worst-case time complexity of the inner loop is $O(k^2)$. Overall, the algorithm's time complexity is $O(N * k^2)$.

4 Experiments

In this section, we conduct extensive experiments on benchmark datasets to evaluate the effectiveness of our proposed framework *LogicED*.

Table 1. Statistics of ED datasets.

Dataset	AIDA	MSNBC	ACE2004	AQUAINT	CWEB	WIKI
# mentions	18,448	656	257	727	11,154	6,821
# documents	946	20	36	50	320	320
# mentions per doc	19.5	32.8	7.1	14.5	34.8	21.3

4.1 Experimental Setup

Datasets. We validate the effectiveness of our framework *LogicED* on six publicly available ED benchmark datasets. Specifically, for the in-domain scenario, we train and fine-tune using the AIDA dataset [12]. For the out-of-domain scenario, we evaluate the performance on five test sets: MSNBC, ACE2004, AQUAINT, WNED-CWEB (CWEB) and WNED-WIKI (WIKI) [11]. The statistics of the datasets are summarized in Table 1.

Entity Abstract Acquisition. In order to extend the textual semantics of the candidate entities, we crawl the abstract of all candidate entities from Wikipedia. In detail, the title of each candidate entity is linked to the Wikidata index. Then, we use the index and MediaWiki tool[1] to retrieve the detailed description of the candidate entity. Note that we only extract the first paragraph of the retrieved description as the candidate's abstract due to the token length limit of PLMs.

Baselines. To demonstrate the superiority of *LogicED*, we compare our system to several state-of-the-art ED models:

- **Local Model.** Shahbazi et al. [23] learn an entity aware extension of Embedding for Language Model (ELMo).
- **Global.** Based on the semantic similarity, the information between random walks on the disambiguation graph induced by choice of entities for each mention [11]. Le and Titov [15], Barba et al. [1], and Yamada et al. [29] approach the ED problem as a relation prediction task, extractive task, and mask prediction task, respectively. These models utilize transformer-based pretrained models to capture semantic features. The global model for entity disambiguation is constructed by employing conditional random fields based on local neural attention [10]. By contrast, Yang et al. [32] uses the structured gradient tree boosting algorithm. Fang et al. [9] converts global linking into a sequence decision problem and proposes a reinforcement learning model. Additionally, Yang et al. [31] proposes a dynamic context augmentation method to fuse the global signal.
- **Auto-regressive approach.** De Cao et al. [4] retrieves entities by generating their unique names in an autoregressive fashion.

Evaluation. In accordance with the established conventions of entity disambiguation evaluation, we employ the *Micro-F1* metric for assessing performance.

---

[1] https://github.com/wikimedia/mediawiki.

Table 2. Results (Micro-F1) on the in-domain and out-of-domain datasets. Bold and underline texts indicate the best and the second-best scores.

Model	In-domain	Out-of-domain					Avgs	
	AIDA	MSNBC	AQUAINT	ACE2004	CWEB	WIKI	Avg	Avg_{OOD}
Ganea and Hofmann [10]	92.2	93.7	88.5	77.5	77.9	77.5	86.4	85.2
Guo and Barbosa [11]	89.0	92.0	87.0	88.0	77.0	84.5	86.2	85.7
Yang et al. [32]	**95.9**	92.6	89.9	88.5	<u>81.8</u>	79.2	88.0	86.4
Shahbazi et al. [23]	93.5	92.3	90.1	88.7	78.4	79.8	87.1	85.9
Yang et al. [31]	93.7	93.8	88.2	90.1	75.6	78.8	86.7	85.3
Le and Titov [15]	89.6	92.2	90.7	88.1	78.2	81.7	86.8	86.2
Fang et al. [9]	94.3	92.8	87.5	91.2	78.5	82.8	87.9	86.6
De Cao et al. [4]	93.3	94.3	89.9	90.1	77.3	87.4	88.8	87.8
Barba et al. [1]	92.6	94.7	91.6	91.8	77.7	<u>88.8</u>	89.5	88.9
Yamada et al. [30]	92.4	<u>96.3</u>	<u>93.6</u>	<u>91.9</u>	78.9	**89.1**	<u>90.3</u>	<u>89.9</u>
LogicED	<u>95.1</u>	**96.6**	**95.5**	**95.5**	**86.7**	82.3	**92.0**	**91.3**

Furthermore, we provide two separate evaluations to emphasize the average performance across all datasets: the average score across out-of-domain datasets and AIDA (Avg), and the average score on out-of-domain datasets only (Avg_{OOD}), which excludes the result on AIDA.

4.2   Experimental Results

Table 2 shows our experimental results where we report the *Micro-F1* score for all six benchmark datasets and calculate the average score with and without the AIDA dataset. As observed, *LogicED* significantly improves the performance on MSNBC (+0.3%), AQUAINT (+1.9%), ACE2004 (+3.6%), and CWEB (+7.8%). On average, *LogicED* improves the performance by 1.4% in Avg_{OOD} and 1.7% in Avg, indicating the effectiveness of our proposed system and its suitability for the ED task.

Compared to the Local and global without graph-based approaches, our model performs better because of the incorporation of the global semantic graph module, which better captures inter-entity dependency features and global coherence. Moreover, compared to the global with graph-based approach, our model performs better because we define the problem as a supervised contrastive learning task and incorporate the CASA mechanism in our local contextual module, which captures richer semantic features. In conclusion, the comparative results demonstrate the superiority of our *LogicED* framework. However, we observe that our model's performance is relatively low on the WIKI dataset. When comparing with other state-of-the-art models, we observe a significant increase in the score of the WIKI dataset since 2019. Notably, the four models that achieved higher scores all utilize a substantial amount of extra training data, obtained through either data augmentation or external resources.

4.3   Hyper-parameter Analysis

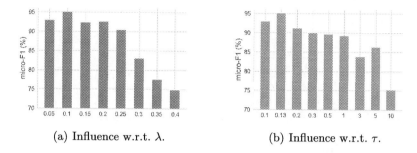

(a) Influence w.r.t. λ.   (b) Influence w.r.t. τ.

Fig. 4. Influence of Hyper-parameters in *LogicED*.

We tune the temperature τ of *InfoNCE* loss and λ used in the MSF algorithm. The results are presented in Fig. 4, the *Micro-F1* score decreases gradually as lambda increases as Fig. 4(a) shows, with the highest score achieved at a lambda of 0.1. As for temperature, the *Micro-F1* score in Fig. 4(b) exhibits a similar trend of decreasing performance as the temperature increases, and the drop in performance is more drastic as the temperature τ increases. Therefore, in our system, we adopt the setting where $\lambda = 0.1$ and $\tau = 0.13$.

4.4   Ablation Study

We conduct ablation study to further demonstrate the contribution of individual components, i.e., CASA, MSF and contrastive learning strategy. We study the impact of removing each module, denoted as "w/o MSF", "w/o CASA", "w/o MSF and w/o CASA", and "w/o InfoNCE", the results are presented in Table 3. For removing the CASA module, we replace it with the original Self-attention layer. Moreover, we evaluate the effectiveness of *InfoNCE* loss with the commonly adopted binary Cross-entropy loss.

According to the results, our design outperforms all variants on six datasets, validating the effectiveness of each module. Overall, our CASA module forces the

Table 3. Experimental results (Micro-F1) of ablation study.

Model	In-domain	Out-of-domain					Avgs	
	AIDA	MSNBC	AQUAINT	ACE2004	CWEB	WIKI	Avg	Avg_{OOD}
LogicED	95.1	96.6	95.5	95.5	86.7	82.3	92.0	91.3
w/o MSF	94.9	96.6	95.3	94.8	86.5	82.1	91.7	91.1
w/o CASA	91.7	96.3	95.5	95.3	86.3	82.0	91.2	91.1
w/o MSF and CASA	93.1	96.4	93.2	95.3	85.6	81.1	90.8	90.3
w/o InfoNCE	94.0	96.4	95.5	95.1	86.3	82.0	91.6	91.0

model to learn more mutual information and therefore enhanced the performance of MSF, and *InfoNCE* loss significantly improves the capacity of the local model.

5   Conclusion

Our work proposes a novel approach for entity disambiguation by framing it as a contrastive learning problem. We introduce *LogicED*, which leverages both local and global semantics for disambiguation. Our model incorporates a CASA module to improve mutual comprehension between mentions and entities and uses an MSF greedy algorithm to integrate mention-entity compatibility and entity-entity coherence. Experimental results show that *LogicED* achieves the state-of-the-art performance.

Acknowledgement. This work was supported in part by the Guangdong Provincial Key Laboratory of Interdisciplinary Research and Application for Data Science, BNU-HKBU United International College (2022B1212010006) and in part by Guangdong Higher Education Upgrading Plan (2021–2025) (UICR0400001-22, UICR0400017-21, UICR0400003-21).

References

1. Barba, E., Procopio, L., Navigli, R.: Extend: extractive entity disambiguation. In: ACL (1), pp. 2478–2488 (2022)
2. Bevilacqua, M., Blloshmi, R., Navigli, R.: One SPRING to rule them both: symmetric AMR semantic parsing and generation without a complex pipeline. In: AAAI, pp. 12564–12573 (2021)
3. Botha, J.A., Shan, Z., Gillick, D.: Entity linking in 100 languages. In: EMNLP (1), pp. 7833–7845 (2020)
4. Cao, N.D., Izacard, G., Riedel, S., Petroni, F.: Autoregressive entity retrieval. In: ICLR (2021)
5. Cao, N.D., et al.: Multilingual autoregressive entity linking. Trans. Assoc. Comput. Linguistics **10**, 274–290 (2022)
6. Chen, L., Varoquaux, G., Suchanek, F.M.: A lightweight neural model for biomedical entity linking. In: AAAI, pp. 12657–12665 (2021)
7. Devlin, J., Chang, M., Lee, K., Toutanova, K.: BERT: pre-training of deep bidirectional transformers for language understanding. In: NAACL-HLT (1), pp. 4171–4186 (2019)
8. Fang, W., Zhang, J., Wang, D., Chen, Z., Li, M.: Entity disambiguation by knowledge and text jointly embedding. In: CoNLL, pp. 260–269 (2016)
9. Fang, Z., Cao, Y., Li, Q., Zhang, D., Zhang, Z., Liu, Y.: Joint entity linking with deep reinforcement learning. In: WWW, pp. 438–447 (2019)
10. Ganea, O., Hofmann, T.: Deep joint entity disambiguation with local neural attention. In: EMNLP, pp. 2619–2629 (2017)
11. Guo, Z., Barbosa, D.: Robust named entity disambiguation with random walks. Semant. Web **9**(4), 459–479 (2018)
12. Hoffart, J., et al.: Robust disambiguation of named entities in text. In: EMNLP, pp. 782–792 (2011)

13. Humeau, S., Shuster, K., Lachaux, M., Weston, J.: Poly-encoders: architectures and pre-training strategies for fast and accurate multi-sentence scoring. In: ICLR (2020)
14. Kolitsas, N., Ganea, O., Hofmann, T.: End-to-end neural entity linking. In: CoNLL, pp. 519–529 (2018)
15. Le, P., Titov, I.: Improving entity linking by modeling latent relations between mentions. In: ACL (1), pp. 1595–1604 (2018)
16. Liu, Y., et al.: RoBERTa: a robustly optimized BERT pretraining approach. arXiv preprint arXiv:1907.11692 (2019)
17. Mikolov, T., Chen, K., Corrado, G., Dean, J.: Efficient estimation of word representations in vector space. In: ICLR (Workshop Poster) (2013)
18. Onoe, Y., Durrett, G.: Fine-grained entity typing for domain independent entity linking. In: AAAI, pp. 8576–8583 (2020)
19. Pennington, J., Socher, R., Manning, C.D.: GloVe: global vectors for word representation. In: EMNLP, pp. 1532–1543 (2014)
20. Pershina, M., He, Y., Grishman, R.: Personalized page rank for named entity disambiguation. In: HLT-NAACL, pp. 238–243 (2015)
21. Puduppully, R., Dong, L., Lapata, M.: Data-to-text generation with entity modeling. In: ACL (1), pp. 2023–2035 (2019)
22. Rao, D., McNamee, P., Dredze, M.: Entity linking: finding extracted entities in a knowledge base. In: Multi-source, Multilingual Information Extraction and Summarization, pp. 93–115. Theory and Applications of Natural Language Processing (2013)
23. Shahbazi, H., Fern, X.Z., Ghaeini, R., Obeidat, R., Tadepalli, P.: Entity-aware ELMo: learning contextual entity representation for entity disambiguation. arXiv preprint arXiv:1908.05762 (2019)
24. Sil, A., Yates, A.: Re-ranking for joint named-entity recognition and linking. In: CIKM, pp. 2369–2374 (2013)
25. Sun, Y., Lin, L., Tang, D., Yang, N., Ji, Z., Wang, X.: Modeling mention, context and entity with neural networks for entity disambiguation. In: IJCAI, pp. 1333–1339 (2015)
26. Tedeschi, S., Conia, S., Cecconi, F., Navigli, R.: Named entity recognition for entity linking: what works and what's next. In: EMNLP (Findings), pp. 2584–2596 (2021)
27. Vaswani, A., et al.: Attention is all you need. In: NIPS, pp. 5998–6008 (2017)
28. Xue, M., et al.: Neural collective entity linking based on recurrent random walk network learning. In: IJCAI, pp. 5327–5333 (2019)
29. Yamada, I., Asai, A., Shindo, H., Takeda, H., Matsumoto, Y.: LUKE: deep contextualized entity representations with entity-aware self-attention. In: EMNLP (1), pp. 6442–6454 (2020)
30. Yamada, I., Washio, K., Shindo, H., Matsumoto, Y.: Global entity disambiguation with BERT. In: NAACL-HLT, pp. 3264–3271 (2022)
31. Yang, X., et al.: Learning dynamic context augmentation for global entity linking. In: EMNLP/IJCNLP (1), pp. 271–281 (2019)
32. Yang, Y., Irsoy, O., Rahman, K.S.: Collective entity disambiguation with structured gradient tree boosting. In: NAACL-HLT, pp. 777–786 (2018)
33. Yin, W., Yu, M., Xiang, B., Zhou, B., Schütze, H.: Simple question answering by attentive convolutional neural network. In: COLING, pp. 1746–1756 (2016)

KFEA: Fine-Grained Review Analysis Using BERT with Attention: A Categorical and Rating-Based Approach

Liting Huang, Yongyue Yang, Xingli Tang, Hui Zhou[✉], and Chunyang Ye

Hainan University, Haikou, Hainan, China
{litinghuang,yyy,txl,zhouhui,cyye}@hainanu.edu.cn

Abstract. User reviews contain many key phrases that are crucial for business understanding, but they are often obscured by the sheer volume of reviews. Extracting key phrases from user reviews could help to understand what users are concerned about and provide timely improvement suggestions. Current pattern-based methods for target phrase extraction usually analyze reviews at a coarse-grained level, making the extracted topics unfocused and useless. Hence, in order to address this issue, we proposed a fine-grained analysis approach (KFEA) to extract, cluster, and visualize key phrases from e-commerce reviews. In order to fully utilize the relevant information from comments, KFEA fuses the information like categories and ratings from a large volume of user reviews, and then extracts key phrases with the help of a pre-trained model. A method is also designed to cluster and visualize the extracted key phrases for business understanding. Our evaluation on 6,088 reviews from 6 products shows that KFEA can effectively extract key phrases and perform clustering and visualization. In particular, KFEA achieved an precision of 76.6% and a recall of 81.8% in extracting key phrases from manually annotated data. KFEA's cross-categories effectiveness is also validated on 16,772 reviews from products like mobile phones, laptops, and furniture.

Keywords: Consumer reviews · Key phrases extraction · Deep mixture model

1 Introduction

The act of users commenting and expressing their opinions in cyberspace is known as e-WOM (electronic word-of-mouth). Growing consumers rely on e-WOM when making purchasing decisions [4]. Therefore, such importance of e-WOM has motivated businesses to strive to understand the users' opinions in reviews to improve user satisfaction with businesses' products. However, with the large number of user reviews, manual analysis is labor-intensive and inefficient.

In recent years, many approaches have been proposed to analyze user reviews automatically. For example, Hong et al. employed a CNN model to analyze user

X. Yang et al. (Eds.): ADMA 2023, LNAI 14176, pp. 260–274, 2023.
https://doi.org/10.1007/978-3-031-46661-8_18

Fig. 1. An example of Jingdong Mall reviews

reviews in the fresh agricultural produce industry and rank the importance of various dimensions such as convenience, communication, reliability, and responsiveness [9]. Xu used latent semantic analysis to extract hidden semantic structures from user reviews in the catering industry and rank the importance of factors such as food, food-related services, and restaurant services [27]. However, these feature extraction methods can only extract coarse-grained topics, from which businesses still cannot accurately understand which aspects of their products users are complaining about.

As illustrated in Fig. 1, user reviews contain key phrases, such as "the fill light is not on" that provide valuable insights. Current pattern-based methods for extracting target phrases have limitations in retrieving fine-grained phrases and often only retrieve coarse-grained phrases, such as "the fill light" which do not provide specific insights into user dissatisfaction. Additionally, the creation of text patterns in these methods is done manually, making it inefficient for generating patterns for a large number of different product categories.

The primary challenge in extracting key phrases from a large volume of unstructured consumer reviews is that reviews are context-dependent, making it difficult to capture the semantic characteristics of the reviews. To address this issue, we utilize a BERT [6] pre-trained model to learn the semantic information within the reviews. In order to fully utilize the relevant information in the reviews. Our method introduces Transformer [25] that combines semantic and rating information from reviews and incorporates category information. This enables the model to effectively extract key phrases from reviews of different categories and ratings.

We used the collected dataset to evaluate the effectiveness of KFEA. The experimental results demonstrated that KFEA outperforms existing baselines in extracting key phrases and clustering. Moreover, it also showed that the features that users cared about vary across different products. The findings of this paper could help businesses prioritize features of different products.

The contribution of this paper has mainly three aspects: First, we create a multi-category dataset from JingDong Mall product reviews for key phrases extraction. This dataset will be made public on GitHub to facilitate related research in this field. Second, we proposed a group-aware semantic-aware analysis methodology of product reviews in a fine-grained way. We design a BERT-Attn-CRF model to extract key phrases and utilize the BERTopic [8] model to cluster the extracted key phrases into topics, and finally visualize the topics for understanding. Third, we evaluate KFEA on 6088 user reviews from six product-sand extensive experimental results showed that KFEA outperforms the existing methods and has broader application scope.

The rest of this paper is organized as follows: Sect. 2 reviews related work in the area of user review analysis. Section 3 presents the details of KFEA model. Section 4 evaluates our proposal using extensive experiments. Section 5 concludes the work and highlights some future research directions.

2   Related Work

Previous work has mainly focused on using text mining techniques to extract high-level topics features from text and quantifying the decisive topics for products [3]. Chatterjee et al. used a hybrid model of text mining, deep learning, and econometrics to predict the high-level topics that affect user satisfaction [3]. Xu et al. used text mining methods to model food user reviews and identify the high-level topics that affect user satisfaction [12,27]. These studies on reviews mainly classify and summarize high-level topics of reviews, such as the value and restaurant service features.

Explicit features mentioned by users can be easily extracted from reviews. However, fine-grained features are more difficult to identify than explicit features. Park used an association rule-based method to extract sub-features from user reviews and achieved good results [18]. Li et al. proposed a model based on semantic and emotional similarity to extract implicit features [13]. Wang et al. proposed a BERT-CRF model to concatenate categories and ratings to obtain problem features in English app store reviews [26]. However, the personal information of the review publisher (such as emotional state) and product information have not been fully utilized.

In similar studies, Tan et al. used review information and rating information as inputs to the MK model to predict user preferences and attributes [24]. Babak extracted features from reviews and used them for similarity analysis, and finally used a deep neural network to generate recommendations [21]. Nimesh et al. extracted and identified key aspects of customer concerns through a rule-based model to determine customer preferences and establish personalized product recommendation methods [28]. However, these methods only use reviews for personalized recommendations and are not suitable for improving user satisfaction.

Other related research areas are sentiment analysis and opinion mining, which focus on analyzing and identifying users' emotional states and subjective information from reviews [2]. Su used a Convolutional Attention Long Short-Term Memory (CA-LSTM) model to classify emotions based on expert knowledge and fuzzy mathematics [22]. Some researchers have applied deep learning to sentiment analysis [1]. As BERT is a pre-trained model that can more easily obtain semantic information, some scholars have also used BERT for research. Munikar et al. used BERT for fine-grained sentiment classification tasks and achieved good results [16]. However, the above-mentioned methods can only judge emotional attitudes at a coarse-grained level and cannot extract fine-grained features. To address above issue, this paper focuses on user reviews and incorporates a state-of-the-art technique (i.e., BERT, Attention) for the semantic-aware learning, and the results show its effectiveness.

3   KFEA Model

3.1   Overview

Figure 2 shows the main architecture of KFEA. KFEA has three main components. The first part of the study focuses on comprehensively understanding the semantics and category, rate information of reviews. A BERT-Attn-CRF model is then constructed using the acquired encoded vectors to extract key phrases. In the second part, KFEA employs the BERTopic method to cluster the extracted key phrases. The topics of key phrases are clustered to summarize their common aspects. In the third part, KFEA visualizes the topics obtained in the second part to facilitate an understanding of which aspects of the product users are dissatisfied with.

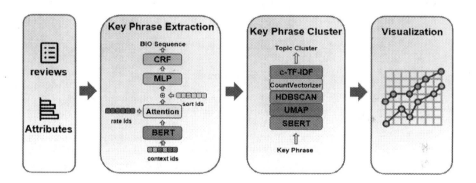

Fig. 2. The overall workflow of the KFEA approach.

3.2   Review Information Collection

The input of category information and rating information can improve the performance of KFEA in extracting key phrases. Users pay attention to different aspects of feature for different categories of products, so this paper uses category as an input. In addition, compared to high-rated reviews, low-rated reviews contain more key phrases, Therefore, this paper also uses rating information as an input of the model. When obtaining user reviews from JingDong Mall, category information and rating information can be directly obtained.

3.3 Key Phrase Extraction

This paper uses Named Entity Recognition (NER) to extract key phrases [17]. The BERT-Attn model is used to encode and learn state features from reviews, and the CRF model is used to learn transition features that represent the relationship between states [10].

In this paper, each sentence is tagged using the BIO format [5], where

- B-label (Beginning) indicates that the current word is the beginning of a key phrase.
- I-label (Inside) indicates that the current word is within a key phrase but not at the beginning position.
- O-label (Outside) indicates that the current word is outside a key phrase.

Sentences annotated with BIO tags are used as input to the BERT-Attn-CRF model.

Figure 3 illustrates the proposed BERT-Attn-CRF model. Specifically, BERT-Attn consists of two parts: a category encoder that maps discrete data to category vector (x_s) and rating vector (x_r), and a rating encoder that integrate textual vector (x_t) and rating vector into fused rating vector via fine-grained fusion method (x_f). Finally, the two fused vectors $(x_s$ and $x_f)$ are concatenated to obtain the final vector (x_i).

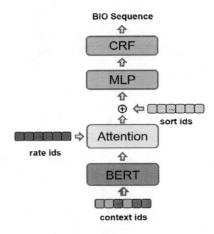

Fig. 3. Structure of BERT-Attn-CRF.

To obtain x_f, we designed the FFR (Fine-Grained Fusion Rate Embedding) submodule in BERT-Attn. FFR submodule introduce cross-attention to obtain x_f. The textual information x_t is used as the query, and the rating information x_r is used as the key and value. The specific formula is given as follows:

$$Attn(Q, K, V) = softmax(\frac{QK^T}{\sqrt{d_k}})V \tag{1}$$

where d_k represents the dimension of the word vector embedding.

Subsequently, x_s and x_f are concatenated to obtain a vector (x_i) representing each input. The vector then enters a multi-layer perceptron (MLP), which computes the probability vector (denoted as p) for each BIO label:

$$x_i = W[x_s : x_f] \tag{2}$$

$$p = f(x_i) \tag{3}$$

f serves as the activation function, and W represents the training parameters.

The probability vector p is input into the CRF layer to determine the label sequence. Let x represent the input p and y represent the output, which has a one-to-one correspondence with p. The feature function is denoted by f and W represents the corresponding weight of the feature function.

The formula is as follows:

$$p(y \mid x) \propto exp\left[\sum_{k=1}^{K} W_k f_k(y, x)\right] \tag{4}$$

3.4 Key Phrase Clustering and Visualization

To better understand key phrases for businesses, this section clusters the obtained key phrases with similar semantics. In this paper, we use a pre-trained word embedding-based method BERTopic for clustering. BERTopic works in the following four sequential steps: sentences are embedded using sentence-transformers [19], then each embedding is dimensionally reduced using UMAP. The results obtained from UMAP are clustered using HDBSCAN [15], clustering similar key phrases together. Finally, topics are derived from key phrases using c-TF-IDF, and the top four keywords (or all keywords if less than four) are selected based on c-TF-IDF scoring.

$$c - TF - IDF_i = \frac{k_l}{o_l} \times \frac{p}{\sum_j^q k_j} \tag{5}$$

The frequency of each keyword k in topic l is divided by the frequency of keyword o. The total number of unincorporated key phrases p is divided by the total frequency q of keyword k across all topics.

With Bertopic topic clustering, KFEA can group features with similar meanings on the same topic into the same topic, and we implement this clustering method based on Bertopic's open source library.

We use line graphs to visualize the clustering results of key phrases, as well as the topics obtained and summarized manually in the topic clustering as the clustering id.

For e-WOM, negative reviews are more reliable and useful than positive ones. Under the influence of negative feedback, product sales may decline [14]. Customers usually express their opinions more specifically in the reviews [7]. However, the large number of reviews makes it impossible for customers to access them all. Therefore, users generally use the obvious feedback as a basis, and we take the number of medium and poor reviews and the average rating as visualization metrics. To show the attention gap between each product with different cluster id, this paper only visualizes the clustering id that all products possess in a same category.

4   Evaluation

4.1   Experimental Setup

In our experiment, we obtained reviews of products from three categories (two in each category) from JingDong Mall. JingDong Mall is a popular electronic product platform in China, so we chose it as the source of our dataset. These products all contain a large number of user reviews, which can be used to extract key phrases. To reduce the training time, we deleted reviews longer than 128 characters. In the end, we collected a dataset containing the latest 6088 reviews (with a ratio of good, medium and poor reviews of 4:3:3), covering categories such as mobile phones, laptops and home furnishings. We manually labeled these reviews. These labels can test the ability to obtain key phrases in real-world scenarios.

The user reviews are manually labeled by three authors as the basis for validating KFEA's effectiveness. The first two authors make independent labels first which will be applied once the results are identical. If there were any discrepancies, all three authors shall negotiate and decide the final label.

To better study the generalization of KFEA, we also uses an English app store review dataset to evaluate the KFEA model [26].

Metrics for Key Feature Extraction. We evaluated the performance of the model using three metrics: precision, recall, and F1-score. Pattern matching-based event extraction often achieves good detection results in specific domains but has poor cross-domain and portability. For different categories of products, template construction requires a significant amount of resources and expertise. Therefore, we only compared our proposed model with the mainstream baseline method BERT-CRF and the latest method SIRA.

- *Precision*: Precision is the ratio of correctly predicted positive instances to all predicted positive instances. It expresses the accuracy of predictions among positive results.
- *Recall*: Recall is the ratio of correctly predicted positive instances to all actual positive instances.
- *F1 − score*: F1-score is a measure that balances precision and recall. It is the harmonic mean of precision and recall.

Metrics for Key Feature Clustering. For topic clustering, we used two metrics to measure the ability of the model and compared our proposed model with the latest method SIRA.

The Rand Index (RI) views clustering as a series of decision-making processes and calculates the ratio of correct decisions.

$$RI = \frac{TP + TN}{TP + FP + TF + FN} \tag{6}$$

TP, FP, FN, and TN represent true positives, false positives, false negatives, and true negatives in the confusion matrix, respectivelt. A higher RI value indicates better clustering performance.

Topic Coherence (TC) [20] is used to measure the coherence of words within a topic.

$$v(\vec{W'}) = \left\{ \sum_{w_i \in W'} NPMI(w_i, w_j)^{\gamma} \right\}_{j=1,\dots,|W|} \tag{7}$$

$$NPMI(w_i, w_j)^{\gamma} = \left(\frac{log \frac{P(w_i, w_j) + \varepsilon}{P(w_i), P(w_j)}}{-log \left(P\left(w_i, w_j \right) + \varepsilon \right)} \right)^{\gamma} \tag{8}$$

$$\phi s_i \left(\vec{u}, \vec{w} \right) = \frac{\sum_{i=1}^{|W|} u_i \cdot w_i}{\| \vec{u} \|_2 \cdot \| \vec{w} \|_2} \tag{9}$$

where C_v is based on a sliding window, top single-group segmented words, and normalized indirect confirmation measures normalized pointwise mutual information (NPMI) and cosine similarity. W represents the topics in topic clustering, and the vector $v(W')$ calculates the consistency of pairing word w_i in W' with all words w_j in W. Additionally, ε is used to interpret zeros and the logarithm of γ to give more weight to higher NPMI values [23]. $P(w_i, w_j)$ calculates the joint probability of two words based on the document. For each $s_i = (W', W^*)$, we calculate the confirmation measure ϕ for the similarity of W' and W^* relative to all words in W.

4.2   Results and Analysis

Evaluation of Key Phrase Extraction. In this experiment, we used a dataset from Jingdong Mall for comparison. When reimplementing the baseline, we followed the baseline steps and hyperparameters to ensure the fairness of the experiment. To verify the effectiveness of key feature extraction, we performed nested cross-validation on the dataset [11].

Table 1 shows the performance of KFEA in extracting key phrases. The precision, recall, and F1-score were 77.30%, 82.58%, and 79.85%, respectively. We could see that KFEA achieved good results, better than the latest baseline. Afterward, we observed the extracted key phrases and found that there were fewer key phrases in high-rated reviews. Because the KFEA model combines ratings and text, while SIRA simply concatenates ratings and text, KFEA can achieve better results. The following experimental results further illustrate the advantages of our method.

Table 1. The results of key phrase extraction in Jingdong Mall.

Product		Method		
		BERT -CRF	SIRA	KFEA
Oppo	P	0.707	0.724	**0.728**
	R	0.746	**0.764**	0.746
	F1	0.726	**0.743**	0.737
Honor	P	0.825	0.769	**0.875**
	R	0.846	0.769	**0.897**
	F1	0.835	0.769	**0.886**
Huawei	P	0.707	0.694	**0.738**
	R	0.732	0.768	**0.804**
	F1	0.719	0.729	**0.769**
Xiaomi	P	0.814	**0.850**	0.809
	R	0.875	0.850	**0.950**
	F1	0.843	0.850	**0.874**
Nanjiren	P	0.735	0.750	**0.755**
	R	0.800	0.800	**0.822**
	F1	0.766	0.774	**0.787**
Kawashi	P	0.727	**0.528**	0.786
	R	0.828	**0.655**	0.759
	F1	0.774	0.585	**0.772**
Overall	P	0.747	0.721	**0.773**
	R	0.796	0.773	**0.826**
	F1	0.771	0.746	**0.799**

Table 2. The results of key phrase extraction in app store.

APP		Method	
		SIRA	KFEA
BPI Mobile	P	0.723	**0.781**
	R	0.825	**0.877**
	F1	0.771	**0.826**
Chase Mobile	P	**0.674**	0.660
	R	0.763	**0.868**
	F1	0.716	**0.750**
Yahoo Mail	P	0.729	**0.736**
	R	**0.843**	0.726
	F1	**0.782**	0.731
Gmail	P	0.750	**0.818**
	R	0.716	**0.806**
	F1	0.733	**0.812**
Snapchat	P	**0.837**	0.784
	R	0.817	**0.826**
	F1	**0.827**	0.805
Instagram	P	0.786	**0.817**
	R	0.746	**0.831**
	F1	0.765	**0.824**
Overall	P	0.750	**0.769**
	R	0.785	**0.819**
	F1	0.767	**0.793**

Among the six categories in JingDong Mall, Honor phones achieved the highest precision of 87.50%, and Xiaomi laptops achieved the highest recall of 95.00%. Huawei laptops had the lowest precision of 73.77%, and Kawashi had the lowest recall of 75.86%. Surely, we also studied some cases where correct conclusions could not be drawn. In some cases, there were omissions or additions of words in the key phrases extracted by KFEA. For example, a review stating "Personally, I suggest better to go to a physical mobile phone store to buy. This time I was very dissatisfied with my purchase. They said they would give headphones but didn't. Can you still trust the merchant based on this?" would be incorrectly extracted into "give headphones" instead of "give headphones but didn't", which would reduce the precision and degrade the model performance.

Table 2 shows the result, In six categories in the app store, KFEA's precision, recall, and F1-score reached 76.90%, 81.93%, and 79.35%, respectively. Compared to SIRA's precision, recall, and F1-score of 75.00%, 78.50%, and 76.71%, KFEA performs better. Subsequently, we examined the key phrases extracted from the app store and found significant differences in the key phrases contained in different categories of apps. KFEA achieved better results by integrating category and rating information in different ways, while SIRA simply concatenated them.

Table 1 shows the performance of KFEA and two baselines. The result shows that KFEA outperforms both baselines on most metrics. This indicates that there is room for improvement in the aforementioned deep learning-based baselines. This demonstrates that fine-grained feature fusion can improve the performance. In KFEA, both the category information and rating information are given sufficient attention to more fully characterize review information and better serve the key phrases extraction.

Table 3. The results of key phrases clustering.

Product		Method	
		SIRA	KFEA
Oppo	**RI**	0.213	**0.958**
	TC	0.874	**0.989**
Honor	**RI**	0.400	**0.936**
	TC	0.906	**0.996**
Huawei	**RI**	0.084	**0.949**
	TC	0.902	**0.993**
Xiaomi	**RI**	0.099	**0.963**
	TC	0.881	**0.983**
Nanjiren	**RI**	0.456	**0.884**
	TC	0.947	**0.973**
Kawashi	**RI**	0.134	**0.903**
	TC	0.911	**0.982**
all	**RI**	0.231	**0.932**
	TC	0.904	**0.986**

Evaluation of Key Phrase Clustering. Table 3 shows the performance of KFEA in topic clustering of key phrases and its performance compared to the SIRA baseline. KFEA outperforms the baseline in all aspects, with an average RI and TC of 0.932 and 0.986, respectively, higher than SIRA's 0.231 and 0.904.

As can be seen from the table, KFEA's RI performance is improved more significantly compared to SIRA. This indicates that KFEA have divided similar problem phrases into the same clusters or placed dissimilar problem phrases into different cluster. According to the experimental results, KFEA is better at evenly dividing key phrases compared to SIRA, and the distribution of key phrases is closer to reality. Because the topics in topic coherence are obtained based on the number of topics, words in the same problem phrase generally appear in the topic, so the gap in topic coherence is not large. However, it can still be seen that KFEA's performance is stronger than SIRA's. The reason for this performance is that KFEA uses the BERT pre-training model to fully exploit the semantic-aware of key phrases. In contrast, SIRA is simply a clustering method that uses cosine similarity to construct graphs.

Analysis and Visualization. Merchants evaluate user behavior activity based on their reviews and ratings of products. Users place greater emphasis on high-quality reviews rather than those that are ranked higher. An excessive number of medium-to-poor reviews can reduce users' favorability towards a product and decrease its sales volume. Thus, this study counts the number of medium-to-poor reviews and visualizes them.

Average rating is an independent parameter and also a critical factor affecting product sales. Therefore, in this paper, we will also average the extracted topics and determine which topics have the lowest average scores and visualize them.

This study employs line graphs to visualize the results, as shown in Figs. 4 and 5. In Fig. 4, the x-axis represents the cluster id , while the y-axis represents the score for each cluster id. The majority of key phrases extracted from reviews are distributed within the 1–3 range, as evidenced by the final topic scores falling within this range. We subtract the corresponding score from 3 to obtain the final result. Higher scores then indicate a greater need for attention, making it easier for merchants to observe. In Fig. 5, the x-axis represents the cluster id, while the y-axis represents the rate for each cluster id. Each product has varying quantities key phrases, so comparing their quantities is not reasonable. Therefore, this study uses the ratio for comparison. By comparing the heights of the line graphs, we can directly understand the distribution of key topics in reviews. It can be seen that Huawei and Xiaomi belong to the same category, but there is a gap between their ratio for each cluster. With these two methods to select different features that influence e-WOM, merchants can improve various aspects of their products based on the selected features to achieve better e-WOM (Table 4).

As illustrated in the figure, the issue with C7 for Lenovo laptops has the most significant impact on product ratings and should be addressed as a priority. The figure also indicates that Huawei laptops received the highest number of negative reviews for C2, suggesting that C2 should also be prioritized. Different products

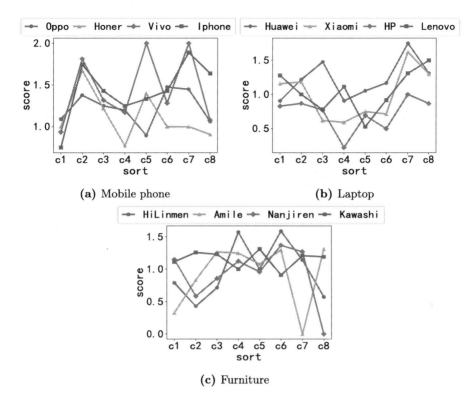

Fig. 4. The Results Of Key Phrases Clustering

Table 4. The cluster name of the category.

#	Mobile phone	Laptop	Furniture
c1	Endurance	Black Screen	Scent
c2	Price reduction	Giveaways	Size
c3	Black Screen	Keyboard	Packaging
c4	Signal	Noise	Logistics
c5	Logistics	Heat	Fabrics
c6	Caton	Packaging	Installation
c7	Charging	Start up	Alignment
c8	Camera	Price reduction	Material

have unique features and feature distributions, highlighting the necessity of fine-grained mining and analysis of reviews. Through visualization, merchants can gain a clear understanding of the features associated with their products and take subsequent actions. This information can also provide merchants with a general

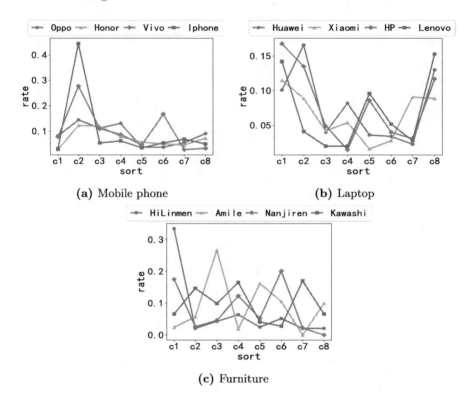

(a) Mobile phone   (b) Laptop

(c) Furniture

Fig. 5. Proportion the number of key phrases in different categories.

understanding of similar products, aiding in competitive analysis. Additionally, priorities can be determined based on the line chart.

5   Conclusion

The key phrases contained in these reviews greatly affect the e-WOM of the product. Therefore, businesses should pay attention to these features that are closely related to user ratings. This paper proposes a semantic-aware fine-grained electronic comment analysis method KFEA, which first uses BERT-Attn-CRF to obtain key phrases that have a great correlation with the rating of the product. Then BERTopic is used for clustering to analyze the priority of topics. Finally, we use a dataset containing 6 categories to evaluate the model and validate the effectiveness of KFEA.

In the future, businesses can collect more comprehensive data for experimentation so that the experiments in the article can be practiced in the real world scenarios.

Acknowledgement. This work is supported by Hainan Provincial Key Research and Development Program (No. ZDYF2022GXJS230), and National Natural Science Foundation of China (No.61962017).

References

1. Alamoudi, E.S., Alghamdi, N.S.: Sentiment classification and aspect-based sentiment analysis on yelp reviews using deep learning and word embeddings. J. Decis. Syst. **30**, 259–281 (2021)
2. Cambria, E., Schuller, B., Xia, Y., Havasi, C.: New avenues in opinion mining and sentiment analysis. IEEE Intell. Syst. **28**, 15–21 (2013)
3. Chatterjee, S., Goyal, D., Prakash, A., Sharma, J.: Exploring healthcare/health-product ecommerce satisfaction: a text mining and machine learning application. J. Bus. Res. **131**, 815–825 (2021)
4. Cheng, X., Zhou, M.: Study on effect of ewom: a literature review and suggestions for future research. In: 2010 International Conference on Management and Service Science, pp. 1–4 (2010)
5. Dai, H., Lai, P.T., Chang, Y.C., Tsai, R.T.H.: Enhancing of chemical compound and drug name recognition using representative tag scheme and fine-grained tokenization. J. Cheminform, **7**, S14–S14 (2015)
6. Devlin, J., Chang, M.W., Lee, K., Toutanova, K.: Bert: Pre-training of deep bidirectional transformers for language understanding. arXiv preprint arXiv:1810.04805 (2018)
7. Ghose, A., Ipeirotis, P.G.: Estimating the helpfulness and economic impact of product reviews: mining text and reviewer characteristics. IEEE Trans. Knowl. Data Eng. **23**, 1498–1512 (2010)
8. Grootendorst, M.: Bertopic: Neural topic modeling with a class-based TF-IDF procedure. CoRR abs/ arXiv: 2203.05794 (2022)
9. Hong, W., Zheng, C., Wu, L., Pu, X.: Analyzing the relationship between consumer satisfaction and fresh e-commerce logistics service using text mining techniques. Sustainability **11**, 3570–3586 (2019)
10. Huang, Z., Xu, W., Yu, K.: Bidirectional lstm-crf models for sequence tagging. arXiv preprint arXiv:1508.01991 abs/1508.01991 (2015)
11. Kohavi, R., et al.: A study of cross-validation and bootstrap for accuracy estimation and model selection. In: IJCAI, vol. 14, pp. 1137–1145 (1995)
12. Li, S., Liu, F., Zhang, Y., Zhu, B., Zhu, H., Yu, Z.: Text mining of user-generated content (ugc) for business applications in e-commerce: A systematic review. Mathematics **10**, 1–27 (2022)
13. Li, S., Zhang, Y., Li, Y., Yu, Z.: The user preference identification for product improvement based on online comment patch. Electron. Commer. Res. **21**, 423–444 (2021)
14. Mandal, S., Maiti, A.: Network promoter score (neps): an indicator of product sales in e-commerce retailing sector. Electron. Mark. **32**, 1327–1349 (2022)
15. McInnes, L., Healy, J., Astels, S.: hdbscan: hierarchical density based clustering. J. Open Source Softw. **2**, 205–206 (2017)
16. Munikar, M., Shakya, S., Shrestha, A.: Fine-grained sentiment classification using bert. In: 2019 Artificial Intelligence for Transforming Business and Society (AITB), vol. 1, pp. 1–5 (2019)
17. Nadeau, D., Sekine, S.: A survey of named entity recognition and classification. Lingvisticae Investigationes **30**, 3–26 (2007)

18. Park, S., Kim, H.M.: Phrase embedding and clustering for sub-feature extraction from online data. J. Mech. Design **144**, 054501-1-054501-10 (2022)
19. Reimers, N., Gurevych, I.: Sentence-bert: sentence embeddings using siamese bert-networks. In: Proceedings of the 2019 Conference on Empirical Methods in Natural Language Processing and the 9th International Joint Conference on Natural Language Processing (EMNLP-IJCNLP), pp. 3982–3992 (2019)
20. Röder, M., Both, A., Hinneburg, A.: Exploring the space of topic coherence measures. In: Proceedings of the Eighth ACM International Conference on Web Search and Data Mining, pp. 399–408 (2015)
21. Shoja, B.M., Tabrizi, N.: Customer reviews analysis with deep neural networks for e-commerce recommender systems. IEEE Access **7**, 1–1 (2019)
22. Su, Y., Shen, Y.: A deep learning-based sentiment classification model for real online consumption. Front. Psychol. **13**, 886982–886991 (2022)
23. Syed, S., Spruit, M.: Full-text or abstract? examining topic coherence scores using latent dirichlet allocation. In: 2017 IEEE International Conference on Data Science and Advanced Analytics (DSAA), pp. 165–174 (2017)
24. Tan, Y., Zhang, M., Liu, Y., Ma, S.: Rating-boosted latent topics: understanding users and items with ratings and reviews. In: IJCAI, vol. 16, pp. 2640–2646 (2016)
25. Vaswani, A., et al.: Attention is all you need. Adv. Neural. Inf. Process. Syst. **30**, 5998–6008 (2017)
26. Wang, Y., Wang, J., Zhang, H., Ming, X., Shi, L., Wang, Q.: Where is your app frustrating users? In: Proceedings of the 44th International Conference on Software Engineering, pp. 2427–2439 (2022)
27. Xu, X.: What are customers commenting on, and how is their satisfaction affected? examining online reviews in the on-demand food service context. Decis. Support Syst. **142**, 113467–113479 (2021)
28. Yadav, N.B.: Harnessing customer feedback for product recommendations: an aspect-level sentiment analysis framework. In: Human-Centric Intelligent Systems, pp. 1–11 (2023)

Emotional Analysis

Discovery of Emotion Implicit Causes in Products Based on Commonsense Reasoning

Qiutong Guo[1,2,3,4], Jianxing Yu[1,2,3,4]([✉]), Yufeng Zhang[1,2,3,4], Haowei Jiang[1,2,3,4], Wei Liu[1,2,3,4], and Jian Yin[1,2,3,4]

[1] School of Artificial Intelligence, Sun Yat-sen University, Guangzhou, China
{guoqt6,zhangyf283,jianghw9}@mail2.sysu.edu.cn,
{liuw259,issjyin}@mail.sysu.edu.cn
[2] Guangdong Key Laboratory of Big Data Analysis and Processing, Guangzhou, China
[3] Pazhou Lab, Guangzhou 510330, China
[4] Key Laboratory of Sustainable Tourism Smart Assessment Technology, Ministry of Culture and Tourism, Guangzhou, China
yujx26@mail.sysu.edu.cn

Abstract. This paper focuses on the task of product emotion cause analysis which aims to find essential causes of certain emotions from product reviews. Current works only study the explicit causes which are some spans in the given text. However, some crucial and useful causes may be expressed vaguely. They may not mention but can be inferred from the text semantics. They can capture the deeper reasons to explain some unknown phenomena, which can well support the applications like market research and product optimization. To address this problem, we in this paper propose a new task of Emotion Implicit Cause Discovery (EICD). We develop a novel method that can deduce the implicit causes based on the contexts and commonsense knowledge. Our method first retrieves related knowledge from the large language model to construct reasoning graphs for the emotions and potential causes. We then encode the structural knowledge in the graph and infer the implicit cause by deductive reasoning. To evaluate our method, we construct a large dataset called EICDset, based on Amazon product reviews. Experiments on it demonstrate the effectiveness of our model.

Keywords: Implicit emotion cause discovery · Commonsense reasoning · External knowledge

1 Introduction

With the rapid growth of the Internet, a huge amount of subjective text is emerging, such as e-commerce reviews and psychological counseling conversa-

Supplementary Information The online version contains supplementary material available at https://doi.org/10.1007/978-3-031-46661-8_19.

tion data. These contents contain plentiful users' emotions which consist of complex feelings, perceptions, thoughts, and behaviors. Such emotional texts have rich subjective knowledge, reflecting people's habits and preferences. Therefore, exploring the knowledge embodied in these subjective sources is highly valuable. For instance, in product or service management, companies can effectively grasp the genuine needs of users by analyzing their emotions and underlying causes in the reviews. This analysis can help to clarify the direction of improvement and optimization, leading to the launch of products and services that align with consumer expectations.

Emotion cause analysis has attracted extensive attention recently. Previous works mostly focus on emotion cause extraction (ECE) and emotion-cause pair extraction (ECPE). ECE aims to extract cause clauses of emotion from the given text, while ECPE is used to extract all pairs of emotion clauses and cause clauses. In real applications, emotional expressions are often complex. The implicit causes of emotions may not be directly mentioned in the text, but humans can conveniently understand them by commonsense reasoning over text semantics. As shown in Fig. 1, the user expresses the emotion of 'anger'. While previous work on ECE can extract the explicit emotion cause of 'bought this miscellaneous brand mobile phone' it fails to discover the more essential cause of the 'anger', which is that 'The cell phone signal is poor' This cause is hidden and does not appear in the text. To discover it, we need to reason over complex content with commonsense knowledge. Since this cause is not a span in the given text, traditional extraction-based methods may be failed. Due to the lack of commonsense reasoning ability, it is difficult for these methods to infer the complex contexts and the background knowledge behind the text. Thus, this topic has great research value. Implicit causes provide a deeper understanding of the users' opinions, leading to more accurate decision-making for products operation.

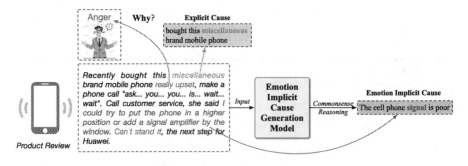

Fig. 1. An example of emotion implicit cause. The implicit cause 'The cell phone signal is poor' for the emotion 'anger' is implied but unmentioned in the text.

To address this problem, this paper proposes a novel topic and practical method to derive the implicit emotion cause. Unlike traditional extraction-based methods, our approach generates implicit emotion causes directly. This generative way provides greater flexibility and can uncover a wider range of cause types.

Specifically, we first identify potential emotion and cause clauses in the text. To capture the background knowledge, we retrieve the related commonsense contents from the large language model (LLM) GPT [16] and the knowledge graph (KG) ATOMIC [17]. The knowledge coverage of LLM is huge but there may exist many irrelevant noises, while the KG covers small-scale high-quality knowledge. We then apply graph networks to integrate these contents and learn their structural contexts. To filter the irrelevant noises, we apply the heuristic rules and a multi-headed attention mechanism, which can derive strong associations of emotion and implicit causes. Finally, we decode the implicit cause with its corresponding emotion. To evaluate our model, we create a large-scale corpus, called EICDset. Experiments on it demonstrate the effectiveness of our model. In summary, the main contributions of this paper include:

- We propose a new task of Emotion Implicit Cause Discovery (EICD), which aims to discover implicit causes that underlie explicit cause events in texts.
- We build a high-quality dataset EICDset for the task and propose a novel generative model which can yield implicit emotion causes directly.
- We conduct extensive experiments to fully evaluate our proposed model.

2 Related Work

Our work relates to emotion cause analysis and knowledge augmentation. Next, we survey the works on these topics.

2.1 Emotion Cause Analysis

Lee et al. [11] first proposed the concept of Emotion Cause Extraction and attracted attention from researchers. It aims to extract events, clauses, phrases, or words that describe the causes of emotions from the text that contains emotions. In the early stages, emotional cause extraction tasks were dominated by the rule-based approach. Researchers [12,13] developed rule-based systems by analyzing the linguistic features of the corpus with empirical guidance and used these rules to extract causal events that lead to emotional changes [7,10]. With the introduction of machine learning methods and deep learning methods, these models' performance has significantly improved [5,22]. Gui et al. [9] constructed an emotion cause extraction corpus using the W3C Emotion Markup Language scheme on SINA city news, which has garnered wide attention and is now considered a benchmark dataset for ECE research. Ding et al. [3] and Xia et al. [21] utilized positional embeddings to model cause clauses that occur near emotion clauses. Recognizing that emotions and causes are highly correlated and can inform each other, Turcan et al. [19] proposed a multi-task learning framework that simultaneously performs emotion recognition and cause extraction.

The requirement of emotion annotation prior to cause extraction in ECE tasks limits their potential application scenarios. To address this limitation, Xia et al. [20] proposed a novel task called emotion-cause pair extraction (ECPE) that jointly extracts all possible pairs of emotion and cause clauses. However, the two-step pipeline approach they used may lead to error transfer and other defects. Therefore, recent research has shifted towards developing end-to-end deep models [15,23]. Ding et al. [4] proposed the ECPE-2D model that uses a 2D Transformer to directly model clause pairs. However, this model has a high time and space complexity due to a large number of parameters. Fan et al. [6] analogized it to a directed graph generation problem and proposed TransECPE, a transition-based model for emotion cause extraction. Bao et al. [1] presented a multi-granularity semantic perceptual graph model that leverages both fine-grained and coarse-grained semantic features to capture causal cues in context.

2.2   Commonsense Knowledge Utilization

Commonsense knowledge utilization can bring rational external knowledge for models. In the study of emotion cause extraction, some works [8,14,24] have studied how to use external knowledge bases to introduce commonsense knowledge. Yan et al. [22] leveraged ConceptNet, a common sense knowledge graph, to establish edges connecting candidate cause clauses to emotion clauses, in order to identify cause clauses more accurately by establishing causal paths. Turcan et al. [19] and Ghosal et al. [8] incorporated common sense knowledge into the model input by generating knowledge under specific relationships for the text using COMET. They applied this technique to span-level emotion cause extraction and conversational causal emotion entailment tasks, respectively.

By retrieving the external knowledge, we are able to obtain targeted commonsense knowledge for the clause under some specific relations. However, this knowledge is usually confined to fixed types of relations with limited coverage. Prompt learning has recently received increasing attention as a new paradigm for pre-trained language models. It is a technique for extracting implicit knowledge from large language models, which can be designed with prompts to guide the large language model to generate content that matches a specific task. In their study on commonsense question answering, Sun et al. [18] proposed a framework based on prompts to enhance the model's commonsense reasoning abilities by utilizing the implicit knowledge contained within the large language model. In this paper, we propose to leverage a large language model with vast amounts of human-shared knowledge to obtain more comprehensive information. By designing prompt statements, we allow the large language model GPT [16] to flexibly generate commonsense related to cause events in clauses.

3   Construction of Corpus

In the work, we construct a new corpus for the EICD task by augmenting the Multilingual Amazon Reviews Corpus with additional annotations. The original dataset comprises product reviews from Amazon, where each record contains the review text, review title, star rating, anonymous reviewer ID, anonymous product ID, and coarse-grained product category. The dataset includes reviews in six languages (English, Japanese, German, French, Chinese, and Spanish) collected between November 1, 2015, and November 1, 2019. Our study focuses on the English subset of the dataset and enhances it with new annotations to support the task of EICD. Figure 2 provides an overview of the corpus construction process.

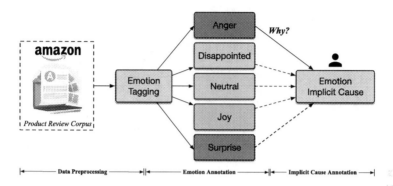

Fig. 2. Process of the Corpus Construction

Data Preprocessing: To ensure that our model is exposed to diverse sentiments, we employ a random sampling strategy based on the star rating of each review to select data from the original dataset. As the star rating conveys the attitude of the reviewer to some extent, this approach allows us to provide the model with data that captures a range of potential attitudes. Specifically, we randomly selected 10,000 samples from each of the five-star ratings (ranging from 1–5) in the corpus. To facilitate the discovery of causes for emotions, we excluded reviews where no emotion was evident.

Emotion Annotation: To indicate cause discovery by emotion in the implicit emotion cause Discovery task, we annotate reviews in the dataset with emotion categories. Specifically, we establish five emotion categories, namely anger, disappointment, neutral, joy, and surprise, through analyzing the dominant emotions conveyed by users in product reviews. Our strategy is based on Ekman's well-established study of emotion categories and takes into account the original classification of the reviews into five stars. In this process, we employ three annotators who perform annotation simultaneously. Two of the annotators independently assess the emotion of each instance. In cases where there is a disagreement between the two annotators, the third annotator acts as an arbiter to resolve the disagreement.

Implicit Cause Annotation: The identification of implicit causes is a crucial factor in understanding emotions expressed in language. To ensure accurate labeling of these causes, we adhere to three fundamental principles: (1) Implicit causes, as distinct from explicit cause events, are usually not explicitly stated in the given text. Thus, we avoid having too many consecutive texts overlapping with the given text in our annotation. Specifically, we set the length of the overlapping consecutive texts to no more than 3; (2) we specify a standardized labeling format for simplicity and ease of reading, which begins with the phrase 'because of'; and (3) we limit the length of the reason text to no more than 30 words. These principles are designed to facilitate the effective identification of implicit causes of emotions in the given text.

Table 1. Samples from our corpus

Review Texts	Emotion	Cause
But the plumper made my lips actually sting!	anger	because of the physical discomfort caused by the product.
Good looking but the steam power is weak. I assumed I could use it for the bathroom too but seems its more suitable for a jewelry cleaning and maybe the kitchen.	disappointed	because of a mismatch between expectations and reality of this steam cleaner, leading to feelings of disappointment.
It has a calming affect, but if you're looking to help with sleep issues, it didn't knock me out like other melatonin brands.	neutral	because of the user's expectation of the product's effects on sleep, they were disappointed with its performance, despite its calming effect.
The product looks exactly as pictured. Once assembled it is a very nice addition to have.	joy	because of the satisfaction of seeing a product that meets expectations and adds value to one's life.
Very sturdy! I was so surprised at the high quality of these. They dont wobble, slip, or bend.	surprise	because of the expectation of poor quality from similar products.

After preprocessing and annotating the data, we obtained a corpus of 50,000 product reviews. This corpus comprises Amazon product review text, review emotions, and their corresponding implicit causes. Table 1 illustrates some examples of the corpus, with one sample for each emotion category.

4   Methodology

In this work, we propose a novel emotion implicit cause generation model for EICD task. Given a document $D = \{c_1, c_2, ...c_{|D|}\}$ containing several clauses and the emotion e corresponding to it. We denote the implicit cause discovery of emotions as a text generation task, using the model based on commonsense reasoning to generate implicit causes that trigger the emotion. In this section, the model is described in detail.

4.1   Overall Architecture

The overall architecture of our model is shown in Fig. 3, which comprises three main components: the Commonsense-based Semantic Graph construction module (CSG), the Emotion and Cause Graphs representation module (ECG), and the Implicit Cause Generation module (ICG). We first take all the clauses in document D as the input of our model. Subsequently, the CSG module constructs semantic graphs of causes clauses and potential emotion clauses by analyzing the semantic relations and relevant commonsense knowledge in the given document. The ECG module characterizes the structural features of the graphs while incorporating commonsense knowledge and contextual semantics. The ICG module first identifies significant candidate combinations of emotions and causes, computes causal combinations with strong associations between them, and finally decodes the implicit causes. The structure of each module is described below.

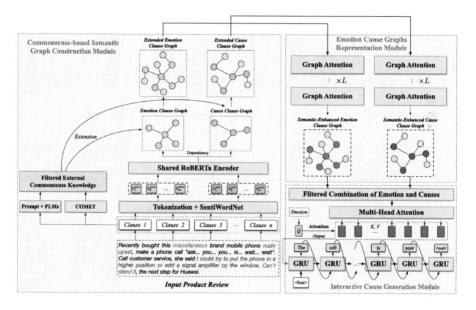

Fig. 3. The overall architecture of the approach.

4.2 Commonsense-Based Semantic Graph Construction

This module is responsible for constructing the emotion clause graph and causes clause graph. To bridge the knowledge gap between explicit cause events and implicit essential causes, we incorporate external commonsense knowledge to obtain clues for potential causal reasoning. By constructing graphs, this unit portrays the logical relationships between clauses at a fine-grained level.

Constructing Clause Graphs. Aiming at a more refined utilization of emotional information to support causal reasoning, we first extract potential emotion clauses and cause clauses from the documents. To identify the potential emotion clauses, we utilize SentiWordNet to obtain the sentiment score of each word in the clause. When a clause's average positive or negative score is greater than

0.5, it is considered a potential emotion clause. We obtain the set of potential emotion clauses $C^{emo} = \{c_1^{emo}, c_2^{emo}, ..., c_j^{emo}\}$. Regarding the cause clause, it should be noted that it may be either the emotion clause itself or other clauses within the document. Therefore, we obtain the set of potential cause clauses $C^{cau} = \{c_1^{cau}, c_2^{cau}, ..., c_{|D|}^{cau}\}$ by considering all clauses in the ground of document D.

Subsequently, we conduct representation learning of clauses. To accomplish this, we encode clauses of document D with the pre-training model RoBERTa, resulting in a representation v of each clause. Finally, we construct the emotion and cause clause graphs by utilizing the clauses as nodes and the dependencies between clauses as edges.

Introducing Commonsense Knowledge. Incorporating commonsense knowledge is essential for bridging the knowledge gap between explicit cause events and implicit causes. One of the most direct sources of commonsense knowledge is the Knowledge Graph (KG), but its coverage is limited as knowledge retrieval is typically restricted to fixed types of relations. To obtain more comprehensive knowledge, we propose to use the Large Language Models (LLMs) that contain a significant amount of common sense. By combining these two sources of knowledge, we can better bridge the knowledge gap between the explicit cause events and the implicit causes. To enrich emotional, causal, and commonsense information, we extend the original clause graph by adding knowledge as neighboring nodes of the corresponding clause nodes. The specific method is as follows.

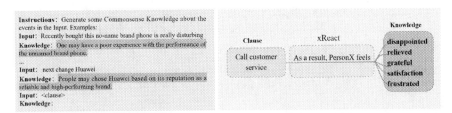

(a) Knowledge Generation Prompt   (b) Example of COMET generation

Fig. 4. The strategy of introducing commonsense knowledge consists of two parts: 1) LLMs, using GPT to generate based on the knowledge generation prompt. 2) KG, leveraging COMET to generate knowledge via specific relations.

textbfLarge Language Model. LLMs are models that have learned a vast amount of general language knowledge and world knowledge by pre-training on massive amounts of text data. Prompt learning is a technique used to extract the knowledge built into LLMs by providing prompts to guide them to generate content that matches a specific task. This flexibility allows LLMs to generate more general and wider coverage knowledge for clauses. We first use knowledge generation prompts to let GPT [16] generate knowledge related to the events in the clauses.

Figure 4a shows the design of the knowledge generation prompt for our task. The knowledge generation prompt consists of instructions, some demos, and a new clause placeholder. When generating knowledge for a new clause, we insert the clause v into a placeholder and repeatedly sample the generated continuation of that prompt to obtain a set of knowledge statements $K_v = \{k_1, k_2, ..., k_{10}\}$. Considering that these contents may contain noise, we propose to use point-wise mutual information (PMI) to quantify the correlation between clause v and each generated knowledge k. Since both v and k are sentences, we use a PrLM to estimate the probabilities. Conditional probabilities are calculated by treating conditional sentences as prefixes. Therefore, we use clause v as the condition to calculate the mutual information between the knowledge and the question.

$$\text{PMI}\,(k; v) = \log \frac{\text{PrLM}\,(k \mid v)}{\text{PrLM}\,(v)} \tag{1}$$

PMI measures the correlation between clauses and knowledge, with higher values indicating greater dependency. A positive score indicates a positive correlation between the clause and the corresponding knowledge, while a score of zero indicates independence and a negative score indicates potential contradiction. We only consider knowledge with positive PMI scores for selection.

Knowledge Graph. Our approach leverages social commonsense knowledge from the Knowledge Graph ATOMIC-2020 [17], using the COMET generating model trained by Bosselut et al. [2] to infer texts outside the graph's scope. While ATOMIC-2020 contains nine social-interaction relations, not all of them apply to our condition.

For the cause clause graph, we use four cause-effect related types in ATOMIC: xIntend, xWant, xNeed, and xEffect. Specifically, xIntend describes the intention or purpose behind the behavior; xEffect denotes the effect caused by the event or behavior; xWant represents a goal or desire that a person wants to achieve; and xNeed represents someone's need or requirement. For the emotion clause graph, we use three types related to emotion expressions: xReact, oReact, and xAttr. The xReact and oReact types describe one's own or others' feelings and reactions to the event, while xAttr represents attributes or traits. An illustration of generating commonsense knowledge through xReact relations is shown in Fig. 4b.

To filter the noise in the content generated by COMET, we propose a sentiment-based approach. We use SentiWordNet to map emotions to sentimental polarity and group knowledge into positive, negative, and neutral sentiment categories. For the emotion of the document, we map the joy and surprise to positive, and anger and disappointment to negative sentiment. Based on this, we can select the matching knowledge based on the sentiment of the input document.

4.3   Emotion and Cause Graphs Representation

To effectively discover implicit causes of emotions, it is crucial to not only incorporate external commonsense but also utilize this knowledge to enhance the model's capacity to comprehend the semantics of text and reason about causal logic. Accordingly, this module employs the graph attention mechanism to model the internal structure of potential emotion and cause clauses graphs and capture possible commonsense associations. Specifically, we leverage multiple graph attention layers to extract the structural information of the graphs and compute the degree of relevance between emotions and causes, thereby incorporating both common sense and contextual knowledge into the graph representation.

We model inter-clause relationships by stacking multiple layers of graph attention. Taking the cause graph as an example, the number of nodes n of the cause graph after knowledge enhancement is the sum of the number of original document clauses and the new neighboring knowledge nodes. We assume that $\{h_1^{t-1}, h_2^{t-1}, ..., h_n^{t-1}\}$ denote the input node representation of the t-th layer graph attention, and h_i^{t-1} is the $t-1$-th layer output representation of node B. The input of the first layer graph attention is $\{h_1^0, h_2^0, ..., h_n^0\}$. Each node in the graph is updated by the following aggregation, as shown in Eq. 2.

$$h_i^{(t)} = \text{ReLU} \left(\sum_{j \in N(i)} \alpha_{ij}^{(t)} W^{(t)} h_j^{(t-1)} + b^{(t)} \right) \tag{2}$$

where $h_i^{(t)}$ is the output of the hidden layer, $W^{(t)}$ and $b^{(t)}$ are the learned parameters, and $N(i)$ denotes the neighbors connected to node v_i. Then, the attention weight $\alpha_{ij}^{(t)}$ can be calculated to represent the relevance of node v_i and node v_j.

$$e_{ij}^t = w^{t^T} \tanh \left([W^t h_i^{t-1}; W^t h_j^{t-1}] \right) \tag{3}$$

$$\alpha_{ij}^{(t)} = \frac{\exp \left(\text{LeakyReLU} \left(e_{ij}^{(t)} \right) \right)}{\sum_{k \in N(i)} \exp \left(\text{LeakyReLU} \left(e_{ik}^{(t)} \right) \right)} \tag{4}$$

where $[, ; ,]$ is the concatenation operation. By stacking T graph attention layers to model inter-clause dependencies, we obtain the output clauses representation of the last layer, which can be denoted as $h^T = (h_1, h_2, ..., h_n)^T$.

This module updates the graph nodes by computing multiple layers of graph attention layers, which incorporate both contextual information from the clauses and external common sense knowledge into the node representations. Moreover, the externally introduced common sense knowledge is fully integrated with the original clause, enabling knowledge transfer in the graph during node attention computation, and helping the model to better understand the semantics. In the subsequent interactive inference unit, the model is equipped with sufficient background common sense and contextual information to reason about the implicit causes of emotions.

4.4   Implicit Cause Generation

This module filters the candidate combinations of emotion-cause matches based on the designed rules, and then reasons and generates the implied causes of emotions on this basis. More specifically, we first design a set of combination filtering rules for the emotions and their corresponding causes, which allows us to filter out explicit causes. Then, using an interactive attention mechanism, we select the combination with the strongest causal association among the filtered candidates. Finally, we decode the cause node representations within this combination to output the implied causes.

Combination and Filtering. We first combine the nodes in the emotion and cause graphs. Considering that there exists much noise in the candidate emotion-cause combinations, we design the filtering rules as follows.

- a. Considering that implicit causes generally do not appear in the given text, for the cause graph, the nodes of the original document clauses in the graph are not considered, and only the extended knowledge nodes are kept as a way to avoid finding some explicit causes.
- b. Considering that the emotion corresponding to the implicit cause is generally reflected in the document, only the original clauses nodes of the document in the emotion graph are considered, and no extended knowledge nodes are considered to ensure the accuracy of the emotional expression of the original document.
- c. Considering that the relative positions of emotion and cause are often relatively close to each other, we constrain the relative positions d of emotion and cause sentences. Only the extended neighboring knowledge nodes of cause clauses with a relative position ≤ 2 to the emotion clause are considered.

After obtaining a candidate combination of sentiment and cause, we concatenate the hidden layer vector representations of the emotion clause h_i^{emo} and the corresponding cause h_j^{cau} to obtain the joint representation vector $H_i = \left[h_i^{emo}, h_j^{cau}\right]$ for the candidate combination.

Interaction and Generation. This module employs the multi-headed attention mechanism to interactively compute candidate emotion and cause combinations, aiming to identify combinations with strong causal relationships. We represent all candidate combinations as a matrix $H = [H_1, H_2, ...H_n] \in R^{t \times 2d}$, where t is the number of all candidate combinations that satisfy the condition. In order to fully obtain the global information, this module uses a multi-headed attention mechanism on H, as shown in Eqs. 5 and 6.

$$H^i = \text{Attention}\left(Q^i, K^i, V^i\right) = \text{softmax}\left(\frac{Q^i K^{i^T}}{\sqrt{d_k}}\right) V^i \tag{5}$$

$$\hat{H} = \text{concat}_{i \in N_h}\left(H^i\right) \tag{6}$$

where $Q^i = HW_Q^i, K^i = HW_K^i, V^i = HW_V^i, W_Q^i \in R^{2d \times d_k}, W_K^i \in R^{2d \times d_k}, W_V^i \in R^{2d \times d_k}$, N_h represents the number of attention heads and d_k represents the dimension of the hidden vector in the multi-headed attention mechanism. To explicitly introduce relative position information, this module uses a relative position representation r_{j-i}. For each candidate emotion and cause combination $\langle c_i, c_j \rangle$, its representation is concatenated from the previous emotion cause combination representation $\hat{h}_{ij} \in \hat{H}$ with the relative position vector r_{j-i}, as shown in Eq. 7.

$$p_{i,j} = \text{ReLU}\left(W_p\left[\widehat{h_{ij}}; r_{j-i}\right] + b_p\right) \tag{7}$$

where W_p and b_p are the learned parameters.

After that, we use the attention mechanism to aggregate all candidate representations $p_{i,j}$ according to the emotion type of the document. Specifically, the emotion of the document will get the corresponding representation e through the embedding layer. We use q as query and all the candidate representations $p_{i,j}$ as key and value to input into the attention module to get the aggregated cause representation p. We utilize the gated recurrent units (GRUs) to build the decoder. We take cause p as the initial hidden state of the GRU, and the symbol ¡bos¿ is fed into the GRU as the initial input. Finally, the GRU model will generate the implicit emotion causes in a sequential manner. To address the problem of repetitive text generation, we introduce the Coverage mechanism, which keeps track of the generated content and penalizes the duplicate output during decoding.

5   Experiments and Results

We conduct experiments on the implicit emotion cause dataset described in Sect. 3 and randomly select 90% of the data for training and the remaining 10% for testing. We implement our experiments using PyTorch. We train our proposed models on a single NVIDIA GeForce RTX 3090 and compare the performance with the following baselines to demonstrate the effectiveness:

- **S2SA**: A Sequence-to-Sequence framework with attention and copy mechanisms, where the input sequence is review text and the output sequence is the corresponding answer.
- **S2SAE**: An extension of S2SA that incorporates sentiment information during answer generation.
- **PGN**: An abstractive summarization model employing a pointer network for word copying from reviews and an encoder-decoder network for generating new words.
- **PGN-E**: An extension of PGN that incorporates sentiment information during answer generation.
- **BART-base**: A denoising autoencoder for pretraining sequence-to-sequence models, and it is particularly effective when fine-tuned for text generation.(The bart-base model is with 6 layers for encoder and decoder, the number of attention head is 12, and the hidden size is 768.)

The evaluation of the models is conducted using three well-established automatic metrics: BLEU, ROUGE-L, and METEOR.

Table 2. Automatic metrics comparison between baselines.

Method	BLEU	ROUGE-L	METEOR
S2SA	12.30	16.43	0.17
S2SAE	13.26	17.02	0.18
PGN	12.37	17.43	0.17
PGN-E	12.87	18.01	0.19
BART-base	16.07	20.71	0.20
Our Method	**17.33**	**21.51**	**0.22**

As can be seen from Table 2, our method outperforms all the baseline methods in terms of all three metrics: BLEU, ROUGE-L, and METEOR. Specifically, our method achieves a BLEU score of 17.33, a ROUGE-L score of 21.51, and a METEOR score of 0.22, which are the highest among all the methods compared.

Comparing the baseline methods, BART-base shows superior performance with a BLEU score of 16.07, a ROUGE-L score of 20.71, and a METEOR score of 0.20. The other baseline methods, namely S2SA, S2SAE, PGN, and PGN-E, exhibit relatively lower performance, with the least effective method being S2SA, which has a BLEU score of 12.30, a ROUGE-L score of 16.43, and a METEOR score of 0.17 (Table 3).

Table 3. Effective of using CSG module

Method	BLEU	ROUGE-L	METEOR
Our Method/wo CSG	16.58	21.02	0.21
Our Method	**17.33**	**21.51**	**0.22**

Moreover, the ablation study performed with our method, by removing the CSG module (Our Method/wo CSG), reveals that the performance of our method without CSG is slightly lower than with the CSG module. Our Method/wo CSG obtains a BLEU score of 16.58, a ROUGE-L score of 21.02, and a METEOR score of 0.21. This indicates that the CSG module contributes to the overall improvement in the performance of our method.

The human evaluation and case study are given in Appendix. The results demonstrates that our outperformance is significant over the compared baselines. The ablation study further confirms the importance of the CSG module in our method, as its removal leads to a slight degradation in performance. These results substantiate the effectiveness of our proposed method for the given task.

6   Conclusion

In this paper, we proposed a new task of Emotion Implicit Cause Discovery (EICD). It aimed to identify implicit causes that trigger emotions from the reviews. Existing works mostly focused on explicit causes and overlooked the importance of unexpressed causes implied in the text semantics. Also, traditional methods based on extraction lack the flexibility to deal with non-spans cause identification. To address this problem, we developed a novel approach, which can derive implicit causes based on context and human-shared knowledge. Specifically, we retrieved related commonsense contents from large language models and knowledge bases to bridge the knowledge gap between cause events and implied causes. We then constructed reasoning paths from emotions to potential implicit causes. We used graph neural networks to encode the structural context of clauses. To evaluate the effectiveness of our approach, we constructed a large corpus based on Amazon product reviews. We conducted experiments on it to demonstrate the effectiveness of our model. In future work, we will further improve the performance and reliability of our model. Besides, we will extend its application scenarios to enhance its practical usefulness.

Acknowledgements. This work is supported by the Key-Area Research and Development Program of Guangdong Province (2020B0101100001), National Natural Science Foundation of China (62276279), the Tencent WeChat Rhino-Bird Focused Research Program (WXG-FR-2023-06), and Zhuhai Industry-University-Research Cooperation Project(2220004002549).

References

1. Bao, Y., Ma, Q., Wei, L., Zhou, W., Hu, S.: Multi-granularity semantic aware graph model for reducing position bias in emotion cause pair extraction. In: Findings of the Association for Computational Linguistics: ACL 2022, Dublin, pp. 1203–1213 (2022)
2. Bosselut, A., Rashkin, H., Sap, M., Malaviya, C., Celikyilmaz, A., Choi, Y.: COMET: commonsense transformers for automatic knowledge graph construction. In: Proceedings of the 57th Annual Meeting of the Association for Computational Linguistics, Florence, pp. 4762–4779 (2019)
3. Ding, Z., He, H., Zhang, M., Xia, R.: From independent prediction to re-ordered prediction: integrating relative position and global label information to emotion cause identification. In: AAAI Conference on Artificial Intelligence (2019)
4. Ding, Z., Xia, R., Yu, J.: ECPE-2D: emotion-cause pair extraction based on joint two-dimensional representation, interaction and prediction. In: Proceedings of the 58th Annual Meeting of the Association for Computational Linguistics, pp. 3161–3170 (2020)
5. Fan, C., et al.: A knowledge regularized hierarchical approach for emotion cause analysis. In: Proceedings of the 2019 Conference on Empirical Methods in Natural Language Processing and the 9th International Joint Conference on Natural Language Processing (EMNLP-IJCNLP), Hong Kong, pp. 5614–5624 (2019)

6. Fan, C., Yuan, C., Du, J., Gui, L., Yang, M., Xu, R.: Transition-based directed graph construction for emotion-cause pair extraction. In: Proceedings of the 58th Annual Meeting of the Association for Computational Linguistics, pp. 3707–3717 (2020)
7. Gao, K., Xu, H., Wang, J.: A rule-based approach to emotion cause detection for Chinese micro-blogs. Exp. Syst. Appl. **42**(9), 4517–4528 (2015)
8. Ghosal, D., Majumder, N., Gelbukh, A., Mihalcea, R., Poria, S.: COSMIC: COmmonSense knowledge for eMotion identification in conversations. In: Findings of the Association for Computational Linguistics: EMNLP 2020, pp. 2470–2481 (2020)
9. Gui, L., Wu, D., Xu, R., Lu, Q., Zhou, Y.: Event-driven emotion cause extraction with corpus construction. In: Proceedings of the 2016 Conference on Empirical Methods in Natural Language Processing, Austin, Texas, pp. 1639–1649 (2016)
10. Gui, L., Yuan, L., Xu, R., Liu, B., Lu, Q., Zhou, Y.: Emotion cause detection with linguistic construction in chinese weibo text. In: Zong, C., Nie, J.-Y., Zhao, D., Feng, Y. (eds.) NLPCC 2014. CCIS, vol. 496, pp. 457–464. Springer, Heidelberg (2014). https://doi.org/10.1007/978-3-662-45924-9_42
11. Lee, S.Y.M., Chen, Y., Huang, C.R.: A text-driven rule-based system for emotion cause detection. In: Proceedings of the NAACL HLT 2010 Workshop on Computational Approaches to Analysis and Generation of Emotion in Text, Los Angeles, pp. 45–53 (2010)
12. Lee, S.Y.M., Chen, Y., Huang, C.R., Li, S.: Detecting emotion causes with a linguistic rule-based approach 1. Comput. Intell. **29**(3), 390–416 (2013)
13. Lee, S.Y.M., Chen, Y., Li, S., Huang, C.R.: Emotion cause events: corpus construction and analysis. In: Proceedings of the Seventh International Conference on Language Resources and Evaluation (LREC'10). European Language Resources Association (ELRA), Valletta (2010)
14. Li, J., Lin, Z., Fu, P., Wang, W.: Past, present, and future: Conversational emotion recognition through structural modeling of psychological knowledge. In: Findings of the Association for Computational Linguistics: EMNLP 2021, pp. 1204–1214 (2021)
15. Li, W., Li, Y., Pandelea, V., Ge, M., Zhu, L., Cambria, E.: Ecpec: emotion-cause pair extraction in conversations. In: IEEE Transactions on Affective Computing, pp. 1–12 (2022)
16. Radford, A., et al.: Improving language understanding by generative pre-training (2018)
17. Sap, M., et al.: Atomic: an atlas of machine commonsense for if-then reasoning. In: Proceedings of the AAAI Conference on Artificial Intelligence, vol. 33, pp. 3027–3035 (2019)
18. Sun, Y., Zhang, Y., Qi, L., Shi, Q.: TSGP: two-stage generative prompting for unsupervised commonsense question answering. In: Findings of the Association for Computational Linguistics: EMNLP 2022, Abu Dhabi, pp. 968–980 (2022)
19. Turcan, E., Wang, S., Anubhai, R., Bhattacharjee, K., Al-Onaizan, Y., Muresan, S.: Multi-task learning and adapted knowledge models for emotion-cause extraction. In: Findings of the Association for Computational Linguistics: ACL-IJCNLP 2021, pp. 3975–3989 (2021)
20. Xia, R., Ding, Z.: Emotion-cause pair extraction: a new task to emotion analysis in texts. In: Proceedings of the 57th Annual Meeting of the Association for Computational Linguistics, Florence, pp. 1003–1012 (2019)
21. Xia, R., Zhang, M., Ding, Z.: Rthn: a RNN-transformer hierarchical network for emotion cause extraction. In: International Joint Conference on Artificial Intelligence (2019)

22. Yan, H., Gui, L., Pergola, G., He, Y.: Position bias mitigation: a knowledge-aware graph model for emotion cause extraction. arXiv preprint arXiv:2106.03518 (2021)
23. Wu, Z., Dai, X., Xia, R.: Pairwise tagging framework for end-to-end emotion-cause pair extraction. Front. Comput. Sci. **17**(2), 172314 (2023). https://doi.org/10.1007/s11704-022-1409-x
24. Zhong, P., Wang, D., Miao, C.: Knowledge-enriched transformer for emotion detection in textual conversations. In: Proceedings of the 2019 Conference on Empirical Methods in Natural Language Processing and the 9th International Joint Conference on Natural Language Processing (EMNLP-IJCNLP), Hong Kong, pp. 165–176 (2019)

Multi-modal Multi-emotion Emotional Support Conversation

Guangya Liu[1,2], Xiao Dong[1,2], Meng-xiang Wang[5], Jianxing Yu[1,2,3,4(✉)], Mengjiao Gan[1,2], Wei Liu[1,2], and Jian Yin[1,2]

[1] School of Artificial Intelligence, Sun Yat-sen University, Zhuhai, China
{liugy28,dongx55,ganmj}@mail2.sysu.edu.cn,
{yujx26,liuw259,issjyin}@mail.sysu.edu.cn
[2] Guangdong Key Laboratory of Big Data Analysis and Processing, Guangzhou, China
[3] Pazhou Lab, Guangzhou 510330, China
[4] Key Laboratory of Sustainable Tourism Smart Assessment Technology, Ministry of Culture and Tourism, Zhuhai, China
[5] China National Institute of Standardization, 100088 Beijing, China
wangmx@cnis.ac.cn

Abstract. This paper proposes a new task of *Multi-modal Multi-emotion Emotional Support Conversation* (**MMESC**), which has great value in various applications, such as counseling, daily chatting, and elderly company. This task aims to fully perceive the users' emotional states from multiple modalities and generate appropriate responses to provide comfort for improving their feelings. Traditional works mainly focus on textual conversation, while a single-modal cannot accurately reflect the users' emotions, such as saying fine with an inconsistent disgusting feeling. To address this problem, we propose a new task on multi-modalities and exploit a new method called *FEAT* for this new task. *FEAT* can integrate fine-grained emotional knowledge from multiple modalities. It first recognizes the users' mental states based on an emotion-aware transformer. It then generates supportive responses using a hybrid method with multiple comfort strategies. To evaluate our method, we construct a large-scale dataset named *MMESConv*. It is almost two times larger than existing single-modal datasets. There are three modalities in this dataset (text, audio, and video) with fine-grained emotion annotations and strategy labels. Extensive experiments on this dataset demonstrate the advantages of our proposed framework.

Keywords: Multi-modal · Multi-emotion · Emotional Support Conversation

1 Introduction

Emotional Support Conversation (*ESC*) is a hot research topic that aims to reduce the users' emotional distress by gradually chatting with psychological

X. Yang et al. (Eds.): ADMA 2023, LNAI 14176, pp. 293–308, 2023.
https://doi.org/10.1007/978-3-031-46661-8_20

care skills [9]. This task has great value in various applications. For example, in the field of social interactions, it can encourage the upset person to stay positive; for school counseling, it can aid an anxious student to cope with academic pressure, and in customer service, it can calm down an irate customer and provide problem-shooting suggestions.

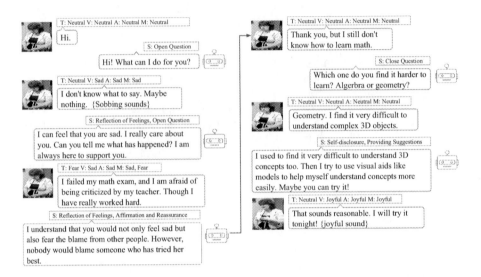

Fig. 1. An example of an effective Multi-modal Multi-emotion Emotional Support Conversation. T: Textual Emotion, V: Visual Emotion, A: Acoustic Emotion, M: Multimodal Emotion, S: Support Strategy.

Previous studies [3,19] have shown that using multiple modalities can provide complementary information. That can enhance the representational capability, leading to improved performance in downstream tasks, such as classification [3], retrieval [19], and text generation [10]. Existing researches [12,16,17] on *ESC* mainly focus on textual conversations. They often ignore the potential value of additional complementary contexts from other modalities, such as video and audio. In real-world conversations, individuals may intentionally or unintentionally express their genuine emotions through various modalities, including tone of voice, facial expressions, body gestures, and more [18]. As illustrated in the red content of Fig. 1, while the text of the help-seeker may indicate a neutral emotional state, the audio and video modalities reveal that she is actually sad due to her sobbing sounds and morose body gesture. If the model only considers the textual information, it may respond with an indifferent statement, such as 'OK, goodbye!' instead of expressing empathy and comforting the sad individual, as shown in Fig. 1. Moreover, as illustrated in the blue content of Fig. 1, individuals commonly express multiple emotions within a single utterance [21]. However, if the model only considers the textual modality, it may only recognize one emotion, such as 'fear', while ignoring other emotions expressed through audio and video modalities.

In this paper, we propose and investigate the *Multi-modal Multi-emotion Emotional Support Conversation* (**MMESC**) task, which involves recognizing the potentially blended emotions of help-seekers, selecting suitable strategies, and generating supportive responses using multi-modal data.

We propose a novel model called *FEAT*, a *Fine-grained Multi-emotion **A**ware **T**ransformer* framework, to fully utilize the multi-modality data and enhance the effectiveness of generated supportive responses as depicted in Fig. 3. Previous research [18] has suggested that there will be more complementary information for inter-modality fusion when there is a larger divergence between distinct modalities. Nonetheless, it can be exceedingly challenging for models to distinguish between nuanced and intricate emotions expressed in different modalities. In *FEAT*, we integrate fine-grained emotional knowledge of both uni-modality and multi-modality to help the model to learn complementary information from different modalities.

To evaluate our method, we collect a large-scale *Multi-modal Multi-emotion Emotional Support Conversation* (*MMESConv*) dataset from YouTube[1] videos. The *MMESConv* dataset is almost twice the size of previous *Emotional Support Conversation* datasets, including 1,599 dialogues with annotations for three modalities: text, audio, and video. We have provided fine-grained emotion annotations for each modality and multi-modality at the utterance level. We find that 24.96% of the utterances had multiple emotions. In addition to the emotion annotations, we have also annotated 9 emotional support response strategies.

Extensive experiments on our *MMESConv* dataset demonstrate the superiority of our proposed framework over existing methods in the **MMESC** task. In summary, our contributions are as follows:

- We introduce a new task called *Multi-modal Multi-emotion Emotional Support Conversation* (**MMESC**), which aims to model real-world *ESC* more accurately. The system needs to recognize the emotion of the help-seeker at the utterance level, predict appropriate support strategies, and generate supportive responses.
- Our proposed method, *FEAT*, integrates emotional knowledge from different modalities in an explicit way, enhancing the model's ability to recognize the help-seeker's emotion and select appropriate strategies. Based on the blended emotion and strategy knowledge, the model is able to produce more supportive responses.
- To evaluate our method, we provide the largest dataset *MMESConv* with fine-grained emotion and strategy annotations at the utterance level. We compare *FEAT* to a comprehensive set of existing methods and demonstrate its superior performance on the *MMESConv* dataset.

We hope that *MMESConv*, *FEAT*, and our solid baselines will advance future research on real-world emotional support dialogue systems.

---

[1] https://www.youtube.com/.

2   Related Work

2.1   Emotion Support Conversations

In an effective *Emotional Support Conversation* (*ESC*), supporters should use diverse support strategies to improve the mental state of help-seekers. Liu et al. [9] recently released a benchmark dataset of uni-modal (text) *ESC*, which highlights the significance of utilizing appropriate strategies to alleviate emotional distress in help-seekers. Several researchers have attempted to develop robust models based on the mentioned benchmark dataset. For instance, Tu et al. [16] focused on the nuanced emotions of the help-seeker and attempted to use a combination of strategies in a single utterance based on the predicted emotional label. Notwithstanding earlier research that emphasized the importance of choosing effective support strategies based on the help-seeker's emotions, the dataset being limited to a uni-modal text presents difficulties in capturing the precise emotion, which hinders the progress of current models. To better understand their complex and changing emotions as the conversation progresses, additional multimodal data such as images and audio are also required. In response, we created a more extensive multi-modal (text, audio, and videos) dataset than the previous one with utterance-level emotion annotations.

2.2   Emotion Recognition in Conversation

The goal of the *Emotion Recognition in Conversation* (*ERC*) task is to automatically identify the emotion of each speaker in a conversation, which plays a vital role in creating a more engaging dialogue system. The *Interactive Emotional Dyadic Motion Capture Database* (*IEMOCAP*) [2] is one of the earliest datasets for this task, including uni-modal textual conversations across ten hours of various dialogue scenarios. Subsequently, researchers expanded this task to a multimodal setting by introducing the *Multimodal Emotion Lines Dataset* [13]. To model various interactions such as global, local, intra-speaker, and inter-speaker interactions, Zhao et al. [21] suggested a dialog-aware interaction module. While identifying the emotion of the speakers is a crucial aspect of our work, we aim to go beyond that by not only recognizing emotions but also selecting appropriate emotional support strategies and generating effective support responses.

2.3   Multi-modal Dialogue Systems

Multi-modal dialogue systems aim to comprehend the context of multiple modalities and provide responses to users. These systems come in various forms based on task definitions [8]. One branch of multi-modal dialogue systems is focused on generating accurate answers to specific questions. *Visual Question Answering* (*VQA*) [1] was the earliest example of this kind, where an agent provides an accurate answer to a question about the content of an image. Task-oriented multi-modal dialogue systems assist humans in completing a particular task, such as shopping and ticket-booking. *Multimodal Dialogs Dataset* (*MMD*) [15] is a benchmark dataset for this type of system, focusing on dialogues related to the fashion e-commerce domain. Despite the popularity of multi-modal dialogue systems, few studies have investigated multi-modal emotional dialogue systems due to the absence of benchmark datasets. Specifically, there has been no previous research on multi-modal emotional support dialogue systems. Therefore, we created a high-quality, large-scale multi-modal emotional support conversation dataset, including three modalities to support further research in this area. Our study is the first to explore multi-modal *ESC* and construct a high-quality multi-modal benchmark.

3 Task Definition

When a user experiences negative emotions, they may seek help to improve their emotional state, often due to a specific problem. The supporter, whether human or system, must possess the necessary skills to comfort the user through a conversation and reduce their emotional distress. During this process, the system collects information from multiple modalities (text, audio, and video) to generate supportive responses. To address the challenges of developing a multi-modal multi-emotion emotional support dialogue system, we propose three benchmark tasks, each with corresponding evaluation metrics. These tasks capture various obstacles that must be overcome to create an effective system as elaborated in the following sections.

In the task of **MMESC**, a dialogue U can be defined as a sequence of utterances exchanged between a help-seeker and a supporter $U = \{u_1, u_1, \cdots, u_{|U|}\}$. The help-seeker's utterances contain three modalities: textual(t), acoustic(a), and visual(v). The **MMESC** task consists of three sub-tasks. First, given a context $C = \{u_1, u_2, \cdots, u_p\}$, the model identifies the emotion expressed in the current utterance of the help-seeker, u_p. Second, the model predicts appropriate support strategies for the response. Finally, the model generates the most likely sequence of textual response $Y = \{y_1, y_2, \cdots, y_n\}$, where n is the response length.

4 MMESConv Dataset

We will detail our efforts to create a high-quality *Multi-modal Multi-emotion Emotional Support Conversation* (*MMESConv*) dataset in the following sections.

4.1 Dialogue Selection

In mental counseling, psychologists often utilize emotional support techniques to comfort those seeking help. [20] Therefore, we collected counseling-related videos from the YouTube website to ensure the quality of our dataset. We instructed our human workers to follow strict rules when selecting videos. We present those rules in appendix A.1 in detail. Our adherence to these rules ensured the creation of a high-quality *MMESConv* dataset.

4.2   Annotation

We introduce the annotation details of *MMESConv* in the following content.

Text and Speaker Annotation. We utilize a professional transcription[2] service to transcribe the audio into text and create timestamps. This process allows us to acquire the utterances and label the speakers as either speaker 1 or speaker 2. Workers are then asked to label the help-seeker and the supporter.

Video Annotation. Instead of relying on face detection, we manually use the professional video annotation tool $CVAT$[3] to label the gestures and faces of the speakers, resulting in a more accurate extraction of video features.

Emotion Annotation. Following previous researches [21], we annotate seven emotions for the help-seeker utterances, namely disgust, happiness, surprise, anger, sadness, fear, and neutrality. Unlike previous studies in the field of *ERC* that only labeled one multi-modal emotion per utterance, we annotate three uni-modal emotions (text, audio, and video) for each utterance. Workers are allowed to select multiple emotion labels for utterances with multiple emotions. We assign three workers to annotate the emotion and strategy for each dialogue and apply the same majority voting strategy as in the previous work [21]. To prevent the mutual inference of different modalities, workers could only view the current modality while labeling emotions. We instruct workers to label the text, audio, video, and multi-modal emotions in that order. We remove the audio and text content from the video clips when labeling the video modality.

Strategy Annotation. We select nine support strategies following previous studies [9], except that we divide the *Question* strategy into *Open Question* and *Close Question* strategies. The human Workers will label multiple strategies for utterances with blended strategies, such as *Question* and *Self-Disclosure*.

4.3   Dataset Statistics

The statistics presented in Table 1 provide a summary of our dataset, which contains 1,599 dialogues and 57,457 utterances. Compared to the previous dataset, this dataset represents a nearly twofold increase in size. It is also noteworthy that 24.96% and 36.42% of the utterances contain multiple emotions and multiple strategies, respectively, indicating that it is common for individuals to express blended emotions and use multiple support strategies in a single utterance. As for the distribution of strategies in Table 2, the most commonly employed support strategies by supporters are *Open Question* and *Restate or Paraphrase*.

The statistics for the distribution of emotions are presented on the left side of Fig. 2, where one multi-modal annotation and three uni-modal annotations (Text, Audio, and Video) are compared. Our dataset shows that fear and sadness are the most commonly expressed emotions in the conversations between help-seekers and supporters, which is expected since most help-seekers tend to

---

[2] We use paid audio transcription service from Otter.ai (https://otter.ai/) .

[3] https://www.cvat.ai/.

Table 1. Statistics of MMESConv.

Category	Num
#dialogues	1,599
#Utterances	57,457
Avg.length of dialogues	35.9
Avg.length of utterances	36.9
#help-seeker's utterances	28261
#supporter's utterances	29196
#multiple emotion utterances	7054
#multiple strategy utterances	10634

Table 2. Statistics of Support Strategy annotations

Support Strategy	Num	Proportion
Open Question	12,150	30.50%
Close Question	6,501	16.32%
Restate or Paraphrase	7,472	18.76%
Reflection of Feelings	2,816	7.07%
Self Disclosure	204	0.51%
Affirmation and Reassurance	1,901	4.78%
Providing Suggestions	2,748	6.90%
Information	3,195	8.02%
Other	2,843	7.14%
Overall	39,830	100%

have negative emotional states. Surprise is the least frequently expressed emotion since supporters usually comfort the help-seeker and help reduce their emotional distress. The right side of Fig. 2 displays the confusion matrix [18] of the annotation differences among various modalities. Notably, the difference between M and A is more significant than that between V and T, which can be explained by the fact that audio contains text information that is closer to the multi-modal, while the connection between video and text is sparse. Moreover, we can observe that acoustic annotation is the most similar to multi-modal annotations, while textual annotation is the least similar. These observations demonstrate that non-verbal cues can play an essential role in better recognizing the emotions of help-seekers.

4.4   Quality Control

Testing Programs for Annotators. We instruct the workers on the definitions of different emotions or strategies before they could annotate them. We then test

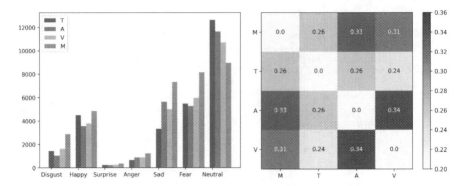

Fig. 2. Left: the distribution of emotion in one Multi-modal annotation and three uni-modal (Text, Audio, and Video) annotations. Right: the confusion matrix shows the emotion annotation difference between different modalities, where a larger number indicates a greater difference.

the workers using 100 utterances of various emotions or strategies to ensure their ability to classify the utterances based on their training.

Validation Testing. We analyze the inter-annotator agreement through validation testing. If there is a poor agreement, all the annotators should review the annotation and make corrections. Our final Fleiss's Kappa score [4] is 0.61, which is higher than that of previous datasets.

5   Methodology

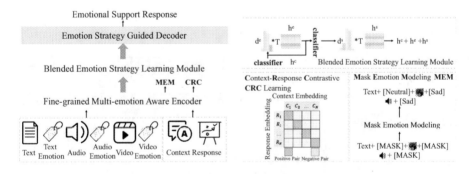

Fig. 3. Left: the overall architecture of our *FEAT* framework. Right: the illustration of the *MEM*, *CRC*, and *Blended Emotion Strategy Learning Module*

Our framework *FEAT* comprises three main components, namely the *Fine-grained Multi-emotion Aware Encoder*, the *Blended Emotion Strategy Learning Module*, and the *Emotion Strategy Guided Decoder*, as shown in Fig. 3. In the lower part of the framework, we concatenate the extracted textual, emotional, acoustic, and visual features and feed them into the encoder of *FEAT*. To enhance the model's ability to model the complex emotions of the help-seeker and choose appropriate support strategies, we utilize the *Masked Emotion Modeling (MEM)* pretext task and the *Context-Response Contrastive (CRC) Learning* task in the *Fine-grained Multi-emotion Aware Encoder*. In the *Blended Emotion Strategy Learning Module*, we use the multi-modal context representation learned from the *Fine-grained Multi-emotion Aware Encoder* to obtain the blended emotion distribution, which is then combined with the multi-modal context representation to obtain the blended strategy distribution. Finally, in the *Emotion Strategy Guided Decoder*, we generate supportive responses based on the multi-modal context representation, the blended emotion distribution, and the blended strategy distribution. We will provide further details on our model in the subsequent sections.

5.1 Fine-Grained Emotion-Aware Encoder

The preceding research [18] suggests that if there is a significant distinction between various modalities, there could be more complementary information for inter-modality fusion. However, since the emotion conveyed through different modalities can often be nuanced and intricate, it can be challenging to directly evaluate the semantic differences between modalities. To address this issue, we present a *Fine-grained Multi-emotion Aware Encoder* that enables the model to explicitly use emotion knowledge from different uni-modalities to recognize comprehensively the help-seeker's multi-modal emotion in the current utterance of a given multi-modal context $C = \{u_1, u_2, \cdots, u_p\}$. Specifically, our *Fine-grained Multi-emotion Aware Encoder* takes different modality features and their corresponding emotional tags as inputs and extracts fused feature embeddings to generate reasonable emotional responses. Before entering the encoder, the different modalities are pre-processed as follows.

The textual feature of the current utterance u_p, denoted as h_p^t, is extracted using the BlenderBot encoder [14]. We use the same BlenderBot encoder to extract the features of uni-modal emotion labels e_t, e_a, and e_v, denoted as $h_p^{e_t}$, $h_p^{e_a}$, and $h_p^{e_v}$, respectively, as emotional features.

For the audio modality, we extract MFCC features from the audio, denoted as h_p^a, while for the video modality, we feed annotated ordinal frames sampled from the video into the transformer encoder, denoted as h_p^v. After being processed via different feature extractors, the four types of features are concatenated and fed into a transformer encoder to capture the token relationships between different modalities as shown in Eq. 1:

$$h_p^c = f\left(cls_p + h_p^t + h_p^{e_t} + h_p^a + h_p^{e_a} + h_p^v + h_p^{e_v}\right), \tag{1}$$

where $f(\cdot)$ denotes the Transformer encoder and h_p^c is the representation of the context. To integrate the dialog history of previous rounds, we initialize the current cls_p by using the representation of the previous round cls_{p-1}.

Masked Emotion Modeling. We then adopt *Masked Emotion Modeling* pretext ask (*MEM*) to facilitate fine-grained multi-emotion feature learning. Specifically, unlike common *Maksed Language Modeling* pretext task where 15% of the random individual words are masked, the uni-modal emotion labels are entirely masked out for the *MEM* task. Next, the remaining inputs are used to reconstruct the masked information. The loss function of the ith modality emotion is defined as Eq. 2:

$$\mathcal{L}_{M_i}(\theta) = -M_{t_{msk} \sim \mathbf{t}} \log P_\theta \left(t_{msk} \mid t_{\neg msk}, \mathbf{M}_{\neg i} \right), \tag{2}$$

where $t_{\neg msk}$ denotes the unmasked tokens surrounding the masked token t_{msk}, θ represents the network parameters. M_i and $M_{\neg i}$ are the ith modality and the remaining modalities, respectively.

Previous studies have suggested that there is a discernible pattern in the context and support strategy of a corresponding response [9]. For example, in the initial stages of an *ESC*, the help-seeker often expresses their negative emotions, while the supporter is likely to use the *Question* strategy to understand the root of the help-seeker's feelings. To enhance the model's ability to select appropriate support strategies, we have developed the Context-Response Contrastive (*CRC*) Learning module, which pulls the corresponding context and strategy-guided responses closer in the representation space.

During the training phase, we input the response representation $h_p^{r_t}$ for the given context $C = \{u_1, u_1, \cdots, u_p\}$ and the corresponding strategy representation $h_p^{r_s}$ (both obtained from the *BlenderBot* encoder) into the encoder, as shown in Eq. 3

$$h_{p+1}^r = f \left(cls_p^r + h_p^{r_t} + h_p^s \right). \tag{3}$$

We then construct triples $x_p = \left\{ c_p, r_p^+, r_p^- \right\}$ where c_p is the multi-modal representation of the context h_p^c, r_p^+ is the corresponding strategy-guided response representation $h_p^{r_t}$, r_p^- is the representations of the other responses in the same batch. For each triple X_p, the training objective is to minimize cross-entropy loss with in-batch negatives [7] as Eq. 4:

$$\mathcal{L}_c = -\log \frac{e^{\langle c_p, r_p^+ \rangle}}{\sum_{j=1}^K e^{\langle c_p, r_j^+ \rangle} + e^{\langle c_p, r_p^- \rangle}}, \tag{4}$$

where K is the batch size.

5.2   Blended Emotion Strategy Learning Module

In this module, our aim is to endow the model with the capability to recognize blended multi-modal emotions and to predict appropriate strategies accordingly. Firstly, using the context representation h_p^c, we employ a fully-connected layer as a classifier to obtain the emotion distribution d_e. We define the multi-emotion classification loss, denoted as $\mathcal{L}_e$, as shown in Eq. 5. Inspired by Tu et al. [16], we construct an emotion codebook and weight it with d_e to obtain a blended emotion representation h^e. Next, we concatenate h_p^c and h_p^e and use the resulting vector to predict the strategy distribution d_s. We define the multi-strategy classification loss, denoted as $\mathcal{L}_s$, as shown in Eq. 6. Similar to the construction of the emotion codebook, we construct a strategy codebook to obtain a blended strategy representation h_p^s. The equations for $\mathcal{L}_e$ and $\mathcal{L}_s$ are presented below:

$$\mathcal{L}_e = -\frac{1}{m}\sum_{i=1}^{m}\sum_{j=1}^{7}e_j^{(i)}\log(\hat{e}_j^{(i)} \mid C, e_t, e_a, e_v) \tag{5}$$

$$\mathcal{L}_s = -\frac{1}{m}\sum_{i=1}^{m}\sum_{j=1}^{9}s_j^{(i)}\log(\hat{s}_j^{(i)} \mid C, e_t, e_a, e_v), \tag{6}$$

where C is the context, e_t, e_a, and e_v are the textual emotion, acoustic emotion, and visual emotion respectively, m is the number of instances, $e_j^{(i)}$ is the true emotion label for jth class of the ith instance, $s_j^{(i)}$ is the true strategy label for jth class of the ith instance.

5.3 Emotion Strategy Guided Decoder

After concatenating h_p^c, h_p^e, and h_p^s, we utilize the *BlenderBot* decoder to produce the emotional support response. We define the response generation loss $\mathcal{L}_r$ as Eq. 7

$$\mathcal{L}_r = -\sum_{t=1}^{n}\log\left(p\left(y_t \mid y_{j<t}, C, e_t, e_a, e_v\right)\right), \tag{7}$$

where n is the length of the response, C is the context, and e_t, e_a, e_v are the textual emotion, acoustic emotion, and visual emotion respectively.

Overall, the total loss $\mathcal{L}_{total}$ of the **MMESC** task is the sum of $\mathcal{L}_s$, $\mathcal{L}_e$, $\mathcal{L}_s$, and $\mathcal{L}_r$ as Eq. 8

$$\mathcal{L}_{total} = \mathcal{L}_e + \mathcal{L}_s + \mathcal{L}_r. \tag{8}$$

6 Experiments

We conduct comprehensive experiments and comparison studies on our proposed *MMESConv* dataset. Specifically, we evaluate the model's performance on three tasks: recognizing the utterance-level emotion of the help-seeker (*Emotion* task), predicting support strategies (*Strategy* task), and generating supportive responses (*Response* task). We provide a detailed description of evaluation metrics in Appendix A.2. To train and evaluate our model, we split our *MMESConv* dataset into training, validation, and testing sets in an 8:1:1 ratio. During training, we use two-turn utterances prior to the target responses as context due to the relatively long utterances in our dataset. We use the PyTorch framework for all experiments and tuned the model's parameters on the validation set before reporting its performance on the testing set. We train our *FEAT* using the Adam optimizer with a warm-up learning rate of e^{-4} for 5 epochs and set the batch size to 4 during training. During inference, we use a batch size of 1 and decode the response using Top-k and Top-p sampling with $p = 0.9$, $k = 30$, and temperature $\tau = 0.7$.

6.1   Experimental Results

Table 3 presents the experimental results for the *Strategy* task and the *response* task. In the *strategy* task, our model delivers state-of-the-art performance when compared to existing works. For the *response* task, our model yields promising results for most of the evaluation metrics. Specifically, our model achieves the lowest score on the *PPL* metric, indicating the generation of responses with better overall quality. Conversely, we obtain higher scores on *B-2*, *B-4*, and *R-L* metrics, indicating that our model generates responses with more accurate semantics and grammar. Our model exhibits comparable performance to existing works in terms of the *D-1* and *D-2* scores, which indicates a fair generation diversity.

Table 3. Comparison results on Strategy Task and Response Task

Methods		Strategy	Response					
		Accuracy	PPL ↓	D-1 ↑	D-2 ↑	B-2 ↑	B-4 ↑	R-L ↑
Previous Methods	**Transformer**	–	84.39	1.13	7.01	6.88	1.24	14.53
	BlenderBot [9]	29.38	18.14	2.70	17.33	5.29	1.34	15.34
	MISC [16]	31.25	16.74	4.22	19.26	7.40	2.70	17.91
	PoKE [17]	32.76	15.15	3.77	**22.99**	6.27	1.24	15.93
	FADO [12]	32.29	15.75	3.35	21.96	8.11	2.53	18.99
Ours	**FEAT**	**33.80**	**14.80**	**4.89**	22.86	**8.46**	**3.05**	**19.59**

   In Table 4, we present our model's performance on the emotion task in comparison to works within the *ERC* domain. Our model surpasses all prior models that incorporate input from three modalities, emphasizing the importance of incorporating fine-grained multiple emotion annotations. We also conclude that the absence of a corresponding dataset has impeded the performance improvements of existing models in the *ESC* tasks towards the task of emotion recognition. Therefore, our proposed **MMESC** can offer more options for future research on real-world multi-modal emotional support dialogue systems.

Table 4. Comparison results on Emotion Task

Methods		Emotion			
		L	L,A	L,V	L,A,V
Previous Methods	**DialogueGCN** [5]	46.53	49.04	49.26	49.93
	MMGCN [6]	46.17	49.33	50.42	51.18
	DialogueRNN [11]	48.80	51.87	52.28	51.66
	MDI [21]	49.42	50.24	52.07	50.99
Ours	**FEAT**	**49.97**	**52.21**	**52.43**	**53.44**

6.2 Ablation Study

To analyze the impact of various modalities and components in our model, we present the ablation study results in Tables 5 and 6, respectively. The results in Table 5 clearly show that our model's performance in all three sub-tasks improves with the incorporation of more modality information into the network. Specifically, the emotion recognition task benefits significantly from the addition of complementary multi-modal information into the network. This enables our model to select more appropriate strategies based on emotional information, resulting in the generation of more suitable responses.

Table 5. Model performance under the uni-modal and multi-modal conditions

Modality Combinations	Emotion	Strategy	Response					
	WF1	Accuracy	PPL ↓	D-1 ↑	D-2 ↑	B-2 ↑	B-4 ↑	R-L↑
L	49.97	29.48	16.15	2.71	19.38	5.52	1.29	15.51
L,A	52.21	31.51	15.42	4.73	21.51	7.15	2.48	17.72
L,V	50.43	29.93	16.02	3.10	19.59	7.73	2.65	16.90
L,A,V	**53.44**	**33.80**	**14.80**	**4.89**	**22.86**	**8.46**	**3.05**	**19.59**

As shown in Table 6, we design several variants of *FEAT* by removing some specific parts:1) **w/o** e_t, e_a, e_v, uni-modal emotion input e_t, e_a, and e_v are removed from the input. 2) **w/o MEM**, the *MEM* pretask is removed while keeping the uni-modal emotion input e_t, e_a, and e_v. 3) **w/o CRC**, the *Context-Response Contrastive Learning* task is removed, which means that we only encode the context of a dialog during the training stage. 4) **w/o** $\mathcal{L}_e$, the loss of emotion recognition task is removed from the total loss.

We can observe that removing each component will result in a performance degradation. Specifically, **w/o** e_t,e_a,e_v cause the emotion recognition *WF1* to drop significantly, which proves the fine-grained emotional knowledge's huge contribution to the model's ability to comprehend complex human emotions. Similarly, **w/o MEM** demonstrates the importance of *MEM* pretask for enhancing the model's ability to integrate uni-modal emotion information. **w/o CRC** causes a great accuracy drop for the strategy selection task, which suggests that CRC can improve the model's ability to select proper strategies for a given context. Moreover, **w/o** $\mathcal{L}_e$ causes the strategy prediction accuracy to drop 2.46 % and the PPL to drop 1.22 %. It shows that recognizing the emotion of the help-seeker at the utterance-level is an essential sub-task for producing more supportive responses according to proper strategies.

Table 6. Ablation study

Model	WF1	Accurary	PPL	D-1	B-2	R-L
w/o e_t,e_a,e_v	48.94	32.15	15.93	3.34	6.16	18.09
w/o MEM	49.35	32.82	16.89	3.94	6.93	18.56
w/o CRC	53.21	30.18	16.76	3.75	7.89	18.04
w/o $\mathcal{L}_e$	–	31.34	16.03	2.98	5.89	16.25
Ours	**53.44**	**33.80**	**14.80**	**4.89**	**8.46**	**19.59**

7   Conclusions

In this study, we introduce a novel task, *Multi-modal Multi-emotion Emotional Support Conversation* (**MMESC**), which holds great potential for applications such as counseling, daily chatting, and providing companionship to the elderly. In this task, a help seeker's mixed and fluctuating emotions must be recognized by a supporter, who then selects appropriate support strategies to generate effective support responses and improve the help seeker's mental state. To tackle this task, we propose the *Fine-grained Multi-emotion Aware Transformer* framework (*FEAT*), which uses fine-grained emotional knowledge from various modalities to precisely recognize the help seeker's mixed emotions and generate supportive responses based on selected strategies. To evaluate our proposed approach and promote research in the **MMESC** field, we create the *MMESConv* dataset on a large scale, which is almost twice the size of previous ones and includes fine-grained, utterance-level emotion annotations and strategy labels. Our extensive experiments demonstrate the superior performance of *FEAT* compared to existing models. We believe that our work will inspire the development of more engaging and effective emotional support dialogue systems based on multi-modal input.

Acknowledgements. This work is supported by the Key-Area Research and Development Program of Guangdong Province (2020B0101100001), National Natural Science Foundation of China (62276279), the Tencent WeChat Rhino-Bird Focused Research Program (WXG-FR-2023-06), and Zhuhai Industry-University-Research Cooperation Project (2220004002549). Jianxing Yu is the corresponding author.

A   Appendix

A.1   Dialogue Selection Guidelines

1) Each video must feature only two people, a help-seeker experiencing negative emotions, and a supporter utilizing support strategies to comfort them. 2) The audio must be clear and understandable, and the video quality must be high enough to distinguish the speaker's gestures and expressions. 3) Each conversation should consist of at least five rounds of interaction. 4) Help-seekers with serious mental illnesses should be excluded. 5) Workers must select videos from a variety of topics and annotate the seeker's problem for each dialogue.

A.2   Evaluation Metrics

We utilize a series of automatic evaluation metrics to assess the performance of our models. In the emotion task, we employ the *weighted F-score* metric. (*WF1*) as it can better evaluate the model's performance on unbalanced emotion distribution. A higher *WF1* score indicates better recognition of the help-seeker's emotions at the utterance-level. For the support task, we use the prediction accuracy metric as it evaluates the model's ability to select the best strategies for generating supportive responses. A higher accuracy score indicates better performance in strategy prediction. In the response task, we employ *PPL* (*perplexity*), *B-2* (*BLEU-2*), *B-4* (*BLEU-4*), and *R-L* (*ROUGE-L*) metrics to assess the generation quality in terms of semantics and grammar. Furthermore, we also use *D-1* (*Distinct-1*) and *D-2* (*Distinct-2*) metrics to evaluate the generation diversity.

References

1. Antol, S., et al.: VQA: visual question answering. In: ICCV, pp. 2425–2433 (2015)
2. Busso, C., et al.: Iemocap: interactive emotional dyadic motion capture database. In: Language Resources and Evaluation, pp. 335–359 (2008)
3. Dong, X., et al.: M5product: self-harmonized contrastive learning for e-commercial multi-modal pretraining. In: CVPR, pp. 21252–21262 (2022)
4. Fleiss, J.L., Levin, B., Paik, M.C.: Statistical Methods for Rates and Proportions. John Wiley & Sons (2013)
5. Ghosal, D., Majumder, N., Poria, S., Chhaya, N., Gelbukh, A.: Dialoguegcn: a graph convolutional neural network for emotion recognition in conversation. In: EMNLP-IJCNLP, pp. 154–164 (2019)
6. Hu, J., Liu, Y., Zhao, J., Jin, Q.: MMGCN: multimodal fusion via deep graph convolution network for emotion recognition in conversation. In: ACL, pp. 5666–5675 (2021)
7. Li, J., Zhong, J.: Building an efficient retrieval-based dialogue system with contrastive learning. In: IJCNN, pp. 1–8. IEEE (2022)
8. Liu, G., Wang, S., Yu, J., Yin, J.: A survey on multimodal dialogue systems: recent advances and new frontiers. In: AEMCSE, pp. 845–853. IEEE (2022)

9. Liu, S., et al.: Towards emotional support dialog systems. In: ACL, pp. 3469–3483 (2021)
10. Ma, Z., Li, J., Li, G., Cheng, Y.: Unitranser: a unified transformer semantic representation framework for multimodal task-oriented dialog system. In: ACL, pp. 103–114 (2022)
11. Majumder, N., Poria, S., Hazarika, D., Mihalcea, R., Gelbukh, A., Cambria, E.: Dialoguernn: an attentive RNN for emotion detection in conversations. In: AAAI, vol. 33, pp. 6818–6825 (2019)
12. Peng, W., Qin, Z., Hu, Y., Xie, Y., Li, Y.: FADO: feedback-aware double controlling network for emotional support conversation. Knowl. Based Syst. **264**, 110340 (2023)
13. Poria, S., Hazarika, D., Majumder, N., Naik, G., Cambria, E., Mihalcea, R.: Meld: a multimodal multi-party dataset for emotion recognition in conversations. In: ACL, pp. 527–536 (2019)
14. Roller, S., et al.: Recipes for building an open-domain chatbot. In: EACL: Main Volume, pp. 300–325 (2021)
15. Saha, A., Khapra, M., Sankaranarayanan, K.: Towards building large scale multimodal domain-aware conversation systems. In: AAAI, vol. 32 (2018)
16. Tu, Q., Li, Y., Cui, J., Wang, B., Wen, J.R., Yan, R.: Misc: a mixed strategy-aware model integrating comet for emotional support conversation. In: ACL, pp. 308–319 (2022)
17. Xu, X., Meng, X., Wang, Y.: Poke: prior knowledge enhanced emotional support conversation with latent variable. arXiv preprint arXiv:2210.12640 (2022)
18. Yu, W., et al.: Ch-sims: a Chinese multimodal sentiment analysis dataset with fine-grained annotation of modality. In: ACL, pp. 3718–3727 (2020)
19. Zhan, X., et al.: Product1m: towards weakly supervised instance-level product retrieval via cross-modal pretraining. In: CVPR, pp. 11782–11791 (2021)
20. Zhang, J., Danescu-Niculescu-Mizil, C.: Balancing objectives in counseling conversations: advancing forwards or looking backwards. In: ACL, pp. 5276–5289 (2020)
21. Zhao, J., et al.: M3ed: multi-modal multi-scene multi-label emotional dialogue database. In: ACL, pp. 5699–5710 (2022)

Exploiting Pseudo Future Contexts
for Emotion Recognition in Conversations

Yinyi Wei[1], Shuaipeng Liu[2(✉)], Hailei Yan[2], Wei Ye[3], Tong Mo[1],
and Guanglu Wan[2]

[1] Peking University, Beijing, China
wyyy@pku.edu.cn
[2] Meituan Group, Beijing, China
liushuaipeng@meituan.com
[3] National Engineering Research Center for Software Engineering, Peking University,
Beijing, China

Abstract. With the extensive accumulation of conversational data on
the Internet, emotion recognition in conversations (ERC) has received
increasing attention. Previous efforts of this task mainly focus on leveraging contextual and speaker-specific features, or integrating heterogeneous external commonsense knowledge. Among them, some heavily rely on future contexts, which, however, are not always available in
real-life scenarios. This fact inspires us to generate pseudo future contexts to improve ERC. Specifically, for an utterance, we generate its
future context with pre-trained language models, potentially containing
extra beneficial knowledge in a conversational form homogeneous with
the historical ones. These characteristics make pseudo future contexts
easily fused with historical contexts and historical speaker-specific contexts, yielding a conceptually simple framework systematically integrating multi-contexts. Experimental results on four ERC datasets demonstrate our method's superiority. Further in-depth analyses reveal that
pseudo future contexts can rival real ones to some extent, especially in
relatively context-independent conversations.

Keywords: Emotion Recognition in Conversations · Conversation
Understanding · Pseudo Future Contexts

1 Introduction

Emotion recognition in conversations (ERC), aiming to identify the emotion of
each utterance in a conversation, is an essential part of conversation understanding. It plays a significant role for various downstream tasks, e.g., opinion mining
in social media, building intelligent assistant through empathetic machine, health
care, and detecting fake news in social media [13, 20].

A common solution for ERC is to jointly leverage the contextual and speaker
information of an utterance, which can be roughly divided into sequence-based

© The Author(s), under exclusive license to Springer Nature Switzerland AG 2023
X. Yang et al. (Eds.): ADMA 2023, LNAI 14176, pp. 309–323, 2023.
https://doi.org/10.1007/978-3-031-46661-8_21

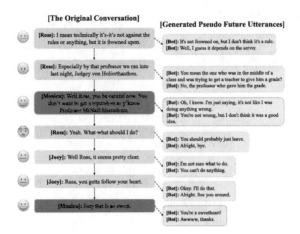

Fig. 1. An conversation of MELD. Original utterances are in dark-colored blocks and generated pseudo future contexts are in light-colored blocks.

methods [11,16] and graph-based methods [6,19]. To further enrich conversational information, another line of research proposes to enhance representations of utterances with knowledge augmentation. These methods are mainly devoted to introducing external commonsense knowledge to facilitate comprehension of utterances, and have amply proved their effectiveness [4,28].

Though many sophisticated designs have been proposed to exploit context information, we found some of them heavily rely on future contexts [6]. The problem is that we may face the unavailability of future contexts. For example, imaging a scenario of detecting the emotion of an utterance just-in-time in an after-sale conversation, we will have no future contexts available. This fact motivates us to simulate unseen future states by generating pseudo future contexts. As shown in Fig. 1, for each utterance, we can use a pre-trained language model (e.g., DialoGPT [26]) to generate its future context, introducing consistent yet extra beneficial knowledge for emotion recognition. Compared with heterogeneous external knowledge used in previous works (e.g., commonsense knowledge), the knowledge extracted is represented in a conversational form homogeneous with historical contexts, and hence can be easily integrated.

We further design a novel context representation mechanism that can be applied indiscriminately to multi-contexts, including historical contexts, historical speaker-specific contexts, and pseudo future contexts. Specifically, for each type of contexts, we employ relative position embeddings to capture the positional relationship between an utterance and its surroundings, obtaining a refined representation of an utterance, as well as a local conversational state. We then utilize GRUs to model how the local state evolves as the conversation progresses, characterizing long-term utterance dependencies from a global view. A final representation, which fuses multi-contexts information from both local and global perspectives, is fed into a classifier to produce the emotion label of an utterance.

We conduct extensive experiments on four ERC datasets, and the empirical results verify the effectiveness and potential of our method.

In summary, our contributions are: (1) We propose improving ERC from a novel perspective of generating pseudo future contexts for utterances, incorporating utterance-consistent yet potentially diversified knowledge from pre-trained language models. (2) We design a simple yet effective ERC framework to integrate pseudo future contexts with historical ones and historical speaker-specific ones, yielding competitive results on four widely-used ERC datasets. (3) We analyze how pseudo future contexts correlate with characteristics of emotion-consistency and context-dependency among utterances in conversations, revealing that pseudo future contexts resemble real ones in some scenarios.

2 Related Works

Works on ERC can be roughly categorized into sequence-based methods and graph-based methods.

(1) **Sequence-Based Methods**. Sequence-based methods treat each utterance as a discrete sequence with RNNs [3] or Transformers [21]. [7,16] utilized RNNs to track context and speaker states. [10] used two levels of Transformers to capture contextual information and seize speaker information with an auxiliary task. [22] treated ERC as an utterance-level sequence tagging task with a conditional random field layer. [11] utilized BART with supervised contrastive learning and an auxiliary response generation task.

(2) **Graph-Based Methods**. Treating an utterance as a node, contextual and speaker relations as edges, ERC can be modelled using graph neural networks. [6] modelled both the context- and speaker-sensitive dependencies with graph convolutional networks. To reflect the relational structure, [8] proposed relational position encodings in graph attention networks. Considering the temporal property, [19] proposed to encode with a directed acyclic graph.

(3) **ERC with External Knowledge**. Some works propose to introduce heterogeneous commonsense knowledge for ERC. [27] combined word and concept embeddings with two knowledge sources. Taking advantage of COMET, a Transformer-based generative model [1] for generating commonsense descriptions, [4,9,28] enhanced the representations of utterances with commonsense knowledge. It is noteworthy that our method distinguishes itself by not relying on the design of intricate instructions or prompts to extract knowledge from the pre-trained language models, setting it apart from other methods that utilize generative models as knowledge bases [14,23].

3 Methodology

3.1 Task Definition

Formally, denote $\mathcal{U}$, $\mathcal{P}$ and $\mathcal{Y}$ as conversation set, speaker set and label set. For a conversation $U \in \mathcal{U}$, $U = (u_0, \cdots, u_{n-1})$, where u_i is the i-th utterance. The

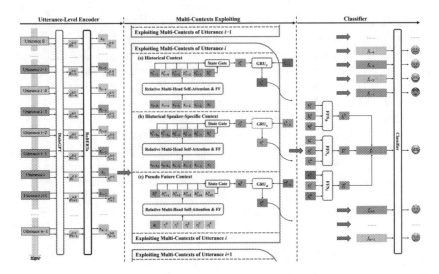

Fig. 2. The overall architecture of our proposed ERCMC framework. Utterances from various speakers are in different coloured blocks (e.g., orange, blue and green). Generated utterances are in yellow blocks. (Color figure online)

speaker of u_i is denoted by function $P(\cdot)$. For example, $P(u_i) = p_j, p_j \in \mathcal{P}$ means that u_i is uttered by p_j. The goal of ERC is to assign an emotion label $y_i \in \mathcal{Y}$ to each u_i, formulated as an utterance-level sequence tagging task in this work. As with most previous works [4,28], for an utterance, we only use its historical utterances, while future utterances are not available to fit real-life scenarios.

3.2 Generation of Pseudo Future Contexts

Knowing that DialoGPT [26] is trained on large-scale conversational data (e.g., a 27GB Reddit dataset) and can generate responses of high quality, we employ it as a knowledge base to introduce homogeneous external knowledge.

Given an utterance u_i, we utilize a DialoGPT $\mathcal{G}$ to recursively generate a pseudo future context $g_i^{0:m}$ by considering at most k historical utterances prior to u_i, where $k = \max(i, k)$, m is the maximum number of generated utterances.

$$g_i^0 = \mathcal{G}(u_{i-k}, \cdots, u_i) \tag{1}$$

$$g_i^j = \mathcal{G}(u_{i-k}, \cdots, u_i, g_i^0, \cdots, g_i^{j-1}) , \ j \in [1, m] \tag{2}$$

3.3 Proposed Framework

Figure 2 shows the overall architecture, which is named ERCMC (**E**motion **R**ecognition in **C**onversations with **M**ulti-**C**ontexts).

Utterance-Level Encoder. Given a conversation $U = (u_0, \cdots, u_{n-1})$, for $u_i \in U$, we first obtain m generated pseudo future utterances $g_i^{0:m}$. Then the utterance and its pseudo future utterances are fed into an encoder $\mathcal{M}$ (RoBERTa [15] in our experiments) to obtain their representations (we represent an utterance with the embedding of the [CLS] token).

$$x_i = \mathcal{M}(u_i) \tag{3}$$

$$r_i^j = \mathcal{M}(g_i^j) \tag{4}$$

where $x_i, r_i^j \in \mathbb{R}^{d_m}$. x_i and r_i^j are the representations of u_i and g_i^j respectively.

Multi-contexts Exploiting. To exploit multi-contexts for an utterance, we consider the three local areas for x_i: the historical context C_i, the historical context of the same speaker S_i and the future pseudo context A_i.

$$C_i = (x_{i-\ell}, \cdots, x_{i-1}, x_i) \tag{5}$$

$$S_i = (x_{b_\ell}, \cdots, x_{b_1}, x_i) \tag{6}$$

$$A_i = (x_i, r_i^0, \cdots, r_i^{m-1}) \tag{7}$$

where $\ell + 1$ is the size of the three local areas. In S_i, $P(u_{b_j}) = P(u_i)$. In C_i and S_i, $\ell = \max(i, \ell)$, and in A_i, m is set to the initial value of ℓ.

For each local area of an utterance, we calculate three representations focusing on local-aware state, local state, and evolution of local states. We take C_i as an example in the following calculation.

We first use a multi-head self-attention layer with relative position embeddings [18] to obtain a local-aware embedding considering C_i:

$$h_j^c = \max(0, h_j'^c W_1^{PF} + b_1^{PF}) W_2^{PF} + b_2^{PF} \tag{8}$$

$$h_j'^c = (head_{1j}^c \| \cdots \| head_{n_h j}^c) W^O \tag{9}$$

$$head_{pj}^c = \sum_{k=i-\ell}^{i} \alpha_{jk}(x_k W_p^V + r p_{pjk}^V) \tag{10}$$

$$\alpha_{jk} = \frac{\exp e_{jk}}{\sum_{o=i-\ell}^{i} \exp e_{jo}} \tag{11}$$

$$e_{jk} = \frac{x_j W_p^Q (x_k W_p^K + r p_{pjk}^K)^\top}{\sqrt{d_k}} \tag{12}$$

where $h_j^c \in \mathbb{R}^{d_m}$, $\|$ is the concatenation of two vectors, $W_1^{PF} \in \mathbb{R}^{d_m \times 4d_m}, b_1^{PF} \in \mathbb{R}^{4d_m}, W_2^{PF} \in \mathbb{R}^{4d_m \times d_m}, b_2^{PF} \in \mathbb{R}^{d_m}$. The number of head is n_h, $W^O \in \mathbb{R}^{d_v n_h \times d_m}$. For each head p, $W_p^Q, W_p^K \in \mathbb{R}^{d_m \times d_k}$, $W_p^V \in \mathbb{R}^{d_m \times d_v}$ are projections, $r p_{pjk}^V \in \mathbb{R}^{d_v}$ and $r p_{pjk}^K \in \mathbb{R}^{d_k}$ are relative position embeddings between x_j and x_k. Similarly, for S_i and A_i, two embeddings h^s and h^a are obtained.

A state gate is then used to get the local state s_i^c:

$$s_i^c = \sum_{j=i-\ell}^{i-1} \beta_j h_j^c \tag{13}$$

$$\beta_j = \frac{\exp st_j}{\sum_{k=i-\ell}^{i-1} \exp st_k} \tag{14}$$

$$st_j = \tanh\left(h_j^c W^S (h_i^c)^\top\right) \tag{15}$$

where $s_i^c \in \mathbb{R}^{d_m}$, $W^S \in \mathbb{R}^{d_m \times d_m}$. Similarly, for S_i and A_i, two local states s_i^s and s_i^a are obtained.

For states in a global view, a GRU unit is utilized to characterize the evolution of local states:

$$t_i^c = \mathrm{GRU}(s_i^c, t_{i-1}^c) \tag{16}$$

where $t_i^c \in \mathbb{R}^{d_m}$ is the tracked global state prior to u_i and t_0^c is initialized with zero. Similarly, for S_i and A_i, two GRU units are used to get t_i^s and t_i^a.

Classifier. For u_i, the exploited outcomes from multi-contexts are integrated into a final representation f_i.

$$f_i = f_i^h \parallel f_i^s \parallel f_i^t \tag{17}$$

$$f_i^h = (h_i^c \parallel h_i^s \parallel h_i^a) W_h^F \tag{18}$$

$$f_i^s = (s_i^c \parallel s_i^s \parallel s_i^a) W_s^F \tag{19}$$

$$f_i^t = (t_i^c \parallel t_i^s \parallel t_i^a) W_t^F \tag{20}$$

where $W_h^F, W_s^F, W_t^F \in \mathbb{R}^{3d_m \times d_m}$.

To obtain the labels $Y = (y_0, \cdots, y_{n-1})$ with the representations $F = (f_0, \cdots, f_{n-1})$, we apply a feed forward network and a softmax layer:

$$y_i = \mathrm{argmax}(P_i) \tag{21}$$

$$P_i = \mathrm{softmax}(F_i W^M + b^M) \tag{22}$$

where $W^M \in \mathbb{R}^{3d_m \times |\mathcal{Y}|}$ and $b^M \in \mathbb{R}^{|\mathcal{Y}|}$.

The model is trained using negative log-likelihood loss.

4   Experimental Setups

4.1   Datasets and Evaluation Metrics

We evaluate our proposed framework on four ERC datasets: IEMOCAP [2], DailyDialog [12], EmoryNLP [25], MELD [17]. IEMOCAP and DailyDialog are two-party datasets, while EmoryNLP and MELD are multi-party datasets. For

Table 1. Statistics of datasets.

Dataset	Conversations			Utterances			Classes
	Train	Dev	Test	Train	Dev	Test	
IEMOCAP	120		31	5,810		1,623	6
DailyDialog	11,118	1,000	1,000	87,170	8,069	7,740	7
EmoryNLP	659	89	79	7,551	954	984	7
MELD	1,038	114	280	9,989	1,109	2,610	7

the four datasets, we only use the textual parts. Statistics of these datasets are shown in Table 1.

Following [4,28], we choose weighted-average F1 for IEMOCAP, EmoryNLP and MELD. Since the *neutral* class constitutes to 83% of the DailyDialog, micro-averaged F1 excluding *neutral* is chosen.

4.2   Baselines

For models without external knowledge, we compare with: (1) **Sequence-based methods**: **DialogueRNN** [16] uses GRUs to track context and speaker states; **HiTrans** [10] utilizes two level Transformers with an auxiliary task to leverage both contextual and speaker information; **CoG-BART** [11] uses BART with supervised contrastive learning and a response generation task. (2) **Graph-based Methods**: **DialogueGCN** [6] models dependencies about context and speaker with graph convolutional networks; **RGAT** [8] uses relational graph attention networks with relational position encodings.

For models with external knowledge, we compare with: **KET** [27] uses a Transformer to combine word and concept embeddings; **COSMIC** [4] is a modified DialogueRNN with commonsense knowledge from COMET; **TODKAT** leverages COMET to integrate commonsense knowledge and a topic model to detect the potential topics of a conversation; **SKAIG** [9] models structural psychological interactions between utterances with commonsense knowledge.

Three variants of ERCMC are compared: (1) ERCMC without future contexts; (2) ERCMC with multi-contexts; (3) ERCMC using real future contexts.

4.3   Implementation Details

We use DialoGPT-medium to generate 5 pseudo future utterances for each utterance (i.e., $m = 5$) with k historical utterances, where $k = 2$ for IEMOCAP and DailyDialog, and $k = 4$ for EmoryNLP and MELD. For backbones, we select RoBERTa-base for IEMOCAP and RoBERTa-large for the other datasets. For each utterance, the max length is set to 128. For multi-contexts exploiting, we set the size of the three local areas to 6 (i.e., $\ell = 5$). The number of head n_h is set to 8. d_m is set to 768 for IEMOCAP and 1,024 for the other datasets. $d_k, d_v = \frac{d_m}{n_h}$. For training setup, the batch size is set to 1 conversation for

Table 2. Overall results. In each part, the highest scores are in boldface. * indicates using future contexts. C, S, PF, and RF denote historical contexts, historical speaker-specific contexts, pseudo future contexts, and real future contexts.

Methods		IEMOCAP	DailyDialog	EmoryNLP	MELD
		Weighted F1	Micro F1	Weighted F1	Weighted F1
Without External Knowledge					
DialogueRNN		62.57	55.95	31.70	57.03
+ RoBERTa		64.76	57.32	37.44	63.61
DialogueGCN*		64.18	–	–	58.10
+ RoBERTa*		64.91	57.52	38.10	63.02
RGAT*		65.22	54.31	34.42	60.91
+RoBERTa*		**66.36**	**59.02**	37.89	62.80
HiTrans*		64.50	–	36.75	61.94
CoG-BART*		66.18	56.29	**39.04**	**64.81**
With External Knowledge					
KET		59.56	53.37	34.39	58.18
COSMIC		65.28	58.48	38.11	65.21
SKAIG*		**66.96**	**59.75**	**38.88**	65.18
TODKAT		61.33	58.47	38.69	**65.47**
Variants of Our Model					
ERCMC	C & S	65.47	59.85	38.71	65.21
	C & S & PF	66.07	59.92	**39.34**	**65.64**
	C & S & RF*	**66.51**	**61.33**	38.90	65.43

IEMOCAP and DailyDialog, and 2 conversations for EmoryNLP and MELD. The gradient accumulation step is set to 16 for DailyDialog and 4 for IEMOCAP, EmoryNLP and MELD. We train the model for 20 epochs for IEMOCAP and 10 epochs for the other datasets with the AdamW optimizer whose learning rate is set to 3e-5 for IEMOCAP and 1e-5 for the other datasets. Each model is trained for max epochs and choose the checkpoint with the best validation performance. The dropout rate is set to 0.1. All the experiments are done on a single NVIDIA Tesla V100. All of our results are the average of 5 runs. We do not implement on larger generative models, such as ChatGPT, due to their significant computational resources requirements and associated costs. Our code is available at https://github.com/Ydongd/ERCMC.

5   Results and Insights

5.1   Overall Results

Overall results are shown in Table 2, from which we have three main observations: (1) Compared with models using future contexts, ERCMC in C&S&RF

Table 3. Various combinations of Multi-contexts. RAW denotes no contex.

Part	IEMOCAP	DailyDialog	EmoryNLP	MELD
RAW	56.48	57.46	37.78	64.06
C	63.95	59.14	37.88	64.20
S	64.39	59.48	37.97	64.43
PF	57.38	58.16	37.84	64.20
C & PF	62.29	59.50	37.90	64.36
S & PF	63.35	59.66	37.98	64.76
C & S	65.47	59.85	38.71	65.21
C & S & PF	66.07	59.92	39.34	65.64

setting achieves competitive performance, demonstrating the superiority of our proposed framework. (2) In stead of leveraging heterogeneous commonsense knowledge as previous works, we use homogeneous conversational knowledge, allowing ERCMC in C&S&PF setting to overtake other models using external knowledge. (3) Additional future contexts provide more information about conversations and thus bring improvements over C&S setting when using pseudo or real future contexts in our model. Compared with using real future contexts, using pseudo future contexts performs better on EmoryNLP and MELD, and underperforms on IEMOCAP and DailyDialog. An in-depth analysis is given in Sect. 5.4.

5.2   Collaboration of Multi-contexts

To reveal the collaborative effect of multi-contexts on ERCMC, we show various combinations of multi-contexts in Table 3, from which we observe that: (1) All combinations improve the performance compared with only using raw information, suggesting that the context information is critical for ERC. (2) Regardless of the number of contexts used, settings with historical speaker-specific contexts produce the best results, indicating that historical contexts uttered by the same speaker is the predominant part. (3) Using C or S alone, and a combination of both, bring more improvements than using PF and its combination. The observation implies that historical contexts and historical speaker-specific contexts are more significant for ERC, while pseudo future contexts serve as a supplementary role to provide extra knowledge.

5.3   Ablation Study

We form the final representations with local-aware embeddings, local states and tracked global states, which are denoted as h, s, and t, respectively. Results with removal of each part are shown in Table 4a, from which we can find that though the calculation of local states and tracked global states contain some

Table 4. Ablation study of ERCMC in C&S&PF setting.

(a) Results with different compositions of the final representations, h, s, and t denote local-aware embedding, local state, and tracked global state, respectively.

Dataset	w/o h	w/o s	w/o t	h, s, t
IEMOCAP	62.88	64.35	64.81	66.07
DailyDialog	59.16	59.59	59.50	59.92
EmoryNLP	20.08	38.65	38.74	39.34
MELD	51.51	65.06	65.30	65.64

(b) Results with different position embeddings. N, S, L, and R denote using no embeddings, sinusoidal, learnable, and relative positon embeddings, respectively.

Dataset	N	S	L	R
IEMOCAP	65.48	64.61	65.28	66.07
DailyDialog	59.89	59.90	59.83	59.92
EmoryNLP	38.64	38.51	38.57	39.34
MELD	64.98	64.83	64.41	65.64

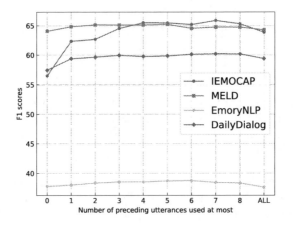

Fig. 3. Effect of number of historical utterances with ERCMC in C&S setting.

information from local-aware embeddings, performances drop dramatically when original embeddings are lost, especially on EmoryNLP and MELD. And performances fall slightly when local states and tracked global states are removed.

Effect of Position Embeddings. Unlike modeling the distances between utterances with sinusoidal position embeddings in previous sequence-based methods [10,27], we utilize relative position embeddings and further compare with three position embeddings in Table 4b: without position embeddings; sinusoidal position embeddings that use sine and cosine functions of different frequencies; position embeddings learned from scratch. It is clear that the distance information exploited by relative position embeddings at utterance-level is more compatible with a sequence-based method and therefore achieves better results.

Effect of Size of Local Areas. In our experiments, we set the size of local areas to 6 (i.e., using five historical utterances at most). We further investigate the size effect of local areas by employing ERCMC in C&S setting. From the results in Fig. 3, we can draw a general trend that performances show an upward tendency when starting using historical utterances and decrease when using all historical contexts due to information redundancy. We can also see

Table 5. Performance and emotion-consistency on four simplified test sets.

(a) Simplified test set of IEMOCAP with 1468 utterances.

Setting	IEMOCAP		
	Performance	WT_1	WT_2
PF	57.81	35.85	38.18
C & S & PF	66.30		
RF	62.81	50.10	50.47
C & S & RF	66.68		

(b) Simplified test set of DailyDialog with 3123 utterances.

Setting	DailyDialog		
	Performance	WT_1	WT_2
PF	51.19	44.77	60.43
C & S & PF	53.80		
RF	53.78	76.79	78.90
C & S & RF	54.53		

(c) Simplified test set of EmoryNLP with 608 utterances.

Setting	EmoryNLP		
	Performance	WT_1	WT_2
PF	40.94	29.95	31.36
C & S & PF	41.86		
RF	40.64	27.11	29.22
C & S & RF	41.73		

(d) Simplified test set of MELD with 1360 utterances.

Setting	MELD		
	Performance	WT1	WT2
PF	64.07	39.52	43.00
C & S & PF	65.68		
RF	63.69	35.62	38.38
C & S & RF	64.97		

that compared with IEMOCAP and DailyDialog, MELD and EmoryNLP have a smoother performance, especially at the beginning and end of the curve.

5.4 Future Context: Pseudo or Real

In Sect. 5.1, we have mentioned that both pseudo and real future contexts improve the performance of ERCMC, and compared with the C&S&RF setting, the C&S&PF setting performs better on EmoryNLP and MELD and worse on IEMOCAP and DailyDialog. This section provides a more in-depth analysis of how pseudo and real future contexts affect the final results.

We first categorize the four datasets concerning several previous observations.

Firstly, in Table 3, C&S setting obtains relative improvements over RAW setting of 15.92% on IEMOCAP, 4.16% on DailyDialog, 2.46% on EmoryNLP and 1.80% on MELD. Secondly, in Table 4a, compared with the full C&S&PF setting, when the original local-aware embeddings are removed, the performances drops by 4.83% on IEMOCAP, 1.27% on DailyDialog, 48.96% on EmoryNLP and 21.53% on MELD. Thirdly, in Fig. 3, the curves of EmoryNLP and MELD are stabler than those of IEMOCAP and DailyDialog.

From them, we can conclude that conversations in IEMOCAP and DailyDialog are more context-dependent, while conversations in EmoryNLP and MELD are relatively context-independent. An alternative explanation for this conclusion is that IEMOCAP and DailyDialog are two-party datasets, and therefore the conversations are more emotion-focused, whereas EmoryNLP and MELD are multi-party datasets, resulting in more diffuse conversations and inconsistent emotions.

To further explore the effect of using pseudo or real future contexts, we attempt to investigate the attributes of future contexts. Previous works [5,24] have reported a common issue with ERC: different emotions in consecutive utterances may confound a model and thereby degrade performance. Inspired by the issue, we define a new concept of "emotion-consistency" as the degree of emotional consistency of the subsequent utterances with the first utterance within a local area. Given a local area $LC = (u_0, \cdots, u_\ell)$, the emotion-consistency is:

$$EC(LC) = 100 \cdot \sum_{i=1}^{\ell} \phi(u_i, u_0) \cdot wt_{i-1} \tag{23}$$

where $WT = (wt_0, \cdots, wt_{\ell-1})$ is the weight in different positions, $\phi(u_i, u_0)$ is a function to indicate whether u_i and u_0 having the same emotion.

For quantitative analysis, we employ ERCMC in C setting to predict emotions and calculate the emotion-consistency scores. To align with pseudo future contexts, we simplify the test sets to ensure each utterance has at least ℓ consecutive utterances ($\ell = 5$ in our experiments). Two kinds of weight are considered, $wt_i^1 = \frac{1}{\ell}, wt_i^1 \in WT_1; wt_i^2 = \frac{\exp e^{\ell-i}}{\sum_{j=1}^{\ell} \exp e^{j-1}}, wt_i^2 \in WT_2$. WT_1 relies uniformly on consecutive utterances, while WT_2 favours utterances near the first utterance.

From Table 5a, 5b, 5c and 5d, we can observe that: (1) Emotion-consistency scores of IEMOCAP and DailyDialog are higher than those of EmoryNLP and MELD. The finding confirms our previous observations that IEMOCAP and DailyDialog are more context-dependent, whereas EmoryNLP and MELD are somewhat context-independent. (2) Performance and emotion-consistency scores are positively correlated. And combined with Table 2, using pseudo future contexts achieves competitive results on IEMOCAP and DailyDialog, and outperforms using real future contexts on EmoryNLP and MELD. The results demonstrate that pseudo future contexts can replace real ones to some extent when the dataset is context-dependent, and serve as more extra beneficial knowledge when the dataset is relatively context-independent.

5.5   Case Study

We illustrate two examples to promote understanding of our method in Fig. 4.

In the first case from IEMOCAP, the emotions of the two speakers are consistent, which greatly helps to distinguish emotions with similar meanings, such as "Excited" and "Happy", while pseudo future contexts mislead the model to some extent.

In the second case from MELD, the emotions of multiple speakers are diverse and a high reliance on real future contexts may influence the model's judgement. More consistent supplementary knowledge from pseudo future contexts could help to identify the correct emotions.

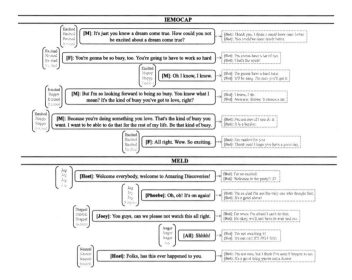

Fig. 4. Two cases from IEMOCAP and MELD. In the boxes on the left, from top to bottom, are: labels, predictions from C&S, C&S&PF, and C&S&RF settings.

6   Conclusion

In this paper, we propose a conceptually simple yet effective method of acquiring external homogeneous knowledge by generating pseudo future contexts that are not always available in real-life scenarios. Furthermore, a novel framework named ERCMC is proposed to jointly exploit historical contexts, historical speaker-specific contexts, and pseudo future contexts. Experimental results on four ERC datasets demonstrate the superiority and potential of our method. Further empirical investigations reveal that pseudo future contexts can rival real ones to some extent, especially when the dataset is less context-dependent. Our research could inspire more future works in conversation understanding. In the future, we plan to generate pseudo future contexts in a more controllable way, and extending our method to more tasks.

Acknowledgements. This work was supported by the National Key R&D Program of China [2022YFF0902703].

References

1. Bosselut, A., Rashkin, H., Sap, M., Malaviya, C., Celikyilmaz, A., Choi, Y.: Comet: commonsense transformers for automatic knowledge graph construction. In: Proceedings of the 57th Annual Meeting of the Association for Computational Linguistics, pp. 4762–4779 (2019)
2. Busso, C., et al.: Iemocap: interactive emotional dyadic motion capture database. Lang. Resour. Eval. **42**(4), 335–359 (2008)

3. Elman, J.L.: Finding structure in time. Cogn. Sci. **14**(2), 179–211 (1990)
4. Ghosal, D., Majumder, N., Gelbukh, A., Mihalcea, R., Poria, S.: Cosmic: common-sense knowledge for emotion identification in conversations. In: Findings of the Association for Computational Linguistics: EMNLP 2020, pp. 2470–2481 (2020)
5. Ghosal, D., Majumder, N., Mihalcea, R., Poria, S.: Exploring the role of context in utterance-level emotion, act and intent classification in conversations: an empirical study. In: Findings of the Association for Computational Linguistics: ACL-IJCNLP 2021, pp. 1435–1449 (2021)
6. Ghosal, D., Majumder, N., Poria, S., Chhaya, N., Gelbukh, A.: Dialoguegcn: a graph convolutional neural network for emotion recognition in conversation. In: Proceedings of the 2019 Conference on Empirical Methods in Natural Language Processing and the 9th International Joint Conference on Natural Language Processing (EMNLP-IJCNLP), pp. 154–164 (2019)
7. Hu, D., Wei, L., Huai, X.: Dialoguecrn: contextual reasoning networks for emotion recognition in conversations. In: Proceedings of the 59th Annual Meeting of the Association for Computational Linguistics and the 11th International Joint Conference on Natural Language Processing (Volume 1: Long Papers), pp. 7042–7052 (2021)
8. Ishiwatari, T., Yasuda, Y., Miyazaki, T., Goto, J.: Relation-aware graph attention networks with relational position encodings for emotion recognition in conversations. In: Proceedings of the 2020 Conference on Empirical Methods in Natural Language Processing (EMNLP), pp. 7360–7370 (2020)
9. Li, J., Lin, Z., Fu, P., Wang, W.: Past, present, and future: conversational emotion recognition through structural modeling of psychological knowledge. In: Findings of the Association for Computational Linguistics: EMNLP 2021, pp. 1204–1214 (2021)
10. Li, J., Ji, D., Li, F., Zhang, M., Liu, Y.: Hitrans: a transformer-based context-and speaker-sensitive model for emotion detection in conversations. In: Proceedings of the 28th International Conference on Computational Linguistics, pp. 4190–4200 (2020)
11. Li, S., Yan, H., Qiu, X.: Contrast and generation make bart a good dialogue emotion recognizer. In: Proceedings of the AAAI Conference on Artificial Intelligence, vol. 36, pp. 11002–11010 (2022)
12. Li, Y., Su, H., Shen, X., Li, W., Cao, Z., Niu, S.: Dailydialog: a manually labelled multi-turn dialogue dataset. In: Proceedings of the Eighth International Joint Conference on Natural Language Processing (Volume 1: Long Papers), pp. 986–995 (2017)
13. Lin, Z., Madotto, A., Shin, J., Xu, P., Fung, P.: Moel: mixture of empathetic listeners. In: Proceedings of the 2019 Conference on Empirical Methods in Natural Language Processing and the 9th International Joint Conference on Natural Language Processing (EMNLP-IJCNLP), pp. 121–132 (2019)
14. Ling, T., Chen, L., Lai, Y., Liu, H.L.: Evolutionary verbalizer search for prompt-based few shot text classification. arXiv preprint arXiv:2306.10514 (2023)
15. Liu, Y., et al.: Roberta: a robustly optimized bert pretraining approach. arXiv preprint arXiv:1907.11692 (2019)
16. Majumder, N., Poria, S., Hazarika, D., Mihalcea, R., Gelbukh, A., Cambria, E.: Dialoguernn: an attentive RNN for emotion detection in conversations. In: Proceedings of the AAAI Conference on Artificial Intelligence, vol. 33, pp. 6818–6825 (2019)

17. Poria, S., Hazarika, D., Majumder, N., Naik, G., Cambria, E., Mihalcea, R.: Meld: a multimodal multi-party dataset for emotion recognition in conversations. In: Proceedings of the 57th Annual Meeting of the Association for Computational Linguistics, pp. 527–536 (2019)

18. Shaw, P., Uszkoreit, J., Vaswani, A.: Self-attention with relative position representations. In: Proceedings of the 2018 Conference of the North American Chapter of the Association for Computational Linguistics: Human Language Technologies, Volume 2 (Short Papers), pp. 464–468 (2018)

19. Shen, W., Wu, S., Yang, Y., Quan, X.: Directed acyclic graph network for conversational emotion recognition. In: Proceedings of the 59th Annual Meeting of the Association for Computational Linguistics and the 11th International Joint Conference on Natural Language Processing (Volume 1: Long Papers), pp. 1551–1560 (2021)

20. Shu, K., Mosallanezhad, A., Liu, H.: Cross-domain fake news detection on social media: a context-aware adversarial approach. In: Khosravy, M., Echizen, I., Babaguchi, N. (eds.) Frontiers in Fake Media Generation and Detection, pp. 215–232. Springer, Singapore (2022). https://doi.org/10.1007/978-981-19-1524-6_9

21. Vaswani, A., et al.: Attention is all you need. Adv. Neural Inf. Process. Syst. **30** (2017)

22. Wang, Y., Zhang, J., Ma, J., Wang, S., Xiao, J.: Contextualized emotion recognition in conversation as sequence tagging. In: Proceedings of the 21th Annual Meeting of the Special Interest Group on Discourse and Dialogue, pp. 186–195 (2020)

23. Wei, Y., Mo, T., Jiang, Y., Li, W., Zhao, W.: Eliciting knowledge from pretrained language models for prototypical prompt verbalizer. In: Artificial Neural Networks and Machine Learning-ICANN 2022: 31st International Conference on Artificial Neural Networks, pp. 222–233 (2022)

24. Yang, L., Shen, Y., Mao, Y., Cai, L.: Hybrid curriculum learning for emotion recognition in conversation. In: Proceedings of the AAAI Conference on Artificial Intelligence, vol. 36, pp. 11595–11603 (2022)

25. Zahiri, S.M., Choi, J.D.: Emotion detection on tv show transcripts with sequence-based convolutional neural networks. In: Workshops at the Thirty-Second AAAI Conference on Artificial Intelligence (2018)

26. Zhang, Y., et al.: Dialogpt: large-scale generative pre-training for conversational response generation. In: Proceedings of the 58th Annual Meeting of the Association for Computational Linguistics: System Demonstrations, pp. 270–278 (2020)

27. Zhong, P., Wang, D., Miao, C.: Knowledge-enriched transformer for emotion detection in textual conversations. In: Proceedings of the 2019 Conference on Empirical Methods in Natural Language Processing and the 9th International Joint Conference on Natural Language Processing (EMNLP-IJCNLP), pp. 165–176 (2019)

28. Zhu, L., Pergola, G., Gui, L., Zhou, D., He, Y.: Topic-driven and knowledge-aware transformer for dialogue emotion detection. In: Proceedings of the 59th Annual Meeting of the Association for Computational Linguistics and the 11th International Joint Conference on Natural Language Processing (Volume 1: Long Papers), pp. 1571–1582 (2021)

Generating Enlightened Suggestions Based on Mental State Evolution for Emotional Support Conversation

Mengjiao Gan[1,2,3,4], Jianxing Yu[1,2,3,4(✉)], Xiao Dong[1,2,3,4], Shuang Qiu[1,2,3,4], Wei Liu[1,2,3,4], and Jian Yin[1,2,3,4]

[1] School of Artificial Intelligence, Sun Yat-sen University, Zhuhai, China
{ganmj,dongx55}@mail2.sysu.edu.cn, qiushuang@gdei.edu.cn,
{liuw259,issjyin,yujx26}@mail.sysu.edu.cn
[2] Guangdong Key Laboratory of Big Data Analysis and Processing, Guangzhou, China
[3] Pazhou Lab, Guangzhou 510330, China
[4] Key Laboratory of Sustainable Tourism Smart Assessment Technology, Ministry of Culture and Tourism, Zhuhai, China

Abstract. Emotional support conversation aims to provide comfort and suggestions to users and gradually reduce their negative emotions such as anxiety. It is a valuable topic for many applications, including mental health support and customer service chats. However, due to the lack of enough expert knowledge, existing methods fail to provide enlightened suggestions to reverse users' worries. Additionally, these methods neglect to grasp the mental state evolution of users. To address these problems, we propose a novel method that considers **M**ental State **E**volution to provide **K**nowledge-grounded **S**uggestions (MEKS). In detail, we first create a suggestion corpus called MentalQA to grasp the psychological knowledge by resorting to the mental health forum. The relevant passages are selected based on both the context and the original response. Then we leverage graph structure to enrich the context with the inferred user's mental state evolution. Furthermore, we introduce a gate to combine textual expert knowledge with the mental state evolution graph, so as to facilitate the generation of supportive responses. Experimental results show that this method can provide reasonable solutions to help the users.

Keywords: Emotional support conversation · dialog system · knowledge enhanced generation

1 Introduction

Emotional support conversation (ESC) [10] aims to provide psychological comfort through dialogue for the users experiencing emotional distress. This task

Supplementary Information The online version contains supplementary material available at https://doi.org/10.1007/978-3-031-46661-8_22.

needs a series of procedures and support skills to help the users identify their problems, provide empathetic responses, and give guiding knowledge to cope with the difficulty. Existing works like emotional conversation [23] only need to express certain emotions in responses, while empathetic conversation [15] solely requires to reply based on the interlocutors' feelings, without caring whether it can reduce their anxiety. Different from these tasks, ESC aims to help users to tackle emotional stress by providing useful suggestions. ESC has great value in various applications, such as providing companionship to the elderly; giving comfort to curing patients with mental health problems; appeasing angry customers and offering solutions in customer service chats.

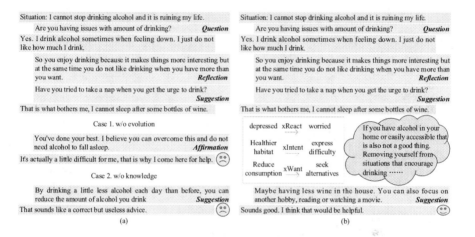

Fig. 1. We present an illustrative example to demonstrate the limitations of existing models in (a) and show an instance of an emotional support conversation in (b). In (a), Case 1 is problematic as it concentrates excessively on the user's present state of insomnia, while overlooking the user's primary intention of reducing alcohol consumption. Meanwhile, Case 2 recognizes the user's preference for seeking an alternative to alcohol abstinence, but it only offers general suggestions that are not tailored to the user's specific needs due to the absence of expert knowledge. In (b), we utilize a white box to illustrate the evolution of the user's fine-grained mental state, which comprises the user's reaction, intention, and desire. Moreover, we use horizontal text to represent domain passages chosen from collected knowledge sources, which are deemed relevant to the conversation.

However, it is challenging to build an ESC model. Existing work mostly focuses on the task of emotional conversation or empathetic conversation. They are not aimed at approaching the users' emotional problems. The current ESC models are severely impeded by two significant limitations. The first limitation stems from the paucity of professional psychological content in the dataset used for training the ESC model. Insufficient content renders the model unable to provide appropriate and effective suggestions to help the users seeking emotional support. Given that the primary goal of an ESC model is to facilitate the resolution of users' emotional concerns, the dearth of professional content can substantially undermine the model's effectiveness. The second limitation arises from the ESC model's inability to accurately capture and simulate the user's mental state and its evolution over time. Consequently, the ESC model may fail to adjust self-care recommendations in accordance with the user's prevailing mental state. We provide a visual demonstration of the aforementioned limitations in ESC modeling in Fig. 1(a). The figure illustrates two typical cases that further underscore the importance of addressing these limitations. In case 1, if the mental state and its evolution of the user are not taken into account, the generated response may overly emphasize the last utterance, disregarding the broader context of the conversation. In case 2, when expert knowledge is absent, the generated response may be overly general and not tailored to the user's specific needs. Therefore, it is crucial to address these limitations. However, existing ESC methods have certain limitations, as they do not consider professional prior knowledge and users' historical mental states. For example, existing methods in the ESC field neglect the importance of suggestions and lack the necessary domain knowledge to provide them effectively [10,13,14]. Some approaches rely solely on partial knowledge derived from historical conversations, which could limit their adaptability to new scenarios and reduce the diversity of generated responses [22]. Additionally, some methods only consider the user's most recent emotional state, failing to capture the full range of mental states and resulting in less adaptive responses [3,19]. These limitations could have a negative impact on the quality of generated responses, underscoring the need for models that can provide reasonable and adaptable advice.

To tackle these problems, we propose a method that considers **M**ental State **E**volution to provide **K**nowledge-grounded **S**uggestions (MEKS). To address the issue of insufficient professional knowledge, we first create the MentalQA dataset from an online mental health forum that contains a vast collection of more than 100k question-answer pairs. The dataset covers a wide range of mental health topics, including depression, anxiety, breakups, and family stress, among others. Each topic comprises multiple questions, and each question has several answers that are labeled with the number of likes from users. The diverse range of topics, numerous questions, and professional answers from therapists underscore the wealth of mental health knowledge embedded in the dataset. More details are given in ??. To identify passages related to the dialog history, we employ a two-stage matching process to filter redundant content. We first use the Dense Passage Retriever (DPR) [6] to select a larger set of passages based on the context. We then employ a refiner that chooses a subset of the passages based on the original response generated by language models (LM). That allows us to retrieve passages based on both contextual information and the parametric knowledge implied in the LM [2]. To capture the dynamic mental states, we resort to a knowledge graph called ATOMIC-2020 [5] which contains plentiful social-emotional relations. We utilize a generative model called COMET [1] to derive how the user reacts, what the user intends, and what the user wants. These aspects form the fine-grained mental state of the user. Our proposed approach, MESK, leverages the prevalent usage of graph models [18] and employs a directed graph to capture the structural information of the components instead of relying on a linear sequence, which has been the case in some previous methods such as [17,19]. By modeling the components and their evolution using a graph structure, MESK can incorporate the user's mental state into the context and capture the relational information between the various components, including the situation and utterances, successive utterances, the user's utterance and corresponding mental state, and the historical and current mental states. As shown in Fig. 1(b), based on this mental state evolution graph and the selected domain knowledge with strategy awareness, our approach generates supportive responses that consider knowledge strategically. In summary, the main contributions of this paper include:

- We propose a new topic on ESC which integrates professional psychological knowledge to provide problem-shooting suggestions. We create a large-scale psychological knowledge corpus called MentalQA for this topic.
- We propose MEKS, a model that captures the evolution of the user's fine-grained mental state to generate contextually relevant suggestions.
- We conduct extensive experiments to fully evaluate our proposed model. Significant outperformance is obtained, especially in terms of the diversity of generated texts and the provision of sound suggestions.

2   Related Work

Emotional support conversation (ESC) represents an extension of empathetic conversation, aimed at reducing stress and helping individuals work through challenges. To ensure the availability of high-quality conversations that provide examples of effective emotional support, ESConv is constructed with rich annotation in a help-seeker and supporter mode [10].

The selection of support strategies and the generation of strategy-constrained responses are the main research areas in ESC. To this end, lookahead heuristics are proposed to estimate future user feedback, enabling the planning of strategies in advance [3]. Peng *et al.* [14] utilize dual-level feedback to schedule strategies and generate strategy-constrained responses, whereas Xu *et al.* [22] model the one-to-many mapping relationship of strategies. To enhance their model, they also incorporate prior knowledge in the form of exemplars and strategy sequences.

Aside from the support strategy selection and strategy-constrained response generation, attention is paid to the user's state, as it plays an important role in dynamic strategy selection and measuring the effectiveness of ESC. Therefore, COMET is introduced to capture the user's instant state, and strategies are formulated as a probability distribution over a codebook [19]. Peng *et al.* [13] argues the necessity of the hierarchical relationships with the global cause and local psychological intention behind conversations.

3 Problem Definition

The main objective of the ESC system is to generate a supportive response y given the situation s and the dialog history $C = [u_1, \cdots, u_k]$. Here, the situation s describes the causes of the emotional problem, and C is a sequence of k utterances. To ensure the response is informed by domain knowledge, we integrate the knowledge source MentalQA as P. Additionally, we construct a graph G based on the situation, dialog history, and inferred mental state to enrich the context with the evolution of the user's mental state and provide adaptive suggestions. Consequently, our **MEKS** generates the response by leveraging the information from both G and P, allowing for more effective and adaptive emotional support.

4 Methodology

The proposed framework, named MEKS and illustrated in Fig. 2, consists of three main modules: mental state evolution, domain knowledge selection, and strategy-aware generation. In the first module, a pretrained reasoning model called COMET is utilized to infer the user's fine-grained mental state. To further improve our model, we model the context and evolution of the user's mental state separately using a graph structure. We then feed the graphs into the CompGCN [20] model to obtain the graph embedding, which helps capture the relationships between different aspects of the user's mental state. Next, domain knowledge selection is performed through a two-stage matching process. Finally, a gate is introduced to combine expert knowledge with the mental state evolution graph, so that facilitates the generation of supportive responses.

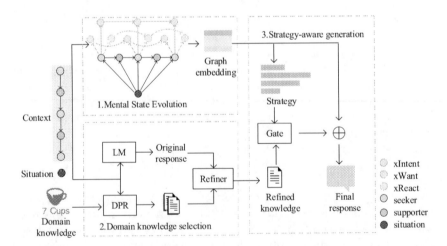

Fig. 2. Framework of MEKS, which consists of three modules: (a) Mental State Evolution infers the user's fine-grained mental state and models the context and the evolution process with a graph, (b) Domain Knowledge Selection retrieves relevant passages from collected mental health knowledge and (c) Strategy-aware Generation combines the evolution graph embedding with retrieved domain knowledge strategically to assist response generation.

4.1 Mental State Evolution

The module aims to enhance the contextualization of the user's mental state and track its evolution process through a three-step approach that includes mental state inference, mental state graph construction, and encoding. Initially, a reasoning model that includes social-interaction commonsense is employed to infer both the historical and current mental state based on the context. Then, a mental state evolution graph is constructed based on the inferred states to capture structured information more effectively. Finally, the graph is encoded to generate a representation that efficiently encapsulates the evolution of the user's mental state.

Mental State Inference. Our model MEKS aims to gain a thorough understanding of the user's mental state by examining three distinct dimensions: the user's reactions, intentions, and desires. By considering not only the user's emotions but also their intentions and desires, we can achieve a more comprehensive understanding that facilitates better strategy selection and response generation. To capture the corresponding relationship state when given the user's utterance and specific relation as input, we utilize a pre-trained generative commonsense reasoning model called COMET [1]. We focus on three types of relations, namely xReact, xIntent, and xWant, to infer how the user reacts, what the user intends, and what the user desires, respectively. We concatenate the user's utterances from each turn and feed them, along with the three types of relations, as input to COMET. This process enables us to obtain the user's mental state for each turn, including their reaction, intention, and desire. The inference process is formulated as follows:

$$\mathcal{M}_u = \bigcup_{r}^{r \in Rel} \mathcal{M}_u^r,$$
$$\mathcal{M}_u^r = \text{COMET}(u, r), \tag{1}$$

where $\mathcal{M}_u$ means the mental state inferences for utterance u, $\mathcal{M}_u^r$ represents the specific state obtained for utterance u and relation r.

Mental State Graph Construction. Upon generating the inferred mental states using the COMET model, we propose a graph-based model to elucidate the structural relationships among the situation, dialogue history, and user's mental states. In this model, the situation, dialogue history, and user's mental states are represented as nodes, while the relationships between them are depicted as edges. Specifically, the inputs, which include the situation, dialogue history, and user's mental states, are denoted as nodes, and the relationships among different nodes are illustrated as edges and incorporated into the graph-based model. A comprehensive definition of our graph model is introduced in the subsequent section.

Nodes: In graph G, the situation, each utterance in the context, and each mental state are represented as nodes. Consequently, the nodes V are denoted as a set comprising the situation s, utterance in dialogue history u, and the specific mental state $\mathcal{M}_u^r$ for utterance, e.g., $V = \{s, u_1, \cdots, u_k, \mathcal{M}_{u_1}^r, \cdots, \mathcal{M}_{u_k}^r\}$. Suppose the graph contains n nodes, the nodes are denoted as $V = \{v_i\}, 1 \leq i \leq n$. **Edges:** The relationships among the situation, each utterance in dialogue history, and each mental state are represented as edges in graph G. These edges can be categorized into the following four types: (1) Connection between the situation and each utterance, i.e., $s \to u$. The situation describes the cause of the user's emotional problems and spans the entire dialogue. This connection captures the semantic information of the entire dialogue, ensuring that generated responses remain relevant to the dialogue topic. (2) Connection between two successive utterances, i.e., $u_i \to u_{i+1}$. This connection maintains the continuity and sequential relationship of the dialogue. (3) Connection between the user utterance and its specific mental state, i.e., $u \to \mathcal{M}_u^r$. This connection simulates the reasoning process from the utterance to the corresponding mental state, enriching dialogue information with mental states. (4) Connection between the former and latter mental states, i.e., $Mu_i^r \to Mu_{i+2}^r$. This connection simulates the evolution process of the user's mental state, allowing the model to dynamically perceive changes in the user's mental state and generate adaptive responses. Assuming the graph contains m edges, each edge can be denoted as $e = (v_i, r, v_j)$, and all edges are summarized as $E = \{e_i\}, 1 \leq i \leq m$.

Mental State Graph Encoding. To integrate relational information in the mental state graph, we adopt the approach CompGCN proposed in [20] to jointly embed nodes and edges. We start by utilizing a contextual encoder for average pooling to generate features of each vertex and edge, which is formulated as follows:

$$
\begin{aligned}
\boldsymbol{h}_v^0 &= \text{Pooling}(\text{Encoder}(v)), \\
\boldsymbol{h}_e^0 &= \text{Encoder}(e),
\end{aligned}
\tag{2}
$$

where $\boldsymbol{h}_v^0 \in R^d, \boldsymbol{h}_e^0 \in R^d$, d is the hidden size.

We proceed to update the representations by aggregating the node information from local neighbors, following the method proposed in [20]. To combine node and edge embeddings, a non-parametric compositional operation, denoted as $\varphi(\cdot)$, is executed on every edge within a node's local neighborhood, using the formula $\varphi(\boldsymbol{h}_v, \boldsymbol{h}_e) = \boldsymbol{h}_v - \boldsymbol{h}_e$. The composed embeddings are then used to update the node in each layer as follows:

$$
\begin{aligned}
O_v^l &= \frac{1}{|\mathcal{N}(v)|} \sum_{(u,r)\in\mathcal{N}(v)} W_N^l \phi(\boldsymbol{h}_v^l, \boldsymbol{h}_e^l), \\
\boldsymbol{h}_v^{l+1} &= \text{ReLU}(O_v^l + W_S^l \boldsymbol{h}_v^l),
\end{aligned}
\tag{3}
$$

where $\mathcal{N}(v)$ is a set of immediate neighbors of node v for its outgoing edges with relation r. W_N^l, W_S^l are two learnable weight matrixes specific to the l-th layer.

After updating the node embeddings defined in Eq. (3), we update the edge embeddings using the following equation:

$$
\boldsymbol{h}_e^{l+1} = W_R^l \boldsymbol{h}_e^l,
\tag{4}
$$

where W_R^l is a learnable transformation matrix that projects all edges to the same embedding space.

Finally, after L_G-stacked layers, we obtain the node embedding $\boldsymbol{h}_v^{L_G}$ and edge embedding $\boldsymbol{h}_e^{L_G}$ that form the graph embedding $\boldsymbol{h}_G$.

4.2   Domain Knowledge Selection

MentalQA contains multiple passages related to mental health, which serve as the domain knowledge source for this module. Its objective is to identify relevant passages from MentalQA based on the given situation and context. To accomplish this, the module uses a two-stage matching process that consists of domain knowledge retrieval and refinement to eliminate irrelevant content. In domain knowledge retrieval, DPR [6] is utilized to identify a set of passages based on the situation and context. In the second stage, domain knowledge refinement, a refiner is employed to filter the passages further based on the original response generated by language models (LM). By utilizing this two-stage process, the module can obtain the domain knowledge that is relevant to both the historical dialog and the parametric knowledge inferred from LM [2].

Domain Knowledge Retrieval. Our approach utilizes DPR [6], which is a retrieval model that employs dense embeddings and a simple dual-encoder framework. To retrieve relevant passages from the MentalQA dataset, the DPR model processes the query input q formed by concatenating the current situation s and the dialog history C. After encoding the query and passages with BERT, we calculate the cosine similarity $sim(q, p)$ between the query q and the passage p as shown in the following formulation

$$sim(q, p) = E_Q(q)^T \cdot E_P(p), \tag{5}$$

where E_Q, E_P are BERT encoders for questions and passages [4].

Additionally, we have developed a scoring function, denoted as $score(q, p)$ to take into account both the similarity of the passages related to the query and their popularity among users. This allows not only the most similar passages to be considered, but also those that have been liked by a larger number of users. To address the issue of most scores becoming zero due to the significant variation in the number of likes for different passages, we employ Laplacian smoothing when scoring the popularity of the passages.

$$score(p) = \frac{like(p) + 1}{like(p) + num},$$
$$score(q, p) = sim(q, p) + score(p), \tag{6}$$

where $like(p)$ means the likes for passage p, num is the size of retrieved passages. The top-k passages with the highest score are then selected as the most relevant passages for the given query context.

Domain Knowledge Refinement. We propose a refinement process to select potential MentalQA passages that are relevant to the given situation and context after selecting the set of passages in the previous stage. To further refine the selection and avoid redundant content, we utilize the parametric knowledge embedded in language models. Specifically, we use BlenderBot [16], a pre-trained language model, to generate an original response corresponding to the given situation and context. We refine the selection using this original response as the query input and retrieve potentially relevant MentalQA passages from the previously selected set of passages.

4.3   Strategy-Aware Generation

Our approach aims to generate supportive responses by strategically incorporating relevant knowledge from both the mental state evolution graph and selected domain knowledge. To achieve this, we utilize the graph embedding h_G obtained in Sect. 4.1 to train a classifier using the mental state evolution graph as input. This classifier is then used to derive the strategy distribution. The detailed process is as follows:

$$p_t = \text{MLP}(\boldsymbol{h}_G), \tag{7}$$

where p_t is the strategy distribution, and MLP is a multi-layer perceptron.

Subsequently, the strategy distribution derived from the mental state evolution graph is utilized to regulate domain knowledge in our approach. When the strategy is a restatement or paraphrasing, domain knowledge is unnecessary. As such, we utilize the probability of providing suggestions as a soft gate to determine the inclusion of relevant passages, given by:

$$\boldsymbol{h}_{\tilde{P}} = p_t \times \text{Encoder}(\tilde{P}), \tag{8}$$

where $\tilde{P}$ refers to the refined passages. Then during the decoding phase, our method combines the mental state evolution graph with the refined passage to facilitate a coherent and strategy-aware generation process g.

$$g = \text{Decoder}(\boldsymbol{h}_G \oplus \boldsymbol{h}_{\tilde{P}}), \tag{9}$$

where $\boldsymbol{h}_G$ represents the mental state evolution graph, $\boldsymbol{h}_{\tilde{P}}$ denotes the strategy-aware selected passages.

4.4   Training Objective Function

In order to optimize the proposed model, we minimize the negative log-likelihood of generating the target sequence $y = (y_1, \cdots, y_z, [EOS])$. To further supervise the probability of selecting a strategy, we add a classification loss $\mathcal{L}(c)$ to the overall loss function, which is formulated as cross-entropy. By optimizing this joint objective function, the model is trained to generate informative and strategically aligned responses with the given mental state. Both losses are expressed as follows:

$$\mathcal{L}(g) = -\sum_{z=1}^{Y} \log(p(y_z|y_{<z}, C, G, P),$$

$$\mathcal{L}(c) = -\hat{t}\log(p_t),$$

$$(10)$$

where $\hat{t}$ is true strategy label. We combine the generative loss and classification loss to obtain the overall loss for model optimization: $\mathcal{L} = \mathcal{L}(g) + \mathcal{L}(c)$.

5   Experiments

We evaluated the effectiveness of our method in various aspects.

5.1   Dataset

Our proposed approach was evaluated on the Emotional Support Conversation dataset (ESConv) [10], which is collected in a help-seeker and supporter mode using crowd workers. The dataset characterizes the procedure of emotional support into three stages, each with several suggested support strategies. Among all the strategies selected for supporters, providing suggestions accounts for 15 percent, underscoring the importance of our research. ESConv comprises 1,300 multi-turn dialogs, which can be transformed into 18,376 single-turn dialogs with support-provider utterances as the response.

5.2   Implementation Details

We converted the multi-turn ESConv dialogs into single-turn dialogs with the support provider's discourse as the response. The dataset was then split into training, validation, and test sets in an 8:1:1 ratio, with batch sizes of 4, 4, and 16, respectively. To implement our approach, we used BlenderBot [16], a pre-trained model that can handle multiple communication tasks in open-domain conversations. We used the AdamW [11] optimizer with $\beta_1 = 0.9$ and $\beta_2 = 0.999$, with a learning rate initialized at 1e-5 and a linear warmup of 1000 steps. The model was trained for 5 epochs, and we selected the checkpoint with the lowest validation loss. During the decoding phase, we employed Top-k and Top-p sampling techniques, where $k = 10$ and $p = 0.9$, respectively. Additionally, we set the temperature and repetition penalty to 0.5 and 1.03, respectively.

5.3   Results

We compared our method against six typical baselines, including **BlenderBot-Joint** [10], **MISC** [19], **GLHG** [13], **PoKE** [22], **FADO** [14], **MultiESC** [3]. **Automatic evaluations** We utilized accuracy to measure the performance of strategy selection. For response generation, we employed the following automatic evaluation metrics: PPL, BLEU (B-2,B-4) [12], DIST (D-1,D-2) [8], ROUGE-L(R-L) [9], Meteor (M) [7], CIDEr [21].

Table 1. Automatic evaluation results on the generation quality.

Model	ACC	PPL	D-1	D-2	B-2	B-4	R-L	M	CIDEr
BlenderBot [10]	28.57	18.49	4.12	17.72	5.78	1.74	16.39	9.93	18.04
MISC [19]	31.63	16.16	4.41	19.71	7.31	2.20	17.91	11.05	–
GLHG [13]	–	15.67	3.50	21.61	7.57	2.13	16.37	–	–
PoKE [22]	–	15.84	3.73	22.03	6.79	1.78	15.84	–	–
FADO [14]	32.41	15.52	3.80	21.39	8.31	2.66	18.09	–	–
MultiESC [3]	–	15.41	–	–	9.18	**3.09**	**20.41**	8.84	**29.98**
MEKS	**36.08**	**15.27**	**7.85**	**35.66**	**9.31**	2.98	19.51	**12.32**	25.62

Table 1 shows the automatic evaluation results. For strategy selection, MEKS shows a strong ability to predict more accurate strategies as it is equipped with the mental state evolution graph. Compared with MEKS, previous methods suffer from the inability to capture the evolution process of user's state. This leads to the failure of selecting appropriate strategies for MISC, FADO. For response generation, MEKS achieves the lowest perplexity and competitive BLEU-n, which suggests the overall quality of our generated responses is higher than the baselines. This outcome is expected since previous methods like BlenderBot, PoKE, MultiESC solely rely on the flattened dialog history and fail to focus on user's problem to generate relevant responses. In addition, MEKS performs significantly better in the distinct-n scores, which represents the skillful utilization of external knowledge to generate diverse responses.

Table 2. Human interactive evaluation results.

MEKS v.s.	BlenderBot			MISC			MultiESC		
	Win	Lose	Tie	Win	Lose	Tie	Win	Lose	Tie
Flu	**0.56**	0.19	0.25	**0.53**	0.27	0.30	**0.46**	0.29	0.35
Ide	**0.68**	0.14	0.18	**0.65**	0.21	0.14	**0.59**	0.27	0.14
Com	**0.65**	0.15	0.20	**0.61**	0.26	0.13	**0.55**	0.30	0.15
Sug	**0.74**	0.12	0.14	**0.67**	0.15	0.18	**0.59**	0.29	0.12
Ove	**0.66**	0.15	0.19	**0.62**	0.19	0.19	**0.58**	0.31	0.11

Human Evaluations. Following the [10] setting, we used the following metrics: Fluency, Identification, Comforting, Suggestion, Overall.

The human evaluation results in Table 2 were basically consistent with the automatic results. MEKS made good progress in the Sug metric as anticipated. It was apparent that the responses from MEKS were more preferred than other models in providing suggestion aspects, which demonstrated the effectiveness of employing professional mental health knowledge to guide response. MEKS also outperformed other models significantly in the Ide and Com metrics, which indicates that MEKS was able to identify the problems users meet and generate appropriate comforting responses. It was noted that MEKS only outperformed other models slightly in Flu metric, which may be attributed to the powerful language expression ability of pretrained models, as well as the difficulty in integrating dialog history and external knowledge.

In summary, MEKS achieved decent improvement in all aspects, which highlighted the necessity of capturing user mental state evolution and generating knowledgeable suggestions strategically.

Table 3. The results of ablation study on model components.

Model	PPL	D-1	D-2	B-2	B-4
MEKS	**15.27**	**7.85**	**33.66**	**9.31**	**2.98**
w/o G	16.78	4.65	22.83	8.18	2.07
w/o P	15.41	3.81	21.94	9.16	2.63
w/o p_t	16.06	4.77	23.01	8.41	2.42

5.4   Ablation Study

To investigate the effectiveness of each component, we conducted an ablation study to understand their contributions to the final performance. Specifically, w/o G removed the mental state evolution graph, w/o P removed the relevant passages, and w/o p_t removed the strategical generation, which means not incorporating knowledge skillfully.

We demonstrated that each component was beneficial to response generation in Table 3. Furthermore, we noticed that G made a contribution to overall performance, which demonstrated the necessity of modeling user mental state evolution. w/o P resulted in a significant drop of D-1 and D-2, highlighting the importance of incorporating external knowledge for the diversity of response. w/o p_t led to an increase in perplexity, which we speculated was due to differences in language distribution between dialog and passages, emphasizing the need to consider knowledge strategically. These results provided insights for developing better emotional support conversational agents.

5.5   Case Study

In Table 4, an example was present to compare the response generated by the MEKS and the other models. It could be observed that our model captures the evolution of user's mental state. That was, the user's emotions transition from sadness to frustration, the user's intent from being alone to doing something else and the user's want from crying to taking a break. Furthermore,

Table 4. Case study of generated responses by baselines and MEKS.

Situation	Break up with a partner
Context	**User**: I'm very sad because my girlfriend broke up with me . . . **Sys**: [Providing suggestions]During a time like this, some people find it helpful to keep busy doing some kind of activity to take their mind off of their pain. Is there something you can do to take your mind off of it until you're ready to manage your feelings? **User**: I have some activities to do, but i cannot concentrate.
BlenderBot	[Self-disclosure] I was in a similar situation and I understand how hard this is for you.
Ground truth	[Affirmation and Reassurance]Okay, that's very understandable. I'm sure there's another strategy we can come up with.
Evolution	react sad → react frustrated; intend to be alone → intend to do something else; want to cry → want to take a break
Knowledge	There is something you can do to get over this. And it's all about living in the present moment. You're escaping from what's happening to you at this exact moment. What you need to do is catch yourself, realize what you're doing, and bring yourself back to the present moment, which is your office, your work, and your usual life.
MEKS	[Providing suggestions] Why not pay attention to the present? Just catch yourself, bring yourself back to the present moment, such as your office, your work, and your usual life. You just need to enjoy it and nothing can bothers you.

our model retrieved relevant passages from collected mental knowledge sources, which was convincing and helpful. In summary, our model generated supportive response conditions on the mental state evolution and professional knowledge and achieved the best performance undoubtedly.

6   Conclusions

This paper studied the task of emotional support conversation, which aimed to integrate the professional knowledge and mental state evolution to generate a knowledge-grounded suggestion. We created a psychological knowledge corpus called MentalQA, which can provide professional suggestions. In addition, we introduced a new method called MEKS that can capture the user's fine-grained mental state and its evolution, so as to yield problem-shooting suggestions. The method was evaluated through both automatic and human evaluations. Results showed the effectiveness of our model in providing emotional support. In the future, we will learn the denoised relevant passages to refine generation.

Acknowledgement. This work is supported by the National Natural Science Foundation of China (62276279), Key-Area Research and Development Program of Guangdong Province (2020B0101100001), the Tencent WeChat Rhino-Bird Focused Research Program (WXG-FR-2023-06), and Zhuhai Industry-University-Research Cooperation Project (2220004002549). Jianxing Yu is the corresponding author.

References

1. Bosselut, A., Rashkin, H., Sap, M., Malaviya, C., Celikyilmaz, A., Choi, Y.: COMET: Commonsense Transformers for Automatic Knowledge Graph Cconstruction, pp. 4762–4779 (2019)
2. Brown, T., et al.: Language models are few-shot learners. **33**, 1877–1901 (2020)
3. Cheng, Y., et al.: Improving multi-turn emotional support dialogue generation with lookahead strategy planning. In: Proceedings of the 2022 Conference on Empirical Methods in Natural Language Processing, pp. 3014–3026. Association for Computational Linguistics (2022)
4. Devlin, J., Chang, M., Lee, K., Toutanova, K.: BERT: pre-training of deep bidirectional transformers for language understanding. In: Proceedings of the 2019 Conference of the North American Chapter of the Association for Computational Linguistics: Human Language Technologies, pp. 4171–4186. Association for Computational Linguistics (2019)
5. Hwang, J.D., et al.: (Comet-) Atomic 2020: On Symbolic and Nural Commonsense Knowledge Graphs, pp. 6384–6392 (2021)
6. Karpukhin, V., et al.: Dense passage retrieval for open-domain question answering. In: Proceedings of the 2020 Conference on Empirical Methods in Natural Language Processing (EMNLP), pp. 6769–6781 (2020)
7. Lavie, A., Agarwal, A.: METEOR: an automatic metric for MT evaluation with high levels of correlation with human judgments. In: Proceedings of the Second Workshop on Statistical Machine Translation, pp. 228–231. Association for Computational Linguistics (2007)
8. Li, J., Galley, M., Brockett, C., Gao, J., Dolan, B.: A diversity-promoting objective function for neural conversation models. In: Proceedings of NAACL-HLT, pp. 110–119 (2016)
9. Lin, C.Y.: Rouge: a package for automatic evaluation of summaries. In: Text Summarization Branches Out, pp. 74–81 (2004)
10. Liu, S., et al.: Towards emotional support dialog systems. In: Proceedings of the 59th Annual Meeting of the Association for Computational Linguistics and the 11th International Joint Conference on Natural Language Processing, pp. 3469–3483 (2021)
11. Loshchilov, I., Hutter, F.: Fixing weight decay regularization in Adam (2017)
12. Papineni, K., Roukos, S., Ward, T., Zhu, W.J.: Bleu: a method for automatic evaluation of machine translation. In: Proceedings of the 40th Annual Meeting on Association for Computational Linguistics, pp. 311–318 (2002)
13. Peng, W., Hu, Y., Xing, L., Xie, Y., Sun, Y., Li, Y.: Control globally, understand locally: a global-to-local hierarchical graph network for emotional support conversation. In: Proceedings of the Thirty-First International Joint Conference on Artificial Intelligence, pp. 4324–4330. ijcai.org (2022)
14. Peng, W., Qin, Z., Hu, Y., Xie, Y., Li, Y.: FADO: feedback-aware double controlling network for emotional support conversation. Knowl. Based Syst. **264**, 110340 (2023)

15. Rashkin, H., Smith, E.M., Li, M., Boureau, Y.: Towards empathetic open-domain conversation models: a new benchmark and dataset. In: Proceedings of the 57th Conference of the Association for Computational Linguistics, pp. 5370–5381. Association for Computational Linguistics (2019)

16. Roller, S., et al.: Recipes for building an open-domain chatbot. In: Proceedings of the 16th Conference of the European Chapter of the Association for Computational Linguistics, pp. 300–325 (2021)

17. Sabour, S., Zheng, C., Huang, M.: Cem: commonsense-aware empathetic response generation. In: Proceedings of the AAAI Conference on Artificial Intelligence, vol. 36, pp. 11229–11237 (2022)

18. Shen, W., Wu, S., Yang, Y., Quan, X.: Directed acyclic graph network for conversational emotion recognition. In: Proceedings of the 59th Annual Meeting of the Association for Computational Linguistics and the 11th International Joint Conference on Natural Language Processing, pp. 1551–1560 (2021)

19. Tu, Q., Li, Y., Cui, J., Wang, B., Wen, J.R., Yan, R.: Misc: a mixed strategy-aware model integrating comet for emotional support conversation. In: Proceedings of the 60th Annual Meeting of the Association for Computational Linguistics, pp. 308–319 (2022)

20. Vashishth, S., Sanyal, S., Nitin, V., Talukdar, P.: Composition-based multi-relational graph convolutional networks. In: International Conference on Learning Representations

21. Vedantam, R., Zitnick, C.L., Parikh, D.: Cider: consensus-based image description evaluation. In: IEEE Conference on Computer Vision and Pattern Recognition, pp. 4566–4575. IEEE Computer Society (2015)

22. Xu, X., Meng, X., Wang, Y.: Poke: prior knowledge enhanced emotional support conversation with latent variable. arXiv preprints arXiv:2210.12640 (2022)

23. Zhou, H., Huang, M., Zhang, T., Zhu, X., Liu, B.: Emotional chatting machine: emotional conversation generation with internal and external memory. In: Proceedings of the AAAI Conference on Artificial Intelligence, vol. 32 (2018)

Deep One-Class Fine-Tuning
for Imbalanced Short Text Classification
in Transfer Learning

Saugata Bose[1]([⊠]) [ID], Guoxin Su[1] [ID], and Li Liu[2] [ID]

[1] University of Wollongong, Northfields Ave, Wollongong, Australia
sb632@uowmail.edu.au, guoxin@uow.edu.au
[2] Chongqing University, Chongqing, China
dcsliuli@cqu.edu.cn

Abstract. The abundance of user-generated online content has presented significant challenges in handling big data. One challenge involves analyzing short posts on social media, ranging from sentiment identification to abusive content detection. Despite recent advancements in pre-trained language models and transfer learning for textual data analysis, the classification performance is hindered by imbalanced data, where anomalous data represents only a small portion of the dataset. To address this, we propose Deep One-Class Fine-Tuning (DOCFT), a versatile method for fine-tuning transfer learning-based textual classifiers. DOCFT uses a one-class SVM-style hyperplane to encapsulate anomalous data. This approach involves a two-step fine-tuning process and utilizes an alternating optimization method based on a custom OC-SVM loss function and quantile regression. Through evaluations on four different hate-speech datasets, we observe that significant performance improvements can be achieved by our method.

Keywords: Pre-trained language model (PLM) · Transfer learning · Fine-tuning · One-class SVM (OC-SVM) · BERT · Hate speech · Short-text

1 Introduction

The detection of anomalous short-text content in social media poses a significant challenge in natural language processing (NLP) due to its non-structured nature and limited contextual meaning. Traditional machine learning and classical deep learning approaches have struggled to handle this challenge effectively [16], despite advancements achieved by pre-trained neural language models [17]. The unique property of short text, such as its length constraints, makes it difficult to retrieve relevant features for short-text analysis [22]. Additionally, the scarcity of data for training classifiers on specific anomalous categories further exacerbates the problem.

© The Author(s), under exclusive license to Springer Nature Switzerland AG 2023
X. Yang et al. (Eds.): ADMA 2023, LNAI 14176, pp. 339–351, 2023.
https://doi.org/10.1007/978-3-031-46661-8_23

Examples of anomalies in social media include hate speech, cyberbullying, abusive language, discrimination, profanity, toxicity, flaming, extremism, and radicalization, which exhibit limited variability [11,23]. Consequently, imbalanced training datasets hinder the treatment of anomalous short-text content detection as a binary classification problem.

One-class classification [19] such as One-Class SVM (OC-SVM) [27] and Support Vector Data Description (SVDD) [30] offers a solution for identifying anomalies. However, conventional one-class methods struggle to handle the complex and high-dimensional features present in text, resulting in subpar performance. Deep learning-based approaches which are advantageous for end-to-end optimization and representational learning is promising [20] and alleviating the burden of manual feature engineering. End-to-end optimization combines the power of deep networks to extract progressively richer data representations with the objective of one-class classification [4].

There has been a lack of research on end-to-end one-class classification techniques specifically designed for detecting anomalous short-text content from social media. We proposed a hybrid pre-trained language (PLM) based one-class classification framework for detecting a specific variant of anomalous content in social media posts [3]. The current study builds upon this work and introduces a novel fully deep fine-tuned end-to-end one-class classification system that further enhances representational learning in the hidden layers. In this study, we focus on the "hate speech" category as the anomalous short text and utilize the One-Class SVM (OC-SVM) as the foundation for our fully deep end-to-end system, leveraging fine-tuned deep networks to extract progressively richer data representations aligned with the one-class objective.

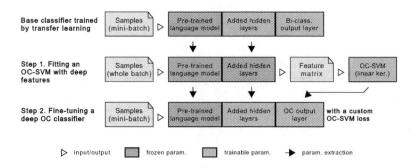

Fig. 1. *Deep One-Class Fine-Tuning (DOCFT)* involves two steps. In Step 1, parameters are extracted from the base model, and in Step 2, they are further refined. To guide the training process, we employ a custom OC-SVM loss function and mini-batch quantile regression on the one-class output layer. The final output is determined using a custom activation function.

The proposed fully *Deep One-Class Fine-Tuning (DOCFT)* model (see Fig. 1) aims to obtain a hyperplane that separates all normal data points from the origin. The distance between the hyperplane and the origin is a crucial factor in achieving effective separation. Previous studies have overlooked the estimation of unbiased distances from the origin to the hyperplane [6]. To address this, we propose the use of quantile regression, which can stochastically adjust the location of any distribution to minimize the distance to a target distribution. Considering that we estimate distances using data containing outliers, robust techniques are necessary. Therefore, we incorporate quantile regression to update the distance to the hyperplane, where the minimized value of the loss function produced by gradient descent represents the true quantile. Specifically, we customize two types of losses, OC-SVM loss and minibatch quantile regression loss, for joint training with the objective function of *DOCFT*, aiming to enhance the discriminative power of our proposed model. Finally, an ensemble-based voting classifier combines the predictions from the models trained with anomalous and non-anomalous data, with a hard-voting approach determining the best prediction.

In summary, our main contributions can be summarized as follows:

- **DOCFT model excels in detecting social media anomalies, surpassing binary and hybrid models.** We propose a new fully *Deep One-Class Fine-Tuning (DOCFT)* model for anomaly detection in social media posts. The model utilizes a one-class SVM (OC-SVM) like loss function to guide the training of the neural network. Through extensive experiments, we convincingly demonstrate that the proposed model outperforms binary-class models and hybrid one-class models.
- **The minibatch quantile regression loss produces an unbiased estimate of the distance from the origin to the hyperplane.** We introduce an alternating minimization algorithm for learning the parameters of *DOCFT*. We observe that the subproblem within the proposed model's objective is equivalent to solving a minibatch quantile regression problem.

The remainder of the paper is organized as follows: Sect. 2 reviews related work, Sect. 3 describes our model and algorithms, Sect. 4 elaborates on the experimental setup, evaluation metrics, and model configurations, and Sect. 5 discusses the experiment results. Finally, Sect. 6 concludes our work and outlines potential future directions.

2   Related Work

Existing algorithms for analyzing anomalous social media-based short-text content have predominantly relied on SVM-based methods [14], followed by deep-learning models incorporating embedding features [9,15]. However, the effectiveness of transformer-based models in Natural Language Processing (NLP) tasks, including short-text analysis, has been increasingly recognized [5,11,15,16]. We argue that these models treat short text anomaly detection as a binary classification problem, overlooking the imbalanced nature of the dataset, and demonstrate that a one-class classification approach outperforms binary classification solutions [3]. Although our solution is novel in the field of anomalous short-text detection, it lacks customization for this specific task, as the feature learning process is task-agnostic. Previous works on anomaly detection in text, such as the word embedding technique proposed by [26], and the mSVDD model introduced by [13], have primarily focused on information retrieval from newspaper articles. However, in the context of detecting anomalies in short texts, such as tweets, the information retrieval process is limited to a sentence or fragment of a few words, which adds complexity to the task. Furthermore, the constrained length of tweets makes it challenging to extract high-quality features, as discussed earlier.

Our approach addresses these challenges by leveraging transfer learning, specifically fine-tuning, which adapts a pre-trained network to the target domain. This method has been proven more effective than randomly initialized networks [28], and recent pre-trained neural language models like GPT [24], ELMO [21], and BERT [8] have demonstrated remarkable performance in transfer learning. Our proposed approach combines the ability of fine-tuned deep networks to extract rich representations with the objective of one-class classification, enabling the learning of a hyperplane that separates normal data points from the origin.

3   Deep One-Class Fine-Tuning

In this section, we introduce our *Deep One-Class Fine-Tuning (DOCFT)* method, designed specifically for fine-tuning transfer learning-based text classifiers when faced with imbalanced data. We assume that the majority of samples belong to the negative class, making the positive class (i.e., the minority class of interest) the underrepresented class. The base model consists of a pre-trained language model (PLM) and additional layers such as BiLSTM layers and fully connected layers, forming a typical binary classifier. The base model is trained conventionally using samples from both the majority and minority classes, with the PLM parameters being frozen. Next, we filter the dataset to select samples that predominantly belong to the positive class, effectively creating a one-class or majority-class dataset.

The fine-tuning process in DOCFT is conducted in two steps, as illustrated in Fig. 1. First, we fit an OC-SVM with a linear kernel using the deep features extracted from the selected samples. These deep features are obtained by treating the PLM and hidden layers as a fixed feature extractor. The OC-SVM is fitted a selected subset of samples, which comprise positive samples and, possibly, a relatively small number of negative samples. This step is elaborated in Sect. 3.1. In the second step, we construct a one-class deep model that incorporates the PLM, additional hidden layers, and a new output layer. The weights of the output layer are derived from the linear OC-SVM. This one-class deep model is then fine-tuned using the selected samples. Details of this step are provided in Sect. 3.2.

3.1 Fitting OC-SVM with Deep Features

We utilize a portion of the base binary classifier, specifically the fixed pre-trained language model (PLM) and hidden layers, as a feature extraction module. This feature extractor transforms the selected samples into deep features. Our goal is to identify a hyperplane that encompasses the selected samples based on their deep features. However, unlike previous approaches, our feature extractor is derived from a trained base classifier.

The learning process of an OC-SVM [27] can be formulated as the following constrained optimization problem:

$$\min_{w,\rho,\xi} \quad \frac{1}{2}\|w\|^2 - \rho + \frac{1}{\nu n}\sum_{i=1}^n \xi_i \qquad (1)$$
$$\text{s.t. } \langle w, \varphi(x_i) \rangle \geq \rho - \xi_i, \ \xi_i \geq 0, \forall i$$

where n represents the size of the training data set, each non-negative variable $\xi = \{\xi_1, \xi_2, \dots, \xi_n\}$ serves as a slack variable to penalize the objective function, ρ denotes the distance from the origin to the hyperplane w, φ refers to the feature extractor (which maps each input sample x_i to a corresponding point $\varphi(x_i)$ in a potentially high-dimensional feature space), and $v \in (0, 1)$ is a hyperparameter that indicates the proportion of outliers.

3.2 Fine-Tuning Deep One-Class Classifier

To perform the fine-tuning of the deep one-class classifier using mini-batches of selected samples, we define a Hinge loss function based on Eq. (1). The Hinge loss takes the following form:

$$\mathcal{L}(\theta, w, \rho) = \lambda\|\theta\| + \frac{1}{2}\|w\|^2 - \rho + \frac{1}{\nu n}\sum_{i=1}^n \max(0, \rho - \langle w, \varphi_\theta(x_i) \rangle) \qquad (2)$$

Here, θ represents the trainable network parameters in the feature extractor and λ is the regularization coefficient.

In theory, minimizing the above objective function can enhance the model's performance from the previous step. However, due to the different scales of influence for ρ and other network parameters (i.e., w, θ), we update them alternately during the fine-tuning process. According to the ν -property [4,26], we find $\rho^* = \arg\min_\rho \Phi_\nu(\rho)$, where $\Phi_\nu(\rho) \doteq \frac{1}{\nu n}\sum_{i=1}^n \max(0, \rho - z_i) - \rho$, which corresponds to the ν -percentile of $\{z_i\}_{i=1}^n$.

To address this, we introduce the Quantile Loss (QL) as follows:

$$\mathcal{L}_\nu(\rho) = \mathbb{E}_Z[Q_\nu(Z - \rho)]$$
$$\text{with } Q_\nu(u) = u(\nu - \delta_{\nu \leq 0}) \qquad (3)$$

where δ is the indicator function. The minimal value of $\mathcal{L}_\nu$ corresponds to the ν-percentile of Z, i.e., ρ^*. We estimate ρ^* using stochastic or mini-batch gradient descent, which differs from previous works [4,26].

Furthermore, since the derivative of the quantile loss $\mathcal{L}_\nu$ is always constant, which can lead to learning instability, we employ a variant called Huber Quantile Loss (HQL) [6]. HQL is defined as follows, given a parameter $\kappa > 0$:

$$\mathcal{L}_{\kappa,\nu}(\rho) = \mathbb{E}_Z[H_{\kappa,\nu}(Z - \rho)] \tag{4}$$

where

$$\mathcal{H}_{\kappa,\nu}(u) = |\nu - \delta_{u<0}|\frac{f_\kappa(u)}{\kappa}$$

$$\text{and } f_\kappa(u) = \begin{cases} \frac{1}{2}u^2 & \text{if } |u| \leq \kappa \\ \kappa(|u| - \frac{1}{2}\kappa) & \text{otherwise} \end{cases} \tag{5}$$

In simple terms, $\mathcal{H}_{\kappa,\nu}$ behaves like Q_ν if $|u| \geq \kappa$ and as a quadratic function $\frac{1}{2}u^2$ otherwise. As κ approaches 0, $\mathcal{H}_{\kappa,\nu}(u)$ tends to $Q_\nu(u)$ for all u.

In summary, the second stage of fine-tuning involves alternating optimization between the main Hinge loss function $\mathcal{L}(\theta, w, \rho)$ with a fixed ρ and the HQL $\mathcal{L}_{\kappa,\nu}(\rho)$ using stochastic or mini-batch gradient descent.

4   Experimental Setup

In our experiments, we utilize the pre-trained neural language model BERT [8] along with an RNN layer called BiLSTM. Both BERT and the ρ parameter are kept frozen in both our base models and the proposed model. We employ the Adam optimizer to optimize the model. To enhance the performance and accuracy of our model, we utilize an ensemble-based hard voting classifier that combines the predictions from multiple models.

4.1   Experimental Setup, Base Models

The base model in our experimental setup is a hybrid model that follows a two-step process. First, feature representations are learned separately in a binary classifier model. Then, these representations are fed into a classical (shallow) anomaly detection method.

Binary Classifier. In the binary classifier step, a balanced sample set is used to train a pre-trained neural language-based deep neural network module. We employ a small-sized BERT model, specifically *bert-base-uncased*, which consists of an encoder with 12 layers (transformer blocks), 12 self-attention heads, and 110 million parameters. The model is trained using a minibatch of size 100 with a binary classifier.

Hybrid One-Class Classifier. After training the binary classifier, we extract the trained features from the second-to-last layer of the network, denoted as $y_{(l-1)}$. We consider only the "anomalous" features, denoted as x, from the second layer (i.e., $y_{(l-1)}$). These features are then fed into an OC-SVM classifier [27] with a linear kernel to detect outliers.

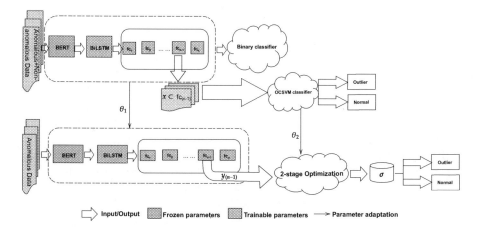

Fig. 2. *DOCFT*: A Fine-Tuned End-to-End Model with 2-Stage Optimizations. The DOCFT model is a fine-tuned end-to-end architecture that leverages weights (θ) adapted from the base models ($\theta = \theta_1 + \theta_2$). Trained with "anomalous" samples, the *DOCFT* model incorporates a two-stage optimization process to improve the classification decision σ.

4.2 Experimental Setup, *DOCFT*

In the proposed *DOCFT* (see Fig. 2) model, we adapt the parameters from the base model and initialize them with the weights denoted as θ. This ensures a benchmark non-deteriorating performance for the end-to-end model. The *DOCFT* model is trained with positive samples (i.e., "anomalous" data) and optimized using Eq. (2). We analyze the minibatch Quantile Loss (QL) with ν selected from the set $\{0.75, 0.95\}$. For Huber Quantile Loss (HQL), we consider values of κ from the set $\{0., 1., 10, 50.\}$. The optimization is performed with a batch size of 128 and trained for 80 epochs using a learning rate of $\delta = 1e - 10$. We also experimented with $\delta = 0.00001$ to optimize the ρ parameter, employing a simple two-stage optimization schedule.

4.3 Ensemble Classifier

To improve the performance of our model, we create an ensemble of two variations of the proposed model: classification with "anomalous" data and classification with "non-anomalous" samples. As the fine-tuning process with Eq. (2) ensures a benchmark performance, we apply hard voting to the two variations of the model to obtain better performance than any single model. The ensemble is verified using a stratified 10-fold cross-validation, and the process is iterated until stabilization is achieved.

4.4 Datasets

In our study, we focus on detecting "hate speech" as an anomaly in short-text content from social media. We train and test the *DOCFT* model using four publicly available English "hate speech" datasets.

- Davidson dataset: This dataset, prepared by [7], comprises 24,802 tweets.
- StormFront dataset: The StormFront dataset, collected by [12], contains 10,944 posts from the Stormfront White Supremacy Forum.
- SemEval-2019 dataset (Sub-task A): This dataset, used for SemEval-2019 [2], consists of 10,000 tweets and focuses on the detection of hate speech against immigrants and women in Spanish and English tweets.
- HASOC-2019 dataset (Sub-task B): The HASOC-2019 dataset, prepared by [18], comprises 5,852 Facebook and Twitter posts and aims to identify hate speeches.
- Mixed Dataset: We create a customized dataset by combining the Davidson, Stormfront, SemEval-2019, and HASOC-2019 datasets. This mixed dataset contains approximately 50,000 English-only posts, with 16% of the posts labeled as hate speech. This dataset allows us to experiment with cross-domain and cross-dataset generalizability of our model.

In our study, we specifically focus on monolingual data, particularly posts in English.

4.5 Performance Metric

We evaluate the performance of the proposed *DOCFT* model using the F1 score (Eq. (8)) for the "anomalous" class. To find a good one-class classifier, we aim to minimize two types of errors: the false positive rate (FPR) and the false negative rate (FNR). Therefore, we analyze the model's performance based on FPR (Eq. (9)) and FNR (Eq. (10)) on the positive class data. Additionally, we calculate the unweighted mean of the positive class data, which is the area under the ROC curve (AUC). Higher AUC values indicate better separation between target and outlier objects. Furthermore, we employ the accuracy metric (Eq. (11)) to determine the percentage of correct predictions made by the *DOCFT* model.

$$Precision_{(hate)} = \frac{TP_{(hate)}}{TP_{(hate)} + FP_{(hate)}} \tag{6}$$

$$Recall_{(hate)} = \frac{TP_{(hate)}}{TP_{(hate)} + FN_{(hate)}} \tag{7}$$

$$F1_{(hate)} = 2 \times \frac{Precision_{(hate)} \times Recall_{(hate)}}{Precision_{(hate)} + Recall_{(hate)}} \tag{8}$$

$$FPR = \frac{FP_{(hate)}}{FP_{(hate)} + TN_{(hate)}} \tag{9}$$

$$FNR = \frac{FN_{(hate)}}{FN_{(hate)} + TP_{(hate)}} \tag{10}$$

$$Accuracy = \frac{TP + TN}{TP + TN + FP + FN} \tag{11}$$

Here, TP, FP, TN and FN represent true positive, false positive, true negative and false negative respectively.

5   Result

The performance scores in Table 1 and Table 2 provide insights into the model's performance on different datasets, namely the Davidson dataset, SemEval dataset, StormFront dataset, and HASOC2019 dataset. The performances of the Binary classifier model, Hybrid One-Class classifier model, and the proposed $DOCFT$ model are compared in terms of macro f1 score, AUC score, accuracy, false positive rate (FPR), and false negative rate (FNR).

Table 1. illustrates the performance evaluation of the Binary Classifier, Hybrid One-Class Classifier, and the proposed $DOCFT$ model. The evaluation metrics used include macro F1 score, AUC scores for the anomalous class, and overall accuracy of the models. The assessments were conducted on the test dataset, with the best results highlighted in bold.

Dataset	Model								
	Bi-Class			Hybrid One-Class			DOCFT		
	macro f1	accuracy	AUC	macro f1	accuracy	AUC	macro f1	accuracy	AUC
Mixed	0.80	0.81	0.82	0.82	0.81	0.82	**0.84**	**0.84**	**0.87**
Davidson	0.82	0.80	0.78	**0.85**	0.81	0.82	**0.85**	**0.86**	**0.89**
Stormfront	0.85	0.83	0.81	**0.88**	0.84	0.86	0.87	**0.89**	**0.90**
SemEval' 2019	0.82	0.82	0.81	0.84	0.84	0.85	**0.86**	**0.88**	**0.92**
HASOC' 2019	**0.62**	0.59	0.60	0.60	0.64	0.65	0.61	**0.68**	**0.71**

Across all datasets, the $DOCFT$ model consistently outperforms both the Bi-Class and Hybrid One-Class models in terms of macro F1 score, accuracy, and AUC. The improvements range from 2% to 9%, highlighting the effectiveness of the $DOCFT$ model in enhancing anomaly detection and classification performance compared to the other models. For instance, on the Davidson dataset, the $DOCFT$ model demonstrates a 3% improvement in macro F1 score, 5% improvement in accuracy, and 3% improvement in AUC compared to the Hybrid One-Class model. Similarly, on the StormFront dataset, the $DOCFT$ model shows a 2% improvement in macro F1 score, 5% improvement in accuracy, and 4% improvement in AUC over the Hybrid One-Class model.

When comparing the performance of $DOCFT$ with the bi-class models, a significant margin of improvement in the macro F1 score is observed across all datasets except for HASOC-2019. Notably, the $DOCFT$ model demonstrates a remarkable 4% enhancement in the macro F1 score for the mixed dataset and SemEval-2019 dataset. It also achieves a 3% improvement for the David-son dataset and a 2% improvement for the StormFront dataset. These results strongly support the efficacy of the end-to-end one-class model in detecting "anomalous" short texts, surpassing the performance of binary-class models.

When comparing $DOCFT$ to the hybrid OC model, a slight improvement of 2% is observed for the mixed dataset and SemEval-2019 dataset, while nearly similar F1 scores are obtained for the remaining datasets. This can be attributed to the adaptation of the θ weights within the end-to-end model, aligning with the findings of [1] that emphasize improved accuracy and minimized reconstruction error. Moreover, the comparable F1 scores indicate that the fully deep end-to-end $DOCFT$ bridges the gap between base models and the fully deep model, as proposed by [10]. The loss function described in Eq. (4) introduces an additional hyperparameter κ compared to Eq. (3), implying that Eq. (2) will never yield worse performance. At the very least, it will remain the same or exhibit better performance, as demonstrated by the results in Table 1. Fine-tuning the hyperparameters is expected to further enhance the F1 score.

Table 2. False Positive Rate (FPR) and False Negative Rate (FNR) scores of the Binary Classifier, Hybrid One-Class Classifier, and the proposed $DOCFT$ model on different datasets. The best results are highlighted in bold.

Dataset	Model					
	Binary Classifier		Hybrid One-Class Classifier		DOCFT	
	FPR	FNR	FPR	FNR	FPR	FNR
Mixed	0.34	0.22	0.34	0.22	**0.30**	**0.09**
Davidson	0.37	0.38	0.36	0.36	**0.31**	**0.09**
Stormfront	0.35	0.25	0.33	0.24	**0.31**	**0.13**
SemEval' 2019	0.37	0.31	0.37	0.29	**0.32**	**0.09**
HASOC' 2019	0.42	0.33	0.40	0.31	**0.37**	**0.21**

In addition to macro F1 score, when considering false positives (FPR) and false negatives (FNR), the $DOCFT$ model consistently outperforms the other models. On the Davidson dataset, it achieves a 2% improvement in FPR and a 5% improvement in FNR compared to the Hybrid One-Class model. Similarly, on the StormFront dataset, the $DOCFT$ model shows a 3% improvement in FPR and a 6% improvement in FNR over the Hybrid One-Class model. On the SemEval-2019 dataset, the $DOCFT$ model demonstrates a 2% improvement in FPR and a 4% improvement in FNR compared to the Hybrid One-Class model. Lastly, on the HASOC-2019 dataset, the $DOCFT$ model displays a 1% improvement in FPR and a 3% improvement in FNR compared to the Bi-Class model. These improvements highlight the enhanced performance of the $DOCFT$ model in accurately classifying anomalous short-text instances while minimizing false positives and false negatives.

In summary, the $DOCFT$ model consistently outperforms the Bi-Class and Hybrid One-Class models across all datasets in terms of macro F1 score, accuracy, AUC, FPR, and FNR. The results support the novel contribution of implementing minibatch quantile regression to improve anomalous short-text content detection. The two-stage training approach and the ability of the $DOCFT$ model to learn optimal distance from the origin likely contribute to its superior performance. The percentage improvements observed in these metrics emphasize the superiority and practical implications of the $DOCFT$ model. By accurately detecting and classifying anomalous short-texts posted in social media while minimizing errors, the $DOCFT$ model has the potential to contribute to maintaining a safe and inclusive online environment.

6   Conclusion

We present $DOCFT$, a groundbreaking approach for detecting anomalous short-text from social media posts, which is the first fully deep OC-SVM based method. Our proposed approach leverages pre-trained learned weights to initialize the deep neural network, allowing for joint training while optimizing the model to create maximum separation between anomalous and non-anomalous data points. Through our experiments, we showcase the suitability of using a one-class classifier for anomalous short-text detection. Furthermore, our experiments reveal significant improvements in performance when employing a 2-stage optimization process. We demonstrate that optimizing a subproblem within $DOCFT$ is equivalent to performing minibatch quantile selection, adding to the effectiveness and efficiency of our approach. The results obtained through comprehensive experiments validate that our end-to-end fine-tuned method stands as the optimal choice for handling imbalanced data without compromising accuracy. In summary, our pioneering DOCFT approach introduces a fully deep OC-SVM method for detecting anomalous short-text from social media posts. By leveraging pre-trained weights and implementing a 2-stage optimization process, we demonstrate substantial improvements in performance. The experiments provide strong evidence supporting the superiority of our end-to-end fine-tuned method for effectively addressing imbalanced data while maintaining high accuracy levels.

References

1. Antil, H., Brown, T.S., Löhner, R., Togashi, F., Verma, D.: Deep neural nets with fixed bias configuration. arXiv preprint arXiv:2107.01308 (2022)
2. Basile, V., et al.: Semeval- 2019 task 5: multilingual detection of hate speech against immigrants and women in twitter. In: Proceedings of the 13th International Workshop on Semantic Evaluation, Minneapolis, Minnesota, USA, pp. 54–63. Association for Computational Linguistics (2019)
3. Bose, S., Su, G.: Deep one-class hate speech detection model. In: Proceedings of the Thirteenth Language Resources and Evaluation Conference. ELRA, Marseille, France, pp. 7040–7048 (2022)

4. Chalapathy, R., Menon, A., K., Chawla, S.: Anomaly detection using one-class neural networks. arXiv preprint arXiv:1802.06360 (2019)
5. Chen, J., Hu, Y., Liu, J., Xiao, Y., Jiang, H.: Deep short text classification with knowledge powered attention. In: Proceedings of AAAI 2019/IAAI 2019/EAAI 2019, Honolulu, Hawaii, USA, pp. 6252–6259. AAAI Press (2019)
6. Dabney, W., Rowland, M., Bellemare, M. G., Munos, R.: Distributional reinforcement learning with quantile regression. In: Proceedings of the AAAI 2018/IAAI 2018/EAAI 2018, New Orleans, Louisiana, USA, pp. 2892–2901. AAAI Press (2018)
7. Davidson, T., Warmsley, D., Macy, M., Weber, I.: Automated hate speech detection and the problem of offensive language. arXiv preprint arXiv:1703.04009 (2017)
8. Devlin, J., Chang, M., Lee, K., Toutanova, K.: BERT: pre-training of deep bidirectional transformers for language understanding, In: Proceedings of NAACL-HLT 2019, Minneapolis, Minnesota, pp. 4171–4186. Association for Computational Linguistics (2019)
9. Duarte, J.M., Berton, L.: A review of semisupervised learning for text classification. Artif. Intell. Rev. (2023). https://doi.org/10.1007/s10462-023-10393-8
10. Evci, U., Dumoulin, V., Larochelle, H., Mozer, M. C.: Head2toe: utilizing intermediate representations for better transfer learning. arXiv preprint arXiv:2201.03529 (2022)
11. Fortuna, P., Nunes, S.: A survey on automatic detection of hate speech in text. ACM Comput. Surv. 51(4), 1–30 (2018)
12. Gibert, O., Perez, N., García-Pablos, A., Cuadros, M.: Hate speech dataset from a white supremacy forum. In: Proceedings of the ALW2, pp. 11–20. Association for Computational Linguistics, Brussels, Belgium (2018)
13. Hu, C., Feng, Y., Kamigaito, H., Takamura, H., Okumura, M.: One-class text classification with multi-modal deep support vector data description. In: Proceedings of the 16th Conference of the European Chapter of the Association for Computational Linguistics, pp. 3378–3390. Association for Computational Linguistics (2021)
14. Jahan, M.S., Oussalah, M.: A systematic review of hate speech automatic detection using natural language processing. arXiv preprint arXiv:2106.00742 (2021)
15. Li, Q., et al.: A survey on text classification: from traditional to deep learning. ACM Trans. Intell. Syst. Technol. 13(2), 1–41 (2022)
16. Luo, X., Yu, Z., Zhao, Z., Zhao, W., Wang, J.: Effective short text classification via the fusion of hybrid features for IoT social data. Digit. Commun. Netw. 8(6), 942–954 (2022)
17. Malik, J. S., Pang, G., Hengel, A.: Deep learning for hate speech detection: a comparative study. arXiv preprint arXiv:2202.09517 (2022)
18. Mandl, T., et al.: Overview of the HASOC track at FIRE 2019: hate speech and offensive content identification in Indo-European languages. In: Proceedings of the FIRE 2019, pp. 14–17. Association for Computing Machinery, New York (2019)
19. Moya, M.M., Koch, M.W., Hostetler, L.D.: One-class classifier networks for target recognition applications. https://www.osti.gov/biblio/6755553. Accessed 8 Apr 2023
20. Pang, G., Shen, C., Cao, L., Hengel, A.V.D.: Deep learning for anomaly detection: a review. ACM Comput. Surv. 54(2), 1–38 (2022)
21. Peters, M.E., et al.: Deep contextualized word representations. In: Proceedings of the 2018 Conference of the North American Chapter of the Association for Computational Linguistics: Human Language Technologies, New Orleans, Louisiana, vol. 1, pp. 2227–2237. Association for Computational Linguistics (2018)

22. Pitsilis, G.K., Ramampiaro, H., Langseth, H.: Effective hate-speech detection in twitter data using recurrent neural networks. Appl. Intell. **48**(12), 4730–4742 (2018)
23. Poletto, F., Basile, V., Sanguinetti, M., Bosco, C., Patti, V.: Resources and benchmark corpora for hate speech detection: a systematic review. Lang. Resour. Eval. **55**(2), 477–523 (2020)
24. Radford, A., Narasimhan, K., Salimans, T., Sutskever, I.: Improving language understanding by generative pre-training. https://s3-us-west-2.amazonaws.com// openai-assets/research-covers/language-unsupervised/. Accessed 8 Apr 2023
25. Ruff, L., et al.: Deep one-class classification. In: Dy, J., Krause, A. (eds.) Proceedings of the 35th International Conference on Machine Learning, vol. 80, pp. 4393–4402. PMLR (2018)
26. Ruff, L., Zemlyanskiy, Y., Vandermeulen, R., Schnake, T., Kloft, M.: Self-attentive, multi-context one-class classification for unsupervised anomaly detection on text. In: Proceedings of the 57th Annual Meeting of the Association for Computational Linguistics, Florence, Italy, pp. 4061–4071. ACL (2019)
27. Schölkopf, B., Alexander, J.S.: Support vector machines, regularization, optimization, and beyond, pp. 656–657. MIT Press (2002)
28. Sun, M., Dou, H., Yan, J.: Efficient transfer learning via joint adaptation of network architecture and weight. In: Vedaldi, A., Bischof, H., Brox, T., Frahm, J.-M. (eds.) ECCV 2020. LNCS, vol. 12358, pp. 463–480. Springer, Cham (2020). https://doi. org/10.1007/978-3-030-58601-0_28
29. Tax, D.M.J.: Data description toolbox. https://homepage.tudelft.nl/n9d04/. Accessed 8 Apr 2023
30. Tax, D.M.J., Duin, R.P.W.: Support vector data description. Mach. Learn. **54**(1), 45–66 (2004)

EmoKnow: Emotion- and Knowledge-Oriented Model for COVID-19 Fake News Detection

Yuchen Zhang[1,2], Xing Su[1], Jia Wu[1], Jian Yang[1], Hao Fan[2(✉)], and Xiaochuan Zheng[2]

[1] School of Computing, Macquarie University, Sdyney, NSW, Australia
{yuchen.zhang3,xing.su2}@students.mq.edu.au, {jia.wu,jian.yang}@mq.edu.au
[2] School of Information Management, Wuhan University, Wuhan, Hubei, China
{hfan,zhengxiaochuan}@whu.edu.cn

Abstract. Content-based methods are inadequate for detecting fake news related to COVID-19 due to the complexity of this domain. Some studies integrate the social context information of the news to improve performance. However, such information is not consistently available and sometimes not helpful regarding COVID-19, as most users lack professional knowledge about it and may be unable to respond accurately. Additionally, fake news often employs emotional manipulation to exploit people's emotions to shape their beliefs and actions. Therefore, we propose EmoKnow, an emotion- and knowledge-oriented model, for detecting fake news about COVID-19. Our proposed method incorporates language modeling, emotion feature extraction, and external knowledge sources to provide an informative representation of news. Experimental results on four COVID-19-related datasets show that EmoKnow significantly outperforms state-of-the-art approaches.

Keywords: COVID-19 · Fake News Detection · Text Embedding · Emotion Features · Knowledge Representation

1 Introduction

The coronavirus disease in 2019 (COVID-19) has caused a major crisis that has seriously affected the global population. Unfortunately, the situation has been further aggravated by the rampant dissemination of a large volume of fake news about COVID-19 through social networks. On the social level, the credibility of public health departments will be undermined by fake news, which could pose a considerable challenge to disease prevention and control not only for COVID-19 but also for future health crises. For example, vaccine misinformation has eroded public trust in vaccines. Over 319,000 unvaccinated people died from COVID-19 after vaccines were open to all adults[1]. On the individual level, false prevention

---

[1] https://www.npr.org/2022/05/16/1099070400/how-vaccine-misinformation-made-the-covid-19-death-toll-worse.

X. Yang et al. (Eds.): ADMA 2023, LNAI 14176, pp. 352–367, 2023.
https://doi.org/10.1007/978-3-031-46661-8_24

or cure of COVID-19 may make people take an irrational response and lead to worse results. Given the severe consequences caused, the significance of detecting fake news about COVID-19 and ensuring accurate information reaches the public cannot be overemphasized.

Although many fake news detection methods have made progress in certain domains, such as politics and celebrities, identifying fake news on COVID-19 presents several unique challenges. Classic Content-based approaches may struggle to discern between real and fake news on COVID-19, as COVID-19 is a complex domain that involves many different aspects [35], *e.g.*, science, medicine, and public health, and requires a more nuanced approach. Some approaches adopt context information such as user's information or comments on the original posts to improve the performance [17,27,30]. However, such auxiliary information is not consistently available and sometimes not helpful regarding COVID-19, as most users lack professional knowledge about it and may not be capable of providing accurate responses. To address this issue, we incorporate external knowledge sources which can provide accurate background information that is hard to obtain from news texts alone. Additionally, recent studies [18,38] have revealed that emotional manipulation is a prevalent tactic employed in fake news by exploiting people's emotions to spread false information and influence their beliefs and actions. Therefore, emotion features are also crucial to consider when identifying the authenticity of news on COVID-19.

In this paper, we propose an **Emo**tion- and **Know**ledge-oriented Model (EmoKnow) for detecting fake news about COVID-19. Specifically, we first utilize a language model to learn the textual representation of news content. Then emotion representation is captured by extracting multiple emotion features from news content. Subsequently, we incorporate a knowledge graph to get knowledge-based representation through knowledge graph representation learning and entity tagging. Finally, we fuse these representations to improve the performance in detecting fake news about COVID-19.

The main contributions of this work can be summarized as follows: 1) We leverage an external knowledge source to furnish our model with knowledge about COVID-19, which is not easily obtainable from news text, leading to enhanced fake news detection; 2) Our model demonstrates the existence of substantial variations in emotion features of fake and real news on COVID-19. With its ability to effectively detect these emotional cues in the news, our model significantly improves the performance of detecting fake news; and 3) By integrating emotion and external knowledge, our proposed method offers an informative representation of news, including text, emotion, and knowledge-based features. We run experiments on four public fake news datasets about COVID-19, and our proposed model, EmoKnow, significantly outperforms state-of-the-art methods.

2   Related Work

2.1   Content-Based Fake News Detection

Content-based methods mainly leverage linguistic features extracted from the news content, including elements such as the title, text, keywords, and overall writing style. Zhang *et al.* [36] proposed to distinguish the authenticity of news articles by examining specific elements within the news content, such as URLs. Dong *et al.* [6] applied BiGRU to extract features from news content and side information, facilitating better fake news detection. Horne *et al.* [11] leverage the differences in structural features, such as text length and the number of different words, to detect fake news on social media.

2.2   Emotion-Based Fake News Detection

There are also studies that aim at identifying specific emotional signals that can distinguish between fake and real news. Studies [34,39] have found that real news may contain a range of emotions based on the subject matter, while fake news tends to be sensational and designed to evoke strong emotions like anger, fear, or excitement. Additionally, fake news is more likely to use manipulative language and emotional appeals, while real news is generally balanced and objective. Giachanou *et al.* [8] proposed an approach that incorporates emotional signals from news content for fake news detection. Guo *et al.* [10] proposed a CNN-based model that can simultaneously exploit content and social emotions for fake news detection. Ajao *et al.* [1] analyzed the relationship between news veracity and the sentimental words in news and proposed an emotion-aware model to detect fake news.

2.3   Knowledge-Based Fake News Detection

Both textual features and emotion features are gained from the provided news content. However, the texts carry limited information and lack related background knowledge, which may provide more details or relations about entities mentioned in the text. Knowledge-based methods enrich the news features by introducing knowledge from external sources, such as a knowledge base (KB) and a knowledge graph (KG). Studies [9,16,26] have shown that external knowledge is essential for boosting fake news detection. Hu *et al.* [12] proposed a graph neural network [31] to compare news entities to knowledge entities, which significantly outperforms other methods. Dun *et al.* [7] implement a method that merges semantic features from news and knowledge features from external knowledge to improve the performance of fake news detection. Wang *et al.* [29] proposed an approach that converts news text and external knowledge from KGs into graphs and utilized a graph convolutional network to learn knowledge-based representation for fake news detection.

In summary, content-based methods for fake news detection are among the earliest approaches that have been proposed. One major drawback of these methods is that they rely solely on the information presented in the news content

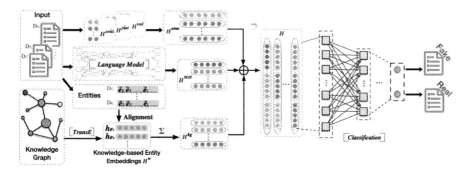

Fig. 1. The framework of EmoKnow for fake news detection. Firstly, the input news set is fed into a language model to learn textual representation H^{text}. Then we extract various emotional features (*i.e.*, sentiment score H^{senti}, Plutchik emotion feature H^{plut} and text VAD score H^{vad}) from news text and concatenate them to get news emotion representation H^{emo}. To leverage external knowledge, we extract entities $\tilde{E}_{D_i}$ mentioned in the news and utilize TransE to learn Knowledge-based entity embeddings H^w. By aligning news entities to KG, we can retrieve news entity embeddings from H^w, and obtain news knowledge-based representation H^{kg}. The news representation H can be obtained by concatenating H^{text}, H^{emo} and H^{kg}. Finally, the news representation H is fed into the classifier, which classifies the news as fake or real.

and can struggle with complex or nuanced situations, like identifying fake news on COVID-19. Knowledge-based methods, on the other hand, show promise in addressing the limitation. Knowledge graphs are ideal for providing auxiliary knowledge for the model to learn from. However, incorporating external knowledge alone may not be sufficient to fully tackle the issue, as identifying emotional manipulations in fake news on COVID-19 is also crucial. Hence, we proposed an approach incorporating language modeling, sentiment analysis, and external knowledge sources to solve the above-mentioned problems.

3   Methodology

We propose an **Emo**tion- and **Know**ledge-oriented Model for COVID-19 Fake News Detection, which is shorted as EmoKnow. The overall framework of Emo-Know is shown in Fig. 1. For each piece of news, we get its representation by extracting features from news text, text emotion, and related knowledge graphs. The whole process consists of four steps. The first step is to get the text-level representation from the news text. We feed the preprocessed news text to the language model to obtain textual representation. Then, we learn news emotion representation by extracting various emotion features of news content. After that, we leverage the knowledge graph to get the knowledge-based representation by entity tagging and knowledge embedding. Finally, we concatenate these representations and get the final news representation, which will be fed to a classifier for fake news detection.

3.1   Textual Representation

For fake news detection tasks, textual features are considered one of the most discriminate features that can be captured [37]. For this purpose, we employ Albert, a lite version of Bert [5], to encode news text and generate text-level representation. Like Bert, Albert is a transformer-based language model. It can achieve state-of-the-art performance while having significantly fewer parameters than BERT. Given a set of news $\mathcal{D}$, it contains n pieces of news $\mathcal{D} = [D_1, D_2, \cdots, D_n]$, where each news contains t words $D_i = [w_1, w_2, \cdots, w_t]$. $\boldsymbol{H}^{text}$ is the learned text representations of news set $\mathcal{D}$ and $\boldsymbol{h}_i^{text}$ is the text representation of news D_i.

3.2   Emotion Representation

Incorporating emotion features allows for a more nuanced understanding of the news content and can aid in differentiating between real and fake news. Emotion features, such as the emotional polarity or sentiment of the news text, can help the model identify manipulative language and sentiment manipulation commonly used in fake news. We denote $\boldsymbol{H}^{emo}$ as emotion representation of news set $\mathcal{D}$. For one piece of news D_i, its emotion representation $\boldsymbol{h}_i^{emo}$ consists of three different emotion features, including sentiment score, Plutchik emotion feature, and text Valence-Arousal-Dominace(VAD) score.

$$\boldsymbol{h}_i^{emo} = \boldsymbol{h}_i^{senti} \oplus \boldsymbol{h}_i^{plut} \oplus \boldsymbol{h}_i^{vad}, \tag{1}$$

where $\oplus$ denotes representation concatenating.

Sentiment Score. The sentiment score is an overall evaluation of the emotional tendency and intensity of the input text, which can provide the macro-level emotion features. Given an input news D_i, with the sentiment analysis toolkits, we can get its sentiment score feature $\boldsymbol{h}_i^{senti}$. In this study, the sentiment score $\boldsymbol{h}_i^{senti}$ of news D_i is calculated by the VADER [13] sentiment analysis tool.

Plutchik Emotion Feature. Except for the above most basic emotion categories, many classic emotion analysis models have been proposed in psychological science. Robert Plutchik created the famous Plutchik's wheel of emotions [25], which consists of eight primary emotions: *joy, sadness, anger, fear, trust, disgust, surprise*, and *anticipation*. We further extracted emotional features $\boldsymbol{h}^{plut}$ based on Plutchik's eight primary emotions.

First, we leverage the pre-trained multiclass emotion classifier [14] to predict the probabilities that the input news content conveys certain emotions. We denote Plutchik's primary emotions as $\boldsymbol{Emo}^{plut} = \{emo_1, emo_2, \cdots, emo_8\}$. For each news D_i, the output of the classifier is $p_i^{plut} = [p_i^{emo_1}, p_i^{emo_k}, ..., p_i^{emo_8}]$, $p_i^{emo_k}$ is the probability that news D_i expresses emo_k ($emo_k \in \boldsymbol{Emo}^{plut}$). Then we calculate the word-level emotion score of each news in the above eight emotions based on emotion lexicons to get nuanced emotion features. Emotion lexicons can provide emotional words representing different categories of emotions and their scores. We use the NRC emotion lexicon [21] and NRC intensity lexicon [23] to calculate the word-level emotion score of each news, as the emotion categories contained in these emotion lexicons are consistent with Plutchik's primary emotion.

Given a piece of news $D_i = [w_1, w_2, \cdots, w_t]$, we calculated its word-level emotion score $s_i^{emo_k}$ in each emo_k ($emo_k \in \boldsymbol{Emo}^{plut}$) separately. For any word w_i in the news D_i, if w_i in the NRC emotion lexicon, we then set a window size and search the negative words and degree words before w_i. The calculation of the word-level emotion score $s_i^{emo_k}$ incorporates emotion words, negative words, and degree words:

$$s_i^{emo_k} = \sum_{j=1}^{t} (-1)^{\boldsymbol{neg}(w_j)} * \boldsymbol{deg}(w_j) * \boldsymbol{emo}(w_j), \tag{2}$$

$$\boldsymbol{emo}(w_j) = \begin{cases} 1, & \text{if } w_j \in \mathscr{E}_e \\ 0, & \text{otherwise} \end{cases}, \tag{3}$$

where $\boldsymbol{neg}(w_j)$ is the number of negative words of w_j in the search window, and $\boldsymbol{deg}(w_j)$ is the values of degree words which are set according to human-annotated degree dictionary.

We calculate $s_i^{emo_k}$ of each $emo_k (emo_k \in Emo^{plut})$ as described above, and can get Plutchik emotion score s_i^{plut} that covers all Plutchik's emotions by concatenate them:

$$s_i^{plut} = s_i^{emo_1} \oplus s_i^{emo_2} \oplus \cdots s_i^{emo_8}. \tag{4}$$

As for emotion intensity features, the extracting process is similar to emotion score, except that each word found in the intensity lexicon will be set to a value according to the intensity lexicon. We denote emotion intensity of news D_i in emo_k ($emo_k \in Emo^{plut}$) as $int_i^{emo_k}$.

$$int_i^{emo} = \sum_{j=1}^{t} (-1)^{\boldsymbol{neg}(w_j)} * \boldsymbol{deg}(w_j) * \boldsymbol{int}(w_j), \tag{5}$$

where $int(w_j)$ is the intensity value of w_j which is set according to NRC intensity lexicon. Thus, the emotion intensity score int_i^{plut} of the news D_i can be extracted:

$$int_i^{plut} = int_i^{emo_1} \oplus int_i^{emo_2} \oplus \cdots int_i^{emo_8}. \tag{6}$$

Finally, the Plutchik Emotion Feature h_i^{plut} can be obtained:

$$h_i^{plut} = p_i^{plut} \oplus s_i^{plut} \oplus int_i^{plut}. \tag{7}$$

Text VAD. We also conducted a lexicon-based Valence-Arousal-Dominance analysis to capture emotion features of news from other dimensions. Text VAD words and their corresponding values for each dimension are obtained from the NRC VAD lexicon [22]. For news $D_i = [w_1, w_2, \cdots, w_t]$, we calculate the text valence score s_i^{val}, text arousal score s_i^{aro} and text dominance score s_i^{dom}, respectively. Then, we concatenate them to get text VAD score $\boldsymbol{h}^{vad}$:

$$\boldsymbol{h}^{vad} = \boldsymbol{s}_i^{val} \oplus \boldsymbol{s}_i^{aro} \oplus \boldsymbol{s}_i^{dom}, \tag{8}$$

where the valence score, arousal score and dominance score are calculated according to Eq. 5, respectively.

3.3   Knowledge-Based Representation

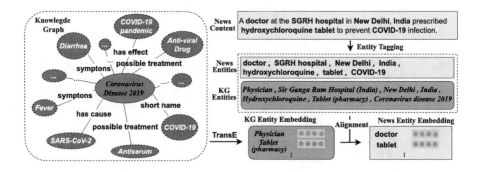

Fig. 2. The Process of Entity Tagging.

In this work, we incorporate the knowledge graph WiKiData5M [28] as an external knowledge source. WiKiData5M is a million-scale knowledge graph that contains over 4.5 million entities, 822 relations, and over 20 million *[entity, relation, entity]* triplets, for example, *["Joe Biden", "position held", "President of the United States"]*. Introducing this knowledge can enrich news representation, leading to better performance of fake news classification.

To get the knowledge-based representation of a certain piece of news, we first extract the entities mentioned in the news and find their aligned entities in the KG. With the entity tagging tool TAGME[2], we can retrieve entities mentioned in the news and align them with their corresponding wiki entities in Wikidata5m. As shown in Fig. 2, for the news *"A doctor at the SGRH hospital in New Delhi India prescribed hydroxychloroquine tablet to prevent COVID-19 infection."*, we can extract entity *"Doctor"* and link it to WikiData entity *"Physican"* and *"SGPH hospital"* is linked to *"Sir Ganga Ram Hospital"*, etc.

The second step is to get the knowledge-based entity embeddings from WiKi-Data5M, where the knowledge graph embedding model TransE [2] is employed.

---

[2] https://sobigdata.d4science.org/web/tagme.

TransE treats the entities and relations in the knowledge graph as two matrices. The entity matrix $N \in \mathbb{R}^{u \times d}$, and the relation matrix $R \in \mathbb{R}^{v \times d}$, where u is the number of entities, v is the number of relations, and d is the dimension of both entities vector embeddings and relations vector embeddings.

Given an existing triplet (p, r, q) in the knowledge graph, $p, q \in N$, $r \in R$, relation r is seen as a translation vector from p to q. The ideal state of the TransE model is $p + q \approx r$ for an existing triplet. While for a corrupted triplet (p', r, q'), a non-existent triplet formed by randomly replacing one entity(head or tail) in an existing triplet with another entity from the knowledge graph, the model expects the opposite. To learn such embeddings, TransE employ Hinge loss function $\mathcal{L}_{kg}$ during model training, which is defined as:

$$\mathcal{L}_{kg} = \sum_{(p,r,q) \in S} \sum_{(p',r,q') \in S'} [\gamma + \mathrm{Dis}(p + r, q) - \mathrm{Dis}(p' + r, q')]_+, \qquad (9)$$

where $[x]+$ represents the positive part of x, $\gamma > 0$ denotes a margin hyperparameter. Dis can be the L1 or the L2-norm to measure the dissimilarity. S is the training set of the triplet.

In practice, for a piece of news D_i, we align its entity set $\tilde{E}_{D_i} = [\tilde{e}_1, \tilde{e}_2, ..., \tilde{e}_m]$ with the corresponding wiki entity set $E_i = [e_1, e_2, ..., e_k]$ in WiKiData5M. Then We feed WiKiData5M entity set $\mathcal{E} = [e_1^w, e_2^w, \cdots]$ and relation set $\mathcal{R} = [r_1^w, r_2^w, \cdots]$ to TransE model. Of particular note is the wiki entity set E_i of each news are extracted from the WiKiData5M entity set $\mathcal{E}$ (i.e., $E_i \in \mathcal{E}$).

After the training process of TransE mentioned above, wiki entity embeddings H^w are learned. With entity tagging and knowledge-based entity embeddings learning, we can obtain Knowledge-based embeddings $h_{\tilde{e}_i}$ of entities $\tilde{e}_i$ by retrieving the embeddings from H^w:

$$h_{\tilde{e}_i} = h_{e_i}, \forall h_{e_i} \in H^w, \qquad (10)$$

where $h_{\tilde{e}_i}$ and h_{e_i} denotes the Knowledge-based embedding of a news entity $\tilde{e}_i$ and its corresponding wiki entity e_i, respectively. Accordingly, we denote h^{kg} as Knowledge-based embedding of one piece of news, and h^{kg} is the sum of $h_{\tilde{e}_i}$:

$$h^{kg} = \sum^{m} h_{\tilde{e}_i}. \qquad (11)$$

3.4   Fake News Detection

To classify the news as fake or real, we concatenate the obtained textual representation, emotion representation, and knowledge-based representation to get the final representation H:

$$H = H^{text} \oplus H^{emo} \oplus H^{kg}. \qquad (12)$$

Then, we feed H to a classifier that consists of multiple fully-connected layers for fake news detection. The first three layers adopt the relu activation function, and the last layer uses the softmax activation function:

$$y' = \text{softmax}(W \cdot H + b). \tag{13}$$

The model training aims to minimize the cross-entropy loss function:

$$\mathcal{L} = -y \log y' - (1 - y) \log(1 - y'), \tag{14}$$

where y' denotes the predicted label while y is the ground truth label.

4   Experiments

This section presents an evaluation of EmoKnow's efficiency based on experimental results. We first outline the experimental setup, including the datasets and baselines used. Subsequently, we present an analysis of the overall results and conduct an ablation study, a case study, and a parameter analysis to demonstrate the superior performance of our approach.

4.1   Experimental Setup

Datasets. To assess the effectiveness of our method, we perform experiments on four public fake news detection datasets about COVID-19, *i.e.*, AAAI 2021 Covid-19 Fake News [24], COVID-19 Fake News Dataset [15], MM COVID [19] and ReCOVery [40]. Table 1 shows the statistic of each dataset.

Table 1. Statistics of Datasets.

Dataset	Fake	Real	Total
AAAI 2021 Covid-19 Fake News	5101	5600	10701
COVID-19 Fake News Datase	1059	2061	3120
MM COVID	2098	1805	3903
ReCOVery	664	1354	2018

Baselines. We compare our model with several representative fake news detection models, which are listed as follows:

- **textCNN** [3] uses multiple kernels in convolutional neural networks(CNNs) to extract the key features in sentences. For text classification, it can extract important features more efficiently and achieve better classification results.
- **textGCN** [33] proposes a text graph convolutional network for text classification. TextGCN can perform well without any external word embeddings or knowledge in text classification tasks.

- **HAN** [32] consists of a hierarchical structure and the attention mechanism. HAN encodes sentence-level and word-level attention for each news content.
- **Bert** [5] is a transformer-based language model that can consider the entire input news content at once to understand the context of news content in both directions and lead to better performance.
- **KGInformed** [16] combines text representation that learned by RoBerta [20] and knowledge graph-based representation from WiKiData5M [28] to improve fake news classification performance.
- **DualEmo** [38] uses Bi-GRU [4] to encode the news text and captures the emotion gap between news and their comments. As our model focuses on the news contents rather than social contexts, we implement a lite version of DualEmo where emotions of comments are excluded.

Experimental Settings and Evaluation. Across all models, the datasets are divided into train, validation, and test sets using a split ratio of 60%-20%-20%. To ensure robustness, all experimental results are averaged over 5-fold cross-validation. For our model, the text-level representation is learned by Albert, and the dimension of text-level representation is 768. The window size for searching negative words and degree words is set to 4 and the dimension of emotion representation is 44. KG-based entity embeddings are obtained by TransE, and the dimensions of both KG-based entity embeddings and KG-based news representation are 512. We adopt the Adam optimizer and set the learning rate to 0.01 across all datasets. For all baseline models, we follow the model architecture presented in their papers. All the experiments are conducted by Python 3.8, 1 NVIDIA Volta GPU with 12G DRAM.

To evaluate the performance of our model and baselines, we adopt Accuracy (Acc), Precision (Pre), Recall (Rec), and F1 score (F1) for evaluation.

4.2   Overall Results

Table 2. Performance Comparison on Four COVID-related Datasets.

	AAAI 2021 Covid-19 Fake News				COVID-19 Fake News Dataset			
	Accuracy	Precision	Recall	F1 Score	Accuracy	Precision	Recall	F1 Score
textCNN	0.810 ± 0.098	0.817 ± 0.085	0.809 ± 0.099	0.858 ± 0.108	0.700 ± 0.023	0.375 ± 0.041	0.498 ± 0.003	0.416 ± 0.005
textGCN	0.826 ± 0.066	0.835 ± 0.058	0.790 ± 0.132	0.780 ± 0.151	0.743 ± 0.065	0.581 ± 0.207	0.589 ± 0.141	0.541 ± 0.169
HAN	0.867 ± 0.005	0.869 ± 0.005	0.868 ± 0.004	0.867 ± 0.005	0.708 ± 0.016	0.604 ± 0.212	0.518 ± 0.020	0.451 ± 0.044
Bert	0.801 ± 0.008	0.801 ± 0.008	0.801 ± 0.008	0.800 ± 0.008	0.706 ± 0.019	0.453 ± 0.204	0.501 ± 0.002	0.416 ± 0.009
KGInformed	0.838 ± 0.010	0.838 ± 0.009	0.838 ± 0.010	0.837 ± 0.010	0.702 ± 0.020	0.352 ± 0.009	0.498 ± 0.002	0.413 ± 0.007
DualEmo	0.906 ± 0.002	0.906 ± 0.002	0.906 ± 0.001	0.906 ± 0.002	0.721 ± 0.024	0.643 ± 0.054	0.597 ± 0.041	0.599 ± 0.051
EmoKnow	**0.908 ± 0.002**	**0.908 ± 0.002**	**0.909 ± 0.002**	**0.908 ± 0.002**	**0.805 ± 0.019**	**0.773 ± 0.020**	**0.722 ± 0.022**	**0.739 ± 0.022**
	MM COVID				ReCOVery			
	Accuracy	Precision	Recall	F1 Score	Accuracy	Precision	Recall	F1 Score
textCNN	0.582 ± 0.035	0.478 ± 0.170	0.547 ± 0.039	0.474 ± 0.101	0.658 ± 0.011	0.460 ± 0.104	0.501 ± 0.020	0.442 ± 0.039
textGCN	0.717 ± 0.156	0.735 ± 0.236	0.685 ± 0.178	0.622 ± 0.241	0.718 ± 0.037	0.691 ± 0.178	0.609 ± 0.102	0.565 ± 0.124
HAN	0.855 ± 0.005	0.854 ± 0.005	0.854 ± 0.006	0.853 ± 0.005	0.722 ± 0.041	0.462 ± 0.197	0.501 ± 0.007	0.425 ± 0.011
Bert	0.730 ± 0.093	0.727 ± 0.094	0.722 ± 0.101	0.720 ± 0.103	0.682 ± 0.030	0.441 ± 0.213	0.506 ± 0.012	0.416 ± 0.032
KGInformed	0.752 ± 0.015	0.759 ± 0.006	0.742 ± 0.017	0.744 ± 0.019	0.697 ± 0.044	0.649 ± 0.193	0.539 ± 0.038	0.488 ± 0.070
DualEmo	0.874 ± 0.015	0.871 ± 0.015	0.874 ± 0.016	0.872 ± 0.016	0.639 ± 0.018	0.471 ± 0.052	0.493 ± 0.021	0.459 ± 0.040
EmoKnow	**0.935 ± 0.007**	**0.935 ± 0.006**	**0.935 ± 0.007**	**0.935 ± 0.007**	**0.787 ± 0.027**	**0.780 ± 0.050**	**0.713 ± 0.028**	**0.728 ± 0.033**

Table 2 shows the results of our proposed model and selected baseline algorithms on four COVID-19-related datasets. The optimal values of each metric are highlighted in bold. Our proposed EmoKnow achieves optimal results for all metrics on all datasets. Such results indicate that emotion features and external knowledge can indeed provide indispensable information for distinguishing the authenticity of the news. They also reveal the effectiveness of our model in capturing these features, which is precisely what other baselines cannot do. Specifically, Our model, EmoKnow, shows exceptional performance in comparison to all baseline models, surpassing them in all evaluation metrics, with a minor exception on AAAI 2021 Covid-19 Fake News dataset, where our results are slightly higher than the sub-optimal. It should be noted that EmoKnow reached a level of recall far exceeding that of other models in most cases. A high recall is notably important in the context of fake news detection because it is often more crucial to ensure that as little fake news as possible is missed and not detected, which is vital for effectively combating the dissemination of fake news. In addition, while some of the baseline algorithms experience quite fluctuations in their results during 5-fold cross-validation, EmoKnow is able to produce dependable and consistent results.

As for the baseline models, textCNN and textGCN have inconsistent performance. They may perform well on certain datasets, but their recall and precision can be quite low on other datasets. Both HAN and Bert utilize attention mechanisms. However, their performance differs. Bert generally gets mediocre performance across all datasets, while HAN obtains good performance on two datasets but its recall and precision fall short on the other two. KGInformed, like Bert, can obtain consistent results, but the performance is insufficient. DualEmo obtains the sub-optimal results on two datasets, but performs the worst on ReCOVery, which indicates that the model is not robust enough.

In summary, EmoKnow shows excellent capability and stability across all datasets. We attribute the superiority of EmoKnow to the integrated emotion features and external knowledge. The incorporation of these features allows for a more informative representation of the news. This is due to the ability to capture sentiment and emotional tone that traditional text-based models lack, and knowledge-level features that cannot be derived solely from the news texts or social context.

4.3   Ablation Study

To demonstrate the effectiveness of the incorporated emotion features and external knowledge, we conduct several ablation experiments (Fig. 3). We feed the identical classifier used in EmoKnow with three different input representations, i.e., text representation H^{text} solely, text representation concatenated with emotion representation $H^{text} \oplus H^{emo}$, and text representation concatenated with knowledge representation $H^{text} \oplus H^{kg}$.

While using only text representation results in mediocre accuracy and subpar performance on other metrics across all datasets, once we concatenated it with emotion representation or knowledge representation, we observed a significant

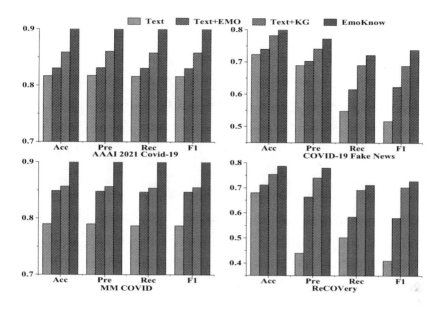

Fig. 3. Ablation study.

boost in performance. Concatenating it with knowledge representation leads to a 3% greater improvement across all metrics compared to concatenating it with emotion representation. Notably, the best performance is always achieved by EmoKnow. On average, EmoKnow yields a 10% higher in accuracy, 17% in precision, 16% in recall, and 20% on f1 score compared to using text representation alone. These results indicate that EmoKnow is effective in utilizing all three representations to improve the performance of fake news detection. It is also worth noting that the improvement seen by concatenating all three representations is substantial, which further solidifies the effectiveness of our model.

4.4 Case Study

To explore how news authenticity and textual, emotion, and knowledge features are related, We conduct a case study on the MM COVID dataset. Figure 4 demonstrates that keywords in real news are quite different from those in fake news. We also investigated emotion variations between fake and real news. In general, over half of real news has a neutral tone, whereas only around 30% of fake news is neutral. The proportion of negative news is much higher in fake news than in real news. Besides, fake news tends to convey higher levels of sadness, fear, anger, and disgust than real news, whereas the levels of anticipation and trust are lower in fake news. These results confirm news text modeling and emotion extraction module in EmoKnow can capture distinctive features that can be utilized for detecting the authenticity of the news (Table 3).

We further analyze how knowledge-based features can be utilized in some cases. Figure 5 shows two news pieces that are misclassified by Albert and KGIn-

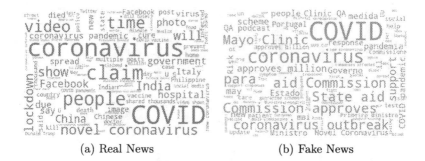

(a) Real News   (b) Fake News

Fig. 4. Word clouds of news in the MM COVID dataset.

Table 3. Emotion Variations of news in the MM COVID dataset.

	Neutral	Positive	Negative	Anger	Anticipation	Disgust	Fear	Joy	Sadness	Surprise	Trust
Fake News	34.08%	18.95	36.91%	9.23%	10.59%	7.27%	23.96%	9.68%	15.95%	4.49%	18.83%
Real News	51.57%	17.33	20.59%	7.36%	14.04%	5.76%	22.86%	7.70%	13.36%	4.82%	24.11%

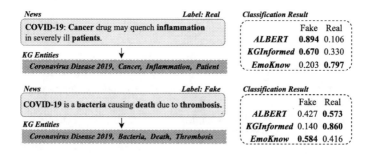

Fig. 5. Classification results of two news pieces on MM COVID dataset.

formed but correctly classified by EmoKnow. It reveals EmoKnow's ability to capture and utilize knowledge-based features to effectively detect fake news.

4.5   Parameter Analysis

We also investigated the sensitivity of EmoKnow to the parameters regarding the number of layers of the classifier and the embedding size of the news texts. The results show that the performance varies slightly with changes in the embedding size, but the performance declines when the number of layers is set to 2. In general, our method can obtain acceptable results in most cases steadily (Fig. 6).

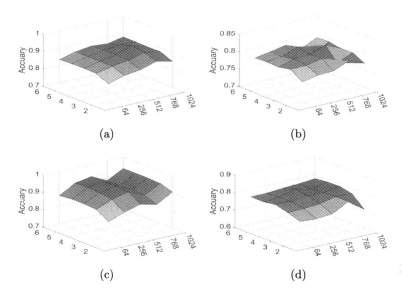

Fig. 6. Parameter analysis of `EmoKnow`. (a) AAAI 2021 Covid-19 Fake News, (b) COVID-19 Fake News, (c) MM COVID, and (d) ReCOVery.

5 Conclusion

In this paper, we proposed an **Emo**tion- and **Know**ledge-oriented model, Emo-Know, that incorporates language modeling, sentiment analysis, and external knowledge sources for COVID-19 fake news detection. Specifically, our approach starts by using a language model to learn the textual representation of news content. Then multiple emotion features are extracted to obtain emotion representation. We also incorporate external knowledge sources to get knowledge-based representation. By concatenating these representations, we enhance the ability to identify fake news on COVID-19. Experiments conducted on four public fake news datasets about COVID-19 show the superior performance of our model. For future work, we will investigate ways to improve the integration of textual representation, emotion representation, and knowledge-based representation to improve the performance of fake news detection.

Acknowledgements. This work was supported by the Australian Research Council (ARC) DP230100899, DE200100964 and LP210301259.

References

1. Ajao, O., Bhowmik, D., Zargari, S.: Sentiment aware fake news detection on online social networks. In: ICASSP, pp. 2507–2511 (2019)
2. Bordes, A., Usunier, N., Garcia-Duran, A., Weston, J., Yakhnenko, O.: Translating embeddings for modeling multi-relational data. In: NIPS (2013)

3. Chen, Y.: Convolutional neural network for sentence classification. Master's thesis, University of Waterloo (2015)
4. Cho, K., Van Merriënboer, B., Bahdanau, D., Bengio, Y.: On the properties of neural machine translation: encoder-decoder approaches. arXiv preprint arXiv:1409.1259 (2014)
5. Devlin, J., Chang, M.W., Lee, K., Toutanova, K.: BERT: pre-training of deep bidirectional transformers for language understanding. arXiv preprint arXiv:1810.04805 (2018)
6. Dong, M., Yao, L., Wang, X., Benatallah, B., Sheng, Q.Z., Huang, H.: Dual: a deep unified attention model with latent relation representations for fake news detection. In: WISE (2018)
7. Dun, Y., Tu, K., Chen, C., Hou, C., Yuan, X.: KAN: knowledge-aware attention network for fake news detection. In: AAAI, pp. 81–89 (2021)
8. Giachanou, A., Rosso, P., Crestani, F.: Leveraging emotional signals for credibility detection. In: SIGIR, pp. 877–880 (2019)
9. Guo, B., Ding, Y., Yao, L., Liang, Y., Yu, Z.: The future of false information detection on social media: new perspectives and trends. CSUR **53**, 1–36 (2020)
10. Guo, C., Cao, J., Zhang, X., Shu, K., Yu, M.: Exploiting emotions for fake news detection on social media. arXiv preprint arXiv:1903.01728 (2019)
11. Horne, B.D., Adali, S.: This just. In: Fake News Packs a Lot in Title, Uses Simpler, Repetitive Content in Text Body, More Similar to Satire Than Real News. In: AAAI (2017)
12. Hu, L., et al.: Compare to the knowledge: graph neural fake news detection with external knowledge. In: ACL, pp. 754–763 (2021)
13. Hutto, C., Gilbert, E.: Vader: a parsimonious rule-based model for sentiment analysis of social media text. In: ICWSM (2015)
14. Kant, N., Puri, R., Yakovenko, N., Catanzaro, B.: Practical text classification with large pre-trained language models. arXiv preprint arXiv:1812.01207 (2018)
15. Koirala, A.: Covid-19 fake news dataset (2021)
16. Koloski, B., Perdih, T.S., Robnik-Šikonja, M., Pollak, S., Škrlj, B.: Knowledge graph informed fake news classification via heterogeneous representation ensembles. Neurocomputing **496**, 208–226 (2022)
17. Kumar, S., Shah, N.: False information on web and social media: a survey. arXiv preprint arXiv:1804.08559 (2018)
18. Li, M.H., Chen, Z., Rao, L.L.: Emotion, analytic thinking and susceptibility to misinformation during the covid-19 outbreak. Comput. Hum. Behav. **133**, 107295 (2022)
19. Li, Y., Jiang, B., Shu, K., Liu, H.: MM-COVID: a multilingual and multimodal data repository for combating COVID-19 disinformation (2020)
20. Liu, Y., et al.: Roberta: a robustly optimized BERT pretraining approach. arXiv preprint arXiv:1907.11692 (2019)
21. Mohammad, S., Turney, P.: Emotions evoked by common words and phrases: using mechanical turk to create an emotion lexicon. In: NAACL HLT Workshop, pp. 26–34 (2010)
22. Mohammad, S.M.: Obtaining reliable human ratings of valence, arousal, and dominance for 20,000 english words. In: ACL (2018)
23. Mohammad, S.M.: Word affect intensities. In: LREC (2018)
24. Patwa, P., et al.: Fighting an infodemic: Covid-19 fake news dataset. In: AAAI Workshop, pp. 21–29 (2021)
25. Plutchik, R.: The multifactor-analytic theory of emotion. J. Psychol. **50**, 153–171 (1960)

26. Qian, S., Hu, J., Fang, Q., Xu, C.: Knowledge-aware multi-modal adaptive graph convolutional networks for fake news detection. TOMM **17**(3), 1–23 (2021)
27. Su, X., Yang, J., Wu, J., Zhang, Y.: Mining user-aware multi-relations for fake news detection in large scale online social networks. In: WSDM, pp. 51–59 (2023)
28. Wang, X., Gao, T., Zhu, Z., Liu, Z., Li, J., Tang, J.: Kepler: a unified model for knowledge embedding and pre-trained language representation. arXiv preprint arXiv:1911.06136 (2019)
29. Wang, Y., Qian, S., Hu, J., Fang, Q., Xu, C.: Fake news detection via knowledge-driven multimodal graph convolutional networks. In: ICMR, pp. 540–547 (2020)
30. Yang, S., Shu, K., Wang, S., Gu, R., Wu, F., Liu, H.: Unsupervised fake news detection on social media: a generative approach. In: AAAI, pp. 5644–5651 (2019)
31. Yang, Z., et al.: State of the art and potentialities of graph-level learning (2023)
32. Yang, Z., Yang, D., Dyer, C., He, X., Smola, A., Hovy, E.: Hierarchical attention networks for document classification. In: NAACL HLT, pp. 1480–1489 (2016)
33. Yao, L., Mao, C., Luo, Y.: Graph convolutional networks for text classification. In: AAAI, pp. 7370–7377 (2019)
34. Yin, H., Song, X., Yang, S., Li, J.: Sentiment analysis and topic modeling for covid-19 vaccine discussions. World Wide Web **25**(3), 1067–1083 (2022)
35. Yin, H., Yang, S., Li, J.: Detecting topic and sentiment dynamics due to COVID-19 pandemic using social media. In: Yang, X., Wang, C.-D., Islam, M.S., Zhang, Z. (eds.) ADMA 2020. LNCS (LNAI), vol. 12447, pp. 610–623. Springer, Cham (2020). https://doi.org/10.1007/978-3-030-65390-3_46
36. Zhang, X., Lashkari, A.H., Ghorbani, A.A.: A lightweight online advertising classification system using lexical-based features. In: SECRYPT (2017)
37. Zhang, X., Ghorbani, A.A.: An overview of online fake news: characterization, detection, and discussion. Inf. Process. Manag. **57**, 102025 (2020)
38. Zhang, X., Cao, J., Li, X., Sheng, Q., Zhong, L., Shu, K.: Mining dual emotion for fake news detection. In: WWW 2021, pp. 3465–3476 (2021)
39. Zhou, L., Tao, J., Zhang, D.: Does fake news in different languages tell the same story? An analysis of multi-level thematic and emotional characteristics of news about covid-19. Inf. Syst. Front. **25**, 1–20 (2022)
40. Zhou, X., Mulay, A., Ferrara, E., Zafarani, R.: Recovery: a multimodal repository for covid-19 news credibility research. In: CIKM 2020, pp. 3205–3212 (2020)

Popular Songs: The Sentiment Surrounding the Conversation

Julian Stefanzick and Xin Zhao[✉]

Seattle University, Seattle, WA 98122, USA
{stenfanzi,xzhao1}@seattleu.edu

Abstract. Music plays an important role in our daily life. It can have a powerful effect on our emotions, mental health, and even the community we live in. Although numerous studies have been conducted to prove the great impact music has on humans, few investigations place an emphasis on the exploration of the relationship between music and listener's sentiment. To this end, we first examined three song demographics: Beats Per Minute, Key, and Length, and six song metrics: Danceability, Energy, Speechiness, Acousticness, Liveness, and Valence of popular songs, and then conducted an empirical study to examine the potential correlation between song demographics/metrics and the sentiment expressed as in written text (such as social media). To accomplish this, we scraped around 20 million tweets referencing the most popular songs from 2018 to 2022 as shown on Spotify's Top Global chart, as well as the immediate surrounding tweets, and performed a double sentiment analysis on the data. Our study concludes that there exists a significant correlation between all the pairs of song metrics. From the sentiment analysis of tweets, our results indicate that there may not be a significant correlation between the sentiment expressed in tweets of a song's listeners and the song itself. Our study provides empirical evidence for a deeper understanding of popular songs using data mining techniques.

Keywords: Data Mining · Sentiment Analysis · Song Metrics · Regression Analysis

1 Introduction

Music is a higher revelation than all wisdom and philosophy. Music is the electrical soil in which the spirit lives, thinks and invents.

— *Ludwig van Beethoven*

Music is integral to humankind. Regardless of age, culture, or race, music is all around—music creation, instrument playing, or listening and singing, are common activities in most people's daily life [14]. These activities not only help us express our personal feelings, positive or negative, but also have an impact on our surrounding environment. Numerous studies have been conducted to offer empirical evidence on the effect that music has on human mood, health, and community [15].

Research communities have seen several investigations into music popularity predictions based on music patterns [16], metrics [17], and deep learning techniques [18]. However, not much academic research has been conducted to analyze the relationship between music and listener sentiment as expressed in digital written text (such as social media). To this end, we first analyzed the five most popular songs on Spotify for each month between March 2018 to February 2022 to gain a better understanding of the **characteristics of popular songs in recent years**. Combining the tweets that reference these songs, as well as the immediate surrounding tweets, we further **explore if a relationship exists between music listened to and written sentiment**.

Obtaining a listener's sentiment is not an easy task. To estimate the sentiment, we focus on the sentiment expressed through the listener's tweets. We chose Twitter[1] in this study due to its enormous user base and the multitude of tools available to access and scrape data. It is estimated that there are around 395 million active daily Twitter users and about 500 million tweets are sent every day[2]. Additionally, Twitter was chosen due to the multitude of available tools we had available to scrape data from the platform.

Sentiment analysis by mining users' tweets is not a new topic in the research community—several investigations have adopted Twitter to analyze sentiment [9–12]. The results of these studies show that using tweets could provide reliable results regarding users' sentiment. To this end, we scraped around 20 million tweets. This number includes direct references to popular songs we selected and the immediate surrounding tweets. Our results show that there may not be a measurable relationship between the sentiment expressed in tweets and the songs Twitter users listened to. Moreover, We hope this paper sparks increased interest in exploring the relationship between the music we listen to and the effect it has on our writing and work. The contribution of the paper is to **gain deeper insights into the relationship between music and humans**. This work could also be of interest to the sentiment analysis community as a new way to explain various online text-based interactions and what leads to a particular sentiment being expressed.

The paper is organized as follows: In Sect. 2, we introduce the background of this paper, along with our research questions. Section 3 explains the methodology of our investigation, and Sect. 4 describes the results of our experiments. Threats to the validity of this paper are discussed in Sect. 5, together with the techniques we undertook to mitigate these threats. Finally, in Sect. 6, we conclude the paper and lay out our future work.

---

[1] https://Twitter.com/home.

[2] https://thesocialshepherd.com/blog/Twitter-statistics.

2   Background and Research Questions

2.1   Song Demographics

The word *Demographics* is originally defined as "the statistical characteristics of human populations[3]." In the context of our paper, we define the song demographics as the following indicators. Our investigation focuses on three dimensions: Beats per minute (BPM), Key, and Length.

BPM. BPM (Beats Per Minute) refers to the number of beats in one minute. It indicates the tempo of a song. Even though the BPM for the music can be of any value theoretically, research has shown that the BPM for the majority of songs is around 120 [5].

Key. In music, a key is "the main group of pitches, or notes, that form the harmonic foundation of a piece of music [13]." In other words, keys identify the tonic tone/chord in a song. The keys in a piece of music are represented using seven English letters: A, B, C, D, E, F, and G. To enrich music pitches, the symbol # is used to differentiate the intervals of different keys. The key notation and representation are shown in Fig. 1.

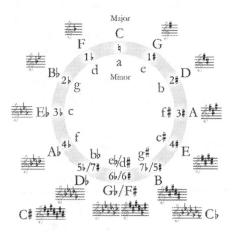

Fig. 1. Music Keys

Length. The length of a song refers to the song's duration. A song may last as short as tens of seconds (for example, "Her Majesty" by The Beatles, 17 s) and as long as over 13 h ("The Rise and Fall of Bossanove" by PC III, 2016) Even though there is no systematic research on the perfect length of a song, the average length of a pop song is around three minutes and thirty-second and the average length of a rock song is four minutes and fifteen seconds[4].

---

[3] https://www.merriam-webster.com/dictionary/demographic.
[4] https://bigtimemusicians.com/how-long-should-a-song-be.

Based on the above discussions, the first research question posed in our study is:

> **Research Question 1 (RQ 1):**
> **What are the demographics of *popular* songs?**

2.2 Song Metrics

Additionally, we are interested in exploring deeper song metrics (features). Spotify internally designates each song a score for each of the following categories:

Danceability. Danceability refers to how favorable a song is for dancing. Danceability ranges from 0.0 (least dance-able) to 1.0 (most dance-able).

Energy. In this paper, we follow the definition provided by Ewer [3]—the energy of the song is "the sense of forward motion in music." Energy is measured from 0.0 to 1.0, where 1.0 indicates that a song is very intense and engaging.

Valence. Music valence can be understood in terms of positive or negative emotions the music expresses [2]. Valence is a measurement scale from 0.0 to 1.0. A higher value indicates more positiveness in a song.

Speechiness. Speechiness refers to whether a track contains spoken words. It ranges from 0.0 to 1.0. Usually, a value below 0.33 indicates that the song does not contain any speech.

Acousticness. In the context of music, acoustic refers to "without electrical amplification [4]". In the scope of this paper, acousticness is a confidence metric from 0.1 to 1.0 of whether the music is acoustic.

Liveness. This characteristic of a song detects the presence of audiences in a song. It is a probability to indicate whether the song is performed live or not. The value of liveness ranges from 0.1 to 1.0.

Following these metrics, we are interested in understanding the metrics in popular songs. Therefore, the second research question in our paper is:

> **Research Question 2 (RQ 2):**
> **Is there a *relationship* among popular songs' metrics?**

2.3 Sentiment Analysis

Sentiment Analysis is the process of the detection of mood in textual data with the help of Natural Language Processing (NPL) techniques. The resulting sentiment usually falls into the following three categories: positive, negative, and neutral.

There are two approaches to analyzing the sentiment expressed in the text. The first approach is *manual examination*. In this approach, a human examines the textual data manually and labels the sentiment for each piece of information. By nature, humans have a better understanding of human language than computers, however, this approach is slow in terms of examination speed. Moreover, personal bias may be introduced in the process of analysis. Therefore, the approaches widely adopted today for sentiment analysis tasks are based on *automatic classification*.

Automatic techniques to analyze sentiment mainly consists of three steps. 1) Data Collection. 2) Data Processing 3) Data Analysis. Data collection is mostly achieved through Web scraping, a process of applying computers to extract information from the Internet. Data processing is the process of converting unstructured textual data to explicit features that are recognizable by computers. This process is also formally known as feature engineering [7]. In the last step—data analysis, different algorithms/models are applied to the data to find the sentiment expressed in the data.

Using sentiment analysis, we explore if a relationship exists between a human's mood, as expressed in text, and the popular songs they listen to. To this end, our research question is:

> **Research Question 3 (RQ 3):**
> **Is there a relationship between a popular song's demographics/metrics and a listener's sentiment as expressed in written text?**

3   Methodology

3.1   Song Selection

Music is produced rapidly. Spotify is "seeing a new track uploaded to its platform every 1.4 s." To this end, we chose the *popularity of the song* as the criteria to select our songs for analysis. To achieve more comprehensive results, We decided to target the most popular **five** songs of each month. The decision to choose only the top five songs each month for our analysis stemmed from a limit of time and resources available. Scraping a broad range of tweets across several years is an intensive task that quickly maxed out our computational power. In future work, we would be interested in greatly expanding the number of songs we use to find tweets to scrape.

1) Popularity of the Songs
One of the first issues of consideration we ran into was how to determine the popularity of songs for each month. We initially examined the Billboard Global 200[5] (a website ranking the top songs based on steaming and sales worldwide) chart which reports the top 200 played songs each week. The reason why we chose Billboard is that Billboard is considered one of the most reputable sources for reporting the top played songs [1]. However, we were only able to find Billboard chart data on a weekly basis instead of monthly. This posed a problem when determining the top five songs for each month because some weeks would start halfway through the end of one month and finish in the next month. For example, Friday, January 28th through Thursday, February 3rd. Without clear numbers,

---

[5] https://www.billboard.com/charts/billboard-global-200/.

it was difficult to accurately determine the top five songs for each month when exclusively viewing the Billboard Global 200.

To solve this, we instead used the Spotify Top 100 Global charts published by Spotify. This increased the accuracy of the determination of popular songs because they were published directly from the source of the song playback. We were also confident that these songs would well represent the most popular songs in the world as Spotify has over 400 million users worldwide, making it the most popular music streaming service in terms of subscribers by far[6]. The Spotify charts website also tracks the top global songs on a weekly basis similar to Billboard, however with the inclusion of streaming numbers, as well as daily streaming reports. This allowed us to easily determine the top five songs streamed on Spotify each month globally.

2) Time Span of the Songs

The time span we selected was March 2018 through February 2022. Once again, the decision to target this time frame stemmed from available data and a limit to our time and computational resources. The top five songs over a four-year time span gave us 240 data points. In total, we ended up with 104 unique songs for analysis, as a song may be on the top five list for multiple months.

3.2   Song Demographics and Metrics

We are interested in two song demographics: the BPM each song plays in, and the run time for each song. These were both obtained from Tunebat[7]. Tunebat is a music database which uses the Spotify Web API. Additionally, we also used Spotify's online Web API[8] to obtain the six metrics of the song. These values are assigned to each song by internal Spotify processes. We were unable to find information on their methods of calculation, as we believe Spotify does not disclose these methods publicly.

To better illustrate the above-mentioned information we retrieved, we present a table (Table 1) to indicate the most streamed five songs in January 2022.

Table 1. The Demographics and Metrics of the Five Most Streamed Songs in January 2022

No.	Artist	Song	BPM	Length	Danceability	Energy	Valence	Speechiness	Acousticness	Liveness
1	GAYLE	abcdefu	122	2:49	0.695	0.54	0.415	0.0493	0.299	0.367
2	The Kid LAROI, Justin Bieber	STAY (with Justin Bieber)	170	2:22	0.591	0.764	0.478	0.0483	0.0383	0.103
3	Glass Animals	Heat Waves	81	3:59	0.761	0.525	0.531	0.0944	0.44	0.0921
4	Imagine Dragons	Enemy	77	2:53	0.728	0.783	0.555	0.266	0.237	0.434
5	Adele	Easy On Me	142	3:45	0.604	0.366	0.13	0.0282	0.578	0.133

---

[6] https://www.businessofapps.com/data/spotify-statistics/.

[7] https://tunebat.com/.

[8] https://developer.spotify.com/console/get-audio-features-track/?id=.

3.3   Tweets Collection

In this section, we detail the tools and search processes used to scrape the song referencing and surrounding tweets.

Tools. We chose a free multipurpose social media scraping package called SNScrape[9]. SNScrape provides a rich collection of services, including scraping based on user profiles, hashtags, or queries. This greatly expanded the customization available for how we searched for tweets.

Query. One of the most important considerations was what the search term should be for Twitter scraping. We wanted to collect the most amount of tweets possible for analysis, however, we had to ensure the tweets were referencing the songs as intended. For example, single-word songs such as "Mood" by 24kGoldn or "Positions" by Ariana Grande would result in tweets that would use the song name in its wording but would obviously not be in reference to the song itself (Fig. 2 shows such an example).

Fig. 2. A Tweet Containing *Word* "Mood"

The solution was to create search terms which included both the song name and the artist name. For example, for "Mood" by 24kGoldn, the search term would be "Mood 24kGoldn". One result after the search with the combination of song name and artist name is shown in Fig. 3.

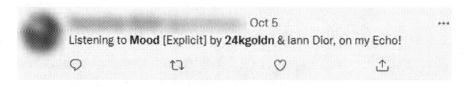

Fig. 3. A Tweet Containing *Song* "Mood"

---

[9] Available at https://github.com/JustAnotherArchivist/snscrape.

Language. We decided to only target tweets written in English. This decision was made to ensure consistency in our sentiment analysis stage. To this end, we employed the use of spaCy[10], an open-source library for language processing and detection. We chose spaCy for its speed and simplicity [2]. We excluded tweets that had less than 90% English to include song titles that may not be traditional English words. This method, while intended to avoid misrepresenting intentions when converted to English, runs the side effect of excluding many tweets that would be of importance to our investigation. In future investigations, we would like to include all tweets in our investigation given enough time and resources.

Scraping Results. The above-mentioned process leads to a tuple as one piece of the scraping result. In this paper, we define this tuple t as:

$$t = \langle d \,|\, u \,|\, c \rangle$$

where d, u, and c represent the date the tweet was sent, the username of the person who sent the tweet, and the tweet itself, respectively. Once SNScrape had determined that it had found all of the possible tweets that included the search term within the month time frame, the list was then converted to a Pandas[11] DataFrame and saved as a CSV file. Next, the CSV file was opened in Excel. Excel's "Remove Duplicates" feature was used to delete any duplicate entries that may have been recorded. Duplicates occurred when multiple search terms were used and the same tweet would be scraped for each search term.

Related Tweets. After scraping all of the tweets that were related to the songs, the next step was to scrape the other tweets sent by the users who had made the song-related tweets. In this paper, we scraped the tweets made the day before, the day of, and the day after the original song-related tweet. This process included similar tools and techniques to when we scraped the original song-related tweets. SNScrape was once again used for the tweet scraping, and spaCy was used again to only select tweets written on Instagram.

3.4   Data Cleaning

Once all of the tweets were collected, a data cleaning process was used to select the most relevant data. Relevant data was defined as data that would be worth recording the sentiment of. Furthermore, relevant data was defined as originating from tweets sent by an individual which contained non-automatically generated information. In particular, tweets that were not relevant and therefore not intended for further analysis included tweets sent by radio stations, tweets that contained links to articles or videos, or tweets which were automatically generated by external services. For example, tweets that included the hashtag #NowPlaying were overwhelmingly sent by radio stations or automatically posted to

---

[10] https://spaCy.io/.

[11] https://pandas.pydata.org/.

users' accounts when connected to services such as Spotify. We did not want to include these types of tweets in our analysis because they would impact our overall scoring in a way that would skew our results away from what the actual human-represented sentiment was in a given month.

At no point did any person write these tweets, so the data from a sentiment analysis on these tweets would not add any substance to our final analysis. For these mentioned reasons, tweets that contained these characteristics were removed from the data pool and from further consideration. This process removed a large amount of automatically computer-generated tweets, something that would have been unfeasible and overly time-consuming to perform manually. While this process saved many hours of work, one downside is that we were not able to guarantee that every tweet of this type was caught and removed from consideration.

Next, all of the tweets which passed this portion underwent a further cleaning process. We used regular expressions in Python to conduct the cleaning. This portion of the data cleaning focused on removing information from the tweets that provided no useful information to our overall sentiment score. Data of this type that was removed included hyperlinks, references to other Twitter users via their handles, the hashtag symbol, and retweet information from retweeted tweets. Once again, it is worth mentioning that these automatic processes removed a large amount of data unintended for analysis, however, because it was automated by the program certain data unintended for analysis may have still found its way into the final pool, just as data which was intended for analysis may have been removed from the final pool as an unintended consequence of the data cleaning process.

3.5   Sentiment Analysis

In this section, we discuss the two sentiment analysis techniques we applied in our investigation: TextBlob [19] and VADER [20]. Both TextBlob and VADER follow a rule-based approach to sentiment analysis. When these scores are combined, an overall sentiment score for the tweet is found. This method of sentiment analysis was chosen over machine learning techniques because we did not need the tools to adapt to changing trends or new rules. We wanted a consistent sentiment scoring method across all of the data.

We decided to run two sentiment analysis tests using different tools to obtain a higher accuracy for our final sentiment scores. While both TextBlob and VADER are regarded as highly accurate in their sentiment scoring, neither tool is perfect. In an effort to capitalize on both their strengths and obtain the most accurate picture of overall sentiment, a further cleansing process was used. After all of the selected tweets were run through both sentiment analysis tools, only the tweets where the tools agreed upon the tweets' sentiment rating of negative, neutral, or positive were kept. For example, if TextBlob rated a tweet as Negative but VADER rated the same tweet as Neutral, then the tweet would be removed from further consideration. However, if TextBlob labeled a tweet as positive and VADER also labeled the same tweet as positive, then that tweet would be included in our final set of data for the month. This ensured a higher accuracy of our final data pool's sentiment labels.

The final step each individual tweet took was to combine the polarity score of TextBlob's sentiment analysis with the compound score of VADER's sentiment analysis. Since both metrics used a range of -1.0 to represent tweets that were labeled as extremely negative, to 1.0 to represent tweets that were labeled as extremely positive, we were able to take a simple average by combining the scores and dividing by two to arrive at the final combined sentiment value from -1.0 to 1.0 for each tweet. The overall sentiment label of negative, neutral, or positive remained unchanged during this averaging process as both tweets started with the same label due to the final previous data cleaning process. An example of sentiment analysis results can be seen in Table 2.

Table 2. The Sentiment of the Most Stream Five Songs in January 2022

No.	Artist	Song	Total Score	Total Tweets	Combined Sentiment Score
1	GAYLE	abcdefu	2163.952128	14680	0.1474
2	The Kid LAROI, Justin Bieber	STAY (with Justin Bieber)	5927.723526	41242	0.1437
3	Glass Animals	Heat Waves	3748.23804	25533	0.1468
4	Imagine Dragons	Enemy	2579.916671	17864	0.1445
5	Adele	Easy On Me	8743.125508	53558	0.1632

Finally, we ended up with 238 collections of tweets for the unique month and song combinations. A collection can be defined as all of the cleaned tweets with their combined sentiment scores for a particular month and of a particular chart rating for that month. For example, one collection may include the cleaned data and sentiment scores for tweets sent in January 2019 which the original tweet used in the data collection process contained the number one most streamed song on Spotify for that month. Each collection underwent an averaging process where the combined sentiment scores for each tweet were added up, with the total value being divided by the number of tweets in each collection. These scores were all compiled into a single spreadsheet which included columns for the Month/Year/ChartRank, song artist, song name, the added-up sentiment score of each individual song for the entire associated collection, the total number of the tweets in the associated collection, and the combined sentiment score average of the entire associated collection.

3.6   Significance Analysis

The final step in this process was a series of regression tests. We used IBM's SPSS Statistics for this process because the software is widely regarded as an accurate and comprehensive statistical tool used by previous sentiment analysis studies [8]. We used a single regression test for each song metric and characteristic against the found sentiment for a collection of song-based tweets. This was intended to pinpoint exactly which characteristic of a song may be behind a particular increase in positive or negative expressed sentiment.

4   Results and Discussions

4.1   Song Demographics

In total, we analyzed 104 unique songs. We conducted some basic statistical tests on the song demographics to learn more about our sample pool. The first song demographic we examined is the song's length. The result shows that among these 104 songs, the shortest song lasts for 2 minutes and 8 s while the longest song lasts for 5 min and 13 s. The average length is 3 min 20 s (200 s).

As to the tempo of these popular songs, we found that the average tempo is 116.99 BPM and the median is 117 BPM. The range spans 75 BPM to 171 BPM. Last, our results also show that 58 (56%) of the songs were written in a Major key while 46 (44%) of the songs were written in a minor key. The most popular keys for songs to be written in were G# Major, C# Major, and C Major, each with 9 songs written in the key. The least popular keys were D minor and D# Major, both with 0 songs.

4.2   Song Metrics

In this subsection, we focus on the question "Are there any correlations among different song metrics?" To this end, we performed a regression analysis between each pair of the song metrics we defined in the previous section. The results are shown in Table 3. However, we must note that due to the fact that the calculation of these metrics is not disclosed by Spotify, we must clarify that it is unclear if these correlations are due to the design of these metrics or not.

Table 3. Absolute p-value for different pairs of song metrics

p-value	Danceability	Energy	Speechiness	Acousticness	Liveness	Valence
Danceability	-	<0.001	<0.001	<0.001	<0.001	<0.001
Energy		-	<0.001	<0.001	<0.001	<0.001
Speechiness			-	<0.001	<0.001	<0.001
Acousticness				-	<0.001	<0.001
Liveness					-	<0.001
Valence						-

Finding 1: There is a correlation between all the pairs of song metrics.

4.3   Relationship Between Sentiment and Music

Our results come from the sentiment analysis performed on the tweets containing references to the songs, the tweets from one day before the tweet was sent, the day the tweet was sent, and the day after the tweet was sent.

The first linear regression test looked to see if there was a relationship between a song's BPM and the combined sentiment score of the month of the song's popularity which the tweets were compiled from. The dependent variable in this test was the combined sentiment score, and the independent variable was the BPM. The results of this test gave us a p-value of .696. This suggests that there is no correlation that results in a change in the sentiment score when there is a change in the BPM.

The next linear regression test looked to see if there was a relationship between a song's length and the combined sentiment score of the month of the song's popularity which the tweets were compiled from. The dependent variable in this test was the combined sentiment score, and the independent variable was the length. The results of this test gave us a p-value of .508. This suggests that there is no correlation that results in a significant change in the sentiment score when there is a change in the length of a song.

Using the same regression analysis for all the pairs, we can obtain the results of the relationship between music demographics and characteristics, and sentiment. Please see Table 4 for a complete list of all tests and their accompanying results.

Table 4. Results: Tests and p-values

Test	p-value
Sentiment vs. BPM	.696
Sentiment vs. Length	.508
Sentiment vs. Danceability	.060
Sentiment vs. Energy	.286
Sentiment vs. Speechiness	.057
Sentiment vs. Acousticness	.428
Sentiment vs. Liveness	.918
Sentiment vs. Valence	.564

All tests resulted in p-values greater than 0.005. The results mean that we could not reject the null hypothesis. This suggests that there is no measurable linear effect on tweet sentiment by the various song features we tested for. In the future we are interested in using a larger data set and including additional tests such as a Pearson Correlation Coefficient analysis to gain additional insight.

Finding 2: There is no significant linear correlation between a listener's written text sentiment and various song characteristics.

5   Threats to Validity

This section examines the threats to the validity of our study. It is organized by the discussion of three different types of threats: internal threats, external threats, and construct threats defined by Parker's work [6]. We also present the means we undertook to mitigate these threats.

5.1   Internal Threats

In this investigation, we have summarized three internal threats. First, the technique we chose for sentiment analysis may have misinterpretations regarding the sentiment. To mitigate this threat, we chose two popular sentiment analysis tools to reduce the possibility of misinterpretation. Second, the tweets we scraped may not reflect the sentiment of the user. However, this threat resides in any sentiment analysis using tweeter because it is not possible that tweets sent by users to capture the sentiment of users every single time. It is also difficult to determine what other factors in a tweeter's life may apply to the sentiment score of their tweet.

Last, the scraping tool we adopted - SNScrape, gave us a slightly different number of total tweets (± 100 tweets) when running the program multiple times using the same conditions (month and song) to scrape tweets. However, based on the large number of total tweets, we believe this difference will not make a significant impact on the result interpretation.

5.2   External Threats

In the context of this study, the external threat to validity comes from the fact that our experimental results are based on 104 unique songs. The results may be not applicable to other songs. However, similar to any other empirical study, it is not feasible to include every piece of music in one study. To mitigate this, we chose the most popular streamed song in the past four years and obtained a large data set from Twitter to explore. In the future, we plan to conduct a follow-up study to further increase the number of songs to reduce external threats.

5.3   Construct Threats

Our construct threat in this study refers to whether the technique we use could accurately explore the relationship between sentiment and music features. In our study, we applied regression analysis to inspect the relationship. Even though there may exist other statistical approaches to explore the relationship between two variables, based on our limited knowledge, regression analysis fits the experiment setting of our study, thus providing us the confidence to draw conclusions based on regression analysis.

6   Conclusion and Future Work

By mining tweets, our paper targets a better understanding of the characterization of popular songs. First, we examined the song demographics and found some common features among popular songs. For example, the tempo for most popular songs is around 117 BPM. Second, we analyzed the metrics of popular songs from six perspectives and conducted a regression analysis among these metrics. We found that these metrics are correlated with each other. However, our experiment did not see any significant relationship between the sentiment of Twitter users' tweets and the various song metrics of songs they had tweeted about.

For future work in the area of Twitter sentiment analysis and song characteristics, we suggest examining a larger sample size of songs. In our experiment, we had 238 data points which included 104 unique songs. Future work may take the entire Spotify top 100 songs for each month into consideration. Additionally, a longer time frame may be established to source data. Both Twitter and Spotify started in 2006, meaning that there are over fifteen years of potential data to use in further analysis, whereas our experiment only examined March 2018 through February 2022. Other future work could look further into Twitter users' sentiments to try to determine a change in sentiment before and after listening to a song. Another direction could be to examine the differences in song metrics of popular music before and after the start of the COVID-19 pandemic. Additionally, using other social media platforms where music is discussed such as YouTube, Instagram, and Reddit would increase data and may lead to more insights.

References

1. Seeger, A.: Music and dance. In: Companion Encyclopedia of Anthropology, pp. 720–739. Routledge (2002)
2. Juslin, P.N., Västfjäll, D.: Emotional responses to music: the need to consider underlying mechanisms. Behav. Brain Sci. **31**(5), 559–575 (2008)
3. Ewer, G.: Making an energy chart for your song (2014). https://www.secretsof songwriting.com/2014/07/16/making-an-energy-chart-for-your-song
4. Beranek, L.: Music, Acoustics, and Architecture. MIT Press, Cambridge (2008)
5. Hurless, N., Mekic, A., Pena, S., Humphries, E., Gentry, H., Nichols, D.: Music genre preference and tempo alter alpha and beta waves in human non-musicians. Impulse **22**(4), 1–11 (2013)
6. Randall, M.P.: Threats to the validity of research. Rehabil. Couns. Bull. **36**(3), 130–138 (1993)
7. Zheng, A., Casari, A.: Feature Engineering for Machine Learning: Principles and Techniques for Data Scientists. O'Reilly Media, Sebastopol (2018)
8. Chakraborty, K., Bhatia, S., Bhattacharyya, S., Platos, J., Bag, R., Hassanien, A.E.: Sentiment analysis of COVID-19 tweets by deep learning classifiers-a study to show how popularity is affecting accuracy in social media. Appl. Softw. Comput. Part A **97**, 106754 (2020)

9. Agarwal, A., Xie, B., Vovsha, I., Rambow, O., Passonneau, R.J.: Sentiment analysis of Twitter data. In: Proceedings of the Workshop on Language in Social Media, pp. 30–38 (2011)

10. Giachanou, A., Crestani, F.: Like it or not: a survey of Twitter sentiment analysis methods. ACM Comput. Surv. **49**(2), 1–41 (2016)

11. Kouloumpis, E., Wilson, T., Moore, J.: Twitter sentiment analysis: the good the bad and the omg! In: Proceedings of the International AAAI Conference on Web and Social Media, vol. 5, no. 1, pp. 538–541 (2011)

12. Neethu, M.S., Rajasree, R.: Sentiment analysis in Twitter using machine learning techniques. In: 2013 Fourth International Conference on Computing, Communications and Networking Technologies, pp. 1–5 (2013)

13. Chase, S.: What Is A Key In Music? A Complete Guide. Online blog (2022). https://hellomusictheory.com/learn/keys/

14. Welch, G.F., Biasutti, M., MacRitchie, J., McPherson, G.E., Himonides, E.: The impact of music on human development and well-being. Front. Psychol. **11**, 1246 (2020)

15. Welch, G.F., McPherson, G.E.: Commentary: music education and the role of music in people's lives. In: Music and Music Education in People's Lives: An Oxford Handbook of Music Education, vol. 1, no. 1 (2018)

16. Lee, J., Lee, J.S.: Predicting music popularity patterns based on musical complexity and early stage popularity. In: Proceedings of the Third Edition Workshop on Speech, Language & Audio in Multimedia, pp. 3–6 (2015)

17. Lee, J., Lee, J.S.: Music popularity: metrics, characteristics, and audio-based prediction. IEEE Trans. Multimedia **20**(11), 3173–3182 (2018)

18. Martín-Gutiérrez, D., Peñaloza, G.H., Belmonte-Hernández, A., García, F.Á.: A multimodal end-to-end deep learning architecture for music popularity prediction. IEEE Access **8**, 39361–39374 (2020)

19. Loria, S.: Textblob Documentation. Release 0.15, vol. 2, no. 8 (2018)

20. Elbagir, S., Yang, J.: Twitter sentiment analysis using natural language toolkit and VADER sentiment. In: Proceedings of the International Multiconference of Engineers and Computer Scientists, vol. 122, pp. 12–16 (2019)

Market Sentiment Analysis Based on Social Media and Trading Volume for Asset Price Movement Prediction

Jiahao Li, Yuyun Gong, Qinghua Zhao, Yufan Xie, Simon Fong, and Jerome Yen[✉]

Faculty of Science and Technology, University of Macau, Macau SAR, China
{mc05504,mc15076,mc05505,mc05384,ccfong,jeromeyen}@um.edu.mo

Abstract. As more and more netizens participate in financial market transactions, online discussions on asset price movements are becoming more comprehensive and timely. Online text, especially from social media, has the potential to be an important data source for financial opinion mining. Market sentiment analysis mainly includes direct analysis methods in the form of text-based surveys and indirect inference methods based on structured data such as price, trading volume, and volatility. In theory, the former is helpful for us to understand investor sentiment earlier, but due to the difficulty of obtaining a sufficient number of objective survey samples, its obtained research attentions are far less than the latter. To combine the advantages and offset the weakness of these two approaches, this paper uses Valence Aware Dictionary and Sentiment Reasoner (VADER) and Fast Fourier Transform (FFT) to construct social media sentiment indexes based on plenty of daily discussion texts about Bitcoin (BTC) and S&P500 (SPX) from Reddit for analyzing their interaction with prices. We also propose a new time series synchronization verification method called Rolling Time-lagged Cross-correlation (RTLCC) surface, and corresponding feature constructing methods, in which RTLCC helps us observe Time-lagged Cross-correlation from the perspective of Rolling Correlation while determining the hyperparameters (Window Size & Time Offset) for features construction. Finally, based on these features, we use four machine learning classifiers for modeling and verify the effectiveness of the proposed market sentiment analysis pipeline, in which on the prediction of 10-day price movements, the best model achieves 89.9% in accuracy (ACC) and 92.5% in AUC.

Keywords: Price Movement Prediction · Market Sentiment Analysis · VADER · Fast Fourier Transform · Time Series Synchronization Verification · Machine Learning

1 Introduction

Behavioral Finance Theory [17] argues that asset price movements align with the prevailing investors' sentiment. Since the text data of social media contains

X. Yang et al. (Eds.): ADMA 2023, LNAI 14176, pp. 383–398, 2023.
https://doi.org/10.1007/978-3-031-46661-8_26

abundant information that can be used to support trading decision-making [21], Market Sentiment Analysis is becoming a research hot spot in predicting financial market movements. The sentiment from social media text directly affects investors' propensity to trade, so it is often considered a direct measurement of market sentiment. Besides, the trading volume is proven to correlate with the market sentiment [15], and it is used to be an indirect measurement of market sentiment because it indirectly reflects the market sentiment. If the above two can be combined, not only the expansion of data sources is realized, but also the convenience from structured data analysis ideas of the latter are introduced while maintaining the timeliness of the former. Thus we attempt to quantify the correlation between market sentiment, trading volume, and price, aiming to construct a general pipeline for predicting price movement based on market sentiment analysis.

To be specific, we summarize market sentiment analysis approaches and introduce related works in Sect. 2. And in third section, a VADER-based sentiment index constructed on text data from Raddit discussion is proposed. Then we design a new synchronization verification indicator, RTLCC surface, and a set of relevant feature construction methods in the feature engineering. For the experimental part in Sect. 4, we respectively evaluate the effectiveness of these features on Support Vector Machine (SVM), Random Forest (RF), XGBoost (XGB), and LightGBM (LGB), while comparing their predicting performances. In short, the overall process has been briefly summarized in Fig. 1.

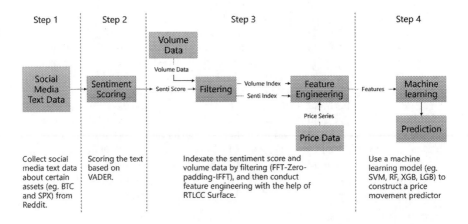

Fig. 1. The overall pipeline of the asset price movement prediction

2   Related Work

2.1   Market Sentiment Analysis

Market Sentiment Analysis applies the NLP technology to quantitative finance and targets to analyze people's attitudes toward assets through computation of subjectivity in texts [22]. The analysis results are often turned into the sentiment index [19], which is a productive tool to quantify sentiment in figures and can be widely applied for Financial Market Predictive Analysis.

For the industry application of sentiment index, back in 1993, the CBOE Volatility Index (VIX) [20] was introduced to measure the market's expectation of 30-day volatility based on the assumption that the trading volume was a good proxy for investors' sentiment [6], and in nowadays, the S&P 500 Twitter Sentiment Index and S&P 500 Twitter Sentiment Select Equal Weight Index are always used to track the performance of the constituents with the most positive sentiment. Furthermore, a growing body of research keeps showing the actual value of the sentiment index: Huang et al. [9] devised an index capable of revealing investors' sentiment and predicting the overall stock market by using the least squares method, which outperformed well-established macroeconomic variables; Da Z Engelberg et al. [5] constructed a sentiment index to predict short-term returns and volatility, which is derived from daily Internet search volumes based on millions of households.

In addition, many studies also show that it is feasible and efficient to utilize the sentiment from social media to enhance financial data mining. Karabulut [11] declared that Facebook's Gross National Happiness (GNH) with the ability to predict changes in both daily returns and trading volume in the US stock market. As one of the most representative methods for sentiment analysis of social media, VADER [10] is an efficient rule-based algorithm that can help calculate a specified set of predetermined sentiment scores by identifying each feature (word, expression, and abbreviation) in a sentence. Toni Pano et al. [14] performed VADER-based sentiment analysis on BTC tweets to identify the role of different text preprocessing strategies in predicting Bitcoin prices. Kim Y B et al. [12] successfully predicted the price fluctuations of cryptocurrencies such as Bitcoin and Ethereum by using VADER to tag user comments in online communities.

2.2 Price Movement Prediction Based on Machine Learning

It is difficult to achieve financial prediction using market sentiment analysis alone, which is often only used as an important factor mining method. To achieve price movement prediction, it is also necessary to build a prediction model based on classification algorithms, and machine learning is a promising method. Nowadays, many scholars convince that some patterns are invisible to traditional financial or economic theories but can be detected and exploited by machine learning. Therefore, they have tried to use different machine-learning models to predict asset price movement.

As early as 2013, Alexander Porshnev et al. [16] used SVM with historical close price and Twitter tweets as input to achieve better results than random prediction in the price movements prediction of the S&P 500 Index and found that the market sentiment derived from text data improves the performance of the predictor. Furthermore, Al Nasseri et al. [1] demonstrate that decision tree algorithms can effectively quantify the relationship between semantic terms on StockTwits and trading behavior, like forecasting the impact of sentiment changes on the Dow Jones Industrial Average (DJIA) index, which has helped us to understand how emotions and language on social media platforms influence financial markets and provides a potential avenue for developing decision-making tools in the investment field. Recently, Guliyev H [8] compared four different machine learning models on predicting the monthly movements of WTI Oil's price, which shows that the XGBoost model made the best result of 91.8% accuracy.

3   Methodology

In this paper, VADER was used for calculating the sentiment score based on the discussions about different assets (e.g., BTC and SPX) on social media (Reddit) every day, which are Fourier transformed into sentiment indexes. Besides, Fourier Transform will also be performed on these assets' volumes to obtain another kind of sentiment index. Then, multiple time series correlation analyses, feature construction, and feature selection are performed on the price time series and the above two sets of sentiment indexes to obtain important features.

Finally, four different machine learning models (e.g., SVM, RF, XGB, LGB) are used to predict the movement of assets' prices with these features, and the prediction is a binary time series (e.g., up and down).

3.1   Sentiment Index Construction

As the information update frequency of the sentiment index is too fast to be matched with the price, it is a suitable way to change sentiment analysis into a sequence prediction task by constructing sentiment indexes. The data of close price and volume can be collected from Yahoo Finance as the price series and volume sequence. The sentiment analysis begins with the collection of daily discussion texts for specific assets from Reddit. The volume sequence can be used to represent the indirect sentiment index (Volume Index), while the direct sentiment index (Senti Index) needs to be constructed on these texts. The construction steps are as follows:

(1) Use API. PRAW to scrape the daily discussion describing a specific asset from Reddit. The statical description of the collected data is shown in Table 1.

Table 1. Data description.

	BTC	SPX
Time Frame	2018.3.1 - 2022.3.1	2018.3.1 - 2022.3.1
Trading Days	1455	1008
Comment Numbers	371477 (77752.881kb)	735289 (156367.301kb)
Missing Values/Days	86 (5.9%)	10 (1.0%)

(2) Calculate the sentiment scores of those text data by VADER and aggregate them by days. An example of how to get a sentiment score of a comment by VADER is shown in Table 2, in which the α is the approximate max sentiment score, and the meanings of the 'pos', 'neg', 'neu', 'total' are respectively the positive, negative, neutral, and compound sentiment score. +1 is to compensate for neutral words.

(3) Interpolate missing values by moving averages. Discussions on social media for certain assets are not present on all trading days, resulting in missing values in the sentiment score sequence. The numbers of various assets' missing values are shown in Table 1, none of which exceeds 5.9% of the respective trading days.

(4) Remove the noise of the sentiment scores by transforming this time-domain sequence into the frequency-domain one by FFT [4] and filtering out relative high-frequency components according to the threshold T while conducting zero-padding, which refers to the ratio of the relative low-frequency part retained after low-pass filter processing to all original components. Then the zero-padded sequences will be turned back into the time domain form by Inverse Fast Fourier Transformation (IFFT) [4]. For ease of representation, FFT_T represents a complete FFT-Zero-padding-IFFT period with $T\%$ as the filtering threshold like 2%, 4%, ..., 100%. Figure 2 describes the construction process of BTC's Senti Index when T is equal to 10%.

Table 2. An Example of VADER-based Sentiment Scoring.

	pos	neg	neu	total
Just			0+1	**Normalizing function:** $\frac{x}{\sqrt{x^2+\alpha}}$ $x = -2.4 - 2.1 = -4.5$
gonna		0+1		
hurt		2.4+1		
the			0+1	
poor		2.1+1		
even			0+1	
more			0+1	
			0+1	
	6.5	6	12.5	
	0/12.5	6.5/12.5	6/12.5	$-4.5/((-4.5^2) + 15)^{0.5}$
	0	0.52	0.48	-0.7579

From top to bottom of Fig. 2, the first and last subplots respectively represent the time-domain sentiment sequence before and after processing. The second and third subplots represent the frequency-domain sentiment sequence before

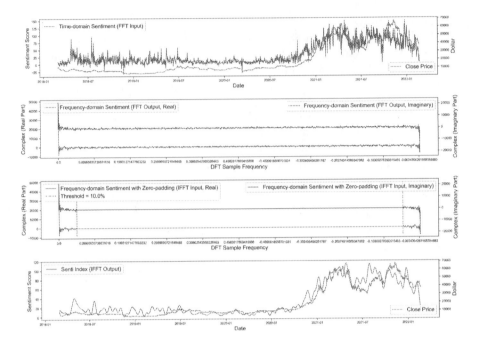

Fig. 2. The FFT-Zero-padding-IFFT period of BTC's Senti Index under FFT_{10}.

and after zero-padding respectively, and their x-axis is an array containing the Discrete Fourier Transform (DFT) sample frequency bin centers in cycles/second of the sample spacing, while their left and right y-axes represent the real and imaginary parts of the complex form of the frequency components.

3.2 Synchronization Verification

For ease of understanding, we declare definitions of universal variables for Sect. 3.2 and Sect. 3.3 uniformly on Table 3.

Pearson Correlation is a global measurement of the time series synchronicity, which calculates correlation by taking a linear relationship as one value between -1 and 1. It is easy to get an intuitive interpretation from Table 4 that there is a high correlation between the price and sentiment indexes. By introducing the WS, Rolling Pearson Correlation can calculate the Pearson Correlation in each rolling window, thus its measurement of the correlation is more comprehensive, but the leading relationships between sequences still cannot be observed. Based on Pearson Correlation, Time-lagged Cross-correlation (TLCC) [18] is used to determine which sequence is the leading sequence by introducing the TO. However, the above methods can not verify the synchronicity in a fine-grained way by observing the relationship among WS, TO, and $\widehat{C}$ at the same time. Windowed Time-lagged Cross-correlation (WTLCC) [3] combines WS and TO to calculate the TLCC in each fixed-size window, but its calculation result will be distorted due to the lack of data for actual calculations when the absolute value of TO is close to the size of the preset window.

This paper proposed a new synchronization verification method in time series analysis for the sentiment indexes designed above and related price time series, called the RTLCC surface, which can determine the values of WS and TO as hyperparameters for the subsequent feature construction while observing TLCC calculated in Rolling Correlation. The RTLCC is designed to calculate and find out the extreme value of the correlation by enumerating (WS, TO) combinations.

The enumerated values of TO construct the x-axis and the ones of WS are for the y-axis, whereas the z-axis is composited by $\widehat{C}$. Then a 3D coordinate map of $(TO, \ WS, \ \widehat{C})$, the RTLCC surface, is built up as Fig. 3.

Table 3. Variable definition declarations.

Name	Description
r_{sp}	The Pearson Correlation Coefficient between Senti Index and the price
r_{vp}	The Pearson Correlation Coefficient between Volume Index and the price
WS	The window size
TO	The time offset
C_{sp}	The Rolling Correlations between Senti Index and the price: $C_{sp} = \{c_{win}, c_{win+1}, \ ..., c_i\}, \ i \leq n$ where n (same below) is the number of trading days and win (same below) is the size of the rolling window.
C_{vp}	The Rolling Correlations between Volume Index and the price
$\widehat{C}$	The average of Rolling Correlations between Senti/Volume Index and the price
c_i	The ith value of C_{sp} or C_{vp}: $$c_i = \frac{\sum_{j=i-win}^{i}(x_j-\hat{x})(y_j-\hat{y})}{\sqrt{\sum_{j=i-win}^{i}(x_j-\hat{x})^2}\sqrt{\sum_{j=i-win}^{i}(y_j-\hat{y})^2}}$$ where x_j, y_j are the jth sample pair in the sequences, while $\hat{x}$, $\hat{y}$ are relevant means.
MAC_{sp}	The moving average of C_{sp}: $MAC_{sp} = \{avg_{2 \times win}, avg_{2 \times win+1}, \ ..., avg_i\}$ where i is less than or equal to the length of C_{sp}, while $avg_i = \frac{1}{win}\sum_{j=i-win}^{i} c_j$.
MAC_{vp}	The moving average of C_{vp}
MSC_{sp}	The moving standard deviation of C_{sp}: $MSC_{sp} = \{std_{2 \times win}, std_{2 \times win+1}, \ ..., std_i\}$ where i is less than or equal to the length of C_{sp}, while $std_i = \sqrt{\frac{\sum_{j=i-win}^{i}(c_j-avg_i)}{win-1}}$.
MAC_{vp}	The moving standard deviation of C_{vp}

Table 4. Pearson correlation snapshot.

r_{sp}, r_{vp}	BTC	SPX
FFT_{10}	0.935, 0.815	0.928, −0.175
FFT_{20}	0.924, 0.794	0.904, −0.170
FFT_{30}	0.915, 0.766	0.892, −0.167
FFT_{40}	0.907, 0.754	0.881, −0.164
FFT_{50}	0.902, 0.743	0.873, −0.162
FFT_{60}	0.896, 0.729	0.862, −0.160
FFT_{70}	0.891, 0.721	0.852, −0.159
FFT_{80}	0.887, 0.716	0.844, −0.156
FFT_{90}	0.883, 0.711	0.838, −0.155
FFT_{100}	0.879, 0.703	0.829, −0.154

The construction processes of the surface are as follows:

(1) Select WS_j from $Range_{WS}$ like 3, 4, ..., 63 as the rolling window size.
(2) Select TO_j from $Range_{TO}$ like −30, −29, ..., 30 as the offset based on the selected WS_i.
(3) Loop through the above two steps in a nested way and compute all

$$\widehat{C}_{ij} = \frac{1}{n - WS_i} \sum_{k=WS_i}^{n-j} c_{kj}. \tag{1}$$

of which every value c_i of corresponding Rolling Correlations is adjusted to be

$$c_{kj} = \frac{\sum_{p=k-WS_i}^{k} (x_p - \hat{x}_k)(y_{p+j} - \hat{y}_k)}{\sqrt{\sum_{p=k-WS_i}^{k} (x_p - \hat{x}_k)^2} \sqrt{\sum_{p=k-WS_i}^{k} (y_{p+j} - \hat{y}_k)^2}}. \tag{2}$$

in which x_p is p^{th} value in Senti Index or Volume Index, while y_{p+j} is $(p+j)^{th}$ value in the relevant price sequence. And $\hat{x}_k = \frac{1}{WS_i}\sum_{p=k-WS_i}^{k} x_p$, while $\hat{y}_k = \frac{1}{WS_i}\sum_{p=k-WS_i}^{k} y_{p+j}$ when the j is fixed.

Each point on the RTLCC surface represents the average of a certain group of Rolling Correlations between the index and price under a specific combination of WS_i and TO_i, and this kind of average is also adopted in studies [2, 7] as an important financial indicator. An ideal coordinate point can be defined in this 3D space when the meaning of each axis has been defined in directions. For example, the ideal point can be (WS_{id}, TO_{id}, $\widehat{C}_{id}$) → (0, −∞, +∞) to make the 'correlation' $\widehat{C}_{ij}$ positively highest possible with the smallest window size and most negative offset (Scene 1). The reasons for pursuing a small window and negative offset are: a smaller window means spending less time looking back at the historical data, so there will be fewer data that has to be used for the model learning; a more negative offset means that the more days the sentiment index leads the price, so there will be longer periods of time can be used to design trading strategies in advance. Moreover, the ideal point also can be (WS_{id}, TO_{id}, $\widehat{C}_{id}$) → (0, −∞, −∞) to make the 'correlation' $\widehat{C}_{ij}$ negatively highest possible with the smallest window size and most negative offset (Scene 2), when there is a potential negative correlation between the sequences to be observed.

This paper adapts the Weighted Euclidean Distance based on Min-Max Normalization to calculate the distance from the ideal point to every point on the surface, which enables the comparability between these three dimensions and the one between surfaces. And in Scene 1, the ideal point is ($WS_{id}, TO_{id}, \widehat{C}_{id}$) = (0, 0, 1), while the ideal point becomes (0, 0, 0) in Scene 2. Denote (WS_i, TO_j, $\widehat{C}_{ij}$) as 3D coordinates of any point on the surface, its distance to the idea point is:

$$Distance\left(WS_i,\ TO_j, \widehat{C}_{ij}\right) = \sqrt{w_1 \times (WS_i - WS_{id})^2 + w_2 \times \left(TO_j - TO_{id}\right)^2 + w_3 \times \left(\widehat{C}_{ij} - \widehat{C}_{id}\right)^2}. \quad (3)$$

in which the w_1, w_2, and w_3 are weights representing the importance of each parameter, subjecting to $w_1 + w_2 + w_3 = 1$. Ultimately, the entire analysis task is reduced to finding the minimum of the 2D matrix. For instance, the RTLCC of Bitcoin under FFT_{10} is shown in Fig. 3, from which three hyperparameters of the actual best point are obtained and recorded in Table 5.

Table 5. Best parameter combination for BTC under FFT_{10} found by RTLCC based on the training set.

	Volume Index VS Bitcoin Price	Senti Index VS Bitcoin Price
Window Size (WS')	55 days	56 days
Time Offset (TO')	2 days	6 days
Rolling Correlations' Average ($\widehat{C}'$)	0.331	0.522

3.3   Feature Construction

In this subsection, we construct features based on the price, Senti Index, and Volume Index. Then hundreds of features for each asset are constructed, which can be summarized into 28 features according to different FFT_T, as shown in Table 6, where price_up_down is the predicting target.

This paper regards price movement prediction as a multivariate time series forecasting task and believes each asset price depends not only on its historical values but also on its relationship with relevant sequences. Thus, features are constructed based on two rules: the construction based on short-term and long-term change value in single time series (Rule 1); the construction based on correlations between various time series (Rule 2).

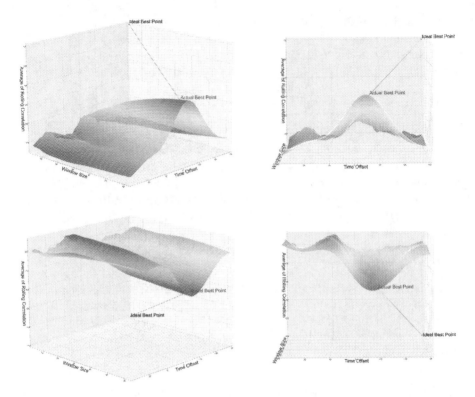

Fig. 3. The RTLCC surfaces of BTC's Senti Index (upper two) and SPX's Volume Index (bottom two) under FFT_{10} based on the training set.

Based on Rule 1, features 1 to 15 are constructed. The short-term trend refers to the difference between two consecutive sequence units, while the long-term trend refers to the trend in a specific time span having more than two units. For the price $P_A = \{p_1, p_2, \ldots, p_i\}$, $i \leq n$, the first order difference, price_diff, is equal to $p_i - p_{i-1}$, and the short-term trend, price_trend, is computed by setting $x = p_i - p_{i-1}$ in this formula:

$$\text{Sign}(x) = \begin{cases} 0, x \leq 0 \\ 1, x > 0 \end{cases}. \tag{4}$$

The long-term change value, his_price_up_down_value, is represented by the tangent of the included angle α between the first-order polynomial fitted straight line $y_1 = A \bullet x + B$ with the horizontal line $y_2 = B$. When the x is equal to TO', this value is:

$$\tan\alpha = \frac{y_1|_{x=TO'} - y_2}{x} = \frac{A \bullet TO' + B - B}{TO'} = A. \tag{5}$$

Meanwhile, the long-term trend, his_price_up_down_trend, is defined as $Sign(\tan\alpha - tan0°)$. For both Senti Index and Volume Index, the definition of

Table 6. Feature description.

	FEATURES	DESCRIPTION
0	price_up_down	Future 1, 5, or 10 days' price movement $(0, 1)$
1	price	Price time series
2	price_diff	The amount of price change between two consecutive days
3	price_trend	Price's short-term trend $Sign(price\_diff)$ $(0, 1)$
4	his_price_up_down_value	The amount of price change in the past TO' days $tan\alpha$
5	his_price_up_down_trend	Price's long-term trend in the past TO' $Sign(tan\alpha)$ days $(0, 1)$
6	senti (FFT_T)	Senti Index under FFT_T
7	senti_diff (FFT_T)	The amount of change in the Senti Index for two consecutive days
8	senti_trend (FFT_T)	Senti Index's short-term trend $Sign(senti\_diff)$ $(0, 1)$
9	his_senti_up_down_value (FFT_T)	The amount of Senti Index's change in the past TO' days
10	his_senti_up_down_trend (FFT_T)	The trend of Senti Index's change in the past TO' days $(0, 1)$
11	volume (FFT_T)	Volume Index under FFT_T
12	volume_diff (FFT_T)	The amount of change in Volume Index between two consecutive days
13	volume_trend (FFT_T)	Volume Index's short-term trend $Sign(senti\_diff)$ $(0, 1)$
14	his_volume_up_down_value (FFT_T)	The amount of Volume Index's change in the past TO' days
15	his_volume_up_down_trend (FFT_T)	The trend of Volume Index's change in the past TO' days $(0, 1)$
16	rolling_corr_senti (FFT_T)	Senti Index's rolling correlations C_{sp}
17	mean_std_diff_senti (FFT_T)	The difference between MAC_{sp}, the moving average of C_{sp}, and MSC_{sp}, the moving standard deviation of C_{sp}
18	mean_std_sum_senti (FFT_T)	The sum of MAC_{sp} and MSC_{sp}
19	corr_up_value_senti (FFT_T)	The difference between C_{sp} and $(MAC_{sp}+MSC_{sp})$
20	corr_down_value_senti (FFT_T)	The difference between C_{sp} and $(MAC_{sp}-MSC_{sp})$
21	corr_up_down_senti (FFT_T)	The state of C_{sp} $(-1, 0, 1)$
22	rolling_corr_volume (FFT_T)	Volume Index's rolling correlations C_{vp}
23	mean_std_diff_volume (FFT_T)	The difference between MAC_{sp}, the moving average of C_{vp}, and MSC_{sp}, the moving standard deviation of C_{vp}
24	mean_std_sum_volume (FFT_T)	The sum of MAC_{vp} and MSC_{vp}
25	corr_up_value_volume (FFT_T)	The difference between C_{vp} and $(MAC_{vp}+MSC_{vp})$
26	corr_down_value_volume (FFT_T)	The difference between C_{vp} and $(MAC_{vp}-MSC_{vp})$
27	corr_up_down_volume (FFT_T)	The state of C_{vp} $(-1, 0, 1)$

short-term or long-term features are similar. For the predicting target, the asset price movements are defined as price_up_down, of which the formula is:

$$Movement = Sign\left(tan\alpha\right) = Sign\left(\frac{y_1|_{x=future} - y_2}{x}\right). \tag{6}$$

where $future$ can be any reasonable integer such as 1, 5, 10, which means the time lag of each predicting step will be 1, 5, or 10 trading days.

Based on Rule 2, features 16 to 27 are constructed. The Bollinger Bands strategy is adopted to observe the correlation between the sentiment index and the price of the same asset based on the assumption that sentiment can reflect future prices in advance. There are five sequences concerned in our Bollinger Bands strategy, namely the normalized price sequence, normalized Senti/Volume index, Rolling Correlations between price and the normalized index, and the upper and lower boundary of the Bollinger Band. By observation, we find that future price movements are often related to the current trends of both the price and sentiment index, as well as the relationship between the rolling correlations and the upper and lower boundary. To quantify the association between these sequences, the following feature construction is performed.

Take Senti Index as an example, the *win* of variables in Table 3 equals to WS' and the state of C_{sp}, corr_up_down_senti, can be formulated as:

$$\text{State}\,(C_{sp}) = \begin{cases} 1, & \text{state1} = 1 \\ 0, & \text{state1} = 0, \text{state2} = 1 \\ -1, & \text{state1} = 0, a\text{state2} = 0 \end{cases} \tag{7}$$

$$\text{state1} = Sign\,[C_{sp} - (MAC_{sp} + MSC_{sp})]\,. \tag{8}$$

$$\text{state2} = Sign\,[C_{sp} - (MAC_{sp} - MSC_{sp})]\,. \tag{9}$$

As a result, the above formula will give out a ternary sequence, and the relationship between the sentiment index and price is summarized into three types of situations. Finally, features [1, 2, 4, 6, 7, 9, 11, 12, 14, 16–20, 22–26] need to be standardized into the zero-mean and unit variance distribution as:

$$v' = \frac{v - \text{mean}(V)}{\sigma}. \tag{10}$$

where v is every single value in a certain feature sequence V, while σ is the standard deviation of V.

4   Evaluation

The evaluation process is carried out according to an experiment comparing the classification performance of four different machine learning models constructed based on two different sets of features as the training data.

4.1   Validation Method and Indicator

Walk-forward Validation method is used in this paper, which adopts the sliding method to split the training set and the test set, and it only takes the part of the accessible historical data closest to the predicted time span as the training set. Moreover, the indicators used here are the 0.5-threshold Accuracy (ACC) and Area under the ROC Curve (AUC).

In the prediction process, the errors will continue to accumulate. The larger the test set is divided, the harder for steps at the end of the test set to be predicted accurately. Thus, the sub-test set with a small size of just one predicting step for each test iteration was designed in the following experiment, and all predicted values are then concatenated in chronological order for comparison with the target values. The overall test set (out-of-sample data) consisting of every sub-test set accounts for 20% of all data and the training set (in-sample data) used for each prediction always accounts for 80% of all data.

4.2 Experiment

In the comparative experiment, four different machine learning models, including SVM, Random Forest, XGB, and LGB are used to be classifiers with the Original or All Features as the training data. Then, by comparing the ACC and AUC of these classifiers, it is proved that the proposed feature construction method is effective, and the difference in the performance of models is further explored with results in Table 7. The 'Indicator' and 'Change' are expressed as percentages, in which the former represents the value of ACC or AUC under a specific combination of training data, model type, and time lag per predicting step, while the latter represents the change of the above indicator values upon their own mean. The 'Mean' is the mean of the four models' performance under the same time lag and training data. In addition, the Average Change represents the average of the Changes under different training data and the same time lag and model type. The Original Features and All Features, respectively, refer to modeling by using only five original features (features 1 to 5) and using all features (features 1 to 27).

The Original (ori_) or All (all_) in Fig. 4 represents that the Original or All Features are training data, and the polyline with a name containing the above two abbreviations is associated with the performance of the corresponding model. Among them, the red line and the blue line represent the moving average of ACC with 10% of the test set's size as the rolling window size, and they are used to observe the change in the predicting accuracy of the model over time. The four parallel lines represent the global metrics of the entire test set, and their specific scores are shown in the title of the figure.

Through the above experiments, under the same training data, whether for ACC or AUC, it can be found that the model based on the Original Feature is inferior to the corresponding model based on All Features, and they all increase with a longer time lag. This shows that our feature construction method is able

Table 7. Comparative experiments results

Trainning Data		Original Features (BTC)		All Features (BTC)		Original Features (SPX)		All Features (SPX)		Average Change
% Indicator (Change)		ACC	AUC	ACC	AUC	ACC	AUC	ACC	AUC	
Time lag/Step	Asset									
1 day	SVM	48.7 (4.5)	47.1 (1.9)	57.4 (0.0)	58.5 (−3.9)	52.6 (6.7)	40.4 (−6.7)	55.0 (3.6)	42.7 (−17.2)	−1.4
	RF	47.5 (1.9)	48.1 (4.1)	54.4 (−5.2)	56.5 (−7.2)	47.4 (−3.9)	42.6 (−1.6)	57.3 (7.9)	56.4 (9.3)	0.7
	XGB	46.0 (−1.3)	46.6 (−3.5)	57.8 (0.7)	62.5 (2.6)	45.6 (−7.5)	44.1 (1.8)	50.3 (−5.3)	52.9 (2.5)	−1.3
	LGB	44.1 (−5.4)	45.1 (−2.4)	60.1 (4.7)	66.1 (8.5)	51.5 (4.5)	45.9 (6.0)	49.7 (−6.4)	54.4 (5.4)	1.9
Mean		46.6	46.2	57.4	60.9	49.3	43.3	53.1	49.7	-
5 days	SVM	49.6 (−13.7)	54.5 (−8.6)	70.2 (−7.6)	78.9 (−4.4)	58.8 (−1.5)	59.2 (−4.4)	68.8 (−10.8)	76.3 (−5.5)	−7.1
	RF	58.0 (0.9)	58.6 (−1.7)	79.8 (5.0)	83.1 (0.7)	58.8 (−1.5)	65.1 (5.2)	80.6 (4.5)	82.9 (2.7)	2.0
	XGB	63.0 (9.6)	63.3 (6.2)	78.6 (3.4)	84.0 (1.8)	59.4 (−0.5)	61.3 (−1.0)	78.8 (2.2)	82.5 (2.2)	3.0
	LGB	59.2 (3.0)	61.9 (3.9)	75.2 (−1.1)	84.1 (1.9)	61.8 (3.5)	61.9 (0.0)	80.0 (3.8)	81.1 (0.5)	1.9
Mean		57.5	59.6	76.0	82.5	59.7	61.9	77.1	80.7	-
10 days	SVM	53.6 (−3.4)	53.5 (−5.8)	78.5 (−10.0)	86.1 (−4.2)	59.2 (−0.2)	59.4 (−1.2)	81.7 (−6.1)	85.8 (−5.3)	−4.5
	RF	52.9 (−4.7)	51.1 (−3.0)	89.7 (2.9)	90.2 (0.3)	58.0 (−2.2)	62.2 (3.5)	86.4 (−0.7)	91.0 (0.4)	−0.4
	XGB	57.5 (3.6)	58.5 (3.0)	89.7 (4.1)	90.2 (2.7)	58.0 (−2.2)	59.9 (−0.3)	89.9 (3.3)	93.0 (2.6)	2.1
	LGB	57.9 (4.3)	60.0 (5.6)	89.7 (2.9)	91.0 (1.2)	62.1 (4.7)	58.7 (−2.3)	89.9 (3.3)	92.5 (2.1)	2.7
Mean		55.5	56.8	87.2	89.9	59.3	60.1	87.0	90.6	-

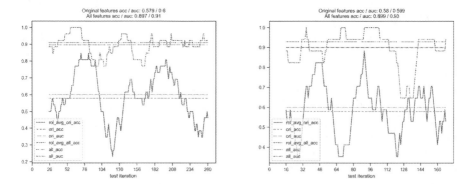

Fig. 4. The model comparison between Original Features and All Features of BTC-LGB under 10-day time lags (left) and the one of SPX-XGB (right).

to increase the classification's performance. For both Bitcoin and S&P 500, LGB and XGB models perform better on comprehensive performance than SVM and RF, no matter what the predicting time lag is equal to. Among them, the XGB model performs best when the time lag is equal to 5 days, with an Average Change of 3.0%, while the LGB model performs best when the time lag equals 10 days, with an Average Change of 2.7%. Although when the time lag is equal to 1 day and the training data is All Features, RF performs better both on ACC and AUC than other models, its performances are lower under any other same conditions. Moreover, by comparing the Average Change, even though XGB achieves the highest average improvement of 3% when the time lag equals 5 days, it performs below the average of all models in a one-day scenario, while only LGB is able to achieve the highest or second high average improvement in all cases, thus LGB is stated to be the best and most robust model founded in this experiment.

When we compare the performance of the models trained on all Features, we can also find that the related models of both BTC and SPX perform well on a 10-day-long time lag (Mean: ACCs' is 87.1%, and AUCs' is about 90.3%). However, the former has a clear advantage over the latter in a 1-day-short time lag. When the latter is almost unpredictable (Mean: ACC's is 53.1%, and AUC's is 49.7%), the former's means of ACC and AUC are respectively as high as 57.4% and 60.9%. It can be seen that the market sentiment of different assets has different effecting time lag on their price movements. In the short term, BTC is more effective than SPX, and as the time lag increases, the gap between the two continues to decrease.

5   Conclusions and Future Work

A novel pipeline of asset price movement prediction based on market sentiment analysis is proposed in this paper, which effectively solves the problem that the cost of text data collection by the direct analysis method of market sentiment is too high. At first, we quantify the market sentiment from social media and trading volume as sentiment indexes by VADER and FFT-Zero-padding-IFFT processing, and then change it into a form of structured data analysis while introducing the convenience of indirect analysis methods. Then, we design a new synchronization verification method, RTLCC, to find out the best observing window size and time offset for feature construction. Finally, classifiers for asset price movement prediction are built and compared based on four different machine learning algorithms for providing an experimental basis for model selection. In short, our research proves that it is feasible to use the public sentiment from social media for price movement prediction, and at the same time provides new ideas for unstructured data analysis for market sentiment, as well as methodological innovations in this field.

In the future, in terms of price movement prediction modeling, we plan to design ensemble models for further comparative experiments and more effective forecasting. For applications, we will try to increase the update frequency of both price and text to 30 or even 15 min among more assets to further explore the potential of this kind of model in high-frequency trading.

Acknowledgment. This work is supported by the Macau SAR Science and Technology Development Fund (No. 0032/2022/A, No. 0091/2020/A2), Guangzhou Development Zone Science and Technology Grant (No. 2021GH10, No. 2020GH10, and No. EF003/FST-FSJ/2019/GSTIC), and Collaborative Research Grant of University of Macau (No. MYRG-GRG2022, No. MYRG-CRG2021-00002-ICI).

References

1. Al Nasseri, A., Tucker, A., De Cesare, S.: Quantifying stocktwits semantic terms' trading behavior in financial markets: an effective application of decision tree algorithms. Expert Syst. Appl. **42**(23), 9192–9210 (2015)
2. Aloui, C., Nguyen, D.K., Njeh, H.: Assessing the impacts of oil price fluctuations on stock returns in emerging markets. Econ. Model. **29**(6), 2686–2695 (2012)
3. Boker, S.M., Rotondo, J.L., Xu, M., King, K.: Windowed cross-correlation and peak picking for the analysis of variability in the association between behavioral time series. Psychol. Methods **7**(3), 338 (2002)
4. Cochran, W.T., et al.: What is the fast fourier transform? Proc. IEEE **55**(10), 1664–1674 (1967)
5. Da, Z., Engelberg, J., Gao, P.: The sum of all fears investor sentiment and asset prices. Rev. Financ. Stud. **28**(1), 1–32 (2015)
6. Gervais, S., Kaniel, R., Mingelgrin, D.H.: The high-volume return premium. J. Financ. **56**(3), 877–919 (2001)
7. Goetzmann, W.N., Li, L., Rouwenhorst, K.G.: Long-term global market correlations (2001)
8. Guliyev, H., Mustafayev, E.: Predicting the changes in the WTI crude oil price dynamics using machine learning models. Resour. Policy **77**, 102664 (2022)
9. Huang, D., Jiang, F., Tu, J., Zhou, G.: Investor sentiment aligned: a powerful predictor of stock returns. Rev. Financ. Stud. **28**(3), 791–837 (2015)

10. Hutto, C., Gilbert, E.: Vader: a parsimonious rule-based model for sentiment analysis of social media text. In: Proceedings of the International AAAI Conference on Web and Social Media, vol. 8, pp. 216–225 (2014)
11. Karabulut, Y.: Can facebook predict stock market activity? In: AFA 2013 San Diego Meetings Paper (2013)
12. Kim, Y.B., et al.: Predicting fluctuations in cryptocurrency transactions based on user comments and replies. PLoS ONE 11(8), e0161197 (2016)
13. Oliveira, N., Cortez, P., Areal, N.: The impact of microblogging data for stock market prediction: using twitter to predict returns, volatility, trading volume and survey sentiment indices. Expert Syst. Appl. 73, 125–144 (2017)
14. Pano, T., Kashef, R.: A complete VADER-based sentiment analysis of bitcoin (BTC) tweets during the ERA of COVID-19. Big Data Cogn. Comput. 4(4), 33 (2020)
15. Pettengill, G.N.: Holiday closings and security returns. J. Financ. Res. 12(1), 57–67 (1989)
16. Porshnev, A., Redkin, I., Shevchenko, A.: Machine learning in prediction of stock market indicators based on historical data and data from twitter sentiment analysis. In: 2013 IEEE 13th International Conference on Data Mining Workshops, pp. 440–444. IEEE (2013)
17. Ritter, J.R.: Behavioral finance. Pacific-Basin Financ. J. 11(4), 429–437 (2003)
18. Shen, C.: Analysis of detrended time-lagged cross-correlation between two nonstationary time series. Phys. Lett. A 379(7), 680–687 (2015)
19. Solt, M.E., Statman, M.: How useful is the sentiment index? Financ. Anal. J. 44(5), 45–55 (1988)
20. Whaley, R.E.: The investor fear gauge (2000)
21. Xing, F.Z., Cambria, E., Welsch, R.E.: Natural language based financial forecasting: a survey. Artif. Intell. Rev. 50(1), 49–73 (2018)
22. Zhang, L., Wang, S., Liu, B.: Deep learning for sentiment analysis: a survey. Wiley Interdiscip. Rev. Data Min. Knowl. Discov. 8(4), e1253 (2018)

Data Mining

Efficient Mining of High Utility Co-location Patterns Based on a Query Strategy

Vanha Tran[1] , Lizhen Wang[2](✉) , Jinpeng Zhang[3,4,5], and Thanhcong Do[1]

[1] FPT University, Hanoi 155514, Vietnam
`hatv14@fe.edu.vn, congdthe150385@fpt.edu.vn`
[2] Dianchi College of Yunnan University, Kunming 650213, China
`lzhwang@ynu.edu.cn`
[3] Yunnan University, Kunming 650091, China
`zjp@ynufe.edu.cn`
[4] Yunnan University of Finance and Economics, Kunming 650221, China
[5] Yunnan Key Laboratory of Service Computing, Kunming 650221, China

Abstract. A high utility co-location pattern (HUCP) is a set of spatial features, which is supported by groups of neighboring spatial instances, and the pattern utility ratios (PUR) of the spatial feature set are greater than a minimal utility threshold assigned by users, can reveal hidden relationships between spatial features in spatial datasets, is one of the most important branches of spatial data mining. The current algorithm for mining HUCPs adopts a level-wise search style. That is, it first generates candidates, then tests these candidates, and finally determines whether the candidates are HUCPs. It performs mining from the smallest size candidate and gradually expands until no more candidates are generated. However, in mining HUCPs, the UPR measurement scale does not hold the downward-closure property. If the level-wise search style is adopted, unnecessary candidates cannot be effectively pruned in advance, and the mining efficiency is extremely low, especially in large-scale and dense spatial datasets. To overcome this, this paper proposes a mining algorithm based on a query strategy. First, the neighboring spatial instances are obtained by enumerating maximal cliques, and then these maximal cliques are stored in a hash map structure. Neighboring spatial instances that support candidates can be quickly queried from the hash structure. Finally, the UPR of the candidate is calculated and a decision is made on it. A series of experiments are implemented on both synthetic and real datasets. Experimental results show that the proposed algorithm gives better mining performance than existing algorithms.

Keywords: Prevalent co-location pattern · High utility co-location pattern · Maximal clique · Query strategy

X. Yang et al. (Eds.): ADMA 2023, LNAI 14176, pp. 401–416, 2023.
https://doi.org/10.1007/978-3-031-46661-8_27

1   Introduction

With the development of sensor equipment, a large amount of data with spatial information can be collected every day from various sources, such as handheld mobile devices, vehicles with navigation systems, location sensing, wireless networks, and so on. These massive spatial datasets contain potential and valuable knowledge. Therefore, spatial data mining has attracted a lot of researchers. Prevalent co-location pattern (PCP) mining, which can efficiently disclose interesting relationships of spatial features in spatial neighbor regions, is an important task of spatial data mining [14].

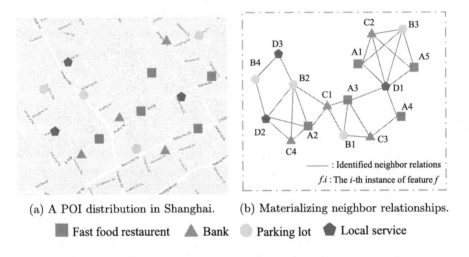

(a) A POI distribution in Shanghai.   (b) Materializing neighbor relationships.

■ Fast food restaurent   ▲ Bank   ● Parking lot   ⬟ Local service

Fig. 1. An illustration of a spatial dataset.

For example, Fig. 1(a) shows a part distribution of a point of interest (POI) dataset of Shanghai City, China. It includes 16 points of interest (spatial instances), e.g., a fast food restaurant on Xinzha Road (red square), a China of bank on Jiaozhou Road (light sea green triangle). These spatial instances are classified into 4 categories (spatial features), i.e., fast food restaurants, banks, parking lots, and local services. When employing co-location pattern mining technology, we can discover that within a radius of 250 m, fast foods, banks, parking lots, and local services frequently appear together [11]. Therefore, {Fast food, Bank, Parking lot, Local service} is a PCP. If a businessman wants to open a new fast food restaurant, he needs to look for a suitable address. Through the information brought by the above PCP, we can recommend to him some suitable addresses such as these addresses that are within 250 m around of parking lots, banks and local services are the best.

PCPs have been proven to be widely used in many fields such as transportation [8], public health [9], ecology [4], criminology [6], environmental management

[1], and so on. The spatial co-location pattern described above is called traditional PCPs. The traditional PCP only reflects the prevalence of a set of spatial features in neighbor geospatial spaces. The importance of all spatial features is considered equally.

However, in reality, each spatial feature has its own importance in the space. For example, the role of a parking lot is completely different from that of a fast food restaurant. Especially, near a residential area, a parking lot is more important than a fast food restaurant. Therefore, if the importance of spatial features is not considered, the mining results may lose their true meaning. In order to overcome the above problem, a concept of high utility co-location patterns is proposed [13]. By setting a utility value for each spatial feature to reflect its importance, we aim to find groups of feature combinations whose utility values are greater than a certain utility threshold assigned by users.

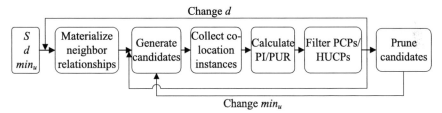

(a) The common PCP/HUCP mining framework using level-wise search style.

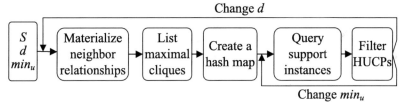

(b) The proposed mining algorithm using query strategy.

Fig. 2. Frameworks of mining PCPs/HUCPs.

A common framework for mining PCPs and HUCPs is to adopt a mining method like the Apriori algorithm, that is a level-wise search style. The common framework is described in Fig. 2(a). Given a spatial dataset S, a distance threshold d to determine neighboring instances, and a prevalent/utility threshold min_u to filter PCPs/HUCPs, the mining process first materializes neighbor relationships between spatial instances. For example, Fig. 1(b) describes the result after materializing neighbor relationships. The uppercase letters, A, B, C, and D represent fast food restaurants, banks, parking lots, and local services, respectively. The number immediately following the uppercase letter indicates the ID of instances belonging to the feature, e.g., A1 is the first instance of feature A, B2 is the second instance of feature B, and C3 is the third instance of feature C.

If the distance between two instances (usually using the Euclidean distance metric) is not larger than d, the two instances are neighbors. As shown in Fig. 1(b), neighboring instances are connected by a solid line. The mining process then generates a set of size 2 candidates (groups of distinguishing spatial features), i.e., {A, B}, {A, C}, {C, D}, and so on. After that, all co-location instances that support these candidates are collected. This step is the most expensive phase in the whole mining process [11] since it needs to verify and ensure that the instances in each co-location instance have a neighbor relationship with each other. Next, it calculates the participation index (PI)/pattern utility ratio (PUR) of the candidates to filter PCPs/HUCPs. Finally, based on the filtering result, it executes pruning strategies to reduce unnecessary candidates. The mining process continues with size 3, size 4, and so on candidates until no more candidates are generated.

Unfortunately, PUR does not satisfy the downward-closure property, unnecessary candidates cannot be effectively pruned in advance. This makes mining less efficient. Although some mining algorithms and pruning strategies have been proposed such as extended pruning algorithm (EPA) [13] and efficient high utility co-location pattern mining algorithm (EHUCM) [7], the performance of these algorithms is extremely low under large-scale and dense spatial datasets.

To improve the performance of mining HUCPs, this paper proposes an efficient algorithm that is shown in Fig. 2(b). Different from the existing algorithms, we avoid the most time-consuming phase, collecting co-location instances. After materializing neighbor relationships in the first phase, the proposed algorithm does not generate candidates, it lists all maximal cliques (a maximal clique is a set of spatial instances in which all the instances are neighbors with each other). The maximal cliques belonging to the same feature sets are then put into a hash map structure. The upper bound size of candidates is the longest keys in the hash structure. Then it utilizes a query strategy to derive all support instances of any candidates from this hash. Finally, HUCPs can be filtered quickly. This mining framework does not need to use any pruning strategies.

Note that the mining method proposed in this paper has a similar idea to the clique-based PCP mining developed by Wang et al. [2]. However, our method has obvious differences. First, we use maximal cliques, while the clique-based employs cliques, the number of cliques is normally much larger than that of maximal cliques, the hash map structure becomes huge. Thus, collecting support instances of a pattern from the huge hash structure requires more execution time. Second, in the clique-based algorithm, a size k pattern is frequent, it is still necessary to calculate the PIs of all its subsets, the upper-lower closure property is not exploited to prune unnecessary candidates.

The rest of this work is organized as follows. Section 2 gives the preliminary of mining HUCPs. The proposed mining algorithm is described in Sect. 3. Section 4 is a detailed report on our experiments. Finally, conclusions and future work are given in Sect. 5.

2   Preliminaries

Given a spatial dataset $S = \{o_1, ..., o_n\}$, each object o_i in S is called a spatial instance, that is represented by <spatial feature, instance identification, location (x, y)>, users set a distance threshold d to define the neighbor relationship between instances, i.e., iff the distance between two instances is not larger than d, they are neighbors. The set of all features that all instances in S belong to is $F = \{f_1, ..., f_m\}$.

Definition 1 (Candidate co-location). *A candidate co-location c is a subset of the spatial feature set F, $c = \{f_1, ..., f_k\} \subseteq F$. The number of features in c is called the size of it, i.e., c is a size k candidate.*

Definition 2 (Co-location instances and table instance). *Given a candidate $c = \{f_1, ..., f_k\}$, its co-location instance is a set of spatial instances $I = \{o_1, ..., o_k\}$ which meets the two conditions: (1) it includes the instances of all features in the candidate; and (2) each spatial instance in I has a neighbor relationship with each other. The set of all co-location instances of c is called the table instance, $T(c)$.*

Definition 3 (External utility). *The external utility is used to reflect the importance of different features. According to the user's application scenario, each feature $f_t \in F$ is assigned a utility value u_{f_t} to reflect its importance.*

For example, as shown in Fig. 1(a), the external utilities of fast food restaurants, banks, parking lots, and local services are assigned as 8, 11, 21, and 7, respectively.

Definition 4 (Support instances and internal utility). *The support instances of a feature $f_t \in c$ are the distinguishing instances of f_t in the table instance of c and is denoted as*

$$SupIns(f_t, c) = \{o_i\}, \forall o_i \in T(c) \wedge \text{ the feature of } o_i \text{ is } f_t \tag{1}$$

The number of support instances is called the internal utility and is denoted as

$$q(f_t, c) = |SupIns(f_t, c)| \tag{2}$$

where $|*|$ represents the cardinality of a set.

Definition 5 (Utility of feature in a candidate). *Given a candidate $c = \{f_1, ..., f_k\}$, the utility of a feature $f_t \in c$ is the product of external utility and internal utility and is denoted as*

$$u(f_t, c) = u_{f_t} \times q(f_t, c) \tag{3}$$

Definition 6 (Pattern utility). *Given a candidate $c = \{f_1, ..., f_k\}$, it pattern utility is the sum of the utilities of all features in c and denoted as*

$$u(c) = \sum_{f_t \in c} u(f_t, c) \tag{4}$$

Definition 7 (Utility of a spatial dataset). *The utility of a spatial dataset S is the sum of all feature utilities in S, denoted as*

$$u(S) = \sum_{f_t \in F, c=\{f_t\}} u(f_t, c) \tag{5}$$

Definition 8 (Pattern utility ratio (PUR)). *Given a spatial dataset S and a candidate c, the pattern utility ratio of c in S is denoted as*

$$PUR(c) = \frac{u(c)}{u(S)} \tag{6}$$

Definition 9 (High utility co-location pattern, HUCP). *If the pattern utility ratio of a candidate c is not smaller than a user-specified utility threshold, min_u, i.e., $PUR(c) \geq min_u$, c is a high utility co-location pattern.*

For example, Fig. 3 illustrates the process of discovering HUCPs from the spatial dataset in Fig. 1. This mining process follows the framework shown in Fig. 2(a). First, size 2 candidates are generated, e.g., {A, B}, {A, C}, {A, B}, and so on. Then, the mining process collects the co-location instances of these candidates, e.g., the table instance of candidate {A, B} is $T(\{A, B\}) = \{\{A1, B3\}, \{A5, B3\}, \{A3, B1\}, \{A2, B2\}\}$. The support instances of features A and B in candidate {A, B} are $SupIns(A, A, B) = \{A1, A3, A5, A2\}$ and $SupIns(B, \{A, B\}) = \{B1, B2, B3\}$, respectively.

Size 2 candidates

A	B	A	C	A	D	B	C	B	D	C	D
A1	B3	A1	C2	A1	D1	B1	C1	B2	D2	C2	D1
A5	B3	A3	C1	A5	D1	B1	C3	B2	D3	C4	D2
A3	B1	A3	C3	A3	D1	B2	C1	B3	D1		
A2	B2	A4	C3	A4	D1	B2	C4			Candidate	
		A2	C1	A2	D2	B3	C2	Co-location instance			
		A2	C4								

Size 3 candidates

A	B	C	A	B	D	A	C	D	B	C	D
A1	B3	C2	A1	B3	D1	A1	C2	D1	B2	C4	D2
A3	B1	C1	A5	B3	D1	A5	C2	D1	B3	C2	D1
A3	B1	C3	A2	B2	D2	A2	C4	D2			
A2	B2	C1									
A2	B2	C3									

Size 4 candidates

A	B	C	D
A1	B3	C2	D1
A5	B3	C2	D1
A2	B2	C4	D2

Fig. 3. An illustration of mining HUCPs using a level-wise search style.

The internal utilities of A and B are $q(A, \{A, B\}) = |\{A1, A3, A5, A2\}| = 4$ and $q(B, \{A, B\}) = 3$. The utilities of the two features in the candidate are $u(A,$

{A, B}) $= u_A \times q$(A, {A, B}) $= 8 \times 4 = 32$ and u(B, {A, B}) $= u_B \times q$(B, {A, B}) $= 11 \times 3 = 33$, respectively. Thus, u({A, B}) $= u$(A, {A, B}) $+ u$(B, {A, B}) $= 32+33 = 65$. Moreover, the utility of the dataset is $u(S) = u$({A}) $+ u$({B}) $+ u$({C}) $+ u$({D}) $= 8 \times 5 + 11 \times 4 + 21 \times 4 + 7 \times 3 = 189$. Therefore, the pattern utility ratio of the candidate is PUR({A, B}) $= u$({A, B})$/u(S) = 65/189 = 0.34$. If a user sets a utility threshold is $min_u = 0.3$, {A, B} is a HUCP since PUR({A, B}) $= 0.34 \geq min_u = 0.3$. The mining process continues with candidates of greater size until no candidates are generated.

Lemma 1. *PUR does not hold the downward-closure property.*

Proof. Similar to the steps in the above example, we can calculate PUR({A, B, D}) $= 0.32$ and PUR({A, B, C, D}) $= 0.54$. It is easy to see that, {A, B} $\subset$ {A, B, D} and PUR({A, B}) $= 0.34 \geq PUR$({A, B, D}) $= 0.32$. However, {A, B, D} $\subset$ {A, B, C, D}, PUR({A, B, D}) $= 0.32 \leq PUR$({A, B, C, D}) $= 0.54$. The UPR of a pattern can be greater or less than the UPRs of its subsets.

Table 1. The enumerated maximal cliques

Maximal cliques	Maximal cliques	Maximal cliques
{A1, B3, C2, D1}	{A3, B1, C3}	{B4, D2}
{A5, B3, C2, D1}	{A3, B1, C1}	{B4, D3}
{A3, D1}	{A2, B2, C1}	{B2, D3}
{A4, D1}	{A2, B2, C4, D2}	

3 The Query Strategy for Mining HUCPs

3.1 Enumerating Maximal Cliques

It can be seen that the most critical step in calculating the PUR of a candidate is to collect all support instances of the candidate. Therefore, if all support instances of a candidate can be quickly collected, the mining efficiency can be improved. Moreover, support instances are obtained from co-location instances. It is not difficult to recognize that a co-location instance is a spatial clique. From the perspective of graph theory, a clique may be a subset of many other cliques. A clique is a maximal clique if it is not a subset of any other clique. Many other subsets can be derived from the maximal clique. If we can get all the maximal cliques first, then we can easily calculate the support instances.

Although enumerating all maximal cliques of a graph is an NP-hard problem, many efficient maximal clique enumeration algorithms have been proposed [3,5, 12,15]. The focus of this paper is not to design a new algorithm for enumerating maximal cliques, we are working on how to use maximal cliques to improve mining HUCP efficiency. Thus, in this work, we employ an efficient maximal clique enumeration algorithm by Yu et al. named MCE. For more details about the MCE algorithm, please refer to [15].

Table 1 lists all the maximal cliques enumerated based on Fig. 1. As can be seen, the neighbor relationship of all instances is stored in a set of maximal cliques.

The next question is how to quickly derive co-location instances of candidates from these maximal cliques. A naive approach is to perform a combination on each maximal clique to generate all subsets, and then gather these subsets as table instances of candidates. But this method will cause all table instances of all candidates to be saved at the same time, the memory consumption is too expensive. A more effective approach is to combine these maximal cliques into another data structure to facilitate the gathering of support instances of candidates.

3.2   Support Instance Hash Map Structure

Definition 10 (Support instance hash map structure). *Support instance hash map structure is a two-level hash structure, the keys and values of the first level are feature sets of the instances in maximal cliques and another hash structure, respectively. The keys and values of the second level are the feature types of instances in maximal cliques and the instances themselves.*

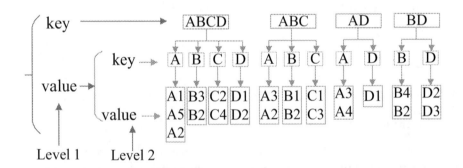

Fig. 4. The constructed support instance hash map structure.

Figure 4 depicts the support instance hash map structure composed of the maximal cliques listed in Table 1. In the whole mining process, we only need to keep this hash structure, so the memory consumption is very small.

Algorithm 1 describes the process of building the support instance hash map structure. First, the algorithm traverses each maximal clique (Step 1), and it first obtains the feature set of all instances belonging to (Step 2). If the feature set is currently a key in the hash, update the hash with the instances in the maximal clique (Steps 3–4). While if it is not a key, a new item is constructed and put into the hash structure (Steps 5–7). Finally the algorithm returns a complete hash structure.

Note that the mining method proposed in this work is different from [2]. The work in [2] enumerates cliques, not maximal cliques. The number of cliques is often much larger than the number of maximal cliques. This makes the hash

Algorithm 1. Constructing the support instance hash map structure

---

    Input: a set of maximal cliques, MCs
    Output: a hash structure, $SupInsHash$
1: **for** $mc \in MCs$ **do**
2:    $c \leftarrow$ Get_feature_set(mc)
3:    **if** c is a key of $SupInsHash$ **then**
4:       $SupInsHash \leftarrow$ Update_value($mc, SupInsHash$)
5:    **else**
6:       $SupInsHash \leftarrow$ Create_new_item(mc)
7:    **end if**
8: **end for**
9: **return** $SupInsHash$

---

structure larger and contains a lot of redundant information. The storage cost of the hash structure is more expensive. In addition, when we desire to obtain the support instances of a pattern, these support instances are repeatedly distributed in multiple items in the hash structure, these same instances have to be taken out and put into a set to calculate the pattern's PI. The mining performance shows less efficiency. In this work, we employ maximal cliques to avoid the above problems.

3.3 Query Support Instances

Lemma 2. *The upper bound on the size of possible candidates is the longest key in the support instance hash map structure.*

Proof. A candidate is considered to calculate its PUR, there must be at least one co-location instance of it. Since a co-location instance is a clique, the longest-size co-location instance corresponds to the longest-size maximal clique. The maximal clique is converted into a form of keys and values in the hash structure. Therefore, the longest key corresponds to the longest feature set, which is also the longest-size possible candidate.

Lemma 3. *The support instances of a candidate $c = \{f_1, ..., f_k\}$ can be queried and gathered from the values of the keys that are supersets of c in the support instance hash map structure.*

Proof. Based on Definition 2, a co-location instance of a candidate c may be a maximal clique or a subset of a maximal clique. From Definition 10, we know that an item of the support instance hash map structure is a set of maximal cliques that belong to the same feature set. Therefore, a co-location instance of candidate c can be dispersed in multiple maximal cliques, and these maximal cliques form different items in the hash structure. When all the items in the hash structure are traversed, those items are supersets of the candidate, their values are exactly support instances of the candidate. Therefore, all co-location instances of c can be collected completely.

Algorithm 2 (called QS-HUCPM) describes the proposed query-based HUCP mining algorithm. Given a spatial dataset S, the external utility of each feature U_F, a distance threshold d, and minimum utility threshold min_u, it first materializes the neighbor relationship of spatial instances (Step 1). Then it employs MCE to enumerate all maximal cliques (Step 2). Next, it constructs the support instance hash map structure (Step 3).

The algorithm uses the key set in the hash structure as an initial set of candidates (Step 4). First, it takes out the set of the longest-size keys (step 5). Then for each candidate, it queries which items in the hash structure are the superset of the candidate and takes the corresponding value as the candidate's supporting instances (Steps 7–10). When all items in the hash structure are traversed completely, all supporting instances of the candidate are collected. The algorithm according to Definitions 2–6 calculates the candidate's PUR and makes a decision (Steps 11–14). When a candidate is calculated, the algorithm generates its direct subsets as new candidates and puts them into the candidate set (Steps 15–18). The algorithm continues to compute until no more candidates are generated.

Algorithm 2. Mining HUCPs based on a query strategy, QS-HUCPM

Input: S; d, min_u, U_F
Output: a set of HUCPs, $HUCPs$
1: $G \leftarrow$ Materialize_neighbor_relationships(S, d, min_u)
2: $MCS \leftarrow$ Enumerate_maximal_cliques(G)   ▷ employ the MCE algorithm [15]
3: $SupInsHash \leftarrow$ Construct_support_instance_hash(MCS)   ▷ use Algorithm 1
4: $Cands \leftarrow$ Get_all_keys(SupInsHash)
5: $l \leftarrow$ Get_longest-size_cands(Cands)
6: **while** $Cands \neq \emptyset$ **do**
7:      $Cands_l \leftarrow$ Get_size-l_cands($Cands, l$)
8:      **if** $Cands_l \neq \emptyset$ **then**
9:          **for** $c \in Cands_l$ **do**
10:             $SupIns \leftarrow$ Get_support_instances($SupInsHash, c$)
11:             $UPR \leftarrow$ Calculate_UPR($c, SupIns, U_F$)
12:             **if** $UPR \geq min_u$ **then**
13:                 $HUCPs \leftarrow HUCPs \cup \{c\}$
14:             **end if**
15:             **if** $l \geq 3$ **then**
16:                 $NewCands \leftarrow$ Generate_direct_subsets(c)
17:                 $Cands \leftarrow Cands \cup NewCands$
18:             **end if**
19:         **end for**
20:     **end if**
21:     $l \leftarrow l - 1$
22: **end while**
23: **return** $\epsilon HUCPS$

It can be seen from Algorithm 2 that the algorithm discovers HUCPs in a top-down direction style. This is totally opposite to the traditional level-wise style which discovers HUCPs from the lowest-size candidates. Because through the maximal clique and hash structure we can find the longest-size possible candidate. However, the traditional mining framework (shown in Fig. 2(a)) cannot determine the upper bound of the size of candidates. If $F = \{f_1, \ldots, f_k\}$ is a feature set, then in the worst case, the longest-size candidate is $|F|$. In fact, the feature set of the longest-size maximal clique is often much smaller than $|F|$, because the candidate search space can be greatly reduced.

4 Experiments

To demonstrate that the proposed algorithm can effectively improve mining efficiency, we performed a series of experiments. We choose two state-of-the-art algorithms used to mine HUCPs, i.e., extended pruning algorithm (EPA) [13] and efficient high utility co-location pattern mining algorithm (EHUCM) [7]. Both two algorithms use a level-wise mining style combined with some pruning strategies. All algorithms are coded in C++ and performed on a Windows 10 with Intel(R) Core i7-3770 and 16 GB of RAM.

Table 2. A summary of the experimental datasets.

Real datasets				
Name	Spatial frame size	# instances	# features	Property
Las Vegas	38,649 × 62,985 (m^2)	22,725	17	Dense, clustering
Toronto	27,583 × 77,312 (m^2)	16,692	16	Dense, uniform
Beijing	138,700 × 227,031 (m^2)	104,909	19	Dense
Vegetation	71,920 × 66,440 (m^2)	408,945	16	Dense
Synthetic datasets				
Figs. 5, 6, 7, 8(a, b)	5,000 × 5,000	50k	15	Dense
Figs. 9, 10(a, b)	5,000 × 5,000	*	15	Dense
k = 1000				

4.1 Datasets

We use a set of synthetic datasets generated by a dataset generator [14]. The generation of feature utility values satisfies the Gauss distribution. In addition, there are four real POI datasets in our experiments. The first two datasets are extracted from the Yelp dataset[1]. The feature utility in both the two datasets (Las Vegas and Toronto) is the average of the number of check-ins in instances. The fourth dataset is the vegetation dataset collected from the Three Parallel Rivers of Yunnan Protected Area [10]. The utility value of its feature set is the

---

[1] https://www.yelp.com/dataset.

price of each type of vegetation. The third dataset is a set of POI collected from Beijing city, China [2]. The feature utility values in it are generated in the same way as the simulated dataset. Table 2 lists the main parameters of these datasets in detail.

4.2   Improvement of Performance on Different Distance Thresholds

Figures 5 and 6 depict the performance of the compared algorithms at different distance thresholds, including execution time and memory consumption, respectively. From the two figures, it can be found that as the distance threshold increases, the more neighboring instances there are, all the algorithms require more execution time and memory space. However, the performance of EPA and EHUCM deteriorates rapidly at large distance thresholds. This is because when the distance threshold is increased, the number of neighboring instances increases, and EPA and EHUCM need more execution time to collect and verify the co-location instances of each candidate. At the same time, these co-location instances also require a larger storage space. While the proposed algorithm can give a good performance. Because for each candidate, it only needs to traverse the hash structure once to query all instances that support that candidate. Thus mining performance can be greatly improved.

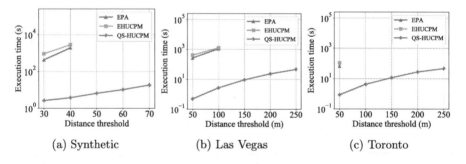

(a) Synthetic   (b) Las Vegas   (c) Toronto

Fig. 5. The execution time of the compared algorithms on different distance thresholds ($min_u = 0.2$ for all cases).

4.3   Improvement of Performance on Different Utility Thresholds

Figures 6 and 7 depict the efficiency of the three algorithms performed at different minimal utility thresholds. Because the performance of the three algorithms is different, if the distance threshold of the synthetic dataset is set to be small, the execution time of EPA and EHUCPM will not change much with the increase of utility thresholds, which shows that the effect of the pruning strategy is very small. On real datasets, when the distance threshold is 200 m, these two algorithms take too much execution time to complete the mining task.

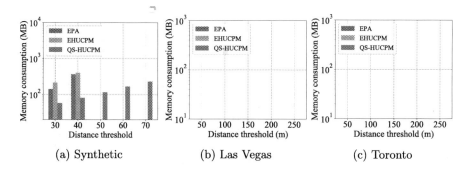

Fig. 6. The memory consumption of the compared algorithms on different distance thresholds ($min_u = 0.2$ for all cases).

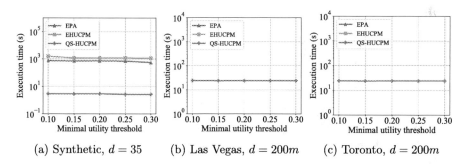

(a) Synthetic, $d = 35$ (b) Las Vegas, $d = 200m$ (c) Toronto, $d = 200m$

Fig. 7. The execution time of the compared algorithms on different minimal utility thresholds.

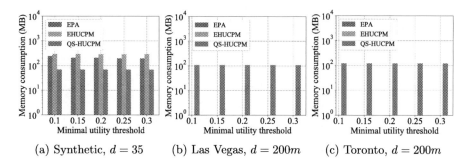

(a) Synthetic, $d = 35$ (b) Las Vegas, $d = 200m$ (c) Toronto, $d = 200m$

Fig. 8. The memory consumption of the compared algorithms on different minimal utility thresholds.

Since our algorithm only needs to keep the hash structure in the mining process, computing the UPR of each candidate only needs to execute a query, thus the execution time and memory consumption are not affected by minimal utility thresholds.

4.4   Improvement of Performance on Different Sizes of Datasets

In the final experiments, we demonstrate that the proposed algorithm is able to get promoted to large-scale spatial datasets. In order to generate different sizes of real datasets, we differently sample the two large-scale datasets, Beijing and vegetation datasets. Figures 9 and 10 show the performance of the compared algorithms as the dataset size increases. It is obvious that the performance of EPA and EHUCPM that adopt the level-wise mining framework deteriorates rapidly and is unusable for large-scale datasets. On the contrary, the proposed algorithm can give good scalability.

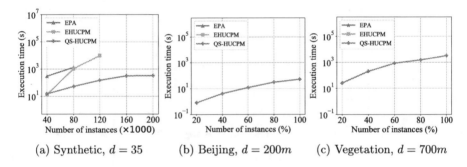

(a) Synthetic, $d = 35$ (b) Beijing, $d = 200m$ (c) Vegetation, $d = 700m$

Fig. 9. The execution time of the compared algorithms on different sizes of spatial datasets ($min_u = 0.2$ for all cases).

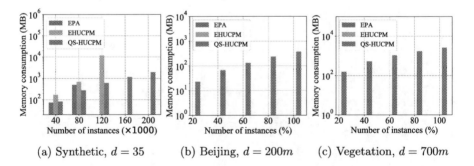

(a) Synthetic, $d = 35$ (b) Beijing, $d = 200m$ (c) Vegetation, $d = 700m$

Fig. 10. The memory consumption of the compared algorithms on different sizes of spatial datasets ($min_u = 0.2$ for all cases).

5   Conclusions and Future Work

This paper proposes a mining HUCP algorithm based on maximal cliques and hash structure. This algorithm only needs to obtain neighbor relationships of spatial instances once, and then store them in a two-level hash structure. Afterward, utilizing the performance of the hash structure, the support instances of each candidate can be quickly obtained by executing a hash query. In addition, the proposed algorithm effectively obtains an upper bound on the size of candidates. This property is especially useful for measurement scales that do not satisfy the downward-closure property since the candidate search space can be greatly reduced. Experimental results on real and simulated datasets demonstrate that the proposed algorithm is capable of large-scale and dense spatial datasets.

In order to further improve the mining efficiency, the future work of this paper is to optimize the maximal clique enumeration algorithm. Besides, from Algorithm 2, it can be easily parallelized to further improve mining efficiency.

Acknowledgements. This work is supported by the National Natural Science Foundation of China (62276227, 61966036), the Project of Innovative Research Team of Yunnan Province (2018HC019), the Yunnan Fundamental Research Projects (202201AS070015), and Yunnan University of Finance and Economics Scientific Research Fund (2022B03).

References

1. Andrzejewski, W., Boinski, P.: Parallel GPU-based plane-sweep algorithm for construction of ICPI-trees. J. Database Manag. (JDM) **26**(3), 1–20 (2015)
2. Bao, X., Wang, L.: A clique-based approach for co-location pattern mining. Inf. Sci. **490**, 244–264 (2019)
3. Cheng, J., Ke, Y., Fu, A.W.C., Yu, J.X., Zhu, L.: Finding maximal cliques in massive networks. ACM Trans. Database Syst. (TODS) **36**(4), 1–34 (2011)
4. Deng, M., Cai, J., Liu, Q., He, Z., Tang, J.: Multi-level method for discovery of regional co-location patterns. Int. J. Geogr. Inf. Sci. **31**(9), 1846–1870 (2017)
5. Eppstein, D., Löffler, M., Strash, D.: Listing all maximal cliques in large sparse real-world graphs. J. Exp. Algorithmics (JEA) **18**, 3-1 (2013)
6. He, Z., Deng, M., Xie, Z., Wu, L., Chen, Z., Pei, T.: Discovering the joint influence of urban facilities on crime occurrence using spatial co-location pattern mining. Cities **99**, 102612 (2020)
7. Li, Y., Wang, L., Yang, P., Li, J.: EHUCM: an efficient algorithm for mining high utility co-location patterns from spatial datasets with feature-specific utilities. In: Strauss, C., Kotsis, G., Tjoa, A.M., Khalil, I. (eds.) DEXA 2021. LNCS, vol. 12923, pp. 185–191. Springer, Cham (2021). https://doi.org/10.1007/978-3-030-86472-9_17
8. Liu, W., Liu, Q., Deng, M., Cai, J., Yang, J.: Discovery of statistically significant regional co-location patterns on urban road networks. Int. J. Geogr. Inf. Sci. **36**(4), 749–772 (2022)

9. Shu, J., Wang, L., Yang, P., Tran, V.: Mining the potential relationships between cancer cases and industrial pollution based on high-influence ordered-pair patterns. In: Chen, W., Yao, L., Cai, T., Pan, S., Shen, T., Li, X. (eds.) ADMA 2022. LNCS, vol. 13725, pp. 27–40. Springer, Cham (2022). https://doi.org/10.1007/978-3-031-22064-7_3

10. Tran, V.: Meta-PCP: a concise representation of prevalent co-location patterns discovered from spatial data. Expert Syst. Appl. **213**, 119255 (2023)

11. Tran, V., Wang, L., Chen, H., Xiao, Q.: MCHT: a maximal clique and hash table-based maximal prevalent co-location pattern mining algorithm. Expert Syst. Appl. **175**, 114830 (2021)

12. Wu, Q., Hao, J.K.: A review on algorithms for maximum clique problems. Eur. J. Oper. Res. **242**(3), 693–709 (2015)

13. Yang, S., Wang, L., Bao, X., Lu, J.: A framework for mining spatial high utility co-location patterns. In: 2015 12th International Conference on Fuzzy Systems and Knowledge Discovery (FSKD), pp. 595–601. IEEE (2015)

14. Yoo, J.S., Shekhar, S.: A joinless approach for mining spatial colocation patterns. IEEE Trans. Knowl. Data Eng. **18**(10), 1323–1337 (2006)

15. Yu, T., Liu, M.: A linear time algorithm for maximal clique enumeration in large sparse graphs. Inf. Process. Lett. **125**, 35–40 (2017)

Point-Level Label-Free Segmentation Framework for 3D Point Cloud Semantic Mining

Anan Du[1] , Shuchao Pang[2,3(✉)] , and Mehmet Orgun[3]

[1] University of New South Wales, Canberra, Australia
anan.du@unsw.edu.au
[2] Nanjing University of Science and Technology, Nanjing, China
[3] Macquarie University, Sydney, Australia
pangshuchao@njust.edu.cn, mehmet.orgun@mq.edu.au

Abstract. 3D point cloud data semantic mining plays a key role in 3D scene understanding. Although recent point cloud semantic mining methods have achieved great success, they require large amounts of expensive manual annotated data. More importantly, the lack of large-scale annotated datasets limits those approaches in many real-world applications, especially for point-level semantic mining tasks such as point cloud semantic segmentation. In this work, we propose a novel point cloud segmentation framework, called Point-level Label-free Segmentation framework (PLS), that does not require point-level annotations. In this framework, the point cloud semantic mining task is formulated as a clustering problem based on mutual information. Meanwhile, our method can directly predict clusters that correspond to the given semantic classes in a single feed-forward pass of a neural network. We apply the proposed PLS to the shape part segmentation task. Experiments on the benchmark ShapeNetPart dataset demonstrate that our method has the ability to discover clusters that match semantic classes, and it can produce comparable results with methods using incomplete labels on several categories.

Keywords: Data mining · Semantic mining · 3D point cloud · Mutual information · Unsupervised clustering

1 Introduction

With the increasing ability to obtain 3D point cloud data, 3D point cloud data mining, especially point cloud semantic mining, plays a vital role in intelligent 3D scene understanding. As one of the point-level point cloud semantic mining tasks, point cloud semantic segmentation has attracted much attention recently, and many impressive results have been achieved [2,8,11,12,25,30,32]. The success of these approaches is attributed to deep network design and the availability of large amounts of point-level annotated training data. However, preparing such fully-annotated datasets involves a significant effort in data cleansing and manual labeling, especially for point-level annotations.

The existing studies have made some efforts to reduce the annotation cost. Some of these studies [7,13,16,18] focus on developing self-supervised learning methods to learn general feature representations without using manual annotations, and then use a small amount of fully-annotated data to train a classifier or use a traditional clustering algorithm such as K-means for the downstream segmentation task. However, the traditional clustering algorithms are not suitable for large-scale datasets. Besides, both types of methods separate the representation learning and clustering/classification into two stages, which results in suboptimal feature representation for the downstream tasks. More recently, a few studies [9,15,31] propose weakly supervised methods for point cloud segmentation, which require a fraction of point-level labels per sample and produce promising results to compete with their fully-supervised counterparts. Although these methods largely reduce the cost of manual annotations, point-level labels are still indispensable. To mitigate the requirement of manual annotations, a reasonable way is to design unsupervised methods. The feasibility of such an approach has been practically proved in the image segmentation field [10], but rarely studied in the field of point cloud semantic segmentation. To our knowledge, [1,14,17] can perform the 3D shape part segmentation task in an unsupervised manner. But these methods are category-specific, which need to learn one specific model for each 3D shape (or object category). Therefore, it is hard to adapt them directly to other application scenes, such as 3D scene semantic segmentation.

In this work, for efficient point cloud semantic mining, we investigate *whether it is possible to develop a general unsupervised method for point cloud semantic segmentation without using point-level annotations.* Intuitively, we require a clustering algorithm that can group data points into classes without using manual labels. At the same time, unlike traditional clustering methods that lack extensibility, the trained model should directly output class identities like the general segmentation models. The model is built upon the state-of-the-art deep neural networks for point cloud feature embedding and is better to be trained in an end-to-end manner to avoid degenerate solutions.

With these motivations, we propose a point-level label-free framework for point cloud semantic segmentation inspired by the study by Ji, Henriques and Vedaldi [10]. We combine representation learning and clustering into a unified framework without using any point-level labels. The advantages of our method are as follows: 1) Our method directly trains a randomly initialized neural network into a classification function in an end-to-end fashion and without any point-level labels. The training objective of our method is to maximize the mutual information of the predictions from different constructed views of the input pairs. 2) Furthermore, the same data can be clustered in many equally good ways. To reduce this ambiguity, we introduce the accessible object-level annotations for further data mining and build our multi-head model, where each head is related to one object. As there is potentially a hierarchical structure in the labels of 3D shape part segmentation, exploiting this hierarchical information helps to reduce the solution space and improve the robustness. In specific, our multi-head model uses a shared embedding network and learns a private classifier for each shape category. The object-level annotations work as a probe that indicates which classifier head should be used for prediction. 3) To further reduce the effect of noisy data and ambiguity, we add an overclustering auxiliary task, which is parallel to the main task but outputs more clusters. Adding the auxiliary task is a general technique that could be useful for other algorithms, such as the study of domain adaptive semantic segmentation by Zheng and Yang [34].

Our key contributions are summarized as follows:

- We propose an unsupervised segmentation framework for 3D point cloud semantic mining without using any point-level annotations, which highly reduces the annotating cost.
- Our method simultaneously learns point-level representation and clustering in an end-to-end fashion, which benefits from the strong representative ability of deep neural networks and avoids the degenerate solutions that are prevalent in those methods which trivially combine clustering and representation learning.
- We apply the proposed framework to the 3D shape part segmentation task. Experiments demonstrate that our method has the ability to discover clusters that match semantic classes in an end-to-end fashion, which in turn might encourage future research on unsupervised point cloud segmentation without point-level annotations.

2 Related Work

2.1 Deep Architectures for Point Cloud Semantic Mining

Various architectures have been recently proposed for 3D semantic mining. Pioneered by PointNet [19] and PointNet++ [20], *point-based networks* take raw point clouds as input to learn per-point features. Most of the following works focus on exploring local structure [25,28] or spatial context [5,30] to enhance the learned feature representations. In general, point-based networks require subsampling a large-scale point cloud to small patches due to the limitation of computational resources. *Projection-based networks* first convert the raw point clouds into regular grids, such as multiple 2D images [24,32] and 3D voxels [2,6], and then apply 2D or 3D convolutions to learn feature representations. Along this line of works, Choy, Gwak and Savarese [2] developes MinkowskiEngine to implement 3D deep architectures efficiently, and achieves outstanding performance on several vision tasks like point cloud segmentation and detection. Therefore, we leverage the 3D MinkUNet presented in [2] as the backbone of our approach.

2.2   Point Cloud Semantic Mining with Less Annotations

Self-supervised Pre-training Plus Semi-supervised Finetuning. Self-supervised learning methods aim to learn general feature representations without manual annotations. Recent works [4,7,13,16,18,29] focus on designing self-supervised tasks on unlabeled 3D datasets to pre-train the networks to learn initial per-point representations. And then, they finetune all or parts of layers with a few annotated task data. This type of method has shown promising results on many 3D vision tasks.

Weakly-Supervised Point Cloud Semantic Mining. Weakly-supervised methods for point cloud segmentation only use *inexact annotations* or *incomplete point-level annotations*. Along this line of methods, Wei et al. [27] use scene-level annotations and subcloud-level annotations to predict pseudo point-level labels, followed by supervised training with pseudo labels to learn the final segmentation model. Xu and Li [31] use subcloud-level labels and a fraction of point-level labels to develop a four-branches framework for weakly-supervised point cloud segmentation. A most recent work [9] further reduces the amount of used point-level labels to 0.1% and achieves state-of-the-art performance. Liu, Qi and Fu [15] propose to use one labeled point per object and applies an iterative strategy to train the network, which achieves state-of-the-art performance on the weakly-supervised configuration and comparable performance with its fully-supervised counterpart.

Unsupervised Point Cloud Semantic Mining. Most of the traditional clustering approaches are generative and do not require labels. Popular clustering approaches like Kmeans, Ncut [22] can be used for unsupervised point cloud semantic mining based on the spatial or color affinities. However, clustering based on low-level features is unstable. Several studies [1,14,17,26] propose to first apply self-supervised methods to learn point feature representations and correspondences with strong generalization ability, and then apply unsupervised methods, such as K-means, for point cloud shape part segmentation task. But these methods are category-specific, which need to learn one specific model for each 3D shape. Therefore, they are not flexible enough to be applied to other point cloud semantic segmentation tasks.

3   PLS: Point-Level Label-Free Segmentation Framework

To avoid the requirement of point-level annotations, we formulate the point-level point cloud semantic mining task into a clustering problem based on mutual information (MI) and propose the **P**oint-level **L**abel-free point cloud **S**egmentation framework, called PLS. As shown in Fig. 1, PLS contains a Siamese network structure with two shared-parameter deep neural networks (SegNet). Given a paired input $(\mathbf{x}, \mathbf{x}')$, where $\mathbf{x}$ and $\mathbf{x}'$ are two point clouds with M points, SegNet, together with a softmax layer, generates paired dense predictions, which are denoted as $(P(\mathbf{y}_u|\mathbf{x}_u), P(\mathbf{y}'_u|\mathbf{x}'_u))$, where $u = \{1, ..., M\}$ is a point location. The objective of PLS is to maximize the MI, denoted as $I(\mathbf{P}_\delta)$, between the output predictions from $\mathbf{x}$ and $\mathbf{x}'$. In fact, the dense predictions can be interpreted as the cluster assignment probabilities, and maximizing $I(\mathbf{P}_\delta)$ forces SegNet to map inputs into semantic consistent representation space.

In the following, we first introduce the principle of clustering using mutual information. Secondly, we introduce the formulation and explanation of our PLS. Thirdly, we apply PLS to the shape part segmentation task.

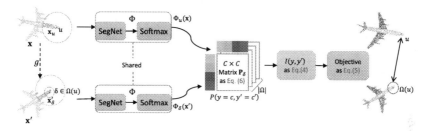

Fig. 1. An overview of our proposed point-level label-free point cloud segmentation framework (PLS). $(\mathbf{x}, \mathbf{x}')$ are a pair of point clouds, g denotes a random transformation, red point u denotes an arbitrary point and $\mathbf{x}_u$ denotes the patch (shown in dark grey) centered at u of point cloud $\mathbf{x}$. $\Omega(u)$ denotes the set of local neighbors of point u and the red solid circle denotes the local range. Green point $\delta \in \Omega(u)$ is an arbitrary neighbor of point u and $\mathbf{x}'_\delta$ denotes the patch centering at δ of point cloud $\mathbf{x}'$. Φ denotes a deep neural network that consists of a network SegNet, followed by a softmax layer. Dashed lines denote shared parameters. Take $(\mathbf{x}, \mathbf{x}')$ as input, then the outputs of $\Phi(\mathbf{x}_u)$ and $\Phi(\mathbf{x}'_\delta)$ can be read off as the feature vector corresponding to point u of tensor $\Phi(\mathbf{x})$ and the feature vector corresponding to point δ of tensor $\Phi(\mathbf{x}')$ respectively, given by $\Phi(\mathbf{x}_u) = \Phi_u(\mathbf{x})$ and $\Phi(\mathbf{x}'_\delta) = \Phi_\delta(\mathbf{x}')$. y and y' are two cluster assignment random variables over C classes for $\mathbf{x}_u$ and $\mathbf{x}'_\delta$ respectively. The $C \times C$ matrix $\mathbf{P}_\delta$ denotes the joint probability distribution $P(y = c, y' = c')$ after marginalization over a batch of patch pairs $(\mathbf{x}_u, \mathbf{x}'_\delta)$, given by Eq. (6). The mutual information $I(y, y') = I$ is computed as Eq. (4). Considering all $|\Omega|$ neighbors of point u together, we get the final objective as Eq. (5).

3.1 Clustering Using Mutual Information

Clustering is an important research topic in data mining. In information theory, the mutual information (MI) of two random variables $I(X; Y)$ measures the information that X and Y share, in other words, $I(X; Y)$ measures how much knowing X reduces uncertainty about Y and vice-versa. If the state of X/Y is deterministic when the state Y/X is revealed, the MI is maximized. Our goal is to gain a representation with a learned deep neural network that preserves the common features between a pair of data samples while discarding the instance-specific details, which can be achieved by maximizing the mutual information between the representations of the input pairs.

Formally, let $\mathbf{x} \sim \mathcal{X}$, $\mathbf{x}' \sim \mathcal{X}$ be a pair of data samples from a joint probability distribution $P(\mathbf{x}, \mathbf{x}')$, and $\varPhi : \mathcal{X} \to \mathcal{Y}$ be a deep neural network that maps the inputs to a representation space $\mathcal{Y}$. The objective function is:

$$\max_{\varPhi} I\left(\varPhi(\mathbf{x}), \varPhi\left(\mathbf{x}'\right)\right) \tag{1}$$

In a clustering task, $\mathcal{Y} \in [0,1]^C$ is the cluster assignment corresponding to the semantic classes with C possible clusters, and the network $\varPhi$ is terminated with a C-way softmax layer. Let $\mathbf{y} \sim \mathcal{Y}$ and $\mathbf{y}' \sim \mathcal{Y}$ be the cluster assignment variables for the paired inputs $\mathbf{x}$ and $\mathbf{x}'$ respectively, then the outputs $\varPhi(\mathbf{x})$, $\varPhi(\mathbf{x}')$ can be interpreted as the probability distribution of $\mathbf{y}$ and $\mathbf{y}'$ over the C clusters: $P(\mathbf{y}|\mathbf{x}) = \varPhi(\mathbf{x})$, $P(\mathbf{y}'|\mathbf{x}') = \varPhi(\mathbf{x}')$. Therefore, the conditional joint distribution $P(\mathbf{y}, \mathbf{y}'|\mathbf{x}, \mathbf{x}')$ is computed as follows:

$$P(\mathbf{y}, \mathbf{y}'|\mathbf{x}, \mathbf{x}') = \varPhi(\mathbf{x})^{\top} \varPhi(\mathbf{x}') \tag{2}$$

After marginalization over the batch, the joint probability distribution is:

$$P(\mathbf{y}, \mathbf{y}') = \mathbb{E}_{\mathbf{x} \sim \mathcal{X}}\left[P(\mathbf{y}, \mathbf{y}'|\mathbf{x}, \mathbf{x}')\right] = \frac{1}{N} \sum_{i=1}^{N} \varPhi(\mathbf{x}_i) \varPhi(\mathbf{x}_i') \tag{3}$$

where N is the number of pairs in a batch, and the joint probability distribution is a $C \times C$ matrix. For simplicity, we denote this matrix as $\mathbf{P}$. The marginals $\mathbf{P}_c = P(\mathbf{y})$ and $\mathbf{P}_c' = P(\mathbf{y}')$ can then be obtained by summing over the rows and columns of $\mathbf{P}$. As for each $(\mathbf{x}_i, \mathbf{x}_i')$ we also have $(\mathbf{x}_i', \mathbf{x}_i)$, we symmetrize $P(\mathbf{y}, \mathbf{y}')$ by $[P(\mathbf{y}, \mathbf{y}') + P(\mathbf{y}, \mathbf{y}')^{\top}]/2$ to maximize the MI in both directions.

At last, the MI in Eq. (1) is derived as follow:

$$\begin{aligned} I(\mathbf{y}, \mathbf{y}') &= D_{\mathrm{KL}}\left(P\left(\mathbf{y}, \mathbf{y}'\right) \| P(\mathbf{y})P\left(\mathbf{y}'\right)\right) \\ &= \sum_{c=1}^{C} \sum_{c'=1}^{C} \mathbf{P}_{cc'} \cdot \ln \frac{\mathbf{P}_{cc'}}{\mathbf{P}_c \cdot \mathbf{P}_{c'}} \end{aligned} \tag{4}$$

3.2   PLS

Our PLS formulates point cloud segmentation as a clustering problem using mutual information. As presented in Sect. 3.1, the formulation of clustering requires paired samples as inputs. In the case of point cloud segmentation, we propose to use point cloud pairs generated from the different views of a point cloud. Here, the term 'views' refers to the versions of a point cloud after different transformations, such as rotation, jitter, flipping and so on. The motivation lies in our assumption that the prediction for any point is *view-invariant*. This assumption, in particular, holds true for 3D CAD shapes and indoor scenes with rotations in the XoY plane, e.g. the semantic label should not change with different views.

As for the training objective, in the case of point cloud segmentation, the neural network $\varPhi$ is an encoder-decoder type network, outputting $\mathbf{y} \in [0,1]^{C \times M}$

of the same number of points as the input. This can be seen as clustering point cloud patches defined by the receptive field of the neural network for each output point.

Formally, let $\mathbf{x} \in \mathcal{R}^{M \times 3}$ be a given point cloud, u denote a point location, $\mathbf{x}_u$ be a point cloud patch centered at u. The paired data for training is generated by patch $\mathbf{x}_u$ and its random views $\mathbf{x}'_u = g\mathbf{x}$. Let Φ denote the segmentation network, which is also terminated with a softmax layer; the cluster probability vectors for all patches $\mathbf{x}_u$ can be read off as the column vectors $\Phi_u(\mathbf{x}) = \Phi(\mathbf{x}_u) \in [0,1]^C$ of the tensor $\Phi(\mathbf{x}) \in [0,1]^{C \times M}$, where C is the number of clusters and M is the number of points in a point cloud. Different from the general clustering formulation, we also consider the spatial smoothness constraint for segmentation because a point and its neighbors generally belong to the same semantic class. Specifically, we generate the input pairs of patches by $(\mathbf{x}_u, \mathbf{x}'_{u+\delta})$, where δ is a small displacement. Finally, the objective function of the segmentation method derived from Eq. (1) is as follows:

$$\max_{\Phi} \frac{1}{|\Omega|} \sum_{\delta \in \Omega} I\left(\mathbf{P}_\delta\right) \tag{5}$$

Similar to Eq. (4),

$$\mathbf{P}_\delta = \frac{1}{N \cdot G \cdot M} \sum_{i=1}^{N} \sum_{g \in G} \sum_{u=1}^{M} \Phi_u\left(\mathbf{x}_i\right) \Phi_{u+\delta}^\top \left(g\mathbf{x}_i\right) \tag{6}$$

Auxiliary Head. Intuitively, the same data can often be clustered in many equally good ways. For example, visual objects can be clustered by color, size, typology, viewpoint, and many other attributes. Inspired by the work of Ji, Henriques and Vedaldi [10], we add an overclustering auxiliary task, which shares the same embedding network (SegNet) with the main task, but outputs more clusters. In this way, our model can potentially capture different and complementary clustering axes.

3.3   PLS for Shape Part Segmentation

To verify our method, we apply the proposed PLS on the shape part segmentation task. Shape part segmentation is a particular semantic segmentation task, which contains a hierarchical structure in the labels. Specifically, there are 16 different shapes in the dataset, and each shape contains several different parts. The parts that belong to different shapes do not have overlaps. To reduce unnecessary ambiguity, we exploit this hierarchical information and introduce the object-level information to SegNet. The main idea is to segment each shape by its private classifier, but using a shared embedding network.

The used SegNet is as shown in the below sub-fig of Fig. 1. The SegNet contains an encoder, a decoder and a multi-head classifier. Given a point cloud shape i, SegNet applies the encoder and decoder to obtain the feature vectors

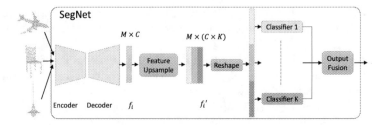

Fig. 2. The architecture of SegNet for 3D shape part segmentation. M denotes the number of points, C denotes the number of classes, K denotes the number of classifiers, f_i and f'_i are feature vectors of input point cloud i.

$f_i \in \mathcal{R}^C$ of each point. Then a Feature Upsampling layer raises the features f_i to f'_i. After that, based on the shape category, the related classifier head is used to produce the prediction output. And finally, the Output Fusion module combines and organizes all outputs to form the final output.

4   Experiments

4.1   Experimental Setup

Datasets. The ShapeNet [33] part segmentation dataset contains 16,881 shapes from 16 categories, represented as point clouds. Most shapes are labeled with two to five parts, and in total, there are 50 parts in the dataset. ShapeNet is widely used as the benchmark for point cloud segmentation evaluation.

Architecture. We use the Minkowski Engine [2] and build a sparse residual U-Net architecture as the embedding network. It follows the architecture ResUNet [3] but with the first convolutional kernel size of 3 and a wider output channel of 96.

Training. We train the networks for 150 epochs using Stochastic Gradient Descent starting with a learning rate of 0.1 with an exponential learning rate schedule with $\gamma = 0.95$. The batch size is set to 48 for all experiments and analysis. The 3D coordinates of the data are normalized to zero mean inside a unit ball with a radius of 1.0. The voxel size is set to 0.02 and the input feature of the point cloud is set to all ones. The transformations for generating paired data we apply include random scaling $\in [0.8, 1.2]$ to a pair, random mirror flipping along x or z axis, different random rotation $\in [0°, 360°]$ along the y-axis (the elevation direction) for both scans in a pair, random shifting voxelized coordinates to make the sparse convolution invariant to even and odds coordinates and adding a small gaussian noise to the feature. The main and auxiliary heads are trained by maximizing Eq. (1) in alternate epochs. Following the training/validation/testing splits, we train our model on the training split data without point-level annotations and use the validation data to compute the label alignment mapping between our cluster assignment and the given labels. All metrics are computed after label alignment.

4.2 Qualitative and Quantitative Results

In this subsection, we compare the performance of our model against three sub-categories of methods. The quantitative results are shown in Table 1. We report the overall per-category mean intersection of union (Cmiou), per-instance mean intersection of union (Imiou) and miou of each category. The qualitative results are shown in Fig. 3.

Comparisons with Traditional Unsupervised Approaches. Traditional unsupervised approaches do not use any annotations, but instead, directly infer clusters with spatial or color affinities. In particular, we compare with Kmeans and normalized cut spectral clustering (Ncut). Both methods are provided with the ground-truth number of clusters. The clustering is applied to each shape, and the produced clusters of each shape are aligned to the given labels to compute the metrics.

As shown in Table 1, our model (PLS) outperforms Kmeans with significant margins of 7.6% on Cmiou and 5.6% on Imiou. As for the miou of each category, PLS achieves significant improvements on most categories, e.g., 23.1% on laptop, 19.2% on lamp and 13.9% on earphone. Ncut performs better than Kmeans, but PLS still outperforms Ncut on Cmiou, Imiou and mious of more than half of all categories (9/16). Besides, here we should also mention that the traditional unsupervised methods Kmeans and Ncut lack extensibility, which limits their application on various real scenes. On the other hand, the clustering here is applied to each shape, so the clusters do not have semantic consistency across shapes. In contrast, our learned model is able to directly output the cluster assignment in a single feed-forward pass of a neural network and capture semantics across instances.

Comparisons with Self-supervised Approaches. The self-supervised approaches aim to learn feature representations by the pretext task, not using manual annotations. In recent studies [21,29], it has been demonstrated that the point features obtained through a self-supervised manner can capture high-level semantics and the self-supervised pre-training improves the performance of multiple downstream supervised tasks. Therefore, to compare with self-supervised approaches, we can extract features by the self-supervised methods and then apply unsupervised clustering approaches to the extracted features to generate clusters. Following this strategy, we implement FCGF by following FCGF [3] to extract per-point features and using Kmeans to generate clusters. While our method aims to directly address the point cloud segmentation task by simultaneously learning the point-level representation and clustering without any point-level annotations. For a fair comparison, we also build our baseline model with the same feature embedding network with FCGF. Because the embedding network of the baseline is too thin to capture complex semantic information, we build our final model with a more wider embedding network.

To compare FCGF with our baseline model (PLS-b), PLS-b increases about 4% on both Cmiou and Imiou, which demonstrates that the proposed method has the ability to simultaneously learn the point-level representation and clustering while avoiding degenerated solutions. It also verifies the benefit of combining task objective and representation learning into a joint framework. Comparing PLS-b with PLS, PLS gains a significant increase by about 10% Cmiou and 11.8% Imiou. This demonstrates that the proposed method has the potential to further improve the performance of the point cloud semantic segmentation task with better feature embedding networks.

Table 1. mIoU(%) evaluation on the ShapeNet dataset. The Labels column (Yes/No) indicates whether using point-level labels. Bold denotes the best results in the related subtable.

Labels	Methods	Cmiou	Imiou	plane	bag	cap	car	chair	ear-	guitar	knife	lamp	laptop	motor-	mug	pistol	rocket	skate-	table
Yes	MT [23]	68.6	72.2	71.6	60.0	79.3	57.1	86.6	48.4	87.9	80.0	73.7	94.0	43.3	79.8	74.0	45.9	56.9	59.8
	WS [31]	74.4	75.5	75.6	74.4	79.2	66.3	87.3	63.3	89.4	84.4	78.7	94.5	49.7	90.3	76.7	47.8	71.0	62.6
No	Kmeans	39.4	39.6	36.3	34.0	49.7	18.0	48.0	37.5	47.3	75.6	42.0	69.7	16.6	30.3	43.3	33.1	17.4	31.7
	Ncut [22]	43.5	43.2	**41.0**	38.0	53.4	20.0	52.1	41.1	**52.1**	**83.5**	46.1	77.5	18.0	33.5	48.0	**36.5**	19.6	35.0
	FCGF [3]	31.2	30.4	20.8	37.0	46.0	19.9	31.4	40.2	35.8	36.1	42.0	34.5	18.4	39.0	21.9	26.0	20.6	30.2
	PLS-b	35.1	34.7	24.7	37.2	36.4	25.9	31.1	**67.4**	23.3	26.3	57.8	34.6	21.5	48.6	21.7	22.9	**43.8**	38.6
	PLS	45.0	46.5	32.8	27.8	**59.2**	25.1	51.9	47.3	37.4	78.3	64.6	92.1	**24.3**	32.0	50.6	30.6	20.8	45.8
	PLS*	**49.9**	**58.5**	19.9	**44.8**	44.6	**29.1**	**73.9**	42.8	49.8	75.1	**66.2**	**93.1**	19.3	**48.6**	**53.8**	31.0	40.1	**65.7**

Fig. 3. Visual results of our method. The odd columns are the ground truth and the even columns are our results. Each color denotes a class.

Comparisons with Semi-supervised Approaches. The semi-supervised approaches use incomplete point-level annotations. Although our method does not use the point-level annotations, to further analyze the gap with methods that use point-level annotations, we also report the results of MT [23] and WS [31], which use a fraction of point-level labels.

From Table 1, we can observe that there is a certain gap between WS and PLS. One important reason may be the same data can be clustered in many equally good ways, but the proposed method does not provide any supervision to direct the clustering ways. Therefore, there is a mismatch between the clusters generated by our method and the given labels, which further causes our models to be worse than WS in terms of the evaluation metrics. While the WS provides the supervision information, even a small amount. Besides, the below overclustering analysis in Sect. 4.3 and visualization in Fig. 4 also support this assumption. Therefore, we re-align the clusters with the ground truth by allowing repeated matching and report the results as PLS*. Comparing PLS* with WS, we observe that the gap is reduced a lot, and on several categories such as laptop and table, PLS* achieves comparable even better results with WS.

4.3   More Analysis and Discussions

Data Augmentations. The proposed method takes paired data as input and learns transformation-invariant features. We investigate the impact of using different transformations, such as rotations, jitter and flips. As shown in Table 2, the useful transformations include jitter, scale, z-axis flips and y-axis rotation. We observe that when using random rotation along an arbitrary axis, the trained model is easy to be confused by parts with similar shapes, such as the screen and keyboard of a laptop, which impacts the performance.

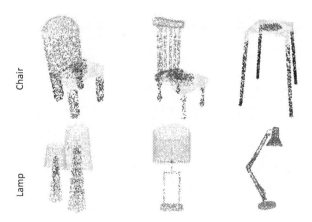

Fig. 4. Visual results of overclustering. Each color denotes a class.

Overclustering. To understand why the overclustering auxiliary task is helpful, we visualize the results of the overclustering task in Fig. 4. We observe that the overclustering outputs fine-grained clusters based on different attributes; meanwhile, the clusters contain consistency across instances. For example, the seat part of the three different chairs are all segmented into several fine-grained parts, and the center dark purple fine-grained parts remain consistent. The yellow corners of the first two chairs with backs are clustered with the four yellow corners of the third chair without a back. The shade parts of the first two lamps (green color) and the third lamp (red color) are identified as two fine-grained classes.

Table 2. Results of applying different transformations.

jitter	scaling	z-symmetry	3Axis- Rotation	y-rotation	Cmiou	Imiou
✓	✓				37.07	32.86
✓	✓	✓			38.50	39.72
✓	✓	✓	✓		34.04	31.64
✓	✓	✓		✓	45.04	46.53

5   Conclusion and Future Work

In this paper, we have presented a point-level label-free framework for 3D point cloud semantic mining. The objective is to optimize mutual information between related pairs. We have shown that the resulting models had the ability to discover clusters that match semantic classes and produced comparable results on several categories with methods using point-level labels.

Since the proposed method is essentially a clustering approach and the same data can be clustered in many equally good ways, the generated clusters may not correspond with the ground truth. In the future, we plan to explore more weak supervision to guide the clustering.

Acknowledgement. This work is supported in part by the National Natural Science Foundation of China under Grants No. 62206128 and in part by the National Computational Infrastructure (NCI) through Australian Government.

References

1. Chen, Z., Yin, K., Fisher, M., Chaudhuri, S., Zhang, H.: Bae-net: branched autoencoder for shape co-segmentation. In: International Conference on Computer Vision, pp. 8490–8499 (2019)
2. Choy, C., Gwak, J., Savarese, S.: 4d spatio-temporal convnets: Minkowski convolutional neural networks. In: IEEE Conference on Computer Vision and Pattern Recognition, pp. 3075–3084 (2019)
3. Choy, C., Park, J., Koltun, V.: Fully convolutional geometric features. In: International Conference on Computer Vision, pp. 8958–8966 (2019)

4. Deng, H., Birdal, T., Ilic, S.: PPF-foldnet: unsupervised learning of rotation invariant 3d local descriptors. In: European Conference on Computer Vision, pp. 602–618 (2018)

5. Du, A., Pang, S., Huang, X., Zhang, J., Wu, Q.: Exploring long-short-term context for point cloud semantic segmentation. In: IEEE International Conference on Image Processing, pp. 2755–2759. IEEE (2020)

6. Graham, B., Engelcke, M., van der Maaten, L.: 3d semantic segmentation with submanifold sparse convolutional networks. In: IEEE Conference on Computer Vision and Pattern Recognition, pp. 9224–9232 (2018)

7. Hou, J., Graham, B., Nießner, M.: Exploring data-efficient 3D scene understanding with contrastive scene contexts. In: IEEE Conference on Computer Vision and Pattern Recognition, pp. 15587–15597 (2021)

8. Hou, Y., Zhu, X., Ma, Y., Loy, C.C., Li, Y.: Point-to-voxel knowledge distillation for lidar semantic segmentation. In: IEEE Conference on Computer Vision and Pattern Recognition, pp. 8479–8488 (2022)

9. Hu, Q., et al.: SQN: weakly-supervised semantic segmentation of large-scale 3D point clouds. In: Avidan, S., Brostow, G., Cissé, M., Farinella, G.M., Hassner, T. (eds.) Computer Vision – ECCV 2022: 17th European Conference, Tel Aviv, 23–27 October 2022, Proceedings, Part XXVII, pp. 600–619. Springer, Cham (2022). https://doi.org/10.1007/978-3-031-19812-0_35

10. Ji, X., Henriques, J.F., Vedaldi, A.: Invariant information clustering for unsupervised image classification and segmentation. In: International Conference on Computer Vision, pp. 9865–9874 (2019)

11. Lai, X., et al.: Stratified transformer for 3d point cloud segmentation. In: IEEE Conference on Computer Vision and Pattern Recognition, pp. 8500–8509 (2022)

12. Li, J., Dai, H., Han, H., Ding, Y.: Mseg3d: multi-modal 3d semantic segmentation for autonomous driving. arXiv preprint arXiv:2303.08600 (2023)

13. Li, J., Chen, B.M., Hee Lee, G.: So-net: self-organizing network for point cloud analysis. In: IEEE Conference on Computer Vision and Pattern Recognition, pp. 9397–9406 (2018)

14. Liu, F., Liu, X.: Learning implicit functions for topology-varying dense 3d shape correspondence. In: Advances Neural Information Processing Systems, vol. 33, pp. 4823–4834 (2020)

15. Liu, Z., Qi, X., Fu, C.W.: One thing one click: a self-training approach for weakly supervised 3d semantic segmentation. In: IEEE Conference on Computer Vision and Pattern Recognition, pp. 1726–1736 (2021)

16. Mei, G., et al.: Data augmentation-free unsupervised learning for 3d point cloud understanding. In: Britain Machine Visual Conference (2022)

17. Niu, C., Li, M., Xu, K., Zhang, H.: Rim-net: recursive implicit fields for unsupervised learning of hierarchical shape structures. In: Proceedings of the IEEE/CVF Conference on Computer Vision and Pattern Recognition, pp. 11779–11788 (2022)

18. Pang, Y., Wang, W., Tay, F.E.H., Liu, W., Tian, Y., Yuan, L.: Masked autoencoders for point cloud self-supervised learning. In: Avidan, S., Brostow, G., Cissé, M., Farinella, G.M., Hassner, T. (eds.) Computer Vision – ECCV 2022: 17th European Conference, Tel Aviv, 23–27 October 2022, Proceedings, Part II, pp. 604–621. Springer, Cham (2022). https://doi.org/10.1007/978-3-031-20086-1_35

19. Qi, C.R., Su, H., Mo, K., Guibas, L.J.: Pointnet: deep learning on point sets for 3d classification and segmentation. In: IEEE Conference on Computer Vision Pattern Recognition, pp. 652–660 (2017)

20. Qi, C.R., Yi, L., Su, H., Guibas, L.J.: Pointnet++: deep hierarchical feature learning on point sets in a metric space. In: Advances Neural Information Processing Systems, pp. 5099–5108 (2017)
21. Sauder, J., Sievers, B.: Self-supervised deep learning on point clouds by reconstructing space. In: Advances Neural Information Processing Systems, pp. 12962–12972 (2019)
22. Shi, J., Malik, J.: Normalized cuts and image segmentation. IEEE Trans. Pattern Anal. Mach. Intell. **22**(8), 888–905 (2000)
23. Tarvainen, A., Valpola, H.: Mean teachers are better role models: weight-averaged consistency targets improve semi-supervised deep learning results. In: Advances Neural Information Processing Systems, vol. 30, pp. 1195–1204 (2017)
24. Tatarchenko, M., Park, J., Koltun, V., Zhou, Q.Y.: Tangent convolutions for dense prediction in 3d. In: IEEE Conference on Computer Vision and Pattern Recognition, pp. 3887–3896 (2018). https://doi.org/10.1109/CVPR.2018.00409
25. Thomas, H., Qi, C.R., Deschaud, J.E., Marcotegui, B., Goulette, F., Guibas, L.J.: Kpconv: flexible and deformable convolution for point clouds. In: International Conference on Computer Vision, pp. 6411–6420 (2019)
26. Wang, P.S., Yang, Y.Q., Zou, Q.F., Wu, Z., Liu, Y., Tong, X.: Unsupervised 3d learning for shape analysis via multiresolution instance discrimination. In: AAAI, vol. 35, pp. 2773–2781 (2021)
27. Wei, J., Lin, G., Yap, K.H., Hung, T.Y., Xie, L.: Multi-path region mining for weakly supervised 3d semantic segmentation on point clouds. In: IEEE Conference on Computer Vision and Pattern Recognition, pp. 4384–4393 (2020)
28. Wu, W., Qi, Z., Fuxin, L.: Pointconv: deep convolutional networks on 3d point clouds. In: IEEE Conference on Computer Vision and Pattern Recognition, pp. 9621–9630 (2019)
29. Xie, S., Gu, J., Guo, D., Qi, C.R., Guibas, L., Litany, O.: PointContrast: unsupervised pre-training for 3D point cloud understanding. In: Vedaldi, A., Bischof, H., Brox, T., Frahm, J.-M. (eds.) ECCV 2020. LNCS, vol. 12348, pp. 574–591. Springer, Cham (2020). https://doi.org/10.1007/978-3-030-58580-8_34
30. Xu, C., et al.: You only group once: efficient point-cloud processing with token representation and relation inference module. In: IEEE International Conference on Intelligent Robots and Systems, pp. 4589–4596 (2021). https://doi.org/10.1109/IROS51168.2021.9636858
31. Xu, X., Lee, G.H.: Weakly supervised semantic point cloud segmentation: towards 10x fewer labels. In: IEEE Conference on Computer Vision and Pattern Recognition, pp. 13706–13715 (2020)
32. Yang, J., Lee, C., Ahn, P., Lee, H., Yi, E., Kim, J.: Pbp-net: point projection and back-projection network for 3d point cloud segmentation. In: IEEE International Conference on Intelligent Robots and Systems, pp. 8469–8475 (2020). https://doi.org/10.1109/IROS45743.2020.9341776
33. Yi, L., et al.: A scalable active framework for region annotation in 3d shape collections. ACM Trans. Graph. **35**(6), 1–12 (2016)
34. Zheng, Z., Yang, Y.: Rectifying pseudo label learning via uncertainty estimation for domain adaptive semantic segmentation. Int. J. Comput. Vis. **129**(4), 1106–1120 (2021)

CD-BNN: Causal Discovery
with Bayesian Neural Network

Huaxu Han, Shuliang Wang[(⊠)], Hanning Yuan, and Sijie Ruan

Beijing Institute of Technology, Beijing, China
{slwang2011,hhx7}@bit.edu.cn

Abstract. Causal discovery involves learning Directed Acyclic Graphs (DAGs) from observational data and has widespread applications in various fields. Recent advancements in the structural equation model (SEM) have successfully applied continuous optimization techniques to causal discovery. These methods introduce acyclicity constraints to tackle the challenge of exploring the exponentially large search space that arises as the number of graph nodes increases. However, these methods often rely on point estimates that fail to fully account for the inherent uncertainty present in the data. This limitation can lead to inaccurate causal graph inference. In this paper, we propose a novel method for causal discovery with Bayesian Neural Networks (CD-BNN). CD-BNN incorporates a Bayesian Neural Network to explicitly model and quantify uncertainty in the data while reducing the influence of noise through model averaging. Moreover, we explore the extraction of the final DAG from the BNN using partial derivatives. We conduct a comprehensive set of experiments on both real-world and synthetic data to evaluate the performance of our approach. The results demonstrate that our proposed method surpasses related baselines in accurately identifying causal graphs, particularly when faced with data uncertainty.

Keywords: causal discovery · bayesian neural network

1 Introduction

Causal discovery can automatically learn causal relations in a data-driven way, which has been widely applied in many applications, such as economic, biology, healthcare, genetics and stability [8,11,14,19,22]. Causal discovery methods aim to model complex problems with Directed Acyclic Graphs(DAGs), which is used in the Bayesian network proposed by Pearl in 1988 [14]. DAGs provide a compact and flexible way to decompose the joint distribution that can be used to represent the causal relation of variables. It is possible to infer causal relationships from observational data alone, which becomes particularly valuable when conducting randomized control trials is not feasible.

Traditional causal discovery methods are constraint- and score-based methods. Constraint-based methods, such as PC and SGS [16,17], build the DAG

X. Yang et al. (Eds.): ADMA 2023, LNAI 14176, pp. 431–446, 2023.
https://doi.org/10.1007/978-3-031-46661-8_29

structure by testing the conditional independence of different variables, which have a good performance on a small dataset. On the other hand, score-based methods need to specify the score functions to evaluate the searched DAGs [2,5,7,10,18,25]. Enumerating and scoring all possible DAGs is feasible when the graph has only a few nodes. However, in real-world scenarios, datasets often consist of hundreds of features, posing a significant challenge for these algorithms to exhaustively enumerate all possible DAGs. Recently, [26] proposed NOTEARS method which transformed the search problem in causal discovery into a continuous optimization task by leveraging structural equation model and some graph theory knowledge. Subsequently, an increasing number of researchers have explored the integration of various neural network models with causal discovery, following the direction of continuous optimization. These models include Multilayer Perceptrons (MLPs), Graph Neural Networks (GNNs), Autoencoders, Reinforcement Learning (RL), and Score Matching [6,9,21,24,28].

While recent methods based on SEM have shown improved performance compared to traditional approaches, these causal discovery algorithms rely on point estimates to recover DAG using Maximum Likelihood Estimation (MLE). However, point estimates do not account for the uncertainty inherent in causal relationships. Real-world datasets often contain unknown noise, leading to uncertainty in the data. Ignoring this existing noise can result in inaccuracies in the learned causal structure. Consequently, many methods become unstable and susceptible to vulnerabilities due to the omission of noise consideration.

For solving the problem, we introduce a new causal discovery framework named *Causal Discovery with Bayesian Neural Network(CD-BNN)* based on the structural equation model, enabling continuous optimization with the Bayesian neural network. CD-BNN can learn a robust nonlinear mapping function between variables with a Bayesian neural network. Then, we extract the DAG structure from learned function parameters based on partial derivatives. So unlike other methods, CD-BNN is capable of capturing uncertainty in data, improving the accuracy of DAG structure learning.

Our main contributions can be summarized as follows:

- We study the uncertainty in data when inferring causal relations under the Gaussian Additive Model assumption. The uncertainty maybe leads to miscorrect graph learning. However, the uncertainty not only depends on noise but also on the unknown function between variables, which improves the difficulty of causal discovery because the function f is independent of noise. For solving the problem, we introduced a new view to causal discovery by considering noise and unknown function f together.
- We propose a new method named CD-BNN for causal discovery based on Bayesian Neural Network. With BNN, our method can model the uncertainty in data and build a framework considering the common influence of noise and the unknown function at the same time, which is a more reasonable design than the point estimating methods.
- We finally evaluate our proposed algorithm on both synthetic and real-world data and show competitive results compared to baseline methods.

2 Background

2.1 Casual Graphical Model

A Casual Graphical Model(CGM) $\mathcal{M}(\mathcal{G}, \theta)$ models the joint density $\mathcal{P}(X)$ of a set of d variables $X = \{x_1, x_2, \cdots, x_d\}$ using a directed acyclic graph $\mathcal{G}(\mathcal{V}, \mathcal{E})$. The graph $\mathcal{G}(\mathcal{V}, \mathcal{E})$ consists of a finite set of nodes $\mathcal{V}$ corresponding to random variables X and directed edges set $\mathcal{E} \subset \mathcal{V} \times \mathcal{V}$. Each variable $x_j \in X$ corresponds to exactly one node $n_j \in \mathcal{V}$. If $i, j \in \mathcal{V}$ are two distinct nodes, the ordered pair $(i, j) \in \mathcal{E}$ denotes a directed edge from i to j. In a CGM, we suppose there are no hidden variables. The graph $\mathcal{G}$ and the density $\mathcal{P}(X)$ satisfy the Markov condition [17]. $\mathcal{M}(\mathcal{G}, \theta)$ encodes the conditional independencies of variables. The parameters θ defines the local conditional distributions $p(x_j | X_{\pi_j^{\mathcal{G}}})$ of each variable j given its parents $\pi_j^{\mathcal{G}}$ in the graph $\mathcal{G}$. So the density $\mathcal{P}(x)$ factorizes as $p(x) = \prod_{j=1}^{d} p(x_j | X_{\pi_j^{\mathcal{G}}})$ according to the graph $\mathcal{G}$. When modeling $\mathcal{P}(X)$ using a CGM, each variable is assumed to be independent of its non-descendants given its parents, thus allowing for a compact factorization of the joint $\mathcal{P}(X)$ into a product of local conditional distributions for each variable and its parents in $\mathcal{G}$.

2.2 Structural Equation Model

Structural equation model(SEM) was applied to genetics at the earliest, which can quantitatively represent causal information in data [20]. Given d dimension data $X = \{X_1, X_2, \cdots, X_d\}$, SEM can be formulated as

$$X_j = f_j(X_{\pi_j^{\mathcal{G}}}, \epsilon_j), \ j = 1, \cdots, d, \tag{1}$$

where f_i is nonlinear function, which has two input $X_{\pi_j^{\mathcal{G}}}$ and ϵ_j. $X_{\pi_j^{\mathcal{G}}}$ is the parent nodes set of node j in the causal graph, which means there exists a directed edge from its parent to node j. ϵ_j is a noise variable modeling the unknown influence in the environment. For each ϵ_j, we have $\epsilon_i \perp \epsilon_j$ if $i \neq j$.

Structure identifiability is inevitable when come to SEM. It's impossible to recover the true causal graph from the Markov class if we don't make further assumptions. There may exit two different CGM $\mathcal{M}_1$ and $\mathcal{M}_2$ leading to same distribution $\mathcal{P}_{\mathcal{M}_1}(X) = \mathcal{P}_{\mathcal{M}_2}(X)$. To solve the problem, many specific parametric models have been proposed to ensure structure identifiability [12,15,23]. Throughout the paper, we assume that the distribution $\mathcal{P}(X)$ satisfies the Gaussian additive model and the corresponding SEM write as

$$X_j = f_j(X_{\pi_j^{\mathcal{G}}}) + \epsilon_j, \ j = 1, \cdots, d, \tag{2}$$

where $\epsilon_j \sim \mathcal{N}(0, \sigma_j^2)$. It's known to be identified from observational data [12] and it is possible to recover the DAG from the knowledge of the joint probability distribution $\mathcal{P}(X)$.

2.3   Bayesian Neural Network

Bayesian Neural Network(BNN) can generate probability interpretability for deep learning, and quantify uncertainty. In the Bayesian framework, the network parameters are random variables following certain distributions rather than having a single fixed value, enabling modeling uncertainty about data and model. We can choose a useful prior that reflects our beliefs about parameters before seeing any data [4]. Given datasets $\mathcal{D}$ and new observations $\hat{x}, \hat{y}$, BNN solves the task of inferring a full posterior density over network parameters that model the observations.

$$P(\hat{y}|\hat{x}) = \mathbb{E}_{p(x|\mathcal{D})}\left[p(\hat{y}|\hat{x}, w)\right]. \tag{3}$$

In general, exact Bayesian inference on the weights of a neural network is intractable as the number of parameters is very large and the functional form of a neural network does not lend itself to exact integration. To solve the problem, a variational approximation was proposed, which aims to find the variational parameters θ of a distribution on weights $q(\mathbf{w}|\theta)$ that minimizes the Kullback-Leibler(KL) divergence with the true Bayesian posterior on the weights. The KL divergence is given by

$$\begin{aligned} \theta^* &= \arg\min_{\theta} \mathbf{KL}[q(\mathbf{w}|\theta)\|P(\mathbf{w}|\mathcal{D}))] \\ &= \arg\min_{\theta} \mathbf{KL}[q(\mathbf{w}|\theta)\|P(\mathbf{w})] - \mathbb{E}_{q(\mathbf{w}|\theta)}[\log P(\mathcal{D}|\mathbf{w})]. \end{aligned} \tag{4}$$

3   Causal Discovery with Bayesian Neural Network

We will show our method, named *Causal Discovery with Bayesian Neural Network*. From the Bayesian view, our method can model uncertainty in data, which can recover causal structure approaching ground-truth relation.

3.1   Uncertainty in Data

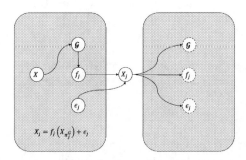

Fig. 1. The structural equation model $X_j = f_j(X_{\pi_j^{\mathcal{G}}}) + \epsilon_j$ has three unknown variables. And it's difficult to recover $\mathcal{G}$ only from observational data X.

In Fig. 1, we draw the true data generation process, in which the random variable X is known, but f_j, $\mathcal{G}$ and ϵ_j need to be inferred from X. What we know is the assumption satisfying the structural equation model. That is $X_j = f_j(X_{\pi_j^{\mathcal{G}}}) + \epsilon_j$, where $\epsilon_i \perp \epsilon_j$ if $i \neq j$. It's difficult to learn causal relations when only known variable X due to uncertainty in data.

In the structural equation model, ϵ_j mainly leads to the noise and uncertainty of data when generating data. However, in the reverse process of inferring $\mathcal{G}$, uncertainty does not only depend on ϵ_j but also the unknown f_j. And because f_j is independent of ϵ, we can't reduce search space by merging the two variables further. But it's still possible to infer $\mathcal{G}$ if we can model uncertainty with other variables properly. BNNs are particularly well-suited for modeling in noisy environments. So we choose BNNs to fit nonlinear functions and eliminate the influence of noise variables. We can model f_j with Bayesian neural network f_{bnn} without considering noise ϵ_j, which reduces the number of variables.

After solving uncertainty problems with BNNs, what we need to deal with is recovering ground-truth causal graph from functions f_{bnn}. It should be noted that $\mathcal{G}$ is invariant and unique to the generation mechanism of data X whatever X changes.

3.2 General Approach

To solve uncertainty when learning causal relations from data. We propose using the Bayesian method to model $\mathcal{G}$, which is a random variable in our framework. Given samples $X^1, X^2, \cdots, X^n$, our framework can represent as

$$\log p(X^1, X^2 \cdots, X^n; \mathcal{G}) = \sum_{k=1}^{n} \log \int_{\mathcal{G}} p(\mathcal{G})p(X^k|\mathcal{G})d\mathcal{G}. \tag{5}$$

However, the Eq. 5 is intractable and it's impossible to average on all graphs $\mathcal{G}$. Instead of computing directly, we use an approximate solution by applying a variational posterior $q(\mathcal{G})$ to approach the actual posterior $p(\mathcal{G})$. Finally, we derive the evidence lower bound(ELBO) as a lower bound on marginal likelihood avoiding optimizing the log-likelihood. This is given by

$$\text{ELBO}(\phi) = \underbrace{-D_{\text{KL}}(q_\phi(\mathcal{G}|\mathbf{X})\|p(\mathcal{G}))}_{\text{model complexity}} + \underbrace{\mathbb{E}_{\mathcal{G} \sim q_\phi(\mathcal{G})}[\log p(\mathbf{X}|\mathcal{G})]}_{\text{reconstruction error}}$$

$$= -\mathbb{E}_{\mathcal{G} \sim q_\phi(\mathcal{G}|\mathbf{X})}[\log q_\phi(\mathcal{G}|\mathbf{X}) - \log p(\mathcal{G})] \tag{6}$$

$$+ \mathbb{E}_{\mathcal{G} \sim q_\phi(\mathcal{G})}[\log p(\mathbf{X}|\mathcal{G})].$$

The formula above consists of three components: the prior distribution $p(\mathcal{G})$, the likelihood distribution $p(\mathbf{X}|\mathcal{G})$, and the variational distribution $q_\phi(\mathcal{G}|\mathbf{X})$. In Sect. 3.3, we will explain how to incorporate the acyclicity constraint into the prior distribution $p(\mathcal{G})$. Then, in Sect. 3.4, we will present the complete architecture of the Bayesian neural network used to compute the likelihood distribution $p(\mathbf{X}|\mathcal{G})$. Finally, in Sect. 3.5, we will describe the variational solution for estimating the posterior distribution $q_\phi(\mathcal{G}|\mathbf{X})$.

3.3   Prior over Graphs

The prior $p(\mathcal{G})$ encodes information about the degree of DAG. [26] explicitly constructs a smooth function with computable derivatives that encode the acyclicity constraint. The graph $\mathcal{G}$ is DAG when satisfying the below condition

$$h(\mathcal{G}) = tr(e^{\mathcal{G} \circ \mathcal{G}}) - d = 0, \tag{7}$$

where $\circ$ is the Hadamard product and d is the number of nodes. In Bayesian framework, we replace $h(\mathcal{G})$ with the prior $p(\mathcal{G})$, same as [3]. So $p(\mathcal{G})$ write as

$$p(\mathcal{G}) \propto exp(-ph(\mathcal{G})^2 - ah(\mathcal{G})), \tag{8}$$

where p and α as penalty terms are gradually increased to favour DAGs. $\mathcal{G} \in [0,1]^{D \times D}$ is a weighted adjacency matrix, with zero entries encouraging sparser graphs.

3.4   Likelihood over Graphs

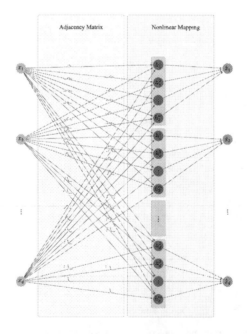

Fig. 2. Bayesian Neural Network Architecture. On the left is the adjacency matrix module, with elements following a certain distribution. On the right is the nonlinear mapping module consisting of d independent multilayer perceptrons.

The primary challenge lies in computing the distribution $p(X_n|\mathcal{G})$, as there is no direct mapping between the variable X_n and the causal graph $\mathcal{G}$. Despite the lack of explicit information about the function f and noise ϵ, we can still establish an approximate connection between them. To tackle the estimation of the likelihood $p(X_n|\mathcal{G})$, we employ Bayesian Neural Networks. BNNs enable us to model the uncertainty associated with the relationship between X_n and $\mathcal{G}$. Through the inherent stochasticity of BNNs, we can capture the uncertainty and provide a probabilistic estimate of the likelihood $p(X_n|\mathcal{G})$. This approach enables us to approximate the likelihood distribution and effectively handle the estimation task in the presence of uncertainty

Figure 2 illustrates the architecture of the entire framework, which comprises two key components: the adjacency matrix module and the nonlinear mapping module. By incorporating the adjacency matrix module and the nonlinear mapping module into our framework, we aim to leverage the power of neural networks not only to learn the function f_{bnn} but also to infer the causal dependencies represented by the graph $\mathcal{G}$. This allows us to shift our focus from solely learning the function to recovering the essential causal structure that governs the relationships among the variables.

Adjacency Matrix Module. The adjacency matrix module as a single module can model uncertainty in data with Bayesian technology, and encode information about causal relations when our neural network converges on a good loss. We will formulate the process to infer a causal graph from a learned function.

It is indeed possible to infer the underlying causal graph $\mathcal{G}$ from a given function f, as demonstrated in the works of [27] and [9]. In these studies, the gradient of the function plays a crucial role in quantifying the magnitude and direction of function changes. By leveraging this gradient information, causal relationships can be derived. Specifically, if the L^2 norm of the partial derivative $\partial_k f_j$ is zero, denoted as $\|\partial_k f_j\|_{L^2} = 0$, it implies that the variable f_j remains unaffected by changes in variable X_k. Consequently, we can infer that X_j and X_k are independent, denoted as $X_j \perp X_k$. This principle forms the basis for deriving causal relations using gradient values. Building upon this idea, we adopt a neural network framework similar to the approach presented in [27]. In this framework, we utilize an adjacency matrix module as the equivalent of the gradient. By effectively learning the function, we can compute the adjacency matrix $W(f) = W(f_1, \cdots, f_d) \in \mathbb{R}^{d \times d}$ from the neural network. The element value $[W(f)]_{kj}$ of $W(f)$ is non-zero when node k is the parent of node j in the causal graph. By adopting this approach, we aim to leverage the power of neural networks to learn and infer causal relationships. This allows us to construct the adjacency matrix and ultimately unravel the underlying causal structure from the given function f. The $[W(f)]_{kj}$ can write as

$$[W(f)]_{kj} := \|\partial_k f_j\|_{L^2}. \tag{9}$$

Nonlinear Mapping Module. For every feature X_i, a full connection layer is used to fit mapping from X_i to X_{-i}, where X_{-i} is a set including all other features excluding X_i. The parameter of L layers of MLP is

$W_{(i)} = \{W_{(i)}^{(L)}, \cdots, W_{(i)}^{(2)}, W_{(i)}^{(1)}\}$. So the output can write as $\hat{X}_i = W_i^{(L)} g(\cdots W_{(i)}^{(2)} g(W_i^{(1)} X_{\pi_i}) \cdots)$ where g is activation function ReLU and X_{π_i} is the output of adjacency matrix module. The zero elements of $W_{i:}^{(1)}$ as mask can filter X_j if $X_j \notin X_{\pi_i}$.

3.5   Variational Distribution over Graphs

An important design choice in variational methods is to specify the variational family. A variational distribution q_ϕ over latent variable z used in ELBO optimization contains two operations: drawing a sample $z \sim q_\phi$, and computing $\log q_\phi(z) - \log p(z)$. Depending on the prior distribution p, it may be additionally possible to compute the term $\mathbb{E}_{z \sim q_\phi}[\log q_\phi(z) - \log p(z)]$ in closed form, possibly reducing variance compared to Monte Carlo estimates [43]. It is also desirable that sampling z can be written as $g(\gamma), \gamma \sim q_0$, i.e. that z is obtained by sampling from a fixed distribution q_0 and transformed through a parameterized, differentiable sampling path g_ϕ. This lets us use pathwise gradient estimators [1].

In our case, the variational posterior $q_\phi(\mathcal{G}|X)$ can factorize as $q_\phi(\mathcal{G}|X) = \prod_{i,j}^d q_\phi(\mathcal{G}_{ij}|X)$, where $q_\phi(\mathcal{G}_{ij}|X)$ is a diagonal Gaussian distribution, then a sample of the weight $\mathcal{G}_{ij}$ can be obtained by sampling from unit Gaussian $\mathcal{N}(\mu, \sigma)$. We parameterize the standard deviation pointwise as $\sigma = \log(1 + \exp(\rho))$ and so σ is always non-negative. The variational posterior parameters are $\theta = (\mu, \rho)$. Thus the transformation from a sample of parameter-free noise and the variational posterior parameters that yields a posterior sample of the weights $\mathcal{G}_{ij} = \mu + \log(1 + \exp(\rho)) \circ \epsilon$ is where $\circ$ is pointwise multiplication.

3.6   Loss Function and Optimization

Next, we will give our loss function considering uncertainty and acyclicity.

$$\mathcal{L}(\mathcal{G}; X) = \frac{1}{2n}\|X - f_{bnn}(\mathcal{G}; X)\|_F^2 + \mathcal{C}_{dag}(\mathcal{G}) + \mathcal{C}_{sparse}(\mathcal{G}), \tag{10}$$

where $\mathcal{G} \sim q_\phi(\mathcal{G}|X)$. $\mathcal{C}_{dag}(\mathcal{G})$ is the acyclic constraint term and $\mathcal{C}_{sparse}(\mathcal{G})$ controls the complexity of whole model. Our optimization aim is to solve

$$\min_{\mathcal{G} \in \mathbb{R}^{d \times d}} \mathcal{L}(\mathcal{G}; X). \tag{11}$$

The optimization is a nonlinear equality-constrained problem, and we use the augmented Lagrangian method to solve it as in [26]. The formula 11 can be decomposed of a series of unconstrained problems

$$\min_{\mathcal{G} \in \mathbb{R}^{d \times d}} \mathcal{L}(\mathcal{G}; X) = f(\mathcal{G}; X) + \lambda_1\|\mathcal{G}\|_1$$

$$\text{s.t.} \quad f(\mathcal{G}; X) = \frac{1}{2n}\|X - f_{bnn}(\mathcal{G}; X)\|_F^2 + \frac{p}{2}h(\mathcal{G})^2 + ah(\mathcal{G}) \tag{12}$$
$$+ \lambda_s\|\mathcal{G}\|_2^2 + \lambda_1\|\mathcal{G}\|_1,$$

where p is a penalty parameter and a is a dual variable. when $p \to \infty$, we have $h(\mathcal{G}) = 0$ and learn f_{bnn} well enough.

4   Experiments

In this section, we apply our method to synthetic and real-world datasets and compare the performance to various baselines, such as DAG-GNN, notears, Grad-DAG, cam, notears-mlp, random and PC. In all cases, We evaluate the learned DAG structure using the following metrics: structural Hamming distance(SHD) and structural intervention distance (SID). The SHD can count the number of edge insertions, deletions or flips to transform one graph to another graph. The lower the better for SHD. The SID is based on a graphical criterion only and quantifies the closeness between two DAGs in terms of their corresponding causal inference statements. It is therefore well-suited for evaluating graphs that are used for computing interventions.

4.1   Synthetic Data

Each dataset corresponds to two kinds of graph, named Erdős-Rényi(ER) graph and scale-free(SF) graph. The ER graph proposed by Erdős and Rényi can control graph density with a probability of whether an existing edge is between any two nodes. The SF graph can model real community structure by restricting the number of edges. We sample ER and SF graphs with average degrees equal to d(ER1/SF1) or 4d(ER4/SF4) by setting the number of graph nodes to $\{10, 20, 50\}$ for verifying the effectiveness of our method when encountering different size of the graph. The synthetic data are generated in two steps. Firstly we sample a random DAG with different graph models. Then the random data X is generated from SEM $X_j = f_j(X_{\pi_j^{\mathcal{G}}}) + \epsilon_j$ with mutually independent noises $\epsilon_j \sim \mathcal{N}(0, \sigma_j^2) \; \forall j$ on the given graph, where functions f_j are independently sampled from a Gaussian process with a unit bandwidth RBF kernel and with σ_j^2 sampled uniformly in $[0.5, 2]$.

We give detailed experiments results in Table 1, 2 and 3. Overall, the CD-BNN algorithm demonstrates remarkable performance in SHD metrics on the 10 nodes graph, ranking second only to the cam algorithm in the ER1-10 and SF1-10 graphs. Additionally, it outperforms the DAG-GNN, notears, random, and PC algorithms in the ER4-10 graph. On the SF4-10 graph, it is only surpassed by the cam and PC algorithms, suggesting that CD-BNN is better suited for different graph types with a smaller number of nodes. As the number of graph nodes increases, Table 2 reveals that even on the 20 nodes dataset, CD-BNN remains second only to the cam algorithm in terms of performance on the ER1-20 and SF1-20 graphs. Moreover, in the case of the ER4-20 graph type, CD-BNN has also shown improvement and is ranked second, just behind the cam algorithm. This indicates that as the number of nodes increases, the recognition ability of CD-BNN improves. However, its performance decreases on the SF4 graph type, suggesting that it may not be suitable for graphs with extreme distribution characteristics of this type. Finally, according to Table 2, on the 50 nodes dataset, the performance on ER1 graph has decreased, but CD-BNN remains second only to the cam algorithm on the ER4 graph. Additionally, it achieves a moderate level of performance on the SF1 and SF4 graphs. Therefore, overall, CD-BNN is better suited for ER graphs and complex graphs with a larger number of nodes.

Table 1. SHD and SID for ER and SF graphs of 10 nodes.

	ER1(8)		ER4(39)		SF1(9)		SF4(23)	
	SHD	SID	SHD	SID	SHD	SID	SHD	SID
DAG-GNN	6	**13**	36	85	8	**15**	23	48
notears	6	16	35	84	8	24	23	63
Grad-DAG	34	**6**	**6**	**10**	30	9	28	**0**
cam	**3**	11	11	**17**	1	**6**	10	18
notears-mlp	**5**	14	**24**	55	9	49	20	77
random	26	14	31	81	22	40	28	51
PC	7	21	34	74	4	23	**15**	41
CD-BNN	**5**	14	**29**	67	4	22	**17**	47

Table 2. SHD and SID for ER and SF graphs of 20 nodes.

	ER1(21)		ER4(76)		SF1(19)		SF4(48)	
	SHD	SID	SHD	SID	SHD	SID	SHD	SID
DAG-GNN	20	106	74	288	17	**49**	46	**101**
notears	**19**	95	75	311	18	59	46	114
Grad-DAG	170	**21**	114	**76**	155	**25**	151	144
cam	2	12	29	97	5	28	34	53
notears-mlp	20	145	**62**	336	**13**	104	46	108
random	92	139	108	**324**	97	75	101	**97**
PC	25	181	70	328	22	187	47	154
CD-BNN	**18**	104	**61**	339	**12**	75	51	150

Next, lets examine more detailed information about various baselines. In Tables 4 and 5, we give details results about the count number of edge insertions, deletions, flips and correctness. In the ER1 graph, it can be observed that under the SHD metric, the results of random and Grad-DAG algorithms are slightly worse. Both algorithms predict more pseudo edges, while other algorithms achieve similar SHD scores. The performance of notears-mlp and CD-BNN are the same, surpassing other algorithms. Under the SID metric, Grad-DAG achieved the optimal score because the algorithm predicted a large number

Table 3. SHD and SID for ER and SF graphs of 50 nodes.

	ER1(51)		ER4(202)		SF1(49)		SF4(136)	
	SHD	SID	SHD	SID	SHD	SID	SHD	SID
DAG-GNN	44	281	203	2230	**47**	**147**	133	**356**
notears	**41**	321	**197**	2230	44	148	130	372
Grad-DAG	997	**208**	935	**1751**	1044	360	996	700
cam	**14**	**46**	**87**	**101**	**16**	**53**	**96**	**280**
notears-mlp	48	311	195	2138	49	153	134	476
random	495	422	535	**2080**	480	246	528	1369
PC	**43**	**240**	201	2362	55	321	**130**	537
CD-BNN	52	372	**188**	2106	49	263	133	464

of correct edges. Although there were many false edges, it did not affect the calculation of SID. However, the PC algorithm, which also predicted 6 correct edges and fewer false edges, but its interference effect was more significant. It can be seen that the factor affecting SID is not only the number of predicted correct edges but also the predicted false edges. Except for the random algorithm, the number of correct edges predicted by other algorithms is similar, and the number of false edges predicted by the notears, notars-mlp, and DAG-GNN algorithms is similar, so they have similar SID.

Table 4. Edges results for ER1 and SF1 of 10 nodes.

	fn		fp		rev		correct	
	ER1	SF1	ER1	SF1	ER1	SF1	ER1	SF1
DAG-GNN	5	8	0	0	1	0	2	1
notears	3	6	0	0	3	2	2	1
Grad-DAG	0	0	32	30	2	0	6	9
cam	1	0	1	0	1	1	6	8
notears-mlp	2	2	0	1	3	6	3	1
random	6	5	19	17	1	0	1	4
PC	2	1	5	2	0	1	6	7
CD-BNN	2	2	0	0	3	2	3	5

The ER graph measures the presence of edges between nodes based on density, while the SF (Scale-Free) graph exhibits a more uneven distribution of edges. In the SF1 graph, it is observed that both CD-BNN and PC algorithms perform well, achieving the best Structural Hamming Distance scores. DAG-GNN and notears-mlp exhibit similar performance in this graph. On the other hand, the Grad-DAG algorithm, which performed well in the ER1 graph, obtains the worst score in the SF1 graph. Although Grad-DAG correctly identifies all the true edges, it also identifies a large number of pseudo-edges, indicating that it is not well-suited for SF graphs with uneven edge distributions. Conversely, CD-BNN demonstrates its ability to handle such extreme scenarios and effectively reduces the number of spurious edges compared to notears-mlp. These findings suggest that the CD-BNN algorithm is more adaptable to graphs with uneven edge distributions, such as the SF1 graph, while Grad-DAG is better suited for graphs with uniform density, like the ER1 graph. By leveraging the Bayesian Neural Network approach, CD-BNN can effectively handle the challenges posed by SF graphs and achieve improved performance, both in terms of correctly identifying edges and minimizing the number of false connections.

Table 5. Edges results for ER4 and SF4 of 10 nodes.

	fn		fp		rev		correct	
	ER4	SF4	ER4	SF4	ER4	SF4	ER4	SF4
DAG-GNN	32	22	1	0	3	1	4	0
notears	29	20	1	0	5	3	5	0
Grad-DAG	0	0	4	28	2	0	37	23
cam	10	9	1	1	0	0	29	13
notears-mlp	20	12	2	1	2	7	17	4
random	19	13	1	11	11	4	9	6
PC	28	14	2	1	4	0	7	9
CD-BNN	25	13	2	0	2	4	12	6

In Table 5, we give the results on ER4 and SF4 graph with 10 nodes. Compared to the ER1 graph, the ER4 graph adds more edges, which is approximately four times than the ER1 graph. Under the SHD metric, Grad-DAG has the best score, followed by the cam and notars-mlp algorithms. DAG-GNN, notears, and PC algorithms have similar scores, while CD-BNN has the worst score, identifying more reverse edges and fewer correct edges, which is relatively normal in terms of the number of false edges. Therefore, this algorithm needs to reduce the number of reverse edges or increase the number of correct edge recognition on ER4 graphs, CD-BNN can consider increasing the number of correct edges. Under the SID metrics, Grad-DAG achieved the optimal score due to identifying more edges, while notears-mlp and cam achieved suboptimal scores, while other algorithms had similar SID.

4.2 Real Data

We consider using the pseudo data set generated by SynTReN generator and the real protein signal data set proposed by Sachs [13]. SynTReN can generate simulated gene expression data similar to the experimental data, and the statistical properties of the generated network topology are closer to the statistical properties of the real biological network. In this experiment, the SynTReN dataset has 20 nodes and 24 edges, while the Sachs dataset has 11 nodes and 17 edges, commonly used in probabilistic graphic model literature.

Table 6. SHD and SID for SynTReN-24 and Sachs-17.

	SynTReN(24)		Sachs(17)	
	SHD	SID	SHD	SID
DAG-GNN	39	148	15	58
notears	39	141	13	49
Grad-DAG	139	**59**	44	38
cam	39	148	**8**	58
notears-mlp	39	184	14	56
random	99	210	29	45
PC	42	185	17	62
CD-BNN	**27**	165	44	**35**

Overall, in Table 6, CD-BNN has the best performance in SynTreN dataset when considering SHD metric and in Sachs dataset when considering SID metric. In the results of SynTReN dataset shown in Table 7, notears-mlp identified more pseudo edges, with the SHD score being the worst. CD-BNN effectively reduced the number of pseudo edges, outperforming all other algorithms. Grad-DAG can recognize the most correct edges, but it also recognizes the most false edges, resulting in poor SHD scores. The random algorithm only recognizes one correct edge, and it also recognizes a large number of false edges, resulting in low SHD scores. PC algorithm is not suitable for complex data, and also recognizes a large number of false edges. Grad-DAG performs well on SID metrics. cam did not outperform other algorithms. notars-mlp achieved better SID scores due to identifying more correct edges, while notears slightly outperformed DAG-GNN and notears algorithms.

Table 7 also shows the results on real dataset of proteins. DAG-GNN has a lower SHD score due to identifying fewer reverse and false edges. Grad-DAG and notears-mlp both recognize more false edges. In terms of SID metric, notears-mlp has the best SID score. Grad-DAG still performs well with a better SID score. DAG-GNN and PC algorithms have better SHD, but their SID scores are not as good as other algorithms. It can be seen that these algorithms can only balance one metric and perform better on one metric while performing worse on the other.

Table 7. Edges results for SynTReN-24 and Sachs-17.

	fn		fp		rev		correct	
	SynTreN	Sachs	SynTreN	Sachs	SynTreN	Sachs	SynTreN	Sachs
DAG-GNN	18	128	10	1	2	4	4	4
notears	19	124	19	0	1	6	4	6
Grad-DAG	4	14	132	945	3	37	17	85
cam	32	32	1	1	3	3	4	4
notears-mlp	6	121	139	2	8	11	10	4
random	15	85	76	420	8	23	1	28
PC	16	107	40	24	1	2	7	27
CD-BNN	24	117	3	4	0	12	0	7

5   Limitations and Feature Work

While the algorithm showcased favorable performance on a majority of the datasets, the experimental results also unveiled certain areas that require improvement. Specifically, as the number of nodes escalated to 50, it failed to surpass the performance of the baseline algorithm. To address this limitation, future endeavors will focus on exploring more intricate network designs that can better learn and represent the causal relationships between the nodes.

6   Conclusion

In this paper, we solve the problem of uncertainty in causal discovery by adopting a Bayesian perspective. We propose a novel causal discovery method called CD-BNN, which incorporates Bayesian Neural Networks into the continuous optimization framework. By integrating BNNs, our method offers a more comprehensive treatment of uncertainty in causal relationships. We conducted extensive experiments from multiple perspectives. The results demonstrate that our algorithm performs admirably on a wide range of datasets.

References

1. Blundell, C., Cornebise, J., Kavukcuoglu, K., Wierstra, D.: Weight uncertainty in neural network. In: International Conference on Machine Learning, pp. 1613–1622. PMLR (2015)
2. Daly, R., Shen, Q.: Learning Bayesian network equivalence classes with ant colony optimization. J. Artif. Intell. Res. **35**, 391–447 (2009)
3. Geffner, T., et al.: Deep end-to-end causal inference. In: NeurIPS 2022 Workshop on Causality for Real-world Impact (2022). https://openreview.net/forum?id=6DPVXzjnbDK

4. Gelman, A.: Bayesian model-building by pure thought: some principles and examples. Statistica Sinica **6**(1), 215–232 (1996)
5. Gheisari, S., Meybodi, M.R.: BNC-PSO: structure learning of Bayesian networks by particle swarm optimization. Inf. Sci. **348**, 272–289 (2016)
6. He, Y., Cui, P., Shen, Z., Xu, R., Liu, F., Jiang, Y.: Daring: differentiable causal discovery with residual independence. In: Proceedings of the 27th ACM SIGKDD Conference on Knowledge Discovery and Data Mining, pp. 596–605 (2021)
7. Heng, X.C., Qin, Z., Wang, X.H., Shao, L.P.: Research on learning Bayesian networks by particle swarm optimization. Inf. Technol. J. **5**(3), 118–121 (2006)
8. Kuang, K., Cui, P., Athey, S., Xiong, R., Li, B.: Stable prediction across unknown environments. In: proceedings of the 24th ACM SIGKDD International Conference on Knowledge Discovery and Data Mining, pp. 1617–1626 (2018)
9. Lachapelle, S., Brouillard, P., Deleu, T., Lacoste-Julien, S.: Gradient-based neural dag learning. arXiv preprint arXiv:1906.02226 (2019)
10. Li, X.L., Wang, S.C., He, X.D.: Learning Bayesian networks structures based on memory binary particle swarm optimization. In: International Conference on Simulated Evolution and Learning (2006)
11. Opgen-Rhein, R., Strimmer, K.: From correlation to causation networks: a simple approximate learning algorithm and its application to high-dimensional plant gene expression data. BMC Syst. Biol. **1**(1), 1–10 (2007)
12. Peters, J., Mooij, J.M., Janzing, D., Schölkopf, B.: Causal discovery with continuous additive noise models (2014)
13. Sachs, K., Perez, O., Pe'er, D., Lauffenburger, D.A., Nolan, G.P.: Causal protein-signaling networks derived from multiparameter single-cell data. Science **308**(5721), 523–529 (2005)
14. Shanmugam, R.: Causality: models, reasoning, and inference-judea pearl; Cambridge University Press, Cambridge, UK, 2000, pp. 384. ISBN 0-521-77362-8. Neurocomputing **1**(41), 189–190 (2001)
15. Shimizu, S., Hoyer, P.O., Hyvärinen, A., Kerminen, A., Jordan, M.: A linear non-gaussian acyclic model for causal discovery. J. Mach. Learn. Res. **7**(10) (2006)
16. Spirtes, P., Glymour, C., Scheines, R.: Causality from probability (1989),
17. Spirtes, P., Glymour, C.N., Scheines, R., Heckerman, D.: Causation, prediction, and search. MIT press (2000)
18. Tong, W., Yang, J.: A heuristic method for learning Bayesian networks using discrete particle swarm optimization. Knowl. Inform. Syst. **24**(2), 269–281 (2010)
19. Varian, H.R.: Causal inference in economics and marketing. Proc. Natl. Acad. Sci. **113**(27), 7310–7315 (2016)
20. Wright, S.: The relative importance of heredity and environment in determining the piebald pattern of guinea-pigs. Proc. Natl. Acad. Sci. U.S.A. **6**(6), 320 (1920)
21. Yu, Y., Chen, J., Gao, T., Yu, M.: DAG-GNN: dag structure learning with graph neural networks. In: International Conference on Machine Learning, pp. 7154–7163. PMLR (2019)
22. Zhang, B., et al.: Integrated systems approach identifies genetic nodes and networks in late-onset Alzheimer's disease. Cell **153**(3), 707–720 (2013)
23. Zhang, K., Hyvarinen, A.: On the identifiability of the post-nonlinear causal model. arXiv preprint arXiv:1205.2599 (2012)
24. Zhang, M., Jiang, S., Cui, Z., Garnett, R., Chen, Y.: D-VAE: a variational autoencoder for directed acyclic graphs. In: Advances in Neural Information Processing Systems 32 (2019)

25. Zhang, X., Jia, S., Li, X., Cong, G.: Learning the bayesian networks structure based on ant colony optimization and differential evolution. In: 2018 4th International Conference on Control, Automation and Robotics (ICCAR) (2018)
26. Zheng, X., Aragam, B., Ravikumar, P.K., Xing, E.P.: Dags with no tears: Continuous optimization for structure learning. In: Advances in Neural Information Processing Systems 31 (2018)
27. Zheng, X., Dan, C., Aragam, B., Ravikumar, P., Xing, E.: Learning sparse nonparametric DAGs. In: International Conference on Artificial Intelligence and Statistics, pp. 3414–3425. PMLR (2020)
28. Zhu, S., Ng, I., Chen, Z.: Causal discovery with reinforcement learning. arXiv preprint arXiv:1906.04477 (2019)

A Preference-Based Indicator Selection Hyper-Heuristic for Optimization Problems

Adeem Ali Anwar[1]([⊠]) [ID], Irfan Younas[2], Guanfeng Liu[1] [ID], and Xuyun Zhang[1]

[1] School of Computing, Faculty of Science and Engineering, Macquarie University, Sydney, NSW, Australia
adeem.anwar@students.mq.edu.au , xuyun.zhang@mq.edu.au
[2] Odyssey Analytics, 5757 Woodway Drive, Houston, TX 77057, USA

Abstract. Heuristics have been effective in solving computationally difficult optimization issues, but because they are often created for certain problem domains, they perform poorly when the challenges are significantly altered. The currently available techniques are either designed to address single- or multi-objective optimization issues solely, or they perform poorly with the same parameters. The multi-domain approach known as hyper-heuristics (HHs) can be used to solve optimization issues with minor variations. Motivated by the notion of utilizing the benefits of low-level heuristics (LLHs) in order to obtain well-distributed and convergent optimum solutions along with taking into account the shortcomings of the work completed in many-objective HHs. For many-objective optimization problems, this paper develops a high-level selection approach that employs indicators by preference and offers a unique selection hyper-heuristic called Preference-based Indicator Selection Hyperheuristic (PBI-HH). In order to establish fairness between exploration and exploitation, the method makes use of a randomization mechanism and a greedy strategy to address a significant problem faced by HHs. Three well-known many-objective evolutionary algorithms are combined in the unique technique that is being proposed. The efficacy of the proposed strategy is assessed by contrasting it with cutting-edge HHs. PBI-HH performs better or equal to the state-of-the-art HHs on 155 out of 160 cases employing the HV indicator and has the optimal μ norm mean values across all datasets.

Keywords: Selection Hyper-Heuristic · Many-objective · Optimization problems

1 Introduction

Over a set of input variables, an algorithm to solve an optimization problem produces an output based on objective function/s (maximize or to minimize) [10]. The problems can be divided into single-, multi-, or many-objective optimisation

© The Author(s), under exclusive license to Springer Nature Switzerland AG 2023
X. Yang et al. (Eds.): ADMA 2023, LNAI 14176, pp. 447–462, 2023.
https://doi.org/10.1007/978-3-031-46661-8_30

depending on the quantity of objectives. Multi-objective optimisation problems (MOOPs) have fewer than or equal to three objectives, whereas single-objective optimisation problems (SOOPs) have one objective to be optimised. Moreover, if there are more than three objectives, then it belongs to many-objective optimization problems (MaOOPs) [2,10]. Evolutionary algorithms are often employed to solve these optimization problems. For SOOP, evolutionary algorithms usually give a single solution, while they can give more than one solutions that are represented by Pareto-front (PF) [12].

Heuristics have successfully addressed the above mentioned optimization problems which are often computationally hard, but they suffer from poor performance when the problems change slightly because they are usually designed for specific problem domains. To address the issue, HHs as a multi-domain technique can be applied to optimization problems that have slight differences [6]. It is a method for automatically creating or choosing low-level heuristics (LLHs) to address computationally challenging issues. As a result, HH may be divided into selection hyper-heuristics and generation hyper-heuristics depending on the characteristics of the search space. Selection hyper-heuristics and generation hyper-heuristics deal with the automation of selecting and generating heuristics for a given optimization problem respectively. In selection hyper-heuristics, high-level methodologies are used to select LLHs [11]. LLHs can be heuristics, operators, meta-heuristics, and even HHs, whereas high-level methodologies can be any selection methodology i.e., random selection, roulette wheel, etc. [6,11,24,25,27].

For MaOOPs, acquiring a solution set that is well-distributed and converged toward optimal solutions is one of the most crucial issues. Recently, researchers have proposed various mechanisms to address this issue. However, the existing mechanisms are either problem-specific and might not generalize well over multiple domains, or they fail to produce good results with the same parameter settings over multiple domains, or they have mostly been proposed to SOOPs or MOOPs. Recent literature advocates that researchers have focused on the selection hyper-heuristics to solve SOOPs and MOOPs, and the work which has been done in many-objective hyper-heuristics (MaOHHs) is very limited.

Inspired by the idea of exploiting the power of low-level heuristics (LLHs) to get well-distributed and converged optimal solutions and considering the limitations of work done in many-objective HHs. This paper proposes a novel selection hyper-heuristic for many-objective optimization problems named Preference-based Indicator Selection Hyper-heuristic (PBI-HH) by developing a high-level selection technique that uses indicators by preference. In PBI-HH, the randomness mechanism and greedy strategy are used to achieve exploration and exploitation respectively. To assess the efficiency of the proposed approach, eight datasets have been taken from two benchmark problems i.e., WFG and DTLZ. Our proposed approach uses the combination of three well-known many-objective evolutionary algorithms (MaOEAs) i.e., NSGA-III [7], I-DBEA [4], and MOEA/D [31] as LLHs. The best that we can tell, the combination of LLHs along with the proposed selection approach has not been studied in the literature. Furthermore, the proposed PBI-HH is compared with the state-of-the-art HH known as

Adaptive epsilon greedy selection hyper-heuristics (HH_EG) [30], Random HH, NSGA-III [7], I-DBEA [4], and MOEA/D [31]. The results have been compared using the multi-domain evaluation technique known as μ norm [1]. Research shows that the proposed algorithm outperformed the state-of-the-art HHs and have the best μ norm mean values across all datasets.

To sum up, a HH for MaOOPs is proposed. PBI-HH has produced the best multi-domain performance across benchmark problems against state-of-the-art algorithms. Moreover, it is shown that the proposed PBI-HH has also produced the best results across different objectives using Hypervolume (HV) [5,23] and Additive Epsilon Indicator (AEI) [18] values except for 7 objectives, where it is the second-best algorithm and performing better than state-of-the-art HHs.

The rest of the paper is as follows. In Sect. 2, the literature review is discussed. In Sect. 3, the proposed PBI-HH is explained. Empirical studies are explained in Sect. 4. Lastly, Sect. 5 describes the conclusion and future work.

2 Related Work

Researchers have worked in the field of HHs for SOOPs and MOOPs extensively and focused on selection hyper-heuristics as well as generation hyper-heuristics. Most of the researchers have focused on single or multi-objective HHs and the work done in the field of MaOHH is very limited. A few recent papers in the field of HHs for MaOHH are given as follows: [3,13–15,19,26,28,30].

MaOHH for MaOOP was introduced [14]. The algorithm is based on the cooperation of different MOEAs and is named as cooperative hyper-heuristic (HH-CO). It is a population-based HH and data is being exchanged between them. In the experiments, it was shown that HH-CO produced good results for complex and MaOPs while using different quality evaluators. [15] extended their work [14] and applied the HH-CO to a many-objective wind turbine design problem. Eight different MOEAs were used along with their population and swapped with each other during the iterations. In the experiments, HH-CO outperformed MOEAs when compared using HV. Sandra et. al proposed a selection HH for many-objective numerical optimization using MOEA and found good results on benchmark and real-life datasets [28]. Atiya et. al proposed genetic programming-based HHs for job-shop scheduling while considering many objectives [19]. Bianca et. al proposed a HH for many-objective quadratic assignment problem using NSGA-III [26]. Based on novel adaptive epsilon-greedy selection [30] suggested a multi-objective HH, named HH_EG. The novel algorithm changes the LLH. The results showed that HH_EG produced better results when compared with other MOEAs and MaOEAs on DTLZ, IMOP, and MaF. Soroush et al. [13] used NSGA-II and k-means to solve a pickup and delivery problem effectively.

In conclusion, researchers have focused on the selection hyper-heuristics with either constructive or perturbative approaches along with different feedback mechanisms and move-accepting techniques to solve SOOPs and MOOPs, and the work which has been done in the field of MaOHHs is very limited. Inspired by the idea of exploiting the strengths of LLHs to get converged and well-distributed optimal solutions and considering the limitations of work done in MaOHHs, this paper proposes a selection HH based on perturbation and reinforcement learning for MaOOPs named PBI-HH. The proposed approach uses the combination of three well-known MaOEAs i.e., NSGA-III [7], I-DBEA [4] and MOEA/D [31] as LLH. To the best of our knowledge, the combination of LLH along with the proposed selection approach has not been studied in the literature.

3   Preference-Based Indicator Selection Hyper-heuristic (PBI-HH)

The proposed selection approach (high-level methodology) is based on the perturbative method along with reinforcement learning as an online feedback mechanism as explained in Algorithm 1 and 2. It uses indicators by preference to develop the high-level methodology to drive the search between LLHs. Moreover, the MaOEAs have been considered as the LLHs. The proposed selection approach chooses the best LLH for a certain generation. In order to do so, it applies all the evaluation measures i.e. indicators (HV [5,23], Inverted generational distance (IGD) [23,29], AEI [18] and Spacing (Sp) [23]) on all the LLHs and then gives the preference to the best one among them according to the fitness value. Then the preferred LLH is being selected and applied on the current generation, which helps in achieving the exploitation. Additionally, to cater to the exploration, randomness has been used by using the threshold (r). A random LLH is applied instead of the best fitness LLH if the threshold is achieved. Furthermore, reinforcement learning is applied based on the HV value of the current and previous generations. Hence, depending on HV values the LLH is changed or sustained.

Algorithm 1. Framework of PBI-HH (Part A)

$P_{current}$: set of input population, P_{new}: set of new population, g: generations, G_{max}: maximum number of generations, a: set of LLHs $\{a_1, ..., a_m\}$, R: reference set, em: set of evaluation measures $\{em_1, ..., em_m\}$, n: number of iterations, o: number of objectives, p: benchmark problem v: set of em values on LLHs $\{emv_{(1,1)}, emv_{(1,2)}, ..., emv_{(i,j)}\}$, f: fitness of LLHs $\{f_1, ..., f_i\}$

Input: G_{max}, o, p, a, em, n

Output: $y_{(G_{max}, hv)}$ and $y_{(G_{max}, aei)}$ (HV and AEI values) // Creation of reference set for calculating evaluation measures

for $i \leftarrow 1$ *to* m **do**

    while $g \leq G_{max}$ **do**

        R ++ ApplyLLH $(a_i, n, P_{current}, o, p)$;

        $g = g + 1$;

    end

end

Algorithm 2. Framework of PBI-HH (Part B)

// Until termination condition
while $g \leq G_{max}$ **do**
  // Calculate em values for all LLHs
  if $g == 1 || g == 2 || flag == 1$ **then**
    for $i \leftarrow 1$ *to* m **do**
      $P_{new} \leftarrow$ ApplyLLH $(a_i, n, P_{current}, o, p)$;
      for $j \leftarrow 1$ *to* m **do**
        $emv_{(i,j)} \leftarrow$ ApplyEvaluationMeasure (P_{new}, em_j, R);
      end
    end
  end
  // Calculate fitness values for all LLHs
  for $i \leftarrow 1$ *to* m **do**
    $f_i \leftarrow (emv_{(i,1)} * -1 + emv_{(i,2)} * 1 + emv_{(i,3)} * 1 + emv_{(i,4)} * 0.5)$;
  end
  r $\leftarrow$ randomValue $(min = 0, max = 10)$;
  x $\leftarrow$ randomValue $(min = 1, max = 3)$;
  if $r > 7$ **then**
    // Apply randomness
    $P_{new} \leftarrow$ ApplyLLH $(a_x, n, P_{current}, o, p)$;
    $y_{(g,hv)} \leftarrow$ ApplyEvaluationMeasure (P_{new}, em_{hv}, R);
    $y_{(g,aei)} \leftarrow$ ApplyEvaluationMeasure (P_{new}, em_{aei}, R);
    $P_{current} \leftarrow P_{new}$;
  else
    // Select min fitness LLH
    x $\leftarrow$ min (f_i);
    $P_{new} \leftarrow$ ApplyLLH $(a_x, n, P_{current}, o, p)$;
    $y_{(g,hv)} \leftarrow$ ApplyEvaluationMeasure (P_{new}, em_{hv}, R);
    $y_{(g,aei)} \leftarrow$ ApplyEvaluationMeasure (a_x, em_{aei}, R);
    $P_{current} \leftarrow P_{new}$;
    // Decision point for changing LLH
    if $y_{(g,hv)} <= y_{(g-1,hv)}$ **then**
      // Calculate em values for all LLHs
      $flag \leftarrow 1$;
    else
      // Use the previous generation LLH
      $flag \leftarrow 0$;
    end
  end
  $g = g + 1$;
end
// Display the fitness values
$Display \leftarrow y_{(G_{max}, hv)}$;
$Display \leftarrow y_{(G_{max}, aei)}$;

3.1   Method

The following section discusses the modules of the proposed framework.

Initialization. On the first generation, a random population is created. The population is generated for subsequent generations based on the non-dominated solutions from the previous generation, as shown in (1).

$$f(\mathcal{P}_{new}(g+1)) = f(\mathcal{P}_{current}(g)),$$
$$\forall P_{current} \in P_{nondominated}$$

(1)

The calculations used to determine the objectives values, and evaluation metrics for the following generation are based on the most recent changes to the solutions. For a fair comparison, the reference set for the evaluation measures is created before the execution of HHs and other algorithms.

Low-Level Heuristic Selection Method. The proposed selection approach chooses the best LLH among MOEA/D, NSGA-III, and I-DBEA for a certain generation as expressed by $a = \{a_1, a_2, a_3, \ldots, a_m\}$. As expressed by $em = \{em_1, em_2, em_3, \ldots, em_m\}$, it applies the evaluation measures (HV, IGD, AEI, and Sp) on all LLHs and then chooses the best one among them according to the fitness value as expressed by $f = \{f_1, f_2, f_3, \ldots, f_m\}$ or chooses a random LLH based on the threshold (r) value. The fitness value of a LLH is calculated by applying all evaluation measures on it and multiplying the values with the constants, and then adding the values to create a fitness value for that certain LLH as expressed by (2) and explained in Algorithm 1 and 2.

$$f_i = (emv_{(i,1)} * -1 + emv_{(i,2)} * 1 + emv_{(i,3)} * 1 + emv_{(i,4)} * 0.5),$$
$$\text{where } i \in a$$

(2)

The LLH improves as the fitness value decreases. The best fitness value is expressed as $x = \operatorname{argmin} f(f_i), \forall i \in \{1, \ldots, m\}$. For HV the constant value is -1, for IGD and AEI it is 1, and for Sp, it is 0.5. For HV the higher value is considered better and for IGD, AEI, and Sp the lesser value is considered better. So, in order to symmetrize the fitness value, HV is inversed by multiplying it with -1. The reason for choosing 1 and 0.5 as constant values are based on the aspects. As HV, AEI, and IGD consider both diversity and convergence of PF, whereas Sp only considers diversity [5,18,23,29]. Furthermore, to cater to the exploration, randomness has been used. The threshold (r) is set to 7, if a randomly generated number (r) is greater than 7, a random LLH is applied instead of the best fitness LLH. r value 7 means that there is a 70% chance of the best LLH being selected while a 30% chance of a random LLH being selected.

Reinforcement Learning Scheme. The reinforcement learning scheme is used as an online feedback mechanism. More specifically, the deterministic policy-based reinforcement learning approach has been used as value-based approaches have disadvantages of random policy and limited problem description [22]. Furthermore, the Markov Decision process has been used as a learning model. The LLH behaves as an agent which interacts with the environment, the state of the agent changes and a reward or punishment has been given to the LLH based on the state, and action is taken accordingly i.e., with better HV it gets rewards, and punishment has been given otherwise. If the HV of the present generation is better than the previous generation, then in the next generation the same LLH is used as the previous generation LLH while considering the old fitness value as expressed in (3). If the HV of the present generation is worse than the previous generation, then in the next generation a new LLH is used. The new LLH is found based on the Low-level Heuristic selection method as explained previously and as expressed in (3).

$$a_x = \begin{cases} a_{old}, & \text{if } y_{(g,hv)} > y_{(g-1,hv)} \\ a_{new}, & \text{if } y_{(g,hv)} <= y_{(g-1,hv)} \end{cases} \qquad (3)$$

3.2 Analysis of PBI-HH

Convergence and diversity of the computed solutions are two important factors in achieving the optimality of meta-heuristics [21]. Given the large number of objectives and optimal solutions, the chance of losing optimal solutions is increased which affects the overall optimality of algorithms [21]. Hence PBI-HH uses MaOEAs as LLHs to tackle this issue. Moreover, PBI-HH is handling the environmental selection and offspring generation efficiently by considering a novel selection mechanism and non-dominated solutions from the last generation respectively. Furthermore, in order to get optimal solutions, exploitation and exploration of an algorithm play an important role as well [21]. The balance of these two components is important to find the global optimum. The proposed PBI-HH demonstrates both attributes. In order to get exploration, a randomness mechanism is applied and in order to achieve exploitation, the greedy approach is used. In the greedy approach, the best fitness LLH is selected during the iterations. The exploitation and exploration in the proposed algorithm are regulated by two factors: reinforcement learning and randomness mechanism. The reason for the proposed algorithm producing better results is also based on the balance of exploration and exploitation. In every generation, a threshold is set, and if the value exceeds the threshold value, then the random LLH is chosen instead of the minimum fitness LLH which helps in the exploration. On the other hand, the minimum fitness LLH is chosen if the threshold value is not met, and reinforcement learning is applied which helps in the exploitation.

3.3   Complexity of One Generation of PBI-HH

Initially, the reference set is created, so the cost for that is $O(LG(N^2M))$, assuming the cost of applying the LLH is $O(N^2M)$ [7], where L is the number of LLHs, N is the population size, G is the number of generations, and M is the dimension of the objectives. The second next step is to calculate the evaluation measures values after applying all LLHs. The Computational cost for that is $O(AE(N^2M))$, where E is the number of evaluation measures. The third step is to calculate the fitness of all LLHs, which takes $O(A)$. The final step is to apply the LLH to the current generation, in the worst case it takes $O(N^2M)$. So overall, the cost of one generation of PBI-HH is $O(AE(N^2M))$.

4   Empirical Studies

Three MaOEAs named NSGA-III [7], MOEA/D [31], and I-DBEA [4] have been considered as LLHs. The proposed PBI-HH is compared with these three algorithms as well as Random HH (R-HH) and a recently proposed HH known as adaptive epsilon greedy selection hyper-heuristics (HH_EG) [30]. The problem is implemented in Java using MOEA framework [20].

4.1   Experimental Settings

Test Problems. The benchmark datasets, DTLZ7, DTLZ4, DTLZ3, DTLZ2, WFG7, WFG5, WFG4, and WFG1, are used in this study [9,16]. DTLZ2 is used to check the algorithms' performance in many objectives, DTLZ3 is used to check the algorithms' ability to converge, DTLZ4 is used to check the algorithms' ability to find diverse solutions and DTLZ7 is used to check the algorithms' ability to uphold sub-populations in various areas of PF [9]. WFG1, WFG4, WFG5, and WFG7 are separable, and their modality is uni-modal except WFG5 and WFG4 which are deceptive and multi-modality respectively [16]. WFG1 geometry is convex and mixed, whereas WFG4, WFG5, and WFG7 have concave geometry [16].

Parameters Settings. The NSGA-III, MOEA/D and I-DBEA, PBI-HH, and R-HH algorithms' HV and AEI values are determined using 25 generations and 4000 iterations. The benchmark datasets DTLZ 2, 3, 4, 7, and WFG 1, 4, 5, 7 have been considered with 3, 4, 5, and 7 objectives and 100 population size. The reference set is created with the generation value 25 and iteration value 10000. Each experiment is done 5 times (5 seeds) to give a fair result. The threshold (r) is set to 7. In all WFG experiments, the number of position- and distance-related variables is fixed to five. The parameters have been taken from [17]. Moreover, the significance levels of algorithms over each other are illustrated using a one-tailed t-test at 0.05 alpha value.

Comparative Study Algorithms. In this study, three MaOEAs (NSGA-III, MOEA/D, I-DBEA) and two HHs (R-HH, HH EG) are employed. NSGA-III, a Non-dominated Sorting Genetic Algorithm, is presented for MaOOPs by [7]. Moreover, it is a refinement of the NSGA-II [8]. An algorithm based on decomposition (MOEA/D) is presented by [31]. It can be applied to both multi- and MaOOPs problems. Another evolutionary method that performs well for MaOOPs is I-DBEA [4]. The problems are addressed using a reference-point-based method in NSGA-III and I-DBEA. The R-HH is a HH in which NSGA-III, MOEA/D, and I-DBEA have been used randomly during the iterations. Whereas HH_EG [30] is a multi-objective HH and is based on a novel adaptive epsilon-greedy selection method.

Performance Metrics. The algorithms' multi-domain performances are compared using several performance metrics. With 0 being the worst result and 1 being the greatest value, HV calculates the distance between the acquired PF and reference point to produce a number between 0 and 1 [5,23]. IGD calculates the distance between every solution of ideal PF to obtained PF and gives a value [23,29]. Less value is considered good. AEI is based on an additive factor. [18]. Less value is considered good. Sp determines the evenness of the solutions of an approximation set. The approximation set's solutions are all uniformly distributed, if the Sp value is 0, [23]. Furthermore, in order to assess the multi-domain efficacy of the proposed PBI-HH, the μ norm is determined [1].

4.2 Experimental Results

Results of the experiment have been discussed in this section. Mean HV values and mean AEI values have been calculated for different datasets over different metaheuristics (LLH), random hyper-heuristic (R-HH), HH_EG, and the proposed PBI-HH considering 3, 4, 5, and 7 objectives.

Figure 1 shows the mean HV and AEI values of DTLZ7, DTLZ4, DTLZ3, DTLZ2, WFG7, WFG5, WFG4, and WFG1 for 3, 4, 5, 7 objectives while comparing NSGA-III, I-DBEA, MOEA/D, HH_EG, R-HH, and PBI-HH.

The proposed PBI-HH for DTLZ2 provides the best HV values across all objectives. For DTLZ3, the proposed PBI-HH has the best HV values on objectives 3 and 4, whereas objectives 5 and 7 are best served by I-DBEA and R-HH, respectively. In DTLZ4, the proposed PBI-HH has the greatest HV values for objectives 3 and 5, whereas MOEA/D has the best outcomes for objectives 4 and 7. PBI-HH has the greatest HV values for DTLZ7 on objectives 3 and 7, whereas I-DBEA performs best on objectives 4 and 5.

The best HV values for WFG1's 3, 4, and 5 objectives are produced by the proposed PBI-HH, whereas the best HV values for WFG1's 7 objectives are created by MOEA/D. For WFG4, NSGA-III has the best HV values on objectives 3 and 4, whereas I-DBEA performs best on objectives 5 and 7. For WFG5, NSGA-III has the best HV values on objectives 3 and 4, whereas I-DBEA performs best on objectives 5 and 7. For WFG7, NSGA-III has the best results on 3 objectives, whereas on 4 and 5 objectives, the proposed PBI-HH has the best HV values and I-DBEA provides the best results on 7 objectives.

For DTLZ2, the proposed PBI-HH has the best AEI values on the 3 and 5 objectives. On 4 objectives, I-DBEA provides the best results. And On 7 objectives, MOEA/D outperformed the other algorithms. In addition, for DTLZ3, NSGA-III has the greatest AEI values for objectives 3 and 4, while I-DBEA and MOEA/D have the best AEI values for objectives 5 and 7, respectively. The best AEI values on 3 objectives for DTLZ4 are found in NSGA-III. MOEA/D outperformed the other algorithms on 4 objectives. On 5 objectives, the proposed PBI-HH has better AEI values. On the 7 objectives, the HH_EG has the better AEI values. Moreover, MOEA/D has the best AEI values for DTLZ7 on objectives 3 and 4. On the 5 and 7 objectives, NSGA-III outperformed the other algorithms.

For WFG1, the proposed PBI-HH has the best AEI values on 3, 4, and 5 objectives and R-HH performed best on 7 objectives. Moreover, while considering WFG4, the proposed PBI-HH has the best results on 3 objectives, whereas I-DBEA outperformed other algorithms on the rest of the objectives. For WFG5, the NSGA-III offers the greatest AEI values on objectives 3 and 4, while I-DBEA has the best outcomes on objectives 5 and 7. Moreover, NSGA-III for WFG7 offers the best AEI values across 3 objectives. On the 4 and 5 objectives, the proposed PBI-HH has outperformed the other algorithms. While HH_EG performed better on 7 objectives.

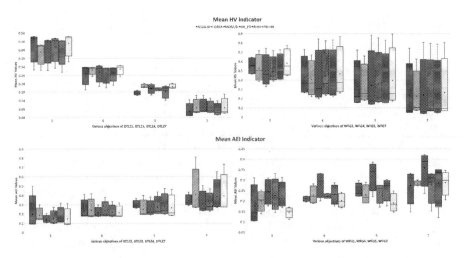

Fig. 1. Mean AEI, HV values on WFG and DTLZ while considering 3, 4, 5, and 7 objectives

HV and AEI data have been used to produce μ norm values across several objectives and algorithms in order to assess the multi-domain performance of PBI-HH. The proposed PBI-HH has the best μ norm mean value on the DTLZ

Table 1. Using HV mean values, μ norm values of datasets over variety of algorithms and objectives

μ norm mean values	μ norm values over variety of objectives and algorithms					
	NSGA-III	I-DBEA	MOEA/D	HH_EG	R-HH	PBI-HH
μ **norm mean on DTLZ**	0.307577	0.629547	0.570742	0.525570	0.408105	**0.927683**
μ **norm mean on WFG**	0.544814	0.615246	0.254058	0.588245	0.44082	**0.884036**
μ **norm mean on DTLZ and WFG combined**	0.426196	0.622396	0.412400	0.527330	0.434642	**0.910523**
Algorithm Ranking	5th	2nd	6th	3rd	4th	1st

and WFG when taking HV values into account, as shown in Table 1. By taking AEI values into account, Table 2 reveals that MOEA/D and PBI-HH both have the best μ norm mean values on the DTLZ benchmark datasets and WFG benchmark datasets, respectively. PBI-HH is the best multi-domain algorithm since it has the best μ norm mean value on both datasets.

Table 2. Using AEI mean values, μ norm values of datasets over variety of algorithms and objectives

μ norm mean values	μ norm values over variety of objectives and algorithms					
	NSGA-III	I-DBEA	MOEA/D	HH_EG	R-HH	PBI-HH
μ **norm mean on DTLZ**	0.550447	0.507966	**0.285180**	0.378909	0.539324	0.309011
μ **norm mean on WFG**	0.295345	0.336753	0.855387	0.392627	0.469936	**0.168668**
μ **norm mean on DTLZ and WFG combined**	0.422896	0.422360	0.570284	0.385768	0.504630	**0.238840**
Algorithm Ranking	4th	3rd	6th	2nd	5th	1st

Table 3 shows the significance levels of algorithms over each other using a one-tailed t-test at 0.05 alpha value. The significance levels have been computed

Table 3. Comparison of HV values on all datasets and objectives over different algorithms. Findings in form of significantly better (+), equal and significantly worse (-)

Algorithms	Significance of algorithms using t-test					
	PBI-HH	NSGA-III	I-DBEA	MOEA/D	HH_EG	R-HH
PBI-HH	-	+18/13/-1	+13/17/-2	+17/14/-1	+13/18/-1	+18/14/-0
NSGA-III	+1/13/-18	-	+9/8/-15	+13/8/-11	+7/14/-11	+6/19/-7
I-DBEA	+2/17/-13	+15/8/-9	-	+13/10/-9	+7/17/-8	+9/17/-6
MOEA/D	+1/14/-17	+11/8/-13	+9/10/-13	-	+2/22/-8	+5/22/-5
HH_EG	+1/18/-13	+11/14/-7	+8/17/-7	+8/22/-2	-	+5/25/-2
R-HH	+0/14/-18	+7/19/-6	+6/17/-9	+5/22/-5	+2/25/-5	-

using HV values on all datasets while considering the 3, 4, 5, and 7 objectives over different algorithms. The proposed PBI-HH is performing significantly better or equal than all other algorithms on 155 out of 160 instances using HV indicator.

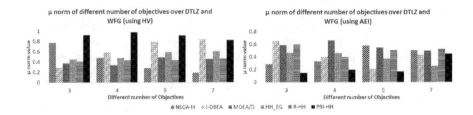

Fig. 2. Considering WFG and DTLZ, μ norm values over different number of objectives

Figure 2 provides the analysis of previously mentioned μ norm values across the objective values to show the dominance of the proposed PBI-HH. It demonstrates that when employing HV values and multi-domain problems, PBI-HH delivers the best μ norm mean values across all objectives with the exception of 7 objectives. It performs better than cutting-edge HHs and is the second-best algorithm in 7 objectives, only behind I-DBEA. Moreover, it exhibits that PBI-HH provides the best μ norm mean values across all the objectives except 7 objectives in multi-domain problems while using AEI values.

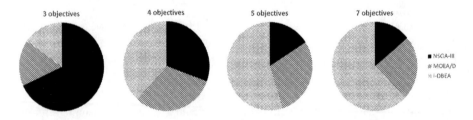

Fig. 3. Utilization of different algorithms over iterations on PBI-HH

Figure 3 shows the utilization of algorithms over 25 iterations while using PBI-HH. NSGA-III is the most used algorithm during the iterations by PBI-HH over 3 objectives, while I-DBEA is being used more during the iterations by PBI-HH over 5 and 7 objectives. Moreover, all algorithms are being equally used during the iterations by PBI-HH over 4 objectives.

Parametrical Analysis. DLTZ2 and WFG7 have been taken to do the parametrical analysis. α, and β show original and changed parameters respectively. In α, the generations are 25, iterations are 4000, population size is 100, reference generations are 25, reference iterations are 10000, seeds are 5, and r is 7.

Whereas in β, the generations are 35, iterations are 8000, population size is 150, reference generations are 35, reference iterations are 20000, seeds are 10, and r is 8. Table 4 illustrates the optimal algorithms for different objectives and the even after changed parameters, the optimal algorithms remains the same except, WFG7 on 3 objective, where PBI-HH now performs optimal.

Table 4. Parametrical analysis of WFG7 and DTLZ2 over different objectives while considering HV. α, β shows original and changed parameters respectively

	DTLZ2		WFG7	
Objectives	α	β	α	β
3	PBI-HH	PBI-HH	NSGA-III	PBI-HH
4	PBI-HH	PBI-HH	PBI-HH	PBI-HH
5	PBI-HH	PBI-HH	PBI-HH	PBI-HH
7	PBI-HH	PBI-HH	I-DBEA	I-DBEA

Convergence Analysis. Two datasets have been taken i.e. DTLZ2 and WFG7 to do the convergence analysis over the proposed PBI-HH and state-of-the-art HH_EG over mean HV values as illustrated in Fig. 4. Generations are set to 25 and two different number of objectives i.e. 4 and 5 are considered.

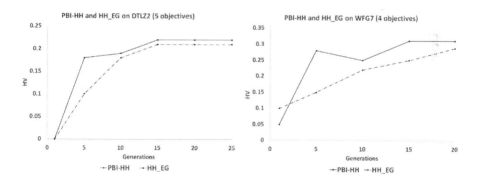

Fig. 4. Convergence analysis of WFG7 and DLTZ2 over EG_HH and PBI-HH

5 Conclusion and Future Work

Preference-based indicator selection hyper-heuristic (PBI-HH), a novel method of choosing LLHs, is proposed. For eight distinct benchmark problems spanning objectives, the proposed approach outperformed existing meta-heuristics

(NSGA-III, I-DBEA, MOEA/D), and hyper-heuristics (R-HH and HH_EG). Moreover, it is shown that the proposed PBI-HH has produced the best results across different objectives using HV and AEI values except for 7 objectives. When it comes to 7 objectives, it comes in second place behind I-DBEA but outperforms cutting-edge HHs, using the HV. Additionally, it produces the best results on 3 out of 4 considered objectives, i.e., producing 75% better results. Furthermore, the proposed PBI-HH performs significantly better than or on an equal level with the state-of-the-art HHs on 155 out of 160 instances utilizing the HV indicator and has provided the best multi-domain performance using μ norm mean values across DLTZ and WFG vs state-of-the-art algorithms.

Future studies can take use of the proposed framework by using the proposed algorithm to solve practical problems like vehicle routing problem and job shop scheduling problems, among others.

Acknowledgments. Adeem Ali Anwar is the recipient of an iMQRES funded by Macquarie University, NSW (allocation No. 20213183) and Dr. Xuyun Zhang is the recipient of an ARC DECRA (project No. DE210101458) funded by the Australian Government.

References

1. Adriaensen, S., Ochoa, G., Nowé, A.: A benchmark set extension and comparative study for the hyflex framework. In: 2015 IEEE Congress on Evolutionary Computation (CEC), pp. 784–791. IEEE (2015)
2. Anwar, A.A., Younas, I.: Optimization of many objective pickup and delivery problem with delay time of vehicle using memetic decomposition based evolutionary algorithm. Int. J. Artif. Intell. Tools **29**(01), 2050003 (2020)
3. Anwar, A.A., Younas, I., Liu, G., Beheshti, A., Zhang, X.: A cricket-based selection hyper-heuristic for many-objective optimization problems. In: Chen, W., Yao, L., Cai, T., Pan, S., Shen, T., Li, X. (eds.) Advanced Data Mining and Applications: 18th International Conference, ADMA 2022, Brisbane, QLD, Australia, November 28–30, 2022, Proceedings, Part II, pp. 310–324. Springer Nature Switzerland, Cham (2022). https://doi.org/10.1007/978-3-031-22137-8_23
4. Asafuddoula, M., Ray, T., Sarker, R.: A decomposition-based evolutionary algorithm for many objective optimization. IEEE Trans. Evol. Comput. **19**(3), 445–460 (2014)
5. Auger, A., Bader, J., Brockhoff, D., Zitzler, E.: Theory of the hypervolume indicator: optimal μ-distributions and the choice of the reference point. In: Proceedings of the tenth ACM SIGEVO workshop on Foundations of genetic algorithms, pp. 87–102 (2009)
6. Burke, E.K., Hyde, M., Kendall, G., Ochoa, G., Özcan, E., Woodward, J.R.: A classification of hyper-heuristic approaches. In: Gendreau, M., Potvin, J.-Y. (eds.) Handbook of Metaheuristics, pp. 449–468. Springer US, Boston, MA (2010). https://doi.org/10.1007/978-1-4419-1665-5_15
7. Deb, K., Jain, H.: An evolutionary many-objective optimization algorithm using reference-point-based nondominated sorting approach, part i: solving problems with box constraints. IEEE Trans. Evol. Comput. **18**(4), 577–601 (2013)

8. Deb, K., Pratap, A., Agarwal, S., Meyarivan, T.: A fast and elitist multiobjective genetic algorithm: NSGA-II. IEEE Trans. Evol. Comput. **6**(2), 182–197 (2002)
9. Deb, K., Thiele, L., Laumanns, M., Zitzler, E.: Scalable test problems for evolutionary multiobjective optimization. In: Abraham, A., Jain, L., Goldberg, R. (eds.) Evolutionary Multiobjective Optimization, pp. 105–145. Springer-Verlag, London (2005). https://doi.org/10.1007/1-84628-137-7_6
10. Derigs, U.: Optimization and operations research-Volume IV. EOLSS Publications (2009)
11. Drake, J.H., Kheiri, A., Özcan, E., Burke, E.K.: Recent advances in selection hyper-heuristics. Eur. J. Oper. Res. **285**(2), 405–428 (2020)
12. Eiben, A.E., Smith, J.E.: Introduction to Evolutionary Computing. Springer Berlin Heidelberg, Berlin, Heidelberg (2003)
13. Fatemi-Anaraki, S., Mokhtarzadeh, M., Rabbani, M., Abdolhamidi, D.: A hybrid of k-means and genetic algorithm to solve a bi-objective green delivery and pick-up problem. J. Ind. Prod. Eng. **39**(2), 146–157 (2022)
14. Fritsche, G., Pozo, A.: Cooperative based hyper-heuristic for many-objective optimization. In: Proceedings of the Genetic and Evolutionary Computation Conference, pp. 550–558 (2019)
15. Fritsche, G., Pozo, A.: The analysis of a cooperative hyper-heuristic on a constrained real-world many-objective continuous problem. In: 2020 IEEE Congress on Evolutionary Computation (CEC), pp. 1–8. IEEE (2020)
16. Huband, S., Barone, L., While, L., Hingston, P.: A scalable multi-objective test problem toolkit. In: Coello Coello, C.A., Hernández Aguirre, A., Zitzler, E. (eds.) Evolutionary Multi-Criterion Optimization, pp. 280–295. Springer Berlin Heidelberg, Berlin, Heidelberg (2005). https://doi.org/10.1007/978-3-540-31880-4_20
17. Li, W., Özcan, E., John, R.: A learning automata-based multiobjective hyper-heuristic. IEEE Trans. Evol. Comput. **23**(1), 59–73 (2017)
18. Liefooghe, A., Derbel, B.: A correlation analysis of set quality indicator values in multiobjective optimization. In: Proceedings of the Genetic and Evolutionary Computation Conference 2016, pp. 581–588 (2016)
19. Masood, A., Chen, G., Zhang, M.: Feature selection for evolving many-objective job shop scheduling dispatching rules with genetic programming. In: 2021 IEEE Congress on Evolutionary Computation (CEC), pp. 644–651. IEEE (2021)
20. Moea framework. http://moeaframework.org/. Accessed 25 Mar 2023
21. Perwaiz, U., Younas, I., Anwar, A.A.: Many-objective bat algorithm. Plos one **15**(6), e0234625 (2020)
22. Qin, W., Zhuang, Z., Huang, Z., Huang, H.: A novel reinforcement learning-based hyper-heuristic for heterogeneous vehicle routing problem. Comput. Indust. Eng. **156**, 107252 (2021)
23. Riquelme, N., Von Lücken, C., Baran, B.: Performance metrics in multi-objective optimization. In: 2015 Latin American Computing Conference (CLEI), pp. 1–11. IEEE (2015)
24. Ross, P.: Hyper-heuristics. In: Burke, E.K., Kendall, G. (eds.) Search Methodologies, pp. 529–556. Springer US, Boston, MA (2005). https://doi.org/10.1007/0-387-28356-0_17
25. Sánchez, M., Cruz-Duarte, J.M., carlos Ortíz-Bayliss, J., Ceballos, H., Terashima-Marin, H., Amaya, I.: A systematic review of hyper-heuristics on combinatorial optimization problems. IEEE Access **8**, 128068–128095 (2020)
26. Senzaki, B.N.K., Venske, S.M., Almeida, C.P.: Hyper-heuristic based NSGA-III for the many-objective quadratic assignment problem. In: Britto, A., Valdivia Delgado,

K. (eds.) Intelligent Systems: 10th Brazilian Conference, BRACIS 2021, Virtual Event, November 29 – December 3, 2021, Proceedings, Part I, pp. 170–185. Springer International Publishing, Cham (2021). https://doi.org/10.1007/978-3-030-91702-9_12

27. S. S., V.C., H. S., A.: Nature inspired meta heuristic algorithms for optimization problems. Computing **104**, 251–269 (2021). https://doi.org/10.1007/s00607-021-00955-5

28. Venske, S.M., Almeida, C.P., Delgado, M.R.: Comparing selection hyper-heuristics for many-objective numerical optimization. In: 2021 IEEE Congress on Evolutionary Computation (CEC), pp. 1921–1928. IEEE (2021)

29. Wang, L., Ng, A.H.C., Deb, K. (eds.): Multi-objective Evolutionary Optimisation for Product Design and Manufacturing. Springer, London (2011). https://doi.org/10.1007/978-0-85729-652-8

30. Yang, T., Zhang, S., Li, C.: A multi-objective hyper-heuristic algorithm based on adaptive epsilon-greedy selection. Complex Intell. Syst. **7**(2), 765–780 (2021)

31. Zhang, Q., Li, H.: MOEA/D: a multiobjective evolutionary algorithm based on decomposition. IEEE Trans. Evol. Comput. **11**(6), 712–731 (2007)

An Elastic Scalable Grouping for Stateful Operators in Stream Computing Systems

Si Lei[1], Dawei Sun[1(✉)], and Atul Sajjanhar[2]

[1] School of Information Engineering, China University of Geosciences,
Beijing 100083, China
leisi@email.cugb.edu.cn, sundaweicn@cugb.edu.cn
[2] School of Information Technology, Deakin University, Victoria 3216, Australia
Atul.sajjanhar@deakin.edu.au

Abstract. In distributed stream computing systems, dynamic data skew and cluster heterogeneity can lead to major load imbalance among multiple instances of stateful operators. Existing stream grouping schemes mainly focus on data load balancing for stateful operators, but they are not considered to be sufficiently elastic scalable, which directly affects the latency and throughput. We propose an elastic scalable grouping (called Es-Stream) for stateful operators. This paper discusses the following aspects: (1) Investigating the dynamic grouping of real-time data stream, proposing a general data stream graph model and a data stream grouping model, as well as formalizing the problem of load balancing optimization and data stream grouping. (2) Utilizing key splitting to solve the bottleneck problem caused by high-frequency keys in the data streams, and lightweight weight adjustment strategy to dynamically change the data tuple allocation probability of the instance according to the network cost, data stream rate and processing rate. (3) Implementing Es-Stream in Apache Storm platform and evaluating the system using metrics such as latency, throughput and load imbalance. Experimental results showed that Es-Stream reduces latency by up to 72%, increases throughput by up to 44% and reduces load imbalance by up to 75%, compared with existing state-of-the-art grouping schemes.

Keywords: Data grouping · Data skew · Load balancing · Stream computing · Heterogeneous cluster

1 Introduction

With the increasing demands for complex processing and analysis of real-time data, distributed stream computing systems, such as Storm [1], Flink [2], Spark [3], and Heron [4], have been widely adopted, e.g., social network analyse [5], real-time event detection [6], and internet of things [7].

Stream processing application is usually modeled as a directed acyclic graph (DAG), which consists of a set of vertices and a set of edges. Each vertex is an operator which contains the processing logic of the application. Tuple is the basic unit of messaging in topology and the edges between vertices indicate how tuples should be transmitted between operators. To improve the system elasticity, distributed stream computing systems create multiple instances of an operator to process data in parallel. Upstream operator use different grouping schemes to divide the output stream into substreams and assign them to multiple downstream instances.

Shuffle Grouping and Key Grouping are the most representative grouping schemes in distributed stream computing systems. Shuffle Grouping is based on polling rules to randomly distribute each tuple to downstream parallel instances, ensuring that the number of tuples processing by each instance is basically the same. Key Grouping uses the specified field as the key and assigns tuples to downstream instances based on a hash function. However, for stateful operators, although Shuffle Grouping can effectively achieve data level load balancing, its cost is too high and it is not easy to scale. Key Grouping can simply store state, but it can easily lead to load imbalance among multiple instances [8].

Elastic stream computing systems are expected to achieve low latency and high throughput over long periods of runtime. This requires it to dynamically adjust the workload of instances distributed on heterogeneous nodes when facing changes in data content and rate. Traditional load balancing methods attempt to perform dynamic rescaling from resource scheduling [9,10], parallelism scaling [11,12] and load migration [13,14]. However, if the data stream changes frequently, these heavyweight online adjustment strategies may cause poor performance.

Therefore, our aim is to improve system performance with a lightweight elastic scalable load balancing strategy from the perspective of data stream grouping. It provides a dynamic strategy to deal with content changes and rate fluctuations in real-time data streams. Contributions of our work are the following:

- We investigate the dynamic grouping of real-time data stream, then propose a general data stream graph mode and a data stream grouping model, as well as formalizing the problem of load balancing optimization and data stream grouping.
- We utilize key splitting to solve the bottleneck problem caused by high-frequency keys in the data stream, and lightweight weight adjustment strategy to dynamically change the data tuple allocation probability of the instance according to the network cost, data stream rate and processing rate.
- We implement Es-Stream in Apache Storm platform and evaluate the system using metrics such as latency, throughput and load imbalance. Experimental results showed that Es-Stream reduces latency by up to 72%, increases throughput by up to 44% and reduces load imbalance by up to 75%, compared with existing state-of-the-art grouping schemes.

The rest of the paper is structured as follows: Sect. 2 reviews the related work. Section 3 describes the system model and formalizes problems. Section 4 focuses on the system architecture and algorithms of grouping. Section 5 discusses the experimental environment and analyzes performance evaluation results of Es-Stream. Finally, conclusions and future work are given in Sect. 6.

2 Related Work

In recent years, there has been a great deal of interest in how stream computing systems maintain a balance when dealing with skewed streaming data.

In order to deal with the uneven load caused by data skew, Partial Key Grouping (PKG) [8] use two new technologies: key splitting and local load estimation, to adapt the classic "power of two choices" to distributed stream settings. When recognizing only two candidates are not enough, D-Choices [15] assign hot keys to $d \geq 2$ choices according to the frequency.

Similarly, in order to solve the problem of hot keys changing with time in the data stream, a new load balancing mechanism (FISH) [16] recently proposed a recent hot key identification based on epoch with inter-epoch hotness decaying and accurate information of remote workers being heuristically inferred through calculation.

Subsequently, a popular-aware differentiated distributed stream processing system (PStream) [17] assigns the hot keys using shuffle grouping while it assigns rare ones using key grouping. PStream utilizes a light-weight probabilistic counting scheme to identify the current hot keys and designs an adaptive threshold configuration scheme to adapt to the dynamical popularity changes in real-time stream. A pre-filter partition is proposed based on Sketch in [18], which uses a heavy hitter algorithm to dynamically monitor items in the stream.

Many studies have considered locality to improve system performance by reducing network cost. In [19], correlated keys are assigned to instances hosted on the same computing node. In [20], a stochastic locality-aware stream partitioning (SLSP) method is proposed that considers both task locality and downstream state. In [21], a network aware grouping is proposed that set a dynamic weight and priority for each downstream instance based on the network location and load between instances.

To summarize, the solutions above provide valuable insight for data stream grouping, but they are not sufficiently elastic scalable. Compared with them, our proposed scheme is more elastic scalable; it not only considers data skew, network cost, and heterogeneity, but also adapts to variable data rates and content changes.

3 Problem Statement

In this section, we formalize the system model to accurately describe the grouping problem in distributed stream computing systems, which includes data stream graph model and data stream grouping model.

3.1 Data Stream Graph Model

The logical topology of a stream application can be viewed as a directed acyclic logical graph $G = (V(G), E(G))$, where $V(G) = \{v_i | i \in 1, ..., n\}$ is a finite set of n vertices. Each vertex $v_i \in V(G)$ is an operator with a logical function $fun(v)$.

it can create multiple instances $v_i = \{v_{i1}, v_{i2}, ..., v_{ik}\}$, $k \in \{1, 2...\}$, that can be deployed on different nodes to complete work in parallel. $E(G) = \{e_{v_{ik}, v_{jm}} | v_{ik} \in v_i, v_{jm} \in v_j\}$ is a finite set of directed edges, and an edge $e_{v_{ik}, v_{jm}} \in E(G)$) indicates communication between instances v_{ik} and v_{jm}, where v_{ik} and v_{jm} are the upstream and downstream instance of $e_{v_{ik}, v_{jm}}$.

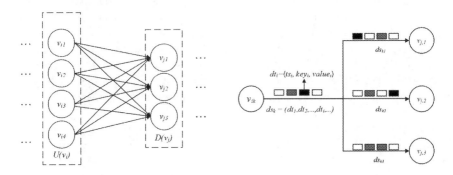

Fig. 1. Instance sets of vertices.　　　　**Fig. 2.** Partitioning data stream.

We give an upstream vertex v_i data to send to the downstream vertex v_j in G. We use $U(v_i)$ and $D(v_j)$ to denote the set of instances within upstream vertex v_i and downstream vertex v_j, where $n_{v_i} = | U(v_i) |$ and $n_{v_j} = | D(v_j) |$ are the number of instances in $U(v_i)$ and $D(v_j)$, respectively. As shown in Fig. 1, $U(v_i) = \{v_{i1}, v_{i2}, v_{i3}, v_{i4}\}$ is the upstream vertex set of v_i and $n_{v_i} = 4$. Correspondingly, $D(v_j) = \{v_{j1}, v_{j2}, v_{j3}\}$ is the downstream vertex set of v_j and $n_{v_j} = 3$.

A data stream ds is an unbounded sequence composed of data tuples with infinite time and number, denoted as $ds = (dt_1, dt_2, \ldots, dt_i, \ldots)$. A data tuple dt_i is a record of attributes $\langle ts_i, key_i, value_i \rangle$, where $ts_i, key_i, value_i$ is representing the key,value and timestamp of the ith data tuple dt_i. Keys are not unique, but are used to group tuples to maintain state.

After the vertex is processed, new tuples will be generated. All output tuples generated by the same vertex form a new data stream.

3.2　Data Stream Grouping Model

For each upstream instance v_i, if the instance number of downstream vertex v_j is greater than one, the output data stream needs to be partitioned among all instances of v_j by a grouping strategy.

We will utilize the partition function $P(ds)$ to map tuples in the data stream output from upstream vertices to downstream vertices. The partition function $P(ds)$ realizes different functions according to vertex function and data characteristics. A tuple dt_i is sent to the instance $v_{jm} \in D(v_j)$ through the function $P(ds)$. As shown in Fig. 2, the data stream $ds_k = \{ds_{k1}, ds_{k1}, ds_{k3}\}$ to 3 downstream instances $v_j = \{v_{j1}, v_{j2}, v_{j3}\}$.

Therefore, the output data stream ds_k of the upstream instance $v_{ik} \in U(v_i)$ is divided into n_{v_j} independent substreams. The input stream of downstream instance v_{jm} is the sum of the substreams sent to it by all upstream instances of v_i. It can be described by (1).

$$ds_k^{out} = \bigcup_{m=1}^{n_{v_j}} ds_{km}. \tag{1}$$

The input rate of downstream instance v_{jm} is the sum of the substreams rates assigned to v_{jm} by all instance of $U(v_i)$. It can be described by (2).

$$\lambda_{v_{jm}} = \sum_{k=1}^{n_{v_i}} \lambda(ds_{km}). \tag{2}$$

For a instance $v_{jm} \in v_j$, the load can be evaluated by (3), which means that more data will simultaneously increase the consumption of CPU, memory and bandwidth.

$$L_{(v_{jm})} = num_{v_{jm}} \times \mu_{v_{jm}}, \tag{3}$$

where $num_{v_{jm}}$ is number of accepted tuples, and $\mu_{v_{jm}}$ is processing rate of v_{jm}.

For a vertex $v_j \in V(G)$, the load balancing deviation of all the instances of v_j can be evaluated by (4).

$$LIB_{v_j} = \frac{1}{n_{v_j}} \sum_{m=1}^{n_{v_j}} \frac{|L(v_{jm}) - L_{avg}(v_j)|}{(L_{avg}(v_j))}, \tag{4}$$

where $L_{avg}(v_j)$ is average load of all n_{v_j} instance. Usually, LIB_{v_j} should satisfy the constraint $LIB_{v_j} \in [LIB_{min}, LIB_{max}]$. This is because achieving absolutely balance in the system is hard and unnecessary, as it can tolerate certain load imbalance.

A directed path from v_{ik} to v_{jm} can be described as $p_{v_{ik}, v_{jm}}$. The latency of $p_{v_{ik}, v_{jm}}$ is calculated by (5).

$$LA(p_{v_{ik}, v_{jm}}) = LA(e_{v_{ik}, v_{jm}}) + LA(v_{jm}), \tag{5}$$

where $LA(e_{v_{ik}, v_{jm}})$ is the network latency between v_{ik} and v_{jm}, $LA(v_{jm})$ is the processing latency of v_j which includes processing time and queuing time.

The average latency between instance v_{ik} and vertex v_j is calculated by (6).

$$LA(p_{v_{ik}, v_j}) = \sum_{m=1}^{n_{v_j}} w_m \times LA(p_{v_{ik}, v_{jm}}), \tag{6}$$

where w_m is weight that v_{ik} assigns tuples to v_{jm}.

An effective stream grouping can often achieve goood load balancing among instances, helping maximize the throughput and minimize the latency of G. The data stream grouping optimization problem can be formalized by (7):

$$\begin{cases} min(LA(p_{v_{ik}, v_j})) \text{ and } min(LIB_{v_j}), \\ subject\ to\ LIB_{min} \le LIB_{v_j} \le LIB_{max}. \end{cases} \tag{7}$$

4   Es-Stream: Architecture and Algorithms

Based on the above theoretical analysis, we have proposed and developed Es-Stream, an elastic Scalable grouping for stateful operators in distributed stream computing systems. To provide an overview of Es-Stream, this section discusses its overall structure, including the system architecture, the key frequency statistic, key frequency classification, and instance assignment.

4.1   System Architecture

The system architecture mainly includes grouper installed on upstream instances and monitor on downstream instances as shown in Fig. 3.

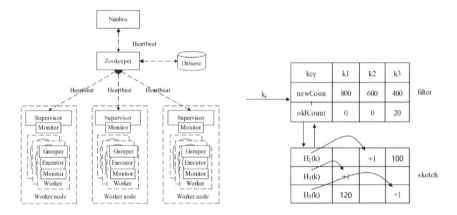

Fig. 3. System architecture.   **Fig. 4.** Key frequency statistics.

Monitor collects and analyzes runtime information stored in the database, such as network latency, data stream rate and instance processing rate. Grouper identifies high-frequency keys in real-time data streams, determines candidate instances of key, and dynamically adjusts the tuple allocation weight of the downstream instances based on runtime information to assign instances.

4.2   Key Frequency Statistics

We are going to use frequency statistics to predict the frequency of keys in real-time data streams [22]. At the same time, effectively identifying key frequencies in stream data that change over time has also become very necessary and important [16].

A pre-filtering method is proposed to filter out high-frequency keys in real-time data streams using a calculator, and process the data overflow using a sketch as shown in Fig. 4, improving the frequency estimation accuracy of high-frequency keys while reducing sketch errors. Filter can store n items, which consists of key, newCount and oldCount. Sketch uses m independent hash functions. In order to adapt to changes in the content of the data stream, we adopt a time-aware decay method that multiplies all counts by α at the end of each time period T to reflect the effect of time decay. It is calculated by (8).

$$C_k(T) = C_k(T-1) \times \alpha, \tag{8}$$

where $C_k(T-1)$ is the count of (T-1)th period in statistics, and α is decaying factor. The frequency of key is the ratio of the count of key to the total count.

Algorithm 1. Key Frequency Statistics

Input: data stream ds_k of upstream v_{ik}; period time T.
Output: frequency $f(k_i)$ of key_i
 1: Get the current time t_{cur}.
 2: **if** $t_{cur} - t_{last} \geq T$ **then**
 3:     Update the count C_k and C_{total} as (8).
 4:     Update t_{last} to the current time t_{cur}.
 5: **end if**
 6: **for** each key_i in ds_k **do**
 7:     **if** find the key_i in the filter **then**
 8:         Update $C_k = C_k + 1$.
 9:     **else**
10:         **if** filter not full **then**
11:             Add key_i to filter, and set count C_k to 1 and count O_k to 0.
12:         **else**
13:             **for** Each line i in the sketch **do**
14:                 Use the hash function $h_i(k)$ to obtain the column j of the key_i.
15:                 Update the count $C_k = C_k + 1$ of column j
16:             **end for**
17:             the count C_k of key_i is the smallest of all line in the Sketch.
18:         **end if**
19:         Get the minimum count C'_k in filter.
20:         **if** $C'_k \leq C_k$ **then**
21:             update sketch with $(k', C_k - O_k)$.
22:             Add key_i to filter and set count C_k to C_k and count O_k to O_k.
23:         **end if**
24:     **end if**
25:     Update $C_{total} = C_{total} + 1$;
26: **end for**
27: return C_k/C_{total}.

The input of this algorithm includes the output data stream ds_k of upstream v_{ik}, period time T. The output is frequency $f(k_i)$ of key_i. Step 1 to step 5 decay the count C_k with α when the time period T ends. Step 6 to step 18 updates the key count C_k when the key arrives. Step 19 to step 23 is the exchange strategy between filter and sketch when the estimated count C_k of k_i in sketch is higher than the smallest count C'_k in the filter. The time complexity of Algorithm 1 is $O(n)$, where n is the size of filter.

4.3   Key Frequency Classification

Consider the skewed distribution of keys, we differentiate between high-frequency and low-frequency key values in data stream. For high-frequency keys, all downstream instances work together to complete its workload. For low-frequency keys, only two instances process it, which can have good load efficiency and only occupy a small amount of memory space to maintain state.

A classification threshold θ is set to classify key based on frequency. When the low-frequency key is assigned to two candidate instances, an appropriate threshold θ is between $2/n$ and $1/5n$ [8], where n is the number of downstream instances. So the ideal average load of the system is $1/n$. When the frequency of key values exceeds the ideal load, it will inevitably lead to load imbalance. Therefore, the classification threshold θ is initialized to $1/n$. At the same time, dynamically adjust the classification threshold θ based on the downstream instance load.

Algorithm 2. Key Frequency Classification

Input: frequency $f(k_i)$ of key_i; Threshold θ; number of downstream instances n; the maximum and minimum load imbalance LIB_{max}, LIB_{min}.
Output: the number d of candidate instances for key_i.
  1: **if** $f(k_i) \geq \theta$ **then**
  2:      Key_i can be distributed to all downstream instances.
  3:      $d = n$.
  4: **else**
  5:      Key_i can be distributed to two downstream instances.
  6:      $d = 2$.
  7: **end if**
  8: Get the load imbalance LIB_{v_j} of the system.
  9: **if** $LIB_{v_j} > LIB_{max}$ **then**
 10:      Update $\theta = \theta/2$.
 11: **end if**
 12: **if** $LIB_{v_j} < LIB_{min}$ **then**
 13:      Update $\theta = \theta + \delta$.
 14: **end if**
 15: **return** d;

The input of this algorithm includes frequency $f(k_i)$ of key_i, Threshold θ, number of downstream instances n, the maximum and minimum load imbalance LIB_{max} and LIB_{min}. The output is the number d of candidate instances for key_i. Step 1 to step 7 categorize keys into high and low frequency keys and distribute different numbers of candidate instances. Step 9 to step 11 get the load imbalance of the instance LIB_{v_j} and dynamically adjust the classification threshold θ. If LIB_{v_j} is greater than LIB_{max}, the algorithm reduces θ by a multiple, hoping to restore system balance as quickly as possible. If LIB_{v_j} is less than LIB_{min}, the algorithm linearly increases θ, making small adjustments to reduce system overhead. The time complexity of Algorithm 2 is $O(1)$.

4.4   Instance Assignment

Downstream instances receive tuples and place them in the message queue, usually using first come first served (FCFS). We adopt the widely used M/M/1 queuing theory model, which allows for estimation of processing time and queuing time.

According to the Erlang formula, the total time of a tuple in an instance follows an exponential distribution. The average processing latency of a tuple in v_{jm} includes the queuing time and processing time of the tuple. Therefore, the average processing latency of a tuple in instance V_{jm} is $LA(V_{jm})$, which is calculated by (9).

$$LA(v_{jm}) = \begin{cases} \frac{1}{\mu_{v_{jm}} - \lambda_{v_{jm}}}, \mu_{v_{jm}} > \lambda_{v_{jm}} \\ +\infty, \quad \mu_{v_{jm}} \leq \lambda_{v_{jm}} \end{cases} \tag{9}$$

where $\mu_{v_{jm}}$ and $\lambda_{v_{jm}}$ are the input rate and processing rate of instance v_{jm}. When $\mu_{v_{jm}} \leq \lambda_{v_{jm}}$, the processing rate of the instance cannot keep up with the input rate of the data, and the number of tuples in the queue increases over time.

In distributed stream computing systems, there are significant differences in network latency between instances at different network distances. Es-Stream comprehensively considers network latency and processing latency to determine instance assignment. For the instance v_{jm}, its weight $weight_{v_{jm}}$ of being assigned a data tuple by instance v_{ik} can be calculated by (10).

$$weight_{v_{jm}} = \frac{(LA(e_{v_{ik},v_{jm}}) + LA(v_{jm}))^{-1}}{\sum_{l=1}^{n_{v_j}}(LA(e_{v_{il},v_{jl}}) + LA(v_{jl}))^{-1}}, \tag{10}$$

where the sum of $LA(e_{v_{ik},v_{jm}})$ and $LA(v_{jm})$ is the total latency from v_{ik} to v_{jm}. The smaller latency of v_{jm}, the higher the probability of v_{jm} being assigned tuples.

Algorithm 3. Instance Assignment

Input: upstream instances set U_{v_i}; downstream instances set D_{v_j}.
Output: target instance v_{jm} of each tuple.
1: **for** each instance v_{ik} in U_{v_i} **do**
2:      Get the network time $LA(e_{v_{ik},v_{jm}})$.
3:      Initialize weights $w_{v_{jm}} = \frac{1}{LA(e_{v_{ik},v_{jm}})}$.
4:      **if** received monitoring information **then**
5:          **for** each instance v_{jm} in D_{v_j} **do**
6:              Get the input rate $\lambda_{v_{jm}}$ and processing rate $\mu_{v_{jm}}$.
7:              Update the weight $w_{v_{jm}}$ from v_{ik} to v_{jm} by (10).

```
 8:          end for
 9:       end if
10:       Sort all $n_{v_j}$ instances by weight in descending order.
11: end for
12: for out tuple $dt_i$ of $v_{ik}$ do
13:       Get the number $d$ of candidate instances based on algorithms 1 and 2.
14:       if $d == n$ then
15:             Candidate instances are all downstream instances.
16:       else
17:             for $i = 1...d$ do
18:                   Use hash function $H_d(k)$ to determine candidate instances $v_{j,H_d(k)}$.
19:             end for
20:       end if
21:       Randomly select a number from 0 to sum of candidate instance weights.
22:       Get target instance $v_{jm}$ of tuple $dt_i$ by weights to randomly.
23: end for
24: return target instance $v_{jm}$ of each tuple;
```

The input of this algorithm includes upstream and downstream instances set U_{v_i} anD_{v_j}. The output is target instance v_{jm} of each tuple. Step 1 to step 10 dynamically adjust instance weights based on monitoring information. Step 1 to step 7 determine the final target instance of the tuple based on random weight grouping. Firstly, select a random number from 0 to the sum of candidate instance weights and traverse the weight table. If the random number is included in the weight range of an instance, that instance is the target instance for tuple. The time complexity of Algorithm is $O(n)$, where n is the number of downstream instances.

5   Performance Evaluation

This section focuses on the evaluation of the proposed Es-Stream, discussing the experimental environment and parameter settings, and providing a performance analysis on the results.

5.1   Experimental Environment

The Es-Stream is implemented on a cluster in the Computer Laboratory of China University of Geosciences, Beijing. The cluster consists of 9 computing nodes, being divided into two types and detailed configuration is shown in Table 1.

One of them is used as a Nimbus node, and 8 are used as Supervisor nodes for task processing. The Zookeeper cluster is established on three nodes within it, responsible for coordinating between Nimbus nodes and Supervisors. Detailed environment configuration is shown in Table 2.

Table 1. Hardware configuration of computing nodes in the cluster.

Type	CPU cores	Memory	Bandwidth	Disk
1	1 vCPU	1 GB	100 Mbps	20 GB
2	2 vCPUs	2 GB	100 Mbps	20 GB

Table 2. Software configuration of the cluster.

Software	Version	Software	Version
OS	Ubuntu 20.04.1 64bit	JDK	jdk1.8 64bit
Python	python 2.7.2	Zookeeper	zookeeper-3.4.14
Storm	apache-Storm-2.1.0	MySQL	MySQL-5.1.73

(a) WordCount.   (b) Top N.

Fig. 5. The logical graph of applications.

We use Amazon Review as a real world dataset, which provides product reviews. Stream application WordCount and Top_N which are classic application for benchmarking distributed stream computing systems, runs to evaluate the system latency, throughput and load imbalance.

The topology of WordCount is composed of three vertices v_{read}, v_{split} and v_{count} as shown in Fig. 5(a). The data stream grouping strategy from v_{read} to v_{split} and v_{split} to v_{count} are ShuffleGrouping, and CustomGrouping P. The CustomGroup P can be replaced with Es-stream, Key Grouping abbreviated as KG, and PKG for comparative evaluation.

The topology of Top_N is composed of four vertices v_{read}, v_{count}, v_{rank} and v_{merge} as shown in Fig. 5(b). The data stream grouping strategy from left to right are CustomGrouping P, Key Grouping and GlobalGrouping.

5.2 Latency

The average latency (AL) is one of the key performance metrics for an elastic stream computing system. On Storm platform, AL can be retrieved through the Storm UI. The shorter the average latency AL, the better the real-time performance.

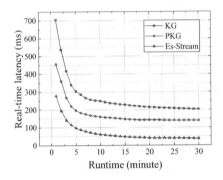

Fig. 6. Real-time latency at stable input rate and parallelism levels.

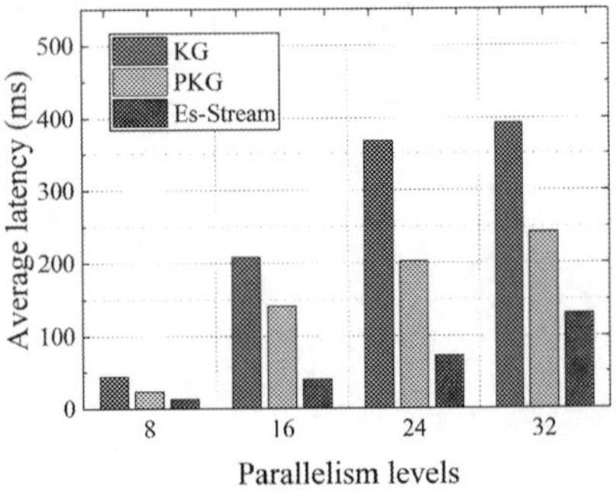

Fig. 7. Average latency at different parallelism levels.

The real-time latency of Es-Stream, KG and PKG can reach a relatively stable state. In the stable phase, Es-Stream has a shorter real-time latency as compared to KG and PKG. As shown in Fig. 6, when the rate of input data stream is 4000 tuples/s and parallelism level is 16, the average latency of Es-Stream, KG and PKG are gauged at 39.52ms, 208.21ms and 140.99ms, Es-Stream reduces it by 81% and 72%.

With the increase of parallelism level or input rate, the average latency increases under three grouping schemes, but Es-Stream has a shorter average latency than KG and PKG. As shown in Fig. 7, when the parallelism is too high or too low, the average latency of Es-Stream is reduced by 68% and 45% compared to KG and PK. When the parallelism is 16 and 24, the average latency of Es-Stream is reduced by 81% and 68% compared to KG and PKG.

When the input rate is 2000 tuples/s as shown in Fig. 8, the average latency difference is not much. When the input rate is 8000 tuples/s, Es-Stream continued to work normally with 36% and 14% decrease in latency compared to KG and PKG, which have already reached the bottleneck.

5.3   Throughput

The average throughput AT is the average rates of successfully processed tuples. The greater the system throughput, the stronger the data processing capability of the stream computing system.

Given the input rate is set to 4000 tuples/s and parallelism level is 16, the real-time throughput of Es-Stream is greater than that of KG and PKG. As shown in Fig. 9, the average throughput of Es-Stream, KG and PKG are gauged at 3766 tuples/s, 3099 tuples/s and 3360 tuples/s, Es-Stream increases it by 21% and 12% and achieves consistent throughput in each second.

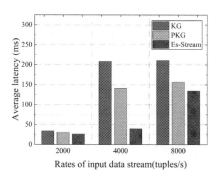

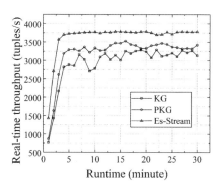

Fig. 8. Average latency at different input rate.

Fig. 9. Real-time throughput at stable input rate and parallelism levels.

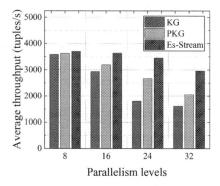

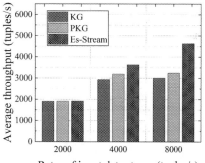

Fig. 10. Average throughput at different parallelism levels.

Fig. 11. Average throughput at different input rate.

With increase in parallelism as shown in Fig. 10, the average throughput under three grouping schemes displays a decreasing trend. And when parallelism level is 32, the average throughput of Es-Stream compared to KG and PKG increases by up to 84% and 44%.

with input rate increases as shown in Fig. 11, average throughput shows a increasing trend. When the data rate is low, the improvement of Es-Stream is not significant. However, as the rate increases, the average throughput of Es-Stream is significantly higher than that of KG and PKG, especially when the rate is 8000 tuple/s, the average throughput of Es-Stream increases by 54% and 43% compared to KG and PKG. This is because Es-Stream assigns more tuples to instances with remaining processing power.

5.4 Load Imbalance

In distributed systems, load imbalance can reflect the load differences in the system. We evaluated the real-time load imbalance of Es-Stream and the average load imbalance at different levels of parallelism.

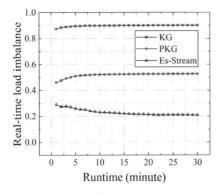
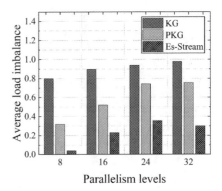

Fig. 12. Real-time load imbalance at stable input rate and parallelism levels.

Fig. 13. Average load imbalance at different parallelism levels.

In different situations, the real-time load imbalance of three grouping schemes can reach a relatively stable state. In the stable phase, Es-Stream has a less real-time load imbalance as compared to KG and PKG. As shown in Fig. 12, when the rate of input data stream is 4000 tuples/s and parallelism level is 16, the average imbalance of Es-Stream is about 0.22, which is reduced by about 74% and 56% compared to KG and PKG.This is because Es-stream distributes the data that should be assigned to the high-load instance to other instances for common processing.

The load imbalance varies with the adjustment of parallelism as shown in Fig. 13. Due to the impact of inter instance communication in the system, the average load imbalance of Es-Stream has been reduced by an average of 75% and 64% compared to KG and PKG.

6 Conclusions and Future Work

In this paper, a light-weight elastic scalable grouping for stateful operators load balance is proposed. Key splitting is used to solve the bottleneck problem caused by high-frequency keys in dynamic real-time stream. Lightweight weight adjustment strategy is adopted to dynamically change the data tuple allocation weight of the instances according to the network cost, data stream rate and processing rate. Our future work will be focusing on integrating the state management as a part of Es-Stream and further improve the efficiency of Es-Stream.

Acknowledgements. This work is supported by the National Natural Science Foundation of China under Grant No. 61972364; the Fundamental Research Funds for the Central Universities under Grant No. 265QZ2021001.

References

1. Toshniwal, A., et al.: Storm@ twitter. In: Proceedings of the 2014 ACM SIGMOD International Conference on Management of Data, pp. 147–156 (2014)
2. Carbone, P., Ewen, S., Fóra, G., Haridi, S., Richter, S., Tzoumas, K.: State management in apache flink®: consistent stateful distributed stream processing. Proc. VLDB Endowment **10**(12), 1718–1729 (2017)
3. Zaharia, M., et al.: Apache spark: a unified engine for big data processing. Commun. ACM **59**(11), 56–65 (2016)
4. Kulkarni, S., et al.: Twitter heron: stream processing at scale. In: Proceedings of the 2015 ACM SIGMOD International Conference on Management of Data, pp. 239–250 (2015)
5. Martí, P., Serrano-Estrada, L., Nolasco-Cirugeda, A.: Social media data: Challenges, opportunities and limitations in urban studies. Comput. Environ. Urban Syst. **74**, 161–174 (2019)
6. George, Y., Karunasekera, S., Harwood, A., Lim, K.H.: Real-time spatio-temporal event detection on geotagged social media. J. Big Data **8**(1), 1–28 (2021). https://doi.org/10.1186/s40537-021-00482-2
7. Ullah, W., et al.: Artificial intelligence of things-assisted two-stream neural network for anomaly detection in surveillance big video data. Futur. Gener. Comput. Syst. **129**, 286–297 (2022)
8. Nasir, M.A.U., Morales, G.D.F., Garcia-Soriano, D., Kourtellis, N., Serafini, M.: The power of both choices: practical load balancing for distributed stream processing engines. In: 2015 IEEE 31st International Conference on Data Engineering, pp. 137–148. IEEE (2015)
9. Jin, H., et al.: Towards low-latency batched stream processing by pre-scheduling. IEEE Trans. Parallel Distrib. Syst. **30**(3), 710–722 (2018)
10. Huang, J., Li, R., Jiao, X., Jiang, Y., Chang, W.: Dynamic DAG scheduling on multiprocessor systems: reliability, energy, and makespan. IEEE Trans. Comput. Aided Des. Integr. Circuits Syst. **39**(11), 3336–3347 (2020)
11. Fu, T.Z., Ding, J., Ma, R.T., Winslett, M., Yang, Y., Zhang, Z.: DRS: auto-scaling for real-time stream analytics. IEEE/ACM Trans. Netw. **25**(6), 3338–3352 (2017)
12. Kalavri, V., Liagouris, J., Hoffmann, M., Dimitrova, D., Forshaw, M., Roscoe, T.: Three steps is all you need: fast, accurate, automatic scaling decisions for distributed streaming dataflows. In: 13th USENIX Symposium on Operating Systems Design and Implementation (OSDI 18), pp. 783–798 (2018)
13. Fang, J., Chao, P., Zhang, R., Zhou, X.: Integrating workload balancing and fault tolerance in distributed stream processing system. World Wide Web **22**(6), 2471–2496 (2019)
14. Mirtaheri, S.L., Grandinetti, L.: Dynamic load balancing in distributed exascale computing systems. Clust. Comput. **20**(4), 3677–3689 (2017)
15. Nasir, M.A.U., Morales, G.D.F., Kourtellis, N., Serafini, M.: When two choices are not enough: Balancing at scale in distributed stream processing. In: 2016 IEEE 32nd International Conference on Data Engineering (ICDE), pp. 589–600. IEEE (2016)

16. Liao, X., Huang, Y., Zheng, L., Jin, H.: Efficient time-evolving stream processing at scale. IEEE Trans. Parallel Distrib. Syst. **30**(10), 2165–2178 (2019)
17. Chen, H., Zhang, F., Jin, H.: Pstream: a popularity-aware differentiated distributed stream processing system. IEEE Trans. Comput. **70**(10), 1582–1597 (2020)
18. Aslam, A., Chen, H., Jin, H.: Pre-filtering based summarization for data partitioning in distributed stream processing. Concurrency Comput. Pract. Experience **33**(20), e6338 (2021)
19. Caneill, M., El Rheddane, A., Leroy, V., De Palma, N.: Locality-aware routing in stateful streaming applications. In: Proceedings of the 17th International Middleware Conference, pp. 1–13 (2016)
20. Son, S., Im, H., Moon, Y.S.: Stochastic distributed data stream partitioning using task locality: design, implementation, and optimization. J. Supercomput. **77**, 11353–11389 (2021)
21. Chen, F., Wu, S., Jin, H.: Network-aware grouping in distributed stream processing systems. In: Vaidya, J., Li, J. (eds.) ICA3PP 2018. LNCS, vol. 11334, pp. 3–18. Springer, Cham (2018). https://doi.org/10.1007/978-3-030-05051-1_1
22. Roy, P., Khan, A., Alonso, G.: Augmented sketch: faster and more accurate stream processing. In: Proceedings of the 2016 International Conference on Management of Data, pp. 1449–1463 (2016)

Incremental Natural Gradient Boosting for Probabilistic Regression

Weiwen Wu, Hui Zhang$^{(\boxtimes)}$, Chunming Yang, Bo Li, and Xujian Zhao

School of Computer Science and Technology, Southwest University of Science and Technology, Mianyang, China
zhanghui@swust.edu.cn

Abstract. The natural gradient boosting method for probabilistic regression (**NGBoost**) is capable of predicting not only point estimates but also target distributions under sample conditions, thereby quantifying prediction uncertainty. However, NGBoost is designed only for batch settings, which are not well-suited for data stream learning. In this paper, we present an incremental natural gradient boosting method for probabilistic regression (**INGBoost**). The proposed method employs scoring rule reduction as a metric and applies the Hoeffding inequality incrementally to construct decision trees that fit the natural gradient, thus achieving incremental natural gradient boosting. Experimental results demonstrate that INGBoost performs well in both point regression and probabilistic regression tasks while maintaining the interpretability of the tree model. Furthermore, the model size of INGBoost is significantly smaller than that of NGBoost.

Keywords: Incremental learning · Natural gradient boosting · Probabilistic regression

1 Introduction

Probabilistic regression is a trending topic in modern machine learning. It involves returning the target distribution based on sample conditions to accurately quantify the uncertainty of prediction [3,11]. Figure 1 illustrates an example of a distribution curve for temperature probabilistic regression where each temperature value in the plot corresponds to a probabilistic density. NGBoost is a natural gradient boosting-based method for probabilistic regression [8], which performs exceptionally well on such problems. However, it can only be used in a batch setting when dealing with large data streams, leading to high memory and time consumption. Therefore, exploring an incremental probabilistic regression method based on natural gradient boosting is necessary.

Incremental decision trees are a type of decision tree that can be constructed incrementally based on data streams. The Hoeffding tree, as the most classical model in incremental decision trees, is constructed by applying the Hoeffding inequality [7]. Most of the existing incremental decision trees are extensions of

X. Yang et al. (Eds.): ADMA 2023, LNAI 14176, pp. 479–493, 2023.
https://doi.org/10.1007/978-3-031-46661-8_32

the Hoeffding tree. This paper proposes a scalable incremental natural gradient boosting method for probabilistic regression. The scalability is due to using scoring rule reduction to incrementally build decision trees instead of adding new decision trees to the collection.

In summary, our contributions are threefold: (i) To the best of our knowledge, for the first time, we propose an incremental natural gradient boosting method for probabilistic regression. (ii) We use the scoring rule reduction as a split metric and apply Hoeffding inequality to incrementally build decision trees to fit the natural gradient. (iii) We demonstrate experimentally that INGBoost performs well in point regression and probabilistic regression, and its model size is smaller than that of NGBoost.

The remainder of this paper is organized as follows: Sect. 2 reviews related work, beginning with incremental decision tree algorithms and progressing to gradient boosting. In Sect. 3, we present our method INGBoost. Section 4 presents an analysis and discussion of the proposed method. In Sect. 5, we experimentally demonstrate that the proposed method is more competitive than existing incremental decision trees in point estimation, close to NGBoost in point estimation and uncertainty estimation. In addition, the size of INGBoost is also smaller than that of NGBoost. Conclusions are given in Sect. 6, where we also discuss possible directions for immediate further work.

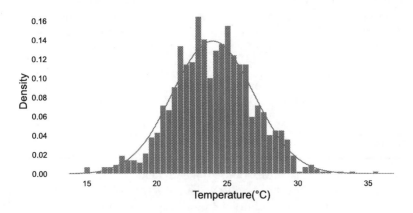

Fig. 1. Probabilistic regression for temperature

2   Related Work

2.1   Incremental Decision Trees

VFDT is a highly influential algorithm in the field of incremental decision trees [7]. Its innovation lies in the application of Hoeffding's inequality to determine the number of samples for leaf node splitting. However, it doesn't handle conceptual drift effectively. CVFDT extends VFDT to enable incremental construction of decision trees on conceptually drifting data streams [13]. It achieves this by periodically validating historical splits and constructing a new decision tree on the nodes where validation fails. VFDTc [10], which is based on Hoeffding trees, stores numerical attributes in a binary search tree to reduce processing time. Furthermore, it incorporates a Bayesian classifier into the leaf nodes to improve classification accuracy. HAT extends the Hoeffding tree algorithm by using the drift detector ADWINP to evaluate splitting performance. Its main advantage is that it doesn't require consideration of the speed and frequency of changes in the data stream [4,5]. While all of the aforementioned incremental decision trees are designed for classification problems, FIMT-DD [14] is a variant of the Hoeffding tree that focuses on data stream regression. It uses standard deviation reduction as a splitting metric and a linear model at the leaf nodes to fit statistical sample information. FIMT-DD uses a threshold mechanism to handle cases where multiple attributes are highly differentiated, but this often results in a delay in splitting. In contrast, ORTO introduces option nodes that allow instances to arrive at multiple leaf nodes and uses a greedy strategy to speed up the split process [16]. Read et al. [20] and Mastelini et al. [19] have both applied the Hoeffding bound to multi-target decision trees, making it possible to solve the problems of multi-target classification and multi-target regression on the data stream. Ikonomovska [15] et al. proposed two integrated regression methods based on incremental decision trees, OBag and ORF, where OBag is integrated by incremental model trees and ORF is integrated by stochastic model trees. There is a main difference between the method presented in this paper and the incremental decision tree. Incremental decision trees can't solve the probabilistic regression problem, but our approach does.

2.2   Gradient Boosting

Gradient boosting is a learning technology that builds multiple classifiers to fit the natural gradient to minimize a differentiable loss function [9]. Most gradient-boosting frameworks are extended based on this technique. Chen et al. [6] proposed XGBoost, which utilizes second-order approximation and regularization to define the loss function, thereby improving accuracy and avoiding overfitting. In addition, XGBoost uses several techniques to improve efficiency and reduce calculations. LightGBM [17] is also a variant designed based on the idea of gradient boosting, which improves the computational efficiency in boosting. All of the gradient boosting models mentioned above can only perform point regression, not probabilistic regression. This is because probabilistic regression often requires simultaneous boosting of multiple parameters, making it a challenge to implement probabilistic regression using gradient boosting. The NGBoost proposed by Duan solves this problem well by using natural gradients and scaling factors to find locally optimal solutions [8]. Although NGBoost performs well in probabilistic regression, it cannot be built incrementally in the face of large-scale data streams. In this paper, we introduce the concept of incremental decision trees into NGBoost to realize INGBoost. When new data arrives, the weak learner in the ensemble uses a split or updates the predicted values of the leaf nodes to achieve the update of INGBoost.

3   INGboost

This section describes the incremental natural gradient boosting method for probabilistic regression. The first key element is the use of scoring rule reduction as a metric to evaluate the split quality and construct decision trees incrementally by using Hoeffding's inequality. Secondly, we fit the natural gradient using the above method of incrementally constructing decision trees.

3.1   Evaluation Metric

In supervised incremental learning, it is not possible to store data in memory and learn from it because they are not finite. Therefore, we need a metric that can quickly evaluate the split quality based on the information already collected. We propose using the Scoring Rule Reduction (**SRR**) as an incremental metric. Before describing SRR, we need to introduce the concepts of scoring rules and natural gradients.

The scoring rule [11] can effectively compare the estimated distribution with the true distribution, and therefore it is used as a loss function in probabilistic regression. A proper scoring rule should satisfy:

$$\mathrm{E}_{y \sim V} \left\{ S \left[V \left(\theta \right), y \right] \right\} \leq \mathrm{E}_{y \sim V} \left\{ S \left[P \left(\theta \right), y \right] \right\} \qquad \forall V, P, \tag{1}$$

where V is the true distribution of outcomes y, P is any other distribution, S is the scoring rule, and θ is the parameter in the distribution.

Logarithmic scoring is the most commonly used proper scoring rule that accurately describes how well the distribution fits the sample. It is defined as:

$$\psi \left(\theta, y \right) = -log P \left(\theta \right) \tag{2}$$

When the standard gradient is used to optimize an objective function, the results of training are greatly influenced by the form of the parameters, which is detrimental to the optimization. The natural gradient is proposed to solve this problem [1]. Unlike the standard gradient, the natural gradient considers the distance between the distributions identified by the parameters which are invariant to parametrization. It represents the direction of the steepest ascent in Riemannian space. From the optimization point of view, the natural gradient is defined as:

$$\tilde{\nabla} S \left[P \left(\theta \right), y \right] \propto \sigma_S \left(\theta \right)^{-1} \nabla S \left[P \left(\theta \right), y \right] \tag{3}$$

where σ_S is the Riemannian metric of the statistical manifold at θ, which is induced by the scoring rule S.

Suppose there is a data set sample D of size N, where the value h_A of attribute A divides the data set D into two subsets, D_L and D_R, and the size is also divided into N_L and N_R. The SRR metric formula for the partition h_A is given below.

$$\text{SRR}(h_A) = sr(D) - \frac{N_L}{N} sr(D_L) - \frac{N_R}{N} sr(D_R) \tag{4}$$

$$sr(D) = \frac{1}{N} \sqrt{\sum_{i=1}^{N} S[P(\theta_i - \eta\bar{g}), y_i]} \tag{5}$$

where S is the scoring rule, $\bar{g}$ is the mean of the natural gradient of the sample, y is the target value, θ is the distribution parameter, and η is the learning rate. This metric formula can evaluate all potential splits and find the best split that minimizes the overall scoring rule.

3.2 Splitting Criterion

To incrementally build the decision tree, we introduce Hoeffding inequality to determine the splitting, which can be well adapted to the data stream. The pseudo-code for building the decision tree is shown in Algorithm 1.

Algorithm 1. Construction of the incremental decision tree

Input: Number of splitting judgments N_{min}, Threshold Index ξ, instance of arrival at
        time step $t\ e\,(x_t, y_t)$.
Output: Decision tree.
1: Instance e arrives at a leaf node.
2: Update statistics (e).
3: **if** $N = N_{min}$ **then**
4:    Let h_A be the attribute with highest SRR (h_A).
5:    Let h_B be the attribute with second-highestt SRR (h_B).
6:    **if** SRR $(h_A) \geq 0$ **then**
7:        Compute u using Equation 7 and ε using Equation 8.
8:        **if** $u + \varepsilon \leq 1$ or $\varepsilon \leq \xi$ **then**
9:            Make split with attribute h_A
10:       **end if**
11:   **end if**
12: **end if**

The Hoeffding inequality allows us to select the best attributes with a certain level of confidence, but this requires that the random variable we define take values with a range restriction. Therefore, we need to construct a bounded random variable that can be used to determine the split.

We rank the potential split points based on the scoring rule reduction. The real-valued random variable u is defined as:

$$z(x) = \frac{1}{1 + e^{-x}} \tag{6}$$

$$u = \frac{z(\text{SRR}(h_B))}{z(\text{SRR}(h_A))} \tag{7}$$

where h_A is the best split and h_B is the second best split. Equation 6 is used to constrain the range of the scoring rule reduction, ensuring that the value of the random variable u falls between 0 and 1. This random variable represents the ratio of the two best splits and can be a good basis for splitting.

The Hoeffdding bound allows a certain confidence level to obtain the true mean of a sequence of random variables [12, 18]. Given a variable u whose range is R. The number of samples observed is N. The Hoeffding bound states that the true mean of the variables is at least $\bar{u} - \varepsilon$ and at most $\bar{u} + \varepsilon$ with probability $1 - \delta$. The value of ε is defined as:

$$\varepsilon = \sqrt{\frac{R^2 \ln(1/\delta)}{2N}} \tag{8}$$

The advantage of the Hoeffding bound is that it gives the exponential decay of the probability that the mean of the observed variable deviates from its true expected value which defines the bound on the true expectation of a random variable as:

$$\bar{u} - \varepsilon \le u_{real} \le \bar{u} + \varepsilon \tag{9}$$

where u_{real} is the true expectation of the random variable.

The upper bound in Eq. 9 is used to determine the splitting. Since the real-valued variable u should be between 0 and 1, the splitting point h_A, with probability $1 - \delta$, is considered as the best splitting point when $u + \varepsilon < 1$. In this case, the split point h_A is also considered to be the point in the global that minimizes the loss of scoring rules and can be applied.

The method of using the Hoeffding bound to determine the split may never make the split when there are multiple attributes with similar distinguishability, which will result in the decision tree not being able to expand further. To solve this problem, we adopted the threshold mechanism used by most incremental decision trees. The threshold mechanism can choose the optimal attribute to split when the Hoeffding bound cannot decide. When ε is less than a certain threshold, the threshold mechanism selects the best split in the current arrival sample, which ensures that the tree continues to expand after a certain number of samples have been observed.

3.3   Prediction of Leaf Nodes

In incremental learning, the data stream generates a new unit of data arriving at a leaf node in the form of a time step. To fit the natural gradient of the data in the node, leaf nodes use the following two ways to determine the output value.

The first way is to take the average as the output value, which is defined as follows:

$$\frac{\sum_{i=1}^{N} y_i}{N} \tag{10}$$

where N is the number of samples counted by the leaf nodes at time step t and y is the target value of the sample. When a node splits, the branch node cannot calculate the output value without a new sample arriving in a short period. To avoid this situation, the split node passes its target value to the child nodes as temporary output until a new sample arrives.

The second way is to add a linear model to the leaf nodes to fit the natural gradient. The linear model is defined as follows:

$$o = v_0 + v_1 x_1 + v_2 x_2 + v_j x_j + \ldots v_n x_n \tag{11}$$

where v_0 represents the constant term, each feature corresponds to a weight v, In order to better adapt the learning model under data stream, we use the stochastic gradient descent method to update the weights, and use the mean squared error as the loss function, which can better simulate the gradual expansion of the model with the arrival of samples.

When the leaf node accepts a new sample, the linear model is equivalent to a fine-tuning in the instance space, and each weight in the model is updated. The updated step size is defined as follows:

$$v_j = v_j + l \left(o - y \right) x_j \tag{12}$$

where o is the predicted value, y is the true value, and l is the learning rate. In the construction of the decision tree, the splitting of leaf nodes and the process of linear model update do not conflict. When the leaf node is split once, the linear model is passed to the child node, avoiding retraining and resulting in a loss of accuracy. The passed linear model continues to update in a new direction based on the instance space that changes subsequently.

3.4   Incremental Natural Gradient Boosting

To achieve incremental natural gradient boosting, we use the above method of building decision trees to fit the natural gradient. The pseudo-code of the algorithm is shown in Algorithm 2 and the training process based on the normal distribution is shown in Fig. 2.

Before natural gradient boosting, the initial parameter θ_0 is selected by minimizing the overall scoring rule for all samples. In an incremental learning setup, obtaining all samples at a certain point in time is impossible, so we only compute θ_0 based on the first arrival sample. For different distributions, there are different ways to calculate θ_0. For example, when the distribution is normal and the first arrival sample is (x_1, y_1), the initial parameter θ_0 is set to $(y_1, 1)$, as shown in line 1 of Algorithm 2.

Algorithm 2. Incremental natural gradient boosting

Input: Boost iterations M, Learning rate η, Parameters of the probability distribution
θ, Scoring rule S, Weak learners f, Sample instance of arrival at time step t e (x_t, y_t)

Output: Weak learners $\left\{ f^{(m)} \right\}_{m=1}^{M}$

1: **if** $t = 1$ **then**
2:     $\theta^{(0)} =$ initialize to marginal(x_t, y_t)
3: **end if**
4: **for** each $m \in [1, M]$ **do**
5:     $g_t^{(m)} \leftarrow \sigma_S \left(\theta_t^{(m-1)} \right)^{-1} \nabla_\theta S \left(\theta_t^{(m-1)}, y_t \right)$
6:     $f^{(m)} \leftarrow learn \left(\left\{ x_t, y_t, g_t^{(m)}, \theta^{(m-1)}, \right\} \right)$
7:     $\theta_t^{(m)} \leftarrow \theta_t^{(m-1)} - \eta \cdot f^{(m)} (x_t)$
8: **end for**

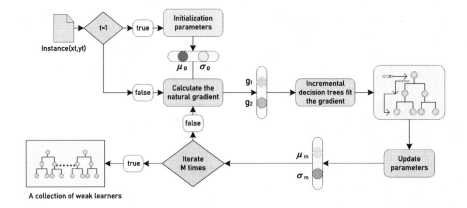

Fig. 2. Incremental natural gradient boosting process based on normal distribution

The proposed method is to use incremental settings. When a new sample arrives, the method does not create new decision trees but rather updates the existing decision trees to fit the natural gradient of that sample, as shown in line 5 of Algorithm 2.

4   Analysis and Discussion

Multi-parameters Boosting. NGBoost achieves multi-parameter gradient boosting by finding a scaling factor that minimizes the scoring rule. However, this approach is only applicable to batch settings. To address this limitation, we propose scoring rule reduction as a metric that can consider the effect of the natural gradient of multiple parameters on the loss function at the same time.

Computational Complexity. The computational variability between ING-Boost and NGBoost focuses on two main aspects. The first is that while NGBoost requires the construction of a series of weak learners for each parameter, ING-Boost only requires the construction of a set of weak learners to fit the natural gradients of all parameters. Therefore, the relative decrease in computational cost in this part is linear in the number of distribution parameters (p). Second, since our weak learners are constructed incrementally, the leaf nodes collect information about the arrival samples. To facilitate scoring rule loss, when a new sample is received, we use a fast sort to sort the array of each attribute. If the number of attributes is j and the number of collected samples is n, the cost of receiving a new sample is $O(j \cdot nlogn)$.

5 Experiments

This section demonstrates the effectiveness of INGBoost in point regression and probabilistic regression tasks on the data stream, as well as the size of the model. In addition, the prediction accuracy under different δ is also checked in this experiment. The mean squared error and mean negative log-likelihood are used to evaluate the model's predictive performance. The number of model nodes is also used to assess the model size.

In all experiments, the common parameters are set to the following values. The minimum number of examples for leaf node split is set to $n_{min} = 150$ and the confidence parameter is set to $\delta = 0.01$. INGBoost and NGBoost are configured with a normal distribution and a logarithmic scoring rule. The boosting iteration is set to $M = 200$ and the learning rate is set to $\eta = 0.01$. In OBag, the maximum number of trees is 10. The maximum depth of the decision tree for the above models is set to 4.

Our experiments use datasets from the UCI Machine Learning Repository and Kaggle Machine Learning Repository [2]. Information about the datasets used in the experiments is shown in Table 1.

Table 1. Details of the dataset used for data stream regression

Datasets	Features	Instances
Abalone	8	4177
Wine	11	4898
Crab Age	8	3893
Heart disease	15	4238
Wind	8	6574
HR comma sep	9	14999
Temperature	23	7752
Churn telecom	20	3333

5.1   Comparison with Existing Incremental Decision Trees

In the experiments in this subsection, we use a tenfold cross-validation procedure for each dataset to evaluate the mean squared error of the models. As baselines, we use three models: **FIMT-DD** [14], **ORTO** [16], and **OBag** [15], where OBag is an online bagging model integrated with FIMT-DD.

The learning curves of the experiments are shown in Fig. 3. At the beginning of the learning phase, the mean squared error of our method is significantly lower than the mean squared error of the baseline, especially on the Wind dataset. This can be attributed to our model's use of scoring rule reduction as the metric for fitting the natural gradient, demonstrating good performance even with small samples. This effect diminishes as more samples are learned, indicating that the model is approaching convergence. In the final stage of learning, INGBoost still outperforms the baseline, although it does not show the same advantage as it did at the beginning of learning. Qualitatively, INGBoost exhibits better convergence properties compared to the baselines.

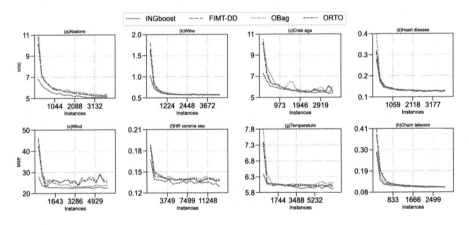

Fig. 3. Learning curves of INGBoost and baseline models on data stream regression

The mean square error of the comparison experiments with the existing incremental decision tree is shown in Table 2. The results demonstrate that our method achieves a lower mean squared error than the baseline across all eight datasets, indicating strong prediction performance. When comparing INGBoost with the best incremental decision tree, ORTO, shown in the table, our method reduces the error by approximately 2.6%. This can be attributed to the fact that the update of each decision tree in INGBoost is based on scoring rule loss, which reduces the error from a global perspective.

Table 2. Results of the final mean square error of INGBoost with the baseline models

Datasets	INGBoost	FIMT-DD	ORTO	OBag
Abalone	5.11 ± 0.71	5.24 ± 0.68	5.23 ± 0.72	5.39 ± 0.83
Wine	0.56 ± 0.07	0.57 ± 0.06	0.57 ± 0.06	0.58 ± 0.05
Crab age	5.44 ± 0.70	5.64 ± 0.73	5.54 ± 0.64	6.00 ± 0.84
Heart disease	1.26E-1 ± 0.16E-1	1.28E-1 ± 0.13E-1	1.29E-1 ± 0.11E-1	1.27E-1 ± 0.12E-1
Wind	2.25E+1 ± 1.22	2.59E+1 ± 1.24	2.49E+1 ± 1.24	2.39E+1 ± 2.47
HR comma sep	1.28E-1 ± 0.11E-1	1.38E-1 ± 0.11E-1	1.37E-1 ± 0.09E-1	1.39E-1 ± 0.12E-1
Temperature	5.90 ± 0.30	6.01 ± 0.33	5.98 ± 0.33	6.07 ± 0.32
Churn telecom	1.04E-1 ± 0.73E-2	1.05E-1 ± 0.80E-2	1.05E-1 ± 0.76E-2	1.05E-1 ± 0.79E-2

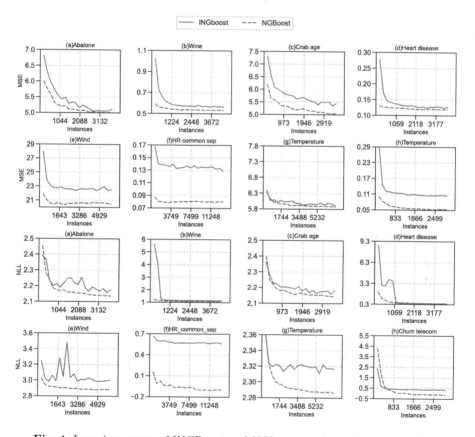

Fig. 4. Learning curves of INGBoost and NGboost on data stream regression

5.2   Comparison with NGBoost

In the experiment in this subsection, we use a tenfold cross-validation procedure for each dataset to evaluate the performance of **INGBoost** and **NGBoost** [8] from three perspectives: learning curve, final prediction, and model size.

Table 3. Final mean square error and mean negative log likelihood of INGBoost and NGBoost

Datasets	INGBoost		NGBoost	
	MSE	NLL	MSE	NLL
Abalone	5.11 ± 0.71	2.17 ± 0.08	5.01 ± 0.49	2.13 ± 0.05
Wine	0.56 ± 0.07	1.13 ± 0.05	0.53 ± 0.06	1.08 ± 0.05
Crab age	5.44 ± 0.70	2.18 ± 0.04	5.04 ± 0.53	2.14 ± 0.04
Heart disease	1.26E-1 ± 0.16E-1	0.43 ± 0.05	1.19E-1 ± 0.14E-1	0.34 ± 0.09
Wind	2.25E+1±1.22	3.00 ± 0.04	2.04E+1 ± 0.74	2.88 ± 0.02
HR comma sep	1.28E-1 ± 0.11E-1	5.62E-1 ± 0.14E-1	0.800E-1 ± 0.62E-2	-1.00E-1 ± 0.11
Temperature	5.90 ± 0.30	2.31 ± 0.02	5.87 ± 0.30	2.28 ± 0.02
Churn telecom	1.04E-1 ± 0.73E-2	3.57E-1 ± 0.9E-2	5.26E-2 ± 0.48E-1	-1.21E-1 ± 0.23

To evaluate the performance of NGBoost on data streams, we combine both old and new data to generate the learning curve, as shown in Fig. 4. Initially, the mean squared error of INGBoost is higher than that of NGBoost. However, as the number of learning samples increases, the difference in mean squared error between the two models becomes smaller. This outcome demonstrates the efficacy of INGBoost in handling data streams and its ability to achieve comparable performance to NGBoost in certain situations.

The mean square error and mean negative log-likelihood of the comparison experiments with NGBoost are shown in Table 3. Except for telecom churn and HR common sep datasets, INGBoost's final mean squared error and mean negative logarithmic likelihood are very close to NGBoost. This indicates that INGBoost is a promising model that can achieve both efficiency and precision in most cases.

The number of nodes in INGBoost and NGBoost on the eight datasets is shown in Fig. 5. With the same number and depth of decision trees, the number of nodes in INGBoost is reduced by about 32.2% compared to NGBoost. This is because the scoring rule reduction in INGBoost can evaluate the natural gradient effect of multiple parameters at the same time, so only one decision tree needs to be constructed to fit the natural gradient at each iteration. As a result, the model size is significantly reduced compared to NGBoost.

5.3   The Impact of Tolerance δ on the Predicted Performance of INGBoost

The experiments in this subsection examine the effect of tolerance δ on the prediction performance of INGBoost. Figure 6 plots the learning curve under different values of δ on two datasets. Since the $1 - \delta$ represents the confidence probability that the Hoeffding boundary determines splitting, the parameter δ indirectly implies the number of samples that need to be collected for decision tree splitting. At the beginning of learning, models with larger δ will split nodes faster, thereby temporarily improving accuracy. But as the number of learning samples increases, the models of smaller δ also gradually split the nodes, and

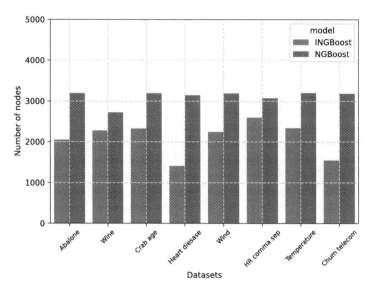

Fig. 5. Results of INGBoost and NGBoost for the number of nodes on data stream regression

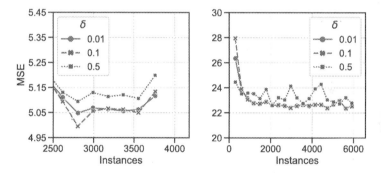

Fig. 6. The prediction performance of INGBoost with different values of δ on two datasets

the quality of the split can be guaranteed with higher confidence, resulting in better accuracy.

6   Conclusion

In this work, we present an incremental natural gradient boosting method for probabilistic regression. The proposed method uses scoring rule reduction as a metric and applies Hoeffding inequality incrementally to construct decision trees that fit the natural gradient, thereby achieving incremental natural gradient boosting. The extensive experimental results demonstrate that the proposed method performs well in data stream regression tasks and reduces the model size.

In future work, we plan to explore the way of storing the numerical attributes of the leaf node samples, which reduces the time complexity of the method.

Acknowments. provincial scientific research institutes' achievement transformation project of the science and technology department of Sichuan Province, China (2023JDZH0011).

References

1. Amari, S.I.: Natural gradient works efficiently in learning. Neural Comput. **10**(2), 251–276 (1998)
2. Asuncion, A., Newman, D.: UCI machine learning repository (2007)
3. Avati, A., Duan, T., Zhou, S., Jung, K., Shah, N.H., Ng, A.Y.: Countdown regression: sharp and calibrated survival predictions. In: Uncertainty in Artificial Intelligence, pp. 145–155. PMLR (2020)
4. Bifet, A., Gavaldà, R.: Learning from time-changing data with adaptive windowing. In: Proceedings of the 2007 SIAM International Conference on Data Mining, pp. 443–448. Society for Industrial and Applied Mathematics (2007). https://doi.org/10.1137/1.9781611972771.42
5. Bifet, A., Gavaldà, R.: Adaptive learning from evolving data streams. In: Adams, N.M., Robardet, C., Siebes, A., Boulicaut, J.-F. (eds.) IDA 2009. LNCS, vol. 5772, pp. 249–260. Springer, Heidelberg (2009). https://doi.org/10.1007/978-3-642-03915-7_22
6. Chen, T., Guestrin, C.: Xgboost: A scalable tree boosting system. In: Proceedings of the 22nd ACM SIGKDD International Conference on Knowledge Discovery and Data Mining, pp. 785–794 (2016)
7. Domingos, P., Hulten, G.: Mining high-speed data streams. In: Proceedings of the sixth ACM SIGKDD International Conference on Knowledge Discovery and Data Mining, pp. 71–80. ACM (2000). https://doi.org/10.1145/347090.347107
8. Duan, T., et al.: NGBoost: natural gradient boosting for probabilistic prediction. In: International Conference on Machine Learning, pp. 2690–2700. PMLR (2020)
9. Friedman, J.H.: Greedy function approximation: a gradient boosting machine. Ann. Stat. **29**(5), 1189–1232 (2001). https://doi.org/10.1214/aos/1013203451
10. Gama, J., Rocha, R., Medas, P.: Accurate decision trees for mining high-speed data streams. In: Proceedings of the ninth ACM SIGKDD International Conference on Knowledge Discovery and Data Mining, pp. 523–528. ACM (2003). https://doi.org/10.1145/956750.956813
11. Gneiting, T., Raftery, A.E.: Strictly proper scoring rules, prediction, and estimation. J. Am. stat. Assoc. **102**(477), 359–378 (2007). https://doi.org/10.1198/016214506000001437
12. Hoeffding, W.: Probability inequalities for sums of bounded random variables. In: Fisher, N.I., Sen, P.K. (eds.) The Collected Works of Wassily Hoeffding, pp. 409–426. Springer, New York (1994). https://doi.org/10.1007/978-1-4612-0865-5_26
13. Hulten, G., Spencer, L., Domingos, P.: Mining time-changing data streams. In: Proceedings of the seventh ACM SIGKDD International Conference on Knowledge Discovery and Data Mining, pp. 97–106. ACM (2001). https://doi.org/10.1145/502512.502529
14. Ikonomovska, E., Gama, J., Džeroski, S.: Learning model trees from evolving data streams. Data Min. Knowl. Disc. **23**(1), 128–168 (2011). https://doi.org/10.1007/s10618-010-0201-y

15. Ikonomovska, E., Gama, J., Džeroski, S.: Online tree-based ensembles and option trees for regression on evolving data streams. Neurocomputing **150**, 458–470 (2015). https://doi.org/10.1016/j.neucom.2014.04.076

16. Ikonomovska, E., Gama, J., Zenko, B., Dzeroski, S.: Speeding-up hoeffding-based regression trees with options. In: Proceedings of the 28th International Conference on Machine Learning (ICML-11), pp. 537–544 (2011)

17. Ke, G., et al.: LightGBM: a highly efficient gradient boosting decision tree. In: Advances in Neural Information Processing Systems 30 (2017)

18. Maron, O., Moore, A.: Hoeffding races: accelerating model selection search for classification and function approximation. In: Advances in Neural Information Processing Systems 6 (1993)

19. Mastelini, S.M., Barbon Jr, S., de Carvalho, A.C.P.d.L.F.: Online multi-target regression trees with stacked leaf models. arXiv preprint arXiv:1903.12483 (2019)

20. Read, J., Bifet, A., Holmes, G., Pfahringer, B.: Scalable and efficient multi-label classification for evolving data streams. Mach. Learn. **88**(1–2), 243–272 (2012). https://doi.org/10.1007/s10994-012-5279-6

Discovering Skyline Periodic Itemset Patterns in Transaction Sequences

Guisheng Chen[1,2] and Zhanshan Li[1,2(✉)]

[1] College of Computer Science and Technology, Jilin University, Changchun 130012, China
[2] Key Laboratory of Symbolic Computation and Knowledge Engineering of Ministry of Education, Jilin University, Changchun 130012, China
lizs@jlu.edu.cn

Abstract. As an extended version of frequent itemset patterns, periodic itemset patterns concern both the frequency and periodicity of itemsets at the same time, so they contain more information than frequent itemset patterns, which only concern the frequency. With further research, we found that, in some cases, the periodic itemset patterns with higher frequency, or with optimal periodicity, or with both higher frequency and optimal periodicity have higher application value. However, there is currently no work focusing on such a kind of periodic itemset patterns. In view of this, this paper first proposes a new concept of skyline periodic itemset patterns, and states the problem of skyline periodic itemset pattern mining, then presents an algorithm called SLPIM (SkyLine Periodic Itemset pattern Miner) for skyline periodic itemset pattern mining. SLPIM first adopts the well-known FP-Growth algorithm to mine all frequent itemset patterns, and then uses an effective judgment strategy to determine which frequent itemset patterns are skyline periodic itemset patterns. Finally, experiments are conducted on two real-world and two simulated datasets. The results show that SLPIM is competent for mining skyline periodic itemset patterns.

Keywords: Pattern mining · Frequent itemset pattern · Periodic itemset pattern · Skyline pattern

1 Introduction

Periodic patterns have a wide range of applications [1], and the periodic pattern mining problem has been an active research topic for decades. There are many kinds of periodic patterns, such as periodic itemset patterns [2], periodic spatial patterns [3], local periodic patterns [4], periodic event patterns [5], and periodic cluster patterns [5]. Among them, periodic itemset patterns are the most common ones. The definitions of periodic itemset patterns are not unique, since different periodicity measures will lead to different definitions. For instance, Tanbeer et al. [2] employed the maximum period as the periodicity measure, and Rashid et al. [6] adopted the variance of periods as the periodicity measure.

X. Yang et al. (Eds.): ADMA 2023, LNAI 14176, pp. 494–508, 2023.
https://doi.org/10.1007/978-3-031-46661-8_33

Chen et al. [7] defined periodic itemset patterns by a novel periodicity measure based on coefficient of variation, and proposed a probability model for periodic itemset pattern prediction. In the definition, an itemset is a periodic itemset pattern if its frequency is not less than a user-defined threshold, and the coefficient of variation for its periods is not greater than a user-defined threshold. In the probability model, periods are supposed to follow a normal distribution. The work has the following limitation: if an item is contained in several patterns, then it will be predicted several times, leading to redundancies in workload. The larger the number of patterns, the more redundancies. So, in such a case, or in other cases that there are too many patterns to cope with, we need to reduce the number of patterns.

One way to reduce the number of periodic patterns is to heighten the frequency threshold or strengthen the constraints on periodicity measures. However, some patterns, which regularly but rarely occur, may be abandoned when the frequency threshold is set a higher value. Fortunately, there is another way that we can cut down the number of periodic patterns by preserving only the special ones according to practical need, without heightening the frequency threshold or strengthening the constraints on periodicity measures. That motivates us to define special types of periodic patterns.

In view of this, we propose a new concept of skyline periodic itemset patterns in this paper, where coefficient of variation is employed as the periodicity measure. Roughly speaking, periodic itemset pattern p is called a skyline periodic itemset pattern if no other periodic itemset patterns in the database have higher frequency and lower coefficient of variation than those of p. So, skyline periodic itemset patterns are a special type of periodic itemset patterns in fact.

The concept of skyline periodic itemset patterns is based on the consideration that both the frequency and periodicity have an impact on the prediction accuracy of the probability model proposed by Chen et al. [7]. On the one hand, higher frequency of an itemset means larger sample size of its periods. On the other hand, if the periods have a lower value for the coefficient of variation, they have a smaller fluctuation. So, theoretically, periodic itemset patterns with either higher frequency or lower coefficient of variation or both can make more accurate predictions, and skyline periodic itemset patterns happen to be such a kind of periodic itemset patterns. So, by preserving only skyline periodic itemset patterns, we can significantly reduce the number of patterns, and, at the same time, ensure the accuracy of predictions as much as possible. In summary, the aim of skyline periodic itemset patterns is to free the users from the overwhelming great number of patterns.

After giving the definition of skyline periodic itemset patterns, we then state the problem of skyline periodic itemset pattern mining, and propose a new algorithm named SLPIM for skyline periodic itemset pattern mining. The SLPIM algorithm is basically consisted of two phases: i) generation of all frequent itemset patterns using off-the-shelf FP-Growth algorithm [8], and, ii) identification of skyline periodic itemset patterns from the frequent itemset patterns.

The contributions of our work are as follows:

1. We present a new concept of skyline periodic itemset patterns, and state the problem of skyline periodic itemset pattern mining in transaction sequences.
2. Propose an algorithm named SLPIM for skyline periodic itemset pattern mining.
3. Empirically, we show that our SLPIM algorithm is competent for skyline periodic itemset pattern mining on two real-world and two synthetic transactional datasets.

The remainder of this paper is organized as follows: Sect. 2 reviews preliminaries and related works. Section 3 includes the new concept and problem statement. We introduce our skyline periodic itemset pattern mining algorithm in Sect. 4. Then, the experiment analysis is reported in Sect. 5. Finally, we make a conclusion in Sect. 6.

2   Preliminaries and Related Works

2.1   Preliminaries

In the following sections, we use $I = \{i_1, ..., i_m\}$ to denote a set of distinct items (products, events, or others). Both transaction b and itemset p are nonempty subsets of I. A transactional database is simply a set of transactions regardless of the order of its transactions, and denoted by $B = \{b_1, ..., b_n\}$. However, a transaction sequence is a set of transactions sorted in ascending order by the indices or time stamps of its transactions, and denoted by $B = \langle b_1, ..., b_n \rangle$. When I represents a set of products sold in a market, then b_i, $i \in \{1, ..., n\}$, is also called a basket, which contains a set of products purchased by a customer sometime, and B is the shopping record of the customer, as Table 1 shows.

Table 1. Transaction sequence of a customer.

Tran. ID	Basket	Tran. ID	Basket
b_1	$\{a, b, c, f\}$	b_9	$\{a, b, g, h\}$
b_2	$\{b, c, d\}$	b_{10}	$\{b, c, d\}$
b_3	$\{a, b, e, f, h\}$	b_{11}	$\{a, b, f, h\}$
b_4	$\{c, d, f, h\}$	b_{12}	$\{c, d, f, g, h\}$
b_5	$\{a, d, e, f, g\}$	b_{13}	$\{a, b, c\}$
b_6	$\{c, e, g, h\}$	b_{14}	$\{b, e, g\}$
b_7	$\{a, b, c, g\}$	b_{15}	$\{a, c, d, e\}$
b_8	$\{a, c, f, g\}$	b_{16}	$\{c, f\}$

Definition 1 (Support). *Given a transaction sequence $B = \langle b_1, ..., b_n \rangle$, itemset p, $p \subseteq I$. If $\exists b_i \in B$, such that $p \subseteq b_i$, then we call b_i a support for p. Let $Sup(p)$ denote all supports of p in B, then $Sup(p) = \{b_i | \forall b_i \in B, p \subseteq b_i\}$.*

Definition 2 (Frequency). *The absolute frequency of itemset p in transaction sequence B is defined as $Freq(p) = |Sup(p)|$, and the relative frequency is $Freq(p) = |Sup(p)|/|B|$.*

Property 1 (Anti-Monotonicity). *Let $p' \subseteq p$, then we have $Freq(p') \geq Freq(p)$.*

For convenience, let $r = |Sup(p)|$ in the following.

Definition 3 (Occurrence List). *The occurrence list of itemset p in transaction sequence B is denoted as $Occur(p)$, and defined as $Occur(p) = \{k | \forall b_k \in B, p \subseteq b_k\}$. All elements of $Occur(p)$ are sorted in ascending order, that means if $Occur(p) = \{w_1, w_2, ..., w_r\}$, then w_1 is the index of the first transaction containing p, ..., w_r is the index of the last transaction containing p.*

Definition 4 (Frequent Itemset Pattern). *Given a transaction sequence $B = \langle b_1, ..., b_n \rangle$, itemset p, frequency threshold θ. If $Freq(p) \geq \theta$, then we call p a frequent itemset pattern, frequent itemset for short.*

Definition 5 (Period List). *The period list, containing all the periods of itemset p in transaction sequence B, is denoted as $Per(p)$, and defined as $Per(p) = \{per_k = w_{k+1} - w_k | \forall k \in \{1, ..., r-1\}, w_k \in Occur(p)\}$.*

Definition 6 (Extended Period List). *The extended period list of p in B is denoted as $ExtPer(p)$, and defined as $ExtPer(p) = \{per_k = w_{k+1} - w_k | \forall k \in \{0, ..., r\}, w_k \in ExtOccur(p)\}$, where $ExtOccur(p) = \{0\} \cup Occur(p) \cup \{n\}$, and $n = |B|$.*

At present, there is no uniform definition of periodic itemset patterns, and different definitions will be obtained by adopting different periodicity measures. The general definition of periodic itemset patterns is: Given a transaction sequence $B = \langle b_1, ..., b_n \rangle$, itemset p, frequency threshold θ, and several periodicity measures $M_1, ..., M_k$, where $M_1, ..., M_k$ are functions of $Per(p)$ or $ExtPer(p)$. If $Freq(p) \geq \theta$ and all the periodicity measures $M_1, ..., M_k$ satisfy the given constraints, then we call p a periodic itemset pattern.

Tanbeer et al. [2] introduced the maximum period as the periodicity measure, and defined periodic itemset patterns as follows:

Definition 7 (Periodic Itemset Pattern). *Given a transaction sequence $B = \langle b_1, ..., b_n \rangle$, itemset p, frequency threshold θ, and maximum period threshold η. If $Freq(p) \geq \theta$ and $max(ExtPer(p)) \leq \eta$, then p is a periodic itemset pattern.*

Chen et al. [7] adopted the coefficient of variation as the periodicity measure, and presented the definition as follows:

Definition 8 (Periodic Itemset Pattern). *Given a transaction sequence $B = \langle b_1, ..., b_n \rangle$, itemset p, frequency threshold θ, and threshold δ for coefficient of variation. If $Freq(p) \geq \theta$ and $Coefvar(p) \leq \delta$, then p is a periodic itemset pattern, where, $Coefvar(p)$ denotes the coefficient of variation for itemset p and $Coefvar(p) = std(Per(p))/mean(Per(p))$, $std(*)$ and $mean(*)$ are the standard deviation and mean, respectively.*

Certainly, there are many other definitions that are based on different periodicity measures, too many to mention here, see related works for more.

2.2   Related Works

Frequent Pattern Mining: Agrawal et al. [9] proposed the first algorithm for frequent itemset pattern mining and named it Apriori, which adopts a breadth-first strategy to explore the search space. Thereafter, some efficient algorithms were proposed, such as H-Mine [10], FP-Growth [8] and LCM [11], those algorithms apply a *pattern-growth* approach to reduce the times of scanning the database.

The FP-Growth algorithm, proposed by Han et al. [8], compresses the database in a tree structure called FP-Tree, which consists of a header table and a prefix-tree. The algorithm finds out all frequent items in the first scan of the database, and constructs FP-Tree in the second scan, then mines frequent itemset patterns from FP-Tree. FP-Growth is efficient since it only scans the database two times, and needs not to generate any candidates.

Periodic Pattern Mining: Tanbeer et al. [2] employed the maximum period as the periodicity measure to define periodic patterns, and then proposed a pattern growth-based mining algorithm PFP-growth (Periodic Frequent Pattern-growth); Their algorithm was implemented by a highly compact tree structure, called PF-tree (Periodic Frequent pattern-tree), which is extended from the FP-Tree. Rashid et al. [6] adopted the variance of periods, which they referred to as patterns' regularity, as the periodicity measure, and used a new type of data structure named RF-tree (Regularly Frequent Pattern tree) to mine the regularly frequent patterns by a *pattern-growth* approach. Kiran et al. [13] presented a new kind of periodic-frequent patterns, which is extended from [2], to discover the patterns that have exhibited partial periodic behaviour in a dataset; In their later work [14], they proposed a new concept known as *local-periodicity* to achieve two pruning techniques, and their mining algorithm, called PFP-growth++, was implemented based on the PFP-growth algorithm by introducing the two pruning techniques. Fournier-Vigera et al. [15] introduced the minimum periodicity, the maximum periodicity and the average periodicity of the periods as periodicity measures. In addition, two efficient algorithms were proposed in their work [16] to discover periodic patterns common to multiple sequences, where periodic patterns were defined by two novel measures: the standard deviation of periods and the sequence periodic ratio; in their later work [4], they presented a novel type of periodic patterns called Local Periodic Patterns (LPPs), which are patterns that have a periodic behavior in some non-predefined time-intervals, and, by extending the Apriori-Tid [9], Eclat [12] and FP-Growth [8] algorithms, three mining algorithms named LPPM$_{breadth}$, LPPM$_{depth}$ and LPP-Growth were proposed to discover the complete set of local periodic patterns; their work [17] proposed a novel function for stability measure to identify stable periodic patterns and proposed an algorithm to discover the complete set of top-k stable periodic itemsets. Nofong [18] introduced an efficient solution to calculate the periodicity; the solution evaluates the periodicity of patterns directly from the occurrence list without calculating the period list. Chen et al. [7] employed coefficient of variation as the periodicity measure to define periodic patterns.

3   New Concept and Problem Statement

In the definition of skyline periodic itemset patterns, we follow Chen et al. [7] to adopt the coefficient of variation as the periodicity measure, and comply with Definition 8 to define periodic itemset patterns.

Definition 9 (Domination). *Assume two itemset p and p' in B. We say p' dominates p, or, p is dominated by p', iff $Freq(p) < Freq(p')$ and $CoefVar(p) \geq CoefVar(p')$, or, $Freq(p) \leq Freq(p')$ and $CoefVar(p) > CoefVar(p')$.*

Definition 10 (Skyline Periodic Itemset Pattern). *Given a transaction sequence $B = \langle b_1, ..., b_n \rangle$, itemset p, threshold θ for frequency, and threshold δ for coefficient of variation. If $Freq(p) \geq \theta$, $Coefvar(p) \leq \delta$, and it is not dominated by any other itemsets in B, then p is a skyline periodic itemset pattern.*

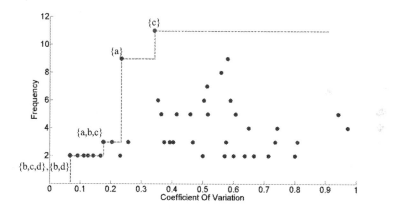

Fig. 1. Diagram of skyline periodic itemset patterns.

Problem Statement: Assume a transaction sequence $B = \langle b_1, ..., b_n \rangle$, frequency threshold θ, and threshold δ for coefficient of variation. The aim of skyline periodic itemset pattern mining is to find out all skyline periodic itemset patterns in transaction sequence B.

Example 1. For the given transaction sequence in Table 1, $\{c\}$, $\{a\}$, $\{a, b, c\}$, $\{b, d\}$ and $\{b, c, d\}$ are skyline periodic itemset patterns when we set $\theta = 2$ and $\delta = 1.0$, where $\{b, d\}$ and $\{b, c, d\}$ have the same frequency and coefficient of variation, as shown in Fig. 1

4   The Proposed Method

Most of the classical frequent itemset pattern mining algorithms use the anti-monotonicity of frequency to prune the search space, and such pruning strategy based on frequency constraint is also applicable to periodic itemset pattern mining algorithms. However, Lemma 1 shows that coefficient of variation does not have the property of monotonicity or anti-monotonicity.

Lemma 1. *Let p' be a subset of p, that is $p' \subseteq p$, then we have $CoefVar(p') < CoefVar(p)$, or $CoefVar(p') > CoefVar(p)$, or $CoefVar(p') = CoefVar(p)$.*

Proof. For the transaction sequence B given in Table 1, the occurrence list of itemset $\{f\}$ is $Occur(\{f\}) = \{1, 3, 4, 5, 8, 11, 12, 16\}$ and the coefficient of variation is $CoefVar(\{f\}) = 0.52$. However, (1) the occurrence list of itemset $\{f, h\}$ is $Occur(\{f, h\}) = \{3, 4, 11, 12\}$ and the coefficient of variation is $CoefVar(\{f, h\}) = 0.94$, so, we have $\{f\} \subset \{f, h\}$ and $CoefVar(\{f\}) < CoefVar(\{f, h\})$; (2) the occurrence list of itemset $\{c, f\}$ is $Occur(\{c, f\}) = \{1, 4, 8, 12\}$ and the coefficient of variation is $CoefVar(\{c, f\}) = 0.13$, so, we have $\{f\} \subset \{c, f\}$ and $CoefVar(\{f\}) > CoefVar(\{c, f\})$; (3) obviously, if $Occur(p') = Occur(p)$, we have $CoefVar(p') = CoefVar(p)$.

That makes it difficult to implement the pruning strategies based on the constraint for coefficient of variation, so we do not consider such kind of pruning strategies in our SLPIM algorithm. SLPIM first uses FP-Growth to mine all frequent itemset patterns, and then identifies which frequent itemset patterns are skyline periodic itemset patterns by an effective strategy.

4.1   Data Structure

Since the FP-Growth algorithm does not preserve any information about the occurrences of itemsets, so, we need additional data structure to preserve the occurrence list for the calculation of the period list. In the implementation of SLPIM, each frequent item i, $\forall i \in I$ s.t. $Freq(i) \geq \theta$, is assigned a bit vector $BitVec(i)$, which has a length of $|B|$. If the kth bit in the bit vector of item i is flipped, then it means that the kth transaction of B contains item i, as shown in Fig. 2(a) and (b). The bit vector of itemset p is obtained by bitwise AND for each bit vector of all items in p, that is, $BitVec(p) = \&_{i \in p} BitVec(i)$. The indices of all the flipped bit in $BitVec(p)$ form the occurrence list of p, as shown in Fig. 2(c).

4.2   Efficient Calculation for Coefficient of Variation

For a given itemset p, when we calculate $CoefVar(p)$, according to the definition of mean and standard deviation, we need to first calculate the period list of p, then calculate $mean(Per(p))$ and $std(Per(p))$. However, inspired by Nofong [18], we can calculate $mean(Per(p))$ and $std(Per(p))$ directly from the occurrence list of p without calculating $Per(p)$ based on the following two lemmas.

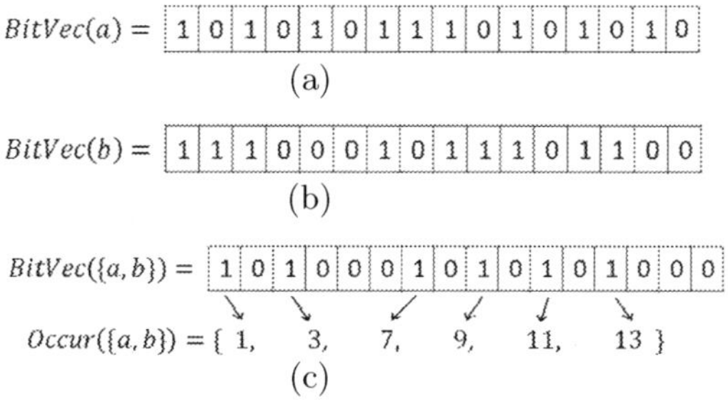

$BitVec(a) =$ | 1 | 0 | 1 | 0 | 1 | 0 | 1 | 1 | 1 | 0 | 1 | 0 | 1 | 0 | 1 | 0 |

(a)

$BitVec(b) =$ | 1 | 1 | 1 | 0 | 0 | 0 | 1 | 0 | 1 | 1 | 1 | 0 | 1 | 1 | 0 | 0 |

(b)

$BitVec(\{a, b\}) =$ | 1 | 0 | 1 | 0 | 0 | 0 | 1 | 0 | 1 | 0 | 1 | 0 | 1 | 0 | 0 | 0 |

$Occur(\{a, b\}) = \{ 1, \quad 3, \quad 7, \quad 9, \quad 11, \quad 13 \}$

(c)

Fig. 2. Diagram of bit vectors. (a), (b) and (c) show the bit vectors of item a, item b and itemset $\{a, b\}$ respectively corresponding to the transaction sequence given in Table 1.

Lemma 2. *Given a transaction sequence* $B = \langle b_1, ..., b_n \rangle$, *and frequent itemset* p *of* B, $Occur(p) = \{w_1, w_2, ..., w_r\}$. *The mean of* $Per(p)$ *can be expressed as* $mean(Per(p)) = (w_r - w_1)/(r - 1)$.

Proof. According to the definition of mean, we have

$$mean(Per(p)) = \frac{per_1 + per_2 + \cdots + per_{r-1}}{|Per(p)|} \tag{1}$$

Since $per_1 = w_2 - w_1$, $per_2 = w_3 - w_2$, $\cdots$, $per_{r-1} = w_r - w_{r-1}$, and $|Per(p)| = |Freq(p)| - 1 = r - 1$, we have

$$mean(Per(p)) = \frac{(w_2 - w_1) + (w_3 - w_2) + \cdots + (w_r - w_{r-1})}{r - 1} \tag{2}$$

This equation is simplified as $mean(Per(p)) = (w_r - w_1)/(r - 1)$.

Lemma 3. *Given a transaction sequence* $B = \langle b_1, ..., b_n \rangle$, *and frequent itemset* p *of* B, $Occur(p) = \{w_1, w_2, ..., w_r\}$. *For convenience, let* $\overline{per} = mean(Per(p))$. *The standard deviation of* $Per(p)$ *can be calculated by*

$$std(Per(p)) = \sqrt{\frac{\sum_{k=2}^{r}(w_k - w_{k-1})^2 + (w_1 - w_r)\overline{per}}{r - 1}} \tag{3}$$

Proof. Let $var(Per(p))$ denotes the variance of $Per(p)$. According to the definition of variance, we have

$$var(Per(p)) = \frac{(per_1 - \overline{per})^2 + (per_2 - \overline{per})^2 + \cdots + (per_{r-1} - \overline{per})^2}{|Per(p)|}$$
$$= \frac{((w_2 - w_1) - \overline{per})^2 + ((w_3 - w_2) - \overline{per})^2 + \cdots + ((w_r - w_{r-1}) - \overline{per})^2}{|Per(p)|} \tag{4}$$

Expanding the expressions in the numerator gives:
$((w_2 - w_1) - \overline{per})^2 = (w_2 - w_1)^2 - 2w_2\overline{per} + 2w_1\overline{per} + \overline{per}^2$.

$((w_3 - w_2) - \overline{per})^2 = (w_3 - w_2)^2 - 2w_3\overline{per} + 2w_2\overline{per} + \overline{per}^2.$

...

$((w_r - w_{r-1}) - \overline{per})^2 = (w_r - w_{r-1})^2 - 2w_r\overline{per} + 2w_{r-1}\overline{per} + \overline{per}^2.$

Summing the above expansion, as a result, the numerator of formula (4) is equal to:

$((w_2 - w_1) - \overline{per})^2 + ((w_3 - w_2) - \overline{per})^2 + \cdots + ((w_r - w_{r-1}) - \overline{per})^2$
$= \sum_{k=2}^{r}(w_k - w_{k-1})^2 - 2w_r\overline{per} + 2w_1\overline{per} + (r - 1)\overline{per}^2$
$= \sum_{k=2}^{r}(w_k - w_{k-1})^2 + (2w_1 - 2w_r + (r - 1)\overline{per})\overline{per}$
$= \sum_{k=2}^{r}(w_k - w_{k-1})^2 + (w_1 - w_r)\overline{per}$, since $\overline{per} = (w_r - w_1)/(r - 1)$.

So, the variance of $Per(p)$, $var(Per(p)) = (\sum_{k=2}^{r}(w_k - w_{k-1})^2 + (w_1 - w_r)\overline{per})/(r - 1)$, and

$$std(Per(p)) = \sqrt{\frac{\sum_{k=2}^{r}(w_k - w_{k-1})^2 + (w_1 - w_r)\overline{per}}{r - 1}} \tag{5}$$

Our expressions in Lemma 2 and Lemma 3 are different from that of Nofong [18], since we define the periodicity measure by $Per(p)$ instead of $ExtPer(p)$.

4.3 Implementation of the SLPIM Algorithm

Algorithm 1 shows the pseudo-code of SLPIM. $PipSet$ and $SlpipSet$ are used for preserving the set of periodic itemset patterns and the set of skyline periodic itemset patterns that have been found. The variables $MaxFreq$ and $MinCoefvar$ keep the maximum frequency and the minimum coefficient of variation of the skyline periodic itemset patterns in $SlpipSet$, respectively. The algorithm first mines all frequent itemsets by FP-Growth and preserves the results in $FpSet$ (Line 3). Then creates bit vector $BitVec(i)$ for each frequent item, where $i \in I$ and $Freq(i) \geq \theta$ (Line 4). For each frequent itemset $fp \in FpSet$, the algorithm needs to judge whether fp is a skyline periodic itemset pattern (Line 5–35). The judgment is as follows: First, calculates the bit vector $BitVec(fp)$, and gets the indices of all non-zero bits in $BitVec(fp)$ to form the occurrence list $Occur(fp)$. Then, calculates the coefficient of variation $CoefVar(fp)$, where $mean(Per(p))$ and $std(Per(p))$ are calculated according to Lemma 2 and Lemma 3. If $CoefVar(fp)$ is not greater than δ, then fp is a periodic itemset pattern (Line 12–15), and there will be four situations. Situation one: If $MaxFreq < Freq(fp)$ and $MinCoefvar > CoefVar(fp)$, that means all the itemsets in $SlpipSet$ are dominated by fp, then $MaxFreq$ and $MinCoefvar$ are replaced by $Freq(fp)$ and $CoefVar(fp)$ respectively, $SlpipSet$ is emptied, and fp is pushed in $SlpipSet$ (Line 16–19). Situation two: If $MaxFreq < Freq(fp)$ and $MinCoefvar \leq CoefVar(fp)$, that means fp can not be dominated by any itemsets in $SlpipSet$ and some itemsets in $SlpipSet$ can be dominated by fp, then $MaxFreq$ is replaced by $Freq(fp)$, the itemsets which are dominated by fp are deleted from $SlpipSet$ by Algorithm 2, and fp is pushed in $SlpipSet$ (Line 20–23). Situation three: If $MaxFreq \geq Freq(fp)$ and $MinCoefvar > CoefVar(fp)$, this situation is the same as Situation two, but $MinCoefvar$ is

---

Algorithm 1: SLPIM(B, θ, δ)

---

1 $PipSet = Null$, $SlpipSet = Null$;
2 $MaxFreq = 0$, $MinCoefvar = +\infty$;
3 $FpSet \leftarrow FP\text{-}Growth(B, \theta)$;
4 Creates bit vector $BitVec(i)$ for each frequent item, $i \in I$ and $Freq(i) \geq \theta$;
5 **for** *all* $fp \in FpSet$ **do**
6   Creates bit vector $BitVec(fp)$ and sets all its components to 1;
7   **for** *all* $i \in fp$ **do**
8           $BitVec(fp) = BitVec(fp) \ \& \ BitVec(i)$;
9   **end**
10      $Occur(fp) \leftarrow$ gets the indices of all non-zero bits in $BitVec(fp)$;
11      $CoefVar(fp) \leftarrow$ calculates the coefficient of variation;
12   **if** $CoefVar(fp) > \delta$ **then**
13   *continue*;
14   **end**
15      $PipSet = PipSet \cup \{fp\}$;
16   **if** $MaxFreq < Freq(fp) \wedge MinCoefvar > CoefVar(fp)$ **then**
17          $MaxFreq = Freq(fp)$, $MinCoefvar = CoefVar(fp)$;
18   Removes all elements from $SlpipSet$;
19          $SlpipSet = SlpipSet \cup \{fp\}$;
20   **else if** $MaxFreq < Freq(fp)$ **then**
21          $MaxFreq = Freq(fp)$;
22          $SlpipSet =$ RemoveNonSLP($SlpipSet, fp$);
23          $SlpipSet = SlpipSet \cup \{fp\}$;
24   **else if** $MinCoefvar > CoefVar(fp)$ **then**
25          $MinCoefvar = CoefVar(fp)$;
26          $SlpipSet =$ RemoveNonSLP($SlpipSet, fp$);
27          $SlpipSet = SlpipSet \cup \{fp\}$;
28   **else**
29   **if** $\exists slp \in SlpipSet$ *such that* slp *dominates* fp **then**
30   *continue*;
31   **end**
32          $SlpipSet =$ RemoveNonSLP($SlpipSet, fp$);
33          $SlpipSet = SlpipSet \cup \{fp\}$;
34   **end**
35 **end**
36 **return** $PipSet$, $SlpipSet$;

Algorithm 2: RemoveNonSLP($SlpipSet, fp$)

---

1 **for** *all* $slp \in SlpipSet$ **do**
2   **if** slp *is dominated by* fp **then**
3   Removes slp from $SlpipSet$;
4   **end**
5 **end**
6 **return** $SlpipSet$;

---

replaced by $CoefVar(fp)$ instead of $MaxFreq$ is replaced by $Freq(fp)$. Situation four: If $MaxFreq \geq Freq(fp)$ and $MinCoefvar \leq CoefVar(fp)$, that means fp can be dominated by some itemsets in $SlpipSet$ and vice versa. In this situation, if fp is dominated by any itemsets in $SlpipSet$, then the judgment of fp is over since fp is not a skyline periodic itemset pattern (Line 29–31), otherwise, the itemsets which are dominated by fp are deleted from $SlpipSet$ by Algorithm 2, and fp is pushed in $SlpipSet$ (Line 32–33).

4.4   Complexity of SLPIM

SLPIM is implemented in two stages: mining stage and judgment stage. In the mining stage, the complexity is determined by FP-Growth.

In the judgment stage, each frequent itemset in $FpSet$ is judged whether it is a skyline periodic itemset pattern. For a frequent itemset fp, $fp \in FpSet$, the calculation of $CoefVar(fp)$ is $O(n)$ in the worst case by exploiting Lemma 2 and Lemma 3, where $n = |B|$, since r is bounded by n. Let N be the number of frequent items of B, then the length of frequent itemset fp has a upper bound of N, that is, $\forall fp \in FpSet$, $|fp| \leq N$, the number of frequent itemsets in $FpSet$ has a upper bound of 2^N, that is, $|FpSet| \leq 2^N$, and $SlpipSet$ is the same, $|SlpipSet| \leq 2^N$. So, in the worst case, the time complexity is $O(2^N * (2^N + N + n))$, and the space complexity is $O(2 * N * n)$, where N depends on the value of θ, a lower value of θ results in a larger number of frequent items.

5   Experiment

SLPIM is the first algorithm devoting to the skyline periodic itemset pattern mining problem. Its performance is not compared with other algorithms since there is currently no method proposed in the literature for mining such a kind of patterns. To evaluate the performance of the proposed SLPIM algorithm, we compared its performance with FP-Growth. The FP-Growth algorithm was chosen for comparisons as the SLPIM algorithm is based on FP-Growth.

The experiments were conducted on a computer with an Intel Core I7-8550U 1.8 GHz processor and 8 GB of RAM, running Windows 10 (64-bit version). SLPIM is implemented in Java.

Table 2. The characteristics of the running datasets.

Dataset	Transaction count	Item count	Ave. item count per trans	Density
Foodmart	4,141	1,559	4.42	0.28%
OnlineRetail	541,909	2,603	4.37	0.17%
C20d10k	10,000	192	20	10.42%
T20i6d100k	99,922	893	19.9	2.23%

5.1   Datasets

Two real-world transactional datasets, Foodmart and OnlineRetail, and two synthetic transactional datasets, C20d10k and T20i6d100k, are selected for the evaluation of SLPIM. They are regarded as transaction sequences by the default order. Table 2 shows the characteristics of these datasets, where column 4 is the average item count per transaction. All the datasets are downloaded from the website of SPMF[1].

5.2   Number of Patterns

Table 3 shows the number of periodic itemset patterns (denoted by $\#PipSet$) and the number of skyline periodic itemset patterns (denoted by $\#SlpipSet$) under different values of θ when δ is set to 1.0, 2.0 and infinity, respectively. Actually, the number of periodic itemset patterns is the same as that of frequent itemset patterns under the same values of θ when δ is set to infinity.

From the table we can see that, for a given value of δ, when θ rises, both the total number of periodic itemset patterns and skyline periodic itemset patterns decrease in all cases of the four datasets. And similarly, for a given value of θ, when δ rises, we get the opposite conclusion. The number of skyline periodic itemset patterns is far more less than that of periodic itemset patterns when θ has a lower value. The number of patterns of the case when $\delta = 2.0$ is very close to that of the case when $\delta = +\infty$, that means few patterns have a coefficient of variation greater than 2.0, except the C20d10k dataset.

5.3   Running Time

Figure 3 shows the comparisons of the running time of SLPIM and FP-Growth under different values of θ when δ is set to infinity. Since SLPIM needs to identify the skyline periodic itemset patterns from all the frequent itemset patterns mined by FP-Growth, it is naturally that SLPIM has longer running time than FP-Growth. From the figures we can see that, when θ has a lower value, the running time difference between SLPIM and FP-Growth is very large, especially in the cases of OnlineRetail and C20d10k since they have the greatest number of frequent itemset patterns. When θ rises, the running time of both the methods decrease in all cases of the four datasets, and the gap of running time between them shrinks as the number of frequent itemset patterns decreases.

5.4   Memory Consumption

Table 4 shows the comparisons of memory consumptions under different values of θ when δ is set to infinity. In the implementation of the SLPIM algorithm, additional data structures are required to preserve the occurrence information

---

[1] http://www.philippe-fournier-viger.com/spmf/index.php?link=datasets.php.

Table 3. Number of patterns.

		$\theta =$	0.0005	0.0015	0.0025	0.0035	0.0045	0.0055	0.0065	
Foodmart	$\delta = 1.0$	#PipSet	1295	1140	754	232	34	3	0	
		#SlpipSet	8	7	6	6	4	1	0	
	$\delta = 2.0$	#PipSet	1644	1477	980	318	48	4	0	
		#SlpipSet	9	8	7	7	5	2	0	
	$\delta = +\infty$	#PipSet	1644	1477	980	318	48	4	0	
		#SlpipSet	9	8	7	7	5	2	0	
		$\theta =$	0.001	0.003	0.005	0.007	0.009	0.011	0.013	0.015
OnlineRetail	$\delta = 1.0$	#PipSet	565	0	0	0	0	0	0	0
		#SlpipSet	24	0	0	0	0	0	0	0
	$\delta = 2.0$	#PipSet	4925	710	350	223	147	112	79	63
		#SlpipSet	44	20	12	7	7	7	6	6
	$\delta = +\infty$	#PipSet	5692	808	396	244	164	126	90	74
		#SlpipSet	44	20	12	7	7	7	6	6
		$\theta =$	0.3	0.4	0.5	0.6	0.7	0.8	0.9	1.0
C20d10k	$\delta = 1.0$	#PipSet	127	127	127	127	127	127	63	0
		#SlpipSet	3	3	3	3	3	3	3	0
	$\delta = 2.0$	#PipSet	2199	383	255	255	191	127	63	0
		#SlpipSet	3	3	3	3	3	3	3	0
	$\delta = +\infty$	#PipSet	5303	2175	1823	959	447	127	63	0
		#SlpipSet	3	3	3	3	3	3	3	0
		$\theta =$	0.015	0.025	0.035	0.045	0.055	0.065	0.075	0.085
T20i6d100k	$\delta = 1.0$	#PipSet	0	0	0	0	0	0	0	0
		#SlpipSet	0	0	0	0	0	0	0	0
	$\delta = 2.0$	#PipSet	421	231	157	97	62	31	23	12
		#SlpipSet	9	8	5	5	4	4	3	2
	$\delta = +\infty$	#PipSet	535	289	193	119	75	39	27	16
		#SlpipSet	9	8	5	5	4	4	3	2

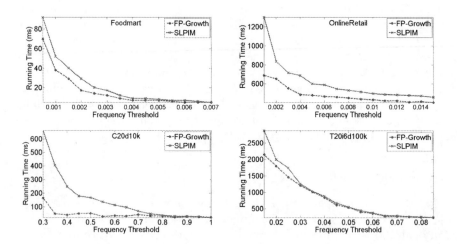

Fig. 3. Running time under different values of θ.

for each frequent item. So, it is naturally that SLPIM has higher memory consumptions. For all cases of the four datasets, when θ rises, the memory consumptions of both SLPIM and FP-Growth reduce as the number of frequent itemset patterns decreases, and the gap of memory consumptions between SLPIM and FP-Growth shrinks as the number of frequent items decreases.

Table 4. Memory consumption (MB).

	$\theta =$	0.0005	0.0015	0.0025	0.0035	0.0045	0.0055	0.0065	
Foodmart	FP-Growth	14.3	13.6	11.1	7.8	7.2	6.5	6.5	
	SLPIM	16.9	16.3	13.0	8.5	7.2	7.2	7.2	
	$\theta =$	0.002	0.004	0.006	0.008	0.01	0.012	0.014	
OnlineRetail	FP-Growth	110.5	105.2	105.1	105.0	105.0	99.8	99.7	
	SLPIM	282.1	250.2	202.5	176.2	164.7	148.3	141.4	
	$\theta =$	0.3	0.4	0.5	0.6	0.7	0.8	0.9	1.0
C20d10k	FP-Growth	15.2	13.1	12.4	11.9	11.8	11.2	11.1	10.5
	SLPIM	33.7	31.2	28.8	27.9	26.6	19.4	17.7	11.9
	$\theta =$	0.015	0.025	0.035	0.045	0.055	0.065	0.075	0.085
T20i6d100k	FP-Growth	329.2	288.3	240.9	201.6	70.4	47.8	21.0	9.4
	SLPIM	370.9	368.8	286.9	229.5	85.9	59.0	29.9	15.0

6 Conclusion

In this paper, we first present the new concept of skyline periodic itemset patterns. On the one hand, skyline periodic itemset patterns have optimal frequency or optimal periodicity or both, so it can meet some special needs of the users. On the other hand, since the number of skyline periodic itemset patterns is far less than that of periodic itemset patterns, our new concept frees the users from the overwhelming great number of patterns. Secondly, we propose a new algorithm named SLPIM for skyline periodic itemset pattern mining. The SLPIM algorithm uses FP-Growth to mine all frequent itemset patterns and then identifies all the skyline periodic itemset patterns from the set of frequent itemset patterns by an effective strategy. The experimentation shows that our SLPIM algorithm is competent for skyline periodic itemset pattern mining.

Acknowledgements. This work was supported in part by the National Natural Science Foundation of China under Grant 61672261 and Grant 62276060, and in part by the Industrial Technology Research and Development Project of Jilin Development and Reform Commission under Grant 2019C053-9.

References

1. Uday Kiran, R., Toyoda, M., Zettsu, K.: Real-world applications of periodic patterns. In: Kiran, R.U., Fournier-Viger, P., Luna, J.M., Lin, J.C.-W., Mondal, A. (eds.) Periodic Pattern Mining, pp. 229–235. Springer, Singapore (2021). https://doi.org/10.1007/978-981-16-3964-7_13

2. Tanbeer, S.K., Ahmed, C.F., Jeong, B.S., Lee, Y.: Discovering periodic-frequent patterns in transactional databases. In: PAKDD, pp. 242–253 (2009)
3. Kiran, R.U., Saideep, C., Zettsu, K., Toyoda, M., Kitsuregawa, M., Reddy, P.K.: Discovering partial periodic spatial patterns in spatiotemporal databases. In: IEEE International Conference on Big Data, pp. 233–238 (2019)
4. Fournier-Viger, P., Yang, P., Kiran, R.U., Ventura, S., Luna, J.M.: Mining local periodic patterns in a discrete sequence. Inf. Sci. **544**, 519–548 (2021)
5. Chen, G., Li, Z.: Discovering periodic cluster patterns in event sequence databases. Appl. Intell. **52**(13), 15387–15404 (2022)
6. Rashid, M.M., Karim, M.R., Jeong, B.S., Choi, H.J.: Efficient mining regularly frequent patterns in transactional databases. In: DASFAA, no. 1, pp. 258–271 (2012)
7. Chen, G., Li, Z.: A new method combining pattern prediction and preference prediction for next basket recommendation. Entropy **23**(11), 1430 (2021)
8. Han, J., Pei, J., Yin, Y., Mao, R.: Mining frequent patterns without candidate generation: a frequent-pattern tree approach. Data Min. Knowl. Discov. **8**(1), 53–87 (2004)
9. Agrawal, R., Imielinski, T., Swami, A.N.: Mining association rules between sets of items in large databases. ACM SIGMOD Rec. **22**(2), 207–216 (1993)
10. Pei, J., Han, J., Lu, H., Nishio, S., Tang, S., Yang, D.: H-mine: fast and space-preserving frequent pattern mining in large databases. IIE Trans. **39**(6), 593–605 (2007)
11. Minato S., Uno T., Arimura H.: LCM over ZBDDs: fast generation of very large-scale frequent itemsets using a compact graph-based representation. In: PAKDD, pp. 234–246 (2008)
12. Zaki, M.J.: Scalable algorithms for association mining. IEEE Trans. Knowl. Data Eng. **12**(3), 372–390 (2000)
13. Kiran, R.U., Reddy, P.K.: An alternative interestingness measure for mining periodic-frequent patterns. In: DASFAA, no. 1, pp. 183–192 (2011)
14. Kiran, R.U., Kitsuregawa, M., Reddy, P.K.: Efficient discovery of periodic-frequent patterns in very large databases. J. Syst. Softw. **112**, 110–121 (2016)
15. Fournier-Viger, P., Lin, C.W., Duong, Q.H., Dam, T.L., Voznak, M.: PFPM: discovering periodic frequent patterns with novel periodicity measures. In: Proceedings of the 2nd Czech-China Scientific Conference 2016 (2016)
16. Fournier-Viger, P., Li, Z., Lin, C.W., Kiran, R.U., Fujita, H.: Efficient algorithms to identify periodic patterns in multiple sequences. Inf. Sci. **489**, 205–226 (2019)
17. Fournier-Viger, P., Wang, Y., Yang, P., Lin, C.W., Yun, U., Kiran, R.U.: TSPIN: mining top-k stable periodic patterns. Appl. Intell. **52**(6), 6917–6938 (2022)
18. Nofong, V.M., Wondoh, J.: Towards fast and memory efficient discovery of periodic frequent patterns. J. Inf. Telecommun. **3**(4), 480–493 (2019)

Double-Optimized CS-BP Anomaly Prediction for Control Operation Data

Ming Wan$^{(\boxtimes)}$, Xueqing Liu, and Yang Li

School of Information, Liaoning University, Shenyang 110036, China
wanming@lnu.edu.cn

Abstract. Automation control, which is one functional core of industrial control system, has become the prime attack target due to its vulnerabilities. Furthermore, many industrial cyber threats can disturb or destroy the correctness of control operation data to cause industrial accidents, when one normal production process is running smoothly and orderly. In order to effectively identify abnormal activities in various control operation data, this paper proposes one BP (Back Propagation) neural network anomaly prediction model based on the double-optimized CS (Cuckoo Search) algorithm. By using the exponential decline strategy and Gaussian perturbation to improve the traditional CS algorithm, this model can obtain one effective anomaly prediction engine based on the optimized BP neural network: for one thing, it can quickly enter the local search through the exponential decline strategy; for another, the information exchange between all local positions and the global optimal positions is realized by Gaussian perturbation. Moreover, the double-optimized CS algorithm not only solves the problem that the traditional BP neural network is prone to fall into local optimal solution, but also eliminates the defect of low vitality in the traditional CS algorithm. Consequently, this model can realize the high-precision prediction of abnormal control operation data. The experimental results show that, compared with other approaches, this model has better prediction performance under both normal and attack states, and can ensure the security of automation control in industrial production.

Keywords: Exponential Decline Strategy · Gaussian Perturbation · CS-BP Prediction · Control Operation Data

1 Introduction

Industrial control systems, such as SCADA (Supervisory Control and Data Acquisition) and DCS (Distributed Control System), are regarded as an integrated platform of all computing devices, networks and control components used for automation control and operation in industrial production processes [1]. However, as one critical infrastructure related to the national economy and people's livelihood, various industrial control systems are facing increasingly severe security risks with the rapid development of ICT (Information Communication Technology) [2, 3]. In recent years, the number of security vulnerabilities in industrial control systems has generally shown an upward trend, and

© The Author(s), under exclusive license to Springer Nature Switzerland AG 2023
X. Yang et al. (Eds.): ADMA 2023, LNAI 14176, pp. 509–523, 2023.
https://doi.org/10.1007/978-3-031-46661-8_34

all kinds of industrial security incidents occur frequently. Especially, APTs (Advanced Persistent Threats), which always launch a multi-step attack by exploiting zero-day vulnerabilities, have become a major attack mode which cannot be ignored in current industrial control systems [4].

The massive emergence of information security vulnerabilities undoubtedly increases many security risks to industrial control systems, which can cause some malicious attacks to destroy normal industrial production activities. Among them, the threats to disturb control operation modes have been considered as one destructive attack type, for example, malicious attackers can falsify or forge control operation data to directly change the automatic control logic through some multi-step and targeted attacks. In practice, the main task of industrial control systems is to complete the automatic control and data acquisition according to some specific technology flows, which always follows a systematic process of continuous production and manufacturing. That is, the control operation data can present some periodic and sequential characteristics to a certain extent, and the above threats may break this regularity when malicious attackers falsify or forge control operation data [5]. From this perspective, anomaly detection and prediction of control operation data has become one area of special focus and research in the field of industrial cyber security, and researchers have achieved some periodic achievements in different ways. Koay et al. provide insights on current advancements of machine learning in industrial control system security, and discuss the performance of some ML-based prediction approaches by comparing different operational datasets [6]. By combining Bloom filter-based classifier and instance-based classifier, Khan et al. propose a hybrid-multilevel anomaly intrusion approach to predict the anomalies from the regular operation data in the gas control network, and more demonstrative feature mining can be further embraced to enhance its detection performance [7]. Wan et al. analyze the weighted function code correlation from industrial control operations, and propose an improved ABC-SVM anomaly detection model to identify abnormal behaviors in industrial control communications [8]. Although this model can intuitively explain the importance and correlation between different control operations, its detection accuracy still needs to be enhanced. Hannon et al. use the probabilistic predictions to principally define an efficient anomaly detection tool for the PMU (Phasor Measurement Unit) data, which can provide fast-sampled operational data to aid control and decision-making in the electric grid [9].

In this paper, we focus on the change of function codes in industrial Modbus/TCP communication, and predict the attack possibility and severity of control operation data under the normal technology flow. Moreover, we propose a BP (Back Propagation) neural network anomaly prediction model based on the double-optimized CS (Cuckoo Search) algorithm, and realize the high-precision prediction of abnormal control operation data. Different from the traditional CS algorithm, this model further improves its slow convergence speed and insufficient vitality in the later stage. On the one hand, the exponential decline strategy is adopted in the speed update formula to quickly enter local optimization, and increases the searching accuracy for the optimal solution; on the other hand, Gaussian perturbation is introduced into the generation process of new local solutions, and it not only improves the convergence speed, but also prevents the search process from falling into local optimal solution by realizing the information exchange between all local optimal solutions and the global optimal solution. In order to evaluate this prediction model, we build a small-scale control system based on Modbus/TCP communication protocol, which simulates the partial control operation process of material synthesis. The experimental results show that, compared with other approaches, our model has stronger prediction performance under both normal and attack states, and effectively detects anomalies in industrial control operations: for one thing, this model has relatively higher prediction accuracy and fewer false positives under normal industrial control operation processes; for another, this model also maintains a high level prediction accuracy, recall and F1 score under the single attack (forging, falsifying or discarding) or multi-step fusion attack.

2 Control Operation Data Preprocessing

In Modbus/TCP communication protocol, each function code can represent a control operation [10], for example, function codes 01 and 06 can respectively represent the control operations "read coil" and "write single hold register". Generally, industrial control systems based on Modbus/TCP use a series of function codes to execute one specific control operation process, and multiple control operation processes are orderly executed to complete the entire technology flow. In the above process, malicious attackers can interfere with the regular control operation by forging, falsifying or discarding function codes, and achieve the goal to destroy normal industrial production activities [11]. From this point, we firstly perform the feature extraction and normalization processing on Modbus/TCP communication data, and obtain the control operation feature for the prediction model. The data preprocessing process is listed as follows:

Step 1: construct the control operation data sequence.

we extract all function codes and relative time from Modbus/TCP communication data, and divide them into m control operation data sequences according to the fixed time interval. Here, we suppose the i - th control operation data sequence as $A_i = a_1^i a_2^i \cdot \cdot \cdot a_n^i, \forall i \in [1, m]$, and all m sequences can form one control operation sequence set $A = \{A_1, A_2, A_3, \cdot \cdot \cdot, A_m\}$.

Step 2: use Box-Cox transformation for data normalization processing.

The Box-Cox transformation belongs to a generalized power transformation method, which is regarded as one commonly used data transformation in current statistical modeling. Furthermore, this method can reduce unobservable errors and the correlation of predicted data to some extent, and its general transformation form is shown in Formula (1). we perform the Box-Cox transformation for all control operation data sequences, and accordingly obtain the control operation feature sequence set $Y = \{Y_1, Y_2, Y_3, \cdot \cdot \cdot, Y_m\}$.

$$Y^{(\lambda)} = \begin{cases} \dfrac{A^\lambda - 1}{\lambda}, & \lambda \neq 0. \\ \ln A, & \lambda = 0. \end{cases} \tag{1}$$

Here, λ is a transformation parameter.

3 Double-Optimized CS-BP Anomaly Prediction Model

Based on the obtained control operation feature sequences, the double-optimized CS-BP anomaly prediction model can train BP neural network as one feasible anomaly prediction engine. However, the traditional BP neural network cannot be directly applied due to its uncertain parameter setting, which may result in its performance degradation. In order to build one fine prediction engine, the parameter optimization should be a key point during the designing of our model. At this stage, many researchers have carried out a whole range of researches on the optimization algorithms [12–14], and we select the CS algorithm as one simple and efficient way to optimize the main parameters of BP neural network. In order to resolve the problems of slow convergence speed and insufficient vitality in the later stage, we further present one double-optimized CS algorithm by using the exponential decline strategy and Gaussian perturbation, which not only significantly enhances the iterative vitality of CS algorithm, but also prevents the prediction model from falling into local optimization.

3.1 BP Neural Network Prediction Engine

Firstly, the phase space is reconstructed by using the delay coordinate method, and each control operation feature sequence Y_i is converted into one n-dimensional state vector. The reconstruction formula is as follows:

$$T(i) = \left\{ Y_{i-(n-1)\tau}, \cdots, Y_{i-\tau}, Y_i \right\} \tag{2}$$

Here, τ is the network time delay; n is the embedded dimension after reconstruction; $T(i)$ represent the reconstruction results.

From the above formula, we can see that the results of phase space reconstruction are directly related to different τ and n. In this model, we also use the mutual information method to determine τ and n, because this method is always used to calculated the delay time in order to obtain the high precision.

Secondly, the above results are input to BP neural network, and the output of its input layer is $y(i) = T(i+1)$. At this point, the number of input nodes and output nodes in BP neural network are n and 1, respectively. Therefore, the number of hidden layer nodes is

$$h = \sqrt{(n+1)} + a \tag{3}$$

Here, a is the adjustment constant between 1 and 10. In effect, a is one important parameter to determine the number of hidden layer nodes, which should be selected based on the following basic principle: the number of hidden layer nodes should be as small as possible to obtain one compact enough network structure under the premise of high prediction accuracy.

When the data is transmitted to the hidden layer, the input of each node in the hidden layer is

$$S_j = \sum_{i=1}^{n} W_{ij} T(i) - \theta_j \tag{4}$$

Here, W_{ij} represents the connection weight between the input layer to the hidden layer; θ_i is the threshold of hidden layer node.

After that, the output of each node in the hidden layer is

$$b_j = \frac{1}{(1 + \exp(\sum_{i=1}^{n} W_{ij}T(i) - \theta_j))} \quad (j = 1, 2, \cdots, p) \tag{5}$$

Finally, the inputs and outputs of the output layer are respectively obtained by

$$L = \sum_{j=1}^{p} W_{jk}b_j - \theta_k \tag{6}$$

$$T(i+1) = \frac{1}{(1 + \exp(\sum_{j=1}^{p} v_j b_j - \gamma))} \tag{7}$$

Here, v_j represents the node weight between the hidden layer and the output layer; γ is the threshold of output layer.

In practice, BP neural network not only randomly initializes the weights and thresholds of each connection before training, but also reduces the error through the reverse iteration to approximate the expected output. As a result, it may take longer time on searching or easily running into the local optimal solution [15], which results in the decrease of prediction performance.

3.2 Traditional Cuckoo Search Algorithm

The typical CS [15, 16], GA (Genetic Algorithm) [17], PSO (Particle Swarm Optimization) [13] and ACO (Ant Colony Optimization) [18] belong to the swarm intelligence optimization algorithms, and all of them can optimize the probability search through an iterative manner to improve BP neural network. Specifically, GA mainly focuses on the problems of combinatorial optimization and continuous optimization, and has the advantages of fast convergence and good versatility. However, it is more fatal to be prone to premature convergence and local optimal solution. Also, PSO aims at solving continuous optimization problems, and offers superior conveniences of fast convergence, few parameters and simple operation. However, it also has some problems such as weak capability in the later iteration and low quality of search solutions. Differently, ACO has advantages in solving discrete optimization problems and small-scale problems, and it has a strong global search ability and robustness by using the positive feedback mechanism, but it also appears the shortcomings of slow convergence speed, easy stagnation and weak solution to continuous optimization problems. In contrast, as a rising star, CS has received widespread attention due to its advantages of fewer parameters, easier coupling with other algorithms and stronger global search ability. However, its slow convergence speed and insufficient vitality in the iteration process still cannot be ignored. In specific applications, the CS algorithm needs to comply with the following three rules: (1) the cuckoo bird should lay one egg at a time, and incubate it in a random nest; (2) each nest may contain either high-quality or low-quality eggs, and only the nests containing high-quality eggs can be left to next generation; (3) once the cuckoo egg is discovered by one host bird, this host bird should discard the nest and search for a new one to avoid affecting the search for the optimal solution.

Based on the above rules, we assume $X_i^{(t)}$ is the position of the i - th nest in the t - th iteration and $L(\lambda)$ is a random search path. The formula for updating the path and position in the nest search is

$$X_i^{(t+1)} = X_i^{(t)} + \partial \oplus L(\lambda) \quad (i = 1, 2, \cdots, n) \tag{8}$$

Here, ∂ represents the step size control, and $\oplus$ represents the point-to-point multiplication operation.

3.3 Double-Optimized CS Algorithm

To solve the above problems of traditional CS algorithm, we further apply the exponential decline strategy in the speed and position updating formula, which can fully leverage its advantages in quickly finding local optimal solutions. Simultaneously, Gaussian perturbation is also used to exchange information between all local optimal solutions and the global optimal solution, and completes the update of global optimal position.

By using the exponential decreasing strategy, we define the search frequency f_i is

$$f_i = f_{\min} + (f_{\max} - f_{\min})\beta \tag{9}$$

Here, β is a random variable between [0,1].

The search frequency is applied in the search speed of traditional CS algorithm, and the new search speed can be calculated by

$$V_i^{(t)} = V_i^{(t-1)} + (X_i^{(t-1)} - X^*)f_i \tag{10}$$

Here, X^* represents the current global optimal position.

According to the new search speed, the position can be updated by

$$X_i^{(t)} = X_i^{(t-1)} + V_i^{(t)} \tag{11}$$

On this basis, Gaussian perturbation is applied to the optimal position update in the t - th iteration. We suppose the matrix $P = [X_1^{(t)}, X_2^{(t)}, \cdots, X_n^{(t)}]^T$, and Gaussian perturbation can be calculated by

$$P' = P + a \oplus \varepsilon \tag{12}$$

Here, a is one constant; ε represents the same order matrix of P.

From the above formula we can see that a can determine the search scope of ε, and it not only stimulates the vitality in the nest search process, but also achieves the information exchange between local and global optimal positions.

3.4   Double-Optimized CS-BP Anomaly Prediction Process

In order to construct a high-precision anomaly prediction model, the double-optimized CS algorithm is further applied to the optimization process of BP neural network. Moreover, the optimization process of CS-BP prediction model is depicted in Fig. 1, and the corresponding pseudocode of optimization algorithm is shown in Table 1. The specific execution steps are as follows:

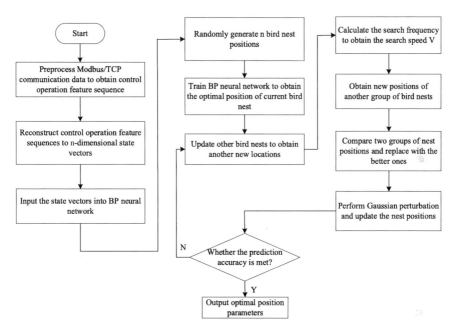

Fig. 1.   Optimization flow chart of CS-BP anomaly prediction model

Step 1: we reconstruct each preprocessed control operation feature sequence into one n-dimensional state vector by using the delay coordinate method, which serves as the input sample to train the BP neural network prediction engine;

Step 2: we randomly generate n bird nest positions, and each position corresponds to one initial weight and threshold of BP neural network. According to the prediction accuracy of each group of bird nests, BP neural network can be trained to obtain the optimal position of current bird nest.

Step 3: based on the obtained optimal position, we further update other bird nests according to Formula (8) to obtain new positions of another group of bird nests. By comparing these two groups of nest positions, we replace the inferior nest positions with the better nest positions, and obtain X_i and $P = [X_1^{(t)}, X_2^{(t)}, \cdots, X_n^{(t)}]^T$;

Step 4: by applying the exponential decline strategy to current nest positions, we can adjust the search speed and update the search location to obtain X_i^*, which is used to update P;

Step 5: Gaussian perturbation is performed on current nest positions to obtain new nest positions P', which are used as the new P in each iteration;

Step 6: according to the anomaly prediction accuracy, we test the new positions in Step 5: if one predefined prediction accuracy is met, we can stop the search and output the optimal position; otherwise, return to Step 3;

Step 7: by using the optimal position parameters as the initial weights and thresholds of BP neural network, we can establish the final double-optimized CS-BP anomaly prediction model to analyze and predict the observed control operation data in real time.

Table 1. Pseudocode of optimization algorithm for CS-BP anomaly prediction model

Algorithm 1: Double optimization of CS-BP model
Input : Control operation feature sequence
Output : Optimal position parameter
1 Initialize Y
2 **for** Y_i in Y
3   $T(i) \leftarrow Y_i$
4   $T(i+1) \leftarrow T(i)$
5   Compute a group of optimal positions
6             $X_i^{(t)} \leftarrow X_i^{(t-1)} + V_i^{(t)}$
7   Compute another group of optimal positions
8   Compare the optimal positions
9             $P' \leftarrow P + a \oplus \varepsilon$
10 **if** the stopping criterion is met **then**
11   Output the optimal position parameter **break**
12 **else**
13   Compute another set of optimal positions
14 **end if**
15 **end for**

4   Experimental Verification and Analysis

In order to evaluate the prediction performance of our model, we build a small-scale control system based on the Modbus/TCP communication protocol, and its hierarchical structure is shown in Fig. 2. Additionally, this system simulates the partial control operation process of material synthesis, and the whole process is set as a cycle of 60 s. The specific process is listed as follows: in the first stage, after reading current coil status and register values, valve 1 and valve 2 are open to fall materials. During this period, the coil status and register values are continuously monitored to mix material 1 and material 2 in a 2:1 ratio; in the second stage, valve 2 and valve 1 is closed in succession, and the conveyor belt is open to complete the material unloading process; in the third stage, the conveyor belt is closed, and the reaction furnace is started; in the last stage, the reaction furnace shuts down and outputs material 3. In our experiment, this system continuously runs for a total of 5 h and 26 min, and 62891 Modbus/TCP function code data are captured from the industrial switch.

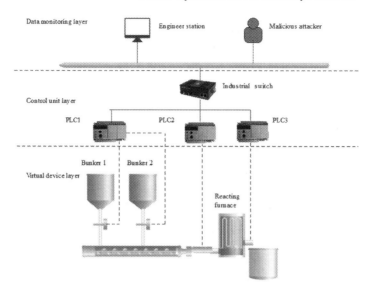

Fig. 2. Small-scale simulation control system based on Modbus/TCP communication protocol

4.1 Prediction Effect Verification Under Attack States

In order to evaluate the prediction effect for abnormal control operation data, we select the prediction accuracy, recall and F1 score as the evaluation indicators under the single attack and multi-step fusion attack, and compare them with the classic LSTM model and GA-BP model to illustrate the advantages and disadvantages of our model.

For the single attack, we design three different attack modes to disturb regular control operations: under the forging attack, we forge control operation 5 to early write coils during the falling process of material 1; under the falsifying attack, we falsify all control operations 3 to 1 during the conveyor belt transportation; under the discarding attack, we discard all control operations on reading coils and registers during the reaction furnace running. Table 2 compares the prediction results of different models under three attack modes. From this table we can draw the following conclusions: firstly, in terms of average prediction accuracy, the prediction accuracy of our model is much higher than the ones of other two models under the forging and discarding attack modes. However, when the falsifying attack occurs, the prediction accuracy of our model is slightly lower than the one of GA-BP model, because the falsifying attack shows the stationary trend with fewer change points of control operation features and the GA algorithm can enhance the global searching ability due to its crossover and mutation operations; secondly, the fitting degrees of our model under three attack modes are 0.91, 0.92 and 0.96, and all of these are closer to 1. That is, the fitting effect of our model is better than the LSTM model and GA-BP model; finally, in terms of prediction time, the running speed of GA-BP model is relatively slow, and our model runs faster against the forgery and discard attacks, but runs slightly slower against the falsifying attack than the LSTM model. In summary, under three different attack modes, although our model may have some shortcomings in individual indicators, the overall prediction effect performs significantly better. Also, we compare our model with the CS-BP model optimized by single exponential decline strategy (CS-BP-SE) and the CS-BP model optimized by single Gaussian perturbation (CS-BP-SG), and the average prediction accuracies under three different attack modes are shown in Table 3. Obviously, the prediction effect of our model is better than the other ones.

Table 2. Prediction effect comparison of different models under three single attacks

Attack Mode	Prediction model	Average prediction accuracy(%)	Fitting degree	Time(s)
	LSTM	84.97	0.74	0.7146
Forging	GA-BP	83.60	0.71	2.3350
	Our model	90.12	0.91	0.6657
	LSTM	88.36	0.80	0.9183
Falsifying	GA-BP	92.49	0.95	1.9540
	Our model	92.17	0.92	1.1822
	LSTM	88.20	0.90	0.5080
Discarding	GA-BP	80.10	0.78	1.0108
	Our model	94.71	0.96	0.3792

Table 3. Prediction accuracy comparison of three CS-BP models

	Prediction model	Attack mode		
		Forging	Falsifying	Discarding
Average prediction accuracy(%)	CS-BP-SE	84.91	86.89	90.18
	CS-BP-SG	84.77	82.60	85.33
	Our model	90.12	92.17	94.71

For the multi-step fusion attack, we logically combine three basic attack modes, and suppose one multi-step fusion attack mode, whose major target is to mix material 1 and material 2 in a wrong 1:1 ratio. More specifically, this fusion attack can be divided into multiple attack steps: firstly, by closing valve 1 in advance to change the mixing ratio of two materials, the control operation 15 is used to close valve 1 and open valve 2; secondly, during a series of operations including opening the conveyor belt, closing the conveyor belt and starting the reactor, the operation to read the register of material 1 is refused; Finally, when the reaction furnace stops, the abnormal material 3 is output. Figure 3 depicts the prediction curves of three models. From the trend of these curves we can see that, on the one hand, the LSTM model and GA-BP model have a greater prediction error of control operation 15 than our model under the first step of fusion attack; on the other hand, the prediction effect of the LSTM model and our model is better than that of the GA-BP model under the second step of fusion attack. Above all, it can be concluded that our model can effectively predict the beginning of supposed multi-step fusion attack, and avoid the following losses caused by this fusion attack.

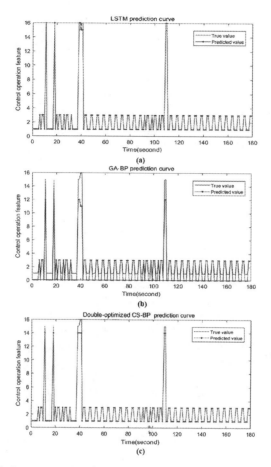

Fig. 3. Prediction curve comparison of three models under multi-step fusion attack

In order to further compare the prediction performance, we perform some experiments to analyze the prediction accuracy, recall and F1 score of each model, and Table 4 shows the compared results of three indicators for different prediction models. From this table we can see that, compared with other two models, our model has higher prediction accuracy and recall, which means that our model can achieve a dynamic balance between prediction accuracy and recall rate. Additionally, F1 score is always used to determine one model is more trustworthy if its value is closer to 1. As shown in this table, the F1 scores of these three models are 0.78, 0.82 and 0.86, respectively. That is, our model has the highest F1 score and is more trustworthy. In summary, our model also exhibits better prediction performance against such multi-step fusion attacks.

Table 4. Prediction indicator comparison of different models under multi-step fusion attack

	LSTM	GA-BP	Double-optimized CS-BP
Accuracy(%)	89.38	88.16	94.35
Recall(%)	71.8	76.71	80.29
F1 score	0.78	0.82	0.86

4.2 Prediction Effect Verification Under Normal States

One effective prediction model can not only detect abnormal control operations when some attacks occurs, but also try to avoid false positives during normal system running, that is, it has sufficient prediction accuracy under normal states. Figure 4 depicts the prediction effect and RMSE (Root Mean Square Error) of three models under normal states. From the prediction effect curves, it can be seen that neither the LSTM model nor the GA-BP model can successfully predict the control operation feature 15 in the 18th second. However, our model can achieve this goal. Additionally, during the stationary period from 140s to 160s, the GA-BP model has the worst prediction effect, while the LSTM model and our model have better prediction ability. According to the RMSE curves, the error occurring probabilities in the overall prediction process of the LSTM model and our model are lower than the one of GA-BP model. In practice, RMSE represents the fitting degree between the test value and the true value, and the smaller RMSE can mean the higher prediction accuracy. From this figure, the RMSE values of three models are 0.21082, 0.78881 and 0.20164, respectively. Therefore, our model has the highest prediction accuracy under normal states.

Table 5 shows the average prediction accuracies of three models under six experiments, and the average prediction accuracies of three models are 92.69%, 94.65% and 97.82%, respectively. Compared with other models, it is not difficult to see that our model has higher prediction accuracy.

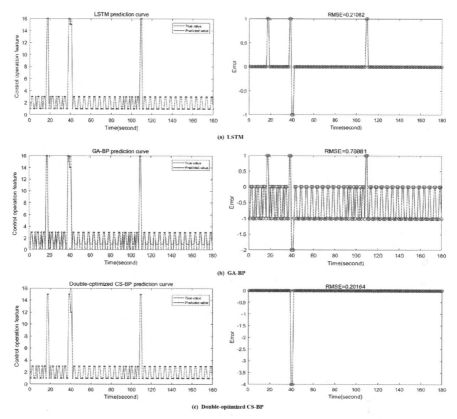

Fig. 4. Prediction effect and RSME of three models under normal states

Table 5. The average prediction accuracies of three models under 6 experiments

	LSTM(%)	GA-BP(%)	Double-optimized CS-BP(%)
1	89.19	88.96	95.87
2	92.90	94.22	96.09
3	96.33	91.71	98.33
4	96.92	97.10	98.17
5	90.30	96.98	99.14
6	90.50	98.92	99.31
Average	92.69	94.65	97.82

5 Conclusion

Focusing on the attack detection problem of control operations in industrial production, this paper proposes a double-optimized CS-BP anomaly prediction model. Moreover, this model uses BP neural network as one feasible anomaly prediction engine, and an improved CS algorithm based on the exponential decline strategy and Gaussian perturbation is designed to optimize the traditional BP neural network. More specially, this optimization not only accelerates the local convergence speed, but also enhances the information exchange between all local optimal solutions and the global optimal solution. By obtaining the optimal parameter setting, the network structure of anomaly prediction engine can be further improved, and the corresponding prediction performance can also be enhanced. In the experimental evaluation, this paper builds a small-scale control system based on Modbus/TCP communication protocol, and select the prediction accuracy, recall and F1 score as the evaluation indicators under the single attack and multi-step fusion attack. By comparing with other approaches, the advantages of this model to detect abnormal control operation are fully demonstrated. Additionally, this paper conducts experimental analysis on the prediction performance during normal system operations, and the experimental results also prove this model has a sufficiently high prediction accuracy to reduce the occurrence of false positives.

Acknowledgments. This work is supported by the Scientific Research Project of Educational Department of Liaoning Province (Grant No. LJKZ0082). The authors are grateful to the anonymous referees for their insightful comments and suggestions.

References

1. Baudouin D., Moalla N., Ouzrout Y.: The challenges, approaches, and used techniques of CPS for manufacturing in Industry 4.0: a literature review. The International J. Adv Manuf. Technol. 113(7–8), 2395–2412 (2021)
2. Wan, M., Li, J., Liu, Y., Zhao, J., Wang, J.: Characteristic insights on industrial cyber security and popular defense mechanisms. Chin. Commun. **18**(1), 130–150 (2021)
3. Tange, K., Donno, M.D., Fafoutis, X., Dragoni, N.: A systematic survey of industrial internet of things security: requirements and fog computing opportunities. IEEE Commun. Surv Tutorials 22(4), 2489–2520 (2020)
4. Yu K., Tan L., Mumtaz S., AI-Rubaye S., AI-Dulaimi A., Bashir A. K., Khan F. A.: Securing critical infrastructures: deep-learning-based threat detection in IIoT. IEEE Commun. Mag. 59(10), 76–82 (2021)
5. Siniosoglou, I., Radoglou-Grammatikis, P., Efstathopoulos, G., Fouliras, P., Sarigiannidis, P.: A unified deep learning anomaly detection and classification approach for smart grid environments. IEEE Trans. Netw. Serv. Manage. 18(2), 1137–1151 (2021)
6. Koay, M.Y.A., Ko, K.L.R., Hettema, H., Radke, K.: Machine learning in industrial control system (ICS) security: current landscape, opportunities and challenges. J. Intell. Inf. Syst. **60**(2), 377–405 (2023)
7. Khan, A.I., Pi, D., Khan, U.Z., Hussain, Y., Nawaz, A.: HML-IDS: a hybrid-multilevel anomaly prediction approach for intrusion detection in SCADA systems. IEEE Access **7**, 89507–89521 (2019)

8. Wan, M., Li, J., Wang, K., Wang, B.: Anomaly detection for industrial control operations with optimized ABC-SVM and weighted function code correlation analysis. J. Ambient. Intell. Humaniz. Comput. **13**(3), 1383–1396 (2022)

9. Hannon C., Deka D., Jin D., Vuffray M., Lokhov Y. A.: Real-time anomaly detection and classification in streaming PMU data. 2021 IEEE Madrid PowerTech, 1–6. IEEE, Madrid, Spain (2021)

10. Cheminod, M., Durante, L., Seno, L.: Valenzano A: performance evaluation and modeling of an industrial application-layer firewall. IEEE Trans. Industr. Inf. **14**(5), 2159–2170 (2018)

11. Kim C., Robinson D: Modbus monitoring for networked control systems of cyber-defensive architecture. In: 2017 Annual IEEE International Systems Conference (SysCon), pp. 1–6, IEEE, Montreal, Canada (2017)

12. Ma, L., Li, N., Guo, Y., Huang, M., Yang, S., Wang, X.: Learning to optimize: reference vector reinforcement learning adaption to constrained many-objective optimization of industrial copper burdening system. IEEE Trans. Cybern. **52**(12), 12698–12711 (2022)

13. Nabaei, A., et al.: Topologies and performance of intelligent algorithms: a comprehensive review. Artif. Intell. Rev. **49**(1), 79–103 (2018)

14. Ma, L., Huang, M., Yang, S., Wang, R., Wang, X.: An adaptive localized decision variable analysis approach to large-scale multi-objective and many-objective optimization. IEEE Trans. Cybern. **52**(7), 6684–6696 (2022)

15. Zhang, W., Han, G., Wang, J., Liu, Y.: A BP neural network prediction model based on dynamic cuckoo search optimization algorithm for industrial equipment fault prediction. IEEE Access **7**, 11736–11746 (2019)

16. Shehab, M., Khader, A.T., Al-Betar, M.A.: A survey on applications and variants of the cuckoo search algorithm. Appl. Soft Comput. **61**(12), 1041–1059 (2017)

17. Katoch, S., Chauhan, S.S., Kumar, V.: A review on genetic algorithm: past, present, and future. Multimedia Tools Appl. **80**(5), 8091–8126 (2021)

18. Karaboga, D., Gorkemli, B., Ozturk, C., Karaboga, N.: A comprehensive survey: artificial bee colony (ABC) algorithm and applications. Artif. Intell. Rev. **42**(1), 21–57 (2012). https://doi.org/10.1007/s10462-012-9328-0

Bridging the Interpretability Gap
in Coupled Neural Dynamical Models

Mingrong Xiang[1], Wei Luo[1(✉)], Jingyu Hou[1], and Wenjing Tao[2]

[1] School of Information Technology, Deakin University, Burwood, VIC 3125,
Australia
mxiang@deakin.edu.au

[2] Key Laboratory of Freshwater Fish Reproduction and Development (Ministry of
Education), Key Laboratory of Aquatic Science of Chongqing, School of Life
Sciences, Southwest University, Chongqing 400715, China

Abstract. Neural ordinary differential equations (NODEs) have
achieved remarkable performance in many data mining applications that
involve multivariate time series data. Its adoption in the data-driven discovery of dynamic systems, however, was hindered by the lack of interpretability due to the black-box nature of neural networks. In this study,
we propose a simple yet effective NODE architecture inspired by the
highly successful generalised additive models. Our proposed model combines linear and nonlinear components to capture interpretable evolution
rules with only a marginal loss of model expressiveness. Experiments
show that our model can effectively recover interactions among variables
in a complex dynamic system from observation data.

Keywords: Neural ODEs · Additive models

1 Introduction

Neural ordinary differential equations (NODEs) have demonstrated significant
results in many fields of dynamic system modelling [3,6,12]. NODEs use neural
networks (NNs) to achieve great expressive power for modelling complex coupling among state variables in the joint evolution rule. However, the lack of
interpretatability of NNs has attracted increasing attention, and the black-box
nature makes it challenging to clarify complex interactions between variables
[7,17,23].

The challenge of interpretability restricts the application of NNs in many
research and application fields, such as drug discovery, medical risk prediction,
and biological network inference [8,25]. These fields require a solid scientific
interpretation of the model rather than focusing on the prediction results for a
particular task. For example, mining interpretable interactions when modelling
dynamic systems of biological networks helps deepen the understanding of organ-
isms at the molecular level and contributes to medical and biological research

X. Yang et al. (Eds.): ADMA 2023, LNAI 14176, pp. 524–534, 2023.
https://doi.org/10.1007/978-3-031-46661-8_35

[15, 19]. However, the interactions in the biological network are difficult to interpret because of the lack of interpretability of NNs. Proper model interpretation is required when modelling dynamic systems with potentially complex interactions, allowing researchers to evaluate whether the model is reliable using their expertise.

In this study, we will focus on the interpretability challenges of mining interactions in modelling dynamic systems using NODE models. For example, we can mine the interpretable interactions between metabolite molecules when modelling the dynamic system of a metabolic network. NNs can easily learn the feature patterns of state variables to create efficient dynamic models. However, functional interactions between variables cannot be explained scientifically, which is regarded as a lack of interpretability. Although NNs are based on simple mathematical operations, the combination of neurons with nonlinear activation in multiple hidden layers makes NNs difficult to evaluate mathematically [7].

In this research, we proposed a novel interpretable NODE model. Inspired by the GAM (Generalized Additive Models) [9], our method is modelling a dynamic system through joint training of linear and nonlinear combinations. Figure 1 shows the core integral calculation functions for recovering the dynamics of a continuous time series dataset. In sub-figure (a) and (b), we use the single linear layer and the three-layer NNs to compute the integral of the dynamic system. (a) and (b) are the baseline methods in our research because a single linear layer is one of the most interpretable methods, and (b) is the most basic form of NNs used for integration in NODE. Sub-figure (c) shows our proposed method. As we can see, each neural network only uses one variable $(x_1, \cdots, x_k)$ as input and trains k neural networks to fit the data simultaneously. All individual outputs will be concatenated and fit into a linear layer, resulting in an interpretable dynamic model.

In particular, the mechanics of linear regression are highly interpretable due to the fairly straightforward calculation of the marginal contribution of the input variables [11, 18]. Therefore, we can use the trained weights of the linear layer as the interaction coefficients among the inputs from x_1 to x_k.

Our main contributions are summarized as follows.

- We extend the highly successful generalised additive model (GAM) in traditional machine learning to the application of the data-driven discovery of ordinary differential equations.
- We proposed a two-part interpretable Neural ODE model inspired by GAM. It consists of nonlinear state variable encoders and a linear matrix that captures the interactions among state variables.
- Our method achieves good performance in terms of both prediction accuracy and relationship identification on a widely used benchmark dataset. To the best of our knowledge, our model is the first NODE model to simultaneously achieve satisfying performance in both criteria.

2   Related Work

Although NNs have achieved great success in many different fields, their potential mechanisms are not fully understood even for a simple structured neural network (e.g. three-layer MLP) [4,8]. The interpretability of the model is essential, referring to the comprehensible logical decision-making mechanism and the interpretable interactions among variables [28]. It is important to know why and how to get the predictions of the model instead of only caring about the accuracy of the results in a specific task. Neural networks are emerging as powerful tools in scientific research fields, especially for data that have complex interactions, such as biological networks and gene expression dynamics [20]. While NNs are able to successfully model complex dynamic systems, existing models have difficulties to interpret the interactions of variables during dynamical evolution processes. As mentioned above, the fundamentals of NNs are comprehensible mathematical operations, but the combination of linear and nonlinear functions in the hidden layers makes exploring the inner workings of the model difficult.

Linear regression is recognized as one of the most interpretable predictive models, and regression coefficients can be directly interpreted [16,18]. Each coefficient describes the effect of a given input on the output change. For example, variables with high coefficients and low variability around them can be considered important. Linear effects are additive, making it easy to separate effects

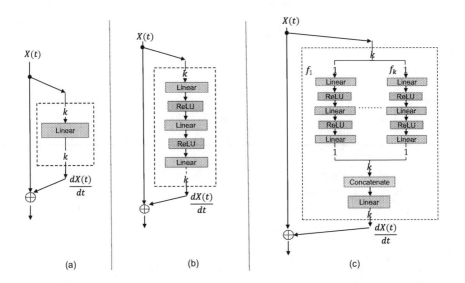

Fig. 1. Three types of Neural ODEs. (a) represents a linear ODE. It is interpretable but only expresses very simple dynamics. (b) represents a nonlinear ODE. It is flexible, but not interpretable. (c) is our proposed architecture. Each input variable goes through a separate nonlinear transformation, and then combined linearly to generate the Jacobian. We can recover interactions among different variables in $X(t)$ from the last linear layer.

so that they are easy to quantify and describe [24]. However, real datasets are often complex, and linear regression is difficult to handle nonlinear associations between input and target values.

Researchers have made great efforts to improve the interpretability of NNs-based models. LIME [22] interprets a single sample using a trained local model. It modifies the feature values of a single data sample of interest and observes the effect on the output. The researchers can then obtain the contribution of each feature to the predictions of a single sample and then interpret them locally. ACD [26] uses hierarchical clustering to stratify the predictions generated by the neural network. Interactions are scored and built into a tree to evaluate the significant feature clusters learned by the network. Network Dissection [29] was interpreted by dividing the input into individual human interpretable labels and assessing the alignment with each corresponding hidden unit in the network. However, these methods are usually complex, require an independent assessment of NNs output or operation, and even require combined previous knowledge (labels) to estimate interactions.

A recently proposed interpretable method, NAM [1], belong to the GAM [9] model family and is trained using a set of networks, each of which processes individual variables. These networks are jointly trained and obtain the final prediction result derived from the weighted average of the independent outputs of all networks. The most important idea of the GAM model family is to use a single smooth function to estimate each additive term in the model to solve the interpretation challenge of the interactions between the independent variable and the dependent variable. Therefore, it can explain how the dependent variable changes with the independent variable in each additive term. Although these methods improve the interpretability of neural network models from different perspectives, methods for modelling dynamical systems and mining interpretable interactions have rarely been published. Inspired by GAM, we propose an interpretable NODE model, which evaluates the interactions between variables when recovering the dynamic system. Our model preserves linear interpretability and considers the powerful modelling capabilities of NNs for variable interactions and nonlinear associations.

3   Method

In this section, we will give the preliminaries of datasets, the descriptions of the competitors, and the details of our method.

3.1   Preliminaries

A continuous time series dataset is usually presented as a matrix:

$$X = \begin{pmatrix} x_{11}, \ x_{12}, \ \ldots, \ x_{1k} \\ x_{21}, \ x_{22}, \ \ldots, \ x_{2k} \\ \vdots \quad \vdots \quad \ddots \quad \vdots \\ x_{n1}, \ x_{n2}, \ \ldots, \ x_{nk} \end{pmatrix} \tag{1}$$

where each row of X contains k variables that corresponding to the n time points t, $x(0) = x(t_1) = (x_{11}, x_{12}, \ldots, x_{1k})$. $x(0)$ indicates the initial values of input when we model a dynamic system.

Our aim is to recover the dynamic system of X and mine for potential interactions among the k variables. In general, the evolution process of all the state variables in $X(t)$ can be described by a differential equation:

$$\hat{X}(t) = \psi(X(t), t) \tag{2}$$

Through the continuously differentiable real vector-valued function ψ, we can recover the state variable at time t in X. In other words, the value of $\hat{X}$ at time t depends on the initial state $x(0)$ and the interactions between the k variables.

3.2   Neural ODEs-Based Models

As a typical combination of a neural network and the ordinary differential equation model, Neural ODE [3] has achieved great success. It can parameterise the continuous dynamics of hidden units using an ODE and a neural network. In case, the Neural ODE equation can be defined as follows:

$$\hat{X}(t) = \frac{dX(t)}{dt} = f(X(t); \theta) \tag{3}$$

where the integration function $f(\cdot; \theta)$ is a neural network with parameter θ. An important reason for the success of neural ODEs and their variants in different dynamic modelling tasks is that they can approximate complex nonlinear dynamics. Theoretically, $f(\cdot; \theta)$ can be any network form, of course, a single linear layer can also be adapted. Therefore, the simplest interpretable NODE model can be implemented by linear regression:

$$X(t + \tau) = X(t) + \int_t^{t+\tau} f_l(X(t), W_l, \beta_l) \tag{4}$$

Each generated state variable in $X(t + \tau)$ depends on the input variable and the result from the integral function $f_l(X(t), W_l, \beta_l)$ at the previous time point. Here, W_l can be used to interpret possible linearity interactions among the k input variables. Additionally, adding more linear layers with nonlinear activation layers to the interaction function can give the NODE model with more powerful nonlinear modelling capabilities.

$$X(t + \tau) = X(t) + \int_t^{t+\tau} f_m(\sigma, X(t), W_m, \beta_m) \tag{5}$$

Thus, f_m can be a three-layer MLP (three linear layers with two nonlinear activation layers), which would be the simplest and standard NNs form in designing a NODE model. However, unlike the single linear layer, we are unable to directly interpret the interactions from the weights of different layers in MLP. One possible solution is to use an additional regression model to approximate the trained model ψ with a linear map A [14]:

$$f_m(X) = AX \tag{6}$$

However, the inferred interactions of a dynamic system are limited by the linear and diagonal assumptions imposed by A. Next, we introduce our design to make the model interpretable and have sufficient learning capacity for nonlinear evolution. Firstly, we assume that our designed network should have an independent linear layer as an output layer. It is important, which allows us directly use the final weights to interpret the interactions among variables. Then, the problem is to guarantee the independence of the input variables to fit into the output layer. That is, each feature retains nonlinearity while evolving independently of the others. Therefore, our idea can be presented as follows:

$$X_h(t) = Concat(f_1(x_{i1}), f_2(x_{i2}), \cdots, f_k(x_{ik})) \tag{7}$$

The key idea is to learn a combination of networks, each function $f_i \in \{f_1, f_2, \cdots, f_k\}$ processes an individual input variable. We design all f_i to be parameterized by the same structure networks (three layers) with a low dimensional hidden layer and nonlinear activation function, then trained jointly. After that, we can concatenate all the single output features from f_i and fit them $X_h(t)$ into the linear output layer f_o.

$$X(t+\tau) = X(t) + \int_t^{t+\tau} f_o(X_h(t), W, \beta) \tag{8}$$

To summarise, all input variables are separately trained by a network using backpropagation and can learn arbitrarily complex functions. The independent outputs are concatenated and fit into linear layers to approximate the state variables at each instant in time t. Thus, we can obtain a weight to interpret the variable interactions when modelling a dynamic system. After that, we use Dormand-Prince solver [2,10] for integration, as it can provide high integration accuracy. Then, we can define a loss function for training the Neural ODE model.

$$L = \frac{\sum_{i=1}^{n} |\hat{X} - X|}{n} \tag{9}$$

4   Experiments

In our experiments, NODEs are used to model the dynamic system, which recovers the $f(;\theta)$ function to approximate the state variables in X. In particular, we consider the model's performance in predicting variables at unseen time points and inferring interactions between variables. We performed the experiments on the datasets of the Metabolic network: yeast glycolysis. We have split the dataset into three sets according to the time t: The first 90% was used to train the dynamic system, following 10% for validation and the remaining 10% for testing the dynamics prediction performance. For the baseline method, we have implemented the Linear ODE and the MLP ODE (refer to Sect. 3.2)

Other Implementation Details. All the NODE methods are implemented using the *torchDyn* Python package [21]. The Linear ODE has been implemented by one linear layer in which the input and output dimensions are equal to the k. The MLP ODE is realised by a 3-layer NN with ReLU activation, so as the individual networks in our proposed method. The hidden dimensions for MLP ODE and our method have been set to 14, and the final output dimensions equal k. The optimal model during training will call back by earlystopping. All the methods have been trained with a 1e–2 learning rate and performed with a single NVIDIA 3090 GPU.

4.1   Datasets

In this research, we performed several experiments on the datasets of the metabolic network: yeast glycolysis. Data were generated from the ODE system with seven state variables (see [13]).

Metabolic networks describe important mechanisms of central cellular metabolism. The study of metabolic networks is beneficial for the research of metabolic diseases and helps to produce high-value products, such as drugs [5,27]. We did the experiments on this metabolic network because it includes many state variables that evolve from complex interactions, which helps us to verify our experiment results more objectively and effectively. The original ODE system is shown below:

$$\dot{s}_1 = 2.5 + \frac{-100s_1s_6}{1 + 13.6769s_6^4}$$

$$\dot{s}_2 = \frac{200s_1s_6}{1 + 13.6769s_6^4} - 6s_2 + 6s_2s_7$$

$$\dot{s}_3 = 6s_2 - 64s_3 + 6s_2s_7 + 16s_3s_6$$

$$\dot{s}_4 = 64s_3 - 13s_4 + 13s_5 - 16s_3s_6 - 100s_4s_7 \tag{10}$$

$$\dot{s}_5 = 1.3s_4 - 3.1s_5$$

$$\dot{s}_6 = \frac{-200s_1s_6}{1 + 13.6769s_6^4} + 128s_3 - 32s_6 - 1.28s_3s_7$$

$$\dot{s}_7 = 6s_2 - 18s_2s_7 - 100s_4s_7$$

In addition to the original equations set, we also derived polynomial functions (eight state variables) from it to simulate the data.

$$
\begin{aligned}
\dot{z}_1 &= 3.5 - 100z_1^2 + 13.6769z_6^4 \\
\dot{z}_2 &= 200z_1^2 + 1 + 13.6769z_6^4 - 6z_2 - 6z_7^2 \\
\dot{z}_3 &= 6z_2 - 64z_3 + 6z_7^2 + 16z_6^2 \\
\dot{z}_4 &= 64z_3 - 13z_4 + 13z_5 - 16z_6^2 - 100z_7^2 \\
\dot{z}_5 &= 1.3z_4 - 3.1z_5 \\
\dot{z}_6 &= -200z_6^2 + 1 + 13.6769z_1^4 + 128z_3 + -32z_6^2 - 1.28z_7^2 \\
\dot{z}_7 &= 6z_2 - 18z_7^2 - 100z_7^2 \\
\dot{z}_8 &= \dot{z}_1 + \alpha
\end{aligned}
\tag{11}
$$

There are several reasons that we did the experiments on derived polynomial functions. Firstly, the polynomial functions contain many linearity interactions, which allow us to check the interpretability of the linear model and compare it with other methods. Secondly, the equations include complex interactions and nonlinear evolution, which makes it better for us to compare the dynamic modelling and prediction performance. Finally, the noisy data z_8 (similar to $\dot{z}_1$, we set $\alpha = 0.1$) is a distractor for mining the interactions. In order to better reflect the performance of different methods, we can generate m datasets for batch training of the model. We set $m = 5$ in our experiments-adding more datasets can improve the accuracy of mining variable interactions and better reflect the gap in model performance.

4.2   Evaluation and Discussion

In this research, we evaluated the methods of both dynamics prediction and interactions mining accuracy. The dynamics prediction was evaluated by MAE, and the interaction mining results were evaluated by AUC. By comparing the ground truth interactions (from those functions in Sect. 4.1) with the models' mining results, it can be found that our model improves performance with better interpretability.

Table 1 records the average dynamic prediction performance. As we can see, the MLP ODE and our method yield much lower MAE than Linear ODE, which means the NNs have a more robust dynamic modelling ability than a single linear layer. It is predictable results that the yeast glycolysis dynamic system contains many non-linear interactions, which NNs are good at. Although our model is more complex than MLP ODE, the dynamic prediction accuracy is not lost and is even better.

Table 1. 10 trials Average MAE of dynamic prediction on two datasets: original (yeast glycolysis) and derived polynomial functions

Dataset	MAE		
	Linear	MLP	Our Method
Original	0.0228	0.0141	**0.0118**
Polynomial	0.0355	0.0280	**0.0257**

On the other hand, our method also achieved significant interaction mining results. As shown in Table 2, the linear model shows a higher interaction mining performance on the original data than on the polynomial, which is 8% higher. Theoretically, a linear model will explore as many linearly dependent interactions as possible. By observing the equations, we can find that the number of linearity interactions in Eq. 10 is more than that in Eq. 11, which is in line with the AUC evaluation results of the linear model. In addition, the AUC of the MLP model is much worse than the linear model. As we mentioned before, the MLP model requires additional regression models to approximate the nonlinear modelling results after the completion of training. Thus, the inference of interactions will be limited by imposed linear and diagonal assumptions. In contrast, our model obtains satisfactory results and performs well on different datasets, yielding 9% and 5% better than the linear model. In order to better illustrate the

Table 2. 10 trials Average AUC of interaction mining on two kinds of simulation data

Dataset	AUC		
	Linear	MLP	Our Method
Original	0.76	0.54	**0.85**
Polynomial	0.68	0.61	**0.73**

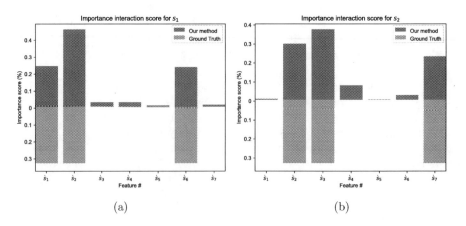

(a)   (b)

Fig. 2. Examples of importance interactions score for variables: the orange bar is the ground truth interactions, and the blue bar is the inferred interactions. (Color figure online)

performance and interpretability of our model, we also visualise the importance score of the variables interaction coefficient inferred by our model and compare it with the ground truth. Figure 2 gives two examples of interaction importance scored results. Here, we assume that variable interactions in the ground truth are equally important. In sub-figure (a), our prediction results are similar to the ground truth, which shows that s_1 is important for $\dot{s}_1$, $\dot{s}_2$ and $\dot{s}_6$ during the evolution processes. In sub-figure (b), we can find that the s_2 is important for the generation of $\dot{s}_2$, $\dot{s}_3$ and $\dot{s}_7$, which is in line with the ground truth interactions. Although it shows a bit deviation in prediction on $\dot{s}_3$, the results for $\dot{s}_1$ and $\dot{s}_5$ are satisfied as we expect, which are close to 0%. In general, our model achieves satisfactory interpretable interaction mining performance without loss of dynamical system modelling accuracy.

5   Conclusion

We proposed an interpretable Neural ODE model for mining the interactions when modelling the dynamic system. The model is easy to implement, but the effect is satisfactory. From the experimental results, we can find that our model obtains significant variables interaction mining performance without sacrificing dynamic prediction accuracy. Our model achieves the independent nonlinear evolution of state variables in the integration process and can directly use the coefficients of the output layer to interpret the model. One promising future direction is to adapt joint training with more inputs by reducing the number of components in the model.

References

1. Agarwal, R., et al.: Neural additive models: interpretable machine learning with neural nets. Adv. Neural. Inf. Process. Syst. **34**, 4699–4711 (2021)
2. Butcher, J.C.: The Numerical Analysis of Ordinary Differential Equations: Runge-Kutta and General Linear Methods. Wiley-Interscience, Hoboken (1987)
3. Chen, R.T., Rubanova, Y., Bettencourt, J., Duvenaud, D.K.: Neural ordinary differential equations. Adv. Neural Inf. Process. Syst. **31**, 1–13 (2018)
4. Doshi-Velez, F., Kim, B.: Towards a rigorous science of interpretable machine learning. arXiv preprint arXiv:1702.08608 (2017)
5. Ducat, D.C., Way, J.C., Silver, P.A.: Engineering cyanobacteria to generate high-value products. Trends Biotechnol. **29**(2), 95–103 (2011)
6. Dupont, E., Doucet, A., Teh, Y.W.: Augmented neural odes. Adv. Neural Inf. Process. Syst. **32**, 1–11 (2019)
7. Fan, F.L., Xiong, J., Li, M., Wang, G.: On interpretability of artificial neural networks: a survey. IEEE Trans. Radiat. Plasma Med. Sci. **5**(6), 741–760 (2021)
8. Gilpin, L.H., Bau, D., Yuan, B.Z., Bajwa, A., Specter, M., Kagal, L.: Explaining explanations: an overview of interpretability of machine learning. In: 2018 IEEE 5th International Conference on data Science and Advanced Analytics (DSAA), pp. 80–89. IEEE (2018)
9. Hastie, T.J.: Generalized additive models. In: Statistical Models in S, pp. 249–307. Routledge (2017)

10. Kimura, T.: On dormand-prince method. Jpn. Malaysia Tech. Instit **40**, 1–9 (2009)
11. Lipton, Z.C.: The mythos of model interpretability: in machine learning, the concept of interpretability is both important and slippery. Queue **16**(3), 31–57 (2018)
12. Lou, A., et al.: Neural manifold ordinary differential equations. Adv. Neural. Inf. Process. Syst. **33**, 17548–17558 (2020)
13. Mangan, N.M., Brunton, S.L., Proctor, J.L., Kutz, J.N.: Inferring biological networks by sparse identification of nonlinear dynamics. IEEE Trans. Molec. Biol. Multi-Scale Commun. **2**(1), 52–63 (2016)
14. Matsumoto, H., et al.: Scode: an efficient regulatory network inference algorithm from single-cell RNA-SEQ during differentiation. Bioinformatics **33**(15), 2314–2321 (2017)
15. Micheletti, C.: Comparing proteins by their internal dynamics: exploring structure-function relationships beyond static structural alignments. Phys. Life Rev. **10**(1), 1–26 (2013)
16. Molnar, C.: Interpretable machine learning. Lulu. com (2020)
17. Montavon, G., Samek, W., Müller, K.R.: Methods for interpreting and understanding deep neural networks. Digital Signal Process. **73**, 1–15 (2018)
18. Munkhdalai, L., Munkhdalai, T., Ryu, K.H.: A locally adaptive interpretable regression. arXiv preprint arXiv:2005.03350 (2020)
19. Nowak, M.A.: Evolutionary Dynamics: Exploring the Equations of Life. Harvard University Press, Cambridge (2006)
20. Park, Y., Kellis, M.: Deep learning for regulatory genomics. Nat. Biotechnol. **33**(8), 825–826 (2015)
21. Poli, M., Massaroli, S., Yamashita, A., Asama, H., Park, J., Ermon, S.: Torchdyn: implicit models and neural numerical methods in pytorch (2021)
22. Ribeiro, M.T., Singh, S., Guestrin, C.: "why should i trust you?" explaining the predictions of any classifier. In: Proceedings of the 22nd ACM SIGKDD International Conference on Knowledge Discovery and Data Mining, pp. 1135–1144 (2016)
23. Roscher, R., Bohn, B., Duarte, M.F., Garcke, J.: Explainable machine learning for scientific insights and discoveries. IEEE Access **8**, 42200–42216 (2020)
24. Schielzeth, H.: Simple means to improve the interpretability of regression coefficients. Methods Ecol. Evol. **1**(2), 103–113 (2010)
25. Sheu, Y.H.: Illuminating the black box: interpreting deep neural network models for psychiatric research. Front. Psychiat. **11**, 551299 (2020)
26. Singh, C., Murdoch, W.J., Yu, B.: Hierarchical interpretations for neural network predictions. arXiv preprint arXiv:1806.05337 (2018)
27. Yarmush, M.L., Banta, S.: Metabolic engineering: advances in modeling and intervention in health and disease. Annu. Rev. Biomed. Eng. **5**(1), 349–381 (2003)
28. Zhang, Y., Tiňo, P., Leonardis, A., Tang, K.: A survey on neural network interpretability. IEEE Trans. Emerg. Topics Comput. Intell. **5**, 726–742 (2021)
29. Zhou, B., Bau, D., Oliva, A., Torralba, A.: Interpreting deep visual representations via network dissection. IEEE Trans. Pattern Anal. Mach. Intell. **41**(9), 2131–2145 (2018)

Multidimensional Adaptative kNN over Tracking Outliers (Makoto)

Jessy Colonval[(✉)] and Fabrice Bouquet

FEMTO-ST Institue, CNRS, Univ. de Franche Comté (UFC), Besançon, France
{jessy.colonval,fabrice.bouquet}@femto-st.fr

Abstract. This paper presents an approach to detect outliers present in a data set, also called aberration. These outliers often cause problems to the learning algorithms by deviating their behavior, which makes them less efficient. It is therefore necessary to identify and remove them during the cleaning data step before the learning process. For this purpose, a method that detects if data is an outlier from its k nearest neighbors is proposed for multidimensional data sets. In order to make the method more accurate, the number of k nearest neighbors chosen is adaptive for each class present in the data set, and each neighbor has a different weight in the decision, depending on their respective proximity. The proposed method is called Makoto for Multidimensional Adaptative kNN Over Tracking Outliers. The effectiveness of this method is compared with four other known methods based on different principles: LOF (Local Outlier Factor), Isolation forest, One Class SVM and Inter Quartile Range (IQR). Thus, on the basis of 406 synthetic data sets and 17 real data sets with distinct characteristics, the Makoto method appears to be more efficient.

Keywords: Machine learning · Data filtering · Spatial Outlier Detection · Kernel Functions · Adaptive kNN · Correlation

1 Introduction

An important principle in the use of artificial intelligence algorithms is the quality of the learning data. The implementation of tools to help filter the data is a critical point [14]. We are interested here in a particular case, that of outliers. These are data that contain information that is incompatible with the rest of the data set. This data can have two effects:

1. Weaken the predictive power of the model obtained at the end of the learning phase.
2. Weaken the score obtained from the model during its validation phase.

It is important to define a few terms that will be used in this paper. A data set can be visualized as a table with rows and columns. Each row represents an individual, often unique, while the columns represent the characteristics of that

© The Author(s), under exclusive license to Springer Nature Switzerland AG 2023
X. Yang et al. (Eds.): ADMA 2023, LNAI 14176, pp. 535–550, 2023.
https://doi.org/10.1007/978-3-031-46661-8_36

individual. There are several terms to designate them, but here, the rows will be called *points* and the columns will be called *attributes*.

In general, there are two types of attributes [1, Chap 11]:

- **Behavioral attributes** is an attribute of interest measured for each point. Points often have only one, but they can have several. These attributes are often non-spatial because it measures a certain quantity. For example, the type of glass, the presence of a heart abnormality or the description of an image. The values of a behavioral attribute are often called class, but here we will call it behavioral value to avoid confusion.
- **Contextual attributes** is an attribute expressing the characteristics of a point is defined on a continuous domain of values, also called spatial attributes. A point often has several of them. For example: the composition of a glass pane, the number of heartbeats per minute, or the color of a pixel.

We are interested in the detection of what we call spatial outliers, i.e. established from a neighborhood. The main criterion of these types of outlier search is the auto-correlation property, i.e. the fact that data in a neighborhood are closely correlated. Thus, an outlier is defined as an abrupt change in behavioral attributes among nearby points according to their contextual attributes.

However, there are two ways to create a neighborhood, depending on the nature of the data used:

- **Multidimensional methods** determine the neighborhood based on a distance between each point.
- **Graph-based methods** determine the neighborhood from the linkage relations between the points. For example, by using edges between nodes, where each node is associated with behavioral attributes.

We are only interested in multidimensional data sets. Thus, the neighborhood will be established from a distance calculation, euclidean for the Makoto method.

Neighborhood-based multidimensional spatial outlier detection methods have already been proposed [5,13]. They differ in the way they establish the neighborhood and how they combine them to make an outlier prediction. The problem is that these techniques consider the neighbors equitably in the final decision, and the majority use all attributes for the creation of the neighborhood.

First, considering all neighbors as equal can be problematic because the neighbors are not equidistant from the starting point. And according to the auto-correlation property, it would make sense to give more importance to the nearest neighbors. For example, when the difference between the nearest and the farthest neighbor is important, then it is intuitive to want to give more importance to the nearest one. However, there are several ways to weight a neighborhood so, depending on the method chosen, it is possible to obtain several outlier predictions. Some methods have decided to generalize this idea and uses different weighting methods of the neighborhood [11,16].

Second, using all contextual attributes in the establishment of the neighborhood can be problematic because of the likely presence of randomly distributed attributes. These attributes are noise and can hinder outlier detection by making some of them undetectable or by considering data as falsely outlier [10]. Moreover, using fewer attributes leads to performance and resource gains, but this is not the main objective

In Sect. 2 we will describe the proposed method in detail and then Sect. 3 will compare this method with others from the state of the art to demonstrate its effectiveness.

2   Principle of the Method Makoto

This section presents the principle of the outlier detection method *Makoto* (**M**ultidimensional **A**daptive **kNN O**ver **T**racking **O**utliers). It can detect outliers only on multidimensional data sets, i.e. with continuous values. This method establishes the outliers using a neighborhood whose number is adaptive according to the distribution of the data set (Sect. 2.3). Each neighbor has a different weight in the decision-making, and this weight is computed using a kernel function (Sect. 2.1 and 2.2). In order to avoid being biased by the choice of a kernel function and to be more confident in the detection, we decide to use several kernel functions and to make a majority vote among their decisions to determine if a point is an outlier or not. Moreover, we use several subsets of the original data set to compute different neighborhoods that will each bring a prediction so that in the end the majority decision will prevail (Sect. 2.4).

2.1   Kernel Functions

To determine the weight of each neighbor, we decide to use the kernel functions [9]. This function gives a weight according to the distance of a neighbor to the origin point. We consider that the distances are between -1 and 1, where 0 is the shortest distance and -1 and 1 are the farthest.

These functions have several properties. They reach their maximum in 0 and decrease as the distance increases. Opposite distances are considered to be equivalent, i.e., their weight is the same. The weighting must be positive or equal to 0. Thus, the following properties must be respected [7]:

- positive, $\forall x \in \mathbb{R}, f(x) \geq 0$;
- max in 0, $\max_{x \in \mathbb{R}} f(x) = f(0)$;
- opposite, $\forall x \in \mathbb{R}, f(x) = f(-x)$;
- continuous by party $\forall x, y \in \mathbb{R}, x < y, f(x) \geq f(y)$.

As mentioned at the beginning of this section, this method use a majority vote on several kernel functions. There are several kernel functions in the literature, only the most common will be studied [9]. However, as shown in Fig. 1, some of the kernel functions are similar. The risk if we use all these functions, in the majority vote, is that a subpart of these functions influence the vote because they often obtain identical results. Thus, some votes will be counted several

(a) Rectangular: $\begin{cases} \frac{1}{2} \text{ if } |x| \leq 1; \\ 0 \text{ otherwise.} \end{cases}$

(b) Triangular: $\begin{cases} (1 - |x|) \text{ if } |x| \leq 1; \\ 0 \qquad\qquad \text{otherwise.} \end{cases}$

(c) Biweight: $\begin{cases} \frac{15}{16}(1 - x^2)^2 \text{ if } |x| \leq 1; \\ 0 \qquad\qquad \text{otherwise.} \end{cases}$

(d) Triweight: $\begin{cases} \frac{35}{32}(1 - x^2)^3 \text{ if } |x| \leq 1; \\ 0 \qquad\qquad \text{otherwise.} \end{cases}$

(e) Tricube: $\begin{cases} \frac{70}{81}(1 - |x|^3)^3 \text{ if } |x| \leq 1; \\ 0 \qquad\qquad \text{otherwise.} \end{cases}$

(f) Epanechnikov: $\begin{cases} \frac{3}{4}(1 - x^2) \text{ if } |x| \leq 1; \\ 0 \qquad\qquad \text{otherwise.} \end{cases}$

(g) Gaussian: $\frac{1}{\sqrt{2\pi}} e^{-\frac{1}{2}x^2}$

(h) Cosine: $\begin{cases} \frac{\pi}{4} \cos(\frac{\pi}{2}x) \text{ if } |x| \leq 1; \\ 0 \qquad\qquad \text{otherwise.} \end{cases}$

(i) Inverse: $\begin{cases} \infty \text{ if } x = 0; \\ \frac{1}{|x|} \text{ otherwise.} \end{cases}$

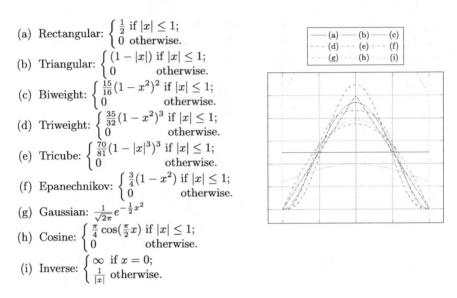

Fig. 1. Kernel functions below in common coordinate system.

times and will have more importance than they should have, which will bias the result. So the decision will be taken only by a specific profile of kernel function and the others will be ignored.

To verify this bias, we calculate the correlation of the votes that each kernel function produces. We calculate this score from the real data sets, presented in the Sect. 3.2, with the Pearson's method. For each of them, the outlier detection is performed with each kernel function to see which points are commonly considered as outliers and then to establish a correlation score to determine which functions often give the same results. Thus, the values in Table 1 are the mean and standard deviation of the correlation scores obtained for a pair of kernel functions over all data sets.

Assuming that correlation scores above 90% give such similar results that it is equivalent to doing the same thing then these functions can be separated into only three different groups:

- Rectangular and Gaussian;
- Triangular, Biweight, Triweight, Tricube, Cosine and Epanechnikov;
- Inverse.

As previously mentioned, if we use simultaneously these functions in the majority vote, then the decision will be made only by the second group. So the final result will have globally the same result as if it had been done with one of the functions of the second group. This vote will be biased and questions the relevance of using all these functions. Thus, we recommend the use of several functions that have different profiles and capture different behaviors.

Table 1. Correlation scores of outlier votes for each kernel function.

	(a)	(b)	(c)	(d)	(e)	(f)	(g)	(h)	(i)
(a)	—	0.68 ± 0.14	0.66 ± 0.14	0.62 ± 0.14	0.66 ± 0.14	0.73 ± 0.12	0.94 ± 0.04	0.71 ± 0.12	0.49 ± 0.17
(b)	0.68 ± 0.14	—	0.96 ± 0.03	0.93 ± 0.05	0.96 ± 0.04	0.92 ± 0.04	0.74 ± 0.12	0.94 ± 0.04	0.71 ± 0.14
(c)	0.66 ± 0.14	0.96 ± 0.03	—	0.94 ± 0.03	0.98 ± 0.02	0.91 ± 0.05	0.72 ± 0.12	0.93 ± 0.04	0.71 ± 0.14
(d)	0.62 ± 0.14	0.93 ± 0.05	0.94 ± 0.03	—	0.93 ± 0.04	0.86 ± 0.07	0.68 ± 0.13	0.87 ± 0.07	0.76 ± 0.13
(e)	0.66 ± 0.14	0.96 ± 0.04	0.98 ± 0.02	0.93 ± 0.04	—	0.92 ± 0.05	0.72 ± 0.12	0.93 ± 0.04	0.70 ± 0.15
(f)	0.73 ± 0.12	0.92 ± 0.04	0.91 ± 0.05	0.86 ± 0.07	0.92 ± 0.05	—	0.79 ± 0.10	0.98 ± 0.01	0.65 ± 0.16
(g)	0.94 ± 0.04	0.74 ± 0.12	0.72 ± 0.12	0.68 ± 0.13	0.72 ± 0.12	0.79 ± 0.10	—	0.78 ± 0.10	0.52 ± 0.16
(h)	0.71 ± 0.12	0.94 ± 0.04	0.93 ± 0.04	0.87 ± 0.07	0.93 ± 0.04	0.98 ± 0.01	0.78 ± 0.10	—	0.66 ± 0.16
(i)	0.49 ± 0.17	0.71 ± 0.14	0.71 ± 0.14	0.76 ± 0.13	0.70 ± 0.15	0.65 ± 0.16	0.52 ± 0.16	0.66 ± 0.16	—

The use of kernel functions, in Makoto method, is based on the assumption that the closer the neighbors are, the more they share common characteristics with the point of origin. However, this neighborhood is established from a Euclidean distance calculation which, by its simplicity, gives only an approximation of the real distance between the points. Thus, it is possible that the points considered to be nearest by the Euclidean calculation are not in reality so, and therefore do not have the characteristics closest to the point of origin. By putting aside this principle, we can imagine other kernel functions which take more care of further points or located on specific distance slices. For example, from the sinusoidal, we get the functions in Fig. 2. Makoto will use these kernel functions with the addition of the Rectangular function.

(a) Sinus: $\dfrac{sin(x\pi + \frac{\pi}{2})}{2}$

(b) Sinus inverse: $\dfrac{sin(x\pi - \frac{\pi}{2})}{2}$

(c) Sinus intermediate: $\dfrac{sin(2x\pi - \frac{\pi}{2})}{2}$

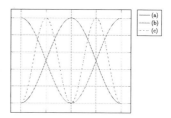

Fig. 2. Kernel functions sinusoidal.

2.2   Weighted k Nearest Neighbor Classification Method

These methods are based on the idea that the closest neighbors should have a higher weight than the farthest neighbors in the prediction. This prediction will be used to determine if a point is an outlier. To do this, the distances on which the search for neighbors is based must be transformed into weight using the functions presented in Sect. 2.1.

The wkNN classification method is used to combine the behavioral attributes of the neighborhood in order to make a prediction. It has been generalized to take the existence of several behavioral attributes within the same point. The prediction algorithm takes the following form:

Step 1 $\rightarrow$ Define the number of neighbors k, the distance function d, usually the Euclidean distance, and the kernel function f.

Step 2 $\rightarrow$ Let $O = \{(X_i, B_i)|i \in [1, n]\}$ the set of n points where X_i is the set of contextual attributes and B_i is the set of behavioral attributes. Steps 3 to 7 are applied for each object, we note the current one (X, B).

Step 3 $\rightarrow$ Find the k nearest neighbors to the current point using the method distance only on contextual attributes, such as $d(X, X_i)$. This set will be denoted $K = \{(X_{k_i}, B_{k_i})|i \in [1, k]\}$.

Step 4 $\rightarrow$ As in the previous step, find the $k + 1$ nearest neighbor to the current object. This neighbor will be denoted (X_{k+1}, B_{k+1}) and will be used to normalize the distances between the point and its neighbors.

Step 5 $\rightarrow$ Normalize the distances of the k nearest neighbors such as: $D_{k_i} = D(X_{k_i}, X_{k+1}) = \frac{d(X, X_{k_i})}{d(X, X_{k+1})}$

Step 6 $\rightarrow$ Transform the normalized distance into weight using the kernel function, $w_{k_i} = f(D_{k_i})$.

Step 7 $\rightarrow$ Let $b \in B$ be a behavioral attribute and C_b the set of possible classes for b then the set of predictions is defined as the class with the highest weighting for each behavioral attribute:

$$\hat{B}_{(X,B)} = \{\forall b \in B | max_b(\forall c \in C_b|\sum_{k_i \in K} \begin{cases} \text{if } c_{k_i} = c, w_{k_i} \\ \text{otherwise}, 0 \end{cases})\} \tag{1}$$

For example, a data set whose contextual attributes are planar coordinates and the behavioral attribute is a colored geometric figure, a *red circle* or a *blue rectangle*. Let a point among this set whose coordinates are center and its unknown geometrical shape will be symbolized by a *green star*. The algorithm presented above is used to determine the shape of the object. To begin, the five closest neighbors are found using the Euclidean distance. To normalize the distances, a sixth neighbor is found, which is not visually represented because it has no influence in the final decision. The weight of each neighbor can be calculated from the normalized distance. Thus, the prediction in the shape of the point can change according to the chosen kernel function. Figure 3 plots the weight of each neighbor and the resulting prediction for the Rectangular, Triangular, Triweight and Gaussian kernel functions. The gradient represents the relative weight according to the relative distance to the $k + 1$ neighbor. The darker it is, the more weight the neighbor will have.

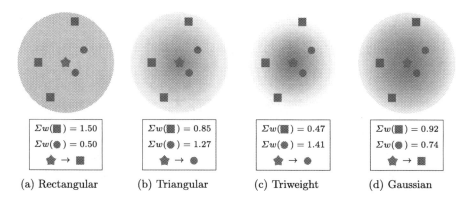

$\Sigma w(\blacksquare) = 1.50$	$\Sigma w(\blacksquare) = 0.85$	$\Sigma w(\blacksquare) = 0.47$	$\Sigma w(\blacksquare) = 0.92$
$\Sigma w(\bullet) = 0.50$	$\Sigma w(\bullet) = 1.27$	$\Sigma w(\bullet) = 1.41$	$\Sigma w(\bullet) = 0.74$
★ → ■	★ → ●	★ → ●	★ → ■
(a) Rectangular	(b) Triangular	(c) Triweight	(d) Gaussian

Fig. 3. Predictions according to the chosen kernel function.

This example illustrates the influence of kernel functions in the prediction. Here, the *blue rectangles* are in the majority but further from the evaluated point than the *red circles*. Thus, the prediction changes according to the difference between the maximum and minimum of the kernel function used (see Fig. 1). In this example, the Triangular and Triweight functions vary enough that the proximity of the red circles counterbalances the numerical advantage of the blue rectangles. In contrast, the Rectangular and Gaussian functions vary little or not at all to allow this.

2.3   Adaptative k per Class

The choice of the number of neighbors k is important because it affects the predictions made. A bad value of k decreases the quality of prediction and thus the quality of outlier detection. If the number of neighbors k is too large, then small groups of points with the same behavioral values may be underrepresented by different neighboring groups and therefore the prediction will not be representative of reality. The presence of this type of such cases when the number of representatives for each behavioral value is unbalanced. For example, in Fig. 3 the number of neighbors k is equal to 5 but if this number is decreased then the weights given by the kernel functions change and the predictions of Rectangular and Gaussian move from *blue rectangle* to *red circle*. And can change if the number k had been higher. Finally, is 5 the right value of k for this example?

Moreover, choosing a unique number of neighbors k is not always relevant, and it is better to choose an adaptive number depending on the point being evaluated. Several methods to establish these numbers have been proposed, and their efficiency compared to the use of a unique k has already been demonstrated [6]. The main problem is the asymmetry of the data sets in the distribution of behavioral values. A number k that is relevant for the majority classes can be destructive for the minority ones. If this number is too high compared to the absolute number of representatives, then we will fall into the case discussed above. This will cause the suppression of these minority classes. Based on this principle, the optimal number k is different for each behavioral value.

The formula for calculating the number of neighbors per class is inspired by the one proposed by Baoli et al. [3]. It has been modified to avoid the use of constants, so the constant α is removed to be replaced by the lower bound equal to 3 to handle cases where the value given by the equation is too low. While the starting value k in the original equation is approximated by the formula $\sqrt{\frac{max(\forall y \in V_b | N(y))}{2}}$.

Let $b \in B$ be a behavioral attribute and V_b the set of behavioral values of b then for each value $x \in V_b$ the number of neighbors is determined according to the following equation:

$$3 \leq \frac{\sqrt{\frac{max(\forall c \in V_b | N(c))}{2}} * N(x)}{max(\forall y \in V_b | N(y))} \leq min(\forall y \in V_b | N(y)) \tag{2}$$

where $N(\_)$ gives the number of elements with the same behavioral value provides as parameter. The number of neighbors must be between 3 and the number of elements of the smallest class. The lower bound is 3 because it is the minimum number of neighbors that we consider reasonable to take to make a decision, less would be absurd. The upper bound is the number of elements of the minority class, so a behavioral value cannot be underrepresented in its neighborhood by a majority class and can influence decisions.

To illustrate the relevance of the upper bound, the *Shuttle* data set (see Sect. 3.2) will be used as an example. The specificity of this data set is the great disparity in the distribution of points for each behavioral value, i.e. between 10 and 45 000. Thus, in the case where a point initially has a behavioral value of **1** when it should be **6**, then the number k used without the bound is equal to 150 (see Table 2). Logically, these nearest neighbors will have a behavioral value equal to **6**, but it only has 10 points in the data set. In the best case where these points were the 10 nearest neighbors, there are still 140 neighbors who will probably be of the majority behavioral value, i.e. **1**. Even with a good neighborhood weighting, the 140 neighbors will have more weight and this hypothetical outlier cannot be detected. It is therefore necessary to limit this number of neighbors in order to ensure that even minority behavioral values can have an influence in the detection.

Table 2. The neighbor number k calculated with and without the upper bound in Eq. 2 for *Shuttle* data set.

Behavioral value		1	2	3	4	5	6	7
Number of members		45 586	50	171	8 903	3 267	10	13
the number k	without bound	150	3	3	29	10	3	3
	with bound	10	3	3	10	10	3	3

2.4   Sub Data Sets and Sub Contextual Attributes and Filtering

In order to make the method more efficient and generalized to avoid **overfitting**, we decide to use this algorithm on a defined number of sub-data-sets n_s inspired by the functioning of **Random Forest** machine learning algorithms. Let n_c be the number of contextual attributes, then $n_s = \sqrt{n_c} * 10$. To do this, we must first decide how to create a subset from a data set. The objective is to have several sub-data-sets, i.e., with only a part of the points, in order to evaluate all the points in slightly different configurations to finally determine that the point is an outlier if it is in the majority of cases.

The subsets should respect several characteristics if they want to use them for this outlier detection method. First, all points must be represented, i.e., a point must be present in at least one subset, so that all points can be evaluated at least once. This feature is most likely to be respected when there are numerous subsets to be generated, however we choose to force it algorithmically in order to ensure it in all circumstances. Second, all subsets must have a distribution of behavioral values similar to the original data set in order to have a representation close to the original data set.

To respect the first condition, an algorithm is used to create the minimum number of subsets that allows the presence of all points at least once. This algorithm only needs the percentage of points present on each subset, p_r. It must not be too low in order to have subsets close from the original set to have reliable decisions, e.g. 70%. Thus, this minimum number of subsets created is equal to $\lfloor \frac{1}{p_r} \rfloor$. Then the rest of the subsets, i.e. $n_s - \lfloor \frac{1}{p_r} \rfloor$, are created randomly. All these subsets always respect the second condition.

Still for the same purpose as the subsets, only a part of the contextual attributes are used. Half of the attributes are kept, and each subset uses a different subgroup of attributes. But before that, a slight filtering of the contextual attributes is performed in order to remove those that are considered useless in establishing the behavioral value because they will influence the creation of neighborhoods, deviating them from reality and making the decision less reliable. Thus, we decide to remove from the process all attributes that are not correctly correlated with at least one of the other attributes. This number must be low enough so that only noisy attributes are removed, e.g. ±0.1. Finally, the correlation matrix is calculated using the **MIC** method, which is more accurate than the **Pearson** and **Spearman** methods which are commonly used.

3   Experimentation

For the experimentation, we study two kinds of data sets. The first is synthetic one and the second is real data sets proposed by the community. Computations have been performed on the supercomputer facilities of the *Mésocentre de calcul de Franche-Comté*. The details of the results on the synthetic data sets and the real data sets used are present in a GitHub directory[1].

---

[1] https://github.com/JessyColonval/Makoto.

During these experiments, Makoto will have 70% of the points of the original data set for each subset and the filtering of contextual attributes is equal to 0.1 of correlation. As comparison, 4 state-of-the-art methods are used: IQR [15], LOF [4], SVM [2] and Isolation Forest [12]. They are chosen because they are often used for outlier detection and have different approaches. However, there are other methods that are more recent as PyOD [17]. But it doesn't use the same logic of detection, in fact they assume that the training subset doesn't contain outliers and the detection is performed only on the test subset.

For validation of certain results, we will use the same 6 *machine learning* (ML) algorithms at three different ways (see Sects. 3.1 and 3.2). They come from the Python library *scikit-learn*[2] and represent different approaches in order to avoid being biased by the results that only one type of *machine learning* would give. These algorithms are: *SVC, KNeighborsClassifier, RandomForestClassifier* (RF), *ExtraTreesClassifier* (ET), *GradientBoostingClassifier* and *LogisticRegression* (LR). All meta-parameters have the default values, except for ET/RF where n_estimator is 200 and LR where max_iter is 1 000.

3.1   On Synthetic Data Sets

These data sets are created with the same algorithm that was designed to generate the *Madelon* data set [8]. They all have only one behavioral attribute whose values are homogeneously distributed. And all contextual attributes are useful for establishing the behavioral value, i.e., there is no noise. The original generation does not contain any outliers, they are added later and are therefore known for further experimentation.

Table 3. Characteristics of synthetic data sets.

samples	attributes	classes
250	10 ... 50, 25	2 ... 5
500	10 ... 70, 25, 75	2 ... 5
1 000	10 ... 100, 25, 75	2 ... 7
2 500	10 ... 100, 25, 75	2 ... 8
5 000	10 ... 100, 25, 75, 125	2 ... 11
10 000, 25 000	10, 50, 75, 100, 250	2, 3, 5, 7, 9, 11

The outliers are created using the 6 ML algorithms described earlier. The idea is to detect the most useful points for predictions in order to change their behavioral value. Thus, the chances of having strong outliers which are harmful to the predictions is increased. While a purely random method wouldn't prevent the selection of points that are not very useful for the prediction. For this purpose, the data sets are randomly divided into a training subset (70%) and a test

---

[2] https://scikit-learn.org/stable/index.html.

subset (30%). These 6 ML algorithms are trained with the training subset, then we see if they are able to correctly predict the points contained in the test subset. These actions are repeated 10 times, and count for each point the number of times they have been correctly predicted. Finally, the points that will change in behavioral value are all those with the lowest scores, and stop once the desired number of outliers is created. In cases where there are more than 2 behavioral values, the new one is chosen randomly.

Thus, with these methods, we create 406 data sets having the characteristics presented in Table 3 and having 5% outliers. These data sets have distinct characteristics to ensure that there is no significant difference in behavior based on the numbers of points, of attributes or of behavioral values. For those with a number of points between 250 and 5 000, the numbers of attributes increase with a step of 10 (except for 25 and 75) and the numbers of behavioral attributes increase with a step of 1. Data sets with 10 000 and 25 000 points are less exhaustive in their number of attributes and behavioral values due to computational time concerns during their generation and outlier detection.

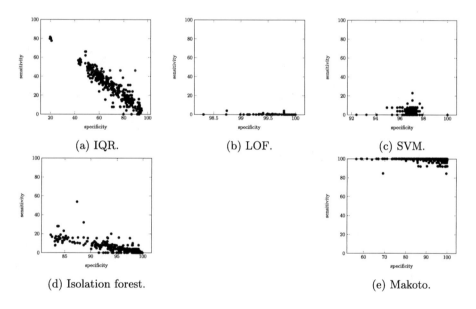

Fig. 4. Specificity and sensitivity of the 5 methods.

Since the generated outliers are known, then it's possible to use the sensitivity and specificity scores to determine which one is the best limit. As a reminder, the calculation of the sensitivity and specificity is based on the numbers of true positive (TP), true negative (TN), false positive (FP) and false negative (FN). The sensitivity measures the ability of a test to give a positive result when a hypothesis is verified, it's calculated by: $\frac{TP}{TP+FN}$. In this case, it is interpreted by the percentage of correctly detected outliers. And the specificity measures the ability of a test to give a negative result when the hypothesis is not verified, it's calculated by: $\frac{TN}{TN+FP}$. In this case, it is interpreted by the percentage of undetected points that are actually healthy. These metrics are interpreted together, a good method must give two scores as close to 100% as possible.

Figure 4 compares the sensitivity and specificity scores obtained with the 5 methods on synthetic data sets. A good method should give sensitivity and specificity scores close to 100%, so visually the points should be at the top right of the plot. **Makoto** is clearly the best method, in most cases it is able to detect numerous outliers ($\geq 80\%$) while keeping well the healthy points ($\geq 80\%$). However, there are some cases where too many healthy points are removed (60–79%) but this is still acceptable compared to the other solutions. While the **LOF**, **SVM** and **Isolation Forest** keep the healthy points well but remove very few outliers ($\leq 20\%$), which makes these methods not very useful. **IQR** is the worst because these solutions are either useless, i.e. few outliers detected with healthy points kept, or harmful, i.e. too many healthy points deleted.

3.2 Real Data Sets

In order to confirm the efficiency of **Makoto**, the comparison continues with 17 real data sets whose contextual attributes have continuous values. They all come from the UCI Machine Learning Repository[3] database, except Mammography, which comes from the BCSC (Breast Cancer Surveillance Consortium) site[4]. The characteristics of these data sets cover several combinations (Table 4), in order to verify that the behavior of a method doesn't change according to these.

Table 4. Characteristics of real data sets.

Data sets	Points	Attributes	Behavioral values	Data sets	Points	Attributes	Behavioral values
annthyroid	7 200	21	3	multiple features	2 000	619	10
breastW	683	9	2				
cardio	2 126	21	10	musk	6 598	166	2
glass	214	9	6	parkinson	756	752	2
ionosphere	351	33	2	pendigits	10 992	16	10
isolet	7 797	617	26	satimage2	6 435	36	6
letter recognition	20 000	16	26	shuttle	58 000	9	7
				wine	178	13	3
mammography	11 183	6	2	wine quality	1 599	11	6

The metrics used to compare the methods studied are different from those used for the synthetic data sets. This is because these data sets come from the

---

[3] https://archive.ics.uci.edu/ml/index.php.
[4] https://www.bcsc-research.org/.

Table 5. Comparison of cross-validation, false positive and class keeping scores of IQR, LOF, SVM, Isolation Forest and Makoto methods on real data sets.

Data sets	IQR CV	FP	CK	LOF CV	FP	CK	SVM CV	FP	CK	IF CV	FP	CK	Makoto CV	FP	CK
annthyroid	Error		✗	98.12%	94.72%	✓	97.12%	97.98%	✓	97.68%	94.22%	✓	Error		✗
				± 0.22	± 0.12		± 0.22	± 0.06		± 0.19	± 0.07				
	53.60%			11.22%			3.01%			5.54%			7.42%		
breastW	97.13%	91.66%	✓	96.73%	95.08%	✓	97.09%	99.67%	✓	98.61%	70.86%	✓	98.8%	6.22%	✓
	± 1.12	± 0.36		± 1.19	± 0.25		± 0.91	± 0.0		± 0.29	± 0.90		± 0.72	± 0.94	
	28.11%			24.60%			7.32%			37.63%			2.20%		
cardio	81.4%	60.35%	✗	82.45%	71.67%	✓	82.06%	85.30%	✓	81.23%	72.71%	✓	90.06%	25.53%	✓
	± 1.94	± 0.19		± 1.04	± 1.14		± 1.5	± 0.3		± 1.26	± 0.35		± 0.97	± 0.36	
	56.35%			1.69%			3.15%			16.04%			14.44%		
glass	73.46%	28.12%	✗	73.49%	40.98%	✓	72.28%	49.17%	✓	74.75%	48.7%	✓	82.72%	21.99%	✓
	± 4.87	± 1.09		± 5.15	± 1.25		± 4.66	± 1.76		± 5.05	± 1.71		± 4.92	± 0.98	
	36.45%			15.89%			1.87%			16.82%			23.83%		
ionosphere	95.12%	59.47%	✓	94.23%	55.61%	✓	91.21%	91.67%	✓	90.81%	86.54%	✓	97.0%	23.76%	✓
	± 2.58	± 0.43		± 2.00	± 0.72		± 1.88	± 0.0		± 1.97	± 0.52		± 1.75	± 1.24	
	48.15%			34.76%			2.85%			32.48%			8.83%		
isolet	Error		✗	93.6%	75.0%	✓	93.53%	90.55%	✓	93.1%	93.4%	✓	95.23%	39.48%	✓
				± 0.5	± 0.0		± 0.44	± 0.25		± 0.49	± 0.14		± 0.34	± 0.61	
	96.63%			0.05%			2.98%			14.21%			3.36%		
letter recognition	90.04%	68.37%	✓	91.48%	94.52%	✓	91.64%	92.75%	✓	91.48%	88.15%	✓	95.03%	49.01%	✓
	± 0.38	± 0.23		± 0.25	± 0.23		± 0.32	± 0.12		± 0.27	± 0.09		± 0.26	± 0.18	
	9.96%			0.19%			3.01%			13.57%			9.61%		
mammography	99.41%	94.33%	✓	98.66%	90.98%	✓	98.69%	91.23%	✓	99.26%	91.06%	✓	99.66%	7.06%	✓
	± 0.04	± 0.03		± 0.09	± 0.19		± 0.12	± 0.16		± 0.03	± 0.2		± 0.09	± 0.13	
	37.61%			1.78%			3.08%			19.88%			1.34%		
multiple features	Error		✗	98.33%	3.33%	✓	98.29%	90.14%	✓	99.0%	91.15%	✓	99.1%	30.36%	✓
				± 0.47	± 10.54		± 0.45	± 0.54		± 0.45	± 0.2		± 0.4	± 2.22	
	97.65%			0.05%			3.00%			35.45%			1.40%		
musk	91.38%	59.22%	✓	96.54%	93.17%	✓	96.45%	99.31%	✓	96.39%	98.78%	✓	98.15%	14.4%	✓
	± 4.61%	± 0.37		± 0.35	± 0.41		± 0.27	± 0.25		± 0.34	± 0.06		± 0.28	± 0.45	
	97.39%			0.32%			2.99%			6.68%			2.39%		
parkinson	Error		✗	86.67%	83.26%	✓	86.45%	87.03%	✓	87.04%	86.19%	✓	87.0%	22.6%	✓
				± 1.67	± 0.79		± 1.9	± 1.08		± 1.78	± 1.14		± 1.74	± 1.69	
	100%			3.04%			3.04%			2.78%			2.12%		
pendigits	98.58%	85.18%	✓	98.67%	76.34%	✓	98.52%	94.99%	✓	98.87%	50.45%	✗	98.91%	44.56%	✓
	± 0.2	± 0.17		± 0.2	± 0.45		± 0.16	± 0.16		± 0.23	± 0.17		± 0.16	± 0.57	
	4.64%			1.56%			2.98%			51.91%			0.99%		
satimage2	89.19%	96.65%	✓	89.89%	76.35%	✓	89.28%	97.56%	✓	88.48%	58.24%	✓	95.25%	10.93%	✓
	± 0.45	± 0.16		± 0.52	± 0.49		± 0.52	± 0.06		± 0.75	± 4.04		± 0.5	± 0.27	
	8.42%			1.54%			2.97%			19.40%			6.96%		
shuttle	99.83%	81.91%	✗	Error		✗	Error		✓	99.78%	60.8%	✗	99.43%	92.26%	✓
	± 0.05	± 0.0								± 0.03	± 0.04		± 0.04	± 0.13	
	79.56%			19.00%			2.79%			15.62%			1.00%		
wine	98.03%	90.2%	✓	98.11%	70.67%	✓	98.3%	88.12%	✓	95.96%	95.83%	✓	98.65%	53.33%	✓
	± 1.66	± 0.0		± 1.54	± 1.41		± 1.45	± 1.01		± 2.81	± 0.0		± 1.22	± 2.64	
	9.55%			2.81%			4.49%			13.48%			2.25%		
winequality	64.42%	51.85%	✓	63.66%	41.31%	✓	63.48%	35.72%	✓	63.99%	46.05%	✓	Error		✗
	± 2.17	± 0.61		± 1.69	± 1.02		± 1.85	± 1.16		± 1.95	± 0.57				
	25.20%			2.94%			2.88%			16.45%			27.39%		

real world, and we're looking for only the outliers contains in these data sets. Thus, it is not possible to measure the ability to detect outliers because we don't know which points are real outliers. To circumvent this problem, the comparison is done using the cross-validation and false positivity scores.

The cross-validation score will measure the performance of the 6 ML algorithms described above after removing outliers. The idea is that the presence of outliers reduces the performance of these algorithms. Thus, by measuring this score for each method, it's possible to determine which removal was the most beneficial for these ML algorithms, therefore which method is the most efficient. To obtain this score, the data set is randomly separate into two subsets: a training subset (70%) and a test subset (30%) which keeps the proportion of the behavioral values of the original set. Then, each of the ML algorithms is trained with the training subset, and we look at their ability to correctly predict the behavioral value of the points contained in the test subset. This operation gives a score as a percentage, the closer it is to 100% the more the algorithm is able to correctly predict this point and the more effective the training, the better the quality of the data. This process is repeated 10 times, i.e. with 10 different separation of the subsets, then averaged in order to have more reliable results.

The false positivity score will measure the relevance of the points that have been removed. It is assumed that a true outlier is a point that will be wrongly predicted, and a false outlier is a point that will be correctly predicted, so that it could have been kept in the training set. However, this way of calculation is imperfect. If too many outliers are removed, then the data set is denatured and those predictions made are not relevant. It works better with few outliers and should be read in addition to the other scores. This score is calculated with the same logic as the cross-validation score. Except that the training set are the healthy points and the test set are the outliers. For the same reasons, this process is repeated 10 times, but only changing the random seeds of the ML algorithms and keeping the same subsets.

The Table 5 gives all the results obtained from the 5 methods on the 17 real data sets. The **CV** columns give, respectively, the averages of the cross-validation scores and the standard deviations, and then the percentage of outliers detected among all points. The **FP** columns give the mean of the false positivity scores and the standard deviations. The **CK** columns indicate if at least one behavioral value was completely considered as an outlier. The cells labeled **Error** appears when a method have denatured the data set too much to be able to compute a part of the metrics. The best methods are those with a high cross-validation score, a low false positivity score, no missing behavioral values and a reasonable number of detected outliers ($\leq 25\%$). Thus, **Makoto** outperforms the other methods on 15 of the 17 data sets.

4   Conclusion

This paper presented a complete outlier detection method, Makoto, and showed its effectiveness by comparing it with 4 other methods in the literature on synthetic and real data sets. However, some of Makoto's results can be improved by changing some of these meta-parameters or the way they are calculated. A possible extension would be to establish a better methodology to choose them. Moreover, this method is similar to a machine learning algorithm and could be used to predict behavioral values.

Acknowledgement. Work supported by the French National Research Agency (contract ANR-18-CE25-0013) and by the EIPHI Graduate School (contract ANR-17-EURE-0002)

References

1. Aggarwal, C.C.: Outlier Analysis. Springer International Publishing, Cham (2015). https://doi.org/10.1007/978-3-319-47578-3
2. Amer, M., Goldstein, M., Abdennadher, S.: Enhancing one-class support vector machines for unsupervised anomaly detection. In: Proceedings of the ACM SIGKDD Workshop on Outlier Detection and Description, ODD 2013, pp. 8–15. Association for Computing Machinery, New York (Aug 2013). https://doi.org/10.1145/2500853.2500857
3. Baoli, L., Qin, L., Shiwen, Y.: An adaptive k-nearest neighbor text categorization strategy. ACM Trans. Asian Lang. Inform. Process. **3**(4), 215–226 (2004). https://doi.org/10.1145/1039621.1039623
4. Breunig, M., Kriegel, H.P., Ng, R., Sander, J.: LOF: identifying density-based local outliers. In: ACM Sigmod Record, vol. 29, pp. 93–104 (Jun 2000). https://doi.org/10.1145/342009.335388
5. Chehreghani, M.H.: K-nearest neighbor search and outlier detection via minimax distances. In: Proceedings of the 2016 SIAM International Conference on Data Mining, p. 9. Society for Industrial and Applied Mathematics (2016). https://doi.org/10.1137/1.9781611974348.46
6. Dietterich, T., Wettschereck, D., Wettschereck, D., Dietterich, T.G.: Locally adaptive nearest neighbor algorithms. In: Advances in Neural Information Processing Systems 6, pp. 184–191. Morgan Kaufmann (1994)
7. Epanechnikov, V.A.: Non-parametric estimation of a multivariate probability density. Theory Probabil. Appli. **14**(1), 153–158 (1969). https://doi.org/10.1137/1114019
8. Guyon, I., Gunn, S., Ben-Hur, A., Dror, G.: Design and Analysis of the NIPS2003 Challenge, vol. 207, pp. 237–263 (Nov 2008). https://doi.org/10.1007/978-3-540-35488-8_10
9. Hechenbichler, K., Schliep, K.: Weighted k-Nearest-Neighbor Techniques and Ordinal Classification. discussion paper 399 (Jan 2004)
10. Keller, F., Muller, E., Bohm, K.: HiCS: high contrast subspaces for density-based outlier ranking. In: 2012 IEEE 28th International Conference on Data Engineering, pp. 1037–1048 (Apr 2012). https://doi.org/10.1109/ICDE.2012.88
11. Kou, Y., Lu, C.T., Chen, D.: Spatial weighted outlier detection. In: Proceedings of the 2006 SIAM International Conference on Data Mining, p. 5. Proceedings, Society for Industrial and Applied Mathematics (Apr 2006). https://doi.org/10.1137/1.9781611972764.71
12. Liu, F.T., Ting, K.M., Zhou, Z.H.: Isolation-Based Anomaly Detection. ACM Trans. Knowl. Dis. Data **6**(1), 3:1–3:39 (2012). https://doi.org/10.1145/2133360.2133363
13. Lu, C., Chen, D., Kou, Y.: Algorithms for spatial outlier detection. In: Third IEEE International Conference on Data Mining, pp. 597–600 (Nov 2003). https://doi.org/10.1109/ICDM.2003.1250986

14. Thung, F., Wang, S., Lo, D., Jiang, L.: An empirical study of bugs in machine learning systems. In: 2012 IEEE 23rd International Symposium on Software Reliability Engineering, pp. 271–280 (Nov 2012). https://doi.org/10.1109/ISSRE.2012. 22
15. Whaley, D.L.: The Interquartile Range: Theory and Estimation (2005)
16. Zhang, S., Wan, J.: Weight-based method for inside outlier detection. Optik **154**, 145–156 (2018). https://doi.org/10.1016/j.ijleo.2017.09.116
17. Zhao, Y., Nasrullah, Z., Li, Z.: Pyod: a python toolbox for scalable outlier detection. J. Mach. Learn. Res. **20**(96), 1–7 (2019). http://jmlr.org/papers/v20/19-011. html

Traffic

MANet: An End-To-End Multiple Attention Network for Extracting Roads Around EHV Transmission Lines from High-Resolution Remote Sensing Images

Yaru Ren⬤, Xiangyu Bai$^{(\boxtimes)}$ ⬤, Yu Han⬤, and Xiaoyu Hu⬤

Inner Mongolia School of Computer Science, Inner Mongolia University, Hohhot 010021, China
bxy@imu.edu.cn

Abstract. Complete and accurate road network information is an important basis in the detection of EHV transmission lines, and regular updates of road distribution near transmission lines are necessary and meaningful. However, no relevant research has been found for this application area, and coupled with the fact that roads themselves are significantly challenging, extracting roads with good connectivity and integrity in remote sensing images remains a problem to be solved. Therefore, in this paper, we develop a new end-to-end road extraction network, Multiple Attention Networks (MANet). Specifically, by fusing convolutional and self-attentive approaches, we focus on global contextual features to obtain an effective feature map. In addition, the Strip Multi-scale Channel Attention (SMCA) module is specifically designed for the line features of roads, focusing on extracting row and column features, while the Edge-aware Module (EAM) is used to extract connected and complete roads, aided by edge information. Meanwhile, in order to enhance the practicality of the study, a Mengxi Transmission Line Road Dataset was constructed independently following the processing process of remote sensing images in industrial production. By conducting relevant quantitative and qualitative experiments on this dataset and the publicly available CHN6-CUG dataset, it is fully verified that the method in this paper is superior to other advanced methods and can still extract roads with strong connectivity in complex backgrounds, which has good potential and outstanding advantages in practical applications.

Keywords: Road extraction · High-resolution remote sensing images · Deep learning · Semantic segmentation

1 Introduction

As one of the major infrastructures, the road network is widely used in various industrial fields and social life. Especially in the detection and inspection of ultra-high voltage transmission lines, regular measurement of their surrounding road networks and timely updating of their detailed information are essential to ensure the safe and smooth operation of the entire power system. With the development of remote sensing technology

© The Author(s), under exclusive license to Springer Nature Switzerland AG 2023
X. Yang et al. (Eds.): ADMA 2023, LNAI 14176, pp. 553–568, 2023.
https://doi.org/10.1007/978-3-031-46661-8_37

in recent years, the extraction of roads from high-resolution remote sensing images and ultra-high resolution remote sensing images has become a popular topic [1].

However, the roads themselves have complex geometric, radiometric, and topological characteristics, such as different widths, directional changes, uniform grayscale, obvious boundaries, and connectivity; they are also in the middle of complex scenes and easily obscured by obstacles such as vehicles, trees, buildings and their shadows, making the task very challenging. In addition, compared with the urban areas where road data are concentrated, the EHV transmission lines are widely distributed and span a large area, and the accessibility of the roads around them is weaker, mainly concentrated in remote areas far from the urban areas, with sparse and disorganized distribution of features and the existence of more third- and fourth-class roads as well as concrete roads and dirt roads, which are more easily integrated into the scene environment and make the extraction more difficult, with obvious differences compared with the former. Therefore, for road extraction around the ultra-high voltage transmission lines, it is important for the field to study a more advanced and suitable method to improve the model performance and further improve the accuracy and quality of road extraction.

In this paper, we propose a Multiple Attention Network (MANet) for road extraction. Even when the road distribution is very hidden and inherently tortuous, the road information can still be captured sensitively for accurate and effective extraction. The main contributions of this paper can be described as follows:

1. A new end-to-end road extraction network, MANet, is proposed, deploying a self-attention mechanism and a channel-attention mechanism, while adding target boundary information to constrain the extracted roads, effectively enhancing road connectivity and reducing disconnections;
2. A Strip Multi-scale Channel Attention (SMCA) mechanism is designed to extract features from two dimensions, row and column, respectively, for the geometric and topological features of winding and narrow roads, and perform multiscale differential fusion to improve the model's ability to perceive roads in complex scenes;
3. To the best of our knowledge, this paper is the first study to perform road extraction in the scenario around EHV transmission lines, for which a road dataset is constructed.

2   Related Work

This section reviews the relevant research methods for road extraction.

The deep learning approach emerging in recent years is data-driven and represented by the semantic segmentation task, which is widely used in the field of road extraction. It mainly relies on the color, geometric and texture features of remote sensing images for feature extraction of images to achieve almost automated road extraction. The first use of deep learning methods for road extraction from remote sensing images in the field of road extraction was by Mnih et al. They used a restricted Boltzmann machine to extract roads from remote sensing images [2]. Subsequently, Wang et al. proposed a neural dynamic tracking framework based on deep convolutional neural networks and finite state machines to extract road networks [3]. However, this method was gradually replaced because it was limited in terms of accuracy and speed, and it was prone to overfitting because the roads themselves accounted for a small proportion of the whole image. Fully

Convolutional Networks (FCN) is regarded as the pioneer of semantic segmentation, and its proposal has led to a significant improvement in semantic segmentation and realized end-to-end image segmentation [4]. Zhong et al. were the first to use FCN for research in the field of road extraction [5]. And then, a series of methods based on encoder-decoder structures were proposed one after another, such as U-Net [6], CasNet [7], D-Linknet [8], etc., which achieved multi-level feature stitching and reuse while extracting features more deeply, all achieving better results at that time. Shi et al. first introduced generative adversarial networks (GAN) into pixel-level remote sensing image classification by acting the basic segmentation network as Generator in GAN [9], and implemented the task of road area detection in Google Earth remote sensing images by GAN model. Recently, with the boom of Transformer structure [10], some researches based on this framework have also emerged, such as RoadFormer [11], HA-RoadFormer [12], etc. Compared with the above-mentioned methods based on convolutional neural networks, they have stronger ability to learn remote features and global modeling, and pay more attention to the global features of images, and all of them have also achieved considerable results.

3   Methodology

3.1   MANet Overall Framework

Figure 1 shows the overall architecture of the proposed MANet model. The whole network is designed with an encoder-decoder structure, using DeepLab V3+ [13] as the semantic segmentation model framework and improving on the original model. In the following, the proposed network structure is briefly described in terms of two components, encoder and decoder.

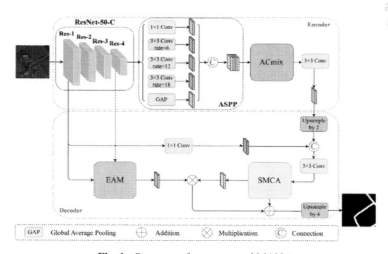

Fig. 1. Structure of our proposed MANet.

The encoder part uses ResNet-50-C [14] as the backbone network. We first use the improved residual network to extract features from four different stages, denoted as

$f_{R_i}(i = 1, 2, 3, 4)$; second, the extracted features f_{R_4} are fed into the ASPP structure to enrich the encoder module in the compiled code network by fusing multi-scale contextual information. Based on this, we further add a self-attention and convolution integration module (ACmix) [15], which effectively integrates the advantages of self-attention and convolution, capturing global contextual information effectively while paying more attention to the acquisition of local information, and at the same time, exploiting the features of the whole image as fully as possible without significantly increasing the computational effort to extract the rich information in it.

In the decoder, the different hierarchical features obtained from the encoder will be convolved and upsampled to obtain the high-resolution segmentation results. We first perform upsampling operation on the multi-scale high-level feature map extracted from the encoder to expand the feature map size by a factor of 2. After compressing the dimensionality of the shallow sub-feature map f_{R_1} in the backbone network as well, we stitch the two together to perform cross-level fusion of high bottom-level features. The feature map is then adjusted by a 3×3 convolution, after which we feed the feature map fused with rich semantic information into our proposed Strip Multi-scale Channel Attention (SMCA), which can fuse features in both row and column directions and is well suited for road feature extraction. At the same time, we consider the importance of edge information, and for the four different layers of features obtained from the initial extraction of the network, we use the shallow features f_{R_1} and the deep features f_{R_4} as the input of the Edge-aware Module (EAM) [16], and get the feature output containing the road boundary information through the detail information in the shallow layer and the semantic information in the deep layer, and then multiply it pixel by pixel with the SMCA output feature map to give the boundary information, and then pixel by pixel Then it is added together to reduce the feature segmentation map from multiple dimensions. Finally, the final road extraction result map is obtained after 4-fold upsampling operation.

3.2 ACmix-Based Encoder

When encoding feature information in the encoder, we deploy the self-attention and convolution integration module(ACmix), which mines the potential connection between convolution and self-attention from a new perspective, decomposing them into two-stage operations - the first stage divides the k × k convolution into k2 independent 1 × 1 convolution operations, in which the self-attention Query, Key, Value mapping is also composed by 1 × 1 convolution; the second stage convolution performs shift and sum operation, and self-attention computes attention weights and aggregates Value values. On this basis, it is found that they rely heavily on the 1 × 1 convolution operation in the first stage, so the first stage operation of both are fused and then the second stage calculation is performed separately, as shown in Fig. 2. The convolution and self-attention mechanisms are effectively integrated while minimizing the computational overhead, so that the local information is complementarily integrated with the global information and the model considers the road itself without forgetting the complex and rich background information in remote sensing images.

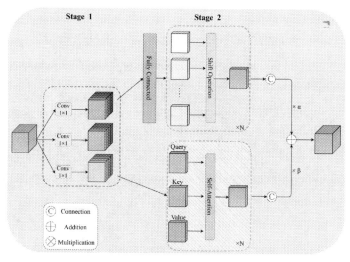

Fig. 2. Self-attention and convolution integration module.

3.3 SMCA and EAM based Decoder

Strip Multi-scale Channel Attention Module. Considering the directional extension of roads, inspired by Strip Attention Networks (SANet) [17], a Strip Multi-scale Channel Attention module is specially designed for road extraction with a three-branch structure that differentially fuses the lineal topological features that focus on roads horizontally and vertically in images from different dimensions, respectively.

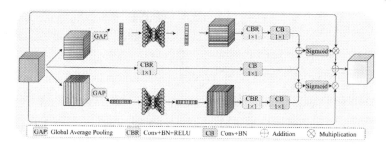

Fig. 3. Strip Multi-scale Channel Attention Module.

The module structure is shown in Fig. 3. SMCA is a three-branch structure, which first divides the feature map into three. The upper branch of the figure is the column pixel feature extraction structure, which firstly performs global average pooling operation to obtain global contextual features from the dimension of each row and performs compression processing of information; then goes through two fully connected layers to learn the attention coefficients of different channels; then the dimensionality is extended to recover the feature map size; and then two 1×1 convolution layers are passed to complete the enhancement of the global channel column direction features. On the other hand, the second branch is directly subjected to two convolution operations to ensure the extraction of local features and avoid losing details. The feature maps of the second branch are summed pixel by pixel with those of the column branch, fused after giving more attention to the multiscale features of the column pixels, fed into the Sigmoid activation function, filtering the miscellaneous terms to a certain extent to obtain the final weights, and multiplying the original feature maps pixel by pixel, the whole process involves the upper and middle branches in the figure, so that the model captures the road distribution information in the vertical direction in space.

The whole process of column pixel feature extraction can be described as follows:

$$\begin{cases} F_1 = \frac{1}{H \cdot C} F_{CB}(F_{CBR}(F_E(F_{Linear-S}(F_{Linear-R}(F_{GAP}(f_m)))))) \\ F_2 = F_{CB}(F_{CBR}(f_m)) \\ F_C = F_S(F_1 \oplus F_2) \otimes f_m \end{cases} \tag{1}$$

where, F_1 and F_2 represent the first and second branch processing respectively, F_C is the whole column pixel branching process. First, for the first branch, the module input feature map f_m goes through the row mapping of $\frac{1}{H \cdot C}$ as well as the global average pooling F_{GAP} to obtain the $W \cdot 1$ dimensional column feature vector, followed by two fully connected layers $F_{Linear-R}$ with $F_{Linear-S}$, $F_{Linear-R}$ representing the fully connected followed by ReLU activation function, $F_{Linear-S}$ representing the post-connected Sigmiod activation function; F_E representing the extended dimensionality, recovering the size, followed by two 1×1 convolutional layers F_{CBR} with F_{CB}, F_{CBR} representing the convolutional layer, BN, ReLU, F_{CB} denotes the convolutional layer with BN layer. The second branch F_2 performs two 1×1 convolutional layer F_{CBR} with F_{CB} pairs only f_m. ... After the with operation, the two are added pixel by pixel ($\oplus$), and then after the Sigmoid function, they are multiplied pixel by pixel ($\otimes$) with the original feature map to obtain the column branch output.

Similar to the column pixel extraction branch, the row pixel extraction branch is the lower branch in the graph, except that the column mapping is done to obtain the $1 \cdot H$ row feature vector. The procedure is as follows:

$$\begin{cases} F_2 = F_{CB}(F_{CBR}(f_m)) \\ F_3 = \frac{1}{W \cdot C} F_{CB}(F_{CBR}(F_E(F_{Linear-S}(F_{Linear-R}(F_{GAP}(f_m)))))) \\ F_R = F_S(F_2 \oplus F_3) \otimes f_m \end{cases} \tag{2}$$

The following is the SMCA module process:

$$F_{SMCA} = F_C \oplus F_R \tag{3}$$

Edge-Aware Module. For the extraction of road features, we not only consider the acquisition of global information that facilitates semantic segmentation. At the same time, we also take into account the importance of edge information. Therefore, we incorporate the Edge-aware Module (EAM) to extract the edge information effectively by using shallow features and deep features as inputs. As shown in Fig. 4.

Specifically, EAM combines the feature maps f_{R_1} and f_{R_4} output from Res-1 and Res-4 in ResNet-50-C, and then changes the number of channels through a 1×1 convolutional layer, and then performs a 2-fold upsampling operation on the feature

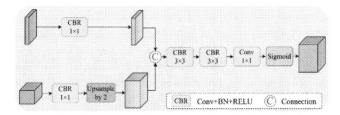

Fig. 4. Edge-aware Module.

map of the f_{R_4} branch, with $\bigcup$ denoting upsampling, so that the feature maps of the two branches are the same size and then stitched together, i.e., $\mathbb{C}$. Then after two 3×3 convolutional layers, BN with Relu, one 1×1 convolutional layer, and finally the edge extraction map is obtained via the Sigmoid function. The procedure is as follows.

$$F_{EAM} = F_S\left(F_{Conv}\left(F_{CBR}\left(F_{CBR}\left(\left(F_{CBR}(f_{R_1})\right)\mathbb{C}\left(\bigcup F_{CBR}(f_{R_4})\right)\right)\right)\right)\right) \qquad (4)$$

4 Results and Discussions

4.1 Dataset Descriptions and Training Details

Dataset Descriptions. To evaluate the effectiveness of our proposed model in detail, we conducted experiments on two sets of road extraction datasets. A detailed description of these two datasets is given below.

The first dataset is our independently constructed transmission line road dataset for the Mengxi power transmission line. The remote sensing images of this dataset are derived from the Gaofen-2 and SuperView-1 satellites, and are local images of some of the areas through which the ultra-high voltage transmission lines are located in the western region of Inner Mongolia Autonomous Region, China in 2022, covering an area size of about 3000 square kilometers, with a total of 18 scenes and a resolution between 0.5 m and 0.8 m.

The images of the public road dataset underwent finer and stricter processing and screening in the production process, with higher image clarity, and the scenes involved were mostly urban areas in developed regions, with roads mainly being highways, and roads in rural scenes were also easier to distinguish; whereas in this study, based on the perspective of industrial production applications, the data came directly from remote sensing satellites, and the image scenes were the environmental images around the channels of ultra-high voltage transmission lines. In order to make this study more practical, we obtained the original remote sensing data, followed the processing process of remote sensing images in industry, and used Arcgis, Qgis and other software to obtain usable remote sensing images through the processes of image mosaic, image color leveling, image correction, image cropping, image slicing and so on. It is undoubtedly a huge workload and inefficient to produce the required datasets for these remote sensing images using traditional manual annotation. In order to reduce the production cost and improve the annotation efficiency, this study selects OSM data combined with professional practitioners for manual review and annotation to construct the dataset. Finally, we obtained a total of 5365 images in the Mengxi transmission line road dataset, with a size of 256 × 256, and divided them into training and testing sets according to the ratio of 8:2, which are 4292 and 1073 images, respectively. However, at present, we cannot make them public due to copyright issues. A partial image of it is shown in Fig. 5.

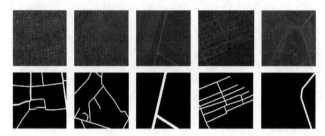

Fig. 5. Mengxi Transmission Line Road Dataset.

The second dataset is the CHN6-CUG road datase t [18], which was produced and shared by the team of Zhu from China University of Geosciences in 2021. Its remote sensing image base map is from Google Earth, and six major cities in China, namely Beijing, Shanghai, Wuhan, Shenzhen, Hong Kong, and Macau, are selected as the study area. The whole dataset is manually labeled and contains 4511 labeled images of 512 × 512 size, of which 3608 are used for model training and 903 are used for testing and result evaluation. A partial image of it is shown in Fig. 6.

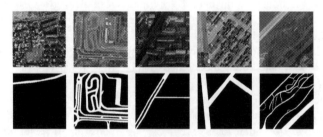

Fig. 6. CHN6-CUG road dataset.

Training Details. All the experiments performed in this paper are based on the Pytorch deep learning framework and are done using the MMSegmentation [19] framework in Open-MMLab. DeepLab V3+ was used as the baseline model. The loss function used in this paper is the joint loss function of the sum of the binary cross-entropy loss function and the dice function. In this paper, Stochastic Gradient Descent (SGD) [20] is used as an optimizer for the model training process and poly learning strategy is used to update the learning rate, i.e $lr = init_{lr} * (1 - iter/maxiter)^{0.9}$. And following the relationship between learning rate and batch size in MMSegmentation, the initial learning rate is set to 5e-3 in this paper. Momentum and weight decay are 0.9 and 0.0001, respectively. The model uses an NVIDIA GPU A100 to accelerate the training. To expand the dataset, we use data augmentation, including random cropping, coefficients from 0.5 to 2, random flipping and multi-scale augmentation. Finally, the original image is used for testing the model.

4.2 Evaluation Metrics

Different evaluation metrics reflect the performance advantages and disadvantages of the developed method from different perspectives. In this paper, six evaluation metrics are mainly used to evaluate the proposed network model, including precise, recall, road Intersection over Union, Mean Intersection over Union, F1-score and aver-age accuracy, all of which are more general evaluation metrics in semantic segmentation of remote sensing images [21].

Road extraction is considered as binary classification tasks, road areas are foreground, i.e., positive samples, and non-road areas are background, i.e., negative samples. Where, TP indicates correctly predicting positive samples as positive, FN indicates incorrectly predicting positive samples as negative, FP indicates incorrectly predicting negative samples as positive, and TN indicates correctly predicting negative samples as negative. The confusion matrix allows the calculation of the above evaluation index values.

$$Precise = \frac{TP}{TP + FP} \tag{5}$$

$$Recall = \frac{TP}{TP + FN} \tag{6}$$

$$IoU = \frac{TP}{TP + FP + FN} \tag{7}$$

$$MIoU = \frac{1}{k + 1} \sum_{i=0}^{k} \frac{TP}{TP + FP + FN} \tag{8}$$

$$F1 - score = 2 \times \frac{Precise \times Recall}{Precise + Recall} \tag{9}$$

$$AverageAccuracy = \frac{1}{k + 1} \sum_{i=0}^{k} \frac{TP + TN}{TP + TN + FP + FN} \tag{10}$$

4.3 Comparative Experiments

Experiments on the Mengxi Dataset. Table 1 shows that the method proposed in this paper significantly outperforms the other methods, where the bolded values represent the best results for quantitative comparison. Specifically, MANet has improved the values of its main indexes to different degrees compared with DeepLab V3 +, which is the most outstanding performance among other methods: recall is improved by 6.37%, F1-score is improved by 4.84%, road IoU is improved by 4.42%, MIoU is improved by 2.31%, and average accuracy is improved by 3.1%. Among them, SANet has the highest accuracy rate of 75.82%, but all other values of this model are low, and the network does not perform well when all values are considered, and the method in this paper still performs best. And it is not difficult to find through the experiment that the accuracy as a single indicator does not effectively evaluate the model performance, and the value is too high even makes the performance degraded.

Table 1. Quantitative comparison between MANet and other methods on the Road Dataset of Mengxi Transmission Line.

Model	Precise	Recall	F1-score	IoU	MIoU	AA
U-Net [6]	37.42	29.55	32.95	19.73	56.5	63.32
UperNet [22]	55.87	15.42	24.17	13.74	54.18	57.35
Swin Transformer [23]	56.83	31.53	40.56	25.44	60.13	65.07
DANet [24]	58.7	34.78	43.68	27.94	61.45	66.68
ViT [25]	64.15	29.37	40.29	25.23	60.18	64.21
Deeplab V3 + [13]	66.37	39.62	49.62	33.0	64.23	69.23
SANet [17]	**75.82**	11.48	19.92	11.06	52.98	55.63
MANet (Ours)	67.42	**45.99**	**54.46**	**37.42**	**66.54**	**72.33**

Figure 7 shows the visualization comparison of the method in this paper with the above seven parties, and the listed images are derived from the Mengxi transmission line road dataset. Compared with other advanced methods, MANet obviously has a better visual effect. Taking the first row as an example, the method designed in this paper has a stronger sensitivity to the roads when the image is weakly illuminated and the road distribution is not obvious. Comparing the whole image, we can see that under the difficult situation of image extraction, the method in this paper can still extract a relatively complete road network and effectively ensure the road connectivity to a certain extent, and the edges are smoother, while other methods have a lot of fractures and burrs on the edges.

Experiments on the CHN6-CUG Dataset. To further verify the superiority of MANet, quantitative evaluation was also conducted on the publicly available CHN6-CUG road dataset in this paper, and the results are shown in Table 2.

Figure 8 shows the visualization comparison between MANet and other methods on the CHN6-CUG road dataset.

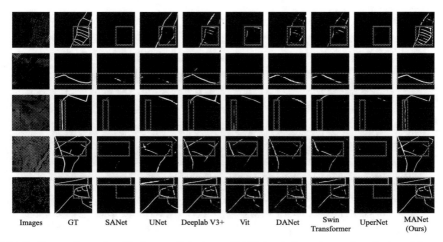

| Images | GT | SANet | UNet | Deeplab V3+ | Vit | DANet | Swin Transformer | UperNet | MANet (Ours) |

Fig. 7. Comparison of qualitative results between MANet and other advanced methods on Mengxi Transmission Line Road Dataset.

Table 2. Quantitative comparison between MANet and other methods on the CHN6-CUG Road Dataset.

Model	Precise	Recall	F1-score	IoU	MIoU	AA
DANet [24]	78.99	69.13	73.73	58.39	77.73	84.01
Deeplab V3 + [13]	**80.23**	70.36	74.97	59.97	78.58	84.65
U-Net [6]	63.51	66.66	65.01	48.2	71.96	82.71
UperNet [22]	76.35	73.46	74.88	59.85	78.45	86.04
Swin Transfomrer [23]	76.34	73.67	74.98	59.97	78.52	86.14
MANet (Ours)	77.06	**74.08**	**75.54**	**60.69**	**78.91**	**86.37**

4.4 Ablation Experiments

Experiments on the Mengxi Transmission Line Road Dataset. To validate the superior performance of the proposed network structure, a series of ablation studies were conducted in this paper to verify the effectiveness of the used and designed modules separately. Table 3 reports the quantitative evaluation of the effectiveness of each module on the Mengxi transmission line road dataset, and B denotes the baseline model Baseline, which is Deeplab V3 + for this method. Figure 9 presents the enhancement of the extracted visual effects after adding each module.

According to the model No. 1 and No. 2 in Table 3, it can be seen that the SMCA module developed in this paper has gained over the baseline model in road extraction except for the precise. From Fig. 9, it is easy to find that the module has a repair function for road breaks.

Table 4 lists several other contemporary advanced attention mechanism modules, namely SE [26], ECA [27], CBAM [28], and also includes the different effects of adding

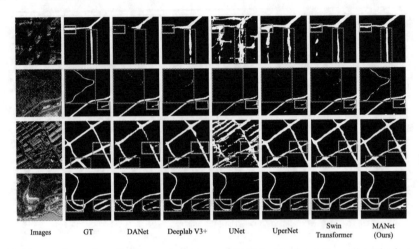

Images   GT   DANet   Deeplab V3+   UNet   UperNet   Swin Transformer   MANet (Ours)

Fig. 8. Comparison of qualitative results between MANet and other advanced methods on the CHN6-CUG road dataset.

Table 3. Quantitative evaluation for key components ablation studies of MANet on the Road Dataset of Mengxi Transmission Line.

Number	Method	Precise	Recall	F1-score	IoU	MIoU	AA
No.1	B	66.37	39.62	49.62	33.0	64.23	69.23
No.2	B + SMCA	64.39	44.23	52.44	35.53	65.5	71.4
No.3	B + SMCA + EAM	65.56	**46.42**	54.35	37.32	66.46	**72.5**
No.4	MANet	**67.42**	45.99	**54.46**	**37.42**	**66.54**	72.33

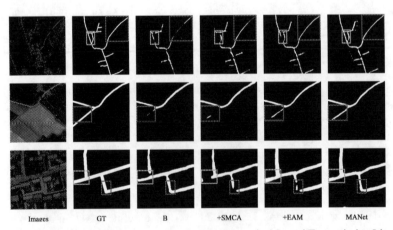

Images   GT   B   +SMCA   +EAM   MANet

Fig. 9. Qualitative results of ablation studies for MANet on the Mengxi Transmission Line Road Dataset.

Table 4. Quantitative research results of different modules and different positions.

Location	Method	Precise	Recall	F1-score	IoU	MIoU	AA
Encoder	B	66.37	39.62	49.62	33.0	64.23	69.23
	B + SA	66.58	38.43	48.47	32.22	63.83	68.65
	B + MSCA	65.75	41.55	50.92	34.15	64.82	70.14
	B + SE	65.58	42.46	51.63	34.8	65.16	70.59
	B + CBAM	65.02	43.06	51.81	34.96	65.22	70.86
	B + ECA	65.48	42.87	51.82	34.97	65.24	70.78
	B + ACmix(Ours)	64.99	**43.19**	**51.89**	**35.04**	**65.26**	**70.92**
	B + SMCA	**68.14**	37.83	48.65	32.14	63.83	68.4
Decoder	B + SA	**66.47**	39.66	49.68	33.05	64.26	69.25
	B + CBAM	66.05	41.16	50.71	33.97	64.73	69.96
	B + ECA	65.77	42.94	51.95	35.09	65.31	70.82
	B + SMCA(Ours)	64.39	**44.23**	**52.44**	**35.53**	**65.5**	**71.4**

two modules, SA and MSCA, to different positions of the baseline network encoder and decoder, respectively, as approximated in this paper. By adding different modules at the encoder locations separately, the SMCA in this paper achieves the highest accuracy rate by 1.77% compared to the baseline, but other values decrease instead of increasing, which is not satisfactory. After analysis, it is concluded that since the features received at the encoder position itself are deep-level features and contain more global information, the feature fusion module in SMCA does not play a role and other geometric information useful for its extraction is lost when focusing on the line features instead, so SMCA is not suitable at the encoder position. By adding each module at the decoder position, our SMCA achieves the highest gain except for accuracy rate, which is better than SA, ECA & CBAM.

Experiments on the CHN6-CUG Dataset. Through ablation experiments on the CHN6-CUG Dataset, it is demonstrated that each module of our approach remains valid with the overall structure of the model on different datasets. As shown in Table 5.

Table 5. Quantitative evaluation for key components ablation studies of MANet on the CHN6-CUG road dataset.

Number	Method	Precise	Recall	F1-score	IoU	MIoU	AA
No.1	B	**80.23**	70.36	74.97	59.97	78.58	84.65
No.2	B + SMCA	79.81	70.73	75.0	60.0	78.59	84.82
No.3	B + SMCA + EAM	77.09	73.61	75.31	60.04	78.76	86.14
No.4	MANet	77.06	**74.08**	**75.54**	**60.69**	**78.91**	**86.37**

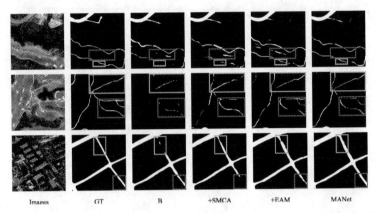

Fig. 10. Qualitative results of ablation studies for MANet on the CHN6-CUG road dataset.

The analysis of the visualization results in the Fig. 10 shows that as the model extracts the road features in depth, it filters out some targets that are extracted incorrectly and corrects the mis-checking phenomenon, making the overall extraction of the model better. At the same time, the line geometric features of road connectivity are used to connect the scattered pixels step by step to identify a more complete road network information.

5   Conclusion

In this paper, an end-to-end extraction method, MANet, is proposed for roads near EHV transmission lines. The model is evaluated in depth on the self-produced road dataset of the Mengxi transmission line and the CHN6-CUG Dataset. The experimental results and comparative analysis show that the algorithm is competitive and advantageous in the road network extraction task. For MANet, the next work can be complemented by incorporating the centerline extraction subtask to improve the network extraction performance. At the same time, the model we designed is focused on solving specific tasks and can be improved in the future to enhance its robustness and expand its applicability.

Acknowledgments. The authors gratefully acknowledge the financial supports by the National Natural Science Foundation of China under Grant No. 62077032, as well as the Inner Mongolia Science and Technology Plan Project under project No. 2021GG0159.

References

1. Hoeser, T., Kuenzer, C.J.R.S.: Object Detection and Image Segmentation with Deep Learning on Earth Observation Data: A Review-Part i: Evolution and Recent Trends. **12**, 1667 (2020)
2. Mnih, V.: Machine Learning for Aerial Image Labeling. University of Toronto (Canada) (2013)
3. Wang, J., Song, J., Chen, M., Yang, Z.J.I.J.o.R.S.: Road Network Extraction: A Neural-Dynamic Framework Based on Deep Learning and a Finite State Machine **36**, 3144-3169 (2015)

4. Long, J., Shelhamer, E., Darrell, T.: Fully Convolutional Networks for Semantic Segmentation (2017)
5. Zhong, Z., Li, J., Cui, W., Han, J.: Fully convolutional networks for building and road extraction: preliminary results. In: 2016 IEEE International Geoscience and Remote Sensing Symposium (IGARSS) (2016)
6. Ronneberger, O., Fischer, P., Brox, T.: U-net: convolutional networks for biomedical image segmentation. In: Medical Image Computing and Computer-Assisted Intervention–MICCAI 2015: 18th International Conference, Munich, Germany, October 5–9, 2015, Proceedings, Part III 18, pp. 234–241. Springer (2015)
7. Cheng, G., Wang, Y., Xu, S., Wang, H., Xiang, S., Pan, C.J.I.T.o.G., Sensing, R.: Automaticroad Detection and Centerline Extraction via Cascaded End-to-end Convolutional Neural Network 55, 3322–3337 (2017)
8. Zhou, L., Zhang, C., Wu, M.: D-LinkNet: LinkNet with pretrained encoder and dilated convolution for high resolution satellite imagery road extraction. In: Proceedings of the IEEE Conference on Computer Vision and Pattern Recognition Workshops, pp. 182–186 (2018)
9. Shi, Q., Liu, X., Li, X.J.I.a.: Road Detection from Remote Sensing Images by Generative Adversarial Networks 6, 25486–25494 (2017)
10. Vaswani, A., et al.: Attention is All You Need 30 (2017)
11. Jiang, X., et al.: Geoinformation: RoadFormer: Pyramidal Deformable Vision Transformers for Road Network Extraction with Remote Sensing Images 113, 102987 (2022)
12. Zhang, Z., Miao, C., Liu, C., Tian, Q., Zhou, Y.J.M.: HA-RoadFormer: Hybrid Attention Transformer with Multi-Branch for Large-Scale High-Resolution Dense Road Segmentation 10, 1915 (2022)
13. Chen, L.-C., Zhu, Y., Papandreou, G., Schroff, F., Adam, H.: Encoder-decoder with atrous separable convolution for semantic image segmentation. In: Proceedings of the European conference on computer vision (ECCV), pp. 801–818 (2018)
14. He, T., Zhang, Z., Zhang, H., Zhang, Z., Xie, J., Li, M.: Bag of tricks for image classification with convolutional neural networks. In: Proceedings of the IEEE/CVF Conference on Computer Vision and Pattern Recognition, pp. 558–567 (2019)
15. Pan, X., et al.: On the integration of self-attention and convolution. In: Proceedings of the IEEE/CVF Conference on Computer Vision and Pattern Recognition, pp. 815–825 (2022)
16. Sun, Y., Chen, G., Zhou, T., Zhang, Y., Liu, N.J.a.p.a.: Context-Aware Cross-Level Fusion Network for Camouflaged Object Detection (2021)
17. Huan, H., Sheng, Y., Zhang, Y., Liu, Y.J.R.S.: Strip Attention Networks for Road Extraction 14, 4516 (2022)
18. Zhu, Q., et al.: A Global Context-Aware and Batch-Independent Network for Road Extraction from VHR Satellite Imagery 175, 353–365 (2021)
19. MMSegmentation contributors. MMSegmentation: Openmmlab Semantic Segmentation Toolbox and Benchmark (2020). https://github.com/openmmlab/mmsegmentation. Accessed 11 Aug 2020
20. Bottou, L.: Large-scale machine learning with stochastic gradient descent. In: Proceedings of COMPSTAT'2010: 19th International Conference on Computational StatisticsParis France, August 22–27, 2010 Keynote, Invited and Contributed Papers, pp. 177–186. Springer (2010)
21. Abdollahi, A., Pradhan, B., Shukla, N., Chakraborty, S., Alamri, A.J.R.S.: Deep Learning Approaches Applied to Remote Sensing Datasets for Road Extraction: A State-of-the-Art Review 12, 1444 (2020)
22. Xiao, T., Liu, Y., Zhou, B., Jiang, Y., Sun, J.: Unified perceptual parsing for scene understanding. In: Proceedings of the European Conference on Computer Vision (ECCV), pp. 418–434 (2018)

23. Liu, Z., et al.: Swin transformer: Hierarchical vision transformer using shifted windows. In: Proceedings of the IEEE/CVF International Conference on Computer Vision, pp. 10012–10022 (2021)

24. Fu, J., et al.: Dual attention network for scene segmentation. In: Proceedings of the IEEE/CVF Conference on Computer Vision and Pattern Recognition, pp. 3146–3154 (2019)

25. Dosovitskiy, A., et al.: An Image is Worth 16x16 Words: Transformers for Image Recognition at Scale. (2020)

26. Hu, J., Shen, L., Sun, G.: Squeeze-and-excitation networks. In: Proceedings of the IEEE Conference on Computer Vision and Pattern Recognition, pp. 7132–7141 (2018)

27. Wang, Q., Wu, B., Zhu, P., Li, P., Zuo, W., Hu, Q.: ECA-Net: Efficient channel attention for deep convolutional neural networks. In: Proceedings of the IEEE/CVF Conference on Computer Vision and Pattern Recognition, pp. 11534–11542 (2020)

28. Woo, S., Park, J., Lee, J.-Y., Kweon, I.S.: Cbam: Convolutional block attention module. In: Proceedings of the European Conference on Computer Vision (ECCV), pp. 3–19 (2018)

Deep Reinforcement Learning for Solving the Trip Planning Query

Changlin Zhao, Ying Zhao, Jiajia Li[(✉)], Na Guo, Rui Zhu, and Tao Qiu

Shenyang Aerospace University, Shenyang, China
{lijiajia,zhurui,qiutao}@sau.edu.cn

Abstract. The Trip Planning Query (TPQ), which returns the optimal path from the starting point to the destination that satisfies multiple types of points of interest (POIs) specified by the user, has attracted more and more attention. The most straightforward approach is to enumerate all the POI combinations that meet the user's needs, and then select the path with the shortest distance. So this problem can be regarded as a combinatorial optimization problem and solved with deep reinforcement learning. Hence, in this paper, we explore the application of deep reinforcement learning in solving TPQ problem. Since the selection of POI can be considered as a sequence decision problem, we model it as a seq2seq problem. Firstly, to help the model reduce the difficulty of selection, we remove POIs that can not be the result, and propose a candidate set generation method. Its nodes are enough to meet the query POIs for the model to select different node sequences. Secondly, we use the encoder-decoder model base on attention mechanism. We concatenate the embedding of the start point, the end point and the selected nodes as the query part of the attention mechanism. We mask the same poi after select so that the model can solve the TPQ problem. Finally, we employ the REINFORCE method for training with a greedy baseline. Our model has a good performance on different maps, different POI densities, and different numbers of required POIs.

Keywords: Trip Planning Query · Deep Reinforcement Learning · Seq2Seq

1 Introduction

Spatial database is more and more widely used in today's society. Among them, the Trip Planning Query(TPQ) problem has attracted more and more attention. Not only should plan the optimal path from start point to end point, but also meet the required POIs. Take Fig. 1 as an example, the uesr wants to start from S to T, and needs pass hospital, mall and hotel. Three paths are given in the figure and they all meet the needs of the user but the path cost is different. We want the path cost to be as samall as possible, which is obviously a combinatorial optimization(CO) problem.

© The Author(s), under exclusive license to Springer Nature Switzerland AG 2023
X. Yang et al. (Eds.): ADMA 2023, LNAI 14176, pp. 569–583, 2023.
https://doi.org/10.1007/978-3-031-46661-8_38

Fig. 1. TPQ on road network

The traditional method to solve the TPQ problem is to design an index structure, which determines a candidate set that meets the query conditions and then combines different solutions from the candidate set. However, the construction of index needs to occupy much more memory space than the model the experiment proves that the efficiency decreases when the required POIs become more and more. In addition, the index needs to be rebuilt when the poi information in the map changes, but the model learned by reinforcement learning can be effective in a unified city with different poi densities and reinforcement learning can solve these problem by learning a policy through the interaction between the agent and the environment to make decisions. It makes the solution of the TPQ problem no longer rely on the algorithm design of experts but lets the machine learn the policy decision by itself. Specifically, there are several challenges:

- How to model the TPQ problem, which is related to what kind of model we design. We model the TPQ problem as a Seq2Seq problem because we can select a candidate set of nodes by querying information to input the model in the form of a sequence, and the model also solves a sequence of points. Based on this, we designed the encoder-decoder model.
- How to determine the input of the model, different from other CO problems with a given scale, the TPQ problem are mostly initiated in cities. For example, if a user wants to visit a hospital, we should consider the starting point and the endpoint of the query initiation and filter out some hospitals to feed to the model for its decision. Based on this, we design the candidate generation strategy to determine the input of the model.
- How to make the model more effective and reasonable in each decision step. We chose to use the attention mechanism because it is calculated between any node and can fully understand the information between nodes. At the same time, we design context_embed as the query part of the attention mechanism in the decoding process. And after a selection, all selected nodes with the same POI information are marked to ensure the reasonable operation of the model.
- How to train the model. There are many ways to train reinforcement learning, but we use REINFORCE because its turn updates allow us to evaluate the

overall quality of a query. So when making a choice, the model takes into account not only the current situation but also the impact of this decision on the overall query, which is an advantage over a single update. At the same time, we introduce the greedy-based baseline to help training.

- How to save the calculation time of the shortest distance between two points. In the process of training and solving, a large number of calculations of the shortest distance between two points are needed, which is complex and time-consuming in a city. We use the H2H index to compute the shortest distance between two points to reduce the time it takes to respond to a query.

2 Related Work

2.1 Traditional Method for the TPQ Problem

In many complex and novel query requirements, TPQ [1] is first proposed and proved to be NP-hard problem, which presents an approximate approximation algorithm that can be solved in polynomial time. The core idea is to find all the same POIs in the whole graph and find the point with the minimum distance from the starting point to the POI point and then to the endpoint. The candidate set is composed of nodes selected by different kinds of POIs(There is only one candidate point for each type of POI), and is selected from the starting point to the endpoint by greedy thinking. The problem of Group TPQ is raised Group TPQ [2] when considering multiple users in the network. The OSR problem is raised [3] when the keywords queried by the user are accessed in a fixed order. However, no matter what type of query, solving the shortest distance between two points is an inevitable problem. H2H index [4] is proposed, with the application of the H2H index significantly reducing the time required for the shortest path query on large-scale real road networks. The IG-TREE index structure proposed by [5] obtained a more effective solution to the TPQ problem. The appropriate candidate set was selected based on this index(There is more than one candidate node for each type of POI), and then the TPQ problem was solved by different combinations in the candidate set.

However, these traditional methods rely on experts to design a reasonable index structure. They have larger storage space requirements and are difficult to ensure efficiency when the query keywords become more, and they cannot adapt to the change of poi density in a road network.

2.2 Reinforcement Learning Method for Routing Problem

The routing problem often involves the combination of different nodes in the road network, so it can be regarded as a combinatorial optimization problem. Mazyavkina et al. [6] introduced many solutions of classical problems of reinforcement learning in combinatorial optimization. Travelling Salesman Problem (TSP) is a classical problem in the field of routing problem, which has a wide range of applications in path planning, gene sequencing and other fields [7]. John

J Hopfield [8]originally thought of using machine learning to solve the TSP problem, and proposed a Hopfield network for solving the small-range TSP problem.

The use of reinforcement learning to solve TSP problems at the present stage can be divided into two ideas, one kind is to consider the topology of the input figure [9,10], and the other is the TSP problem is modeled as a seq2seq problem [11–16]. Petar Velickovic et al. [9]proposed Graph Attention Networks. GAT is utilized by [10], whose trained neural network can output an approximate solution for a new graph instance.

Vinyals et al. [11] proposed a pointer network, which relies on the basic encoding and decoding structure, and introduces the attention mechanism in the encoding part. Since the NLP problem is also a sequential decision problem, the transformer has a good performance on NLP problems [12]. Kaempfer & Wolf [13] trained a transformer-based model to solve the TSP problem, again using beam search to obtain the solution. Deudon et al. [14] also designed a network based on the transformer and trained by policy gradient descent. Wouter Kool et al. [15] used REINFORCE to train the model in the encoder part, drawing on the transformer, and trained the model for other combinatorial optimization problems with the same hyperparameters. Xavier Bresson et al. [16] mainly modified the decoder part of the transformer to calculate attention between each node, and further improved the solution of path length at the cost of solution time.

Although these reinforcement learning methods cleverly solve the routing problem, but they cannot solve the TPQ problem. This is because the TPQ problem involves considering the optimal path planning while selecting between POIs of the same kind. It makes every choice greater influence on the results of the subsequent.

3   Problem Definition

Table 1. Important notation

Notation	Definition
$G = (V, E)$	a road network graph G with n nodes and m edges
$dis(v_1, v_2)$	the shortest distance between v_1 and v_2
P	all kinds of POIs in a graph G
P_Q	some kinds of POIs which need required in a query Q
$Q = \{s, t, P_Q\}$	a query Q obtain start from s to t through P_Q
$range(v_i)$	from s to t through v_i $range(v_i) = dis(s, v_i) + dis(v_i, t)$
V_π	some POIs which selected by model that can meet P_Q
π	π is a path of v which is a solution of Q $\pi = \{s, v_{\pi_1} \ldots v_{\pi_g}, t\}$
$dis(Q)$	the path cost with Q $dis(Q) = dis(s, v_{\pi_1}) + \cdots + dis(v_{\pi_g}, t)$

We give some important notation in Table 1, then we define the TPQ problem as Definition 1.

Definition 1. *TPQ in a road network is defined as: Given a road network graph* $G = (V, E)$. *The user initiates a query* $Q = \{s, t, P_Q\}$ *The model gives a solution sequence* $\pi = \{s, v_{\pi_1} \ldots v_{\pi_g}, t\}$ *that satisfies Q, and we want the path cost* $dis(Q) = dis(s, v_{\pi_1}) + \cdots + dis(v_{\pi_g}, t)$ *to be as small as possible.*

4 Method

The architecture of the network is shown in Fig. 2. When the user initiates a query Q, in order to reduce the calculation of the model, the candidate generate strategy is used to select the appropriate candidate nodes as X_{input} to our model. Then it is encoded by the transformer [12] without position-encode, and the multi-head attention mechanism is also used during decoding. We concatenate the embedding of the start point, the end point, the selected nodes and the mean value of all nodes as the query part in MHA, and mask the selected kinds of POIs . Finally, for faster convergence, we apply REINFORCE based on greedy baselines.

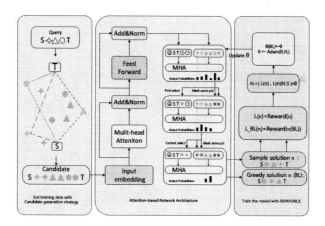

Fig. 2. Overall method architecture for TPQ

4.1 Candidate Generation Strategy

To reduce the computation of the model, we need a strategy to filter out some nodes that are more likely to constitute the optimal solution. We use $range(v_i)$ to measure whether the poi is selected into the candidate set. We select the first k minimum $range(v_i)$ for each poi in the whole graph to the candidate. For example in Fig. 3, the user needs from s to t through $star$, and the graph

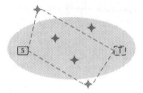

Fig. 3. Candidate generation strategy for TPQ

G has five *stars*. Obviously, the *range(bluestars)* > *range(redstars)*. For our candidate generation strategy, the *readstars* will be selected as the candidate for the star kind. The *candidate set* and s, t make up the input of the model as X_{input}. In the experiments, we let $k = 3$ and show that this works.

4.2   Reinforcement Learning for TPQ

The TPQ problem can be interpreted as players playing a game. It requires the agent to learn a decision policy by constantly interacting with the environment to obtain rewards, and the path distance should be as small as possible when the query requirements are meet, which coincides with the idea of reinforcement learning. Therefore, we use reinforcement learning to model TPQ problems, and the relevant elements are defined as follows:

State: The state represents what the agent needs to describe at time step t. For the TPQ problem, it needs to describe the start and end point of a query, the distance between points, the last selected node, and whether the node was selected.

Action: Action represents a selection made when the model is in the state, and for the TPQ problem, represents the model selecting an unvisited node.

Reward: Since we use REINFORCE method and do round update, we define reward as $dis(Q)$

Policy: We want the model to learn a policy (parameterized by θ) to help the agent make decisions to solve the TPQ problem, denoted as:

$$P\_TPQ_\theta(\pi \mid Q) = P(V_{\pi_1} \mid s,t) \cdot \ldots \cdot P(V_{\pi_g} \mid V_{\pi_g - 1} \cdots V_{\pi_2}, s, t) \qquad (1)$$

4.3   Neural Architecture for TPQ

Encoder. In order to connect nodes to each other, we borrowed the attention-based transformer structure. However, we do not adopt positioning embedding, because different from NLP, the node order of the candidate set we choose for TPQ does not need to be considered when the model is input.

X_{input} is encoded as a 128-dimensional vector by a linear transformation.

$$X_{embed} = W_x \cdot X_{input} + b_x \tag{2}$$

The Attention mechanism first initializes three matrices $W_q \in R^{(d_k,d_h)}, W_k \in R^{(d_k,d_h)}, W_v \in R^{(d_v,d_h)}(d_k = d_v = 16)$,And use$X_embed$ to multiply them separately:

$$q = W_q \cdot X_{embed}, k = W_k \cdot X_{embed}, v = W_v \cdot X_{embed} \tag{3}$$

Then we obtain the similarity by multiplying q and k so that the nodes exchange information.

$$com_{ij} = \begin{cases} \frac{q_i^T \cdot k_j}{\sqrt{d_k}} & \text{if } i \text{ adjust } j \\ -\infty & \text{otherwise} \end{cases} \tag{4}$$

To reduce the variance after softmax, So we're going to divide this by $\sqrt{d_k}$ The attention - weight as defined by the com_{ij} as follow:

$$a_{ij} = \frac{e^{com_{ij}}}{\sum_j e^{com_{ij}}} \tag{5}$$

It is finally obtained by a linear combination of $a_{ij}$and $v$as follows:

$$x_{attention} = \sum_j a_{ij} \cdot v_j \tag{6}$$

But in practice we use Multi-Head Attention, MHA is to the attention mechanism of repeated operation M times(We let M=8)

The results are concatenated and then reduced by linear transformation$W_o \in R^{(d_h,d_v)}$:

$$x_{MHA} = MHA(q,k,v) = Concat(X_{attention_1} \cdots X_{attention_M} \cdot W_o) \tag{7}$$

After X_MHA is obtained, it is added with X_embed to obtain the first sub-layer of the model through BN:

$$x_{sub-layer1} = BN(X_{embed} + x_{MHA}) \tag{8}$$

Feed-Forward results are obtained after ReLu activation function($d_{ff} = 512$) Then the second sub-layer of the model is obtained by BN

$$FF(x_{sub-layer1}) = W_{ff\_1} \cdot ReLu(W_{ff\_0} \cdot x_{sub-layer1} + b_{ff\_0}) + b_{ff\_1} \tag{9}$$

$$x_{sub-layer2} = BN(x_{sub-layer1} + FF(x_{sub-layer1})) \tag{10}$$

In order to take into account the information of all nodes in the whole Query in the decoding part, and our problem is a fully connected graph, we define q_{embed} as:

$$q_{embed} = \frac{1}{n} \sum_{i=1}^{n} x_{sub-layer2_i} \tag{11}$$

Decoder. During decoding, the model selects one node at a time and matches it to the node mask with the same POI in X_{input} to ensure that the model runs correctly The specific design relies on a context_embed to represent the context information of the agent at time step t and uses context_embed to conduct attention operation with other unmasked nodes to obtain the node selection probability. Finally, different search methods for this probability are used to find specific solutions.

Context_embed: We want the model to treat each node "fairly" during input, rather than because the order of input nodes affects subsequent decisions. But when decoding, the model needs to understand the current context information, including the start point of a query, the endpoint of a query, the first selected node, and the currently selected node. And consider the remaining unvisited nodes before making a selection. For this purpose, we design context_embed as follows:

$$X_{context\_embed} = Concat(q_{embed}, X_s, X_{first}, X_{current}, X_t) \qquad (12)$$

When $X_{first}, X_{current}$ is not selected, we use a randomly generated vector instead
Use $X_{context\_embed}$ to conduct attention operations with all the nodes that have not been visited.

$$q = W_q \cdot X_{context\_embed}, k = W_q \cdot X_{sob-layer2_i}, v = W_v \cdot X_{sob-layer2_i} \qquad (13)$$

The dimension reduction of $X_c ontext\_embed$ can be the same as that of $X_{sublayer2_i}$ by W_q. Specifically, we compute the similarity between $X_{context\_embed}$ and all nodes that have not been visited

$$com_{context\_embed,j} = \begin{cases} C \cdot tanh(\frac{q_i^T \cdot k_j}{\sqrt{d_k}}) & if\ j\ without\ mask \\ -\infty\ otherwise \end{cases} \qquad (14)$$

The overall procedure is the same as the encoder procedure above but only MHA with M=1 is used, where q is calculated using $X_{context\_embed}$ and mapped to $[-10,10](C = 10)$using the activation function.
Finally, we compute the output probability for each node:

$$P_i = P\_TPQ_\theta(\pi_t = i \mid Q, \pi_{1:t-1}) = \frac{e^{com_{context\_embed,i}}}{\sum_{j,} e^{com_{context\_embed,j}}} \qquad (15)$$

5   Model Training

For a range of NP-hard problems such as TPQ, supervised learning does not excel because it is difficult to obtain when used as a label for the solution We decided to adopt the REINFORCE algorithm in reinforcement learning. At the same time, we used a baseline based on greedy for reference from [15], and the specific design is as follows:

We define reward as $r(\pi \mid Q) = dis(Q)$, we want it to be smaller and smaller. So we need to find the expectation of:

$$J(\theta|Q) = E_{\pi \sim P\_TPQ_{\theta(\cdot|Q)}}[r(\pi|Q)] \tag{16}$$

And then we need to find the gradient of:

$$\nabla_\theta J(\theta|Q) = E_{\pi \sim P_P Q_{\theta(i)}(Q)}[(r(\pi|Q) - b(Q))\nabla_\theta log(P_T P Q_\theta(\pi \mid Q))] \tag{17}$$

It can help the model update faster by reducing the variance of the gradient. We use Monte-Carlo sampling for it, which can be approximated as follows:

$$\nabla_\theta J(\theta|Q) \approx \frac{1}{B} \sum_{c=1}^{B} (r(\pi_c|Q_c) - b(Q_c))\nabla_\theta log(P\_TPQ_\theta(\pi_c \mid Q_c)) \tag{18}$$

Since we compute the network as well as the trained network during training, each set of networks has its own parameters, which are initially assigned the same random values. When selecting nodes, the network selects the solution in a greedy way, and the main network selects the solution in a sampling way. After solving the gradient, the Adam optimizer is used to optimize the parameters. When the quality of the solution of the main network is higher than the threshold of the network (0.001), we assign the parameters of the main network.

6 Experiment

6.1 Experiment Setup

Datasets: The TPQ problem was conducted in three cities shown in Table 2. According to the normal distribution, parameters shown in Table 3.

Table 2. City information

City	V	E
CAL	21048	43386
NY	264336	733846
COL	435666	1057066

Table 3. Parameters

Parameter	Value
POI density	2%, 4%, 8%, 16%, 32%
query POIs number	3, 5, 7, 9
TOP-K	5, 10, 15

Methods:(1)**Baseline**: We established the selection rule of the candidate set in the previous article. Now we set k=3, find all feasible solutions in the candidate set by permutation and combination, and use the optimal solution as a measure to compare with other algorithms.(2)**TPQ-org***: When k=1, the decision is made by a greedy strategy for the points in the candidate set.(3)**IG-Tree***: It is also solved by designing a candidate set algorithm and performing permutation and combination in it. However, with the help of the ig-tree index structure, it can find out the candidate set that satisfies a query. In Baseline,TPQ-org*, and IG-Tree* algorithms, the algorithm for finding the distance between two points is replaced by the H2H algorithm, which can improve the solution time by at least two orders of magnitude.

Implementaltion Detials: These algorithms are implemented in C++ and compiled with GNN GCC full optimization and conducted on a machine with the hardware is an Intel(R) Xeon(R) W-2245 CPU @3.2GHz with 250GB RAM under Linux(Ubuntu 18.04 LTS, 64bit). Our method all use in PyTorch, and trained using a Quadro RTX5000 GPU. We let sample width =10000 and beam search width =100. For the NY map, we trained only 4%poi density and applied the trained model to the other NY densities. We train different models for different numbers of query POIs.

6.2   Experimental Results and Analysis

As shown in Table 4. We hope that the smaller dis is better, TPQ-org and IG-Tree are worse than Baseline, which indicates that the candidate set rule of our previous article is effective, because Baseline is all feasible solutions in the candidate set and also the upper bound for our model. With the increase of query poi num, the performance of the algorithm becomes worse. The more POIs that are queried, the more choices the model has to make, making it difficult for the model to make a good choice. but the advantages of the algorithm with other algorithms increase. The algorithm gets better as the map POI density decreases. Because the smaller the map density, the more scattered the POI points, and the larger the feedback fluctuation brought by the model's decision, which is more conducive to model learning.

As shown in Table 5, which describes the performance of top-k algorithm on NY, when the query POi is smaller, our model has a greater advantage over IG-Tree, the model performs better with less top-k, and the model performs better with the increase of POI density.

As shown in Table 6. We evaluated the time consumed by different experiments, which is difficult to describe. This is because we deployed it on C++ and used an index to speed up the solution for better performance in comparison experiments. Even so, our time does not have a disadvantage when computing TOP-1 but has a clear advantage compared with IG-Tree when computing top-k.

As shown in Table 7. We concat the poi information of the node in the form of one-hot encoding with the latitude and longitude of the node before entering the model as OUR*. It can be seen that with the increase of query keywords, OUR* performs better. This shows that when there are few query POIs nums, the model performs well enough that additional information interferes with the model's decision, and when there are many query POIs nums, the additional information we provide to the model has a positive effect.

Table 4. Different algorithms performance on NY and CAL

query_poi_num	3		5		7		9	
method	dis	gap	dis	gap	dis	gap	dis	gap
NY-POI-2%								
Baseline	467291	0.00%	494410	0.00%	516625	0.00%	528344	0.00%
TPQ-org*	485646	**3.93%**	549812	11.21%	606071	17.31%	662955	25.48%
IG-Tree*	601186	28.65%	697752	41.13%	777285	50.45%	817364	54.70%
Our(greedy)	504450	7.95%	565092	14.30%	620732	20.15%	692422	31.06%
Our(sample)	488776	4.60%	536021	**8.42%**	569886	**10.31%**	600611	**13.68%**
NY-POI-4%								
Baseline	442899	0.00%	459856	0.00%	474315	0.00%	477530	0.00%
TPQ-org*	454915	**2.71%**	493514	7.32%	533689	12.52%	554711	16.16%
IG-Tree*	546597	23.41%	617756	34.34%	688224	45.10%	713109	49.33%
Our(greedy)	467079	5.46%	509080	10.70%	547863	15.51%	595914	24.79%
Our(sample)	459484	3.74%	490248	**6.61%**	513694	**8.30%**	532407	**11.49%**
NY-POI-8%								
Baseline	427160	0.00%	439092	0.00%	445434	0.00%	447673	0.00%
TPQ-org*	433291	**1.44%**	455748	**3.79%**	475038	6.65%	494407	10.44%
IG-Tree*	505504	18.34%	557663	27.00%	601599	35.06%	623068	39.18%
Our(greedy)	442213	3.52%	468280	6.65%	495705	11.29%	529631	18.31%
Our(sample)	436880	2.28%	457445	4.18%	471822	**5.92%**	483966	**8.11%**
CAL-POI-8%								
Baseline	593527	0.00%	703395	0.00%	799365	0.00%	996654	0.00%
TPQ-org*	614719	**3.57%**	735940	4.63%	856464	7.14%	1115645	11.94%
IG-TREE*	635548	7.08%	758884	7.89%	876654	9.67%	1164848	16.88%
Our(greedy)	620395	4.53%	741034	5.35%	864665	8.17%	1139627	14.35%
Our(sample)	619248	4.33%	732346	**4.12%**	855230	**6.99%**	1100125	**10.38%**
NY-POI-16%								
Baseline	422746	0.00%	432108	0.00%	435482	0.00%	436728	0.00%
TPQ-org*	427316	**1.08%**	442437	**2.39%**	459115	5.43%	464279	6.31%
IG-Tree*	490415	16.01%	533194	23.39%	561393	28.91%	579578	32.71%
Our(greedy)	435718	3.07%	454649	5.22%	476242	9.36%	504332	15.48%
Our(sample)	430503	1.83%	446566	3.35%	458605	**5.31%**	465181	**6.52%**
NY-POI-32%								
Baseline	418552	0.00%	426720	0.00%	428911	0.00%	426718	0.00%
TPQ-org*	420412	**0.44%**	433850	**1.67%**	441476	**2.93%**	444365	**4.14%**
IG-Tree*	469721	12.23%	504905	18.32%	526759	22.81%	537409	25.94%
Our(greedy)	430227	2.79%	445817	4.48%	458839	6.98%	474477	11.19%
Our(sample)	424370	1.39%	436530	2.30%	443357	3.37%	447078	4.77%

Table 5. Different top-k algorithms performance on NY

query_poi_num	3		5		7		9	
method	dis	gap	dis	gap	dis	gap	dis	gap
NY-POI-2% (TOP-5,10,15 Average)								
Baseline	480031	0.00%	501528	0.00%	521552	0.00%	531998	0.00%
IG-Tree*	540425	12.58%	598156	19.27%	656997	25.97%	682894	28.36%
Our(beam search)	521738	8.69%	577832	15.21%	628838	20.57%	697573	31.12%
Baseline	491742	0.00%	507368	0.00%	525440	0.00%	534863	0.00%
IG-Tree*	562630	14.42%	617418	21.69%	672921	28.07%	695523	30.04%
Our(beam search)	544683	10.77%	592831	16.84%	634800	20.81%	702113	31.27%
Baseline	502241	0.00%	511741	0.00%	528324	0.00%	536985	0.00%
IG-Tree*	580465	15.57%	631815	23.46%	684115	29.49%	704519	31.20%
Our(beam search)	566432	12.78%	605417	18.31%	639130	20.97%	704431	31.18%
NY-POI-4% (TOP-5,10,15 Average)								
Baseline	449530	0.00%	463664	0.00%	477218	0.00%	479680	0.00%
IG-Tree*	491886	9.42%	532101	14.76%	573239	20.12%	592792	23.58%
Our(beam search)	478364	6.41%	516980	11.50%	552850	15.85%	600579	25.20%
Baseline	455856	0.00%	466871	0.00%	479554	0.00%	481343	0.00%
IG-Tree*	506642	11.14%	545166	16.77%	585118	22.01%	602079	25.08%
Our(beam search)	494149	8.40%	526759	12.83%	557040	16.16%	602037	25.07%
Baseline	461718	0.00%	469223	0.00%	481234	0.00%	482558	0.00%
IG-Tree*	518210	12.24%	554414	18.16%	593353	23.30%	608608	26.12%
Our(beam search)	512569	11.01%	534468	13.90%	560149	16.40%	603882	25.14%
NY-POI-8% (TOP-5,10,15 Average)								
Baseline	430506	0.00%	440996	0.00%	446904	0.00%	448896	0.00%
IG-Tree*	458909	6.60%	487592	10.57%	509750	14.06%	526375	17.26%
Our(beam search)	448295	4.13%	474931	7.70%	499383	11.74%	532111	18.54%
Baseline	433665	0.00%	442576	0.00%	448099	0.00%	449777	0.00%
IG-TreeE*	467109	7.71%	495715	12.01%	517150	15.41%	532432	18.38%
Our(beam search)	459783	6.02%	481358	8.76%	501295	11.87%	533913	18.71%
Baseline	436623	0.00%	443809	0.00%	448971	0.00%	450431	0.00%
IG-Tree*	473307	8.40%	501445	12.99%	522300	16.33%	536574	19.12%
Our(beam search)	474054	8.57%	487422	9.83%	503865	12.23%	535355	18.85%
NY-POI-16% (TOP-5,10,15 Average)								
Baseline	424813	0.00%	433432	0.00%	436201	0.00%	437757	0.00%
IG-Tree*	446805	5.18%	468511	8.09%	486695	11.58%	493896	12.82%
Our(beam search)	441562	3.94%	462863	6.79%	481648	10.42%	507112	15.84%
Baseline	426824	0.00%	434563	0.00%	436776	0.00%	438575	0.00%
IG-Tree*	452804	6.09%	474607	9.21%	492139	12.68%	498681	13.70%
Our(beam search)	451120	5.69%	469269	7.99%	483312	10.65%	509099	16.08%
Baseline	428781	0.00%	435419	0.00%	437195	0.00%	439181	0.00%
IG-Tree*	457193	6.63%	478876	9.98%	495956	13.44%	501996	14.30%
Our(beam search)	433665	1.14%	474004	8.86%	485332	11.01%	510287	16.19%
NY-POI-32% (TOP-5,10,15 Average)								
Baseline	419835	0.00%	427539	0.00%	429627	0.00%	427166	0.00%
IG-Tree*	436000	3.85%	453526	6.08%	464983	8.23%	470927	10.24%
Our(beam search)	433204	3.18%	449364	5.10%	462256	7.59%	481050	12.61%
Baseline	421031	0.00%	428261	0.00%	430114	0.00%	427517	0.00%
IG-Tree*	439899	4.48%	457626	6.86%	468898	9.02%	474457	10.98%
Our(beam search)	442040	4.99%	454354	6.09%	464033	7.89%	482369	12.83%
Baseline	422169	0.00%	428726	0.00%	430470	0.00%	427773	0.00%
IG-Tree*	442759	4.88%	460525	7.42%	471567	9.55%	476857	11.47%
Our(beam search)	454799	7.73%	458824	7.02%	465190	8.07%	483412	13.01%

Table 6. Time cost with different algorithms

query_poi_num	3	5	7	9
NY-POI-2% time(s)				
Baseline	0.000131	0.000421	0.003986	0.052187
TPQ-org*	0.000127	0.000168	0.000284	0.000345
IG-Tree*(top-k)	1.097157	1.678025	3.993721	12.410499
IG-Tree*	0.020016	0.020564	0.020783	0.021464
Our(greedy)	0.000172	0.000265	0.001256	0.001067
Our(sample)	1.071326	2.089796	2.870463	3.039828
Our(beam search)	0.041838	0.077437	0.105119	0.135248
NY-POI-4% time(s)				
Baseline	0.000248	0.000599	0.004414	0.050006
TPQ-org*	0.000235	0.000445	0.000543	0.000791
IG-Tree*(top-k)	1.282895	2.224972	6.110478	18.404909
IG-Tree*	0.019985	0.020891	0.021162	0.022401
Our(greedy)	0.000353	0.00074	0.001325	0.001895
Our(sample)	1.22083	2.012311	2.731697	3.967486
Our(beam search)	0.04764	0.076903	0.105443	0.138161
NY-POI-8% time(s)				
Baseline	0.000483	0.001015	0.004665	0.050056
TPQ-org*	0.000587	0.000927	0.001358	0.001745
IG-Tree*(top-k)	1.57325	3.454037	10.18716	30.288803
IG-Tree*	0.017453	0.019795	0.017308	0.023776
Our(greedy)	0.000397	0.000697	0.001213	0.000501
Our(sample)	1.26799	1.956981	2.692578	3.272532
Our(beam search)	0.043354	0.076778	0.104875	0.137601
NY-POI-16% time(s)				
Baseline	0.000891	0.001684	0.005545	0.050061
TPQ-org*	0.000774	0.001397	0.001721	0.002444
IG-Tree*(top-k)	1.680179	4.852662	15.72485	45.403581
IG-Tree*	0.014615	0.015341	0.016827	0.019368
Our(greedy)	0.000436	0.000872	0.001634	0.00104
Our(sample)	1.228306	1.98791	2.797348	3.449433
NY-POI-32% time(s)				
Baseline	0.001624	0.003032	0.007553	0.052288
TPQ-org*	0.001427	0.00218	0.003313	0.004427
IG-TREE*(top-k)	2.015365	5.431364	17.572686	71.043175
IG-TREE*	0.01532	0.017298	0.019893	0.023436
Our(greedy)	0.000559	0.001158	0.001547	0.001001
Our(sample)	1.257097	1.664806	2.289789	3.008657
Our(beam search)	0.045972	0.077154	0.10505	0.133948

query_poi_num	3		5		7		9	
method	dis	gap	dis	gap	dis	gap	dis	gap
COL-POI-4%								
Baseline	3971691	0.00%	3923175	0.00%	4137584	0.00%	4184790	0.00%
TPQ-org*	4020777	1.24%	4038049	2.93%	4368701	5.59%	4499382	7.52%
IG-Tree*	4796731	20.77%	5139976	31.02%	5642254	36.37%	6106430	45.92%
Our(greedy)	4118828	3.70%	4122632	5.08%	4425774	6.97%	4695822	12.21%
Our(sample)	4042348	1.78%	4091190	4.28%	4375857	5.76%	4500366	7.54%
Our(greedy)*	4138602	4.20%	4151868	5.83%	4405275	6.47%	4637518	10.82%
Our(sample)*	4069338	2.46%	4099114	4.48%	4361328	5.41%	4455607	6.47%

7   Conclusion

We solve TPQ on road networks using reinforcement learning, where modeling the problem as a seq2seq problem and leveraging transformers is a key step. The correct decision is guaranteed by designing special structures to mask the auxiliary decoding process and the visited nodes simultaneously. Finally, it is trained by REINFORCE. In the future, we will continue to pay attention to the performance of different models on TPQ problems and the possibility that reinforcement learning can be applied to various queries on road networks.

Acknowledgements. The research work was partially supported by the Shenyang Young and Middle-aged Scientific and Technological Innovation Talent Support Plan under Grant No. RC220504; and the Natural Science Foundation of Liaoning Education Department under Grant No. LJKZ0205.

References

1. Li, Y. Cheng, D., Marios, H., George, K., Teng, S.: On trip planning queries in spatial databases. In: SSTD, pp. 273–290 (2005)
2. Tanzima, H., Tahrima, H., Mohammed A., Lars, K.: Group trip planning queries in spatial databases. In: SSTD, pp. 259–276 (2013)
3. Mehdi, S., Mohammad, R., Kolahdouzan, C.S.: The optimal sequenced route query. VLDB J. **17**(4), 765–787 (2008)
4. Ouyang, D., Qin, L., Chang, L., Lin, X., Zhang, Y., Zhu, Q.: When hierarchy meets 2-Hop-labeling: efficient shortest distance queries on road networks. In: SIGMOD, pp. 709–724 (2018)
5. Anasthasia, A., Md. Saiful, I., David, T., Muhammad, A.:IG-Tree: an efficient spatial keyword index for planning best path queries on road networks. In: WWW, vol. 22(4), pp. 1359–1399 (2019)
6. Nina, M., Sergei, S., Sergei, I. Evgeny, B.: Reinforcement learning for combinatorial optimization: A Survey. CoRR abs/ arXiv: 2003.03600 (2020)
7. Applegate, L., Bixby, E., Vaek, C., Cook, J.: The traveling salesman problem: a computational study. In: PUP (2006)
8. Hopfield, J.: Neural computation of decisions in optimization problems. Biol. Cybern. **52**(1985)

9. Petar, V., Guillem, C., Arantxa, C., Adriana, R., Pietro, L., Yoshua B.: Graph attention networks. In: ICLR (Poster) (2018)
10. Iddo, D.A., et al.: Learning to solve combinatorial optimization problems on real-world graphs in linear time. In: ICMLA, pp. 19–24 (2020)
11. Oriol, V., Meire, F., Navdeep, J.: Pointer networks. In: NIPS, pp. 2692–2700 (2015)
12. Ashish, V., et al: Attention is all you need. In: NIPS, pp. 5998–6008 (2017)
13. Yoav, K., Lior, W.: Learning the Multiple Traveling Salesmen Problem with Permutation Invariant Pooling Networks. CoRR abs/ arXiv: 1803.09621 (2018)
14. Michel, D., Pierre, C., Alexandre, L., Yossiri, A., Louis-Martin, R.: Learning heuristics for the TSP by policy gradient. In: CPAIOR, pp. 170–181 (2018)
15. Wouter, K., Herke van, H., Max, W.: Attention, learn to solve routing problems! In: ICLR (Poster) (2019)
16. Xavier, B., Thomas, L.: The Transformer Network for the Traveling Salesman Problem. CoRR abs/ arXiv: 2103.03012 (2021)

MDCN: Multi-scale Dilated Convolutional Enhanced Residual Network for Traffic Sign Detection

Yan Ke[1], Wanghao Mo[1], Zhe Li[2], Ruyi Cao[1], and Wendong Zhang[1(✉)]

[1] XinJiang University, Urumqi, China
{keyan,whm734097706,107552101634}@stu.xju.edu.cn, zwd@xju.edu.cn
[2] The Hong Kong Polytechnic University, Hong Kong SAR, China
lizhe.li@connect.polyu.hk

Abstract. Detecting small, multi-scale, and easily obscured traffic signs in real-world scenarios presents a persistent challenge. This paper proposes an approach that utilizes a multi-scale feature pyramid module to capture hierarchical features, facilitating robust detection of traffic signs across varying viewing angles and scales. To aggregate features at different scales and eliminate background interference, we employ a superposition of null convolution kernels with varying dilation rates, expanding the perceptual field from small to large. This effectively covers the object distribution across multiple scales while enhancing the resolution of the final output feature map for improved small target localization. Our method has demonstrated its effectiveness and superiority over several state-of-the-art approaches through extensive experiments conducted on two public traffic sign detection datasets.

Keywords: Traffic sign detection · Multi-scale feature pyramid · Hierarchical features · Perceptual field expansion · Robust detection

1 Introduction

The rapid advancement of autonomous driving and intelligent driver assistance systems has spurred extensive research on traffic sign detection [4]. However, this task is beset by various challenges, including interference from external factors such as illumination, occlusion, weather conditions, and shooting angles. Furthermore, traffic sign targets are typically small and exhibit variations in scale, exacerbating the difficulty of detection.

To overcome these challenges, researchers have proposed innovative approaches. Wang *et al.* [17] have replaced the original feature pyramid network in YOLOv5, resulting in improved real-time detection performance. Similarly, Yao *et al.* [19] have enhanced the feature fusion method of YOLOv4-Tiny through the introduction of an AFPN (Adaptive Feature Pyramid Network).

Y. Ke and W. Mo—Contribute equally to this work.

X. Yang et al. (Eds.): ADMA 2023, LNAI 14176, pp. 584–597, 2023.
https://doi.org/10.1007/978-3-031-46661-8_39

While single-stage networks are commonly employed in traffic sign detection studies due to computational limitations in real-world applications, this app-roach is not without shortcomings. Environmental changes and occlusions often impact the visibility of traffic signs, leading to a decline in detection performance.

In recent times, researchers have made significant advancements in improv-ing the performance of ATDR (Automatic Traffic Sign Detection and Recog-nition) in real-world scenarios. One such approach involves the utilization of multiscale pre-trained networks, which have shown promising results. In this context, a novel traffic sign detection network called TSingNet has been intro-duced. TSingNet leverages scale-aware and context-rich features to effectively detect and identify small or obscured traffic signs [11]. Furthermore, Shen *et al.* [16] propose a population multiscale attention pyramid network that facilitates optimal feature fusion patterns and the construction of information-rich feature pyramids to detect traffic signs of various sizes. Although these approaches have demonstrated strong performance, they primarily incorporate high-level seman-tic information in the earlier layers. Consequently, they face the challenge of foreground semantics, particularly regarding small traffic signs, which are prone to vanishing at higher levels of the FPN (Feature Pyramid Network).

We introduce MDCN (Multi-Scale Dilated Convolutional Enhanced Residual Network), a novel traffic sign detection network that leverages scale awareness and context-rich feature representation to detect multi-scale and small-object traffic signs efficiently. Our paper makes the following key contributions:

– We employ several novel data augmentation methods to increase the diversity and difficulty of the data, thereby enhancing the generalization ability of the model and effectively addressing the detection of various scales.
– We propose MDRNet(Multi-scale Deep Residual network), a new backbone architecture designed to learn scale-aware and context-rich features for traffic sign detection in outdoor environments. MDRNet aims to narrow the seman-tic gap between multiple scales, leading to improved detection performance.
– To overcome hardware limitations and accommodate high-resolution images, we introduce GN to remove the batch size limitation. Additionally, we utilize WS to further normalize the data from a weight perspective, accelerating model convergence and improving accuracy.

2   Methodology

We developed MDCN based on the Faster R-CNN framework [15]. The archi-tecture of our model is illustrated in Fig. 1, which comprises three primary com-ponents: feature extraction, feature fusion, and detection.

2.1   Multiple Data Augmentation Fusion

In our experiments, we utilized the CTSD and GTSDB datasets, which exhibit an imbalanced distribution of target scales and contain small target samples.

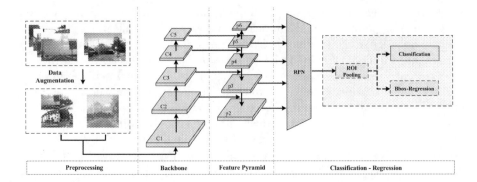

Fig. 1. The Overall architecture of MDCN. The feature maps C1 to C5, obtained from the backbone network, serve as the input for feature extraction. The feature maps P2 to P6 are then generated to extract features at different scales. The RPN (Region Proposal Network) is responsible for extracting regions of interest.

To address this issue and enhance the model's attention towards small targets during training, we expanded and augmented the datasets with additional data specifically focused on small targets.

To achieve this, we employed various data augmentation techniques, including Mixup [22], Mosaic [1], and Random affine [14]. These methods allowed us to synthesize samples and expand the dataset effectively. Fig. 2 provides a visualization of the data augmentation techniques employed.

Mixup involves overlaying two images, resulting in improved image detection accuracy without incurring significant computational overhead. Mosaic, on the other hand, stitches together four images by randomly cropping, scaling, and aligning them. This approach not only enhances the richness of the image background but also increases the diversity of target scales within a single image. These augmentation methods significantly enrich the detection dataset and contribute to the network's robustness.

2.2 MDRNet

To address the challenges posed by small-scale traffic sign image targets, multiple scale levels, and potential occlusion, we propose MDRNet. Our approach involves replacing all 3 × 3 regular convolutions in the conv4-conv5 layers of ResNet50 with dilated convolutions [20]. This modification expands the model's perceptual field without increasing computational complexity or compromising resolution. Additionally, the different perceptual fields obtained from various convolutional layers facilitate the extraction of multi-scale contextual information, thereby enabling the effective detection of traffic sign targets at different scales. Please refer to Fig. 3 for a visual representation of the details.

The regular convolution operation can be represented by Eq. 1, while a dilated convolution is defined as Eq. 2.

(a)   (b)

Fig. 2. Visualization of multiple data augmentation fusion. The ground truth is represented by the green box. (a) demonstrates the application of the Mosaic method, where four images are randomly stitched together. This technique enhances the richness of the image background and increases the diversity of target scales within a single image. (b) depicts the use of Mixup, which involves scaling two sample-label data pairs together to generate a new sample with an adjusted label count. This method effectively improves the accuracy of image detection. (Color figure online)

Let $\mathbf{y}(m,n)$ denote the result of dilated convolution between an input signal $\mathbf{I}(m,n)$ and a filter $\mathbf{F}(i,j)$, where $\mathbf{I}$ has a length and width of M and N, respectively. The formulation of the regular convolution operation is given by Eq. 1:

$$\mathbf{y}(m,n) = \sum_{i=1}^{M}\sum_{j=1}^{N}\mathbf{I}(m+i,n+j) * \mathbf{F}(i,j) \tag{1}$$

In contrast, the dilated convolution introduces a hyperparameter known as the dilation rate, denoted by r. This parameter defines the spacing between values as the convolution kernel processes the data. The formulation of the dilated convolution can be expressed as Eq. 2:

$$\mathbf{y}(m,n) = \sum_{i=1}^{M}\sum_{j=1}^{N}\mathbf{I}(m+r \times i,n+r \times j) * \mathbf{F}(i,j) \tag{2}$$

When the dilation rate r is set to 1, the dilated convolution reduces to a regular convolution.

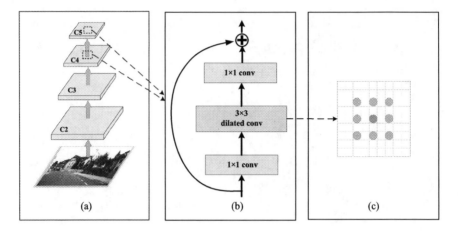

Fig. 3. A detailed design of MDRNet. (a)The network architecture of MDRNet. (b) A residual module is employed in MDRNet to enhance feature extraction and localization. This module consists of two 1×1 regular convolutional modules and a 3×3 dilated convolutional module. (c) The dilated convolutional module showcases the spacing between dots.

2.3   Normalization Methods

Smaller batch sizes have been shown to be more robust to variations between training and test sets [8]. For pixel-level image generation tasks like object detection and image segmentation, memory consumption limits the batch size to be small [6]. Moreover, the large image size of traffic signs and the constraints of general hardware resources further support the use of smaller batch sizes.

However, the effectiveness of BN (Batch Normalization) decreases significantly with small batches, limiting its applicability in micro-batch training. To address this limitation, we employ GN (Group Normalization), which divides channels into groups and calculates the mean and variance within each group for normalization. GN demonstrates stable accuracy across a wide range of batch sizes compared to BN. GN differs from BN in terms of the statistical range over which the mean and variance are calculated. We also introduce WS (Weight Standardization) [13] to further enhance the model's generalization ability and network performance. WS provides regularization without compromising information exchange, thereby improving model generalization. The joint application of GN and WS is illustrated in Fig. 4. The fusion of GN and WS is expressed as follows:

$$\hat{\boldsymbol{x}} = \left[\hat{\boldsymbol{x}}_{i,j} \mid \hat{\boldsymbol{x}}_{i,j} = \frac{1}{\boldsymbol{\sigma}_{i,\cdot}} \left[\sum_{k,l} \boldsymbol{x}_{i-k,j-l} \cdot \left[\frac{1}{\boldsymbol{\sigma}_w} \left(\boldsymbol{\Gamma}_{k,l} - \boldsymbol{\mu}_w \right) \right] - \boldsymbol{\mu}_i \right] \right] \quad (3)$$

where $\boldsymbol{x}$ is the input tensor, $\boldsymbol{\Gamma}$ is the convolution kernel, and $\mathbf{y}$ is the output tensor. i and j denote the indexes of the output tensor, and k and l denote the indexes of the convolution kernel, respectively. In WS, the statistical domain of the mean and standard deviation of the weight parameters is each channel. The $\mu_{i,\cdot}$ and $\sigma_{i,\cdot}$ in Eq. 4 are the mean and standard deviation, calculated as σ_i and μ_i.

$$\mu_{i,\cdot} = \frac{1}{m} \sum_{k \in \mathcal{S}_i} x_k, \quad \sigma_i = \sqrt{\frac{1}{m} \sum_{k \in \mathcal{S}_i} (x_k - \mu_i)^2 + \epsilon} \tag{4}$$

$\mathcal{S}_i$ is the set of pixels for which the mean and variance are calculated, and $\mathcal{S}_i$ of GN is defined as

$$\mathcal{S}_i = \{k \mid k_N = i_N, \left\lfloor \frac{k_c}{C/G} \right\rfloor = \left\lfloor \frac{i_c}{C/G} \right\rfloor\} \tag{5}$$

where G is the number of groups (default value is 32) and C/G is the number of channels per group. $\lfloor - \rfloor$ represents the floor operation. GN computes μ and σ along the (H, W) axes and along a group of C/G channels.

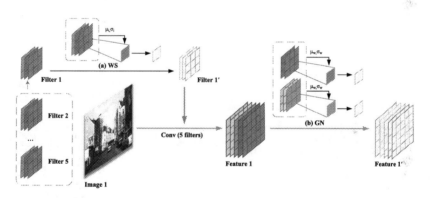

Fig. 4. The joint application of GN and WS involves utilizing μ_i, σ_i, μ_w, and σ_w, which represent the mean and variance of the respective statistical intervals. The normalized convolution kernel of Filter 1 is denoted as Filter 1$'$, while the normalized feature layer of Feature 1 is denoted as Feature 1$'$. It is important to note that the same operation performed on Filter 1 is also repeated for Filter 2 to 5, ensuring consistency across all filters.

3 Experiment

3.1 Implementation Details

Datasets GTSDB [7]: The traffic signs are classified into three categories: triangular warning signs (yellow or red), circular prohibitory signs (red or black), and mandatory signs (blue circles or squares). The GTSDB dataset consists of 900 images with a size of 1360 × 800, including 43 types of traffic signs. Similarly, the CTSD dataset contains 1100 images captured on highways, urban, and rural roads, with sizes of 1024 × 768 and 1280 × 720. It also classifies traffic signs in China into the same three categories as the GTSDB dataset.

Both the CTSD and GTSDB datasets were selected for their inclusion of small targets and multiple scales in complex environments. Figure 5 shows some typical examples from these datasets, where small targets and multiple scales of traffic signs are often present in challenging conditions such as blurred, poorly lit, damaged, and obscured signs. The datasets were divided into training and test sets in a 7:1 ratio.

Evaluation Metrics. In this paper, the evaluation metrics used are AP_{50}, AP_{75}, AP_S, AP_M, AP_L, and AR, which are calculated following the methods described in COCO [10].

Setting. All experiments were conducted on Ubuntu 20.04 using PyTorch 1.9.1. The training process utilized an NVIDIA GeForce GTX 3090 GPU with 24 GB of memory. The SGD optimizer was employed with an initial learning rate of 0.00125, momentum of 0.9, and weight decay of 0.0001.

Fig. 5. Some difficult examples for traffic sign detection.

3.2   Results and Analysis

Comparison with State-of-the-Art Methods on CTSD. We compared our MDCN method with several state-of-the-art approaches on the CTSD dataset. The comparison results are presented in Table 1 and Table 2.

Our proposed MDCN method achieves an impressive 83.0 AP and 87.6 AR, surpassing all other methods in the table. It demonstrates a strong balance between minimizing false detections and reducing missed detections. In terms of detection metrics, MDCN not only exhibits significant improvements over the original Faster R-CNN, but also outperforms both two-stage and single-stage target detection algorithms. It outperforms the second-best DH Faster R-CNN by 1.7 AP and surpasses the lowest AP YOLOF by a remarkable 14.9 AP. MDCN outperforms other methods in all metrics, except for the APL metric. This demonstrates the superiority of our method in multi-scale detection.

Furthermore, our proposed method achieves outstanding results for all three major categories of traffic signs, particularly for the mandatory category, where it outperforms all other detection methods by at least 3 points. This is due to the variable shapes and scales of directional traffic signs, with their rectangular aspect ratios often being more extreme compared to other traffic signs. Consequently, the detection of multi-scale signs in this category poses greater challenges.

Overall, the results validate the effectiveness of our MDCN method in achieving superior performance in multi-scale traffic sign detection.

Table 1. Comparison with the state-of-the-art methods on the CTSD dataset. Set Wa is the abbreviation of Warning, Pr is the abbreviation of Prohibitory, and Ma is the abbreviation of Mandatory.

Method	Backbone	Epo	AP	AP_{50}	AP_{75}	AP_S	AP_M	AP_L	AR	Wa	Pr	Ma
Dynamic rcnn [15]	ResNet-50	1x	81.2	97.9	95.0	60.9	82.8	88.9	85.3	79.9	83.1	80.4
Retinanet [9]		2x	76.5	92.4	90.1	49.6	79.4	82.2	84.3	73.0	78.3	78.1
Cascade rcnn [2]		2x	81.2	96.4	96.0	61.9	83.4	**91.4**	85.4	80.4	83.6	79.6
Baseline [15]		1x	79.0	96.2	94.0	61.0	81.7	85.3	84.1	77.4	81.7	78.1
DH Faster rcnn [18]		1x	81.3	98.3	95.3	64.3	83.1	86.3	84.5	80.8	82.0	81.1
Libra Faster rcnn [12]		1x	80.0	97.3	94.9	63.7	82.4	87.2	85.6	78.7	82.5	78.8
YOLOF [3]		1x	68.1	94.5	81.6	39.4	71.4	77.4	74.7	69.0	64.9	70.2
VfNet [21]		1x	78.4	97.0	93.7	62.5	80.5	88.7	84.7	78.7	80.0	76.4
ATSS [23]		1x	81.1	97.6	96.7	65.5	82.4	87.0	85.6	80.3	81.1	81.9
TOOD [5]		1x	79.8	**98.4**	95.5	63.4	81.5	88.7	83.8	80.5	79.9	79.1
MDCN	MDRNet	1x	**83.0**	98.3	**98.1**	**70.5**	**83.9**	90.8	**87.6**	**81.7**	**84.6**	**82.9**

Comparison with State-of-the-Art Methods on GTSDB. We conducted a comparison between MDCN and several popular detection algorithms on the GTSDB dataset, which is widely recognized as a representative dataset in the field of traffic sign object detection. Table 1 presents the results of this comparison.

Among the two-stage algorithms, DH Faster R-CNN and Cascade R-CNN achieved commendable detection performance with 77.9 AP and 76.6 AP, respectively. However, MDCN surpassed them by achieving an impressive 78.7 AP. Furthermore, MDCN outperformed the baseline by 5.5 AP and surpassed YOLOF, which had the lowest AP, by a significant factor of 1.48. Overall, MDCN exhibited superior detection accuracy compared to other methods, particularly for the warning class of traffic signs that typically occupy a smaller proportion of the image.

Additionally, MDCN demonstrated higher AP scores in the AP_S and AP_L metrics compared to all other methods. Although its AP_M was slightly lower than Dynamic R-CNN, Cascade R-CNN, and DH Faster R-CNN by 0.6 AP_M, 1.2 AP_M, and 2.0 AP_M, respectively, MDCN still outperformed them by significant margins in the AP_S metric (10.1 AP_S, 13.1 AP_S, and 6.8 AP_S, respectively). Moreover, MDCN's AP_L was higher than that of these methods. These results indicate that our method exhibits superior and more stable performance in multi-scale detection.

In conclusion, MDCN demonstrated excellent detection performance on the GTSDB dataset, affirming the effectiveness and generalization capabilities of our proposed model.

Table 2. Comparison with the state-of-the-art methods on the GTSDB dataset. Set Wa is the abbreviation of Warning, Pr is the abbreviation of Prohibitory, and Ma is the abbreviation of Mandatory.

Method	Backbone	Epo	AP	AP_{50}	AP_{75}	AP_S	AP_M	AP_L	AR	Wa	Pr	Ma
Dynamic rcnn [15]	ResNet-50	1x	76.2	96.1	92.5	64.3	82.7	85.6	80.2	76.6	76.3	75.7
Retinanet [9]		2x	67.3	85.5	79.8	44.4	79.5	78.8	76.3	72.3	75.7	53.8
Cascade rcnn [2]		2x	76.6	95.9	91.2	61.3	83.3	82.5	80.6	75.9	78.5	75.4
Baseline [15]		1x	73.2	96.5	92.1	56.9	81.6	85.4	77.8	74.8	77.5	67.3
DH Faster rcnn [18]		1x	77.9	97.8	93.5	67.6	**84.1**	76.3	82.1	73.9	**79.7**	**80.3**
Libra Faster rcnn [12]		1x	75.8	**98.9**	92.6	63.8	80.9	82.9	79.9	75.3	75.5	76.7
YOLOF [3]		1x	53.3	93.3	62.2	39.7	61.7	59.6	64.0	47.4	58.6	53.9
VfNet [21]		1x	72.2	95.6	85.6	56.6	80.6	78.4	78.0	67.1	76.7	73.0
ATSS [23]		1x	75.4	97.1	92.3	57.7	82.0	81.7	79.8	73.2	77.1	75.8
TOOD [5]		1x	76.1	98.0	92.7	67.4	82.0	83.4	82.1	76.5	76.6	75.3
MDCN	MDRNet	1x	**78.7**	98.0	**95.2**	**74.4**	82.1	**85.9**	**83.2**	**81.2**	77.0	78.1

3.3   Ablation Studys

We performed a series of ablation experiments to demonstrate the effectiveness of MDCN for the detection of small traffic sign targets at multiple scales in real-world complex environments.

Component Ablation Studies of MDCN. The effectiveness of various optimization components in improving the performance of the baseline model was evaluated through experiments on the CTSD and GTSDB datasets, and the results are summarized in Table 3 and Table 4.

The addition of GN led to improvements of 1.4 and 3.4 in AP on the CTSD and GTSDB datasets, respectively, highlighting its positive impact on detecting small targets. Combining GN with WS further enhanced the detection performance, resulting in improvements of 1.5 AP and 1.2 AP, as well as 8.9 and 15.4 in APS, respectively, compared to GN alone. This demonstrates that the combination of GN and WS contributes to improved detection of small targets.

The integration of MDRNet significantly improved the AP values by 3.7 AP and 5.3 AP on both datasets compared to the baseline. Moreover, it led to notable improvements in APS, APM, and APL, with gains of 8.9, 2.6, and 4.1, respectively, on both datasets. The largest improvement was observed in APS, indicating the effectiveness of MDRNet in detecting small targets. This improvement can be attributed to the ability of MDRNet to capture more contextual information, which helps reduce the rates of false detections and missed detections for small targets.

Furthermore, the inclusion of the MDEF method resulted in additional improvements of 4.0 AP and 5.5 AP on the two datasets compared to the baseline. MDEF effectively expanded the number of small targets in the dataset and disrupted the regular positioning of traffic signs on the road, thereby preventing overfitting to specific road environments.

Visualizations in Fig. 6 further demonstrate the superior performance of MDCN compared to the baseline. MDCN exhibits greater sensitivity to traffic sign objects with uneven positional distribution and a wide range of scales. Additionally, it demonstrates better detection performance for small objects, as evident from the second row of the visualization.

Table 3. Ablation study on the effectiveness of the various MDCN component modules on CTSD dataset. MDEF is short for Multiple Data augmentation Fusion.

Method	GN	WS	MDRNet	MDEF	AP	AP_{50}	AP_{75}	AP_S	AP_M	AP_L
MDCN	-	-	-	-	79.0	96.2	94.0	61.0	81.7	85.3
	✓	-	-	-	80.4	97.3	96.2	70.2	82.6	84.7
	✓	✓	-	-	81.9	97.1	97.0	68.9	83.6	90.3
	✓	✓	✓	-	82.7	97.6	97.2	69.9	84.3	89.4
	✓	✓	✓	✓	**83.0**	**98.3**	**98.1**	**70.5**	**83.9**	**90.8**

Ablation Experiments of Dilated Convolution Embedding Positions.
In our ablation experiments, we investigated the impact of dilated convolution on the task of traffic sign detection. Dilated convolution is known for its ability to expand the receptive field while preserving resolution. However, it is crucial to carefully select the positions where regular convolutions are replaced with dilated convolutions to achieve optimal results.

We examined the effect of incorporating dilated convolutions from the bottom-up, starting from the conv2 to conv5 layers in the backbone network. The results, as shown in Table 5 and Table 6, revealed that the addition of dilated

Table 4. Ablation study on the effectiveness of the various MDCN component modules on GTSDB dataset. MDEF is short for Multiple Data augmentation Fusion.

Method	GN	WS	MDRNet	MDEF	AP	AP_{50}	AP_{75}	AP_S	AP_M	AP_L
MDCN	-	-	-	-	73.2	96.5	92.1	56.9	81.6	85.4
	✓	-	-	-	76.6	95.9	95.1	69.9	82.0	84.6
	✓	✓	-	-	77.8	97.7	94.3	72.3	81.2	82.5
	✓	✓	✓	-	78.5	97.1	95.6	73.1	82.2	89.6
	✓	✓	✓	✓	**78.7**	**98.0**	**95.2**	**74.4**	**82.1**	**85.9**

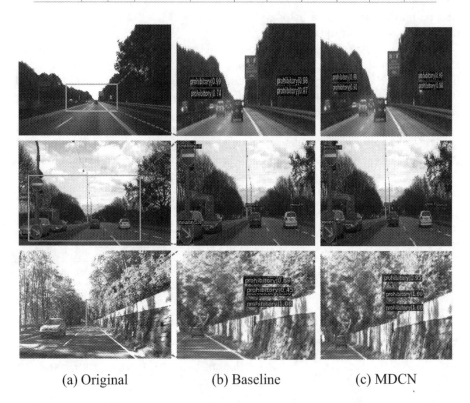

(a) Original   (b) Baseline   (c) MDCN

Fig. 6. Comparison of detection results between baseline and MDCN. (a) is the original image. (b) is the detection result using baseline. (c) is the detection result using MDCN. The images in the last two columns are taken from the green box in the first column. (Color figure online)

convolutions in the last two layers of the convolutional hierarchy yielded the most significant improvements. Specifically, on the CTSD dataset, the APs increased to 81.7 and 81.6, while on the other dataset, they improved to 76.4 and 77.6, respectively.

This observation can be attributed to the fact that deeper convolution layers tend to lose internal data structures, resulting in the loss of some crucial information related to small objects. By replacing the deeper convolutions with dilated convolutions, we were able to improve the localization accuracy of small targets.

Building upon these findings, we gradually added dilated convolution modules in pairs until all 3×3 convolutions between c4 and c5 were replaced. This configuration resulted in the optimal embedding position for MDRNet, achieving the best performance of 83.0 AP and 78.7 AP on both datasets.

These ablation experiments highlight the importance of selecting appropriate positions for dilated convolutions, and they provide valuable insights into the optimal design choices for our MDRNet architecture.

Table 5. Ablation study with gradually dilated modules on CTSD dataset. In the setting column, c2, c3, c4 and c5 stand for applying dilated convolution at c2, c3, c4 and c5 layers.

Method	c2	c3	c4	c5	AP	AP_{50}	AP_{75}	AP_S	AP_M	AP_L	AR
MDRNet	✓	-	-	-	81.3	96.9	95.9	66.0	83.2	91.1	86.7
	-	✓	-	-	81.4	96.8	96.4	66.6	83.5	88.7	86.1
	-	-	✓	-	81.7	97.9	97.8	68.0	82.8	89.4	86.4
	-	-	-	✓	81.6	97.0	96.0	67.1	83.6	90.2	87.4
	✓	✓	-	-	80.5	97.1	96.3	66.6	81.2	89.9	85.0
	-	✓	✓	-	80.4	97.6	97.1	66.5	82.3	87.6	85.9
	-	-	✓	✓	**83.0**	**98.3**	**98.1**	**70.5**	**83.9**	**90.8**	**87.6**

Table 6. Ablation study with gradually dilated modules on GTSDB dataset. In the setting column, c2, c3, c4 and c5 stand for applying dilated convolution at c2, c3, c4 and c5 layers.

Method	c2	c3	c4	c5	AP	AP_{50}	AP_{75}	AP_S	AP_M	AP_L	AR
MDRNet	✓	-	-	-	75.7	95.6	90.3	70.2	79.6	84.2	80.8
	-	✓	-	-	75.9	96.1	93.4	65.0	80.1	84.6	81.5
	-	-	✓	-	76.4	95.9	92.7	66.3	81.8	86.7	80.7
	-	-	-	✓	77.6	97.4	93.1	72.1	81.3	85.4	82.6
	✓	✓	-	-	74.9	95.4	94.0	68.0	78.6	85.0	80.3
	-	✓	✓	-	77.2	96.7	93.7	65.7	81.6	84.2	82.7
	-	-	✓	✓	**78.7**	**98.0**	**95.2**	**74.4**	**82.1**	**85.9**	**83.2**

4 Conclusion

The primary objective of this paper is to enhance the detection performance of small multi-scale traffic signs in complex real-world environments. To achieve this goal, we propose an MDCN traffic sign detection method based on Faster R-CNN. Our approach incorporates MDRNet as the backbone network, which effectively increases the perceptual field and sampling rate. This improvement enhances the feature representation capability for targets of different scales and those that may be partially obscured. Considering the characteristics of the traffic sign dataset, hardware resource limitations, and the dataset's high resolution, we introduce a normalized combination method at the backbone and pooling layers. This technique accelerates the convergence speed and improves the generalization ability of the model. Furthermore, we employ a multi-data augmentation fusion strategy to expand small targets and enhance the network's robustness. The effectiveness and generalization of our proposed MDCN method are validated through comprehensive evaluations on the CTSD and GTSDB datasets. The results demonstrate its superiority in detecting small traffic signs and establish its competitiveness compared to other state-of-the-art methods.

For future research, we intend to explore traffic sign recognition techniques tailored specifically for mobile terminals. By adapting the methodology to mobile devices, we aim to further expand the applicability and practicality of traffic sign detection in real-world scenarios.

Acknowledgment. This work is supported by the Natural Science Foundation of Xinjiang Uygur Autonomous Region (2020D01C33).

References

1. Bochkovskiy, A., Wang, C.Y., Liao, H.Y.M.: Yolov4: optimal speed and accuracy of object detection. arXiv preprint arXiv:2004.10934 (2020)
2. Cai, Z., Vasconcelos, N.: Cascade r-CNN: high quality object detection and instance segmentation. IEEE Trans. Pattern Anal. Mach. Intell. **43**(5), 1483–1498 (2019)
3. Chen, Q., Wang, Y., Yang, T., Zhang, X., Cheng, J., Sun, J.: You only look one-level feature. In: Proceedings of the IEEE/CVF Conference on Computer Vision and Pattern Recognition, pp. 13039–13048 (2021)
4. Elsagheer Mohamed, S.A., AlShalfan, K.A.: Intelligent traffic management system based on the internet of vehicles (IoV). J. Adv. Transp. **2021**, 1–23 (2021)
5. Feng, C., Zhong, Y., Gao, Y., Scott, M.R., Huang, W.: TOOD: task-aligned one-stage object detection. In: 2021 IEEE/CVF International Conference on Computer Vision (ICCV), pp. 3490–3499. IEEE Computer Society (2021)
6. He, K., Gkioxari, G., Dollár, P., Girshick, R.: Mask R-CNN. In: Proceedings of the IEEE International Conference on Computer Vision, pp. 2961–2969 (2017)
7. Houben, S., Stallkamp, J., Salmen, J., Schlipsing, M., Igel, C.: Detection of traffic signs in real-world images: the German traffic sign detection benchmark. In: The 2013 International Joint Conference on Neural Networks (IJCNN), pp. 1–8. IEEE (2013)

8. Keskar, N.S., Mudigere, D., Nocedal, J., Smelyanskiy, M., Tang, P.T.P.: On large-batch training for deep learning: generalization gap and sharp minima. arXiv preprint arXiv:1609.04836 (2016)

9. Lin, T.Y., Goyal, P., Girshick, R., He, K., Dollár, P.: Focal loss for dense object detection. In: Proceedings of The IEEE International Conference on Computer Vision, pp. 2980–2988 (2017)

10. Lin, T.-Y., et al.: Microsoft COCO: common objects in context. In: Fleet, D., Pajdla, T., Schiele, B., Tuytelaars, T. (eds.) ECCV 2014. LNCS, vol. 8693, pp. 740–755. Springer, Cham (2014). https://doi.org/10.1007/978-3-319-10602-1_48

11. Liu, Y., Peng, J., Xue, J.H., Chen, Y., Fu, Z.H.: Tsingnet: scale-aware and context-rich feature learning for traffic sign detection and recognition in the wild. Neurocomputing **447**, 10–22 (2021)

12. Pang, J., Chen, K., Shi, J., Feng, H., Ouyang, W., Lin, D.: Libra R-CNN: towards balanced learning for object detection. In: Proceedings of the IEEE/CVF Conference on Computer Vision and Pattern Recognition, pp. 821–830 (2019)

13. Qiao, S., Wang, H., Liu, C., Shen, W., Yuille, A.: Micro-batch training with batch-channel normalization and weight standardization. arXiv preprint arXiv:1903.10520 (2019)

14. Redmon, J., Farhadi, A.: Yolov3: an incremental improvement. arXiv preprint arXiv:1804.02767 (2018)

15. Ren, S., He, K., Girshick, R., Sun, J.: Faster R-CNN: towards real-time object detection with region proposal networks. In: Advances in Neural Information Processing Systems 28 (2015)

16. Shen, L., You, L., Peng, B., Zhang, C.: Group multi-scale attention pyramid network for traffic sign detection. Neurocomputing **452**, 1–14 (2021)

17. Wang, J., Chen, Y., Dong, Z., Gao, M.: Improved yolov5 network for real-time multi-scale traffic sign detection. Neural Comput. Appl. **35**(10), 7853–7865 (2022)

18. Wu, Y., et al.: Rethinking classification and localization for object detection. In: Proceedings of the IEEE/CVF Conference on Computer Vision and Pattern Recognition, pp. 10186–10195 (2020)

19. Yao, Y., Han, L., Du, C., Xu, X., Jiang, X.: Traffic sign detection algorithm based on improved yolov4-tiny. Signal Process.: Image Commun. **107**, 116783 (2022)

20. Yu, F., Koltun, V.: Multi-scale context aggregation by dilated convolutions. arXiv preprint arXiv:1511.07122 (2015)

21. Zhang, H., Wang, Y., Dayoub, F., Sunderhauf, N.: Varifocalnet: an iou-aware dense object detector. In: Proceedings of the IEEE/CVF Conference on Computer Vision and Pattern Recognition, pp. 8514–8523 (2021)

22. Zhang, H., Cisse, M., Dauphin, Y.N., Lopez-Paz, D.: Mixup: beyond empirical risk minimization. arXiv preprint arXiv:1710.09412 (2017)

23. Zhang, S., Chi, C., Yao, Y., Lei, Z., Li, S.Z.: Bridging the gap between anchor-based and anchor-free detection via adaptive training sample selection. In: Proceedings of the IEEE/CVF Conference on Computer Vision and Pattern Recognition, pp. 9759–9768 (2020)

Identifying Critical Congested Roads Based on Traffic Flow-Aware Road Network Embedding

Jing Zhao[1,2], Peng Cheng[3], Qixiang Ge[2], Xun Zhu[2], Lei Chen[4,5], Xi Guo[6], Jinshan Sun[1], and Yangfang Yang[7(✉)]

[1] State Key Laboratory of Precision Blasting, Jianghan University, Wuhan, China
{jingzhao,sunjinshan}@jhun.edu.cn

[2] School of Artificial Intelligence, Jianghan University, Wuhan, China
geqixiang@stu.jhun.edu.cn, zhuxun@jhun.edu.cn

[3] East China Normal University, Shanghai, China
pcheng@sei.ecnu.edu.cn

[4] The Hong Kong University of Science and Technology (Guangzhou), Guangzhou, China
leichen@cse.ust.hk

[5] The Hong Kong University of Science and Technology, Hong Kong, China

[6] University of Science and Technology Beijing, Beijing, China
xiguo@ustb.edu.cn

[7] China Academy of Transportation Sciences, Beijing, China
yangyf@motcats.ac.cn

Abstract. Traffic congestion occurs frequently and concurrently on urban road networks, and may cause widespread traffic paralysis if not controlled promptly. To relieve traffic congestion and avoid traffic paralysis, it is significant to identify critical congested roads with great propagation influence on others. Existing studies mainly focus on topological measures and statistical approaches to evaluate the criticality of road segments. However, critical congested roads are generated by dynamic changes in traffic flow, so that identifying them involves both the network structure and dynamic traffic flows is required. In this paper, we propose a novel road network embedding model, called *Seg2Vec*, to learn comprehensive features of road segments considering both the road structural information and traffic flow distribution. The Seg2Vec model combines a Markov Chain-based random walk with the *Skip-gram* model. The random walk is conducted on the road network based on the transition probabilities computed from historical trajectory data. Moreover, we define the *propagation influence* of a congested road by a score function based on the learned road representation. The goal is to find the *critical congested roads* with top-K propagation influences. Evaluation experiments are conducted to verify the effectiveness and efficiency of the proposed

Supported by the Project of Hubei Provincial Science and Technology Department(2020BCA084), and Open Foundation of Key Laboratory of Transport Industry of Big Data Application Technologies for Comprehensive Transport (2021B1202).

X. Yang et al. (Eds.): ADMA 2023, LNAI 14176, pp. 598–613, 2023.
https://doi.org/10.1007/978-3-031-46661-8_40

method. A case study of identifying critical congested roads from a congestion cluster is also demonstrated. The identified critical congested roads can facilitate decision-making for traffic management.

Keywords: Road network embedding · Critical congested roads · Congestion propagation · Trajectory data

1   Introduction

Traffic congestion has become a major issue to urban traffic system, since it causes waste of time, energy consumption and air pollution, which further leads to economic cost and damages to citizens' health. In urban road networks, traffic congestion occurs frequently and concurrently, so that it may cause widespread traffic paralysis if not controlled promptly. Therefore, to relieve traffic congestion and avoid traffic paralysis, it is significant to identify *critical congested roads* with great propagation influence on other roads.

Existing studies on identifying critical roads mainly focuses on topological measures and statistical approaches to evaluate the criticality of roads [2,6,9,22]. For instance, a topological graph measure called *betweenness* is defined in [2] to identify critical roads that are located on many shortest paths between other vertices. Guo et al. proposed the weighted degree and the impact distance as the two major measures to identify the most influential locations [6]. Li et al. also studied the percolation process on road networks, and identified bottleneck roads that play a critical role in connecting different functional clusters [9,22].

The above approaches have been proposed to identify fixed critical roads that caused by the essential structure of road networks. However, critical congested roads are generated by dynamic changes of traffic flow or sudden increases of traffic demand, such as traffic accidents, road damage, etc. In order to discover the root cause of traffic congestion, it is necessary to measure the interactions among congested roads and the surrounding roads under real-time conditions. In other words, a congestion occurs on the road segment with great propagation influence on other road segments is considered as the critical congested road. The propagation influence of a road segment is determined by many factors, such as time, location, and properties of road. As a result, identifying critical congested roads involves both the network structure and dynamic traffic flows. How to obtain comprehensive road network features that consider both sides and quantitatively analyze the congestion propagation characteristics of road segments are significant issues studied in this paper.

In order to learn comprehensive features of road segments, road network embedding is extended from graph embedding methods [1]. Graph embedding methods use truncated random walks to map a network to low-dimensional representations, while preserving most of the network information. The learned feature representations are applied widely for network analysis, such as node classification, link prediction and community detection [21]. It has been verified that graph embedding methods are able to capture structural information of road networks [8]. To quantify the traffic interaction between road segments, Road2Vec [10] learns the feature representation of road segments using travel routes. However, straightforwardly using real trajectories to train the model makes the result bias to main roads with high frequencies, and the local structure of road network is ignored. Therefore, training the sequences of graph nodes in an appropriate manner which considers both the road structural information and traffic flow distribution is required.

In this paper, we propose a traffic flow-aware road network embedding model, called *Seg2Vec*, which leverages trajectory data to enhance graph embedding for road representation. The Seg2Vec model combines a Markov Chain-based random walk with the Skip-gram model [14]. The random walk is conducted on the road network based on the transition probabilities computed from historical trajectories. The learned road representation is used to measure the traffic similarity between road segments, which is further used to evaluate the propagation influences of congested roads. Moreover, a critical congested road identification algorithm is proposed to identify critical roads with top-K propagation influences. By monitoring the travel speeds of vehicles, it is able to detect traffic congestion and identify critical congested roads in real time.

Our contributions can be summarized as follows:

1. We propose a road network embedding model, Seg2Vec, to measure the traffic similarity between road segments. The Seg2Vec model learns feature representation of road segments considering both the road network structure and traffic flow distribution.
2. We define the congestion propagation influence of a congested road by a score function based on the learned feature representation of road segments. To the best of our knowledge, it is the first work to measure congestion propagation influence by road embedding.
3. We propose a critical congested road identification algorithm to identify critical roads with top-K propagation influences in real time.
4. We conduct extensive experiments to verify the effectiveness and efficiency of the proposed method using real taxis' trajectories. A case study of identifying critical congested roads for a congestion cluster is also demonstrated.

The rest of this paper is organized as follows. In Sect. 2, we introduce the preliminaries and the problem definition of this paper. The proposed road network embedding model and critical congested roads identification method are introduced in Sect. 3 and Sect. 4, respectively. Experimental results with a case study are shown in Sect. 5. Related work is introduced in Sect. 6. Finally, we conclude the paper in Sect. 7.

2   Problem Statement

2.1   Preliminaries

Definition 1 Road network: *A road network is defined as a directed graph $G = (V, E)$, where V is a set of nodes and E is a set of directed edges on the* road network. Each edge $e_{ij} \in E$ is determined by a source node v_i and a target node v_j, i.e., $e_{ij} = (v_i \rightarrow v_j)$, $v_i, v_j \in V$.

Definition 2 *n-hop neighborhood:* Given a (un)directed graph $G = (V, E)$, the n-hop neighborhood of a node $u \in V$, represented as $N_n(u)$, is defined as nodes within n-hop distance from u on the graph, which means the shortest path from u to $v \in N_n(u)$ or v to u is not large than n, where n is a given parameter.

Note that, we consider both of the upstream and downstream neighborhood of a node as its neighborhood. This is because traffic congestion can propagate both backward and forward, i.e., *bi-directional*.

Definition 3 *Trajectory:* A trajectory s consists of a sequence of location points $\{l_1, l_2, \ldots, l_{|s|}\}$, where each location point $l_i = (x_i, y_i, v_i, t_i)$ corresponds to a location coordinate (x_i, y_i) with a velocity v_i at a time stamp t_i, where $i \in [1, |s|]$.

Traffic congestion generally causes a slowdown in traffic speed on specific roads, which lasts for a short time but happens frequently. Here, we define the congested road by comparing vehicles' speeds with the free-flow speed. The free-flow speed is the average speed that a driver would travel if there is no congestion or other adverse conditions. We use the F percentile of all valid speeds on each road segment as its free-flow speed, as in [20], with a default value of 85. By monitoring vehicles' speeds, traffic congestion is detected rapidly and accurately.

Definition 4 *Road speed:* Given an edge e on a road network, a set of sub trajectories $S_{e,t}$ that matched to edge e during time period t, such that $|S_{e,t}| \geq min\_sup$, where min_sup is a confidence threshold. The road speed $v_{e,t}$ is calculated as follows:

$$v_{e,t} = \frac{1}{|S_{e,t}|} \sum_{s \in S_{e,t}} \sum_{l_i \in s} \frac{v_i}{|s|}, \tag{1}$$

i.e., the average of the speeds of all trajectories in $S_{e,t}$, where the speed of a trajectory $s \in S_{e,t}$ is the average travel speed of all location points $l_i \in s$.

Definition 5 *Congested road:* Given an edge $e \in E$ with road speed $v_{e,t}$, and a free-flow speed v_f. The edge e is defined as a congested road, if $v_{e,t}$ is less than C percentage of v_f, where C is a default parameter.

2.2 Problem Definition

To evaluate the propagation influence of a congested road, we construct a road graph which considers each road segment as a node, and learn feature representation of road segments. Then, we generate congestion clusters from a set of congested roads, and identify critical congested roads for each congestion cluster.

Definition 6 *Road graph:* *Given a directed road network $G = (V, E)$, a road graph is defined as an undirected graph $G_r = (V_r, E_r)$, where V_r contains all road segments in E, i.e., a node $v_i \in V_r$ corresponds to an edge $e_i \in E$. An edge $e_{ij} \in E_r$ between v_i and v_j on G_r exists if there is a path from edge e_i to e_j or from e_j to e_i on G.*

Definition 7 *Road representation:* *Given a road graph $G_r = (V_r, E_r)$, the road representation aims to learn a feature representation of road segments, denoted as $f : V_r \rightarrow \mathbb{R}^d$, which projects each road to a d-dimensional feature vector.*

Definition 8 *Congestion cluster:* *Given a set of congested roads E_t during time period t, a congestion cluster Ec is defined as a subset of E_t, such that each congested road $e \in Ec$ has at least one n-hop neighborhood that is congested, and all of the congested n-hop neighborhood of e belongs to Ec, i.e., $|N_n(e) \cap E_t| \geq 1$ and $N_n(e) \cap E_t \subseteq Ec$. In addition, $|Ec| \geq c$, where c is a default parameter.*

Definition 9 *Propagation influence:* *Given a congestion cluster Ec, and the road representation f, each congested road $u \in Ec$ has a propagation influence with respect to its n-hop neighborhood in Ec, denoted as $Nc(u)$. The propagation influence of u is defined as follows:*

$$PI(u \mid Nc(u)) = w_u \sum_{e \in Nc(u)} Sim(f(u), f(e)), \qquad (2)$$

where, $Sim()$ is a predefined similarity metric, e.g., cosine similarity, and w_u is a normalized occurrence probability of road segment u.

The value of w_u is calculated by historical trajectory data. The intuition is that, the larger the occurrence probability is, the more likely the road segment affects other road segments.

Definition 10 *Critical Congested Roads:* *Given a congestion cluster Ec, and an integer $K \leq |Ec|$. The critical congested roads in Ec are defined as a set of congested roads $E_K \subset Ec$ with maximum size K, such that the sum of propagation influence of each road in E_K is maximal.*

3   Seg2Vec: Traffic Flow-Aware Road Network Embedding

As described above, the propagation influence of a congested road is calculated by the feature representation of road segments. In this paper, we propose Seg2Vec, a traffic flow-aware road network embedding model that learns the structural information of road segments by simulating trajectories using a Markov chain-based random walk. The intuition is that traffic flows on road networks follow certain spatiotemporal distributions, and road segments with high co-occurrences along trajectories indicate high traffic similarities among them. Therefore, we use historical trajectories to capture the transition probability between road segments, then generate neighbor nodes by a sampling strategy based on the precomputed transition matrix, with details as follows.

3.1 Markov Chain-Based Random Walks

The Markov chain is a stochastic process that satisfies the Markov property [4]. The process describes a sequence of possible events in which the probability of each event only depends on the state of the previous event. In this paper, we consider each road segment as a state, the probability of state change between road segments is called transition probability, calculated by historical data.

Given a set of trajectories S mapped to the road graph G_r, each trajectory consists of a sequence of nodes (i.e., road segments). The transition probability distribution can be represented by a transition matrix P, and each element p_{ij}^T represents the probability of changing the state from v_i to v_j during time period T, which is evaluated by the following equation:

$$p_{ij}^T = Pr^T(s_{n+1} = v_j \mid s_n = v_i) = \frac{|S^T(v_i \to v_j)|}{|S^T(v_i)|}, \tag{3}$$

where $|S^T(v_i)|$ is the outflow of road v_i during time period T, and $|S^T(v_i \to v_j)|$ is the traffic flow from road v_i to road v_j during time period T. In addition, the variable $n = 0, 1, ..., l - 1$, where l is the walk length. Note that, the transition probability of current node only depends on the previous node, so the random walk is the first-order Markov chain.

Benefits of Random Walks. The benefits of random walks are reflected in both effectiveness and efficiency. For effectiveness, the traffic distribution of real trajectories essentially has a bias to the main roads with high frequencies, which leads to the learned features bias to these roads. The proposed Seg2Vec model simulates equal number of trajectories for each road using a Markov chain-based random walk, so that it can preserve local network structure with less bias. Moreover, training random walks is more efficient, since the number of walks and walk length of the simulated walks are optional. In addition, random walks can provide sample reuse. For instance, a trajectory with length l will generate k samples for $l - k$ nodes at once, where k is the context size, and $l > k$. Then, for N source nodes each with r random walks, there are $Nrk(l - k)$ samples generated in total.

3.2 Feature Learning Model

Given a road graph $G_r = (V_r, E_r)$, and a set of simulated trajectories S generated by the random walks introduced above. The objective function of the Seg2Vec model is to maximize the log-probability of observing the neighborhood $N(u)$ for a source node $u \in V_r$ conditioned on its feature representation, given by f:

$$\max_f \sum_{u \in V} \log Pr(N(u) \mid f(u)), \tag{4}$$

where, the neighborhood $N(u)$ of road segment u is determined by a sliding window of size k over consecutive road segments of trajectories.

Based on the conditional independence, the probability of observing a neighborhood $n_i \in N(u)$, given node u, is calculated by the softmax function:

$$Pr(n_i \mid f(u)) = \frac{\exp(f(n_i) \cdot f(u))}{\sum_{v \in V_r} \exp(f(v) \cdot f(u))}. \tag{5}$$

Based on the above assumptions, the feature learning model is trained by optimizing the objective function using stochastic gradient ascent (SGD) and back propagation [17].

3.3 The Seg2Vec Algorithm

The pseudo-code of Seg2Vec is given in Algorithm 1. The algorithm consists of three phases, i.e., computing the transition matrix (Line 1), Markov chain-based random walks (Line 2–14) and optimization using SGD (Line 15). The first phase is constructing a transition matrix P by computing the transition probabilities between road segments based on historical trajectories. Since we aim to learn representations for all nodes, we simulate r walks per node with fixed length l starting from each node. The sampling strategy for the random walk is the alias sampling, which is also used in [5,18]. The alias sampling can be done efficiently in $O(1)$ time complexity, with the precomputed transition matrix for the first-order Markov chain. Finally, the generated walks are selected as the input of the training model, and the feature representations of road segments are optimized using SGD. Note that, each phase described above can be conducted in parallel, which contributes to the scalability of the proposed model.

Algorithm 1 The Seg2Vec algorithm.

Input: G, trajectories S, dimensions d, walks per node r, length l, context size k.
Output: Feature representation f.
1: $P =$ComputeTransitionPr(G, S) ▷ Preprocess the transition matrix
2: $walks \leftarrow \emptyset$ ▷ Initialize the walks to empty
3: **for** $iter \leftarrow 1$ to r **do**
4:     **for all** nodes $u \in V$ **do**
5:         $walk \leftarrow [u]$ ▷ Initialize the walk list as the source node u
6:         **for** $n \leftarrow 1$ to $l - 1$ **do**
7:             $v_{cur} = walk[n-1]$
8:             $N(v_{cur}) =$ GetNeighbors(v_{cur}, G) ▷ Get the neighborhoods of v_{cur}
9:             $v_{next} =$AliasSample$(v_{cur}, N(v_{cur}), P)$
10:           Append v_{next} to $walk$
11:        **end for**
12:       Append $walk$ to $walks$
13:     **end for**
14: **end for**
15: $f =$ StochasticGradientDescent$(k, d, walks)$
16: **return** f

---

4   Critical Congested Road Identification

Consider a real-time traffic monitoring system that consecutively collects vehicles' GPS trajectory data, and detects congested roads based on the collected trajectories. In order to identify critical congested roads among a series of detected congested roads, we propose a critical congested roads identification method as shown in Algorithm 2. We first generate several congestion clusters based on the n-hop neighborhoods for each congested road (Line 1). Then, the propagation influence of each congested road is computed by the feature representation of road segments (Line 6). Finally, we select the critical roads with top-K propagation influences in each cluster (Line 9-10).

Algorithm 2 The CriticalRoadIdentification algorithm.

Input: Congested roads E_t, order n, K, feature representation f.
Output: Overall top-K critical roads R.
 1: E_c =GenerateCongestCluster(E_t, n) ▷ Generate a set of congestion clusters for E_t
 2: $R \leftarrow \emptyset$ ▷ Initialize the top-K critical roads for all congestion clusters
 3: **for all** cluster $c \in E_c$ **do**
 4:      $r_c \leftarrow \emptyset$ ▷ Initialize the critical road list for cluster c
 5:      **for all** road $e \in c$ **do**
 6:          $PI(e) = $ ComputePI(e, c, f) ▷ Compute the propagation influence of road e
 7:          Append $(e, PI(e))$ to r_c
 8:      **end for**
 9:      Sort the propagation influences of roads in r_c in descending order
10:      Select the top-K critical roads in r_c and add to result R
11: **end for**
12: **return** R

Figure 1 illustrates an instance of identifying critical congested roads. While Fig. 1a shows a set of congested roads in red, it is difficult to explain the congestion or decide how to relieve them. In this work, we provide an alternative solution to explain the root cause of congestion. Figure 1b shows a congestion cluster, in which each node represents a congested road in Fig. 1a. The value on a node (in red) represents its propagation influence, while the value on an edge (in black) represents the similarity between two roads. The propagation influence is calculated by Eq. 2, assuming the occurrence probability of each node is 1. Here, we consider both the upstream and downstream roads of a congested road as its neighborhoods, e.g., e_2 and e_7 are both e_1's neighborhoods. Moreover, we consider the relationship between any 2-hop neighborhoods, e.g., e_2 and e_7 are mutual neighborhoods. By ranking the propagation influences of congested roads, it is more intuitive to express the criticality of each one. For instance, it is evident that the propagation influence of e_1 is the greatest one, which has strong influence on other roads, followed by e_5, e_2 and e_7.

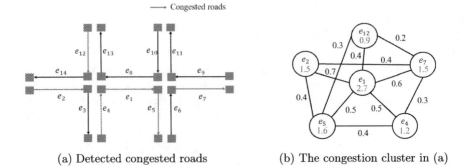

(a) Detected congested roads   (b) The congestion cluster in (a)

Fig. 1. An instance of identifying critical congested roads

5   Experiments

5.1   Datasets and Preprocessing

We use real GPS trajectories generated by taxicabs in Chengdu from October 1st to October 20th, 2016, provided by Didi Chuxing[1] There are about 3,503,276 records of trip orders with 671 millions trajectory points, each of which contains a time stamp, a longitude and latitude, an encrypted driver ID and order ID. We use trajectory data of two weeks for training, and the rest for testing. The road network of Chengdu is obtained from Open Street Map (OSM)[2] with about 3151 nodes and 7336 edges on the road network. Figure 2a shows the original distribution of trajectory data (in red) generated on October 1st, 2016, inside

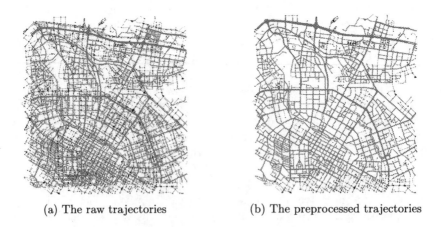

(a) The raw trajectories   (b) The preprocessed trajectories

Fig. 2. The distribution of trajectory data on the road network

[1] Didi chuxing, https://outreach.didichuxing.com/app-vue/dataList.
[2] OSM, https://www.openstreetmap.org/,

the Second Ring Road of Chengdu's road network. Since the raw trajectory data is unordered and full of noise, data preprocessing is necessary. The preprocessing mainly contains three steps, i.e., coordinates transformation, data cleaning and map matching. We utilize an efficient map matching algorithm, called ST-Matching algorithm [12], to map the trajectories to the road network. After data preprocessing phase, the raw trajectory data is transformed into 3,689,336 records of time-sorted trajectories, with about 657 millions of location points mapped to the road network, as shown in Fig. 2b.

5.2   Experimental Setup

We perform extensive experiments and a case study to evaluate the quality and efficiency of the proposed method. The quality of Seg2Vec is evaluated by cosine similarities of the learned road representation, as well as the effectiveness of critical congested road identification, comparing with Road2Vec [10]. The efficiency of our methods is evaluated by the execution time of offline training and online process. In addition, a case study is also demonstrated to show the effectiveness of our proposed method. Our experiments are performed on a 64-bit server running Ubuntu 20.04.4 (OS) with an Intel Xeon Gold 6226R CPU @ 2.90 GHz × 32 and a 256GB RAM.

Time Division. As shown in Fig. 3, the number of traffic flow and traffic congestion on the road network is time-variant. According to the distribution of traffic flow and congestion changing by time, we divide a day into three time periods, i.e., *morning peak*, *normal* and *evening peak*. The morning peak hours of weekdays and weekends are set as 7 am to 10 am and 8 am to 11 am, respectively, while the evening peak hours are set as 5 pm to 8 pm. The remained time periods during daytime are set as normal time.

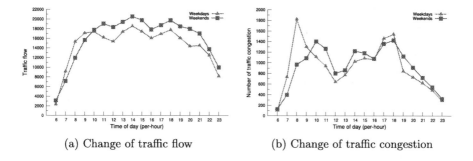

(a) Change of traffic flow   (b) Change of traffic congestion

Fig. 3. Traffic distribution changing over time of day

Effectiveness Evaluation. Given a congestion cluster Ec during time period $[t_s, t_e]$, and a set of critical congested roads E_K with top-K propagation influences, the effectiveness of E_K is measured by the average percentage of traffic volume reduction if deleting the trajectories passing through E_K from the congestion cluster Ec, the calculation function is as follows:

$$Effectiveness(E_K) = \frac{1}{|E_c|} \sum_{e \in Ec} \frac{|V(E_K) \cap V(e)|}{|V(e)|}, \tag{6}$$

where, $V(E_K)$ is the vehicles on the critical congested roads E_K during time period $[t_s, t_e]$, and $V(e)$ is the vehicles passing through road $e \in Ec$ during time period $[t_s, t_e + \delta t]$. In the following, we will investigate the effect of time delay on propagation influence by comparing the effectiveness by varying δt.

5.3   Evaluation Results

Traffic Similarity. We first compare the two learning models by showing the traffic similarity with varying n, i.e., the average cosine similarities of n-hop neighbors for each road segment, where $n \in [1, 5]$. In Fig. 4, the traffic similarities of Road2Vec (in hollow) and Seg2Vec (in solid) are decreased by the increase of n, which indicates the two models can capture the spatial proximity of road segments. Moreover, the traffic similarities of Seg2Vec are larger than the ones of Road2Vec when n equals to 1, while n becomes larger than 2, the traffic similarities of Seg2Vec are smaller than Road2Vec. The result indicates that Seg2Vec is more sensitive to both adjacent and distant neighbors than Road2Vec.

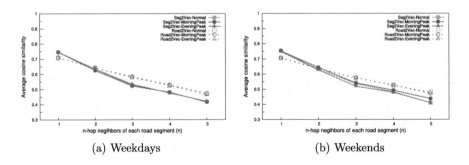

(a) Weekdays   (b) Weekends

Fig. 4. Traffic similarity with varying n

Case Study. We evaluate the effectiveness of our proposed model through a case study of a congestion cluster around Tianfu Square during 9:10-9:20 on October 19th, 2016. As shown in Fig. 5a, there are 15 congested roads (in red) in the congestion cluster. To identify critical congested roads which have strong influences on others, our proposed road embedding model learns the feature representation of road segments. Based on the learned feature vectors and the proposed propagation influence function, we visualize the embedded road segments in Fig. 5b. Each congested road is represented as a node, the size of which represents its propagation influence on others. An edge between two nodes represents the two road segments that are mutual 5-hop neighborhoods, and the thickness of an edge represents the traffic similarity between them. As Fig. 5 shows, the top-2 critical roads are road-1 and road-2, which are both located in the central position of the congestion cluster and have a series of neighborhood nodes. Even though road-3 locates at the marginal position of the cluster, it has strong similarities with its neighborhoods, e.g., road-6 and road-7.

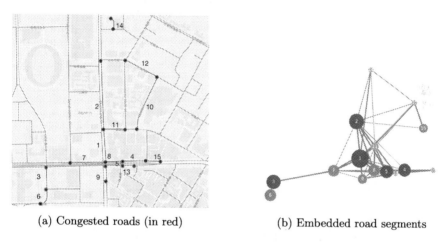

(a) Congested roads (in red)   (b) Embedded road segments

Fig. 5. A case study in Chengdu on weekday morning peak hour

Figure 6 shows the effectiveness comparison of Seg2Vec and Road2Vec. In Fig. 6a, the effectiveness of Seg2Vec for top-K critical roads is better than the ones of Road2Vec, especially for top-1 critical road. This can be explained by the sensitivity of Seg2Vec to adjacent neighbors. The effectiveness increases when K increases, since the more critical congested roads identified, the larger propagation influence on other roads. Figure 6b illustrates that the effectiveness is also affected by time delay δt. As Fig. 6b shows, the effectiveness of Seg2Vec increases when δt changes from 0 to 5 min, while the effectiveness of Road2Vec increases when δt becomes 15 min, which is caused by the time delay of congestion propagation. The result also suggests that the Top-1 critical congested road identified by Seg2Vec spreads more quickly than the one identified by Road2Vec.

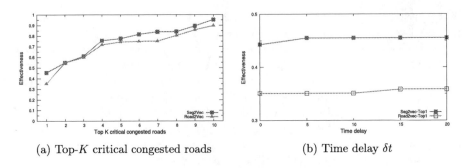

(a) Top-K critical congested roads　　(b) Time delay δt

Fig. 6. Effectiveness with varying K and δt

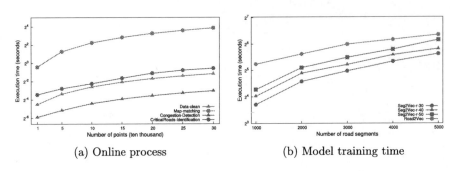

(a) Online process　　(b) Model training time

Fig. 7. Efficiency

Efficiency. We evaluate the efficiency of the online process, which consists of data cleaning, map matching, traffic congestion detection and critical road identification. As shown in Fig. 7a, the execution time of the online process is almost within 1 min with reasonable numbers of trajectory points during 10 min, such as 300 thousand. Therefore, the proposed critical congested road identification method is efficient and suitable for online monitoring scenarios. We also compare the training time of Seg2Vec and Road2Vec. In Fig. 7b, the training time of Seg2Vec is less than Road2Vec when the road segments increases from 1000 to 5000. The result indicates the scalability of our proposed model.

Discussions. The experimental results are summarized as follows:

1. Both of Seg2Vec model and Road2Vec model can capture spatial proximity of road segments, while the Seg2Vec model is more expressive to traffic similarities of adjacent and distant neighborhoods.
2. The effectiveness of Seg2Vec is better than Road2Vec model for top-K critical congested road identification. The effectiveness becomes better when time delay increases, which is because of the delay of congestion propagation.
3. For efficiency, the online process of our proposed method is adaptive to online traffic monitoring scenarios, and the training time of Seg2Vec is also scalable.

6   Related Work

Congestion Propagation : The causal interactions among road segments are explored in [11,15] using pattern mining approaches. Liu et al. [11] proposed a Spatio-Temporal Outlier (STO) method, which divides the road network into regions, and uses a recursive approach to find frequent traffic jam propagation trees in the road network. The disadvantage of the STO method is that the road network is oversimplified to regions and the results are probably imprecise. The Spatio-Temporal Congestion (STC) [15] process is further studied to develop the STO method. Instead of regional division, it models the road network as a directed graph. The STCTree algorithm generates the most frequent subtree from all discovered congestion trees. The limitation of frequent pattern mining is that the latent propagation patterns may be infrequent. Visualization methods that analyze traffic congestion propagation are also studied in [3,20]. Wang et al. [20] provided a system for traffic congestion visualization based on GPS trajectories. Deng et al. [3] proposed a visual analytics system called VisCas, which combines a network inference model with interactive visualizations to infer the latent cascading patterns. They model the congestion cascading network based on the spatial and temporal distance between congestion events. While this kind of hand-crafted features is easy to access and understood, they can hardly reserve the complex spatial and temporal correlations of urban data.

Road Network Embedding. In order to obtain comprehensive features of road networks, road network embedding is extended from graph embedding [21]. Graph embedding generally uses truncated random walks to learn network representations while preserving the network structures [5,16,18]. DeepWalk [16] simulates uniform random walks, which is analogical to a depth-first search (DFS). LINE [18] uses a breadth-first search (BFS), and optimizes a carefully designed objective function that preserves both the local and global network structures. Node2vec [5] proposed a flexibility random walk procedure which combines the DFS and BFS by search bias parameters. Node2vec was extended to learn road segment embeddings in [8], and a case study on the Danish road network was conducted. The results suggested that the network embedding model is able to capture structural information of road networks. To quantify the traffic interaction between road segments, Road2vec [10] learns the feature representation of road segments using travel routes based on the Word2Vec model [14]. Wang et al. [19] proposed RN2Vec to learn intersections and road segments jointly by exploring geo-locality and homogeneity of them. The learned feature vectors of road networks are used for downstream tasks such as road classification and traffic flow prediction [7,13].

7   Conclusion and Future Work

In this paper, we study the problem of identifying critical congested roads for traffic congestion management. We propose a road network embedding model, Seg2Vec, to obtain comprehensive road features that consider both the network structure and traffic flow distribution. We also define a score function to evaluate the propagation influence of congested roads based on the learned road representation. The effectiveness of Seg2Vec is verified by a case study comparing with Road2vec. The scalability of Seg2Vec and the efficiency of the online process are also demonstrated. This paper focuses on measuring the congestion propagation and discovering the critical cause in the local congestion cluster. In future work, we consider to extend the local critical road identification to global critical road identification. In addition, other downstream tasks such as road classification and congestion prediction are also considered.

References

1. Cai, H., Zheng, V.W., Chang, K.C.C.: A comprehensive survey of graph embedding: Problems, techniques, and applications. IEEE Trans. Knowl. Data Eng. **30**(9), 1616–1637 (2018)
2. Demšar, U., Špatenková, O., Virrantaus, K.: Identifying critical locations in a spatial network with graph theory. Trans. GIS **12**(1), 61–82 (2008)
3. Deng, Z., et al.: Visual cascade analytics of large-scale spatiotemporal data. IEEE Trans. Visual Comput. Graphics **28**(6), 2486–2499 (2021)
4. Gagniuc, P.A.: Markov chains: from theory to implementation and experimentation. John Wiley and Sons (2017)
5. Grover, A., Leskovec, J.: Node2vec: scalable feature learning for networks. In: Proceedings of the 22Nd ACM SIGKDD International Conference on Knowledge Discovery and Data Mining, pp. 855–864. KDD '16, ACM, New York, NY, USA (2016). https://doi.org/10.1145/2939672.2939754
6. Guo, S., et al.: Identifying the most influential roads based on traffic correlation networks. EPJ Data Sci. **8**(1), 1–17 (2019). https://doi.org/10.1140/epjds/s13688-019-0207-7
7. Hu, S., et al.: Urban function classification at road segment level using taxi trajectory data: a graph convolutional neural network approach. Comput. Environ. Urban Syst. **87**, 101619 (2021)
8. Jepsen, T.S., Jensen, C.S., Nielsen, T.D., Torp, K.: On network embedding for machine learning on road networks: a case study on the Danish road network. In: 2018 IEEE International Conference on Big Data (Big Data), pp. 3422–3431. IEEE (2018)
9. Li, D., et al.: Percolation transition in dynamical traffic network with evolving critical bottlenecks. Proc. Natl. Acad. Sci. **112**(3), 669–672 (2015)
10. Liu, K., Gao, S., Qiu, P., Liu, X., Yan, B., Lu, F.: Road2vec: measuring traffic interactions in urban road system from massive travel routes. ISPRS Int. J. Geo Inf. **6**(11), 321 (2017)
11. Liu, W., Zheng, Y., Chawla, S., Yuan, J., Xing, X.: Discovering spatio-temporal causal interactions in traffic data streams. In: Proceedings of the 17th ACM SIGKDD International Conference on Knowledge Discovery and Data Mining, pp. 1010–1018 (2011)

12. Lou, Y., Zhang, C., Zheng, Y., Xie, X., Wang, W., Huang, Y.: Map-matching for low-sampling-rate gps trajectories. In: Proceedings of the 17th ACM SIGSPATIAL International Conference on Advances in Geographic Information Systems, pp. 352–361 (2009)
13. Medina-Salgado, B., Sanchez-DelaCruz, E., Pozos-Parra, P., Sierra, J.E.: Urban traffic flow prediction techniques: a review. Sustain. Comput.: Inform. Syst. **35** 100739 (2022)
14. Mikolov, T., Chen, K., Corrado, G., Dean, J.: Efficient estimation of word representations in vector space. CoRR abs/1301.3781 (2013)
15. Nguyen, H., Liu, W., Chen, F.: Discovering congestion propagation patterns in spatio-temporal traffic data. IEEE Trans. Big Data **3**(2), 169–180 (2016)
16. Perozzi, B., Al-Rfou, R., Skiena, S.: Deepwalk: Online learning of social representations. In: Proceedings of the 20th ACM SIGKDD International Conference on Knowledge Discovery and Data Mining, pp. 701–710. KDD '14, ACM, New York, USA (2014). https://doi.org/10.1145/2623330.2623732
17. Rumelhart, D.E., Hinton, G.E., Williams, R.J.: Learning representations by back-propagating errors. Nature **323**, 533–536 (1986)
18. Tang, J., Qu, M., Wang, M., Zhang, M., Yan, J., Mei, Q.: Line: Large-scale information network embedding. In: Proceedings of the 24th International Conference on World Wide Web, pp. 1067–1077 (2015)
19. Wang, M.X., Lee, W.C., Fu, T.Y., Yu, G.: On representation learning for road networks. ACM Trans. Intell. Syst. Technol. (TIST) **12**(1), 1–27 (2020)
20. Wang, Z., Lu, M., Yuan, X., Zhang, J., Van De Wetering, H.: Visual traffic jam analysis based on trajectory data. IEEE Trans. Visual Comput. Graphics **19**(12), 2159–2168 (2013). https://doi.org/10.1109/TVCG.2013.228
21. Xia, F., et al.: Graph learning: a survey. IEEE Trans. Artif. Intell. **2**(2), 109–127 (2021)
22. Zeng, G., et al.: Switch between critical percolation modes in city traffic dynamics. Proc. Natl. Acad. Sci. **116**(1), 23–28 (2019)

A Cross-Region-based Framework for Supporting Car-Sharing

Rui Zhu[1(✉)], Xuexin Zhang[1], Xin Wang[2], Jiajia Li[1], Anzhen Zhang[1], and Chuanyu Zong[1]

[1] Shenyang Aerospace University, Shenyang, China
{zhurui,lijiajia,azzhang,zongcy}@sau.edu.cn
[2] Shenyang Aircraft Corporation, Shenyang, China

Abstract. With the rapid development of mobile Internet and sharing economy, carsharing has attracted a lot of attention around the globe. Many popular taxi-calling service platforms, such as DiDi and Uber, have provided carsharing service to the passengers. Such carpooling service reduces the energy consumption while meeting passengers' convenience and economic benefits. Although numbers of algorithms have been proposed to support carsharing, the computing efficiency and matching quality of these existing algorithms are all sensitive to the distribution of passengers. In many cases, they cannot effectively and efficiently support carsharing in an *on-line* way. Motivated from the aforementioned issues and challenges, in this paper, we propose a novel framework, namely, Cross-Region-based Task Matching (CRTM) for supporting carsharing for smart city. Compared with existing algorithms, CRTM analyzes and monitors regions having multitudes of tasks for car sharing among users. In order to achieve this goal, we first propose a new *machine learning*-based algorithm to find a group of regions which contain many tasks. Then, we propose a novel index, namely, Included Angle Partition-based B-tree (IAPB), for maintaining tasks such as (i)whose pick-up points are contained in these regions, (ii) that may pass this kind of regions. Thirdly, we propose three buffer-based matching algorithms for cross-region-based task matching. Experiment results demonstrate the significant superior performance of the proposed algorithms in terms of energy saving and overall cost minimization.

Keywords: Carsharing · Self-Adaptively Matching · Index · Task Pair

1 Introduction

With the rapid development of Mobile Internet and sharing economy, carsharing has attracted a lot of attention around the globe. At the same time, it generates economic benefits to drivers, passengers and taxi-calling service platforms. Compared with traditional service patterns that allow a taxi service to a single passenger, a carpool taxi can provide service to more than one unrelated passengers at the same time. Take an example in Fig. 1(a). Because the

© The Author(s), under exclusive license to Springer Nature Switzerland AG 2023
X. Yang et al. (Eds.): ADMA 2023, LNAI 14176, pp. 614–629, 2023.
https://doi.org/10.1007/978-3-031-46661-8_41

pick-up/drop-off points of passengers u_1 and u_2 are roughly the same, the service platform can assign u_1 and u_2 to a same taxi. The corresponding path is planned as $p(u_1.s \rightarrow u_2.s \rightarrow u_2.d \rightarrow u_1.d)$. Here, $u_1.s(u_1.d)$ refers to the pick-up(drop-off) point of u_1. Compared with the shortest paths corresponding to u_1 and u_2, the distance difference is small. However, since traveling expenses under sharing mode is much lower, it reduces the self-driving demand of users. Also, the energy consumption can be effectively reduced.

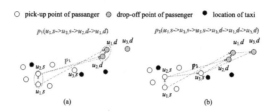

Fig. 1. The Running Example of Car-Sharing

Due to its importance in wide range of applications, many researchers have studied this problem. One type of algorithms use *group nearest neighbour searching algorithm* to support carsharing. Specially, once a passenger u submits a request to the service platform, the platform finds suitable passengers for u via searching its nearest neighbour based on its pick-up and drop-off points. However, in this technique, one problem is some of the proper passengers may be missed. Back to the example in Fig. 1(b). When u_1 travels from $u_1.s$ to $u_1.d$, u_1 may pass the pick-up point of u_3. Obviously, u_1, u_2 and u_3 are suitable for sharing the same taxi. Unfortunately, because the pick-up point of u_1 is far from that of u_3, the system cannot assign the same taxi to u_1, u_2 and u_3.

Another type of algorithms applies *machine learning* for finding high-quality paths in a city [1]. Let C be a city, G_C be the road network corresponding to C. For every two vertexes r and s in G_C, a high-quality path p_{rs} is formed. Here, we call a path p_{rs} as a high-quality path if the number of passengers around p_{rs} is no less than the number of taxis around p_{rs}. Once high-quality paths in a city are learned, the system can provide taxis with high-quality paths so as to guarantee that they can service for, as much as possible, passengers at the same time. However, this type of algorithms have to maintain all high-quality paths. The space cost is high. Also, because the distribution of taxis and passengers are frequently changed, the high-quality paths between every two points is also frequently changed [12,13]. Under such cases, the algorithm has to re-construct the high-quality path set, in which the computational cost is also very high.

Based on the above observations, in this paper, we propose an efficient framework, namely, (**CRTM** Cross-Region-based Task Matching) for supporting *tasks matching*. For simplicity, taxis are denoted as workers, and passengers are denoted as tasks. In a nutshell, let S be the region corresponding to a city C. We partition S into a group of sub-regions. In each sub-region, we evaluate whether it is a b-region. Here, we call a sub-region as a b-region(busy-region in short) if the number of tasks in it is much larger than that of workers. For each b-region, we construct an index to maintain tasks which pass it. In this way, when a passenger u located at a b-region submits a request to the system, the system matches u with a *cross region* task based on their pick-up/drop-off points relationship(u_1 and u_3 in Fig. 1(b)). Compared with the first type of methods, we can find suitable taxis for tasks in a *cross-region* way. Accordingly, the matching quality/amount all can be enhanced a lot. Compared with the second type of algorithms, we need not to maintain all high-quality paths. However, CRTM should overcome the following challenges.

Challenges. First of all, many factors should be considered for evaluating which regions could be considered as b-regions. It is difficult to use score functions and other traditional methods to find b-regions. Secondly, when matching, the system has to consider tasks with pick-up/drop-off points located at different regions. It is also difficult to propose an effective index for supporting the task matching in a cross-region way. Thirdly, since the matching is executed under an *on-line* way, it is difficult to guarantee the matching quality.

Contributions. To deal with the above challenges, the contributions of this paper are as follows.

- The b-region Selection Algorithm: We use an efficiently machine learning algorithm named (ELM Extreme Learning Machine) [7] for finding b-regions. Firstly, we compute the distribution summary of tasks/workers in each sub-region. Based on the computation result, we construct feature vectors, and use ELM-based classifier for identifying b-regions.
- The Index IAPB: It is a two-level index. The first level index mainly maintains position relationship among tasks' drop-off points based on their *relative polar angles*. The second level index mainly maintains tasks' drop-off points based on the their *relative polar radius*. Compared with traditional indexes, IAPB has the ability to compute angle distances among tasks efficiently. Clearly, if the relative polar angle distance between two tasks is small, even the pick-up/-off point distance between them is large. Therefore, IAPB also can match them together. The concept of relative polar angle is explained in Sect. 3.1.
- Buffer-based Self-adaptive Matching Algorithms: In order to guarantee the matching quality, we propose one baseline algorithm, and two optimization algorithms. The key idea is to store a small number of tasks in a buffer, and periodically execute the matching algorithm via using IAPB. If there exists a high-quality matching based on the elements in the buffer, we use the matching result for assignment. Otherwise, we properly enlarge the buffer, and execute the matching until the matching quality is high enough.

The rest of this paper is organized as follows. Section 2 reviews the related work and presents the problem definition. Section 3 discusses the CRTM framework. Section 3.1 discusses the b-region management. Section 3.2 discusses the matching algorithms. Section 4 shows the experimental results. Finally, Sect. 5 concludes the paper.

2 Preliminaries and Definitions

In this section, we first review the algorithms about the problem of spatio-temporal crowdsourcing [3] and ride-sharing [2]. Next, we discuss the problem definition.

2.1 Related Work

Spatio-Temporal Crowdsourcing. Spatio-temporal crowdsourcing becomes important in many smart city applications. It mainly consists of tasks and workers. Tasks and workers first send their information to the platform, then the platform assigns suitable workers to tasks according to their position information.

Task consists of *static offline* and *dynamic online* tasks. Kazemi et al. [10] study the problem of static offline task matching over spatial-temporal crowdsourcing scenarios. They use bipartite graph to model the matching problem. Accordingly, this problem is reduced to the problem of maximum bipartite graph matching. However, their proposed algorithms cannot effectively work under *dynamic online* scenarios.

Compared with *static offline* matching, dynamic online matching allows tasks and workers appear in the platform dynamically. Tong et al. [3] solve this problem by introducing a novel matching model. In addition, they propose greedy algorithms to solve this problem, and carefully analyze the worst case and average case of their proposed algorithms. They find that these algorithms can efficiently and effectively work in most cases. Moreover, they propose a two-step framework. Compared with other efforts, it uses the advantages of *offline prediction* and *online task assignment*. They prove that their proposed algorithm achieves a 0.47 competitive ratio.

Ride-Sharing. This kind of problems is similar with task assignment, which also includes *dynamic sharing* and *static sharing*. According to the survey of Furuhata et al. [4], many algorithms focus on the problem of static ride-sharing. Santi et al. [5] solve this problem via evaluating the potential of ride-sharing based on a graph-based approach. In addition, they prove that ride-sharing problem is NP-hard.

For the problem of dynamic sharing, the bottleneck is to reduce the highly computation cost. The reason behind it is tasks and workers are timely changed. If the system has to spend long time in finding suitable workers, the requirement of real-time cannot be satisfied. Ma et al. propose a novel index for searching suitable workers. Huang et al. [7] propose another index to efficiently answer trip request. Asghari et al. [5] propose the framework APART to solve this problem. The key behind it is that workers and tasks are considered as bidders and goods respectively. When a task t is submitted to the system, the system evaluates which workers could be assigned to t according to the self-defined profiles of riders and passengers. Wang et al. [1] propose an efficient framework named PPVF. Compared with other methods, their key idea is to find a group of high-quality paths from historical transaction record set. Here, a path is regarded as a high-quality path if there are many tasks around it. In this way, they can use these paths for planning. The benefit is since these high-quality paths are crowed with tasks, workers could provide service for as much as possible tasks.

2.2   Problem Definition

Definition 1. *(*TASK*). A task t, denoted by the tuple $\langle u, s, d, p, dl \rangle$, is created when a passenger u submits a requirement to the platform with a pick-up point s and a drop-off point d. Once a task t is generated, the platform assigns a worker w to t, and plans a path p for t so as to transport u from s to d. In addition, dl refers to the deadline of t. In other words, after the task t is generated, a worker must arrive at the pick-up point of t before dl. Otherwise, t is automatically destructed.*

Definition 2. *(*WORKER*). Let w be worker. It is expressed by the tuple $\langle pos, T, c \rangle$. Here, pos refers to the location of w, T refers to a group of tasks that are assigned to w at the same time, and c refers to the maximal number of tasks it can execute at the same time.*

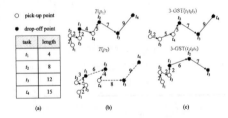

Fig. 2. The Problem Definition($\delta = 1.5$)

Note that, under car-sharing mode, since a group of tasks may be assigned to the same worker, this worker has to pass the pick-up/drop-off points of all tasks, we should evaluate whether these tasks could be matched together. In the following, we first introduce the concept of *group shortest path* and *conditional shortest path* based on the example as stated below. Next, we explain the evaluation method.

Take an example in Fig. 2(a)-(b). There are four tasks, which are $T\{t_1, t_2, t_3, t_4\}$. The length of their corresponding shortest paths are $\{4, 8, 12, 15\}$. $T(p_1)$ and $T(p_2)$ are two paths that pass the drop-off and pick-up points of these tasks. The length of $T(p_1)$, i.e., denoted as $|T(p_1)|$, equals to 31. $|T(p_2)|$ equals to 34. Among all paths that pass the drop-off and pick-up points of tasks in T, $|T(p_1)|$ is the shortest of all. Thus, $T(p_1)$ is the group shortest path of T, i.e., denoted as $\mathsf{GSP}(T)$. The path $p(t_1.s \rightarrow t_2.s \rightarrow t_3.s \rightarrow t_1.d)$ is a sub-path of $\mathsf{GSP}(T)$. It is the conditional shortest path of t_1, i.e., denoted as $\mathsf{CSP}(t_1, T)$.

Let T be a set of tasks serviced by the worker w. We can evaluate whether tasks in T could be serviced by w based on the following 3 facets. Firstly, we use $\mathsf{DR}(t_i, T)$ to evaluate the *detour ratio* of each task $t_i \in T$. It is computed based on Eq. 1. Intuitively, the lower the *detour ratio*, the higher quality service w can provide. We should guarantee that the *detour ratio* of each task should be smaller than a threshold δ. Secondly, the final target of carsharing is reducing the totally travel distance. Thus, we should guarantee that $|\mathsf{GSP}(T)|$ is less than $\sum_{i=1}^{i=|T|} |p(t_i.s \rightarrow t_i.d)|$ with $|T|$ being the number of tasks contained in T. Thirdly, assuming the pick-up order of tasks in T is $\{t_1, t_2, \ldots, t_{|T|}\}$, after w arrives at the pick-up point of t_i, t_i is *reachable* to t_{i+1} if w can arrive at the pick-up point of t_{i+1} before $t_{i+1}.dl$.

We formally explain the concept of *sharable task set*, and lastly explain the problem definition. For simplicity, if a sharable task set contains m tasks, we call it as a m-STS.

$$\mathsf{DR}(t_i, T) = \frac{|\mathsf{CSP}(t_i, T)|}{|p(t_i.s \rightarrow t_i.d)|} \tag{1}$$

Definition 3. *(SHARABLE TASK SET). Given a set of tasks T serviced by the worker w, we call T as a* sharable task set *if*

- $\forall i$, $NR(t_i, T) \leq \delta$
- $\sum_{i=1}^{i=|T|} |p(t_i.s \rightarrow t_d.s)| \leq |GSP(T)|$
- $\forall i$, t_i is reachable to t_{i+1}.

Take an example in Fig. 3(c). Let δ be 1.5. The length of $\mathsf{CSP}(t_1, t_1 \cup t_2 \cup t_3)$, $\mathsf{CSP}(t_2, t_1 \cup t_2 \cup t_3)$, and $\mathsf{CSP}(t_3, t_1 \cup t_2 \cup t_3)$ are 6, 10 and 15 respectively. $\mathsf{NR}(t_1, t_1 \cup t_2 \cup t_3)$, $\mathsf{NR}(t_2, t_1 \cup t_2 \cup t_3)$, $\mathsf{NR}(t_3, t_1 \cup t_2 \cup t_3)$ are $\frac{6}{5}$, $\frac{10}{8}$, and $\frac{15}{12}$ respectively. They are all smaller than 1.5. Thus, $t_1 \cup t_2 \cup t_3$ could be regarded as a 3-STS. Similarity, $t_2 \cup t_3 \cup t_4$ is another 3-STS. However, $\mathsf{NR}(t_1, t_1 \cup t_2 \cup t_3 \cup t_4) \geq$ 1.5. $t_1 \cup t_2 \cup t_3 \cup t_4$ is not a 4-STS.

Definition 4. *(CCRS). Let T be a set of tasks, W be a set of workers. They are partitioned into $\{T_1, T_2, \ldots, T_m\}$ based on their assigned workers, i.e., tasks in $T_i \subset T$ are assigned to the worker $w_i \in W$. The CCRS (short for Conditional based Cross Region Sharing) problem is maximizing $\sum_{i=1}^{i=m} SD(T_i)$ in the premise that each task set T_i is a sharable task set. Here, $SD(T_i)$ is computed based on Eq. 2. It is used for evaluating the benefit of task matching.*

$$SD(T_i) = \sum_{j=1}^{j=|T_i|} |p(t_i.s \rightarrow t_d.s)| - |GSP(T_i)| \tag{2}$$

3   Cross-Region Based Task Matching Framework

In this paper, we propose an efficient framework, named CRTM for supporting CCRS. CRTM is mainly used for assigning tasks to suitable workers based on their position. It maintains two types of regions, which are b-regions and f-regions. Let r be a region with size $s \times s$. If r contains many *unassigned tasks*, but few empty workers, we call r as a b-region. Otherwise, we call it as a f-region. Accordingly, we call a task t as a f-task if the pick-up point of t is located at a f-region, and will pass b-regions, call a task t' as a b-task, if the pick-up point of t' is located at a b-region. In the following, we first discuss the b-region.

3.1   The B-Region Management

In this section, we first discuss the b-region selection algorithm. Next, we explain how to maintain tasks in each b-region.

The B-Region Selection Algorithm. We assume that a city is bounded by a square S. We first use a grid file $\mathcal{G}$ to partition S into a group of cells with side-length b, and then find b-regions based on these cells.

In this section, we use an efficiently *machine learning* algorithm named ELM [9] for identifying b-regions. The reason we introduce *machine learning algorithm* is that many factors should be considered for evaluating which regions could be regarded as b-regions, leading that it is difficult for us to use score functions and other traditional methods to find b-regions. Among all *machine learning algorithms*, ELM is an efficient one. It is based on a generalized single-hidden-layer feed. Compared with neural networks, its hidden-layer nodes are randomly chosen instead of iteratively tuned. The benefit of ELM is it could provide us with good generalization performance at thousands of times faster than traditional learning algorithms.

In order to make ELM effectively work, we first study the features of b-regions. Via deeply studying, we find that *tasks/workers amount, workers' average speed* and *empty worker radio* could be used for identifying b-regions. Specially, for each cell $r \in G$, we use the tuple $\langle r_{tw}, r_s, r_e \rangle$ as the feature vector of c_i. The first element, i.e., denoted as r_{tw}, describes the tasks/workers amount relationship in each region. It equals to $\frac{r_w}{r_t}$. Here, r_w (and r_t) refers to the number of workers (and tasks) contained in the cell r. Intuitively, if $\frac{r_w}{r_t}$ is relatively large, r may be a f-region with high probability. Otherwise, it may be a b-region. Take an example in Fig. 4(a). These two shadow regions are b-regions. They all contain 4 tasks, but each of them only contains one worker. Therefore, these two regions could be used as b-regions.

The second element, i.e., denoted as r_s, describes the average speed of all workers contained in r. If the average speed it high, it means this region contains a small number of vehicles, and it is unlikely to be a b-region. Otherwise, this region may be crowed with many vehicles, and this region may be a b-region. The third element, i.e., denoted as r_e, describes the empty radio of workers contained in r. Intuitively, the lower the r_e, the more chance a task can be assigned to an empty worker, and the higher probability of r being a b-region.

After discussing the feature construction, we then discuss the classifier construction algorithm. Let $\mathcal{H}$ be a set of historical records. We partition $\mathcal{H}$ into a group of subsets $\{H_1, H_2, \ldots, H_{|G|}\}$ based on records' position, i.e., records in H_i are contained in the cell c_i. Next, we construct a group of subsets $\{TH_1, TH_2, \ldots, TH_{|G|}\}$ based on $\{H_1, H_2, \ldots, H_{|G|}\}$, use these subsets as training sets,i.e, $\frac{|H_i|}{3} = |H_i|$. Then, we form a group of feature vectors, and use these vectors for training. After training, we construct feature vectors based on other subsets as the manner discussed above, input these feature vectors into the classifier for classifying.

The Index IAPB In this section, we propose a novel index named (IAPB Included Angle Partition-based B-tree) to maintain b-tasks in each b-region.

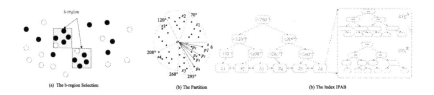

(a) The b-region Selection　　　(b) The Partition　　　(b) The Index IPAB

Fig. 3. The B-region Indexing

The IAPB Construction. We now formally explain the IAPB construction. It is a two-level index. The first level index mainly maintains position relationship among b-tasks' drop-off points based on their *relative polar angles*. The second level index mainly maintains b-tasks' drop-off points based on the their *relative polar radius*. Here, let A, B be the drop-off points of tasks t_A and t_B. O be the center point of the b-region b, V be a virtual point in the horizontal direction. We use V and O as *angle reference point* and *reference origin*. The relative polar angle of t_A equals to $\angle AOV$, and the relative polar radius of t_A equals to $|OA|$(See Fig. 4(b)).

Specially, let C be a city in $[0, 1]^2$ space, b be a b-region with center point O, T be a set of b-tasks contained in b, and V be a virtual point in the horizontal direction. We first compute the *relative polar angle* of all tasks in T. Next, we partition tasks in T into a group of subsets $\{T_1, T_2, \ldots, T_m\}$ based on their relative polar angles. The partition result should satisfy that:

- $|T_i| = \sqrt{|T|}$
- given two task $t_i \in T_i$ and $t_j \in T_j$ with drop-off point A and B respectively, $\angle VOA \leq \angle VOB$.

After partitioning, we use a group of buckets $\{B_1, B_2, \ldots, B_m\}$ to maintain tasks in $\{T_1, T_2, \ldots, T_m\}$. In addition, we use a B$^+$-tree structure to maintain these buckets. Lastly, we use another two B^+-trees to maintain tasks in each bucket respectively (See Fig. 4(c)). We use an example to explain the index IAPB. There are 36 b-tasks contained in b. We partition it into 6 ($=\sqrt{36}$) subsets, i.e., $\{T_1, T_2, \ldots, T_6\}$. Next, we use a B^+-Tree, i.e., the first-level index, to maintain these six subsets based on their relative polar angle lower-bounds. For example, relative polar angles lower-bound of tasks in T_1 is $0°$. Thus, the value of the leaf-node corresponding to T_1 is 0. The relative polar angles of tasks in T_4 is $129°$, the value the leaf-node corresponding to T_4 is 129. Thirdly, we use another two B^+-Trees, i.e., the second-level index, to maintain tasks in each T_i. In Fig. 4(c), the B^+-Tree BT_6^R maintains tasks in T_6 based on tasks' relative polar radius. The B^+-Tree BT_6^A maintains tasks in T_6 based on tasks' relative polar angle.

The IAPB Maintenance. When a task t located at a b-region is generated, if it passes b, we insert t into the index I_b corresponding to b. Next, we find the bucket it should be inserted into via computing its relative polar angle. Assuming it is inserted into the bucket B_i, we insert t into the B^+-Trees B_i^A and B_i^R respectively based on its relative polar angle and relative polar radius respectively. After insertion, if $|B_i| \geq 2\sqrt{|T|}$, we should averagely split the subset T_i into two subset T_i^1 and T_i^2 with size $\sqrt{|T|}$ respectively. Finally, we update the index IAPB based on the splitting result.

We periodically delete meaningless b-tasks. Here, we call a task t' as a meaningless b-task if no worker is assigned to it until $t'.dl$. Here, $t'.dl$ refers to the deadline of t'. Specially, if a task s' turns to meaningless, we first remove it from the index IAPB. After removing, if $|B_i|$ reduces to $\sqrt{|T|}$ and $|B_i| + |B_{i-1}| \leq 2\sqrt{|T|}$, we merge B_i into B_{i-1}. If $|B_i|$ reduces to $\sqrt{|T|}$ and $|B_i| + |B_{i-1}| \geq 2\sqrt{|T|}$, we generate the new B_i and B_{i-1} based on the elements in $B_i \cup B_{i-1}$. Since the algorithm is simple, we skip the details for saving space.

3.2   The Matching Algorithms

First of this section, we propose a novel algorithm named (STPM Sharable Task Pair based self-adaptively Matching), which supports task matching when the capacity of workers equals to 2. Next, we propose an optimization algorithm of STSM, i.e., called as (OSTPM Optimization Sharable Task Pair based self-adaptively Matching).

The Algorithm STPM. Before explaining the algorithm, we first introduce the concept of *Sharable Task Pair*. For simplicity, given two tasks t_1 and t_2, if t_1 and t_2 could construct a sharable task pair, t_1 is *sharable* with t_2. In addition, if t_1 and t_2 are *assigned* to the common worker at the same time, t_1 is *matched* to t_2.

Definition 5. *(SHARABLE TASK PAIR). Let t_f be a task that passes the b-region b, t_b be a pair whose pick-up point is located at b. t_f and t_b are regarded as a sharable task pair if :*

- $|p(t_b.s \rightarrow t_b.d)| + |p(t_f.s \rightarrow t_f.d)| \geq$ GSP$(t_b \cup t_f)$
- NR$(t_b, t_b \cup t_f) \leq \delta$ *and* NR$(t_f, t_b \cup t_f) \leq \delta$;
- t_f *is reachable to t_b.*

After explaining the concept of Sharable Task Pair, we now formally discuss the matching algorithm. It contains two steps, which are *sharable task pair set construction* and *buffer-based self-adaptive matching*. For simplicity, T_f refers to a set of cross-region tasks that pass b, and T_b refers to a set of tasks whose pick-up points are located at b. The first step is to find all sharable task pairs from T_f and T_b. The second step is to find the optimal matching based on these sharable task pairs.

The Sharable Task Pair Set Construction. This step contains two steps, which are *filtering* and *verification*. In the filtering phase, for each task $t_f^i \in T_f$, we filter tasks in T_f which are not sharable with t_b^i based on the their drop-off points. In other words, we assume that tasks in T_b have the same pick-up points with elements in T_f, and we only use drop-off points of tasks for filtering. In the verification step, we evaluate whether the unfiltered task pairs are really sharable task pairs.

In this section, we use Theorem 1 for filtering task pairs that do not satisfy the constrained condition (I) of *sharable task pair*. Specially, let t_f and t_b be two tasks contained in T_f and T_b respectively, the point A be their common pick-up point, C(or B) be the drop-off point of t_f(or t_b). If $|AB| \geq |BC|$, $\mathsf{GSP}(t_b \cup t_f)$ is "$A \rightarrow C \rightarrow B$". Based on the constrained condition (I) of *sharable task pair*, $|AC| + |BC|$ should be smaller than $|AC| + |AB|$. Thus, $|AB|$ should be smaller than $|BC|$. It implies we can use $|AC|$ and $\angle BAC$ to compute a threshold, i.e., $\frac{|AC|}{2cos\angle BAC}$. In other words, if $|AB|$ is larger than $\frac{|AC|}{2cos\angle BAC}$, t_f and t_b could not be regarded as a candidate sharable task pair. Similarity, we use Theorem 2 for filtering task pairs that do not satisfy the constrained condition (II) of *sharable task pair*. Specially, if $|AB| \leq \frac{|AC|(\delta^2-1)}{2(\delta-cos\theta)}$, we have $|AB|+|BC| \geq \delta|AC|$. It implies we also can use $|AC|$ and $\angle BAC$ to compute another threshold. In other words, if $|AB|$ is larger $\frac{|AC|(\delta^2-1)}{2(\delta-cos\angle BAC)}$, t_f and t_b could not be regarded as a candidate sharable task pair.

Theorem 1. *Let $\triangle ABC$ be a triangle in $2-$dimensional space. The side-length of AC equals to $|AC|$. If $|BC| \leq \frac{|AC|}{2cos\angle BAC}$, we have $|BC| \leq |AB|$.*

Theorem 2. *Let $\triangle ABC$ be a triangle in $2-$dimensional space. The side-length of AC equals to $|AC|$. If $|AB| \geq \frac{|AC|(\delta^2-1)}{2(\delta-cos\angle BAC)}$, we have $|AB| + |BC| \leq \delta|AC|$.*

Take an example in Fig. 5(a). Let t_b be a f-task with relative polar radius $|AB|$. t_f be a b-task. If its relative polar radius is smaller than $|AC_4|$, the constrained condition (I) of *sharable task pair* is satisfied. If its relative polar radius is smaller than $|AC_2|$, the constrained condition (II) of *sharable task pair* is satisfied. Since $|AC_4| \geq |AC_2|$, the filtering threshold is set to $|AC_2|$.

We now formally explain the sharable task pair set construction algorithm. Let T_b be a set of tasks whose pick-up points are contained in the b-region b. T_f be a set of tasks that will pass b, and I be the index IAPB that maintains tasks in T_b. For each task $t_f \in T_f$, we first access the index IAPB I to find the bucket B_i satisfying $l(B_i) \leq a(t_f) \leq u(B_i)$. Here,$l(B_i)$(or $u(B_i)$) refers to the relative polar angle lower-bound(or upper-bound) of tasks in B_i, $a(t_f)$ refers to the relative polar angle of t_f. Next, we scan tasks contained in B_i, evaluate whether these tasks are sharable with t_f based on the definition of sharable task pair set. Thirdly, we further access other buckets. In the following, we use B_{i+1} as the example to show how to access other buckets.

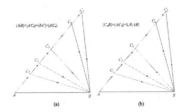

Fig. 4. The Filtering under STPM

Specially, let θ be $u(B_i) - a(t_f)$, we use $r(t_f)$ and θ to compute a threshold η_i via Theorem 1-2. In this way, $\forall t \in B_{i+1}$, if its relative polar radius is larger than this threshold, it could be filtered. For the ones that cannot be filtered by η_i, we use the definition of sharable task pair set for verification. From then on, we repeat the above operations to access other buckets based on t_f. After processing t_f, we use the same method to construct sharable task pairs for the other tasks in T_f. For the limitation of space, we skip the details for saving space.

The Buffer-Based Self-Adaptively Matching Algorithm. Once the sharable task pair set TP is formed, we construct a *bipartite graph* BG based on the pairs in TP. Here, BG has two sub-graphs G_f and G_b. Vertexes in G_b are b-tasks. Vertexes in G_b are f-tasks. Given two vertexes in G_f and G_b, there exists an edge e between them if the corresponding tasks could form a sharable task pair. The weight of e, i.e., denoted as $w(e)$, is computed based on Eq. 2. After constructing BG, we execute *bipartite graph matching* algorithm for finding the current optimal matching.

Next, we evaluate the matching quality. Our goal is to use *rank sum test* for evaluating whether the current matching results and historical off-line matching results obey the same distribution(or having a higher matching quality). Specially, let $\mathcal{P}\{p_1, p_2, \ldots, p_q\}$ be a set of matched task pairs with weights $\{w(p_1), w(p_2), \ldots, w(p_q)\}$, $N(\mu, \sigma^2)$ be the normal distribution function fitted by the historical records. We randomly create q points based on $N(\mu, \sigma^2)$. Next, we use *rank sum test* for evaluating. If we find that the current matching results and historical off-line matching results obey the same distribution(or having a higher matching quality), we assign non-empty workers to tasks in T_f based on the matching result. Otherwise, we only retain task pairs in TP that may turn to meaningless after TM moment, and ignore the other task pairs. Here, TM

refers to the matching algorithm executing period. Lastly, we assign b-tasks to suitable *non-empty workers* based on the retained task pairs in TP.

4   Experimental Evaluation

In this section, we conduct extensive experiments to demonstrate the efficiency of CRTM. The experiments are based on one real data set and one synthetic dataset respectively. In this section, we first explain the settings of our experiments, and then report our findings.

4.1   Experimental Setting

Data Set. In total, two datasets are used in our experiments, one real data set namely DiDi, one synthetic datasets namely Syn-H. DiDi is the taxi-calling data from July 2016 to December 2016 in Beijing. This data set contains 50000 drivers and 500,000 tasks. Each task in the dataset is expressed by the tuple $\langle s, d, r \rangle$. Here, $t.s$ is the pick-up point of t, $t.d$ is the drop-off point of t, and $t.r$ is the request time of t. The data set Syn-H is a synthetic datasets. Tasks/workers in Syn-H obey uniform distribution.

Comparisons. we compare the results of CRTM with two other existing approaches: PPAF and APART. Here, PPAF is proposed by Wang et al. APART is another efficiently algorithm under car-sharing. As we propose three algorithms, OSTSM is more efficient, we use this algorithm for evaluation.

Metrics. In our experiments, we measure the following metrics by varying different parameters of the system, which are *response time*, *empty ratio*, *task weight*, and the *space cost*. Here, *response time* refers to the average time that tasks are assigned to workers after these tasks are submitted to the system. *empty ratio* equals to $\frac{at}{nat+at}$. Here, at refers to the number of assigned tasks, nat refers to the number of unassigned tasks. *space cost* refers to average memory we consume. *sharable task set weight* evaluates the benefit of car-sharing. Obviously, the higher the sharable task set weight, the more benefit we can obtain.

Parameters. We evaluate algorithms differences via the following four parameters. The first parameter is m. It refers to the number of b-regions we should monitor. The second parameter is *vacancy rate* that equals to $\frac{|T|}{|W|}$. Here, $|T|$ and $|W|$ refers to the number of tasks and workers in the system. The third parameter, denoted as W_C, is the capacity of a worker. The last parameter is the number of tasks that are submitted to the system per minute. Table 1 shows the parameter setting, default values are bolded. In implementation, we use 20% tasks in each data set for initiation. After initiation, we submit other tasks to the system based on their submitted time. All the algorithms are implemented with C++, and all the experiments are conducted on a CPU i5 with 16GB memory, running Microsoft Windows 10.

Table 1. Parameter Settings

Parameter	value
Worker Capacity	2, 3, **4**, 5, 6 ,8
Tasks/Workers	2, 3, **4**, 7, 10
Passengers Amount	2000, **4000**, 10000, 20000

4.2   Algorithm Performance

In this section, we compare the response time of our framework named CRTM with two other approaches: APART and PPVF. For the impact of *tasks/workers* to the response time, as is depicted in Fig. 6.(a)(b), CRTM performs the best of all. The reason behind it is, compared with APART and PPVF, CRTM sends tasks to all b-regions that they pass. When $\frac{|T|}{|W|}$ is large enough, CRTM has the ability of providing workers for most part of tasks. It proves that our proposed algorithm could send tasks to suitable b-regions in most cases.

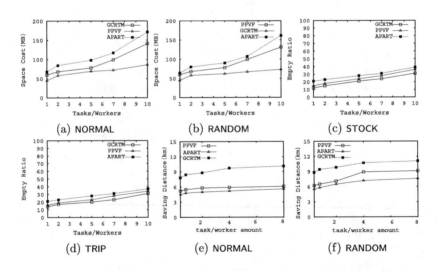

Fig. 5. Comparison of different algorithms under Other Metric.

For the impact of *worker capacity* to the response time, as is depicted in Fig. 6.(c)(d), when the capacity of workers is small, APART performs better than that of CRTM. However, when CRTM is 4, then the response time difference between CRTM and PPVF is small. The reason behind it is that when worker capacity is large enough, PPVF also can provide enough workers with tasks. However, it shows that our proposed framework is not sensitive to the worker capacity. For the impact of *task submitted speed* to the response time, as is depicted in Fig. 6.(e)(f), when the task submitted radio is not large enough, the

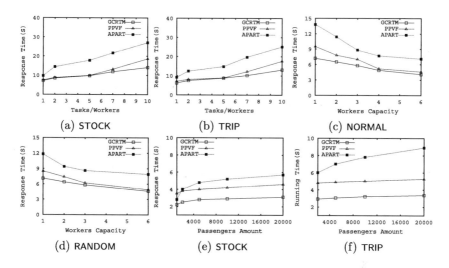

Fig. 6. Running time comparison of different algorithms under different data sets.

response time of CRTM and APART are roughly the same. When *task submitted radio* is much larger than the number of workers in the system, the response time of APART becomes much shorter than that of CRTM. However, we want to highlight that, in most cases, *task submitted radio* may not become so high that the system can not find enough workers for tasks. Next, we compare the space cost of CRTM with two other approaches: APART and PPVF. For the impact of *task/worker* to the space cost, as is depicted in Fig. 5(a)(b), CRTM performs the best of all. The reason behind it is CRTM only selects a small number of regions as b-region. In each b-region, the space cost of the corresponding index is linear to the task scale. Thus, the space cost of CRTM is not high.

Thirdly, we compare the *empty ratio* of CRTM with APART and PPVF. As is depicted in Fig. 5.(c)(d) CRTM performs the best of all. The reason behind it is that, compared with other algorithms, CRTM sends tasks to all b-regions. Thus, when *task/worker* is large, the empty ratio of CRTM is the lowest of all. For the impact of *worker capacity* and *task submitted radio*, CRTM also performs the best of all. Last, we compare the *saving distance* of CRTM with APART and PPVF. As is depicted in Fig. 5.(e)(f), CRTM performs the best of all. The reason behind it is, compared with other algorithms, CRTM sends tasks to all b-regions. Thus, CRTM not only finds task pairs based on nearest neighbour searching, but also has the ability of finding suitable task pairs via a *cross region* way. In other words, the matching under CRTM can find suitable workers from a large scale. Thus, the *saving distance* of CRTM is highest of all.

5   Conclusion

In this paper, we study the problem of tasks matching over smart city. Since the some sub-regions are crowed with multitudes of tasks, the system cannot provide enough workers for tasks contained in these regions. In order to solve this problem, we propose an ELM-based algorithm to identify this type of regions, called as b-region. Based on the identifying result, we construct a group indexes for these b-regions. When new tasks are generated, we insert tasks into these b-regions. The benefit is that when tasks are submitted at these b-regions, we can provide suitable workers located at other regions for them. The results demonstrate the superior performance of our proposed algorithms.

Acknowledgements. This paper is partly supported by the National Key Research and Development Program of China(2020YFB1707901), the National Natural Science Foundation of Liao Ning(2022-MS-303, 2022-MS-302, and 2022-BS-218).

References

1. Bin, W., Rui, Z., Siting, Z., Zheng, Z., Xiaochun, Y., Guoren, W.: PPVF: a novel framework for supporting path planning over carsharing. IEEE Access **7**, 10627–10643 (2019)
2. Kazemi, L., Shahabi, C.: Geocrowd: enabling query answering with spatial crowdsourcing. In: Proceedings of the SIGSPATIAL 2012 International Conference on Advances in Geographic Information Systems, pp. 189–198 (2012)
3. Tong, Y.X., She, J.Y., Ding, B.L.: Online mobile micro-task allocation in spatial crowdsourcing. In: Proceedings of the 32nd International Conference , pp. 49–60 (2016)
4. Furuhata, M., Dessouky, M., Ordez, F.: Ridesharing: the state-of-the-art and future directions. Transp. Res. Part B Methodol. **57**, 28–46 (2013)
5. Santi, P., Resta, G., Szell, M., Sobolevsky, S.: Quantifying the benefits of vehicle pooling with shareability networks. Proc. Nat. Acad. Sci. **111**, 290–294 (2014)
6. Ma, S., Zheng, Y., Wolfson, O.: T-share: a large-scale dynamic taxi ridesharing service. In: 2013 IEEE 29th International Conference on Data Engineering (ICDE), pp. 410–421 (2013)
7. Huang, Y., Bastani, F., Jin, R.: Large scale real-time ridesharing with service guarantee on road networks. In: Proceedings of the VLDB, pp. 2017–2028 (2014)
8. Santos, D.O., Xavier, E.C.: Dynamic taxi and ridesharing: a framework and heuristics for the optimization problem. In: Twenty-Third International Joint Conference on Artificial Intelligence (2013)
9. Guang, B.H., Qin, Y.Z., Chee-Kheong, S.: Extreme learning machine: a new learning scheme of feedforward neural networks. In: International Symposium on Neural Networks (2004)
10. Zhu, R., Wang, B., Yang, X.: SAP: improving continuous top-K queries over streaming data. IEEE Trans. Knowl. Data Eng. **29**, 1310–1328 (2017)
11. Li T., Chen L., Jensen C.S., et al.: Evolutionary clustering of moving objects. In: 2022 IEEE 38th International Conference on Data Engineering (ICDE), pp. 2399–2411. IEEE (2022)

12. Li, T., Huang, R., Chen, L.: Compression of uncertain trajectories in road networks. In: Proceeding of the VLDB, pp. 1050–1063 (2020)
13. Li, T., Chen, L., Jense, C.S.: TRACE: real-time compression of streaming trajectories in road networks. In: Proceeding of the VLDB, pp. 1175–1187(2021)

Attention-Based Spatial-Temporal Graph Convolutional Recurrent Networks for Traffic Forecasting

Haiyang Liu[1], Chunjiang Zhu[2], Detian Zhang[1(✉)], and Qing Li[3]

[1] Institute of Artificial Intelligence, Department of Computer Science and Technology, Soochow University, Suzhou, China
20215227052@stu.suda.edu.cn, detian@suda.edu.cn
[2] Department of Computer Science, UNC Greensboro, Greensboro, NC, USA
chunjiang.zhu@uncg.edu
[3] Department of Computing, The Hong Kong Polytechnic University, Hong Kong, China
qing-prof.li@polyu.edu.hk

Abstract. Traffic forecasting is one of the most fundamental problems in transportation science and artificial intelligence. The key challenge is to effectively model complex spatial-temporal dependencies and correlations in modern traffic data. Existing methods, however, cannot accurately model both long-term and short-term temporal correlations simultaneously, limiting their expressive power on complex spatial-temporal patterns. In this paper, we propose a novel spatial-temporal neural network framework: Attention-based Spatial-Temporal Graph Convolutional Recurrent Network ($ASTGCRN$), which consists of a graph convolutional recurrent module ($GCRN$) and a global attention module. In particular, $GCRN$ integrates gated recurrent units and adaptive graph convolutional networks for dynamically learning graph structures and capturing spatial dependencies and local temporal relationships. To effectively extract global temporal dependencies, we design a temporal attention layer and implement it as three independent modules based on multi-head self-attention, transformer, and informer respectively. Extensive experiments on five real traffic datasets have demonstrated the excellent predictive performance of all our three models with all their average MAE, $RMSE$ and $MAPE$ across the test datasets lower than the baseline methods.

Keywords: Traffic forecasting · Graph convolutional networks · Attention mechanism

1 Introduction

With the development of urbanization, the diversification of transportation modes and the increasing number of transportation vehicles (cabs, electric vehicles, shared bicycles, *etc.*) have put tremendous pressure on urban transportation systems, which has led to large-scale traffic congestions that have become

X. Yang et al. (Eds.): ADMA 2023, LNAI 14176, pp. 630–645, 2023.
https://doi.org/10.1007/978-3-031-46661-8_42

a common phenomenon. Traffic congestion has brought serious economic and environmental impacts to cities in various countries, and early intervention in traffic systems based on traffic forecasting is one of the effective ways to alleviate traffic congestion. By accurately predicting future traffic conditions, it provides a reference basis for urban traffic managers to make proper decisions in advance and improve traffic efficiency.

Traffic forecasting is challenging since traffic data are complex, highly dynamic, and correlated in both spatial and temporal dimensions. Recently, deep learning has dominated the field of traffic forecasting due to their capability to model complex non-linear patterns in traffic data. Many works used different deep learning networks to model dynamic local and global spatial-temporal dependencies and achieve promising prediction performance. On the one hand, they often used Recurrent Neural Networks (RNN) and the variants such as Long Short-Term Memory ($LSTM$) [14] and Gated Recurrent Units (GRU) [6] for *temporal* dependency modeling [1,2,5,22]. Some other studies used Convolutional Neural Networks (CNN) [15,28,30,31] or attention mechanisms [12,32] to efficiently extract temporal features in traffic data. On the other hand, Graph Convolutional Networks ($GCNs$) [21–23,29] are widely used to capture complex *spatial* features and dependencies in traffic road network data.

However, $RNN/LSTM/GRU$-based models can only indirectly model sequential temporal dependencies, and their internal cyclic operations make them difficult to capture long-term global dependencies [20]. To capture global information, CNN-based models [29,30] stack multiple layers of spatial-temporal modules but they may lose local information. The attention mechanism, though effective in capturing global dependencies, is not good at making short-term predictions [32]. Most of the previous attention-based methods [12,13] also have complex structures and thus high computational complexity. For instance in [13], the prediction of architectures built by multi-layer encoder and decoder, though excellent, is much slower than most prediction models by 1 or 2 orders of magnitude. Furthermore, most of the current GCN-based methods [11] need to pre-define a static graph based on inter-node distances or similarity to capture spatial information. However, the constructed graph needs to satisfy the static assumption of the road network, and cannot effectively capture complex dynamic spatial dependencies. Moreover, it is difficult to adapt graph-structure-based spatial modeling in various spatial-temporal prediction domains without prior knowledge (*e.g.*, inter-node distances). Therefore, effectively capturing dynamic spatial-temporal correlations and fully considering long-term and short-term dependencies are crucial to further improve the prediction performance.

To fully capture local and global spatial-temporal dependencies from traffic data, in this paper we propose a novel spatial-temporal networks framework: Attention-based Spatial-temporal Graph Convolutional Recurrent Network ($ASTGCRN$). It consists of a graph convolution recurrent module ($GCRN$) and a global attention module. Our main contributions are summarized as follows:

- We develop a novel spatial-temporal neural network framework, called *AST-GCRN*, that can effectively model dynamic local and global spatial-temporal dependencies in a traffic road network.
- In *ASTGCRN*, we devise an adaptive graph convolutional network with signals at different depths convoluted and then incorporate it into the *GRU*. The obtained *GRU* with adaptive graph convolution can well capture the *dynamic graph structures, spatial features, and local temporal* dependencies.
- We propose a general attention layer that accepts inputs from *GCRN* at different time points and captures the *global temporal* dependencies. We implement the layer using multi-head self-attention, transformer, and informer to generate three respective models.
- Extensive experiments have been performed on five real-world traffic datasets to demonstrate the superior performance of all our three models compared with the current state of the art. In particular, our model with transformer improves the average *MAE*, *RMSE*, and *MAPE* (across the tested datasets) by 0.21, 0.37, and 0.28, respectively. We carry out an additional experiment to show the generalizability of our proposed models to other spatial-temporal learning tasks.

2   Related Work

2.1   Traffic Forecasting

Traffic forecasting originated from univariate time series forecasting. Early statistical methods include Historical Average (*HA*), Vector Auto-Regressive (*VAR*) [34] and Auto-Regressive Integrated Moving Average (*ARIMA*) [19,27], with the *ARIMA* family of models the most popular. However, most of these methods are linear, need to satisfy stationary assumptions, and cannot handle complex non-linear spatial-temporal data.

With the rise of deep learning, it has gradually dominated the field of traffic forecasting by virtue of the ability to capture complex non-linear patterns in spatial-temporal data. RNN-based and CNN-based deep learning methods are the two mainstream directions for modeling temporal dependence. Early RNN-based methods such as *DCRNN* [22] used an encoder-decoder architecture with pre-sampling to capture temporal dependencies, but the autoregressive computation is difficult to focus on long-term correlations effectively. Later, the attention mechanism has been used to improve predictive performance [12,13,26,32]. In CNN-based approaches, the combination of 1-D temporal convolution *TCN* and graph convolution [29,30] are commonly used. But CNN-based models require stacking multiple layers to expand the perceptual field. The emergence of *GCNs* has enabled deep learning models to handle non-Euclidean data and capture implicit spatial dependencies, and they have been widely used for spatial data modeling [15,23]. The static graphs pre-defined according to the distance or similarity between nodes cannot fully reflect the road network information, and cannot make dynamic adjustments during the training process to effectively capture complex spatial dependencies. Current research overcame the limitations of convolutional networks based on static graphs or single graphs, and more adaptive graph or dynamic graph building strategies [2,18,28] were proposed.

In addition to the above methods, differential equations have also been applied to improve traffic forecasting. [10] capture spatial-temporal dynamics through a tensor-based Ordinary Differential Equation (ODE) alternative to graph convolutional networks. [7] introduced Neural Control Differential Equations ($NCDEs$) into traffic prediction, which designed two $NCDEs$ for temporal processing and spatial processing respectively. Although there are dense methods for spatial-temporal modeling, most of them lack the capability to focus on both long-term and short-term temporal correlations, which results in the limitations of capturing temporal dependencies and road network dynamics.

2.2   Graph Convolutional Networks

Graph convolution networks can be separated into spectral domain graph convolution and spatial domain graph convolution. In the field of traffic prediction, spectral domain graph convolution has been widely used to capture the spatial correlation between traffic series. [3] for the first time proposed spectral domain graph convolution based on spectral graph theory. The spatial domain signal is converted to the spectral domain by Fourier transform, and then the convolution result is inverted to the spatial domain after completing the convolution operation. The specific formula is defined as follows:

$$g_\theta \star_G x = g(L)x = U g_\theta(\Lambda) U^{\mathbf{T}} x, \tag{1}$$

In the equation, $\star_G$ denotes the graph convolution operation between the convolution kernel g_θ and the input signal x and $L = D^{-\frac{1}{2}} \mathbf{L} D^{-\frac{1}{2}} = U \Lambda U^{\mathbf{T}} \in \mathbb{R}^{N \times N}$ is the symmetric normalized graph Laplacian matrix, where $\mathbf{L} = D - A$ is the graph Laplacian matrix and $D = diag(\sum_{j=1}^{N} A_{1j}, \cdots, \sum_{j=1}^{N} A_{Nj}) \in \mathbb{R}^{N \times N}$ is the diagonal degree matrix. U is the Fourier basis of G and Λ is the diagonal matrix of $\mathbf{L}$ eigenvalues. However, the eigenvalue decomposition of the Laplacian matrix in Eq. (1) requires expensive computations. For this reason, [8] uses the Chebyshev polynomial to replace the convolution kernel g_θ in the spectral domain:

$$g_\theta \star_G x = g(L)x = \sum_{k=0}^{K-1} \beta_k T_k(\hat{L}) x, \tag{2}$$

where $[\beta_0, \beta_1, \ldots, \beta_{K-1}]$ are the learnable parameters, and $K \geq 1$ is the number of convolution kernels. ChebNet does not require eigenvalue decomposition of Laplacian matrices, but uses Chebyshev polynomials $T_0(\hat{L}) = I_n, T_1(\hat{L}) = \hat{L}$, and $T_{n+1}(\hat{L}) = 2\hat{L}T_n(\hat{L}) - T_{n-1}(\hat{L})$. Here $\hat{L} = \frac{2}{\lambda_{max}} L - I_n$ is the scaled Laplacian matrix, where λ_{max} is the largest eigenvalue and I_n is the identity matrix. When $K = 2$, ChebNet is simplified to GCN [17].

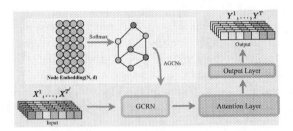

Fig. 1. Detailed framework of the *ASTGCRN* model.

3   Methodology

In this section, we first give the mathematical definition of the traffic prediction problem, and then describe in detail the implementation of the *ASTGCRN* framework (see Fig. 1): *GCRN* and the attention layer.

3.1   Problem Definition

The traffic prediction task can be formulated as a multi-step time series prediction problem that utilizes historical traffic data and prior knowledge of N locations (*e.g.*, traffic sensors) on a road network to predict future traffic conditions. Typically, prior knowledge refers to the road network represented as a graph $G = (V, E, \mathbf{A})$, where V is a set of $N = |V|$ nodes representing different locations on the road network, E is a set of edges, and $\mathbf{A} \in \mathbb{R}^{\mathbf{N} \times \mathbf{N}}$ is the weighted adjacency matrix representing the proximity between nodes (*e.g.*, the road network between nodes). We can formulate the traffic prediction problem as learning a function F to predict the graph signals $Y^{(t+1):(t+T)} \in \mathbb{R}^{T \times N \times C}$ of the next T steps based on the past T' steps graph signals $X^{(t-T'+1):t} \in \mathbb{R}^{T' \times N \times C}$ and G:

$$[X^{(t-T'+1):t}, G] \xrightarrow{F_\Theta} [X^{(t+1):(t+T)}], \tag{3}$$

where Θ denotes all the learnable parameters in the model.

3.2   Adaptive Graph Convolution

For traffic data in a road network, the dependencies between different nodes may change over time, and the pre-defined graph structure cannot contain complete spatial dependency information. Inspired by the adaptive adjacency matrix [2, 5,29], we generate $T_1(\hat{L})$ in Eq. (2) by randomly initializing a learnable node embedding $E_\phi \in \mathbb{R}^{N \times D_e}$, where D_e denotes the size of the node embedding:

$$T_1(\hat{L}) = \hat{L} = softmax(E_\phi \cdot E_\phi^{\mathbf{T}}) \tag{4}$$

To explore the hidden spatial correlations between node domains at different depths, we generalize to high-dimensional graph signals $X \in \mathbb{R}^{N \times C_{in}}$ and concatenate $T_k(\hat{L})$ at different depths as a tensor $\tilde{T}_\phi = [I, T_1(\hat{L}), \ldots, T_{K-1}(\hat{L})]^{\mathbf{T}} \in \mathbb{R}^{K \times N \times N}$.

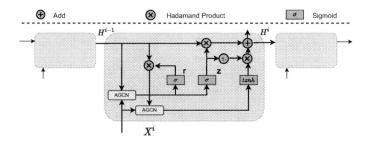

Fig. 2. The Graph Convolution Recurrent Module.

Let C_{in} and C_{out} represent the number of input and output channels, respectively. Then the graph convolution formula in Eq. (2) can be refined as:

$$g_\theta \star_G x = g(L)x = \tilde{T}_\phi X \Psi, \tag{5}$$

where the learnable parameters $\Psi \in \mathbb{R}^{K \times C_{in} \times C_{out}}$. However, the parameters shared by all nodes have limitations in capturing spatial dependencies [2]. Instead, we assign independent parameters to each node to get the parameters $\hat{\Psi} \in \mathbb{R}^{N \times K \times C_{in} \times C_{out}}$, which can more effectively capture the hidden information in different nodes. We further avoid overfitting and high spatial complexity problems by matrix factorization. That is to learn two smaller parameters to generate $\hat{\Psi} = E_\phi W$, where $E_\phi \in \mathbb{R}^{N \times D_e}$ is the node embedding dictionary and $W \in \mathbb{R}^{D_e \times K \times C_{in} \times C_{out}}$ are the learnable weights. Our adaptive graph convolution formula can be expressed as:

$$g_\theta \star_G x = g(L)x = \tilde{T}_\phi X E_\phi W \in \mathbb{R}^{N \times C_{out}} \tag{6}$$

3.3   GRU with Adaptive Graph Convolution

GRU is a simplified version of *LSTM* with multiple *GRUCell* modules and generally provides the same performance as *LSTM* but is significantly faster to compute. To further discover the spatial-temporal correlation between time series, we replace the *MLP* layers in *GRU* with adaptive graph convolution operation, named *GCRN*. The computation of *GCRN* is given as follows:

$$
\begin{aligned}
z^t &= \sigma(\tilde{T}_\phi[X^t, h^{t-1}]E_\phi W_z + E_\phi b_z), \\
r^t &= \sigma(\tilde{T}_\phi[X^t, h^{t-1}]E_\phi W_r + E_\phi b_r), \\
\tilde{h}^t &= tanh(\tilde{T}_\phi[X^t, r^t \odot h^{t-1}]E_\phi W_{\tilde{h}} + E_\phi b_{\tilde{h}}), \\
h^t &= z^t \odot h^{t-1} + (1 - z^t) \odot \tilde{h}^t,
\end{aligned}
\tag{7}
$$

where $W_z, W_r, W_{\tilde{h}}, b_z, b_r$ and $b_{\tilde{h}}$ are learnable parameters, σ and $tanh$ are two activation functions, *i.e.*, the Sigmoid function and the Tanh function. The $[X^t, h^{t-1}]$ and h^t are the input and output at time step t, respectively. The network architecture of *GCRN* is plotted in Fig. 2.

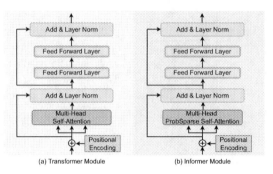

(a) Transformer Module   (b) Informer Module

Fig. 3. The Attention Layer.

3.4 The Attention Layer

GCRN can effectively capture sequential dependencies, but its structural characteristics limit its ability to capture long-distance temporal information. For the traffic prediction task, the global temporal dependence clearly has a significant impact on the learning performance. Self-attention directly connects two time steps through dot product calculation, which greatly shortens the distance between long-distance dependent features, and improves the parallelization of computation, making it easier to capture long-term dependencies in traffic data. Therefore, we propose three independent modules for the self-attention mechanism, namely, the multi-headed self-attention module, the transformer module, and the informer module, in order to directly capture global temporal dependencies. In the following three subsections, we will explain these three modules in detail.

Multi-Head Self-attention Module. Multi-head attention is to learn the dependencies of different patterns in parallel with multiple sets of queries, keys and values (where each set is regarded as an attention head), and then concatenate the learned multiple relationships as the output. We use a self-attentive mechanism to construct the multi-headed attention module. Specifically, $Q_o = H_o W_q$, $K_o = H_o W_k$ and $V_o = H_o W_v$ are derived from the same matrix H_o by linear transformation. Here $H_o \in \mathbb{R}^{N \times T \times C_{out}}$ is the output result of the *GCRN* module, and $W_q \in \mathbb{R}^{C_{out} \times d_q}$, $W_k \in \mathbb{R}^{C_{out} \times d_k}$ and $W_v \in \mathbb{R}^{C_{out} \times d_v}$ are the learnable parameters of the linear projection. For multi-head self-attention mechanism, the formula can be stated as:

$$
\begin{aligned}
MHSelfAtt &= Concat(head_1, \ldots, head_h)W_o, \\
\text{where } head_i &= Att(Q_o, K_o, V_o) \\
&= softmax(\frac{Q_o K_o{}^{\mathbf{T}}}{\sqrt{d}})V_o.
\end{aligned}
\tag{8}
$$

Transformer Module. The Transformer module (see Fig. 3(a)) contains a multi-head self-attention layer and two feed-forward neural networks. For self-attention, each position of the input sequence is equally inner product, which results in the loss of sequential information. We use a fixed position encoding [25] to address this flaw:

$$PE_t(2c) = sin(t/1000^{2c/C_{out}}),$$
$$PE_t(2c+1) = cos(t/1000^{2c/C_{out}}), \tag{9}$$

where t is the relative position of each sequence (time step) of the input and c represents the dimension. In order to better identify the relative positional relationship between sequences, H_o and position encoding are combined to generate $H_o' \in \mathbb{R}^{N \times T \times C_{out}}$:

$$H_o'[:,t,:] = H_o[:,t,:] + PE_t \tag{10}$$

After combining the location encoding, H_o' is fed as input into the multi-headed self-attentive layer for remote relationship capture, and then the output state is passed to the two fully connected layers. Layer normalization and residual connectivity are used in both sub-layers. Finally, Transformer module outputs the result $H_a \in \mathbb{R}^{N \times T \times C_{out}}$.

Informer Module. According to Eq. (8), the traditional self-attention mechanism requires two dot products and $O(T^2)$ space complexity. The sequence lengths of queries and keys are equal in self-attention computation, *i.e.*, $T_q = T_k = T$. After finding that most of the dot products have minimal attention and the main attention is focused on only a few dot products, [33] proposed Prob-Sparse self-attention. ProbSparse self-attention selects only the more important queries to reduce the computational complexity, *i.e.*, by measuring the dilution of the queries and then selecting only the top-u queries with $u = c \cdot lnT$ for constant c. The query dilution evaluation formula is as follows:

$$\bar{M}(q_i, K_o) = \max_j \left\{ \frac{q_i k_j^\mathbf{T}}{\sqrt{d}} \right\} - \frac{1}{T} \sum_{j=1}^T \frac{q_i k_j^\mathbf{T}}{\sqrt{d}} \tag{11}$$

where q_i and k_i represent the i-th row in Q_o and K_o respectively. To compute $\bar{M}$, only $U = T \ln T$ dot product pairs are randomly selected, and the other pairs are filled with zeros. In this way, the time and space complexity are reduced to only $O(T \ln T)$. Therefore, we construct a new module called Informer module (see Fig. 3(b)) by using ProbSparse self-attention to replace the normal self-attention mechanism of Transformer module. It selects top-u query according to $\bar{M}$ to generate sparse matrix Q_o^{spa}. Then the Multi-head ProbSparse self-attention can be expressed as:

$$MHProbSelfAtt = Concat(head_1, \ldots, head_h)W_o,$$
$$where \ head_i = Att(Q_o^{spa}, K_o, V_o)$$
$$= softmax(\frac{Q_o^{spa} K_o^\mathbf{T}}{\sqrt{d}})V_o. \tag{12}$$

Table 1. Statistics of the tested datasets

Datasets	Nodes	Samples	Unit	Time Span
PEMSD3	358	26,208	5 mins	3 months
PEMSD4	307	16,992	5 mins	2 months
PEMSD7	883	28,224	5 mins	4 months
PEMSD8	170	17,856	5 mins	2 months
PEMSD7(M)	228	12,672	5 mins	2 months
DND-US	53	313	1 week	6 years

We choose the $L1$ loss to formulate the objective function and minimize the training error by back propagation. Specifically, the loss function is defined as follows.

$$Loss = \frac{1}{T} \sum_{t=0}^{T-1} (\hat{Y}^t - Y^t) \tag{13}$$

where $\hat{Y}$ is the real traffic data, Y is the predicted data, and T is the total predicted time steps.

4   Experimental Results

In this section, we present the results of the extensive experiments we have performed. We start by describing the experimental setups and then discuss the prediction results obtained in the baseline settings. Finally, the ablation study and the effects of hyperparameter tuning are provided.

Datasets. We evaluate the performance of the developed models on five widely used traffic prediction datasets collected by Caltrans Performance Measure System [4], namely *PEMSD3, PEMSD4, PEMSD7, PEMSD8,* and *PEMSD7(M)* [7,10]. The traffic data are aggregated into 5-minute time intervals, *i.e.*, 288 data points per day. In addition, we construct a new US natural death dataset *DND-US* to study the generalizability of our method to other spatial-temporal data. It contains weekly natural deaths for 53 (autonomous) states in the US for the six years from 2014 to 2020. Following existing works [2], the Z-score normalization method is adopted to normalize the input data to make the training process more stable. Detailed statistics for the tested datasets are summarized in Table 1.

Baseline Methods. We compare our models with the following baseline methods:

- Traditional time series forecasting methods, Historical Average (*HA*), *ARIMA* [27], *VAR* [34], and *SVR* [9];
- *RNN*-based models: *FC-LSTM* [24], *DCRNN* [22], *AGCRN* [2], and *Z-GCNETs* [5];
- *CNN*-based methods: *STGCN* [30], *Graph WaveNet* [29], *MSTGCN, LSGCN* [15], *STSGCN* [23], and *STFGNN* [21];

Table 2. Performance comparison of different models on the tested datasets. Underlined results represent the current best among existing methods. Our three models outperform *almost all* baseline methods, as shown in bold font.

Model	PEMSD3			PEMSD4			PEMSD7			PEMSD8			PEMSD7(M)		
	MAE	RMSE	MAPE	MAE	RMSE	MAPE	MAE	RMSE	MAPE	MAE	RMSE	MAPE	MAE	RMSE	MAPE
HA	31.58	52.39	33.78%	38.03	59.24	27.88%	45.12	65.64	24.51%	34.86	59.24	27.88%	4.59	8.63	14.35%
ARIMA	35.41	47.59	33.78%	33.73	48.80	24.18%	38.17	59.27	19.46%	31.09	44.32	22.73%	7.27	13.20	15.38%
VAR	23.65	38.26	24.51%	24.54	38.61	17.24%	50.22	75.63	32.22%	19.19	29.81	13.10%	4.25	7.61	10.28%
SVR	20.73	34.97	20.63%	27.23	41.82	18.95%	32.49	44.54	19.20%	22.00	33.85	14.23%	3.33	6.63	8.53%
FC-LSTM	21.33	35.11	23.33%	26.77	40.65	18.23%	29.98	45.94	13.20%	23.09	35.17	14.99%	4.16	7.51	10.10%
DCRNN	17.99	30.31	18.34%	21.22	33.44	14.17%	25.22	38.61	11.82%	16.82	26.36	10.92%	3.83	7.18	9.81%
AGCRN	15.98	28.25	15.23%	19.83	32.26	12.97%	22.37	36.55	9.12%	15.95	25.22	10.09%	2.79	5.54	7.02%
Z-GCNETs	16.64	28.15	16.39%	19.50	31.61	12.78%	21.77	35.17	9.25%	15.76	25.11	10.01%	2.75	5.62	6.89%
STGCN	17.55	30.42	17.34%	21.16	34.89	13.83%	25.33	39.34	11.21%	17.50	27.09	11.29%	3.86	6.79	10.06%
Graph WaveNet	19.12	32.77	18.89%	24.89	39.66	17.29%	26.39	41.50	11.97%	18.28	30.05	12.15%	3.19	6.24	8.02%
MSTGCN	19.54	31.93	23.86%	23.96	37.21	14.33%	29.00	43.73	14.30%	19.00	29.15	12.38%	3.54	6.14	9.00%
LSGCN	17.94	29.85	16.98%	21.53	33.86	13.18%	27.31	41.46	11.98%	17.73	26.76	11.20%	3.05	5.98	7.62%
STSGCN	17.48	29.21	16.78%	21.19	33.65	13.90%	24.26	39.03	10.21%	17.13	26.80	10.96%	3.01	5.93	7.55%
STFGNN	16.77	28.34	16.30%	20.48	32.51	16.77%	23.46	36.60	9.21%	16.94	26.25	10.60%	2.90	5.79	7.23%
ASTGCN(r)	17.34	29.56	17.21%	22.93	35.22	16.56%	24.01	37.87	10.73%	18.25	28.06	11.64%	3.14	6.18	8.12%
DSTAGNN	15.57	27.21	14.68%	19.30	31.46	12.70%	21.42	34.51	9.01%	15.67	24.77	9.94%	2.75	5.53	6.93%
STGODE	16.50	27.84	16.69%	20.84	32.82	13.77%	22.59	37.54	10.14%	16.81	25.97	10.62%	2.97	5.66	7.36%
STG-NCDE	15.57	27.09	15.06%	19.21	31.09	12.76%	20.53	33.84	8.80%	15.45	24.81	9.92%	2.68	5.39	6.76%
A-ASTGCRN	**15.06**	**26.71**	**13.83%**	19.30	**30.92**	12.91%	**20.42**	**33.81**	**8.54%**	15.46	24.54	9.89%	2.66	5.36	6.72%
I-ASTGCRN	**15.06**	**26.40**	13.91%	**19.15**	**30.80**	12.89%	20.81	33.83	8.95%	**15.26**	**24.53**	**9.65%**	**2.63**	**5.30**	**6.60%**
T-ASTGCRN	**14.90**	**26.01**	14.17%	19.21	**31.05**	**12.67%**	**20.53**	**33.75**	**8.73%**	**15.14**	**24.24**	**9.63%**	**2.63**	5.32	6.66%

- Attention-based models: *ASTGCN(r)* [12], and *DSTAGNN* [18];
- Other types of models: *STGODE* [10] and *STG-NCDE* [7].

Experimental Settings. All datasets are split into training set, validation set and test set in the ratio of 6:2:2. Our model and all baseline methods use the 12 historical continuous time steps as input to predict the data for the next 12 continuous time steps.

Our models are implemented based on the Pytorch framework, and all the experiments are performed on an NVIDIA GeForce GTX 1080 TI GPU with 11G memory. The following hyperparameters are configured based on the models' performance on the validation dataset: we train the model with 300 epochs at a learning rate of 0.003 using the Adam optimizer [16] and an early stop strategy with a patience number of 15. The code is available at https://github.com/Liuhy-666/ASTGCRN.git.

Three common prediction metrics, Mean Absolute Error (*MAE*), Root Mean Square Error (*RMSE*), and Mean Absolute Percentage Error (*MAPE*), are used to measure the traffic forecasting performance of the tested methods. In the discussions below, we refer to our specific *ASTGCRN* models based on the Multi-head self-attention module, Transformer module, and Informer module as *A-ASTGCRN*, *T-ASTGCRN*, and *I-ASTGCRN*, respectively.

4.1   Experimental Results

Table 2 shows the prediction performance of our three models together with the nineteen baseline methods on the five tested datasets. Remarkably, our three models outperform almost all the baseline methods in prediction on all the datasets. Table 3 lists the training time (s/epoch), inference time (s/epoch) and memory cost (MB) of our models, as well as several recent and best-performing baselines on the *PEMSD4* dataset.

Table 3. Computation time on *PEMSD4*.

Model	Training	Inference	Memory
STGODE	111.77	12.19	8773
Z-GCNETs	63.34	7.40	8597
DSTAGNN	242.57	14.64	10347
STG-NCDE	1318.35	93.77	6091
A-ASTGCRN	45.12	5.18	7167
I-ASTGCRN	58.84	6.51	7527
T-ASTGCRN	54.80	5.62	7319

Table 4. Forecasting performance of several competitive methods on *DND-US*

Dataset	Model	MAE	RMSE	MAPE
DND-US	AGCRN	105.97	325.09	7.49%
	DSTAGNN	47.49	73.37	7.47%
	STG-NCDE	47.70	77.30	6.13%
	A-ASTGCRN	**39.33**	**62.86***	**5.36%**
	I-ASTGCRN	**38.79***	**66.99**	**5.16%***
	T-ASTGCRN	**40.60**	**66.28**	**5.43%**

The overall prediction results of traditional statistical methods (including *HA*, *ARIMA*, *VR*, and *SVR*) are not satisfactory because of its limited ability to handle non-linear data. Their prediction performance is worse than deep learning methods by large margins. *RNN*-based methods such as *DCRNN*, *AGCRN*, and *Z-GCNRTs* suffer from the limitation of *RNNs* that cannot successfully capture long-term temporal dependence and produce worse results than our methods. *CNN*-based models such as *STGCN*, *Graph WaveNet*, *STSGCN*, *STFGCN*, and *STGODE*, have either worse or comparable performance compared to *RNN*-based methods in our empirical study. They get the 1-D *CNN* by temporal information, but the size of the convolutional kernel prevents them from capturing the complete long-term temporal correlation. Although both *ASTGCN* and *DSTAGNN* use temporal attention modules, they ignore local temporal information. *STG-NCDE* achieves currently best performance in multiple datasets. But their temporal *NCDE* using only the fully connected operation cannot pay full attention to the temporal information.

To test the generalizability of our proposed models to other spatial-temporal learning tasks, we perform an additional experiment on the *DND-US* dataset to predict the number of natural deaths in each US state. As shown in Table 4, all our three models again outperform several competitive baseline methods with significant margins.

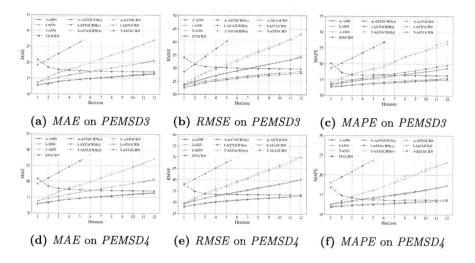

(a) *MAE on PEMSD3*   (b) *RMSE on PEMSD3*   (c) *MAPE on PEMSD3*

(d) *MAE on PEMSD4*   (e) *RMSE on PEMSD4*   (f) *MAPE on PEMSD4*

Fig. 4. Prediction performance at each horizon

4.2   Ablation and Parameter Study

Ablation Study. We refer to the model without an attention layer as *STGCRN*, and the *A-ASTGCRN*, *I-ASTGCRN* and *T-ASTGCRN* without the *GCRN* layer as *A-ANN*, *I-ANN* and *T-ANN*, respectively. Also, *A-ASTGCRN(s)*, *I-ASTGCRN(s)* and *T-ASTGCRN(s)* are variant models that use static graphs for graph convolution. we plot the detailed values of different horizons for our methods on the *PEMSD3* and *PEMSD4* datasets in Fig. 4. It shows that the prediction performance of *STGCRN* become closer to the three models as the predicted horizon increases, and the spatial modeling ability of static graphs is much lower than that of adaptive adjacency matrices. The autoregressive feature of the *GRU* model allows more spatial-temporal information to be pooled in the later time horizons, so that long-term prediction appears to be better than short-term prediction. But the performance of *STGCRN* lags behind the three attention-based models at all time horizons. The attention module is crucial for capturing long-term temporal dependencies in traffic data, further enhancing the modeling of spatio-temporal dependencies. However, only using attention modules to focus on long-term temporal dependencies and removing *GRU* using adaptive graph convolutions hurts prediction performance.

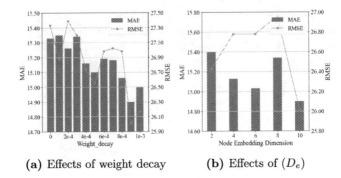

(a) Effects of weight decay   (b) Effects of (D_e)

Fig. 5. Effects of hyperparameter tuning on *T-ASTGCRN* in *PEMSD3*

Table 5. Effect of convolution kernel number K on *T-ASTGCRN*.

Dataset	K	MAE	RMSE	MAPE	Training	Inference	Memory
PEMSD3	1	15.24	26.46	15.10%	**88.94**	**9.95**	**6497**
	2	**14.90**	**26.01**	**14.17%**	95.80	10.22	7555
	3	15.33	27.04	13.92%	121.40	13.39	8535
PEMSD4	1	19.40	31.19	13.00%	**48.72**	**5.24**	**6355**
	2	**19.21**	**31.05**	**12.67%**	54.80	5.62	7319
	3	19.22	31.07	12.84%	66.43	7.12	8137

Parameter Study. To investigate the effects of hyperparameters on the prediction results, we conduct a series of experiments on the main hyperparameters. Figure 5 shows the *MAE* and *RMSE* values of *T-ASTGCRN* in the *PEMSD3* dataset when varying the weight decay and node embedding dimension D_e. It can be seen that fine-tuning the weight decay and node embedding dimension can improve the prediction performance of the model. Meanwhile, Table 5 shows the prediction performance and training cost for varying the number of convolution kernels K. From the experimental results, we can see that with $K = 1$, the graph convolution is simplified to a unit matrix-based implementation, which does not enable effective information transfer between nodes. A larger convolution depth does not improve the prediction performance, but instead incurs longer training time and memory cost. Therefore, for our model and dataset, we set K to 2.

5   Conclusion

In this paper, we design an attention-based spatial-temporal graph convolutional recurrent network framework for traffic prediction. We instantiate the framework with three attention modules based on Multi-head self-attention, Transformer and Informer, all of which, in particular the Transformer-based module, can well capture long-term temporal dependence and incorporate with the spatial and short-term temporal features by the *GCRN* module. Extensive experiments confirm the effectiveness of all our three models in improving the prediction performance. We believe that the design ideas of Transformer and Informer can bring new research thrusts in the field of traffic forecasting.

Acknowledgements. Chunjiang Zhu is supported by UNCG Start-up Funds and Faculty First Award. Detian Zhang is partially supported by the Collaborative Innovation Center of Novel Software Technology and Industrialization, the Priority Academic Program Development of Jiangsu Higher Education Institutions.

References

1. Bai, L., Yao, L., Kanhere, S.S., Yang, Z., Chu, J., Wang, X.: Passenger demand forecasting with multi-task convolutional recurrent neural networks. In: Yang, Q., Zhou, Z.-H., Gong, Z., Zhang, M.-L., Huang, S.-J. (eds.) PAKDD 2019. LNCS (LNAI), vol. 11440, pp. 29–42. Springer, Cham (2019). https://doi.org/10.1007/978-3-030-16145-3_3
2. Bai, L., Yao, L., Li, C., Wang, X., Wang, C.: Adaptive graph convolutional recurrent network for traffic forecasting. Adv. Neural. Inf. Process. Syst. **33**, 17804–17815 (2020)
3. Bruna, J., Zaremba, W., Szlam, A., LeCun, Y.: Spectral networks and locally connected networks on graphs. arXiv preprint arXiv:1312.6203 (2013)
4. Chen, C., Petty, K., Skabardonis, A., Varaiya, P., Jia, Z.: Freeway performance measurement system: mining loop detector data. Transp. Res. Rec. **1748**(1), 96–102 (2001)
5. Chen, Y., Segovia, I., Gel, Y.R.: Z-gcnets: time zigzags at graph convolutional networks for time series forecasting. In: International Conference on Machine Learning, pp. 1684–1694. PMLR (2021)
6. Cho, K., Van Merriënboer, B., Bahdanau, D., Bengio, Y.: On the properties of neural machine translation: Encoder-decoder approaches. arXiv preprint arXiv:1409.1259 (2014)
7. Choi, J., Choi, H., Hwang, J., Park, N.: Graph neural controlled differential equations for traffic forecasting. In: Proceedings of the AAAI Conference on Artificial Intelligence, vol. 36, pp. 6367–6374 (2022)
8. Defferrard, M., Bresson, X., Vandergheynst, P.: Convolutional neural networks on graphs with fast localized spectral filtering. In: Advances in Neural Information Processing Systems 29 (2016)
9. Drucker, H., Burges, C.J., Kaufman, L., Smola, A., Vapnik, V.: Support vector regression machines. In: Advances in Neural Information Processing Systems 9 (1996)
10. Fang, Z., Long, Q., Song, G., Xie, K.: Spatial-temporal graph ode networks for traffic flow forecasting. In: Proceedings of the 27th ACM SIGKDD Conference on Knowledge Discovery & Data Mining, pp. 364–373 (2021)
11. Geng, X., Li, Y., Wang, L., Zhang, L., Yang, Q., Ye, J., Liu, Y.: Spatiotemporal multi-graph convolution network for ride-hailing demand forecasting. In: Proceedings of the AAAI Conference on Artificial Intelligence, vol. 33, pp. 3656–3663 (2019)
12. Guo, S., Lin, Y., Feng, N., Song, C., Wan, H.: Attention based spatial-temporal graph convolutional networks for traffic flow forecasting. In: Proceedings of the AAAI Conference on Artificial Intelligence, vol. 33, pp. 922–929 (2019)

13. Guo, S., Lin, Y., Wan, H., Li, X., Cong, G.: Learning dynamics and heterogeneity of spatial-temporal graph data for traffic forecasting. IEEE Trans. Knowl. Data Eng. (2021)
14. Hochreiter, S., Schmidhuber, J.: Long short-term memory. Neural Comput. **9**(8), 1735–1780 (1997)
15. Huang, R., Huang, C., Liu, Y., Dai, G., Kong, W.: Lsgcn: Long short-term traffic prediction with graph convolutional networks. In: IJCAI, pp. 2355–2361 (2020)
16. Kingma, D.P., Ba, J.: Adam: a method for stochastic optimization. arXiv preprint arXiv:1412.6980 (2014)
17. Kipf, T.N., Welling, M.: Semi-supervised classification with graph convolutional networks. arXiv preprint arXiv:1609.02907 (2016)
18. Lan, S., Ma, Y., Huang, W., Wang, W., Yang, H., Li, P.: Dstagnn: dynamic spatial-temporal aware graph neural network for traffic flow forecasting. In: International Conference on Machine Learning, pp. 11906–11917. PMLR (2022)
19. Lee, S., Fambro, D.B.: Application of subset autoregressive integrated moving average model for short-term freeway traffic volume forecasting. Transp. Res. Rec. **1678**(1), 179–188 (1999)
20. Li, F., et al.: Dynamic graph convolutional recurrent network for traffic prediction: benchmark and solution. ACM Trans. Knowl. Dis. Data (TKDD) (2021)
21. Li, M., Zhu, Z.: Spatial-temporal fusion graph neural networks for traffic flow forecasting. In: Proceedings of the AAAI Conference on Artificial Intelligence, vol. 35, pp. 4189–4196 (2021)
22. Li, Y., Yu, R., Shahabi, C., Liu, Y.: Diffusion convolutional recurrent neural network: data-driven traffic forecasting. In: International Conference on Learning Representations (ICLR 2018) (2018)
23. Song, C., Lin, Y., Guo, S., Wan, H.: Spatial-temporal synchronous graph convolutional networks: a new framework for spatial-temporal network data forecasting. In: Proceedings of the AAAI Conference on Artificial Intelligence, vol. 34, pp. 914–921 (2020)
24. Sutskever, I., Vinyals, O., Le, Q.V.: Sequence to sequence learning with neural networks. In: Advances in Neural Information Processing Systems 27 (2014)
25. Vaswani, A., et al.: Attention is all you need. In: Advances in Neural Information Processing Systems 30 (2017)
26. Wang, X., et al.: Traffic flow prediction via spatial temporal graph neural network. In: Proceedings of The Web Conference 2020, pp. 1082–1092 (2020)
27. Williams, B.M., Hoel, L.A.: Modeling and forecasting vehicular traffic flow as a seasonal arima process: theoretical basis and empirical results. J. Transp. Eng. **129**(6), 664–672 (2003)
28. Wu, Z., Pan, S., Long, G., Jiang, J., Chang, X., Zhang, C.: Connecting the dots: Multivariate time series forecasting with graph neural networks. In: Proceedings of the 26th ACM SIGKDD International Conference on Knowledge Discovery & Data Mining, pp. 753–763 (2020)
29. Wu, Z., Pan, S., Long, G., Jiang, J., Zhang, C.: Graph wavenet for deep spatial-temporal graph modeling. In: Proceedings of the 28th International Joint Conference on Artificial Intelligence, pp. 1907–1913 (2019)
30. Yu, B., Yin, H., Zhu, Z.: Spatio-temporal graph convolutional networks: a deep learning framework for traffic forecasting. In: Proceedings of the 27th International Joint Conference on Artificial Intelligence, pp. 3634–3640 (2018)
31. Zhang, Q., Chang, J., Meng, G., Xiang, S., Pan, C.: Spatio-temporal graph structure learning for traffic forecasting. In: Proceedings of the AAAI Conference on Artificial Intelligence, vol. 34, pp. 1177–1185 (2020)

32. Zheng, C., Fan, X., Wang, C., Qi, J.: Gman: a graph multi-attention network for traffic prediction. In: Proceedings of the AAAI Conference on Artificial Intelligence, vol. 34, pp. 1234–1241 (2020)
33. Zhou, H., Zhang, S., Peng, J., Zhang, S., Li, J., Xiong, H., Zhang, W.: Informer: beyond efficient transformer for long sequence time-series forecasting. In: Proceedings of the AAAI Conference on Artificial Intelligence, vol. 35, pp. 11106–11115 (2021)
34. Zivot, E., Wang, J.: Vector autoregressive models for multivariate time series. Modeling financial time series with S-PLUS®, pp. 385–429 (2006)

Transformer Based Driving Behavior Safety Prediction for New Energy Vehicles

Hao Lin and Junjie Yao[✉]

East China Normal University, Shanghai, China
51205901039@stu.ecnu.edu.cn, junjie.yao@cs.ecnu.edu.cn

Abstract. The classification of driving behavior, with a particular emphasis on discerning safe from unsafe practices, is a task of paramount importance in the appraisal of drivers, and its significance is escalating in the epoch of autonomous driving. Driving behavior classification typically employs an assortment of features, such as velocity, acceleration, pedal pressure, turn signal utilization, and Global Positioning System (GPS) signals, amongst others. Nonetheless, these features exhibit considerable heterogeneity and do not offer comprehensive coverage. The extant literature pertaining to time series classification grapples with efficaciously addressing the high-dimensional nature, voluminous data, and the complexity of scenarios within the safety classification of driving behavior, especially for new energy vehicles. In this study, we have amassed an extensive corpus of sensor data, generated during the operation of new energy vehicles. Our research focused on the classification of driving behaviors concerning safety within the context of new energy vehicles and was predicated upon self-supervised learning. We proffered a time series model that leverages the Transformer architecture, tailored specifically for the aforementioned scenario, and employed a pre-training framework. To ascertain the efficacy of the proposed model, it was subjected to rigorous validation against a dataset comprising driving data from new energy vehicles. The model exhibited commendable performance and was further assessed through a series of downstream tasks.

Keywords: Driving Behavior Analysis · Multivariate Time Series Classification · Pre-Trained Model

1 Introduction

New energy vehicles (NEVs) represent a burgeoning class of automobiles that predominantly employ electric propulsion systems, or, in certain configurations, hybridize the utilization of electric and traditional internal combustion engines. In recent years, there has been a notable acceleration in the adoption and market penetration of NEVs. Concurrently, the widespread incorporation of sensor technologies into NEVs has engendered the acquisition of extensive datasets pertaining to real-world driving dynamics. Figure 1 presents a set of sensor features

X. Yang et al. (Eds.): ADMA 2023, LNAI 14176, pp. 646–660, 2023.
https://doi.org/10.1007/978-3-031-46661-8_43

in a typical NEV. The availability of such data has spurred research into the research of driving patterns and the anomalous driving behaviors.

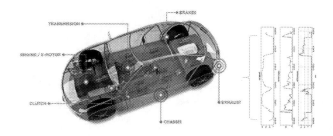

Fig. 1. Rich Sensor Features of New Energy Vehicle

Recently, a lot of work has been conducted in this field [1,3,8,23]. These work are usually based on low-dimensional [5,21], or machine learning methods [28,31]. To effectively detect the driving safety of new energy vehicles, we need to cope with multivariate time series modeling.

The challenge is the sheer volume of the data collected. The brevity of the sampling intervals leads to an accumulation of voluminous datasets, with a single vehicle potentially generating several million data points within the span of a fortnight. Furthermore, the labeling of this data for supervised learning approaches, particularly deep learning which typically requires extensive labeled datasets, is fraught with difficulties. Manual labeling is impractical given that defining labels necessitates the consideration of multifarious features in conjunction. Moreover, the interdependence among features, coupled with the fact that hazardous driving behaviors tend to manifest over a very truncated time scale, exacerbates the complexity of this task.

In this paper, we present a novel self-supervised learning model [32] predicated on a time-series transformer architecture, tailored for the sensor data pertaining to new energy vehicles. Transformer [26,32] is an important and recently developed deep learning paragradim, and one of its key factors to achieve widespread success in NLP is the ability to learn how to express natural language through self-supervised pre-training.

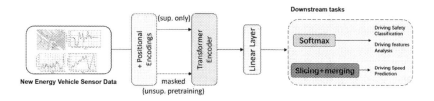

Fig. 2. Pipeline of Ev-transformer

The pipeline of our proposed approach is shown in Fig. 2. It is predicated on the Time Series Transformer [32], inheriting its encoder module. This model introduces modifications to the conventional Transformer architecture, enabling the conversion of word index sequences, originally tailored for discrete use, into multivariate time series. Notably, the Ev-Transformer is endowed with a self-supervised pre-training module, rendering it particularly well-suited for the novel application of driving safety analysis.

Our model was subjected to a scrupulous evaluation process across multiple driving datasets, especially our extensive rich sensor electric vehicle datasets. In this work, we trained a self-supervised model ev-transformer, Fig. 2 to perform the safety risk classification task and other downstream ones such as long-term driving risk prediction and driving risk trend prediction. Transformer model is based on multi-head attention mechanism, which makes it especially suitable for time series data. By considering the context (past - future) of each input sequence element, each input sequence element is represented. This model can model a large number of data in different fields, including driving data.

The findings illustrate that our modeling approach outperforms a plethora of extant monitoring models, particularly in relation to the sensor dataset of new energy vehicles. Notably, our model maintains a considerable edge even under conditions of scant training samples. To summarize, our work has the following main contributions:

1. We undertook the collection and processing of real-world driving data pertaining to new energy vehicles. The data was segmented based on specific indicators, and a subset of the data was subjected to labeling.
2. We conceived and developed a self-supervised model, christened Ev-Transformer, predicated on time series transformer architectures, specifically tailored for the classification of driving behaviors in new energy vehicles. The model yielded commendable results.
3. Leveraging the proposed model, we executed downstream tasks encompassing long-term driving risk prediction and the forecasting of driving risk trends.

2   Related Work

The problem of Driving behavior analysis is related to several directions. We briefly review them as follows.

Multivariate Time Series Classification: There exists a plethora of modeling techniques for multivariate time series. Machine learning algorithms, such as Random Forests [2], XGBoost [4], Support Vector Machines [6], and others [12], have been employed for multivariate time series classification. It is imperative to note that, at present, deep learning has not yet unequivocally established its dominance within the realm of time series analysis. Noteworthy exemplars like TS-CHIEF [25], which is predicated upon tree-structured classifiers, and ROCKET [7], which leverages random convolutional kernel transformations, continue to serve as exemplary benchmarks. However, it is worth mentioning that these methodologies often necessitate extensive manual preprocessing of the raw time series data, which stands at odds with the inherent characteristics of the driving data associated with new energy vehicles.

Deep Learning Methods: At present, the progression of deep learning-based methodologies for time series data prediction and classification can be categorized into several pivotal stages. Initially, the **CNN-LSTM** architecture [10,27] was introduced, which utilizes Convolutional Neural Networks (CNNs) as feature extractors for audio and text data, followed by Long Short-Term Memory (LSTM) networks for further processing. This architecture, originally designed for generating textual annotations for images, has been widely adopted in temporal data analyses. Subsequently, **ConvLSTM** [24], a variant of LSTM that incorporates convolutional operations, emerged as an effective alternative for spatial-temporal data feature extraction. Another notable development is **Deep-ConvLSTM** [22,29], which is a composite framework consisting of convolutional and LSTM recurrent layers, adept at autonomously learning feature representations and modeling temporal dependencies between activations.

A more recent innovation, **LSTM-FCN**, incorporates attention mechanisms in temporal data analysis, significantly boosting the performance of fully convolutional networks with a minimal increase in model size and reduced dependency on data preprocessing. It is evident that models grounded in deep learning can effectively leverage combinations of different networks to robustly capture the characteristics of time series data. Present-day refinements of these methods, such as MLSTM-FCN [13], MCDNN [30], and TapNet [33], predominantly employ an amalgamation of LSTM and multi-layer CNNs to learn low-dimensional feature representations of time series data.

Unsupervised Methods: For unlabeled time series data, such as the driving data of new energy vehicles, extensive data annotation is often prohibitively expensive or impracticable. This is attributable to the substantial time investment, specialized infrastructure, or domain expertise that may be requisite. Contemporary studies on unsupervised learning for multivariate time series have primarily gravitated towards the utilization of autoencoders.

Autoencoders are trained to reconstruct their input, and can be employed as Multi-Layer Perceptrons (MLP) [11,14] or Recurrent Neural Networks (RNN) [18,20] in sequence-to-sequence configurations. More recently, there has been a surge in the adoption of encoder-decoder transformer architectures [15,16,19] for univariate time series prediction, exhibiting competitive performance when juxtaposed with ARIMA, LSTM, and Seq2Seq models. The Time Series Transformer (TST) [32] extends this paradigm by incorporating a comprehensive transformer encoder-decoder structure, initially devised for generative tasks, thereby forging a versatile framework that can be adapted to an array of downstream tasks through the modification of the output layer, while also accommodating self-supervised training.

3 Proposed Approach

3.1 Framework of Ev-Transformer

Figure 3 is the framework of the proposed ev-transformer. $\mathbf{x}_t$ of each timestamp t will be linearly projected to the vector $\mathbf{z}_t$ of same dimensionality as the representation vectors of model and the input of the first attention layer, and then add a positional encoding to form the queries, keys and values.

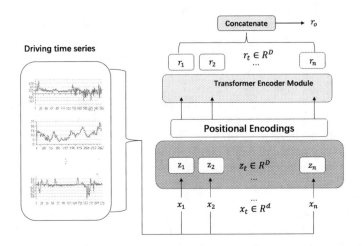

Fig. 3. Ev-transformer Framework.

Diverging from the conventional Transformer models employed in Natural Language Processing (NLP) tasks [26], the Ev-Transformer eschews the use of layer normalization subsequent to the computation of self-attention and the feed-forward segment within each encoder block, opting instead for batch normalization. This modification is predicated on the distinction that, unlike word embeddings in NLP, time series data may encompass outlier values, and layer normalization could exacerbate the influence of these outliers.

Through empirical validation, it was substantiated that, in the context of driving behavior analysis for new energy vehicles, the incorporation of batch normalization yields superior performance.

3.2 Transformer Modules

For the training sample $\mathbf{T} = \{\ \mathbf{I_{n_1}}, \mathbf{I_{n_2}}, \mathbf{I_{n_3}}, ..., \mathbf{I_{n_k}}\ \} \in \mathbf{R^{k \times n \times d}}$, each trip $\mathbf{I_{n_i}} = \{\ \mathbf{x_1}, \mathbf{x_2}, \mathbf{x_3}, ..., \mathbf{x_n}\ \} \in \mathbf{R^{n \times d}}$, is a multivariate time series of length n and d different dimensions, consisting of a sequence of feature vectors $\mathbf{x_t} \in \mathbf{R^d}$, First, the feature vectors $\mathbf{x_t}$ are normalized. For each dimension, the model subtract the average value and divide it by the variance of the training set samples, and then linearly project it to the $\mathbf{D}$ dimension vector space, where $\mathbf{D}$ is the dimension of the transformer model sequence element representations:($\mathbf{W_p}$, $\mathbf{b_p}$ are learnable parameters)

$$\mathbf{z_t} = \mathbf{W_p x_t} + \mathbf{b_p} \tag{1}$$

$\mathbf{z_t} \in \mathbf{R^D}$, $\mathbf{t} = \mathbf{0}, ..., \mathbf{n}$, are the model input vectors,which will become the queries, keys and values of the self-attention layer. Then, because transformer is a feed-forward structure and insensitive to sequence, in order to make it understand the sequence characteristics of time series, the model adds positional encodings $\mathbf{P} \in \mathbf{R^{n \times D}}$ to the input vectors: $\mathbf{Z}' = \mathbf{Z} + \mathbf{P}$. The positional encodings in this model is fully learnable, because it perform better than deterministic encodings through experiments.

3.3   Training Datasets

The datasets employed in this work are an amalgamation of data amassed from various sensors integrated within electric vehicles, encompassing the speedometer, accelerometer, manometer, GPS, and turn signal indicators.

The new energy vehicle driving dataset employed in this study comprises the following:

- **Evs-taxi**[1]: A dataset of sensor readings from new energy taxis, curated and labeled as described in Chap. 4. It encompasses approximately 20,000 data points across 12 dimensions, of which around 7,500 are automatically labeled.
- **Evs-contest**[2]: A dataset derived from a New Energy Vehicle Driving Behavior Analysis Competition, which contains labeled data for 2,000 new energy vehicles. It includes attributes such as speed, acceleration, geolocation information, battery data, and corresponding safety labels.

Dubbed EVs-taxi, this dataset captures the driving patterns of 100 new-energy taxis over a span of 12 d, culminating in an aggregate of approximately 200,000 data points sampled at a frequency of 60 Hz. The dataset is multidimensional, with in excess of 50 dimensions emanating from the various sensors. Given that only a subset of these dimensions is germane to driving behavior, we undertook a pre-processing step to eliminate attributes that bear no ostensible relevance to driving safety, such as the air conditioning mode and indoor and outdoor temperatures. The attributes retained post-filtering were employed as features for the ensuing analyses.

Given that the raw sensor data encompasses the entirety of the vehicle's journey spanning multiple days, and dangerous driving behaviors typically manifest over a duration of mere tens of seconds, it is imperative to segment the original sensor data. The initial segmentation is executed by leveraging the attribute of residual battery power from the sensor data (a single trip is demarcated by a

---

[1] The dataset will be made publicly available following the application of anonymization procedures to ensure data privacy.
[2] https://www.heywhale.com/landing/2022_XGAME.

sequence where the battery level depletes from high to low, invariably indicating a charging cycle). Subsequent to this, a threshold-based segmentation approach predicated on instantaneous speed is employed. Specifically, individual trips are further divided based on instances where the instantaneous speed transitions from 0 to a predetermined threshold and ultimately reverts to 0.

3.4 Pre-training

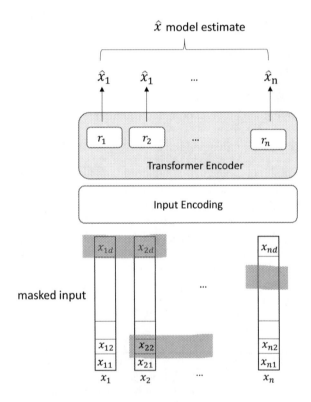

Fig. 4. Self-supervised Module Of Ev-transformer

The vehicle sensor data is of particular interest as it can be leveraged to study the relationship between driving behaviors and vehicle characteristics, thereby facilitating the development of novel approaches to enhance driving safety. Nonetheless, the sensor data emanating from new energy vehicles is not devoid of challenges. A primary concern is the high dimensionality, as onboard data collection systems such as GPS, telematics control units (T-box), and various sensors including speedometers, accelerators, pressure gauges, and signal light controllers often encapsulate in excess of 50 distinct attributes.

The self-supervised pre-training module (Fig. 4) of the model refers to the idea of MLM (masked language model) in BERT [9], specifically, an auto-regressive task of denoising the input: set part of input to 0 then ask the model to give the prediction of the masked values. For each training epoch, a binary noise mask $\mathbf{M} \in \mathbf{R}^{n \times d}$ will be created independently to mask the input by ele-mentwise multiplication: $\tilde{X} = \mathbf{M} \odot \mathbf{X}$, the proportion of each sequence with length of $\mathbf{n}$ (one of the multivariate time series) to be masked is $\mathbf{p}$, the length of each mask segment follows a geometric distribution with mean $\mathbf{l_d}$, in all the proposed experiments $\mathbf{p}$ is set to 0.15, $\mathbf{l_d}$ is set to 3 according to the results.

This masking mode encourages model learning to focus on the previous and subsequent parts in individual variables, as well as the contemporary values of other variables in the time series, so as to learn to model the interdependencies between variables. Through a linear layer on the final output vector represen-tations $\mathbf{r_t}$, the model predicts for each time step $\widehat{\mathbf{x_t}}$:($\mathbf{W_m}$, $\mathbf{b_m}$ are learnable parameters, the MSE loss only consider the predictions on the masked values)

$$\widehat{\mathbf{x_t}} = \mathbf{W_m r_t} + \mathbf{b_m} \tag{2}$$

$$\mathbf{L_{MSE}} = \frac{1}{|\mathbf{M}|} \sum \sum_{(j,i) \in M} (\widehat{\mathbf{x}}(j,i) - \mathbf{x}(j,i))^2 \tag{3}$$

4   Downstream Tasks

Ev-transformer can perform different fine-tuning on the output layer of the model to adapt to different downstream tasks, in this chapter, we will introduce three types of downstream tasks based on Ev-transformer model.

4.1   Downstream Task Tuning

The base model can be used for regression and classification with the following modification: the final representation vectors $\mathbf{r_t}$, $\mathbf{t} = \mathbf{0}, ..., \mathbf{n}$ of all time steps are then concatenated into one vector $\mathbf{r_o}$ as the input of the linear output layer:

$$\widehat{y} = \mathbf{W_o r_o} + \mathbf{b_o} \tag{4}$$

The predictions $\widehat{y}$ then be processed differently according to different down-stream tasks. In regression tasks such as speed prediction will directly calculate the mean square error $\mathbf{L} = ||\widehat{\mathbf{y}} - \mathbf{y}||^2$ where $\mathbf{y} \in \mathbf{R^d}$ are the ground truth values, while in classification tasks such as driving safety classification the predictions $\widehat{\mathbf{y}}$ will pass through a softmax layer and its cross-entropy with the ground truth labels will be the loss:

$$\mathbf{L} = -\frac{1}{batch\_size} \sum_{j=1}^{batch\_size} \sum_{i=1}^{2} \mathbf{y_{ji} \log \widehat{y_{ji}}} \tag{5}$$

where $\mathbf{y_{ji}}$ is the ground truth label of the $\mathbf{j}$th sample in the batch, $\widehat{\mathbf{y_{ji}}}$ is the prediction of the $\mathbf{j}$th sample in the batch for the $\mathbf{i}$th class.

4.2   Classification of Driving Behavior Safety

This model is first used to solve the safety assessment problem of driving behavior of new energy vehicles. Specifically, for the given group of travel trip $\mathbf{T} = \{ \mathbf{I}_{n_1}, \mathbf{I}_{n_2}, \mathbf{I}_{n_3}, ..., \mathbf{I}_{n_k} \} \in \mathbf{R}^{k \times n \times d}$, each trip should have a corresponding label $\mathbf{y}$ representing the trip is safe or dangerous. The label set is represented as $\mathbf{Y} = \{ y_1, y_2, y_3, ..., y_k \}$, the behavior safety prediction problem of new energy vehicles is converted into the two-classification problem of multivariate time series data.

Through the partially labeled data set T, a classifier can predict a certain trip belongs to a safe or dangerous segment. The model predictions $\hat{\mathbf{y}}$ will additionally obtain the distribution on the class(safe or dangerous) through the softmax function, and its cross entropy with the classification basic fact tag will be the sample loss.

4.3   Importance of Driving Behavior Features

On the basis of driving behavior safety classification, this model can also the impact of different features on classification accuracy to find the most important features that affect the safety or risk types of driving behavior of new energy automobiles.

By comparing the degree of performance degradation of the model when no characteristic combination representing certain driving behaviors is input, the impact of this driving behavior on driving safety is reflected. If the representative characteristic of this kind of driving behavior is missing, the degree of performance degradation of the model is greater, indicating that this kind of driving behavior has a greater impact on driving safety of new energy vehicles. On the other hand, if the model performance decreases less when the representative features of this kind of driving behavior are missing, the impact of this kind of driving behavior on driving safety of new energy vehicles is less.

4.4   Prediction of Driving Speed for New Energy Vehicles

Ev-transformer model can also be used for long term prediction of driving features trends, such as speed. The long-term prediction of driving speed is achieved by dividing the driving trip $\mathbf{I}_{n_i}$ into a fixed-length sequence through a fixed-length sliding window and using the model to make continuous regression predictions. In this task, the size of the sliding window is set to 60.

5   Experimental Study

In this section, we will evaluate our method on tasks we mentioned. We will introduce the evaluation settings, the details of the ev driving dataset, metrics and comparison model.

5.1   Experimental Setup

Metrics: For classification tasks, we calculate the *Accuracy, Precision* and *Recall* as evaluation measures. For regression tasks, we use the Root Mean Squared Error($RMAE$) as evaluation measure.

Paramters: Upon finalizing the hyperparameters, the model is retrained on the entirety of the training set and subsequently evaluated on the test set. The corrected Adam optimizer [17] is employed, obviating the necessity for an optimal learning rate.

Datasets: In this paper, the research content of this work is the driving safety for new energy vehicles, we collected sensor data, discussed in Sect. 3.3.

Segmentation results in trips characterized by time lengths ranging approximately between 50 and 100 units. Subsequent to this initial segmentation, it is crucial to ascertain the optimal time window based on empirical analysis, which will inform the execution of the final segmentation.

5.2   Classification Results

Table 1. Accuracy, Precision, Recall of Driving Behavior Safety Classification

Accuracy:	Evs-taxi	Chicago_risk	Evs-contest
XGBoost	0.523	0.549	0.633
MLSTM-FCN	0.547	0.589	0.630
Resnet	0.560	0.603	0.640
Rocket	0.554	0.609	**0.649**
EVT(sup. only)	**0.601**	**0.614**	0.593
Precision:	Evs-taxi	Chicago_risk	Evs-contest
XGBoost	0.554	0.603	**0.678**
MLSTM-FCN	0.557	0.599	0.640
Resnet	0.568	0.614	0.644
Rocket	0.577	**0.622**	0.656
EVT(sup. only)	**0.582**	0.602	0.582
Recall:	Evs-taxi	Chicago_risk	Evs-contest
XGBoost	0.486	0.490	0.580
MLSTM-FCN	0.492	0.570	0.617
Resnet	0.553	0.593	0.632
Rocket	0.563	0.593	**0.641**
EVT(sup. only)	**0.627**	**0.639**	0.611

Table 1 presents the classification accuracy of each method under consideration. The Evs-transformer (employing only the basic model) exhibits superlative accuracy in classifying vehicle driving safety on the EVs-taxi and Chicago_risk datasets (the latter being a benchmark dataset commonly employed for driving accident prediction). However, a noticeable decrement in classification accuracy was observed for Evs-transformer on datasets of lower dimensionality and simplicity. This suggests that the Evs-transformer model may have inherent limitations in handling low-dimensional time series data, possibly attributed to the issues with the attention mechanism in the low-dimensional representation space, compounded by the introduction of positional encoding. To mitigate this issue, incorporating a 1-D convolutional layer for feature extraction from low-dimensional input can be considered.

Furthermore, a discernible observation from the table is that, in juxtaposition with other models, the Evs-transformer's salient improvement, while sustaining commendable precision, is attributed to a substantial augmentation in recall. This suggests that the Evs-transformer is adept at concurrently maintaining a relatively stable proficiency in discerning safe driving behavior and an elevated capacity for identifying dangerous driving behavior, as evinced even in the low-dimensional EVs competition dataset, where classification performance is relatively modest, yet the model's recall remains notably high. This substantiates that the Transformer Encoder architecture is efficacious in capturing the multi-dimensional temporal features characteristic of new energy vehicles' dangerous driving behaviors. This attribute aligns with practical application requisites, where the primacy is accorded to the efficacious identification of dangerous driving behavior over the minimization of false positives in safe driving behavior classification.

Subsequent to the initial analysis, the study delved into evaluating the impact of self-supervised pre-training on the classification of driving behavior safety in new energy vehicles. The experimental results substantiate that the Evs-transformer model, incorporating self-supervised pre-training and subsequent fine-tuning, outperforms a solely supervised model in the task of new energy vehicle driving safety classification. As depicted in Fig. 5, the self-supervised approach utilizing masking and denoising strategies (with 20% of labeled samples deployed for self-supervised pre-training) exhibits a marked improvement in classification accuracy across all three datasets.

5.3   Behavior Features Analysis

The experimental results of driving behavior features are shown in Fig. 6:

The results indicate that omitting any of these three types of features leads to a significant decrement in the model's accuracy. Specifically, the absence of the *w/o turn* feature, which characterizes sharp turns, has a more pronounced impact on the model's accuracy compared to the absence of the *w/o acc* feature for sharp accelerations and decelerations. Conversely, the absence of the *w/o light* feature, which characterizes the use of lights, has a relatively minor impact. These observations align with our initial expectations.

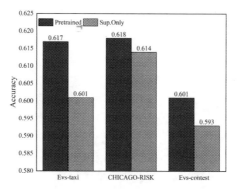

Fig. 5. Comparison of Self-supervised Pre-training Module

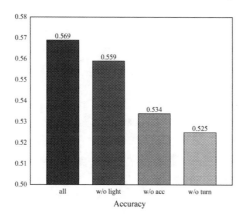

Fig. 6. The Importance of Driving Safety Features

5.4 Behavior Prediction Results

Table 2 shows the root mean square error of some model implementations. Our Evs-transformer model includes the model that uses only the basic module and that uses the same training set on self-supervises pre-training. We compare them with the current best-performing models reported in the archive. Our transformer model ranks first in both new energy vehicles driving datasets.

Subsequently, we investigated the influence of the proportion of labeled data allocated for self-supervised training on prediction outcomes. With the total number of labeled samples held constant, Fig. 7 illustrates that as a larger fraction is appropriated for self-supervised pre-training, the model's root mean square error initially undergoes a rapid decline, followed by a transient increase before diminishing again. The model's performance reaches its zenith when approximately 20% of the labeled data is utilized for self-supervised training.

Table 2. Performance(RMSE) on Driving Speed Prediction

model	dataset	
	Evs-taxi	Evs-contest
SVR	17.091	13.457
XGBoost	16.279	11.310
MLSTM-FCN	14.761	10.014
Resnet	15.213	9.232
Inception	17.781	13.231
EVT(sup. only)	10.357	9.591
EVT(pretrained)	**9.208**	**7.512**

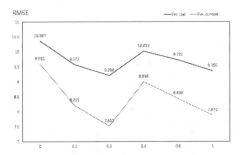

Fig. 7. Use of labeled data for unsupervised pre-training

6   Conclusion

In this study, we introduce a novel self-supervised learning model predicated on a time-series transformer architecture, tailored for the sensor data pertaining to new energy vehicles. Our model was subjected to a scrupulous evaluation process across multiple driving datasets. The findings illustrate that our modeling approach outperforms a plethora of extant monitoring models, particularly in relation to the sensor dataset of new energy vehicles. Notably, our model maintains a considerable edge even under conditions of scant training samples.

Furthermore, we furnish evidence that the unsupervised pre-training of our model, which we designate as EV-Transformer, yields significant performance benefits compared to a fully supervised learning approach. This holds true even in the absence of supplementary unlabeled data. In culmination, the proposed framework exhibits versatility as it can be efficaciously deployed for a range of downstream tasks, including but not limited to, the classification of driving behavior with respect to safety, and the long-term prognostication of trends in driving features.

Acknowledgement. This work is supported by National Natural Science Foundation of China(NSFC), 61972151. We thank the anonymous reviewers for their valuable comments and suggestions.

References

1. Liua, J., Zhong, L., Wickramasuriya, J., Vasudevan, V.: uwave: accelerometer-based personalized gesture recognition and its applications. Pervas. Mobile Comput. **5**(6), 657–675 (2009)
2. Breiman, L.: Random forests, machine learning 45. J. Clin. Microbiol. **2**, 199–228 (2001)
3. Brush, A., Krumm, J., Scott, J., Saponas, S.: Recognizing activities from mobile sensor data: challenges and opportunities (2011)
4. Chen, T., Guestrin, C.: Xgboost: A scalable tree boosting system. ACM (2016)
5. Chen, T., Hu, A., Jiang, Y.: Radio frequency fingerprint-based dsrc intelligent vehicle networking identification mechanism in high mobility environment. Sustainability **14** (2022)
6. Cortes, C., Vapnik, V.: Support-vector networks. Mach. Learn. **20**(3), 273–297 (1995)
7. Dempster, A., Petitjean, F., Webb, G.I.: Rocket: exceptionally fast and accurate time series classification using random convolutional kernels. Data Min. Knowl. Disc. **34**(5), 1454–1495 (2020)
8. Devi, U.S., Harsha, K.S.S., Rao, B.S.: Abnormal driving behaviors detection with smart phones (2018)
9. Devlin, J., Chang, M.-W., Lee, K., Toutanova, K.: Bert: pre-training of deep bidirectional transformers for language understanding. arXiv preprint arXiv:1810.04805 (2018)
10. Donahue, J., et al.: Long-term recurrent convolutional networks for visual recognition and description. Elsevier (2015)
11. Fortuin, V., Hüser, M., Locatello, F., Strathmann, H., Rätsch, G.: Deep self-organization: Interpretable discrete representation learning on time series, arXiv preprint arXiv:1806.02199 (2018)
12. Jason, L., Sarah, T., Anthony, B.: Time series classification with hive-cote. ACM Trans. Knowl. Discov. Data **12**(5), 1–35 (2018)
13. Karim, F., Majumdar, S., Darabi, H., Harford, S.: Multivariate lstm-fcns for time series classification. Neural Netw. **116** (2018)
14. Kopf, A., Fortuin, V., Somnath, V.R., Claassen, M.: Mixture-of-experts variational autoencoder for clustering and generating from similarity-based representations on single cell data. PLoS Comput. Biol. **17**(6), e1009086 (2021)
15. Li, S., et al.: Enhancing the locality and breaking the memory bottleneck of transformer on time series forecasting. In: Advances in Neural Information Processing Systems, 32 (2019)
16. Lim, B., Ark, S., Loeff, N., Pfister, T.: Temporal fusion transformers for interpretable multi-horizon time series forecasting. International J. Forecast. (1) (2021)
17. Liu, L.: On the variance of the adaptive learning rate and beyond. In: International Conference on Learning Representations (2020)
18. Lyu, X., Hueser, M., Hyland, S.L., Zerveas, G., Raetsch, G.: Improving clinical predictions through unsupervised time series representation learning. arXiv preprint arXiv:1812.00490 (2018)

19. J. Ma, Z., Shou, A., Zareian, H., Mansour, A.V., Chang S.-F.: Cdsa: cross-dimensional self-attention for multivariate, geo-tagged time series imputation. arXiv preprint arXiv:1905.09904 (2019)
20. Malhotra, P., TV, V., Vig, L., Agarwal, P., Shroff, G.: Timenet: Pre-trained deep recurrent neural network for time series classification. arXiv preprint arXiv:1706.08838 (2017)
21. Nawaz, S., Mascolo, C.: Mining users' significant driving routes with low-power sensors (2014)
22. Ordóñez, F., Roggen, D.: Deep convolutional and lstm recurrent neural networks for multimodal wearable activity recognition. Sensors **16**, 115 (2016)
23. Ouyang, Z., Niu, J., Guizani, M.: Improved vehicle steering pattern recognition by using selected sensor data. IEEE Trans. Mobile Comput. (2018)
24. Shi, X., Chen, Z., Wang, H., Yeung, D.Y., Wong, W.K., Woo, W.C.: Convolutional lstm network: a machine learning approach for precipitation nowcasting. MIT Press (2015)
25. Shifaz, A., Pelletier, C., Petitjean, F., Webb, G.I.: Ts-chief: a scalable and accurate forest algorithm for time series classification. Data Min. Knowl. Disc. **34**(3), 742–775 (2020)
26. Vaswani, A., et al.: Attention is all you need. In: Guyon, I., et al. (eds.)editors, (n: Advances in Neural Information Processing Systems, vol. 30. Curran Associates Inc (2017)
27. Vinyals, O., Toshev, A., Bengio, S., Erhan, D.: Show and tell: A neural image caption generator. IEEE (2015)
28. Yan, W., Jie, Y., Liu, H., Chen, Y., Martin, R.P.: Sensing vehicle dynamics for determining driver phone use. In: Proceeding of the 11th Annual International Conference on mobile systems, applications, and services (2013)
29. Yang, J.B., Nguyen, M.N., San, P.P., Li, X.L., Shonali, P.K.: Deep convolutional neural networks on multichannel time series for human activity recognition. In Proc, IJCAI (2015)
30. Zheng, Y., Liu, Q., Chen, E., Ge, Y., Zhao, J.L.: Time series classification using multi-channels deep convolutional neural networks. In: Li, F., Li, G., Hwang, S., Yao, B., Zhang, Z. (eds.) WAIM 2014. LNCS, vol. 8485, pp. 298–310. Springer, Cham (2014). https://doi.org/10.1007/978-3-319-08010-9_33
31. Yu, J., Chen, Y., Xu, X.: Sensing Vehicle Conditions for Detecting Driving Behaviors. SECE, Springer, Cham (2018). https://doi.org/10.1007/978-3-319-89770-7
32. Zerveas, G., Jayaraman, S., Patel, D., Bhamidipaty, A., Eickhoff, C.: A transformer-based framework for multivariate time series representation learning. In: Proceedings of the 27th ACM SIGKDD Conference on Knowledge Discovery & Data Mining, pp. 2114–2124 (2021)
33. X. Zhang, Y. Gao, J. Lin, and C. T. Lu. Tapnet: Multivariate time series classification with attentional prototypical network. pages 6845–6852, 2020

Graph Convolution Recurrent Denoising Diffusion Model for Multivariate Probabilistic Temporal Forecasting

Ruikun Li, Xuliang Li, Shiying Gao, S. T. Boris Choy, and Junbin Gao[✉]

The University of Sydney, Camperdown, NSW 2006, Australia
{ruikun.li,xuliang.li,shiying.gao,boris.choy,junbin.gao}@sydney.edu.au

Abstract. The probabilistic estimation for multivariate time series forecasting has recently become a trend in various research fields, such as traffic, climate, and finance. The multivariate time series can be treated as an interrelated system, and it is significant to assume each variable to be independent. However, most existing methods fail to simultaneously consider spatial dependencies and probabilistic temporal dynamics. To address this gap, we introduce the Graph Convolution Recurrent Denoising Diffusion model (GCRDD), a recurrent framework for spatial-temporal forecasting that captures both spatial dependencies and temporal dynamics. Specifically, GCRDD incorporates the structural dependency into a hidden state using the graph-modified gated recurrent unit and samples from the estimated data distribution at each time step by a graph conditional diffusion model. We reveal the comparative experiment performance of state-of-the-art models in two real-world road network traffic datasets to demonstrate it as the competitive probabilistic multivariate temporal forecasting framework.

Keywords: Diffusion Model · Graph Neural Network · Recurrent Neural Network · Multivariate Time-Series Forecasting

1 Introduction

The effectiveness of temporal forecasting methods has been empirically verified in various fields, including finance, healthcare, traffic, etc. [18]. However, the classical univariate methods fail to capture and explain the interactions and co-movements among a group of time series variables [19]. To address that, multivariate temporal forecasting methods have been proposed, with the essential assumption of capturing both inner (auto-correlations) and inter-temporal (temporal correlations among time series) information [11], and the exploration of the correlation among the temporal variables accordingly has become the critical element in making multivariate time series forecasting outstanding the performance of univariate methods.

R. Li and X. Li—These authors contributed equally to this work.

X. Yang et al. (Eds.): ADMA 2023, LNAI 14176, pp. 661–676, 2023.
https://doi.org/10.1007/978-3-031-46661-8_44

Thus, deep-learning-based methods have been prominent in the multivariate time series forecasting field, as auto-adjusted weights for easing effect from exogenous covariates, and significant feature extraction capability. The LSTNet [16] and TPA-LSTM [25] are examples of potent non-linear pattern capture capabilities. They both have demonstrated great performance in the deterministic (non-probabilistic) time series forecasting model. However, the empirical evidence shows that deterministic time series methods have limitations in their prediction reliability [1]. Instead of being stuck where forecast figures are expected to materialize, probabilistic forecasts assume a universe with unequal probability. Many researchers have explored the benefits of probabilistic forecasting and proposed several methods, such as the deep state space model for probabilistic prediction [20], the flexible method for probabilistic modeling with conditional quantile functions using monotonic regression splines [8], and the autoregressive diffusion model sampling from the data distribution at each time step by estimating its gradient [21]. Such approaches, however, have failed to address the pair-wise dependencies among the temporal variables explicitly, which weakens model interpretability.

The graph concept, which naturally contains correlation information among variables, has been introduced more recently to deep learning neural networks, called Graph Neural Networks (GNNs). GNNs aggregate neighborhoods' information to a node and allow it to be aware of the connections between each other by propagating messages in the graph [31]. From the GNN perspective, the multivariate time series can be naturally seen as graph format data, such as the multiple temporal processes viewed as the nodes in a graph and the latent inter-correlations between time series treated as edge links. Many recent studies have proved the significant improvement in the prediction power of considering graph structure as external information, taking both temporal and topology information as input features to conduct forecasting tasks for future values [10,17,24,32]. However, the spatial-temporal graph neural networks still fall short of forecasting tasks due to the following challenges:

- *Challenge 1: Probabilistic Forecasting.* Most temporal prediction studies in GNNs have focused on forecasting value rather than the probability of the expected value range. In most cases, multivariate time series forecasts are commonly applied for real-world decision-making pipelines, and it is beneficial to provide uncertainty estimation for forecasting.
- *Challenge 2: Utilizing Graph Structure Information.* Most of the published probabilistic temporal forecasting studies are limited to considering pair-wise dependencies among the time features. The question is how to combine the graph information and the probabilistic model for time series in an end-to-end framework.

Therefore, we propose an innovative end-to-end forecasting framework, called **Graph Convolution Recurrent Denoising Diffusion Model (GCRDD)**, combining the characteristics of probabilistic diffusion model and recurrent graph neural networks to address these challenges. As shown in Fig. 1, our proposed framework consists of two core components, the graph recurrent module, and the graph diffusion module. To address *Challenge 1*, we apply the recurrent graph module to generate a hidden state feature, which incorporates input graph structure information. Furthermore, given the generated graph-embedded hidden states and temporal features, we modify the classic diffusion model as the graph diffusion module to solve the multivariate probabilistic time series forecasting problem without explicitly making any specific probabilistic assumption. The graph diffusion module is designed not only to train a model with all the inductive biases of probabilistic temporal prediction but also to consider the graph information among time series in the framework we designed (for *Challenge 2*).

The advantage is that our proposed framework generally extends the probabilistic diffusion method to spatial-temporal forecasting with predefined topology information. In summary, our main contributions are as follows:

- Limited studies consider the structure information in the probabilistic multivariate time series forecasting model [4]. To the best of our knowledge, this is one of the pioneer applications of the temporal diffusion model with graph knowledge.
- We propose a modified graph diffusion module that was refined from the TimeGrad [21] to capture hidden spatial dependencies among temporal features. Our method exploits a new field for diffusion models to handle spatial-temporal data.
- Comparing our method with state-of-the-art methods on two spatial-temporal traffic datasets demonstrates its competitive performance.

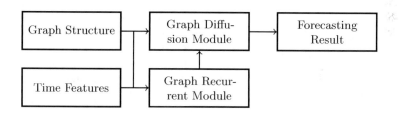

Fig. 1. A concept map of our end-to-end framework.

2   Preliminary

2.1   Problem Formulation

This paper considers the probabilistic spatial-temporal forecasting tasks. In general, spatial time series data consist of a sequence of graphs $\{\mathcal{G}\}_{t=1}^{T} = \{(\mathcal{V}_t, \mathcal{E}_t)\}_{t=1}^{T}$ in which we will assume the graph structure is invariant along the time, i.e., $\mathcal{E} = \mathcal{E}_t$ and the node features $\mathbf{X}_t \in \mathbb{R}^{n \times d}$, where n is the number of temporal variables (nodes) and d represents the dimension of temporal data, will change. Hence we apply a pre-defined static adjacency matrix for spatial time series $\mathbf{A}$. The adjacency matrix is the discrete data that stores the relationships among the network variables, and we denote the time series with graphs as the equivalent form $\{(\mathbf{X}_t, \mathbf{A})\}_{t=1}^T$ or even simply $\{\mathbf{X}_t\}$ without confusion in the context.

To conduct the probabilistic spatial-temporal forecasting task, we aim to estimate a multivariate distribution within a time range from a specific time stamp to the future. We will focus on the basic auto-regressive strategy. At a given time stamp t, we will use the past information back to a length l (called context length) to predict τ steps ahead, that is, to estimate a distribution $p(\mathbf{X}_{t+1}, ..., \mathbf{X}_{t+\tau} \mid \mathbf{X}_t, \mathbf{X}_{t-1}, ..., \mathbf{X}_{t-l+1})$. This setting resembles the seq-to-seq models in the Natural Language Processing (NLP) area [28]. In other words, the time series data will be split into many pairs of the context window sequence on time $[t, t-l+1]$ and the prediction window sequence $[t+1, t+\tau]$. In this paper, we will build models starting at $\tau = 1$.

The main difference between the concept of spatial time series considered in this paper and the general multivariate time series is that we will exploit the connection information among the multivariate temporal variables. In contrast, all the temporal variables inside a general method are naturally assumed independent of each other. Hence, each of the temporal variables of a spatial multivariate time series can be treated as nodes/vertices in a graph, and the edge connections (intercorrelations) between nodes can be represented by a pre-defined adjacency matrix $\mathbf{A}$, either time-invariant (dynamic) or -variant (static).

2.2   Convolution on Graphs

For our purpose, we give a brief review on the widely applied graph convolution.

Given a graph $(\mathbf{X}, \mathbf{A})$ with the node feature $\mathbf{X} \in \mathbb{R}^{n \times d}$ and the adjacency matrix $\mathbf{A} \in \mathbb{R}^{n \times n}$, where n is the number of nodes and d is the feature dimension, then the graph convolution $*_\mathbf{A}$ with a diagonal kernel function $\Theta = \mathrm{diag}(\theta) \in \mathbb{R}^{n \times n}$ is defined as:

$$\Theta *_\mathbf{A} \mathbf{X} := \mathbf{U} \Theta(\Lambda) \mathbf{U}^\mathsf{T} \mathbf{X} \tag{1}$$

where the graph Fourier basis $\mathbf{U} \in \mathbb{R}^{n \times n}$ and spectra Λ (diagonal) come from the eigen-decomposition of the normalized graph Laplacian $\mathbf{L} = \mathbf{I} - \mathbf{D}^{-1/2}\mathbf{A}\mathbf{D}^{-1/2} = \mathbf{U}\Lambda\mathbf{U}^\mathsf{T} \in \mathbb{R}^{n \times n}$. Here $\mathbf{I} \in \mathbb{R}^{n \times n}$ is the identity matrix, $\mathbf{D} \in \mathbb{R}^{n \times n}$ is the diagonal degree matrix with $\mathbf{D}_{ii} = \sum_{j=1}^n \mathbf{A}_{ij}$ [5].

By definition, the graph convolution operation is passing the signal $\mathbf{X}$ to the filter Θ by a multiplication between Θ and graph Fourier transformation $\mathbf{U}^\mathsf{T}\mathbf{X}$ [26]. However, the computation complexity of (1) is $\mathcal{O}(n^3)$ due to the eigen-decomposition, which is expensive for evaluation. To circumvent this problem and reduce the number of parameters, the kernel Θ can be parameterized by a truncated expansion of Chebyshev polynomial $\mathbf{T}_k$ [7]:

$$\Theta(\Lambda) \approx \sum_{k}^{K-1} \theta_k \mathbf{T}_k(\widetilde{\Lambda})$$

where the $\widetilde{\Lambda}$ is the re-scaled eigenvalue matrix, which $\widetilde{\Lambda} = 2\Lambda/\lambda_{\max} - \mathbf{I}$; K represents the kernel size of graph convolution. Then the graph convolution can be rewritten as:

$$\Theta *_{\mathbf{A}} \mathbf{X} = \mathbf{U}\,\Theta(\Lambda)\mathbf{U}^{\mathsf{T}}\mathbf{X} \approx \sum_{k}^{K-1} \theta_k \mathbf{T}_k(\widetilde{\mathbf{L}})\mathbf{X} \tag{2}$$

where $\mathbf{T}_k(\widetilde{\mathbf{L}}) \in \mathbb{R}^{n \times n}$ is the order k Chebyshev polynomial with re-scaled Laplacian $\widetilde{\mathbf{L}} = 2\mathbf{L}/\lambda_{max} - \mathbf{I}$. Helped by Chebyshev polynomial approximation, the computation complexity can be reduced to linear $\mathcal{O}(\mathcal{E})$.

2.3 Diffusion Probabilistic Model

Recently, the so-called diffusion model has attracted much popularity [12,29, 33].This idea has been recently adopted in the so-called TimeGrad models for multivariate time series forecasting [21]. Our research further generalizes the TimeGrad framework to spatial time series forecasting. For our purpose, we summarize the key points of the diffusion model here.

Consider a given data $\mathbf{x}^0$ with diffusion steps s on the superscript, which follows the true data distribution $q_{\text{data}}(\mathbf{x}^0)$ over some data input domain $\mathcal{X} \subset \mathbb{R}^D$, which is to be learned from a set of given training data.

Let $p_\theta(\mathbf{x}^0)$ denote the probability density function which aims to approximate $q_{\text{data}}(\mathbf{x}^0)$ and allows for easy sampling. The point here is that $p_\theta(\mathbf{x}^0)$ is the learning target, not its analytical expression but an easier sampling process.

A diffusion model now assumes a generative model with latent variables $\mathbf{x}^1, \ldots, \mathbf{x}^S \in \mathbb{R}^D$ that hold the form $p_\theta(\mathbf{x}^0) = \int p_\theta(\mathbf{x}^{0:S})\, d\mathbf{x}^{1:S}$ [12]. However, the generative process is unknown and to be learned. The approximate posterior $q(\mathbf{x}^{1:S}|\mathbf{x}^0)$

$$q(\mathbf{x}^{1:S}|\mathbf{x}^0) = \prod_{s=1}^{S} q(\mathbf{x}^s|\mathbf{x}^{s-1})$$

follows a form of Markov chain. This process is generated by adding Gaussian noise into $\mathbf{x}^{s-1}$ gradually, and a name *forward process* or *diffusion process* thence to be called in this stage. The *forward process* follows the form:

$$q(\mathbf{x}^s|\mathbf{x}^{s-1}) \sim \mathcal{N}\left(\mathbf{x}^s; \sqrt{1-\beta_s}\mathbf{x}^{s-1}, \beta_s\mathbf{I}\right) \tag{3}$$

The *forward process* or *diffusion process* applies various variances $\beta_1, \ldots, \beta_S$ with $\beta_s \in (0,1)$. Then a *reverse process*, which is the joint distribution $p_\theta(\mathbf{x}^{0:S})$, is a Markov chain with learned Gaussian transition starting from $p\left(\mathbf{x}^S\right) \sim \mathcal{N}\left(\mathbf{x}^S; 0, \mathbf{I}\right)$ [12]. Then the subsequent transition of:

$$p_\theta\left(\mathbf{x}^{0:S}\right) = p\left(\mathbf{x}^S\right) \prod_{s=1}^{S} p_\theta\left(\mathbf{x}^{s-1}|\mathbf{x}^s\right)$$

is given by the parameterization of denotation:

$$p_\theta\left(\mathbf{x}^{s-1}|\mathbf{x}^s\right) \sim \mathcal{N}\left(\mathbf{x}^{s-1}; \mu_\theta(\mathbf{x}^s, s), \Sigma_\theta(\mathbf{x}^s, s)\mathbf{I}\right) \tag{4}$$

with parameter θ. Both $\mu_\theta : \mathbb{R}^D \times \mathbb{N} \to \mathbb{R}^D$ and $\Sigma_\theta : \mathbb{R}^D \times \mathbb{N} \to \mathbb{R}^+$ are two networks for the Gaussian mean and variance. The purpose of $p_\theta\left(\mathbf{x}^{s-1}|\mathbf{x}^s\right)$ is to extract the Gaussian noise that was added in the $\mathbf{x}^s$. The goal of learned parameters θ is to fit the data distribution $q_{\text{data}}\left(\mathbf{x}^0\right)$ by minimizing the negative log-likelihood.

Define $\alpha_s := 1 - \beta_s$ and the cumulative product of α_s as $\overline{\alpha}_s := \prod_{i=1}^{s} \alpha_i$. From (3), it is easy to see [12]:

$$q(\mathbf{x}^{s-1}|\mathbf{x}^s, \mathbf{x}^0) = \mathcal{N}\left(\mathbf{x}^{s-1}; \widetilde{\mu}_s(\mathbf{x}^s, \mathbf{x}^0), \widetilde{\beta}_s\mathbf{I}\right) \tag{5}$$

where

$$\widetilde{\mu}_s(\mathbf{x}^s, \mathbf{x}^0) = (\mathbf{x}^s - \beta_s\epsilon/\sqrt{1 - \overline{\alpha}_s})/\sqrt{\alpha_s}.$$

and $\widetilde{\beta}_s = ((1 - \overline{\alpha}_{s-1})/(1 - \overline{\alpha}_s))\beta_s$. To make (4) to (5), we can take

$$\mu_\theta(\mathbf{x}^s, \mathbf{x}^0) = (\mathbf{x}^s - \beta_s\widehat{\epsilon}_\theta(\mathbf{x}^s, s)/\sqrt{1 - \overline{\alpha}_s})/\sqrt{\alpha_s}. \tag{6}$$

and $\Sigma_\theta(\mathbf{x}^s, \mathbf{x}^0) = \widetilde{\beta}_s$, where $\widehat{\epsilon}_\theta(\mathbf{x}^s, s)$ is a network to approximate the noise ϵ at diffusion step s.

It turns out that the negative log-likelihood can be bounded by the following objective

$$\sum_{s=2}^{S} D_{\text{KL}}\left(q(\mathbf{x}^{s-1}|\mathbf{x}^s, \mathbf{x}^0)||p_\theta(\mathbf{x}^{s-1}|\mathbf{x}^s)\right)$$
$$- \log p_\theta(\mathbf{x}^0|\mathbf{x}^1) + D_{\text{KL}}\left(q(\mathbf{x}^S|\mathbf{x}^0)||p(\mathbf{x}^S)\right) \tag{7}$$

Given the model assumption (4) to (5), the main components in the objective can be down to the following term:

$$\mathbb{E}_{\mathbf{x}^s, \epsilon}\left[\frac{\beta_s^2}{2\Sigma_\theta\alpha_s(1 - \overline{\alpha}_s)}||\epsilon - \widehat{\epsilon}_\theta(\mathbf{x}^s, s)||^2\right] \tag{8}$$

which resembles the loss in the Noise Conditional Score Networks [27] by using the score matching. Notes that $\mathbf{x}^s = \sqrt{\overline{\alpha}_s}\mathbf{x}^0 + \sqrt{1 - \overline{\alpha}_s}\epsilon$. Once trained, we can compute the following to sample from the reverse $\mathbf{x}^{s-1} \sim p_\theta(\mathbf{x}^{s-1}|\mathbf{x}^s)$ (4)

$$\mathbf{x}^{s-1} = \frac{1}{\sqrt{\alpha_s}}\left(\mathbf{x}^s - \frac{\beta_s}{\sqrt{1 - \overline{\alpha}_s}}\widehat{\epsilon}_\theta(\mathbf{x}^s, s)\right) + \sqrt{\Sigma_\theta}\mathbf{z} \tag{9}$$

where $\mathbf{z} \sim \mathcal{N}(\mathbf{0}, \mathbf{I})$ for $s = S, \ldots, 2$ and $\mathbf{z} = \mathbf{0}$ when $s = 1$.

The *reverse process* is the sampling from the most perturbed distribution x, which is the Gaussian noise, then reduce the noises scale and make the sampling approaching to target distribution $\mathbf{x}^0$, starting from white noise sample $\mathbf{x}^s$.

3   Framework

3.1   Model Architecture

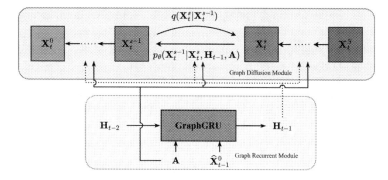

Fig. 2. The GCRDD architecture of the Graph-based Gated Recurrent Unit conditioned diffusion probabilistic model at time $t - 1$

We first introduce the general architecture of our model. As shown in Fig. 2, the GCRDD framework has a bottom-up structure consisting of a graph recurrent module and a diffusion module. The recurrent graph module applies the graph-based gated recurrent unit (GRU) to extract inter-correlated temporal patterns from multivariate time series data. The graph-based GRU is modified by replacing the weight matrix multiplication with the graph convolutional kernel, which captures the spatial dependencies among the time series by neighbor's information aggregation. As our model aims to learn the conditional distribution of future time, we employ a conditional denoising diffusion model as the diffusion module's main body to sample the prediction result. We illustrate the core components of our model GCRDD in the following manner:

3.2   Graph Recurrent Module

The graph recurrent module is designed to adaptively learn the updated hidden state to capture the temporal dynamics of multivariate time series. To achieve this target, we use the autoregressive recurrent neural network (RNN) frameworks [9,28]. RNN frameworks can effectively extract autoregressive features but treat each series independently. Thus, we leverage the graph convolution operations, mentioned in section [2.2], on recurrent architecture to model the temporal dependency. In particular, we use the graph-modified Gated Recurrent Units (GraphGRU). Let $\widehat{\mathbf{X}}_{t-1}^0 = [\mathbf{X}_{t-1}^0 \parallel \mathbf{C}_{t-1}]$ denote the coupled temporal graph signal, where $\mathbf{C}_{t-1}$ is the covariates at time point $t-1$. The covariates $\mathbf{C}_{t-1}$ applied in this model compose two kinds of embedding features, the time-dependent (e.g. date of week, week of year, etc.) and time-independent features. The updated graph hidden state $\mathbf{H}_{t-1}$ can be represented by:

$$\mathbf{H}_{t-1} = \text{GraphGRU}_\theta(\widehat{\mathbf{X}}_{t-1}^0, \mathbf{H}_{t-2}, \mathbf{A}) \tag{10}$$

where GraphGRU is a multi-layers recurrences with shared parameters θ, and $\mathbf{H}_0 = 0$. The graph hidden state $\mathbf{H}_{t-1}$ will accumulate all the past information of the graph time series. The graph-modified GRU takes inputs to update hidden states and accepts adjacency matrix into graph convolution to replace the operation of weight matrix multiplication in classic GRU [6]. Some existing works share similar GRU architectures but with various graph convolution designs, such as T-GCN [34] with GCNConv [14] and DCRNN [17] with the diffusion convolution, etc. To ensure the simplicity of our model, we apply the Graph Chebyshev Convolution GRU defined in GCRN [23]:

$$\mathbf{R}_{t-1} = \text{sigmoid}\left(\Theta_{\mathbf{R}} *_{\mathbf{A}} \left[\widehat{\mathbf{X}}_{t-1}^0 \parallel \mathbf{H}_{t-2}\right]\right)$$

$$\mathbf{U}_{t-1} = \text{sigmoid}\left(\Theta_{\mathbf{U}} *_{\mathbf{A}} \left[\widehat{\mathbf{X}}_{t-1}^0 \parallel \mathbf{H}_{t-2}\right]\right)$$

$$\widetilde{\mathbf{H}} = \tanh\left(\Theta_{\widetilde{\mathbf{H}}} *_{\mathbf{A}} \left[\widehat{\mathbf{X}}_{t-1}^0 \parallel (\mathbf{R}_{t-1} \odot \mathbf{H}_{t-1})\right]\right)$$

$$\mathbf{H}_{t-1} = \mathbf{U}_{t-1} \odot \mathbf{H}_{t-2} + (1 - \mathbf{U}_{t-1}) \odot \widetilde{\mathbf{H}}$$

where $\mathbf{R}_{t-1}$, $\mathbf{U}_{t-1}$, and $\widetilde{\mathbf{H}}$ denote the reset gate, update gate, and candidate activation at time point $t-1$, respectively. $*_{\mathbf{A}}$ is the Graph Chebyshev Convolution defined in (2) and $\Theta_{\mathbf{R}}, \Theta_{\mathbf{U}}, \Theta_{\widetilde{\mathbf{H}}}$ are parameters for the various kernels.

3.3   Diffusion Module

In the original diffusion model, data distribution is learned through a Markov chain $p_\theta(\mathbf{x}^{s-1}|\mathbf{x}^s)$. One strategy to learn this process is to train a noising model $\widehat{\epsilon}_\theta$ to learn the diffusion noise along the *forward process*. This network is shared by all the independent data and their latent data in the training objective. In the case of graph time series forecasting, we need to take into account the time series effect. The Graph Recurrent Module helps carry the intercorrelated temporal information through the recurrent latent variables $\mathbf{H}_{t-1}$, which should be used for the diffusion process.

At a high level, the noise approximation network is defined as:

$$\widehat{\epsilon}_\theta^s(t) = \widehat{\epsilon}_\theta(\mathbf{X}_t^s, \mathbf{H}_{t-1}, \mathbf{A}, s). \tag{11}$$

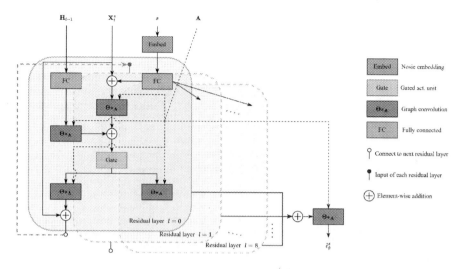

Fig. 3. The $\widehat{\epsilon}_\theta^s(t)$ network architecture

That is, the module $\widehat{\epsilon}_\theta$ will be trained to approximate the diffusion (standard Gaussian) noise $\epsilon_t^s \sim \mathcal{N}(0,\mathbf{I})$ added at time t and diffusion step s. We first calculate $\mathbf{X}_t^s$ as:

$$\mathbf{X}_t^s = \sqrt{\overline{\alpha}_s}\mathbf{X}_t^0 + \sqrt{1-\overline{\alpha}_s}\epsilon_t^s.$$

Then we have the conditional objective variant for the time step t, as:

$$\mathbb{E}_{\mathbf{X}_t^0,\epsilon_t^s,s}\left[\|\epsilon_t^s - \widehat{\epsilon}_\theta\left(\sqrt{\overline{\alpha}_s}\mathbf{X}_t^0 + \sqrt{1-\overline{\alpha}_s}\epsilon_t^s,\mathbf{H}_{t-1},\mathbf{A},s\right)\|^2\right]$$

Using this objective, we can train the network $\widehat{\epsilon}_\theta$ to approach ϵ_t^s by using the procedure in Algorithm 1.

Then we develop the network $\widehat{\epsilon}_\theta(t)$ based on a graph convolution architecture structure (see Fig. 3 for an illustration) that is diverse from DiffWave architecture [15]. In our network architecture, it consists of $l = 0, 1, ..., L - 1$ residual blocks with several components as:

$$\begin{aligned}
\mathbf{Y}_l &= \Theta_{\mathbf{Y}} *_{\mathbf{A}} \left(\mathbf{X}_{t,l}^s + \text{Embed}(s)\right) \\
\mathbf{G}_l &= \text{Gate}\left(\mathbf{Y}_l + \Theta_{\mathbf{H}} *_{\mathbf{A}} \left(\mathbf{H}_{t-1}\right)\right) \\
\mathbf{X}_{t,l+1}^s &= \mathbf{X}_{t,l}^s + \Theta_{\mathbf{G}} *_{\mathbf{A}} \left(\mathbf{G}_l\right) \\
\widehat{\epsilon}_\theta^s(t) &= \Theta_{\mathbf{X}} *_{\mathbf{A}} \left(\mathbf{X}_{t,0}^s + \mathbf{X}_{t,1}^s + \cdots + \mathbf{X}_{t,L-1}^s\right)
\end{aligned} \qquad (12)$$

where $\text{Embed}(\cdot)$ is the embedding of the diffusion step s; $\Theta *_{\mathbf{A}}$ presents an appropriate graph convolution layer like Chebyshev graph layer with different parameters Θs; Gate is defined as $\text{sigmoid}(\cdot) \odot \tanh(\cdot)$ [2]. The initialization value of the $\mathbf{X}_{t,l=0}^s$ is $\mathbf{X}_t^s$.

After the training phase, we estimate values in future steps of all temporal variables and test them with ground truth. In this stage, we apply the sampling procedure in Algorithm 2 to get the next step prediction value $\mathbf{X}_t^0$ and pass the one-step forecasting value into the GraphGRU to get the hidden state $\mathbf{H}_t$ one step ahead. Then we repeat this process until we reach the desired forecasting horizon.

Algorithm 1 Training at time step t

1: **Input:** $\mathbf{H}_{t-1}$ and $\mathbf{A}$
2: **repeat:**
3:   Sample $\mathbf{X}_t^0 \sim q_{\text{data}}$, $\epsilon_t^s \sim \mathcal{N}(0, \mathbf{I})$, and
     $s \sim \text{Uniform}(\{1, ..., S\})$
4:   Take gradient step on
     $\nabla_\theta \| \epsilon_t^s - \widehat{\epsilon}_\theta \left(\sqrt{\overline{\alpha}_s} \mathbf{X}_t^0 + \sqrt{1 - \overline{\alpha}_s} \epsilon_t^s, \mathbf{H}_{t-1}, \mathbf{A}, s \right) \|^2$
5: **until** converged

---

4   Experiments

To evaluate the performance of our proposed framework on multivariate temporal forecasting, we have conducted substantial experiments with seven classic multivariate time series forecasting models on two real-world spatial-temporal datasets on traffic volume. These experiments and studies are expected to answer the following questions:

Algorithm 2 Sampling

1: **Input:** $\mathbf{H}_{t-1}$ and $\mathbf{A}$
2: Sample $\mathbf{X}_t^S \sim \mathcal{N}(0, \mathbf{I})$
3: **for** $s = S, S - 1, ...1$ **do**
4:   **if** $s > 1$ **then**
5:     $\mathbf{Z} \sim \mathcal{N}(0, 1)$
6:   **else**
7:     $\mathbf{Z} = 0$
8:   **end if**
9:   $\mathbf{X}_t^{s-1} = \frac{1}{\sqrt{\alpha_s}} (\mathbf{X}_t^s - \frac{\beta_s}{\sqrt{1 - \overline{\alpha}_s}} \widehat{\epsilon}_\theta(\mathbf{X}_t^s, \mathbf{H}_{t-1}, \mathbf{A}, s)) + \sqrt{\Sigma_\theta} \mathbf{Z}$
10: **end for**
11: **Return:** $\mathbf{X}_t^0$

- **Q1:** Is GCRDD outperforming the classic benchmark approaches regarding multivariate temporal forecasting based on the results from two spatial-temporal datasets?
- **Q2:** Are hyper-parameter spaces of our proposed model performing a difference on two evaluation datasets?

4.1   Experimental Setup

Dataset. To keep fairness, we apply the two most famous real-world spatial-temporal traffic forecasting datasets introduced in work [17], to evaluate the performance of our proposed framework. We keep the same pre-process for each dataset as the setting in [32]. The description of details of the datasets is shown as followings:

- **METR-LA** is a traffic information dataset recorded by 207 loop detectors on Los Angeles County highway from March 1st, 2012, to June 30th, 2012 in 5 min intervals. The total number of observed traffic data points is 6,519,002.
- **PEMS-BAY** is another traffic information dataset recorded by the Performance Measurement System (PeMS) of California Transportation Departments. This dataset contains 325 road sensor records in the Bay Area for six months of data ranging from Jan 1st, 2017, to May 31st, 2017. The total number of observed traffic data points is 16,937,179.

Benchmarks. We compare seven classic multivariate time series forecasting approaches to evaluate our proposed framework performance. The descriptions of benchmarks show as followings:

- **DCRNN*** [17]: A well-known recurrent graph neural network framework that applies bi-directional diffusion convolution on GRU with pre-defined graph information.
- **GCGRU*** [23]: GCGRU is one of the methods under the framework GCRN brought by [23], which applies the pre-defined graph to the classic recurrent neural network structure with graph convolution.
- **TGCN*** [34]: The temporal graph convolutional network model, which combines GCN and the gated recurrent unit, is a novel neural network model for traffic forecasting.
- **A3TGCN*** [3]: A3TGCN is an extension of the TGCN method that uses attention. The spatial aggregation uses GCN, and the temporal aggregation uses GRU. We can pass in the periods to get an embedding for several time-steps, which can be used to predict several steps into the future as the output dimension.
- **GWN** [30]: GWN is developed to capture long-range temporal sequences. It is achieved by an adaptive dependency matrix learned through node embedding. Learned in an end-to-end manner, and with a stacked dilated 1D convolution component, this method captures the hidden spatial dependency.
- **MTGNN*** [32]: A method is developed to specifically analyze multivariate time series data. With a mix-hop propagation layer and a dilated inception layer, MTGNN can automatically extract the uni-directed relationships in the variables.

"*" indicates the model is re-implemented by *PyTorch Geometric Temporal*[1] frameworks [22] in experiment.

---

[1] https://pytorch-geometric-temporal.readthedocs.io/en/latest/index.html.

Evaluation Metrics. We follow the original test metrics setting from [17], which considers three commonly applied metrics in time series forecasting, including (1) Mean Absolute Error (MAE); (2) Mean Absolute Percentage Error (MAPE); and (3) Root Mean Squared Error (RMSE).

Evaluation Method. In the evaluation, we use the data splitting method similar to DCRNN, with which the whole dataset is separated into three parts as training data 70%, validation data 10%, and test data 20%, but without shuffle and normalization process. However, as the graph diffusion model in our framework has a sampling process, it is extremely time consuming when evaluating our model on whole test data. We thus, decide to apply an alternative evaluation method, testing model on ten windows from test data, which is slightly different from the DCRNN setting. To make it a fair evaluation, we randomly pick ten test windows locations without overlap on our test dataset and apply all test metrics for the different models. The aim of this evaluation method is to ensure that all models are tested for different temporal dynamic patterns with limited numbers of test windows, rather than focusing on those overlapped smooth time range. In the end, we report the average of ten test results from different test windows.

4.2   Forecasting Performance (Q1)

Table 1 lists the experiment results over 10 windows, which shows that the GCRDD model sets the new state-of-the-art performance on most of the benchmarks in these two datasets. For short horizon $H = 3$, the proposed model shows a significant out-performance over the other comparison models in these two datasets. As the prediction length increases, we need to increase the number of samples, which leads to the accumulation of errors. Forecast results become more volatile for later data. At the same time, the prediction length also increases the parameters of the RNN, which may have some influence on the experimental results. For $H = 6$, in the PEMS-BAY dataset, the proposed model GCRDD performs best regarding all three evaluation methods. In the METR-LA dataset, the proposed model performs best in the MAE and RMSE. For $H = 12$, the MTGNN model's performance is better than the proposed one in the PEMS-BAY dataset. However, the proposed method GCRDD outperforms MAE and MAPE evaluations in dataset METR-LA. We find that in the 10 windows sampling test dataset, the proposed model stably outperforms in 3 horizons and 6 horizons.

Table 1. Baseline comparison of models on two traffic spatial-temporal datasets.

	Metric	DCRNN	GCGRU	TGCN	A3TGCN	GWN	MTGNN	GCRDD
METR-LA								
3 horizons	MAE	2.91	2.70	2.73	2.75	2.71	2.68	**2.66**
	RMSE	5.32	4.69	4.76	4.85	4.36	4.17	**4.11**
	MAPE	8.7%	8.1%	8.3%	8.6%	8.0%	7.8%	**7.6%**
6 horizons	MAE	3.98	3.94	3.82	3.75	3.85	3.82	**3.73**
	RMSE	6.71	7.24	6.76	6.85	6.67	6.54	**6.07**
	MAPE	11.6%	10.4%	**10.1%**	10.5%	10.9%	10.9%	10.7%
12 horizons	MAE	10.1	10.2	9.99	9.81	9.74	9.69	**9.66**
	RMSE	**13.34**	13.42	14.53	14.56	14.17	13.98	13.9
	MAPE	27.9%	28.3%	26.6%	26.2%	26.1%	25.6%	**25.1%**
PEMS-BAY								
3 horizons	MAE	1.92	1.87	2.38	2.11	1.67	1.53	**1.51**
	RMSE	2.74	2.91	3.44	2.89	2.69	2.57	**2.52**
	MAPE	3.4%	3.2%	4.0%	3.2%	3.0%	2.5%	**2.3%**
6 horizons	MAE	2.87	2.77	2.89	2.62	2.56	2.50	**2.48**
	RMSE	4.71	4.62	4.88	4.56	4.55	4.53	**4.45**
	MAPE	4.9%	4.9%	5.1%	4.7%	4.6%	4.4%	**4.2%**
12 horizons	MAE	4.89	4.73	5.37	5.24	4.73	**4.71**	4.91
	RMSE	8.25	7.62	8.42	8.20	7.32	**6.97**	7.48
	MAPE	12.5%	10.4%	11.4%	10.9%	10.5%	**9.8%**	10.1%

4.3   Model Training Setting (Q2)

To achieve the best performance of our proposed model, we apply the Adam [13] stochastic optimizer to train our graph recurrent diffusion model. After conducting extensive experiments, we have found that our model is sensitive to learning rates and epochs on two datasets. Based on the experiment result, we set the learning rate of our model to 5×10^{-4} on PEMS-BAY and, surprisingly, lock the training epochs number to 2. A subsequent investigation into this abnormal training setting indicates that the PEMS-BAYS dataset has a more smooth dynamic pattern on temporal features. In other words, the PEMS-BAYS data features are easier to capture compared with METR-LA data, and the smaller volume of training settings on the PEMS-BAYS dataset protects our model against the overfitting problem. For the METR-LA dataset, the data features are hard to extract as the Los Angeles metropolitan area has a more complex road situation. We, thus, increase the learning rate to 1×10^{-3} and epochs number to 20. In the training process, we find our model best performed on a linear diffusion schedule, in which the variance increase from $\beta_1 = 1 \times 10^{-4}$ to $\beta_N = 0.1$ in 100 diffusion steps. The batches size of the input is set to 16, which is constructed by taking random windows.

In the evaluation, while the large sampling number results in high precision forecasting results, the model suffers from the computation complexity problem when we set the sampling number over 200. To balance computation cost and efficiency, we set the sampling number to 150 on our model in the evaluation process.

5   Conclusion and Future Work

In this work, we have presented a graph probabilistic multivariate time series forecasting framework GCRDD, which fuses the graph information into a recurrent diffusion architecture to learn and sample from the estimated distribution of temporal variables. In addition, we have conducted extensive experiments to assess estimators regarding MAE, RMSE, and MAPE metrics, and the result demonstrates competitive performance in the spatial-temporal multivariate time series forecasting task.

In designing, we assume an input of static graph information. While the geographical location of road sensors in traffic datasets, like MTER-LA and PEMS-BAY, have limited change, the static graph information in traffic time series can only partially reflect the road connection. The static graph cannot include dynamic information, like temporary road service and reversible road schedules. A possible strategy to improve prediction performance in the future is considering the time-variate graph information in our forecasting framework to make it able to capture the dynamic structural changes.

For some datasets, the temporal features may not explicitly reflect the graph information. Therefore, we could build the adaptive graph learning module to extract structural information dynamically from multivariate time series. As a graph could be treated as a random variable parameterized by a matrix Bernoulli distribution, future work could apply a secondary diffusion module for the graph learning process.

References

1. Afifi, H., Elmahdy, M., El Saban, M., Abu-Elkheir, M.: Probabilistic time series forecasting for unconventional oil and gas producing wells. In: The 2nd Novel Intelligent and Leading Emerging Sciences Conference, pp. 450–455 (2020)
2. an den Oord, A., Kalchbrenner, N., Espeholt, L., kavukcuoglu, K., Vinyals, O., Graves, A.: Conditional image generation with pixel CNN decoders. In: Advances in Neural Information Processing Systems. vol. 29, pp. 4790–4798 (2016)
3. Bai, J., et al.: A3t-gcn: attention temporal graph convolutional network for traffic forecasting. ISPRS Int. J. Geo Inf. **10**(7), 485 (2021)
4. Chen, H., Rossi, R.A., Mahadik, K., Kim, S., Eldardiry, H.: Graph deep factors for probabilistic time-series forecasting. ACM Trans. Knowl. Discov. Data (2022)
5. Chung, F.R.: Spectral Graph Theory, vol. 92. American Mathematical Soc. (1997)
6. Chung, J., Gulcehre, C., Cho, K., Bengio, Y.: Empirical evaluation of gated recurrent neural networks on sequence modeling. In: Neural Information Processing Systems 2014 Workshop on Deep Learning (2014)

7. Defferrard, M., Bresson, X., Vandergheynst, P.: Convolutional neural networks on graphs with fast localized spectral filtering. In: Advances in Neural Information Processing Systems, vol. 29 (2016)
8. Gasthaus, J., et al.: Probabilistic forecasting with spline quantile function RNNs. In: The 22nd International Conference on Artificial Intelligence and Statistics, pp. 1901–1910 (2019)
9. Graves, A.: Generating sequences with recurrent neural networks. arXiv preprint arXiv:1308.0850 (2013)
10. He, H., Zhang, Q., Bai, S., Yi, K., Niu, Z.: CATN: cross attentive tree-aware network for multivariate time series forecasting. Proc. AAAI Conf. Artif. Intell. **36**(4), 4030–4038 (2022)
11. Hmamouche, Y., Przymus, P.M., Alouaoui, H., Casali, A., Lakhal, L.: Large multivariate time series forecasting: survey on methods and scalability. In: Utilizing Big Data Paradigms for Business Intelligence, pp. 170–197. IGI Global (2019)
12. Ho, J., Jain, A., Abbeel, P.: Denoising diffusion probabilistic models. Adv. Neural. Inf. Process. Syst. **33**, 6840–6851 (2020)
13. Kingma, D.P., Ba, J.: Adam: a method for stochastic optimization. In: International Conference on Learning Representations (2015)
14. Kipf, T.N., Welling, M.: Semi-supervised classification with graph convolutional networks. In: International Conference on Learning Representations (2016)
15. Kong, Z., Ping, W., Huang, J., Zhao, K., Catanzaro, B.: Diffwave: a versatile diffusion model for audio synthesis. In: International Conference on Learning Representations (2021)
16. Lai, G., Chang, W.C., Yang, Y., Liu, H.: Modeling long-and short-term temporal patterns with deep neural networks. In: The 41st International Conference on Research & Development in Information Retrieval, pp. 95–104 (2018)
17. Li, Y., Yu, R., Shahabi, C., Liu, Y.: Diffusion convolutional recurrent neural network: data-driven traffic forecasting. In: International Conference on Learning Representations (2017)
18. Lim, B., Zohren, S.: Time-series forecasting with deep learning: a survey. Phil. Trans. R. Soc. A **379**(2194), 20200209 (2021)
19. Patton, A.: Copula methods for forecasting multivariate time series. Handb. Econ. Forecast. **2**, 899–960 (2013)
20. Rangapuram, S.S., Seeger, M.W., Gasthaus, J., Stella, L., Wang, Y., Januschowski, T.: Deep state space models for time series forecasting. In: Advances in Neural Information Processing Systems, vol. 31 (2018)
21. Rasul, K., Seward, C., Schuster, I., Vollgraf, R.: Autoregressive denoising diffusion models for multivariate probabilistic time series forecasting. In: International Conference on Machine Learning, pp. 8857–8868 (2021)
22. Rozemberczki, B., et al.: PyTorch geometric temporal: spatiotemporal signal processing with neural machine learning models. In: Proceedings of the 30th International Conference on Information and Knowledge Management, pp. 4564–4573 (2021)
23. Seo, Y., Defferrard, M., Vandergheynst, P., Bresson, X.: Structured sequence modeling with graph convolutional recurrent networks. In: Cheng, L., Leung, A.C.S., Ozawa, S. (eds.) ICONIP 2018. LNCS, vol. 11301, pp. 362–373. Springer, Cham (2018). https://doi.org/10.1007/978-3-030-04167-0_33
24. Shang, C., Chen, J., Bi, J.: Discrete graph structure learning for forecasting multiple time series. arXiv preprint arXiv:2101.06861 (2021)
25. Shih, S.Y., Sun, F.K., Lee, H.Y.: Temporal pattern attention for multivariate time series forecasting. Mach. Learn. **108**(8), 1421–1441 (2019)

26. Shuman, D.I., Narang, S.K., Frossard, P., Ortega, A., Vandergheynst, P.: The emerging field of signal processing on graphs: extending high-dimensional data analysis to networks and other irregular domains. IEEE Signal Process. Mag. **30**(3), 83–98 (2013)

27. Song, Y., Ermon, S.: Improved techniques for training score-based generative models. Adv. Neural. Inf. Process. Syst. **33**, 12438–12448 (2020)

28. Sutskever, I., Vinyals, O., Le, Q.V.: Sequence to sequence learning with neural networks. In: Advances in Neural Information Processing Systems, vol. 27 (2014)

29. Tashiro, Y., Song, J., Song, Y., Ermon, S.: CSDI: conditional score-based diffusion models for probabilistic time series imputation. Adv. Neural. Inf. Process. Syst. **34**, 24804–24816 (2021)

30. Wu, Z., Pan, S., Long, G., Jiang, J., Zhang, C.: Graph wavenet for deep spatial-temporal graph modeling. In: The 28th International Joint Conference on Artificial Intelligence (IJCAI). International Joint Conferences on Artificial Intelligence Organization (2019)

31. Wu, Z., Pan, S., Chen, F., Long, G., Zhang, C., Philip, S.Y.: A comprehensive survey on graph neural networks. IEEE Trans. Neural Netw. Learn. Syst. **32**(1), 4–24 (2020)

32. Wu, Z., Pan, S., Long, G., Jiang, J., Chang, X., Zhang, C.: Connecting the dots: multivariate time series forecasting with graph neural networks. In: The 26th ACM International Conference on Knowledge Discovery & Data Mining, pp. 753–763 (2020)

33. Yang, L., et al.: Diffusion models: a comprehensive survey of methods and applications. arXiv preprint arXiv:2209.00796 (2022)

34. Zhao, L., et al.: T-GCN: a temporal graph convolutional network for traffic prediction. IEEE Trans. Intell. Transp. Syst. **21**(9), 3848–3858 (2019)

A Bottom-Up Sampling Strategy for Reconstructing Geospatial Data from Ultra Sparse Inputs

Marco Landt-Hayen[1,2]([✉])[ID], Yannick Wölker[1,2][ID], Willi Rath[2][ID], and Martin Claus[1,2][ID]

[1] Christian-Albrechts-Universität zu Kiel, Kiel, Germany
[2] GEOMAR Helmholtz Centre for Ocean Research, Kiel, Germany
mlandt-hayen@geomar.de

Abstract. Working with observational data in the context of geophysics can be challenging, since we often have to deal with missing data. This requires imputation techniques in pre-processing to obtain data-mining-ready samples. Here, we present a convolutional neural network (CNN) approach from the domain of deep learning to reconstruct complete data from sparse inputs. CNN architectures are state-of-the-art for image processing. As data, we use two-dimensional fields of sea level pressure (SLP) and sea surface temperature (SST) anomalies. To have consistent data over a sufficiently long time span, we favor to work with output from control simulations of two Earth System Models (ESMs), namely the Flexible Ocean and Climate Infrastructure and the Community Earth System Model. Our networks can restore complete information from incomplete input samples with varying rates of missing data. Moreover, we present a technique to identify the most relevant grid points of our input samples. Choosing the optimal subset of grid points allows us to successfully reconstruct SLP and SST anomaly fields from ultra sparse inputs. As a proof of concept, the insights obtained from ESMs can be transferred to real world observations to improve reconstruction quality. As uncertainty measure, we compare several climate indices derived from reconstructed versus complete fields.

Keywords: Missing value imputation · Optimal sampling strategy · Convolutional neural networks · Explainable AI

1 Introduction

Geospatial data in the context of ocean and atmosphere are often provided on a two-dimensional latitude-longitude grid. Examples are sea level pressure (SLP) or sea surface temperature (SST). Observational data are obtained e.g., from

This work was supported by the Helmholtz School for Marine Data Science (MarDATA) funded by the Helmholtz Association (Grant HIDSS-0005).

X. Yang et al. (Eds.): ADMA 2023, LNAI 14176, pp. 677–691, 2023.
https://doi.org/10.1007/978-3-031-46661-8_45

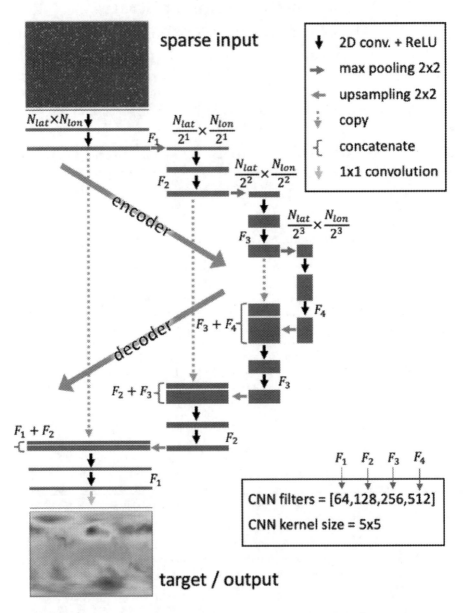

Fig. 1. Encoder-decoder-style U-Net architecture used for reconstructing complete two-dimensional geospatial fields from sparse inputs. In this sketch, we show an input sample for SLP CESM data with 95% missing values and the corresponding complete sample used as target. Blue rectangles symbolize resulting feature maps. Grid dimensions (N_{lat} and N_{lon}) in combination with pooling or upsampling operations determine height and width of feature maps. The depth is specified by the number of CNN filters $F_1..F_4$. (Color figure online)

remote sensing by satellites or from in situ measurements by survey stations, ships and planes. Observations are usually incomplete due to technical, physical or economical reasons. For instance, instrumental errors can lead to missing data. Moreover, infrared radiation is absorbed by clouds.

As prerequisite for advanced data mining, we need consistent and complete data. This requires imputation of missing values. Beckers and Rixen [1] introduced DINEOF as a method to infer missing data from oceanographic data series using empirical orthogonal functions. DINEOF has been and still is a widely used technique in the context of geospatial data. To name a few, Alvera-Azcárate et al. [2] applied the technique to SST data with rates of missing data typically in the range of $40 - 80\%$. DINEOF has also been used for multivariate reconstruction of missing data [3] with a focus on specific regions of interest [4]. The method is powerful, self-consistent and easy to use. But it has its limits in the allowed rates of missing values. For instance, SST samples with cloud coverage exceeding 95% are eliminated in pre-processing, since DINEOF fails in this regime. Chai et al. [5] proposed an encoder-decoder-style U-Net convolutional neural network (CNN) for reconstructing seismic data with regularly and irregularly missing values and found it to be superior compared to Fourier transform interpolation methods. In their work, rates of missing values were also limited to 95%. Similarily, Barth et al. [6] present a convolutional autoencoder to reconstruct missing SST observations and showed that deep neural networks are currently state-of-the art for this task.

In this work, we investigate methods for reconstructing full two-dimensional geospatial fields from incomplete input data on a *global* scale. We introduce a CNN that can handle input data with varying rates of missing values and go beyond existing approaches allowing to have ultra sparse inputs with up to 99.9% missing values. To overcome existing limits, we propose a bottom-up sampling strategy and set the benchmark for the ultra sparse regime. We show how to identify grid points that are essential to reconstruct dominant structures in SLP and SST anomaly fields from only 0.1% of the original data. Furthermore, we assign certain scores that can be visualized as a heat map to give an intuitive understanding of grid points' relevance for *reconstruction*. Thus, our approach differs from existing methods that aim to find feature relevance for *model predictions* like e.g., layer-wise relevance propagation (LRP) [7] or shapley additive explanations (SHAP) [8].

To have consistent data over a sufficiently long time span, we favor to work with output from control simulations of two Earth System Models (ESMs), namely the Flexible Ocean and Climate Infrastructure (FOCI) [9] and the Whole Atmosphere Community Climate Model as extension of the Community Earth System Model (CESM) [10,11]. In particular, we work with SLP and SST anomaly fields over 1,000 and 999 years obtained from FOCI and CESM, respectively. Eventually, we transfer results from ESMs to real world (RW) observations. Our main contributions are as follows:

– We present CNN models that can reconstruct missing data from inputs with varying rates of missing values and clearly outperform DINEOF.

- Our framework is provided as open-source project that can be extended and customized to individual needs, e.g., by including further methods or data.
- We propose a bottom-up sampling strategy to identify grid points in geospatial input fields that are most relevant for successful reconstruction.
- With this optimal subset of grid points, we restore complete information from ultra sparse inputs and set the benchmark in this regime.
- We introduce relative loss reduction maps to give a visual and intuitive understanding of grid points' relevance for reconstruction.
- As a proof of concept, we transfer insights obtained from ESMs to RW observations to improve reconstruction quality.

The rest of this work is structured as follows. In Sect. 2 we provide a short description of ESM and observational data and outline data pre-processing. An overview of our U-Net models is given in Sect. 3. Moreover, we describe our sampling strategy and introduce relative loss reduction maps in Sect. 3. In Sect. 4 we apply our U-Net models and DINEOF to ESM and observational data. Additionally, we evaluate model performance and compute several climate indices on reconstructed versus complete SLP and SST fields. Discussion of all results and a conclusion are found in Sect. 5.

2   Data

In this section, we briefly describe ESM and observational data before we outline data pre-processing. We start with monthly mean SLP and SST fields from FOCI and CESM control runs, respectively. Both simulations were run using pre-industrial external forcing that is representative for the year 1850. FOCI and CESM are based on very different component models (see [9–11] for details) with different strengths and weaknesses in simulating various aspects of the global climate. Additionally, we use reanalysis data provided by the National Oceanic and Atmospheric Administration (NOAA) as RW observations for SLP and SST [12]. Table 1 gives an overview of all data.

Table 1. Overview of CESM, FOCI and RW data used throughout this work. The grid resolution determines the grid dimensions in latitude and longitude. The number of valid grid points excludes permanently missing data, due to land masses. Finally, we show the total number of samples and the time span in years.

Source	Feature	Grid resolution	Grid dimensions $N_{lat} \times N_{lon}$	Valid grid points	Samples (years)
CESM	SLP	$1.8° \times 2.5°$	96×144	13,824	11,988 (999)
	SST			8,276	
FOCI	SLP	$1.8° \times 1.8°$	96×192	18,432	12,000 (1,000)
	SST			12,949	
RW	SLP	$2.5° \times 2.5°$	72×144	10,368	900 (75)
	SST	$2° \times 2°$	80×176	9,913	1,716 (143)

SLP and SST anomaly fields are obtained from raw data by removing the seasonal cycle over a specified climatology period. For RW data, we choose years 1980 through 2009 as climatology period, whereas for ESM data, we use the whole time span. To simulate missing data, we use different settings:

- **Fixed mask**: Randomly choose a subset of grid points as missing data with a discrete rate of missing values. This subset is then identical for *all* samples.
- **Variable discrete mask**: Only fix the discrete rate of missing values. Then, randomly choose subsets of grid points as missing values. Thus, different samples have different subsets of missing values. As extension, we can use each sample multiple times. Here, we use factors 1, 2 and 3 for data augmentation.
- **Variable range mask**: The set of missing values still differs for different samples, as for the variable discrete mask. Additionally, the rate of missing values is randomly drawn from a specified range. Our models allow maximum flexibility. We can set the range from 0 to 100% missing values.
- **Optimal mask**: Once we identified subsets of grid points that are most relevant for successful reconstruction, we create a fixed mask from taking the remaining grid points as missing data. Here, we use optimal masks only for distinct rates of missing values.

Moreover, anomaly fields are scaled to $[0, 1]$ with minimum and maximum values derived from training data. After scaling, missing values are set to *zero*. While SLP data is defined everywhere, we don't have SST data over land masses. Permanently missing values are also set to *zero*. 80% of all samples are used for training, while the remaining 20% are reserved as test data for model evaluation.

3   Models and Methods

We give an overview of our U-Net models in Sect. 3.1. Furthermore, we describe our sampling strategy and introduce relative loss reduction maps in Sect. 3.2. For a detailed technical description of the DINEOF methodology, we refer to Beckers and Rixen [1]. In their work, they introduced DINEOF as a method to infer missing data from oceanographic data series using empirical orthogonal functions.

3.1   U-Nets

In this work, we focus on encoder-decoder-style CNNs, also referred to as U-Nets [13]. A thorough introduction to deep learning in general and CNNs in particular can be found e.g., in the textbook of Goodfellow et al. [14]. A sketch of our U-Net architecture with 4 convolution blocks in the encoder part is shown in Fig. 1. As inputs, we use scaled two-dimensional *sparse* anomaly fields. Scaled *complete* anomaly fields serve as targets. Each convolution block in the encoder path consists of two convolution operations with strides set to one and zero padding to conserve height and width dimensions of the resulting feature maps plus a rectified linear unit (ReLU) activation. The convolution blocks are followed by a 2×2 maximum pooling operation to reduce height and width. Maximum depth of 512 feature maps is reached after 4 convolution blocks. In the decoder path we use skip connections to prevent vanishing gradients [15] and 2×2 upsampling to restore former height and width dimensions. Ultimately, we have a 1×1 convolution operation to obtain a single output channel. The kernel size for convolution operations is set to 5×5. The number of convolution filters F_i determines the depth of the resulting feature maps in the i-th convolution block. For our standard U-Net configuration, we have an increasing number of 64, 128, 256 and 512 filters in the four convolution blocks, respectively. This configuration is used throughout this work. Only for models trained on SST RW data, we add an additional fifth convolution block with $F_5 = 1,024$ filters and reduce kernel size for all convolutions to 4×4 to optimize performance. The U-Net models are trained over 10 epochs using the ADAM optimizer [16] with a mean squared error (mse) loss function. The learning rate is set to 10^{-4} and 10^{-5} for SLP and SST, respectively. The batch size is set to 10. All hyperparameters are the result of a grid search optimization. Therefore, we temporarily used 20% of the training data as validation data.

3.2   Relative Loss Reduction Maps

Assume that we have geospatial input data with a total number v of valid grid points. Furthermore, we assume to have a specific rate $m \in [0,1]$ of missing values. In a first step, we aim to find the subset G_m^s of the most relevant grid points for a single sample s that are essential to restore complete information. Thus, G_m^s contains $g_m \equiv |G_m^s| = (1-m) \cdot v$ grid points, where g_m is rounded to the nearest integer number. Moreover, we assume to have a U-Net model trained on samples over the full range of missing values $[0,1]$ with a variable range mask. We then pick a single sample s and compute the mean state loss (MSL) and the minimum loss (MinL) by using an empty sample of only zeros and the complete sample, respectively, as input for our range model. The difference of MSL and MinL determines the maximum absolute loss reduction (MALR) that we can achieve for a specific sample s, as stated in Eq. 1. Superscript s denotes the sample number.

$$MALR^s = MSL^s - MinL^s \tag{1}$$

To find the optimal subset G_m^s of grid points for a single sample, we start with an empty sample and successively add grid points $g = 1..g_m$. The more information we provide to the U-Net, the lower the reconstruction loss. To add grid point g, we choose the one from all $v - g + 1$ remaining valid grid points, that leads to maximum decline in reconstruction loss. The absolute loss reduction (ALR) for adding a grid point g to an input sample s can be translated into relative loss reduction (RLR) as follows:

$$RLR_g^s = ALR_g^s / MALR^s \tag{2}$$

The relative loss reduction scores $\{RLR_g^s\}_{g=1..g_m}$ are assigned to the corresponding grid points and can be visualized as heat map to give an intuitive understanding of grid points' relevance for reconstruction. The subset of most relevant grid points can vary from sample to sample. Here, we average scores over the first and last 120 training samples to obtain a stable mean relative loss reduction map (MRLRM) for a given geospatial feature. To reduce computational effort, we stop adding grid points when the accumulated relative loss reduction exceeds a threshold t. Here, we set $t = 0.9$ as a tradeoff between computation time and information gain. Finally, we derive optimal masks $G_{m=0.999}$, $G_{m=0.99}$ and $G_{m=0.95}$ that contain most relevant grid points for 99.9%, 99% and 95% missing values, respectively, for a given feature (SLP or SST) and source (CESM or FOCI). Therefore, we fit a set of bivariate Gaussians to each MRLRM using Gaussian Mixture Models (GMMs) [17] and take the grid points with highest RLR as cluster representatives. To summarize the algorithm for deriving optimal masks:

- **Step 1**: Train U-Net with variable range mask.
- **Step 2**: Compute relative loss reduction maps (RLRMs) for a representative subset of training samples.
- **Step 3**: Get MRLRM by averaging over RLRMs obtained in previous step.
- **Step 4**: Derive optimal masks for desired rates of missing values using GMMs.

4   Application and Results

In Sect. 3.2, we introduced the methodology to derive subsets of most relevant grid points from MRLRMs for various rates of missing values. In Sect. 4.1, we transfer these subsets (also referred to as *optimal masks*) obtained from ESM data to RW data. We then apply our U-Net models and DINEOF to RW data in Sect. 4.2 to reconstruct complete information from sparse input data for different types of masks. Additionally, we evaluate model performance and compute several climate indices on reconstructed versus complete SLP and SST fields in Sect. 4.3.

4.1   Derive and Transfer Optimal Masks

Figure 2 shows MRLRMs for ESM data for both features, SLP and SST. From these MRLRMs, we derive optimal masks. In order to transfer optimal masks from ESM to RW data, we need to consider the number of valid grid points for our target grid (RW) from Table 1. Here, we focus on the ultra sparse regime with 99.9%, 99% and 95% missing values. Accordingly, for SLP RW data we aim to find the most relevant 10, 104 and 518 grid points, whereas for SST RW data, we look for the most relevant 10, 99 and 496 grid points, representing 0.1%, 1% and 5% of valid grid points, respectively. As an example, we highlight resulting cluster representatives from GMM for 99.9% missing data (10 grid points) in

Fig. 2. Eventually, we need to deal with the fact that ESM and RW data are on different grids. Therefore, we use nearest neighbour interpolation to transfer optimal masks to the target grid.

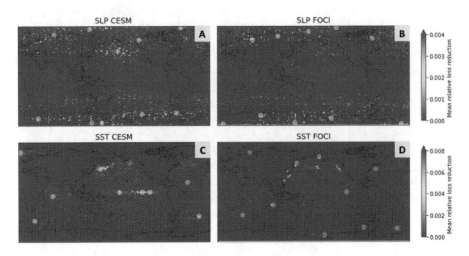

Fig. 2. The upper part shows MRLRMs for SLP obtained from CESM (A) and FOCI (B) data. The lower part shows results for SST CESM (C) and FOCI (D) data, respectively. The maps are averaged over first and last 120 training samples. The 10 most relevant grid points obtained from GMM are highlighted as yellow dots. These grid points represent 0.1% of valid grid points for SLP and SST RW data and hence, form the optimal masks for 99.9% missing values, transfered from ESM to RW data. (Color figure online)

4.2   Reconstruction with U-Nets and DINEOF

In Sect. 2, we introduced different mask types to simulate missing data. In this section, we show results for U-Net models trained on RW data with fixed and variable masks with augmentation factors 1, 2 and 3, respectively, for distinct rates of missing values $m \in \{0.999, 0.99, 0.95, 0.9, 0.75, 0.5\}$. Additionally, we show the test loss for models trained on our optimal masks transferred from ESM to RW data in the ultra sparse regime for $m \in \{0.999, 0.99, 0.95\}$. As a benchmark, we include results from reconstruction with DINEOF for $m \in \{0.99, 0.95, 0.9, 0.75, 0.5\}$.

The main interest of this work lies in the attempt to optimally reconstruct information from ultra sparse inputs. From inspecting loss on test data in Fig. 3, best performance is found for models trained on optimal masks transferred from ESM data. Most competitive are the models trained on variable masks with augmentation factor 3. In Fig. 4, we show reconstruction for an individual SLP RW test sample with 99.9% and 99% missing data, respectively, applying best

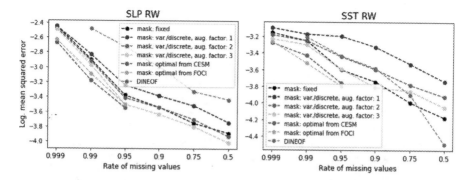

Fig. 3. Loss (mse) for test data on a logarithmic scale for U-Net models trained on SLP (left) and SST (right) RW data with various types of missing masks and distinct rates of missing values. All results are averaged over three runs with random seeds 1, 2 and 3, respectively. For comparison, we show corresponding loss for reconstruction with DINEOF.

performing U-Net models. For comparision, we also show the final reconstruction for DINEOF from 99% missing data. For 99.9% missing data, DINEOF fails.

The target field shows two dominant spots of positive SLP anomaly (red) in the upper center and upper left part of the sample. Moreover, we find alternating positive and negative SLP anomalies in high latitudes (40° to 70° N) and low latitudes (-40° to -70° N), respectively. The reconstruction of the sample from only 0.1% of the data (99.9% missing) using a model trained on a variable mask with augmentation factor 3 appears to be blurry and does not match the corresponding target. Whereas, reconstructions from models trained on optimal masks restore either the most dominant SLP anomaly spot in the upper center or upper left part and we also see a rudimentary replica of the alternating patterns of positive and negative SLP anomalies.

The reconstruction from 1% of the data (99% missing) using a model trained on a variable mask better fits the original target field, as it shows at least the dominant spot of positive SLP anomaly in the upper left part and alternating positive and negative anomalies in low latitudes. Compared to that, the reconstructions of both models trained on optimal masks are far more precise as they restore the correct location and shape for *both* dominant spots of positive SLP anomaly plus the alternating patterns of positive and negative anomalies. The final reconstruction with DINEOF appears to be noisy compared to results from our U-Nets. At least, the alternating positive and negative anomalies are restored in a rudimentary way. Similar results are obtained for reconstruction of SST samples. But for the sake of brevity, we only show results for SLP samples, here.

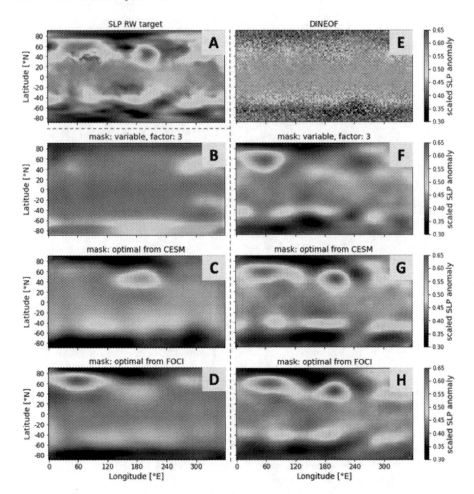

Fig. 4. Reconstruction of a complete SLP RW anomaly field used as target (A) from 0.1% and 1% of all data, hence, with 99.9% (left part) and 99% (right part) missing values, respectively. Here, we show results for best performing U-Net models. In particular, we use models trained on variable masks with augmentation factor 3 (B and F), models trained on optimal masks transferred from CESM (C and G) and FOCI (D and H). For comparision, we show the final reconstruction for DINEOF (E) from 99% missing data.

4.3   Model Evaluation

In this section, we evaluate model performance. The features we use throughout this work (SLP and SST), reflect some of the main dynamics of the Earth system in form of known modes of climate variability, patterns and oscillations, like e.g., the Southern Annular Mode (SAM) [18], the North Altantic Oscillation (NAO) [19], the Atlantic Multidecadal Oscillation (AMO) [20] or the El Niño Southern Oscillation (ENSO) [21].

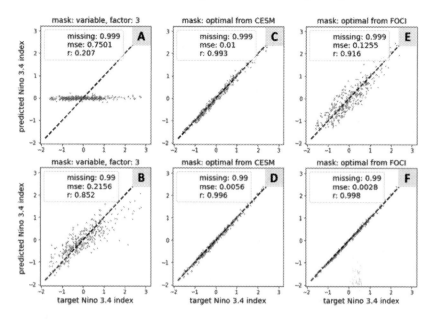

Fig. 5. Compare Niño 3.4 index obtained from reconstructed versus complete SST anomaly fields. Each blue dot represents a RW test sample. We compute indices on reconstructions from 0.1% (upper part) and 1% (lower part) of all data, hence, with 99.9% and 99% missing values, respectively. Here, we show results for best performing U-Net models. In particular, we include models trained on variable masks with augmentation factor 3 (A and B), models trained on optimal masks transferred from CESM (C and D) and FOCI (E and F). Loss (mse) and correlation coefficient are stated for each model and rate of missing values. The dashed black line is the identity.

To describe the Earth System dynamics in a compressed way, two-dimensional geospatial fields can be reduced to specific climate indices that capture the main processes. The SAM index was originally defined by Gong and Wang [18] as the difference of normalized monthly zonal mean SLP at 40°S and 65°S, respectively. The NAO index can be computed from SLP as the normalized difference between Reykjavik (64°9'N, 21°56'W) and Ponta Delgada (37°45'N, 25°40'W) [19]. AMO refers to a natural variability occurring in the SST of the North Atlantic with a multidecadal period of 60 to 80 years. The AMO index is computed from area-weighted SST anomalies of the North Atlantic [22]. ENSO is a complex phenomenon that can be detected as periodic SST fluctuations in the Tropical Pacific. Several indices are defined to compute the current ENSO phase from area-averaged SST anomalies in certain regions. For instance, Morrow et al. [23] define the Niño 3.4 index from the Niño 3.4 region (5°N-5°S, 120°W-170°W) which is often used in the context of ENSO. In Fig. 5, we show the Niño 3.4 index computed on reconstructed versus complete SST RW anomaly fields (test data). We again focus on the ultra sparse regime with 99.9% and 99% of missing data and compare results for models trained on variable masks with augmentation factor 3 with results for models trained on optimal masks derived from CESM and FOCI data. These models perform best, as shown in Sect. 4.2. As metric, we use the mse of predicted versus true Niño 3.4 index and correlation coefficient r. Clearly, highest fidelity in terms of low mse combined with a high degree of correlation is found for models trained on optimal masks. Similar results are obtained for SAM, NAO and AMO indices, as shown in Table 2.

Table 2. Compare climate indices corresponding to SAM, NAO and AMO obtained from reconstructed versus complete SLP and SST fields, respectively. Overview of loss (mse) and correlation coefficient for RW test data with 99.9% and 99% missing values and models trained on different masks to simulate missing values. In particular, we compare results for models trained on variable masks with augmentation factor 3 to models trained on optimal masks transferred from CESM and FOCI data, respectively.

Feature	Index	Mask	Missing rate	Loss (mse)	Correlation
SLP	SAM	var., factor: 3	99.9% | 99%	0.502 | 0.098	0.749 | 0.951
		opt. from CESM		0.230 | 0.050	0.885 | 0.975
		opt. from FOCI		0.288 | 0.048	0.856 | 0.976
SLP	NAO	var., factor: 3	99.9% | 99%	2.171 | 0.677	0.573 | 0.883
		opt. from CESM		1.568 | 0.400	0.691 | 0.931
		opt. from FOCI		1.404 | 0.446	0.726 | 0.923
SST	AMO	var., factor: 3	99.9% | 99%	0.031 | 0.018	0.180 | 0.675
		opt. from CESM		0.018 | 0.007	0.673 | 0.895
		opt. from FOCI		0.025 | 0.007	0.483 | 0.897

5   Discussion and Conclusion

In this work, we presented encoder-decoder-style U-Net models to reconstruct complete information from sparse geospatial input data. To have consistent data over a sufficiently long time span, we favored to work with ESM data since RW data from reanalysis already contains reconstructed information from statistical methods.

To simulate missing data, we used different types of masks. Additionally, we have permanently missing data for SST in terms of land masses. All missing values were set to zero. Our U-Nets showed maximum flexibility in a sense, that different samples can have different sets of missing data plus the rate of missing values can vary over the full range [0,1]. This flexibility allowed us to find most relevant grid points for each sample by using a bottom-up sampling strategy. We successively added and fixed grid points that lead to largest drop in reconstruction loss and assigned the relative loss reduction to the corresponding grid points. Results vary for different samples. Therefore, we averaged over multiple training samples to obtain stable MRLRMs for both features (SLP and SST) and ESMs (CESM and FOCI). MRLRMs have been visualized as heat maps and give an intuitive understanding of grid points' relevance for successful reconstruction. Our MRLRM methodology allows to look inside our U-Nets and adds a new method of explainable artificial intelligence (xAI). For SLP CESM and FOCI data, we find highest values in MRLRMs at high and low latitudes. Whereas, for SST CESM and FOCI, we find concise spots of high values mostly at coastlines of the Northern Pacific and in the Niño 3.4 region, respectively. The location of dominant spots with high values in MRLRMs clearly differs for SST CESM and FOCI. Understanding these differences in detail is a matter of ongoing research and could help to reveal artifacts and biases for certain ESMs.

From the MRLRMs, we then derived optimal masks for distinct rates of missing values in the ultra sparse regime with 99.9%, 99% and 95% missing data. As a proof of concept, we transferred the obtained masks to RW data. For SLP and SST RW data, we have 10,368 and 9,913 valid grid points, respectively. Thus, for SLP RW data we aimed to find the most relevant 10, 104 and 518 grid points, whereas for SST RW data, we looked for the most relevant 10, 99 and 496 grid points, representing 0.1%, 1% and 5% of valid grid points, respectively. Assuming that we only have sparse measurements in practice, the derived masks help to answer the question, how to get most information from a limited number of survey stations and tells us *where* we should measure.

Although we refer to the masks as *optimal* masks, there is no proof for that optimality statement. For instance, finding the optimal 10 grid points from a total number of 10,368 valid grid points, gives us

$$\binom{10,368}{10} \approx 3.9 \cdot 10^{33}$$

possible combinations. Opposed to that, our methods of successively adding and fixing grid points, only requires to search through

$$\sum_{g=1..10} (10,368 - g + 1) \approx 10^5$$

possibilities. This keeps computational effort manageable and still leads to clear outperformance over models trained on fixed and variable masks. We showed that our models trained on the optimal masks succeed in reconstructing dominant large scale structures from only 0.1% and 1% of the original data, whereas all other models fail on this task, including DINEOF. We also found that the reconstructed geospatial fields from models trained on optimal masks reflect the large scale dynamics, as shown for several climate indices obtained from reconstructed versus complete data.

As next steps, we plan to extend our method of identifying most relevant grid points to further geospatial data like e.g., geopotential height, surface air temperature, sea surface salinity or precipitation. We then aim to go beyond reconstructing missing information and will try to predict *future* geospatial fields from (ultra) sparse inputs. This requires additional information in form of multivariate and/or time lagged input features as multiple channels for our U-Net models.

Data Availability Statement. Our framework for reconstructing missing data is hosted on GitHub: https://github.com/MarcoLandtHayen/reconstruct_missing_data. Trained models and all results are stored in a separate Git repository: https://git. geomar.de/marco-landt-hayen/reconstruct_missing_data_results. Observational data used in this work are publicly available [12]. ESM data are stored on Zenodo: https:// doi.org/10.5281/zenodo.7774316.

References

1. Beckers, J.M., Rixen, M.: EOF calculations and data filling from incomplete oceanographic datasets. J. Atmos. Oceanic Tech. **20**(12), 1839–1856 (2003)
2. Alvera-Azcárate, A., Barth, A., Rixen, M., Beckers, J.M.: Reconstruction of incomplete oceanographic data sets using empirical orthogonal functions: application to the Adriatic Sea surface temperature. Ocean Model. **9**(4), 325–346 (2005). https:// doi.org/10.1016/j.ocemod.2004.08.001
3. Alvera-Azcárate, A., Barth, A., Beckers J.M., Weisberg, R.H.: Multivariate reconstruction of missing data in sea surface temperature, chlorophyll, and wind satellite fields. J. Geophys. Res. Oceans **112**(C3) (2007) https://doi.org/10.1029/2006JC003660
4. Yang, Y.C., Lu, C.Y., Huang, S.J., Yang, T.Z., Chang, Y.C., Ho, C.R.: On the reconstruction of missing sea surface temperature data from Himawari-8 in adjacent waters of Taiwan using DINEOF conducted with 25-h data. Remote Sens. **2022**(14), 2818 (2022). https://doi.org/10.3390/rs14122818
5. Chai, X., Gu, H., Li, F., Duan, H., Hu, X., Lin, K.: Deep learning for irregularly and regularly missing data reconstruction. Sci. Rep. **10**, 3302 (2020). https://doi.org/10.1038/s41598-020-59801-x
6. Barth, A., Alvera-Azcárate, A., Troupin, C., Beckers, J.M.: DINCAE 2.0: multivariate convolutional neural network with error estimates to reconstruct sea surface temperature satellite and altimetry observations. Geosci. Model Develop. **15**(5), 2183–2196 (2022). https://doi.org/10.5194/gmd-15-2183-2022
7. Bach, S., Binder, A., Montavon, G., Klauschen, F., Müller, K.-R., Samek, W.: On pixel-wise explanations for non-linear classifier decisions by layer-wise relevance propagation, PLoS ONE **10** (2015)
8. Lundberg, S.M., Lee, S.-I.: A unified approach to interpreting model predictions, In Proceedings of the 31st Conference on Neural Information Processing Systems, Long Beach (2017). https://doi.org/10.48550/arXiv.1705.07874
9. Matthes, K., et al.: The flexible ocean and climate infrastructure version 1 (FOCI1): mean state and variability. Geosci. Model Developm. **13**(6), 2533–2568 (2020)
10. Hurrell, J.W., et al.: The community earth system model: a framework for collaborative research. Bull. Am. Meteor. Soc. **94**, 1339–1360 (2013)
11. Marsh, D.R., Mills, M.J., Kinnison, D.E., Lamarque, J.F., Calvo, N., Polvani, L.M.: Climate change from 1850 to 2005 simulated in CESM1 (WACCM). J. Clim. **26**(19), 7372–7391 (2013)

12. National Oceanic and Atmospheric Administration Download Center. https:// downloads.psl.noaa.gov/Datasets/ncep.reanalysis.derived/surface/pres.sfc. mon.mean.nc, and https://downloads.psl.noaa.gov/Datasets/noaa.ersst.v5/ sst.mnmean.nc
13. Ronneberger, O., Fischer, P., Brox, T.: U-net: convolutional networks for biomedical image segmentation. In: Proceedings of the 18th International Conference on Medical Image Computing and Computer-Assisted Intervention, Munich, vol. 9351, pp. 234–241 (2015). https://doi.org/10.1007/978-3-319-24574-4_28
14. Goodfellow, I., Bengio, Y., Courville, A.: Deep Learning. MIT Press (2016). http:// www.deeplearningbook.org
15. He, K., Zhang, X., Ren, S., Jian, S.: Deep residual learning for image recognition. In: Proceedings of the IEEE Conference on Computer Vision and Pattern Recognition, Las Vegar, pp. 770–778 (2016) https://doi.org/10.1109/CVPR.2016.90
16. Kingma, D.P., Ba, J.: Adam: a method for stochastic optimization. In: Proceedings of the 3rd International Conference for Learning Representations, San Diego (2014). https://arxiv.org/abs/1412.6980
17. Reynolds, D.A., Gaussian Mixture Models, Encyclopedia of Biometrics, pp. 827–832 (2009). https://doi.org/10.1007/978-1-4899-7488-4_196
18. Gong, D., Wang, S.: Definition of antarctic oscillation index. Geophys. Res. Lett. **26**(4), 459–462 (1999)
19. Hurrell, J.W.: Decadal trends in the North Atlantic oscillation: regional temperatures and precipitation. Science **269**(5224), 676–679 (1995)
20. Schlesinger, M.E., Ramankutty, N.: An oscillation in the global climate system of period 65–70 years. Nature **367**, 723–726 (1994)
21. Philander S.G.: El Niño, La Niña, and the Southern Oscillation. Academic Press (1989)
22. Trenberth, K.E., Shea, D.J.: Atlantic hurricanes and natural variability in 2005. Geophys. Res. Lett. **33**(12) (2006)
23. Morrow, R., Ward, M.L., Hogg, A.McC., Pasquet, S.: Eddy response to Southern Ocean climate modes. J. Geophys. Res. **115**(C10) (2010). https://doi.org/10.1029/2009JC005894

Recommendation II

Feature Representation Enhancing by Context Sensitive Information in CTR Prediction

Haibo Liu, Yafang Guo[✉], Liang Wang, and Xin Song

School of Cyber Security and Computer, Hebei University, Baoding 071000, China
Yfguo3080@gmail.com

Abstract. Click-Through-Rate (CTR) is a fundamental metric used to assess the efficacy of recommendation systems. In the past, most CTR prediction approaches focused on modeling the cross feature of various feature fields to improve the accuracy of CTR prediction. But they only learned the fixed representation of feature and neglected the varying significance of different feature fields in distinct contexts - what we refer to as context sensitive information - leading to suboptimal performance. While recent approaches have attempted to leverage linear transformations and feature interactions to capture context sensitive information, they remain inadequate as they overlook the varying importance of original features or different order cross features. In this paper, we propose a new module called Enhancing Feature Network (EFNet). EFNet has two key components: 1) Information Capture Layer (ICL), that dynamically captures explicit and implicit context sensitive information from original embedding features and digs out their corresponding bit-level weights; 2) Enhancing Feature Layer (EFL) that adaptively combine the context sensitive information with original embedding features according to the weights obtained in ICL. It is worth noting that EFNet can be integrated into existing CTR prediction models as a module to boost their overall performance. We conduct comprehensive experiments on four public datasets and the results demonstrate that models incorporating the EFNet module outperform other state-of-the-art models.

Keywords: Feature Interaction · Context Sensitive Information · CTR Prediction · Recommendation System

1 Introduction

CTR prediction is used to estimate the likelihood of a user clicking on items and widely employed in Internet companies [1]. A precise prediction of CTR can provide significant business value while improve user satisfaction [2, 3]. Over the past decades, numerous models [2–7] have been proposed to predict CTR and have achieved considerable success. Following these works, we categorize these CTR prediction models to two types: 1)Traditional methods based models, which aim to model the original features and low-order interaction [3, 4, 8, 9] between various feature fields; 2) Deep learning based models [4, 8–12], which use deep neural networks can capture more abstract features and interactions in a larger feature space.

© The Author(s), under exclusive license to Springer Nature Switzerland AG 2023
X. Yang et al. (Eds.): ADMA 2023, LNAI 14176, pp. 695–708, 2023.
https://doi.org/10.1007/978-3-031-46661-8_46

While these models have considered the modeling different feature fields, they have often overlooked the varying importance of each feature fields and their corresponding cross features in different contexts. For example, consider the following two instances: {*female, hobby, occupation, clothing, festival*} and {*female, hobby, occupation, clothing, workday*}. During the festivals, the cross feature of '*hobby*' and '*festival*' can notably influence clothing choices of women. However, on workdays, the feature '*occupation*' may play a more prominent role. This example highlights how the importance of each feature field can vary in different contexts. Furthermore, the importance of different order interactions is critical to CTR prediction models and has not been adequately considered in previous models.

To address the aforementioned issues, we propose a novel module named Enhancing Feature Network (EFNet), which employs both explicit and implicit methods to dynamically capture the varying significance of each feature field in diverse contexts - we refer to as *context sensitive information* - and adaptively enhance the original features with this information. EFNet consists of two key components: 1) Information Capture Layer (ICL), which using both explicit and implicit methods to dynamically capturing the context sensitive information from the original features and their corresponding bit-level weights; 2) Enhancing Feature Layer (EFL), able to adaptively fuse the context sensitive information into the original features. As an independent module, EFNet can also be integrated into any existing CTR prediction models and notably improve their performance.

The major contributions of this paper are summarized as follows:

- We propose a new module named EFNet, which is the first to simultaneously capture the varying importance hidden in both original features and their different order interactions in different contexts.
- EFNet employs both explicit and implicit methods to capture context sensitive information, which can be employed as a new module in different existing CTR prediction models to improve their performance.
- Our experimental results on four public datasets demonstrate that integrating EFNet into DCN-V2 can outperform other state-of-the-art CTR prediction models. Additionally, our experiments also confirm the compatibility of EFNet with other CTR prediction models to improve their performance.

2   Related Works

Many CTR prediction models have achieved enormous success by modeling different feature fields and their cross features [4, 8–10, 12]. The Factorization Machine (FM) [13] calculates the inner product between feature pairs to capture the explicit feature interactions. However, FM models can only capture explicit second order feature interactions. The attention mechanism [2, 3, 14] considers the correlation between different features and uses key-value pair multiplication to calculate the mutual influence between each feature fields, thereby capturing feature interactions. Models such as [9, 10, 15, 16] use the Cross module to extract the interactions of various feature fields, enabling them to automatically model explicit feature interactions of any finite order. Due to the great success of Deep Neural Network (DNN) in many fields such as computer vision [17], natural language processing [18], the use of DNN in CTR prediction has become a current research trend. In recent years, many CTR prediction models, such as [4, 8–10, 12, 16], utilize DNNs to extract implicit and abstract features interactions while incorporating modules that extract explicit features in order to improve the accuracy of the CTR prediction.

However, none of these models regarded that the importance of each feature field varies in different contexts as context sensitive information. In order to solve this problem, FiBiNet [19] attempts to incorporate the SeNet [20], which is effective in the computer vision containing a Pooling and Multi-Layer Perceptron (MLP) module. However, this method is not suitable for recommendation systems because of the limited ability to extract the context sensitive information. This method only considers using the original features to extract implicit context sensitive information, while ignoring the complementary role played by different orders cross feature interaction in different contexts. FRNet [21] tries to dynamically extract context sensitive information by using self-attention modules with a single-layer MLP. However, it only considers using fixed order feature interactions to represent context sensitive information, ignoring the explicit context sensitive information contained in original features and different order cross features.

Our model draws lessons from the above models, leverages MLP to extract abstract high order features which represent the implicit context sensitive information and Cross module to extract the different orders interactions of various feature fields which represent the explicit context sensitive information. Among them, EFNet add residual terms to reintroduce the original features which are crucial in recommendation systems. Additionally, we dynamically capture the corresponding weights balanced the context sensitive information and original features on bit-level. Ultimately, we adaptively incorporate the bit-level weights to comprehensively fuse context sensitive information and original features, leading to a superior performance.

3 Prilimitery

In this section, we introduce the framework of the most CTR prediction models, which consists of Embedding Layer, Feature Interaction Layer and Prediction Layer. Figure 1 shows an example of our CTR prediction framework.

3.1 Embedding Layer

The input of the CTR prediction task typically comprises categorical and numerical features [10, 15, 16]. Typically, the categorical features are represented as one-hot or multi-hot vectors, while the numerical features are represented as practical significant number. Assuming that an instance X contains m features and can be described as $X \in R^{m \times 1} = [x_1, x_2, \ldots, x_m]^T$. In most cases, the original data of the recommendation systems is extremely sparse [8, 12, 16, 19, 20, 22, 23]. To address this issue, CTR prediction models employ an embedding layer to convert high-dimensional sparse features into low-dimensional dense features, so that the embedding vectors E can effectively alleviate the dimension explosion issue caused by multiple feature vectors [4]. Specifically, for any categorical feature vector $x_c \in R^{c \times 1}$, the Embedding Layer converts it into a fixed dimension h, and flattens it into a column vector $x_c \in R^{h \times 1} = \left[e_{c,1}, e_{c,2}, \ldots, e_{c,h}\right]^T$; for numerical features, we apply a normalization procedure to obtain a new numerical feature $x_n \in R^{n \times 1}$ with standardized values. This mitigates potential training difficulties associated with scaling issues [24] and ensures the overall model is not adversely affected.

Finally, these obtained embedding vectors are concatenated together to form an embedding column vector $E \in R^{f \times 1} = \left[e_1, e_2, \ldots, e_f\right]^T$.

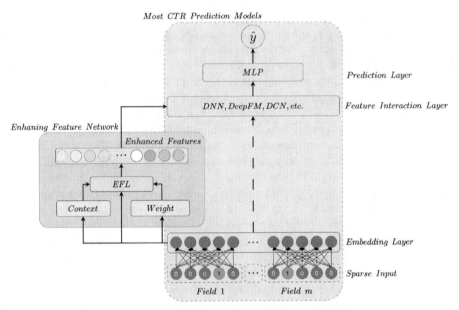

Fig. 1. A example of CTR prediction framework. EFNet should be integrated between Feature Interaction Layer and Embedding Layer.

3.2 Feature Interaction Layer

In previous CTR prediction models, the Feature Interaction Layer is the most critical component. It uses various kinds of operators, such as FM and DNN, to model the cross features. The output of the feature interaction layer $\hat{y}_i$ is a vector with the same dimension as E and typically encompasses all orders of cross features.

3.3 Prediction Layer

To obtain the CTR, a Prediction Layer transforms the output of the Feature Interaction Layer to a probability by function:

$$\hat{y} \in (0, 1) = Sigmoid\left(\sum_i \hat{y}_i\right) \tag{1}$$

The Binary Cross Entropy (BCE) is widely used in the Prediction Layer for loss calculation[13]. Its formula is:

$$Loss(y, \hat{y}) = -[(y \times log\hat{y}) + (1 - y) \times log(1 - \hat{y})], \tag{2}$$

where y represents the true label (0 or 1) and $\hat{y}$ represents the predicted probability of the positive class (clicking the item) obtained by applying the sigmoid function to the output of the Feature Interaction Layer.

4 Proposed Module

In this section, we will describe the architecture of our proposed module EFNet. Specifically, in Sect. 4.1, we introduce the ICL and in Sect. 4.2 we introduce the EFL. The overall structure of EFNet is shown in Fig. 2.

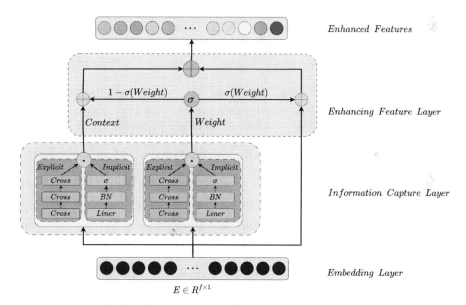

Fig. 2. The architectures of Enhancing Feature Network.

4.1 Information Capture Layer (ICL)

In this section we describe the Information Capture Layer (ICL) in detail. The ICL consists of two key components with distinct functions: the Context Sensitive Information Extractor and the Weight Selection Block. The Context Sensitive Information Extractor extracts explicit and implicit context sensitive information. The Weight Selection Block determines the appropriate bit-level weight for each context-sensitive information that needs to be integrated from a bit-wise perspective.

Context Sensitive Information Extractor. We consider context sensitive information to be present in both explicit and implicit forms hidden in all embedding vectors E obtained from the Embedding Layer. From the perspective of feature crossing, we have realized that the fundamental operator of previous works is merely a simple matrix transformation or pairwise cross of features without considering both original features and the different orders of feature interaction. To tackle this issue, EFNet implements Cross module, a finite order explicit feature cross method [10], to extract explicit context sensitive information. Additionally, EFNet uses the MLP module to extract implicit context sensitive information.

The original features play a significantly more critical role in the recommendation system than the cross features [25]. Balancing the handling of original features and cross features can lead to better performance and enhance the stability of the CTR prediction models. We incorporate the Cross module to capture the context sensitive information, as shown in Fig. 3 (a). The formula is as following:

$$x_{l+1} \in R^{f \times 1} = E \odot (W_l \cdot x_l + b_l) + E, \tag{3}$$

where $\odot$ represents Hadamard product. $E \in R^{f \times 1}$ is the original embedding vector. $x_l \in R^{f \times 1}$ is the result obtained from the l-th cross layer and $x_0 = E$. Both $W_l \in R^{f \times f}$ and $b_l \in R^{f \times 1}$ are the parameter matrices required for the $l + 1$-st order cross. By incorporating the bias matrix b_l and the residual term into the calculation, the Cross Module effectively reintroduces the original embedding vectors into the output x_{Cross}. As a result, x_{Cross} contains both the original embedding features and all finite order feature interaction that contains explicit context sensitive information we extracted.

MLP is the most widely used and effective method in data mining for modeling the implicit and abstract high order features interactions [4, 6]. According to previous works, typically the implicit context sensitive information is not overly complex[21]. Therefore, we employ a single layer MLP with a Batch-Normalization to extract the implicit context sensitive information, followed by a matrix transformation that can map it to the feature space with the same dimension as x_{Cross}.

$$x_{MLP} = \sigma (BN (W_l E + b_l)), \tag{4}$$

where σ is the ReLU function; W_l and b_l are parameter matrices. BN is the Batch-Normalization. The entire MLP is shown in Fig. 3(b).

After obtaining the explicit and implicit context sensitive information using the ICL respectively, we combine them using the Hadamard product method as illustrated in Fig. 2. The formula is as following:

$$x_{Context} = x_{Cross} \odot x_{MLP}, \tag{5}$$

where x_{Cross} and x_{MLP} represent the output of Cross module and MLP module respectively.

Weight Selection Block Note that not all the context sensitive information significantly contributes to the model. In fact, some information may even have a detrimental effect [5, 14]. As such, we employ an adaptive weight selection to determine, from a bit-wise perspective, whether the context sensitive information is beneficial to the overall model.

$$Weight \in R^{f \times 1} = \sigma (x_{Cross} \odot x_{MLP} + E), \tag{6}$$

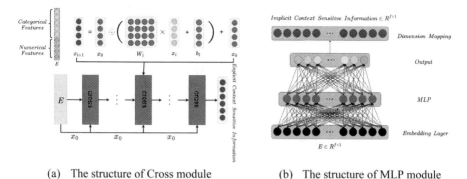

(a)   The structure of Cross module   (b)   The structure of MLP module

Fig. 3. The structure of Cross and MLP module in ICL.

where, x_{Cross} and x_{MLP} represent the output of Cross module and MLP module respectively. E represents the embedding features. σ represents the Sigmoid function to select the context sensitive information incorporating weight. In this method, we reintroduce the original features into the weight, which can ensure that the original features can retain their significance in determining the weight.

4.2 Enhancing Feature Layer (EFL)

The importance of each original feature can vary greatly, and the same case holds of context sensitive information. It is imperative to maintain the importance of original features while integrating context sensitive information. To achieve this, we merge the context sensitive information with the original embedding vector based on their respective bit-level weights. The enhanced features are then obtained by the following formula:

$$EnhancedFea = E \odot Weight + x_{Context} \odot (1 - Weight), \qquad (7)$$

where, E represents the embedding vector. Meanwhile, $x_{Context}$ and $Weight$ represent the context sensitive information and corresponding weight.

First, it is imperative to maintain the significance of original features while integrating context sensitive information. Moreover, in order to make context sensitive information enhancing the original features effectively, we employ $(1 - Weight)$. The weights obtained from the ICL adaptively balance the importance of original features and context sensitive information, which enables us to utilize the useful portion to improve the overall model.

5   Experiment

Following an extensive analysis of numerous CTR prediction models, we opted for DCN-V2 [10] as the feature interaction layer for EFNet due to its straightforward and efficient architecture[1]. To evaluate EFNet module from multiple perspectives, we performed experiments on four public datasets and sought to answer the following questions:

---

[1] Our code is published on https://github.com/Sakuragbb/EFNet.

(RQ1) How does the performance of our proposed module compare to the existing CTR prediction model?

(RQ2) How do hyper-parameters affect the overall performance of EFNet?

(RQ3) Do the individual components of EFNet contribute to the performance of the EFNet?

(RQ4) Can EFNet play a beneficial role in different CTR prediction models?

5.1 Datasets

We utilize the four public datasets for CTR prediction and divide them into three parts with a random seed: 70% is training, 20% is for validation and the remaining 10% for testing.

1. **Criteo**[2]:Criteo Kaggle dataset comprises one month of 45,840,617 ad click instances. It includes 26 categorical feature fields and 13 integer numerical fields.
2. **MovieLens 1M**[3]: MovieLens 1M contains three files: user features, movie features, and user ratings for movies. We use all movie ratings as labels and mark the ratings $> = 4$ as 1, and the other as 0.
3. **Anime**[4]:This dataset contains user preference data from 73,516 users on 12,294 anime. Each user can add anime to their completed list and give it a rating The dataset is a compilation of these ratings and includes 8 categorical feature fields We use all the ratings of user as labels and mark the ratings $> = 8$ as 1.
4. **Criteo_450w**: To evaluate the performance of our module on different data scales, we sampled the Criteo dataset to form a new dataset for 4,584,062 (10%) instances. The dataset is randomly sampled on the Criteo data set with a random seed at a ratio of 0.1.

5.2 Evaluation Metrics and Baseline Methods

In our experiments, we use two widely adopted metrics for CTR prediction tasks: AUC (Area Under the ROC Curve) and Log Loss.

AUC: AUC represents the probability that a positive prediction will be ahead of the negative prediction. So that it is a widely used metric in evaluating classification problems. A value between 0 and 1 is desirable, with a larger AUC indicating better performance. Note that an improvement of 0.1% in AUC is usually considered significant for the CTR prediction as it can result in a substantial increase in revenue for companies with large user bases.

Log Loss: Log Loss, also known as binary cross-entropy loss, is a commonly used metric for binary classification. It measures the performance of classification based on probability. The lower bound of the Log Loss is 0, and a lower value indicates better performance.

We compare EFNet with seven baseline methods, namely: **DNN** [11], **DeepFM** [4], **xDeepFM** [12], **FiBiNet** [19], **InterHAt** [14], **DCN-V2** [24], **FRNet** [21].

---

[2] https://www.kaggle.com/datasets/mrkmakr/criteo-dataset.

[3] http://files.grouplens.org/datasets/movielens/.

[4] https://www.kaggle.com/datasets/CooperUnion/anime-recommendations-database.

5.3   Implementation Details

We implement all the models with Pytorch in our experiments. For the optimization method, we use the Adam with a mini-batch size of 4096. The default learning rate is initially set to 0.001 and reduced by a factor of 2 after every 5 epochs. The embedding size is 10 for Criteo and Criteo_450w, 20 for MovieLens 1M, 10 for Anime. All activation functions in MLP are ReLU unless otherwise specified. To ensure a fair comparison, we take the optimal settings from the original papers for baseline methods.

To ensure fair comparison, we run all experiments five times by 5 random seeds and report the representative results. We observe that the standard deviations of our method are on the order of 1e-4, indicating that our results are stable and repeatable. All models are trained on *RTX 3090 GPUs.*

5.4   Performance Comparison (RQ1)

In this subsection, we summarize the overall performance of related models and EFNet on the four datasets. The experimental results are shown in Table 1.

EFNet model outperforms other models across all 4 datasets. This is because previous works only learned fixed representation of original features and hardly ever considered the varying importance of each original features in different contexts. Meanwhile, it is noteworthy that both FiBiNet and FRNet, which have considered context sensitive information, did not demonstrate superior performance compared to DCN-V2 on the MovieLens 1M and Anime. This can be attributed to the fact that these two datasets contained only 7 or 8 feature fields in our experimental setup, resulting in relatively simple the context sensitive information. Furthermore, the approach employed by FiBiNet for extracting context sensitive information is overly simplistic as it does not consider explicit context sensitive information, while FRNet places excessive emphasis on explicit feature interactions and overlooks the importance of original features. In contrast, our model incorporates both the original features and explicit context sensitive information by modeling different orders of feature interactions, enabling it to perform optimally.

5.5   Hyper-Parameters Study (RQ2)

We perform a hyper-parameter analysis to assess its impact on the performance of EFNet. We keep all other hyper-parameters constant and modify two specific hyper-parameters: the order of cross and the number of neurons in the MLP of ICL. The results shown as Fig. 4 and Fig. 5.

The results show that 'Cross = 2' is the optimal setting. This is because overmining can negatively impact the overall performance of the CTR prediction model by adversely affecting the representation of original features in the explicit context sensitive information. As a result, the optimal cross orders for different datasets are not consistent. This is due to the varying number and structure of feature fields in different datasets, which results in different complexity for explicit context sensitive information and necessitates different cross orders.

In general, the larger the required feature space dimension, the more complex the extracted feature interactions will be. The number of neurons in MLP indirectly represents the complexity of implicit context sensitive information. Similar to the cross

Table 1. Comparison with SOTA CTR prediction methods on four datasets.

	Criteo		MovieLens 1M		Anime		Criteo_450w	
Model	AUC	LogLoss	AUC	LogLoss	AUC	LogLoss	AUC	LogLoss
DNN	0.7920	0.4578	0.8135	0.4911	0.8461	0.3931	0.7825	0.4621
DeepFM	0.8002	0.4578	0.8175	0.4857	0.8475	0.3850	0.7898	0.4524
xDeepFM	0.8026	0.4532	0.8185	0.4798	0.8480	0.3851	0.7926	0.4520
FiBiNet	0.8033	0.4530	0.8182	0.4817	0.8464	0.3826	0.7936	0.4497
InterHAt	0.7998	0.4545	0.8177	0.4896	0.8444	0.3901	0.7901	0.4545
DCN-V2	0.8037	0.4521	0.8182	0.4812	0.8489	0.3894	0.7929	0.4471
FRNet	0.8049	0.4514	0.8179	0.4820	0.8422	0.3982	0.7938	0.4499
EFNet	**0.8069**	**0.4497**	**0.8210**	**0.4785**	**0.8512**	**0.3787**	**0.7954**	**0.4456**

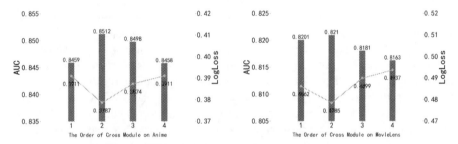

Fig. 4. The performance of different orders of Cross module.

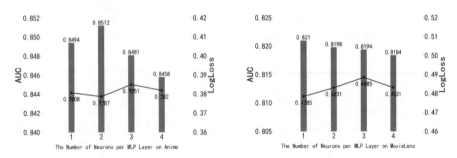

Fig. 5. The performance of different number of neurons in MLP module.

module, different datasets require different number of neurons in MLP layer. It can be seen that a CTR prediction model only needs to use MLP with 200–300 neurons to fully extract implicit context sensitive information.

5.6 Ablation Experiment (RQ3)

To ensure the criticality of every component within EFNet, we design three additional models and named them as Non-Cross, which removes the Cross module to eliminate the explicit context sensitive information; Non-MLP, which removes the MLP module to eliminate implicit context sensitive information; and Non-ICL, which removes both the Cross and MLP module to simultaneously lose both explicit and implicit context sensitive information same as DCN-V2. For simplicity, we perform identical operations on the Weight Selection Block and Feature Interaction Layer of each of these models. The experimental results are presented in Table 2.

Table 2. Comparison the performance of three models to EFNet.

Model	Criteo			Anime		
	AUC	LogLoss	ΔAUC↑	AUC	LogLoss	ΔAUC↑
Non-Cross	0.8052	0.4522	0.19%	0.8497	0.3891	0.09%
Non-MLP	0.8049	0.4534	0.15%	0.8498	0.3889	0.10%
Non-ICL	0.8037	0.4547	-	0.8489	0.3893	-
EFNet	**0.8069**	**0.4497**	0.40%	**0.8512**	**0.3887**	0.27%

As shown in the table, the three models we designed have eliminated part of the ICL, which has a notable impact on performance. This demonstrates that the ICL module we designed is capable of effectively extracting context-sensitive information and plays a crucial role in CTR prediction models. Compared to Non-ICL, Non-Cross only considers and integrates the implicit context sensitive information, resulting in a 0.19% increasing in AUC on Criteo. On the other hand, Non-MLP only considers and integrates the explicit sensitive information, leading a 0.15% increasing in AUC compared to Non-ICL on Criteo. Above cases shows that considering either explicit or implicit context sensitive information can notably improve the CTR prediction performance.

We believe that the impact of explicit and implicit context sensitive information may vary across datasets. For example, on the Criteo dataset, the AUC of Non-Cross decreased by 0.21% compared to EFNet, while on the Anime dataset, the decrease is only 0.18%. Moreover, compared to Non-ICL, which lacks both explicit and implicit context sensitive information, the AUC of Non-Cross increased by 0.19% on the Criteo dataset and 0.09% on the Anime dataset. EFNet, which incorporates both explicit and implicit context sensitive information, experienced an increase in AUC of 0.40% on the Criteo and 0.27% on the Anime when compared to Non-ICL. These examples demonstrate that considering both explicit and implicit context sensitive information can notably improve the performance on different datasets.

5.7 Portability Validation (RQ4)

We integrate EFNet into DNN, DeepFM and DCN-V2, which have the clearest and simplest structure, to avoid being influenced by other model factors. We compare the original models with these models fused with EFNet on Criteo and Anime dataset. The results are shown in Table 3.

Table 3. Comparison of the performance of three different models added EFNet with the original models.

	Criteo			Anime		
Model	AUC	LogLoss	ΔAUC↑	AUC	LogLoss	ΔAUC↑
DNN	0.7920	0.4578	0.63%	0.8461	0.3931	0.30%
DNNEF	0.7970	0.4549	-	0.8486	0.3855	-
DeepFM	0.8002	0.4547	0.52%	0.8475	0.3855	0.14%
DeepFMEF	0.8047	0.4511	-	0.8487	0.3811	-
DCN-V2	0.8037	0.4521	0.40%	0.8489	0.3893	0.27%
DCN-V2EF	0.8069	0.4497	-	0.8512	0.3887	-

The conclusion can be drawn that integrating EFNet into DNN, DeepFM, and DCN-V2 leads to notable AUC improvements in AUC. This suggests that models that take context sensitive information into account overperform the original models. Furthermore, it indicates that considering both the original embedding features and different order cross features is more advantageous for the model than not considering them. Lastly, the results also demonstrate the transferability and effectiveness of our proposed EFNet in CTR prediction models.

Fig. 6. The heat map of two instances. The upper one is the representation of the original features.

Meanwhile, we select two instances in Anime, extracts their original and enhanced features that integrated context sensitive information. We employ the depth of color to represent their numerical values and generate a heat map, as shown in Fig. 6. The distinct color variation represents the differentiated representation of the original features and enhanced features. The color differences indicate that the context sensitive information has significantly altered the distribution of the original features, which is highly beneficial for overall performance improvement. Moreover, a comparison of the two instances highlights the uniqueness of the context sensitive information in each instance, further confirming the dynamic and adaptable nature of our proposed model in capturing and integrating context sensitive information.

6   Conclusion

In this paper, we propose a novel module named EFNet, which can capture the critical context sensitive information and be applied to various CTR prediction models to improve their performance. In EFNet, we design a structure Information Capture Layer to respectively extract explicit and implicit context sensitive information and their corresponding weighs. We also employ an Enhancing Feature Layer to filter and fuse context sensitive information with original features. Experiments on four public datasets demonstrate that the model equipped with EFNet outperform other current CTR prediction models, comprehensively proving the effectiveness and transferability of the EFNet module and the crucial role of each component in EFNet.

Acknowledgements. This research is supported by the Research Found for Talented Scholars of Hebei University under the grant No. 521100221088.

References

1. Fain, D.C., Pedersen, J.: Sponsored search: a brief history. Bull. Am. Soc. Inf. Sci. Technol. **32**(2), 12–13 (2006)
2. Zhou, G., Mou, N., Fan, Y., Pi, Q., Gai, K.: Deep interest evolution network for click-through rate prediction. Proceedings of the AAAI Conference on Artificial Intelligence **33**(2), 5941–5948 (2019)
3. Zhou, G., et al.: Deep interest network for click-through rate prediction. Proceedings of the 24th ACM SIGKDD International Conference on Knowledge Discovery & Data Mining, pp. 1059–1068. Association for Computing Machinery, London, United Kingdom (2018)
4. Guo, H., Tang, R., Ye, Y., Li, Z., He, X.: DeepFM: a factorization-machine based neural network for CTR prediction. Proceedings of the 26th International Joint Conference on Artificial Intelligence, pp. 1725–1731. AAAI Press, Melbourne, Australia (2017)
5. Cheng, Y.: Dynamic explicit embedding representation for numerical features in deep CTR prediction. Proceedings of the 31st ACM International Conference on Information & Knowledge Management, pp. 3888–3892. Association for Computing Machinery, Atlanta, GA, USA (2022)
6. Long, L., Yin, Y., Huang, F.: Hierarchical attention factorization machine for CTR prediction. In: Database Systems for Advanced Applications, pp. 343–358. Springer International Publishing (2022). https://doi.org/10.1007/978-3-031-00126-0_27
7. Li, X., Wang, Z., Wu, X., Yuan, B., Wang, X.: Input enhanced logarithmic factorization network for CTR prediction. In: Advances in Knowledge Discovery and Data Mining, pp. 42–54. Springer International Publishing (2022). https://doi.org/10.1007/978-3-031-05981-0_4
8. Song, W., et al.: AutoInt: automatic feature interaction learning via self-attentive neural networks. Proceedings of the 28th ACM International Conference on Information and Knowledge Management, pp. 1161–1170. Association for Computing Machinery, Beijing, China (2019)

9. Wang, R., Fu, B., Fu, G., Wang, M.: Deep & cross network for ad click predictions. Proceedings of the ADKDD'17, Article 12. Association for Computing Machinery, Halifax, NS, Canada (2017)

10. Wang, R., et al.: DCN V2: improved deep & cross network and practical lessons for web-scale learning to rank systems. Proceedings of the Web Conference 2021, pp. 1785–1797. Association for Computing Machinery, Ljubljana, Slovenia (2021)

11. Karatzoglou, A., Hidasi, B.: Deep learning for recommender systems. Proceedings of the Eleventh ACM Conference on Recommender Systems, pp. 396–397. Association for Computing Machinery, Como, Italy (2017)

12. Lian, J., Zhou, X., Zhang, F., Chen, Z., Xie, X., Sun, G.: xDeepFM: combining explicit and implicit feature interactions for recommender systems. Proceedings of the 24th ACM SIGKDD International Conference on Knowledge Discovery & Data Mining, pp. 1754–1763. Association for Computing Machinery, London, United Kingdom (2018)

13. Rendle, S.: Factorization Machines. 2010 IEEE International Conference on Data Mining, pp. 995–1000 (2010)

14. Li, Z., Cheng, W., Chen, Y., Chen, H., Wang, W.: Interpretable click-through rate prediction through hierarchical attention. Proceedings of the 13th International Conference on Web Search and Data Mining, pp. 313–321. Association for Computing Machinery, Houston, TX, USA (2020)

15. Chen, B., et al.: Enhancing explicit and implicit feature interactions via information sharing for parallel deep CTR models. Proceedings of the 30th ACM International Conference on Information & Knowledge Management, pp. 3757–3766. Association for Computing Machinery, Virtual Event, Queensland, Australia (2021)

16. Yu, R., et al.: XCrossNet: feature structure-oriented learning for click-through rate prediction. In: Advances in Knowledge Discovery and Data Mining, pp. 436–447. Springer International Publishing (2021). https://doi.org/10.1007/978-3-030-75765-6_35

17. Krizhevsky, A., Sutskever, I., Hinton, G.E.: ImageNet classification with deep convolutional neural networks. Commun. ACM. ACM **60**, 84–90 (2017)

18. Mikolov, T., Karafiát, M., Burget, L., Cernock, J., Khudanpur, S.: Recurrent neural network based language model. In: Interspeech, Conference of the International Speech Communication Association, Makuhari, Chiba, Japan (2015)

19. Huang, T., Zhang, Z., Zhang, J.: FiBiNET: combining feature importance and bilinear feature interaction for click-through rate prediction. Proceedings of the 13th ACM Conference on Recommender Systems, pp. 169–177. Association for Computing Machinery, Copenhagen, Denmark (2019)

20. Hu, J., Shen, L., Sun, G.: Squeeze-and-excitation networks. In: 2018 IEEE/CVF Conference on Computer Vision and Pattern Recognition, pp. 7132–7141 (2018)

21. Wang, F., et al.: Enhancing CTR prediction with context-aware feature representation learning. Proceedings of the 45th International ACM SIGIR Conference on Research and Development in Information Retrieval, pp. 343–352 (2022)

22. Pan, Y., Yao, J., Han, B., Jia, K., Zhang, Y., Yang, H.: Click-through Rate Prediction with Auto-Quantized Contrastive Learning. ArXiv abs/2109.13921 (2021)

23. Bayer, I.: FastFM: a library for factorization machines. J. Mach. Learn. Res. **17**, 6393–6397 (2016)

24. Wang, R., Shivanna, R., Cheng, D., Jain, S., Chi, E.: DCN V2: improved deep & cross network and practical lessons for web-scale learning to rank systems. In: WWW'21: The Web Conference 2021 (2021)

25. He, K., Zhang, X., Ren, S., Sun, J.: Deep residual learning for image recognition. In: 2016 IEEE Conference on Computer Vision and Pattern Recognition (CVPR), pp. 770–778 (2016)

ProtoMix: Learnable Data Augmentation on Few-Shot Features with Vector Quantization in CTR Prediction

Haijun Zhao[1,2], Ronghai Xu[1,2], Chang-Dong Wang[1,2], and Ying Jiang[1,2(✉)]

[1] School of Computer Science and Engineering, Sun Yat-sen University, Guangzhou, Guangdong, China
`{zhaohj23,xurh6}@mail2.sysu.edu.cn, jiangy32@mail.sysu.edu.cn`
[2] Key Laboratory of Machine Intelligence and Advanced Computing, Ministry of Education, Guangzhou, China

Abstract. Click-Through Rate (CTR) prediction is a critical problem in recommendation systems since it involves enormous business interest. Most deep CTR model follows an *Embedding & Feature Interaction* paradigm. However, the feature interaction module cannot work well without a good embedding representation of features. Due to the long-tail phenomenon in real scenes, few samples are provided in the dataset for a large proportion of features. In this paper, we present ProtoMix, a model-agnostic framework for learnable data augmentation on few-shot features in CTR prediction. ProtoMix automatically extracts information from co-occurred features within the same instance to assign prototype embedding with vector quantization for few-shot features and further synthesize the embedding representation of the augmented virtual instance for training. Original embedding, feature interaction module, and the embedding generator are jointly trained on a well-designed objective in an end-to-end manner in ProtoMix. We experimentally validate the effectiveness and compatibility of ProtoMix by comparing it with baseline and other data augmentation methods on different deep CTR models and multiple real-world CTR benchmark datasets.

Keywords: Click-Through Rate Prediction · Data Augmentation · Vicinal Risk Minimization

1 Introduction

Click-through Rate (CTR) Prediction plays a crucial role in online advertising and recommendation systems. The CTR task estimates the probability that a user will click on a recommended item under a specific context. Most recommendations are based on the predicted CTR values, so researchers focus on building more expressive CTR prediction models. With the success of deep learning in computer vision and natural language processing, more deep CTR models are proposed and deployed in industrial, such as Wide & Deep in Google Play and DIN in Taobao.

© The Author(s), under exclusive license to Springer Nature Switzerland AG 2023
X. Yang et al. (Eds.): ADMA 2023, LNAI 14176, pp. 709–723, 2023.
https://doi.org/10.1007/978-3-031-46661-8_47

Most deep CTR models follow an *Embedding & Feature Interaction* paradigm [11]. Most of the work focus on designing delicate structures to capture the feature interaction and exploiting the great approximation ability of deep neural network, e.g., wide component and deep component in Wide & Deep, cross layer as parallel component in DCN, attention mechanism for behavioral modeling in DIN. Some of these components are ensembled to avoid undesirable behaviors of deep neural network such as overfitting and improve the generalization performance in real-world scenarios. It implies that we can further unlock the potential of deep learning methods in the CTR task with proper design and tuning.

Embedding is a fundamental technique in recommendation systems, as it converts sparse categorical features into dense embedding vectors. These vectors enable CTR models to generalize well on novel instances with varying feature compositions. However, the long-tail phenomenon in real-world data often results in "head" features with a much higher frequency of occurrence than "tail" features. Consequently, the insufficient number of training samples for "tail" features makes it challenging to obtain an adequate embedding vector representation. Moreover, since the optimization steps for "tail" features are limited, their embedding distributions may differ from those of "head" features, which can further affect the model's performance.

Drawing inspiration from the success of injecting prior knowledge through data augmentation methods in other deep learning applications, we also intend to explore the application of this technique in the field of CTR prediction. However, unlike the semantic invariance of images in computer vision or the synonymous nature of words in natural language processing, it is challenging to derive a fixed rule for data augmentation based on the relationship between input features in CTR prediction. Furthermore, due to the sparsity of feature combinations, the method of generating novel training samples using convex combinations of original training samples in Mixup cannot gain ideal performance in CTR.

We propose ProtoMix, a model-agnostic framework learning data augmentation for few-shot features with vector quantization mechanism. Our framework enhance the deep CTR model in both parts of *Embedding* and *Feature Interaction*. ProtoMix first generates an augmented virtual instance for each training instance in the dataset. The embedding vectors of augmented virtual instance are obtained by mixing the few-shot features embedding with assigned prototype embedding in each field heavily affected by long-tailed phenomenon. By training on both augmented virtual instances and original instances, feature interaction modules in deep CTR models also implicitly improve the generalization and reduce the sensitivity of few-shot feature embeddings.

The contributions of this paper are:

- We propose ProtoMix, a model-agnostic framework for learnable data augmentation with few-shot features. By exploiting the co-occurred features to generate augmented embedding for few-shot features and jointly trained on both original embedding and augmented embedding, ProtoMix allows significant improvement of performance on deep CTR models.

- We first unify the existing approaches for cold-start problem and data augmentation methods from the perspective of vicinal risk minimization. Furthermore, we bridge the inner connection among them, and derives from both of them in our proposed ProtoMix.
- Experiments on multiple datasets and models demonstrates the effectiveness and compatibility of ProtoMix in CTR task. Comparison experiments show the superiority of components in ProtoMix over other data augmentation methods.

2 Related Work

2.1 Embedding

Most of the works in recommendation systems follow the diagram of *Embedding & Feature Interaction* and design the delicate feature interaction module to capture and exploit the feature interaction better [6]. A recent study [11] shows that minor improvement is actually achieved in CTR feature interaction architecture design. Several works turn to the embedding part in different ways [3,4]. These works show significant success without redesigning new architecture for feature interaction, which prompts great potential on enhanced embedding in recommendation systems.

Moreover, extra modification on the embedding part also shows promising results on cold-start problem. Cold-start problem is a long-standing problem in recommendation systems. The problem directly stems from the scarcity of user-item interaction, which makes id embedding of user/item not informative enough for accurate recommendation. The modification of embedding by cold-start method can be summarized into two categories. The first is content-based, which utilizes the attribute information to enhance the id embedding. Another one is DropoutNet, which introduces noise on id embedding in the training procedure, making the downstream feature interaction module more robust to id embedding.

2.2 Data Augmentation

Data augmentation is a series of methods to increase the diversity of training examples without collecting new data. It is a widely used technique in deep learning, which helps a lot in reducing overfitting and increasing the generalization performance. Based on the priors assumed by the data augmentation methods, we categorize these methods into two groups.

The first ones are semantic-based methods, which apply transformations specific on certain modal of data, to generate new samples without affecting semantic information [2,10] For instance, random cropping and flipping applied in training deep learning computer vision models keep the image information unchanged while generating new samples.

The other ones are metric-based methods, which apply simple mathematical operations on data. Most of them are modal-agnostic. Mixup [9] utilizes the element-wise convex combination to generate new training samples with new labels, and dropout masks several input elements to generate new training intermediate features without changing labels.

In the context of click-through rate (CTR) or general recommendation models, most data augmentation techniques [1,7] are often designed and applied to sequence features. However, our proposed method directly affects the feature embeddings themselves, potentially making it applicable to a wider range of scenarios.

3 Methodology

Our goal is to learn data augmentation that leverages information from the co-occurring features within the same instance to generate embeddings to enhance the original few-shot feature embeddings. Specifically, the embedding of a virtual instance is then obtained by concatenating the generated embedding of *long-tailed* fields and the original embedding of the refiging fields. Embedding module, embedding generator, and feature interaction module are jointly trained on synthesized virtual instances and real original instances, using multiple optimization objectives. We have named our proposed framework ProtoMix, and its overview is illustrated in Fig. 1.

In Sect. 3.1, we first review the CTR prediction task and vicinal risk minimization; In Sect. 3.2, we introduce our prototype embedding generator with vector quantization mechanism and adaptive mixing for virtual instance generation; In Sect. 3.3, we introduce our training pipeline with prototype embedding learning as an auxiliary task; In Sect. 3.4, we build up a connection between our method and vicinal risk minimization, a family of learning algorithms.

3.1 Preliminaries

CTR Prediction. CTR prediction task is mostly formulated as a binary classification problem. For each instance $\{\mathbf{x}, y\}$ in dataset $\mathcal{D}$, $y \in \{0, 1\}$ is label observed from the user implicit feedback and $\mathbf{x}$ is a multi-field data:

$$\mathbf{x} = [x_1, x_2, \ldots, x_M], \tag{1}$$

where M is the number of feature fields, $x_i \in \{0, 1\}^{v_i}$ is the one-hot vector of the feature in the i-th categorical field. And the corresponding feature embedding can be obtained by embedding look-up operation on each embedding table and the embedding representation of the multi-field data $\mathbf{x}$ can be constructed as $\mathbf{e} = [e_1, e_2, \ldots, e_M] \in \mathbb{R}^{M \times d}$.

Following the paradigm of *Embedding & Feature Interaction*, the estimated CTR $\hat{y}$ can be obtained by feeding the embedding representation $\mathbf{e}$ into the (parameterized) feature interaction module f_θ,

$$\hat{y} = \sigma(f_\theta(\mathbf{e})), \tag{2}$$

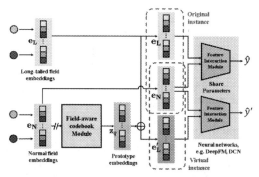

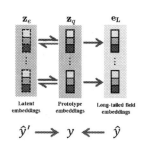

(a) Components of deep CTR models in ProtoMix

(b) Visualization of training objectives in ProtoMix

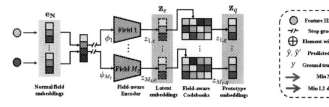

(c) The detail of field-aware prototype embedding generator

Fig. 1. The overview of our proposed framework ProtoMix for data augmentation in CTR task

where $f_\theta(\cdot)$ refers to the delicate feature interaction network parameterized by θ, and $\sigma(\cdot)$ is the sigmoid activation function. Both embedding matrices and feature interaction module parameters are jointly trained end-to-end by minimizing the loss function. Usually, the loss function is the widely-adopted negative log-likelihood:

$$\mathcal{L} = \ell(\hat{y}, y) = -\frac{1}{|\mathcal{D}|} \sum_{\{\mathbf{x},y\} \in \mathcal{D}} y \log(\hat{y}) + (1 - y) \log(1 - \hat{y}). \tag{3}$$

Vicinal Risk Minimization. In statistical learning theory, a supervised learning problem can be formulated as the search of the function $f \in \mathcal{F}$ that minimizes the generalization risk with surrogate loss function $\ell(f(\mathbf{x}, y))$,

$$R(f) = \int \ell(f(\mathbf{x}), y) dP(\mathbf{x}, y). \tag{4}$$

Though the real target distribution $P(\mathbf{x}, y)$ is unknown, generalization risk is unbiasedly approximated by the empirical risk on the given training dataset:

$$R_e(f) = \frac{1}{n} \sum_{i=1}^{n} \ell(f(\mathbf{x}_i), y_i). \tag{5}$$

Therefore, Empirical Risk Minimization (ERM) is minimizing the expectation of the loss function with respect to an empirical distribution $P_e(\mathbf{x}, y)$, which approximates the entire data-distribution space by a finite n number of delta functions located at each $(\mathbf{x}_i, y_i)$:

$$P_e(\mathbf{x}, y) = \frac{1}{n} \sum_{i=1}^{n} \delta_{\mathbf{x}_i}(\mathbf{x}) \delta_{y_i}(y). \tag{6}$$

Vicinal Risk Minimization approximates the generalization risk on a richer distribution, which approximates the entire data-distribution space by a finite n number of vicinity located at each $(\mathbf{x}_i, y_i)$:

$$P_v(\mathbf{x}, y) = \frac{1}{n} \sum_{i=1}^{n} P_{\mathbf{x}_i, y_i}(\mathbf{x}, y), \tag{7}$$

where $P_{\mathbf{x}_i, y_i}(\mathbf{x}, y)$ denotes the user-defined vicinal distribution around the i-th sample. The vicinal risk has an integral form and can be approximated by Monte Carlo estimate with samples.

ERM has been successfully and widely adopted to solve machine learning problems. For instance, optimizing on Eq. (4) is ERM. However, ERM may suffer from the scarcity of sample points, which leads P_e a poor approximation to the true distribution P. Besides, when the function class is extremely rich with very high capacity, undesirable behaviors such as overfitting and memorization limit the performance of the model. VRM alleviates the deficiency above by replacing the empirical distribution with the user-defined vicinal distribution. Training dataset with various data augmentations can be viewed as user-defined vicinal distribution, and optimizing on such training dataset can be viewed as VRM.

3.2   Virtual Instance Generation

The vicinal distribution defined by mixup is simple and effective in other deep learning applications. However, it doesn't work well in deep CTR prediction since such directly linear behavior prior is not suitable in non-linear feature interaction. Our key motivation is to define a more flexible vicinal distribution, which satisfies two requirements. First, the vicinal region should be flexible across different fields, especially should be of wider range for the few-shot features in *long-tailed* field. Second, the vicinality in such distribution is not only metric-based but also semantic-based.

In this part, we introduce virtual instance generation, which defines the vicinal distribution applied in ProtoMix. For each *long-tailed* field, a prototype embedding generator helps define a semantic-based and flexible vicinal distribution by exploiting the co-occurred feature in other fields and generating prototype embedding as a mixing component. The virtual instance is then synthesized by mixing $\mathbf{z}_q$ and $\mathbf{e_L}$ with an adaptive coefficient based on feature frequency.

Prototype Embedding Generator. For j-th long-tailed field, we define a latent embedding space $\mathbf{Z}_j \in \mathbb{R}^{K \times d}$, where K is the size of the discrete latent space. $\mathbf{Z}_j$ consists of K prototype embedding vectors $z_{j,i} \in \mathbb{R}^d, i \in 1, 2, \ldots, K$. We like to choose a matching prototype embedding vector embedding in the codebook $\mathbf{Z}_j$ for augmentation on $\mathbf{e}_j$. To utilize the information from the co-occurred features, the encoder ϕ_j takes $sg[\mathbf{e_N}]^1$ as input, and outputs a latent embedding $z_{j,e} \in \mathbb{R}^d$. Note that $sg(\cdot)$ operation applied here is to avoid the embedding module from receiving the gradient backpropagated via our embedding generator. The encoder can be in various forms, and MLP with only one hidden layer is adopted here.

The chosen prototype embedding for augmentation is obtained by mapping the latent embedding $z_{j,e} \in \mathbb{R}^d$ onto the nearest prototype embedding,

$$z_{j,q} = z_{j,k}, \text{where} \quad k = \text{argmin}_{1 \leq i \leq K} \|z_{j,e} - z_{j,i}\|_2 \qquad (8)$$

Since the nearest assignment operation is not derivable, no gradient is available for updating the encoder ϕ_j. The implementation trick of straight through estimator is used for gradient approximation, which copies gradients from $z_{j,q}$ to $z_{j,e}$.

To learn suitable latent space for data augmentation, extra constraints on optimization objectives are needed. We further introduce them in Sect. 3.3.

Adaptive Mixing for Virtual Instance. Mixing weight is important for Mixup-type methods. Traditional Mixup use random weight sampled from Beta distribution. Since our mixing is proposed for augmenting the few-shot features, we would like to augment more on the few-shot features by an adaptive mixing weight. Inspired by [8], an adaptive mixing weight based on the frequency of the features for mixing embedding is introduced. The generated embedding serves as a vicinal datapoint around the original embedding by mixing the prototype embedding and the original embedding adaptively,

$$e'_j = \sigma(\text{freq}(e_j))e_j + (1 - \sigma(\text{freq}(e_j)))z_{j,q}, \qquad (9)$$

where $\text{freq}(\cdot)$ denotes the accumulated frequency of the feature in the training dataset. The more frequently features appear in the train dataset, the more prior information from normal fields injected into the mixed embedding. Since feature frequence is calculated on only the train dataset, and performance in the test dataset shows the generalization ability of the method.

The generated embedding of *long-tailed* fields $\mathbf{e_{L'}}$ are then obtained by concatenating the generated embedding e'_j of each j-th *long-tailed* field. And we get a virtual instance and its corresponding embedding for data augmentation:

$$\begin{aligned}
\text{original instance} : \{\mathbf{x}, y\} &\xrightarrow{Embedding} \{[\mathbf{e_N}, \mathbf{e_L}], y\}, \\
\text{virtual instance} : \{\mathbf{x}', y\} &\xrightarrow{Embedding} \{[\mathbf{e_N}, \mathbf{e_L}'], y\}.
\end{aligned} \qquad (10)$$

---

[1] sg stand for the stop gradient operator that is defined as identity at forward computation time and has zero partial derivatives.

3.3   Mixed Training

Our training pipeline follows a simple end-to-end manner. For each instance $\{\mathbf{x}, y\}$ in training dataset $\mathcal{D}$, two forward computation processes are executed on both the original instance and virtual instance to obtain the prediction $\hat{y}$ and $\hat{y}'$. And loss functions are summed in the form of convex combination with coefficient γ:

$$\mathcal{L}_{log} = \gamma \ell(\hat{y}, y) + (1 - \gamma)\ell(\hat{y}', y), \tag{11}$$

where $\hat{y}$ and $\hat{y}'$ are the output of deep CTR model with input $[\mathbf{e_N}, \mathbf{e_L}]$ and $[\mathbf{e_N}, \mathbf{e_L}']$.

Following the [5], three extra loss components are needed to learn reasonable dictionary codebooks $\mathbf{Z}$. The first term is align loss which pulls the prototype embedding together with the original embedding for each field in $long - tailed$ fields $\mathbf{L}$:

$$\mathcal{L}_{align} = \sum_{j \in \mathbf{L}} \|z_{j,q} - \text{sg}[e_j]\|_2^2 \tag{12}$$

The vector quantization objective uses the $l2$ error to move the prototype embedding vectors $z_{j,q}$ towards the encoder outputs $z_{j,e}$.

$$\mathcal{L}_{embed} = \sum_{j \in \mathbf{L}} \|z_{j,q} - \text{sg}[z_{j,e}]\|_2^2 \tag{13}$$

Finally, since the volume of the embedding space is dimensionless, it can grow arbitrarily if the prototype embeddings $z_{j,q}$ do not train as fast as the encoder parameters. To ensure the encoder commits to an embedding and its output does not grow, we add a commitment loss,

$$\mathcal{L}_{commit} = \sum_{j \in \mathbf{L}} \|\text{sg}[z_{j,q}] - z_{j,e}\|_2^2 \tag{14}$$

Thus, the total training objective becomes:

$$\mathcal{L} = \mathcal{L}_{log} + \beta_1 \mathcal{L}_{align} + \beta_2 \mathcal{L}_{embed} + \beta_3 \mathcal{L}_{commit}. \tag{15}$$

We set $\beta_2 = 1$ and $\beta_3 = 0.5$ constantly in our implementation as [5]. β_1 is determined by hyperparameter tuning on each dataset, and its specific value are presented Sect. 4.1.

3.4   Connection Between ProtoMix and VRM

As discussed in Sect. 2.2, data augmentation methods can be categorized into two groups: structure-prior-based and simple mathematical operation based. These data augmentation methods turn the ERM in supervised learning into VRM by alternating the approximate target distribution from empirical distribution to vicinal distribution. Structure-prior-based data augmentation methods build a structure-prior-based (semantic-based) vicinal distribution, while simple mathematical operation-based ones build a simple mathematical operation-based (metric-based) vicinal distribution. In this view, ProtoMix generates virtual instances for training, which defines both semantic-based and metric-based vicinal distribution. However, the excessively wide or improper vicinal region may also degrade the performance of the model. (See it in Sect. 4.2.1)

To utilize the vicinal distribution while preventing performance degradation, ProtoMix combines these ideas and trivially balances them. ProtoMix is trained on a hybrid objective, which combines both empirical distribution and vicinal distribution with hyperparameter γ,

$$P_{\mathbf{x},y} = \frac{1}{n} \sum_{i=1}^{n} (\gamma \delta_{\mathbf{x}_i}(\mathbf{x}) \delta_{y_i}(y) + (1 - \gamma) P_{\mathbf{x_i},y_i}(\mathbf{x}, y)), \tag{16}$$

where the vicinal distribution $P_{\mathbf{x_i},y_i}(\mathbf{x}, y) = \mathbb{E}(\delta_{\mathbf{x}'_i}(\mathbf{x}), \delta_{y_i}(y))$ is discrete and dynamically adaptive in the training procedure. This is because the vicinal distribution is defined by the co-learning prototype embeddings codebook.

4 Experiments

We conduct the experiments to answer the following research questions:

- **RQ1**: How do different instance data augmentation methods perform in the CTR prediction?
- **RQ2**: How does ProtoMix perform when plugged into various CTR prediction models?
- **RQ3**: How does each component affect on ProtoMix?

4.1 Experiments Settings

Datasets. We conduct experiments on three public real-world datasets:

- **Taobao Display Ads Click**[2]. The dataset is sampled from the website of Taobao for 8 d of ad display/click logs. It contains 26 million logs generated by 1.14 million users. It is widely used as a benchmark in CTR evaluation.
- **MovieLens-1M**[3]. The dataset is the most widely used benchmark in recommendation systems. It contains 1 million ratings from 6000 users on 4000 movies.
- **Avazu**[4]. The dataset is published for the Kaggle competition and is also widely used as a benchmark in CTR model evaluation. It contains 10 d of 40 million click logs with 23 categorical feature fields.

---

[2] https://tianchi.aliyun.com/dataset/dataDetail?dataId=56.
[3] http://files.grouplens.org/datasets/movielens/1m/.
[4] https://www.kaggle.com/c/avazu-ctr-prediction.

We follow the data splitting setting commonly used in the previous works [6,8, 12]. For Taobao Display Ads Click, the click data are generated from the first seven day logs and used as the training set, and the rest of the data from the last day is used as the test set. For **Avazu** and **Movielens-1M**, data is randomly split into 8:1:1 as the training set, validation set and test set (Table 1).

Table 1. Statistics of the benchmark datasets

Dataset	#instances	#fields	#features	positive ratio
Taobao Ads	26,557,963	15	2,774,595	5.1%
ML-1M	739,015	7	9919	77.8%
Avazu	40,428,970	24	9,449,238	16.9%

Baselines. Since our data augmentation framework is model-agnostic, it can be applied to any deep CTR model following the *Embedding & Feature Interaction* paradigm. We select several representative models to test the general effectiveness of the proposed framework: DNN, Wide & Deep, DCN, DeepFM, xDeepFM, and DCN-v2 [6].

Evaluation Protocols. We adopt **AUC** (Area under the ROC curve) and **Logloss** (binary cross-entropy loss) as the evaluation metrics. All the experiments are repeated five times by changing the random seeds. The unpaired t-test is performed to detect significant differences between each baseline and one with augmentation. Noting that our proposed approach ProtoMix figly affects the few-show features, which occupy less than 10% of instances. The effectiveness of ProtoMix may be underestimated since our experiments are set on the whole dataset to match the real-world scenario. Besides, a **0.001-level** improvement is considered significant in CTR prediction task [6].

Hyper-parameter Settings. We implement all experiments in Pytorch. We optimize all models with mini-batch Adam, where the learning rate is searched from {1e-5, 3e-5, ..., 1e-2} in Taobao Ads and MovieLens-1M, and fixed at 1e-3 in Avazu. The embedding size is set to 16 in all datasets. Besides, the hidden layers in deep CTR models are fixed to 256-256-256 by default in Taobao Ads and MovieLens-1M, and 512-512-512 in Avazu. Explicit feature interaction modules in DCN, xDeepFM, and DCN-v2 are set to 2 layers with the width of 64 neurons. All activation functions applied in MLP are ReLU. All the embedding modules are initialized by $\mathcal{N}(0, 0.0001)$, while linear modules are initialized by Xavier uniform with zero bias term. For our method, the number of prototype embeddings in each long-tailed field is searched in {1, 2, 4,..., 32}, γ is searched in {0.25, 0.5, 0.75}, β_1 is searched in {0, 1e-5, 1e-4, ..., 1}. We select fields *user-id* and *item-id* as *long-tailed* fields for MovieLens-1M and Taobao Ads, *device-id* and *device-ip* for Avazu. γ are selected as 0.25 for all datasets, and β_1 are selected as 0.1 for Taobao Ads, 1 for MovieLens-1M, 0.01 for Avazu.

4.2   Overall Performance (RQ1 and RQ2)

In this section, we first compare the performance of different data augmentation methods on the same baseline CTR model DNN. Then, we evaluate the generalized effectiveness of our proposed framework by showing improvement over six representative deep CTR models.

Compared with Other Data Augmentation Methods. In this study, we applied various data augmentation techniques to Taobao Ads and Movielens-1M datasets. Specifically, we compared the performance of different data augmentation methods based on Mixup. To be more specific, we conducted comparison experiments using two different implementations of Mixup for the CTR task. The first implementation followed the original Mixup, where two training instances were sampled from the same batch and all inputs were mixed after embedding. The second implementation was an alternative version of Mixup, where only embeddings of selected fields were mixed. In our experiments, we tuned the hyperparameter α in Mixup to be 0.1, 0.2, 0.5, 1, and 2. We used the term "Mixup all" to refer to the first implementation and "Mixup u/i" to refer to the second. The performance comparison results are presented in Table 2. Based on our observations, we draw the following conclusions:

- ProtoMix yields robust improvement to the baseline model on different datasets, while other data augmentation methods perform inconsistently. Moreover, standard Mixup may still lead a performance degradation after comprehensive hyperparameters tuning.
- Though other data augmentation methods improve the Logloss metric in most of the experiments, their effects on AUC metric are comparatively weak. We infer that data augmentation methods may increase the fitting of model by introducing more interpolation prior to Mixup training but harm the fine-grained discriminative power of the model, which is a critical deficit in the CTR task.
- In the hyperparameter tuning phase, ProtoMix usually shares the same hyperparameter setting (e.g., *lr*, epochs) as the baseline training methods, while other data augmentation methods may vary a lot. ProtoMix follows a similar training dynamic to baseline training methods, while other Mixup-based data augmentation methods alter the training dynamic.

Compatibility with Different CTR Models. ProtoMix is a model-agnostic framework that can function as a plug-and-play module to enhance the performance of various deep CTR models by training with augmented few-shot features. We conducted extensive experiments by applying ProtoMix to six representative deep CTR baseline models, and the results are presented in Table 3.

ProtoMix yielded significant improvement on Taobao Ads and MovieLens-1M, while the improvement on Avazu was relatively modest. However, it is worth noting that despite the evolution of deep CTR models in recent years, these representative deep CTR models have slight differences in their performance. Additionally, ProtoMix figly operates on few-shot features, which represent a small proportion of samples in the dataset. ProtoMix demonstrated considerable effectiveness within the aforementioned context.

Table 2. Results on model DNN and DCN-v2 with different data augmentation methods

Methods	Taobao Ads		MovieLens-1M	
	AUC	Logloss	AUC	Logloss
DNN	0.63573	0.19483	0.89394	0.32368
w/Mixup all	0.63515	0.19479	**0.89618**	**0.31623**
w/Mixup u/i	0.63033	0.19506	0.89059	0.32777
w/ProtoMix	**0.63996**	**0.19368**	0.89468	0.32170
DCN-v2	0.63588	0.19472	0.89637	0.31907
w/Mixup all	0.63530	0.19480	0.89613	**0.31583**
w/Mixup u/i	0.63286	0.19442	0.89080	0.32286
w/ProtoMix	**0.64022**	**0.19354**	**0.89686**	0.31748

4.3   Ablation Study (RQ3)

To demonstrate the effectiveness of our proposed method, we conduct experiments on Taobao Ads that compare it to several of its ablated variants. We divide our discussion into a few perspectives concerned with our framework.

Vector Quantization. To demonstrate the effectiveness of our embedding generating with vector quantization, we compare our method to two ablated variants. The first drops the codebook assignment procedure, and the second drops the MLP encode procedure. In both variants, only align loss is kept. We name the first variant as **MLPMix** and the second variant as **RandMix**. With the align loss term, variant RandMix and MLPMix can both be regarded as adding identically distributed noise in training, whose coefficient is determined by the frequency of the features. Variant MLPMix shows the effectiveness of the co-occurred feature-based generative mechanism for embedding compared to variant RandMix. Vector quantization mechanism, including commitment loss and embedding loss for learning representative prototype embedding, contributes to the performance gain beyond MLPMix (Table 4).

Coefficient β_1 and γ. The coefficient β_1 is responsible for regulating the strength of the constraint that ensures alignment between the generated embedding and the original embedding. On the other hand, the coefficient γ determines the proportion of original instances versus virtual instances used in training.

Table 3. Results on Taobao Ads, Movielens-1M and Avazu Datasets with six representative deep CTR models trained in both w/ProtoMix and w/o ProtoMix ways. Boldface denotes the highest score on each dataset. $\star$ indicates significance level p-value ≤ 0.05 of comparing w/ProtoMix with w/o ProtoMix in the same model.

Method	Taobao Ads		MovieLens-1M		Avazu	
	AUC	Logloss	AUC	Logloss	AUC	Logloss
DNN	0.63573	0.19483	0.89394	0.32368	0.79416	0.37147
DNN w/ProtoMix	**0.63996***	**0.19368***	**0.89468***	**0.32170***	**0.79440***	**0.37121***
Wide&Deep	0.63573	0.19476	0.89393	0.32333	0.79451	0.37120
Wide&Deep w/ProtoMix	**0.64001***	**0.19365***	**0.89461***	**0.32255**	**0.79465***	**0.37107**
DeepFM	0.63601	0.19453	0.89583	0.31900	0.79466	0.37113
DeepFM w/ProtoMix	**0.64002***	**0.19354***	**0.89646***	**0.31772**	**0.79478**	**0.37097**
DCN	0.63607	0.19478	0.89461	0.32227	0.79450	0.37120
DCN w/ProtoMix	**0.64022***	**0.19354***	**0.89569***	**0.32016***	**0.79473***	**0.37107**
xDeepFM	0.63596	0.19480	0.89445	0.31997	0.79478	0.37110
xDeepFM w/ProtoMix	**0.64022***	**0.19361***	**0.89551***	**0.32058**	**0.79503***	**0.37103**
DCN-v2	0.63588	0.19472	0.89637	0.31907	0.79475	0.37103
DCN-v2 w/ProtoMix	**0.64011***	**0.19358***	**0.89686***	**0.31748***	**0.79483**	**0.37098**

Table 4. Performance of ProtoMix and its variants on Taobao Ads dataset

Method	Taobao Ads	
	AUC	Logloss
DNN	0.63573	0.19483
DNN w/RandMix	0.63603	0.19446
DNN w/MLPMix	0.63682	0.19378
DNN w/ProtoMix	**0.63996**	**0.19368**

In the case of coefficient β_1, we conducted a grid search for hyperparameter tuning and analyzed the results, which are presented in Fig. 2(a). Our findings indicate that higher values of β_1 lead to better performance, indicating the significance of aligning the generated embedding with the original embedding. However, it is noteworthy that the align loss alone, without other components specifically designed in ProtoMix, cannot achieve the same level of performance as the variant RandMix and MLPMix.

To investigate the impact of coefficient γ, we conducted experiments under two different settings, where β_1 was fixed at 1 and 0.01, respectively. The results are presented in Fig. 2(b) and 2(c). When β_1 was set to 1, the performance of ProtoMix was not significantly influenced by different values of γ. However, in the case where β_1 was set to 0.01, a relatively small value of γ yielded better performance. This suggests that the quality of the virtual instances generated by ProtoMix can influence the model's preference bias between original and virtual instances during training.

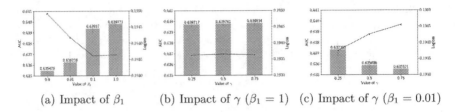

(a) Impact of β_1 (b) Impact of γ ($\beta_1 = 1$)   (c) Impact of γ ($\beta_1 = 0.01$)

Fig. 2. Impact of the coefficient γ and β_1

5   Conclusion

In this paper, we introduce ProtoMix, a model-agnostic framework for learnable data augmentation on few-shot features in click-through rate (CTR) prediction. ProtoMix consists of two key components: (1) *virtual instance generation*: the prototype embedding generator leverages context information to generate prototype embeddings, which are then concatenated with the adaptive mixing of original and augmented embeddings to obtain the embedding of a virtual instance; and (2) *mixed training*: the feature interaction module is trained on both original instances and virtual instances, while the prototype embedding generator is simultaneously trained with additional constraints on the learning objective.

Compared to other commonly used data augmentation methods in CTR prediction, which often show inconsistent performance, ProtoMix consistently and significantly improves performance across different datasets and models, demonstrating its effectiveness and compatibility as a data augmentation framework for CTR tasks. Furthermore, our analysis on unifing existing cold-start solutions and data augmentation methods from the perspective of VRM provide insights that may inspire further improvements in the training of deep CTR models.

Acknowledgements. This work is partially supported by the Special Project on High-performance Computing under the National Key R&D Program (No.2016YFB0200602), and the Natural Science Foundation of Guangdong Province, China (No.2022A1515010831).

References

1. Bian, S., Zhao, W.X., Wang, J., Wen, J.R.: A relevant and diverse retrieval-enhanced data augmentation framework for sequential recommendation. In: Proceedings of the 31st ACM International Conference on Information and Knowledge Management, pp. 2923–2932 (2022)
2. Feng, S.Y., et al.: A survey of data augmentation approaches for MLP. arXiv preprint arXiv:2105.03075 (2021)
3. Guo, W., et al.: Dual graph enhanced embedding neural network for CR prediction. In: Proceedings of the 27th ACM SIGKDD Conference on Knowledge Discovery and Data Mining, pp. 496–504 (2021)

4. Pan, F., Li, S., Ao, X., Tang, P., He, Q.: Warm up cold-start advertisements: Improving CTR predictions via learning to learn id embeddings. In: Proceedings of the 42nd International ACM SIGIR Conference on Research and Development in Information Retrieval, pp. 695–704 (2019)
5. Van Den Oord, A., Vinyals, O., et al.: Neural discrete representation learning. In: Advances in Neural Information Processing Systems 30 (2017)
6. Wang, R., et al.: DCN v2: Improved deep AND cross network and practical lessons for web-scale learning to rank systems. In: Proceedings of the Web Conference 2021, pp. 1785–1797 (2021)
7. Xie, X., Sun, F., Liu, Z., Gao, J., Ding, B., Cui, B.: Contrastive pre-training for sequential recommendation. arXiv preprint arXiv:2010.14395 (2020)
8. Xu, X., et al.: Alleviating cold-start problem in CTR prediction with a variational embedding learning framework. arXiv preprint arXiv:2201.10980 (2022)
9. Zhang, H., Cisse, M., Dauphin, Y.N., Lopez-Paz, D.: mixup: beyond empirical risk minimization. arXiv preprint arXiv:1710.09412 (2017)
10. Zhao, T., Liu, Y., Neves, L., Woodford, O., Jiang, M., Shah, N.: Data augmentation for graph neural networks. In: Proceedings of the AAAI Conference on Artificial Intelligence. vol. 35, pp. 11015–11023 (2021)
11. Zhu, J., Liu, J., Yang, S., Zhang, Q., He, X.: Open benchmarking for click-through rate prediction. In: Proceedings of the 30th ACM International Conference on Information and Knowledge Management, pp. 2759–2769 (2021)
12. Zhu, Y., et al.: Learning to warm up cold item embeddings for cold-start recommendation with meta scaling and shifting networks. In: Proceedings of the 44th International ACM SIGIR Conference on Research and Development in Information Retrieval, pp. 1167–1176 (2021)

When Alignment Makes a Difference: A Content-Based Variational Model for Cold-Start CTR Prediction

Jianyu Ren and Ruoqian Zhang[✉]

School of Computer Science and Technology, Soochow University, Suzhou, China
jyren@stu.suda.edu.cn, rqzhang@suda.edu.cn

Abstract. Click-Through Rate (CTR) prediction is a core task in recommendation systems. Despite VAE-based models have shown promising accuracy performance, they are still weak in supporting cold-start CTR prediction due to limited personal interactions. To this end, this paper proposes a content-based variational CTR model, which jointly models content information and interactions behaviors in a shared probability space via variational inference. Specifically, a three-step scheme is designed to fully utilize content information for the improved ability of preference modeling toward cold-start users. First, our method adopts VAE to model user preferences from personal interactions by probabilistic distributions, instead of a fixed embedding vector for representing the user's interest. Then, we transform content information into variational probabilistic distribution to model the implicit preferences of cold-start users. Finally, a variational alignment strategy is applied to maximize the similarity between variational preference distributions obtained from interactions behaviors and content information respectively, so that the interest of the cold user can be recovered. Besides, we adopt a self-attention mechanism to reasonably balance the importance of latent features for CTR prediction. Experiments on two public real datasets show the effectiveness of the proposed approach.

Keywords: Recommender systems · CTR prediction · Cold start · Variational auto-encoder

1 Introduction

Click-through rate (CTR) prediction is an important task to predict whether users will interact with the recommended items. This task plays a central role in many fields such as social applications [27,31] and e-commerce platforms [5,10]. Recent studies [2,6,15,25] have shown that deep neural networks have achieved promising results in CTR prediction. Most of these CTR models follow a two-stage approach of embedding and prediction. It first vectorizes features of users and items and the interactions between them. Then these vectors will be processed with deep neural networks to output a prediction.

ⓒ The Author(s), under exclusive license to Springer Nature Switzerland AG 2023
X. Yang et al. (Eds.): ADMA 2023, LNAI 14176, pp. 724–739, 2023.
https://doi.org/10.1007/978-3-031-46661-8_48

The paradigm called collaborative filtering played an enormous role within these models. It finds crowds who have similar interactions with the current target user, then provides new recommendations for the target user based on the interactions history of the crowds. This pattern requires massive interaction data to guess users' preferences. Hence, when the users are 'cold', i.e., when they have very little and highly sparse historical interaction data, CTR models tend to perform poorly. An intuitive idea for alleviating this challenge is to take advantage of the content-based filtering paradigm [24,26], where the similarity is calculated by the content information of the user profile rather than the interactions between them. This combination of two paradigms solves the user's cold start problem to a certain extent. However, the fly in the ointment is that they are all based on fixed embedding techniques. This leads to the user's preferences being finally fixed somewhere in the latent space through a series of deterministic processes. Evidence [30] has shown that fully deterministic-based schemes are highly data-sensitive and tend to generate embeddings biased toward hot users. Inspired by the successfulness of variational auto-encoder (VAE) in computer vision [21], natural language processing [32] and recommendation system [28,30], the state-of-the-art method [28] applies VAE to CTR prediction tasks and obtains satisfactory results. By modeling the probability distribution of user preferences in the latent probability space, it can generate more reliable user embeddings for further prediction [7].

Unfortunately, classical VAE-based models are still weak in supporting the cold-start CTR problem. Because the existing VAE framework does not directly input the content information as the variational encoder, that is, the modeling process of the content information is implicit. This makes them unable to integrate the user preferences contained in the content information into the user's embedded representation in a variational probabilistic context. In order to unify content information into the VAE framework, we should model interactions behaviors and content information separately and extract the user preferences from them in the form of probability distributions. Next, we should maximize the similarity of the two user preference distributions in the above probability latent space, that is, align the shapes of the distributions as much as possible so that the user preferences they represent are as convergent as possible. Through this alignment strategy, user preferences from two sources can be organically integrated and contribute to generating better cold user embedding representations.

To tackle the above challenges, we propose a **Content-Based Variational Alignment** Model (CBVA). First, CBVA models user interaction history in the form of probability distributions by employing VAE. Instead of the form of fixed embedding vectors, this probabilistic representation avoids the bias issue caused by deterministic methods. Second, in order to further process the content information under the universal context of probability, user features will also be transformed into variational probabilistic distribution. Then a variational alignment strategy will be applied for knowledge integration of user preference. To be specific, the alignment is done by sharing the decoder and using the frequency of interactions data as a condition of variational inference [17]. The alignment integrates the user preference extracted from the content information into the variational probabilistic distribution modeled from interactions data. Therefore the alignment strategy makes it possible to capture cold-start users' preferences with little personal interaction. After that, reconstructed embeddings that contain more accurate user preferences will be generated and crosses between all embeddings will be calculated. We also design an attention-based weight module to more reasonably fuse crossed latent features before making a prediction.

The main contributions of this paper are summarized as follows:

- We propose the CBVA model, which to the best of our knowledge is the first content-based variational model for user cold-start CTR prediction.
- We utilize a variational alignment strategy to reconstruct cold users' complex preferences. Particularly, we model both users' interactions log and content information into probabilistic distribution and then align these latent probabilistic distributions by sharing a decoder and using frequency as the condition. Cold users' preference extracted from content information is efficiently integrated into their embeddings through this.
- We efficiently integrate the representation of user preferences by the proposed attention-based weight module, which balances the importance of crossed latent features more reasonably.
- Experiments on two public datasets demonstrate that CBVA significantly outperforms the state-of-the-art models.

2   Related Works

2.1   Cold-Start Recommendation

The cold start problem refers to the difficulty of recommender systems to properly recommend users with sparse interaction data, which is very common in CTR prediction. An intuitive idea to solve user cold start is using rich content information of items or users, called content-based methods [24,26,33]. [33] is a representative of these methods, which proposes meta scaling and shifting networks to generate scaling and shifting functions using content information for each item and then warm-up user ID embeddings by the two networks. Another major category is the meta-learning framework [3,9,33]. Meta-learning methods attempt to find good initial values for the deep neural network in order to make good recommendations for new users or items by training in a few iterations [18,19,29]. [9] first introduced the meta-learning method into cold-start CTR prediction. Recently, [28] attempted to employ VAE to overcome the shortcoming that existing methods are all based on point estimate which still has a huge risk to result in isolated and unreliable embedding for cold-start and reaching the state-of-the-art in the field. Like [28] and go a step further, we variationally model historical interactions data and content information jointly. Then a variational alignment strategy will be implemented to integrate user preferences extracted from both sides. Embeddings that fit cold user preferences more accurately can be obtained, which benefits the prediction for cold-start users.

2.2 Variational Auto-Encoder in Recommendation

Variational auto-encoders [8] are auto-encoders that introduce variational inference. Instead of embedding the input as a fixed single point in the latent space,
we transform it into a probability distribution in the latent space in VAE. This
helps to avoid overfitting and ensures that the latent probabilistic distribution
has good implicit semantics of the input, which contributes a lot to the data
generation process. VAE was introduced into recommender systems recently
[1,12,28,30]. [12] introduces a generative model with multinomial likelihood and
uses Bayesian inference for parameter estimation to go beyond the limited modeling capacity of linear factor models. [28] chooses variational inference to handle
the computationally intractable problem in distribution estimate for users and
items embedding by Bayesian inference and apply VAE with the mean-field
assumption in practice. We combine VAE into CTR prediction to simultaneously model the user's history and content information variationally and propose a variational alignment strategy. The strategy is implemented in two ways:
First, Wasserstein distance is used to align probability distributions in the latent
space instead of traditional KL divergence. Then, inspired by [17], we use user
frequency as an explicit condition to directly supervise the variational inference
process.

2.3 Self-Attention Mechanism

The attention mechanism is a way to imitate human attention physiological
mechanisms. It can enhance the weight of some parts in the input data of neural
networks and weaken the weight of other parts [11,20,22]. The attention mechanism calculates the soft weight between sequence tokens by inputting query and
key-value sequences converted from two sequences respectively. Self-attention is
a special case of attention mechanism when query, key-value sequences are converted from the same sequence. [22] first largely applied self-attention to the
NLP translation task for the first time and achieved good results. We introduce
a self-attention mechanism before getting the final prediction to get balanced
weights and fuse latent features more reasonably.

3 Problem Definition

The CTR prediction is a supervised binary classification problem. We denote
the set of all users as U, and the set of all items as V. Each piece of the data
sample is a quadruple in the form of $(u,\ v,\ t,\ y)$, where $u \in U$ denotes a user,
$v \in V$ denotes an item, $y \in \{0, 1\}$ denotes whether the user interacted the item
and $t \in \mathbb{R}$ denotes a timestamp. Each user or item has a distinguishing ID and
other features, such as this user's age, and gender.

Following [23,28,33], we can split the whole classifier into embedder denoted
as $g_\phi(\cdot)$ with learnable parameters ϕ and predictor denoted as $f_\theta(\cdot)$ with learnable parameters θ. $g_\phi(\cdot)$ maps discrete-valued features into unique embeddings

and transforms continuous-valued features into embedding vectors by scaling them. We denote the set of all embeddings by E, and the user ID embedding is denoted as e_{uid}. Other users' feature embeddings are denoted as $e_{u_1}, \ldots, e_{u_n}$, n presents the feature number of each user. e_{vid} and $e_{v_1}, \ldots, e_{v_m}$ represent the same definitions with respect of item. The timestamp t is also embedded and denoted as e_t. Through $g_\phi(\cdot)$ we can get E to provide $f_\theta(\cdot)$ for further prediction. The whole CTR prediction can be formally summarized as follows:

$$E = \{e_{uid}, e_{u_1}, \ldots, e_{u_n}, e_{vid}, e_{v_1}, \ldots, e_{v_m}, e_t\} = g_\phi(u, v, t) \tag{1}$$

where $\hat{y} = f_\theta(E), \hat{y} \in \{0, 1\}$ denotes the predicted result, which indicates whether the user will interact with the item.

The binary cross-entropy also called the log-loss function, is used to judge the performance of classifier part f_θ:

$$\mathcal{L}(\hat{y}, y) = -y \log \hat{y} - (1 - y) \log(1 - \hat{y}) \tag{2}$$

Those users with sparse or even blank historical interaction data lack training process, making their embedding always insufficient [28,33], leading to the cold-start problem.

4 Proposed Method

4.1 Overview

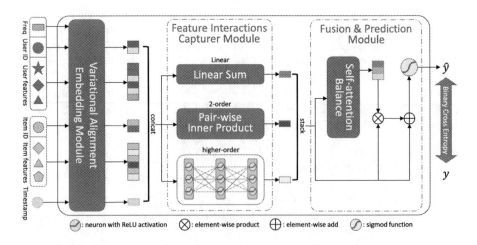

Fig. 1. Architecture figure of CBVA.

CBVA, which is illustrated in Fig. 1, consists of three main steps: First, we will use a variational alignment embedding module which is the embedder in the problem definition. It introduces variational auto-encoders and implements a conditional alignment strategy on variational probability distributions. Through this, we can reconstruct embeddings that adequately express the user's preferences extracted from historical interactions and content information. Second, the generated embeddings will be sent into the feature interactions capture module. It will calculate the low- and high-order interactions between the features and output highly compressed feature crosses. Finally, these crosses will be inputted into the fusion and prediction module for the terminal prediction. We design a self-attention-based balance strategy in this module to refine the weights of feature crosses and fuse them more reasonably. Then the prediction score will be decided by the sigmoid function and converted into a category by comparing it with the set threshold. The last two modules together make up the predictor in the problem definition.

4.2   Variational Alignment Embedding Module

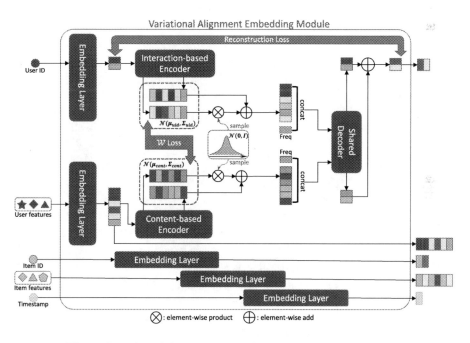

Fig. 2. Specifics of the variational alignment embedding module.

Embedding Layer. The workflow of the entire variational alignment embedding module is expounded in Fig. 2. First, each input including the user ID has its own embedding layer. The parameters of these embedding layers are initially randomly generated and can be optimized by gradient descent algorithms. These embedding layers are simply matrices that store embedded values, and the input takes a row from its corresponding matrix as embedding based on its own values. Therefore we denote all these embedding layers collectively as L_E and the embedding process can be formally represented as follow, where $E = L_E(u, v, t, y)$ is a $\mathbb{R}^k$ vector. We will use hot users whose interaction data is rich to train embedding layers before making predictions for cold users.

It is worth noting that whether the user ID embedding can well represent the user's preference is crucial to the performance of the CTR prediction task [16]. Because When it comes to first-present users, their ID embeddings are only randomly generated. However, such randomly initialized embeddings contain nearly zero useful information. Inspired by many meta-learning ideas [9], we believe that there is common knowledge among all users, i.e., the differences in users' interests are based on more general common interests. Therefore, rather than randomly generating ID embeddings for new users, we prefer to generate initial ID embeddings for new users from the existing ID embeddings of hot users to make the training process easier. So we re-initialize ID embeddings of the cold user in E by taking the average of all existing ID embeddings as the initial ID embedding for new users:

$$e_{uid} = \frac{\sum_{i=1}^{|U|} e_{uid}}{|U|} \tag{3}$$

This simple average aggregation enables us to exploit this commonality of interests among users to enhance the ability to capture the preferences of cold users with sparse interaction data without using heavy tools such as meta-learning.

Interaction-Based Encoder. For reasonable use of user historical interaction data from a probabilistic perspective, we employ VAE for variationally modeling it. The user ID embedding is inputted into the interaction-based encoder to get the variance and expectation of preference probabilistic distribution:

$$\mu_{uid}, \sigma_{uid} = L_{uid}(e_{uid}) \tag{4}$$

where L_{uid} represents the interaction-based information encoder. This encoder infers the latent representation z by $q(z|x)$ based on the observed data, where x corresponds to the id, which saves mainly user preferences extracted from the user's historical interaction records. $\mu_{uid} \in \mathbb{R}^d$ and $\sigma_{uid} \in \mathbb{R}^d$ mean the expectation and the covariance matrix's main diagonal of a Gaussian distribution. Through this, we conclude user preference modeled from interactions into this Gaussian for further alignment in latent probability space.

Content-Based Encoder. We also employ another individual VAE for variationally modeling user content information. The user ID embedding is inputted into the content-based encoder to get the variance and expectation of preference probabilistic distribution:

$$\mu_{cont}, \sigma_{cont} = L_{cont}(e_{u_1}, e_{u_2}, \ldots, e_{u_n}) \tag{5}$$

where L_{cont} represents the content-based information encoder layer. The encoder infers the latent representation z by $q(z|x)$ based on the observed data, where x corresponds to the content information. $\mu_{cont} \in \mathbb{R}^d$ and $\sigma_{cont} \in \mathbb{R}^d$ mean the expectation and the covariance matrix's main diagonal of a Gaussian distribution, which concludes user preference modeled from the content information. Through this step, we successfully model the content information as a distribution and use it as an empirical prior, laying the foundation for the next step to align them in the latent probability space.

Variational Preference Alignment. Next, we will align the distributions in the probability space by minimizing their differences. In general VAE architectures, KL divergence is often used to measure the differences between two distributions. However, there are three fatal drawbacks of KL divergence, making it inapplicable for user cold-start CTR prediction:

- KL divergence is not symmetric, i.e., it is not a well-defined distance function. This means that we must distinguish between primary and secondary distributions and measure the differences only from the perspective of the primary distribution. But we believe that the user preference distribution learned from historical interactions and the preference distribution extracted from content information are equally important, and their knowledge should be integrated equally.
- KL divergence faces a failure problem when the two distributions do not overlap at all. The value of the KL divergence will degenerate to infinite in this situation, which does not give us a good gradient for optimization. Since we are tackling the user cold start problem where the distribution of the data is very sparse and unbalanced, this situation should be taken into consideration.
- KL divergence is only sensitive to the integral of the two distributions, rather than taking specific shape differences between distributions into account. Hence, it is impossible for us to determine directly whether the shapes of the two distributions are close by using the KL divergence value.

Therefore, we use the Wasserstein distance to overcome the above shortcomings of KL divergence. Wasserstein distance is symmetric and shape sensitive, and we can always calculate a usable gradient from it, it is formally defined as:

$$W_p[\mu||\nu] = (\inf_{\gamma \in \Gamma(\mu,\nu)} \mathbb{E}_{(x,y)\sim\gamma}[d(x,y)^p])^{\frac{1}{p}} \tag{6}$$

where μ, ν are two probability distributions and Γ is the set of all coupling of them. In this paper, we choose the Euclidean norm as metric d and let $p = 2$. The 2-order Wasserstein distance has a closed form when μ and ν are multivariate

Gaussian distributions. This matches our assumptions that the family of probability distributions used for approximations is gaussians. We use Wasserstein distance between the distributions as a loss, denoted as $\mathcal{L}_{wd}$:

$$
\begin{aligned}
\mathcal{L}_{wd} &= \mathrm{W}_p[\mathcal{N}(\mu_{uid}, \sigma_{uid}) || \mathcal{N}(\mu_{cont}, \sigma_{cont})] \\
&= ||\mu_{uid} - \mu_{cont}||_2^2 + \mathrm{trace}(\sigma_{uid} + \sigma_{cont} - 2(\sigma_{cont}^{\frac{1}{2}} \Sigma_{uid} \sigma_{cont}^{\frac{1}{2}})^{\frac{1}{2}})
\end{aligned}
\tag{7}
$$

To optimize the entire model by gradient descent algorithm, we use the reparameterization trick to isolate the randomness from sampling. We sample an auxiliary noise variable denoted as ϵ_1, ϵ_2 from standard normal distributions and convert it into a sample from the target distribution. Through this we can obtain latent info of distributions, $z_{uid} \in \mathbb{R}^d$ for user ID and $z_{cont} \in \mathbb{R}^d$ for content information:

$$
\begin{aligned}
z_{uid} &= \epsilon_1 \otimes \mathrm{diag}(\sigma_{uid}^{\frac{1}{2}}) \oplus \mu_{uid}; \quad \epsilon_1 \sim \mathcal{N}(\mathbf{0}, \mathbf{I}) \\
z_{cont} &= \epsilon_2 \otimes \mathrm{diag}(\sigma_{cont}^{\frac{1}{2}}) \oplus \mu_{cont}; \quad \epsilon_2 \sim \mathcal{N}(\mathbf{0}, \mathbf{I})
\end{aligned}
\tag{8}
$$

where $\mathrm{diag}(\cdot)$ means that the elements on the main diagonal are extracted from the square matrix and formed into a vector in order, and $\otimes, \oplus$ mean element-wise multiply, and element-wise plus respectively.

Note that the decoder is shared to force it to decode both probability distributions simultaneously, so its parameters are trained by the dynamic balance. This strategy implements a preliminary pull-in of two distributions in probability space. Therefore, the effective knowledge of the latter can be integrated into the former, and then more robust user ID embeddings will be reconstructed.

Furthermore, we seek a way to explicitly supervise the VAE by the amount of user data to generate more powerful user ID embeddings by learning from user frequency, since frequency has a huge impact on the distribution of ID embeddings. Inspired by the success of conditional VAEs (CVAE) in the field of image generation [4,17], we directly take the user's normalized frequency $f \in [0, 1]$ as a condition for participating in latent variables before decoding:

$$
\begin{aligned}
\hat{e}_{uid} &= L_D(z'_{uid}); \quad z'_{uid} = \mathrm{concat}(z_{uid}, f) \\
\hat{e}_{cont} &= L_D(z'_{cont}); \quad z'_{cont} = \mathrm{concat}(z_{cont}, f)
\end{aligned}
\tag{9}
$$

where $\hat{e}_{uid} \in \mathbb{R}^d, \hat{e}_{cont} \in \mathbb{R}^d$ are reconstructed embeddings and $\mathrm{concat}(\cdots)$ is the concatenate operation between vectors. The explicit condition has two benefits: First, direct input of labels helps guide the decoder to reconstruct to varying degrees according to the user's hotness or coldness. The direct participation of labels will make the gradient more significant, which is beneficial to convergence and easy training. Second, this explicit input compensates for the lack of user frequency information in our model prior, contributing to the aligning between probability distributions.

The reconstruction error will be calculated by the mean square error between the reconstructed and the original use ID embedding. This loss is denoted as

$\mathcal{L}_{recon}$, where the $e^i_{uid}, \hat{e}^i_{uid}$ means the i-th component of the original and recon-structed embedding vector respectively. Finally, we can get the refined user ID embedding $\widetilde{e}_{uid} = \hat{e}_{uid} \oplus \hat{e}_{cont}$.

$$\mathcal{L}_{recon} = \text{MSE}(e_{uid}, \hat{e}_{uid}) = \frac{1}{|e_{uid}|} \sum_i (e^i_{uid} - \hat{e}^i_{uid})^2 \tag{10}$$

4.3 Feature Interactions Capturer Module

Then, following the factorization idea of [6,15,25], we want to consider the different levels of crosses between each feature domain. Therefore, we use three modules to extract 1-order, 2-order, and higher-order features respectively.

The 1-order and 2-order feature crosses denoted as $c_{one} \in \mathbb{R}, c_{two} \in \mathbb{R}$, and the higher-order feature crosses c_{higher} are captured by employing an MLP with an activation function:

$$\begin{aligned}
c_{one} &= \sum \text{concat}(E) \\
c_{two} &= \sum_{e_p \in E} \sum_{e_q \in E; e_p \neq e_q} <V_p, V_q> e_p \cdot e_q \\
a^{(0)} &= \text{concat}(E) \\
a^{(l)} &= \sigma(W^{(l)} a^{(l-1)} + b^{(l)}) \\
c_{higher} &= W^{(out)} a^{(L)} + b^{(out)}
\end{aligned} \tag{11}$$

where $a^{(l)} \in \mathbb{R}^d, W^{(l)} \in \mathbb{R}^{d \times d}, b^{(l)} \in \mathbb{R}^d$ are the output, a ReLU non-linear activation function, the weight and the bias of l-th layer. $W^{(out)} \in \mathbb{R}^{d \times 1}, b^{(out)} \in \mathbb{R}$ are the weight and the bias of MLP output layer. $L, \sigma(\cdot)$ are the hyper-parameter of MLP layers number and the non-linear activation function. $c_{higher} \in \mathbb{R}$ is the captured higher-order feature crosses denoted.

4.4 Fusion and Prediction Module

Before prediction, we need to fuse these extracted feature crosses of each order. [6,25] directly stack or sum each value and then directly enters the sigmoid discriminant function to obtain the prediction. We believe that the imbalance between each order intersection should be considered. Inspired by the use of the self-attention mechanism in NLP [22] tasks to automatically capture importance between sequence elements, we propose a self-attention module to recalculate fusion weights for feature crosses:

$$\begin{aligned}
C &= \text{stack}(c_{one}, c_{two}, c_{higher}) \\
Q &= K = V = C \\
C_b &= \text{softmax}\left(\frac{QK^T}{\sqrt{d_k}}\right)V
\end{aligned} \tag{12}$$

where $C_b \in \mathbb{R}^3$ is the balanced crosses and stack$(\cdots)$ is a stacking operation between tensors to create a new tensor. We will carry out a residual link to constrain the strength of self-attention to avoid overfitting. The prediction score $y' \in [0, 1]$ will be calculated by sigmoid function:

$$\hat{C} = C \oplus C_b$$
$$y' = \text{sigmod}(\sum_{i=1}^{3} \hat{C}^i) \tag{13}$$

where $\hat{C}^i$ means the i-th component of $\hat{C}$. The predicted scores will be classified into positive or negative according to different thresholds, and record the result as $\hat{y}$. We use the binary cross-entropy $\mathcal{L}_{bce}$ as a loss, and the total loss of the entire model to train is the sum of distribution difference loss $\mathcal{L}_{wd}$, reconstruction loss $\mathcal{L}_{recon}$ and binary cross entropy loss $\mathcal{L}_{bce}$:

$$\mathcal{L}_{bce} = -y \log \hat{y} - (1 - y) \log(1 - \hat{y})$$
$$\mathcal{L} = \mathcal{L}_{wd} + \mathcal{L}_{recon} + \mathcal{L}_{bce} \tag{14}$$

5   Experiments

5.1   Datasets

We used the following two public real datasets:

- Movielens-1M[1] is a movie review dataset, containing 1,000,209 anonymous ratings of approximately 3,900 movies made by 6,040 users. Each movie and user has characteristics such as ID, age, gender, movie name, release year, etc.
- Taobao display advertising click dataset[2], TaobaoAD for short, is a click-through rate forecast dataset of Taobao display ads. It randomly sampled 26 million behavior records of 1.14 million users from the Taobao website. Each ad and user has its own features like ID, group ID, gender, and price.

In Movielens-1M, a movie rating ≥ 4 is considered a positive review, otherwise a negative one. We follow the settings of [13,33], splitting the entire dataset into two groups, hot and cold. The hot group only contains hot users defined as those whose number of interactions records is greater than the threshold N, while the cold group contains only cold users. We set $N = 25$ and $N = 5$ for Movielens-1M and TaobaoAD. It is worth noting that the ratio between hot and cold is about 8:2, which is consistent with the definition in the long tail data. Then, the log of cold users in the cold group is sorted by timestamp and divided into four groups denoted as warm-a, -b, -c, and test, following [13,33].

5.2   Baselines

We divide the baselines into two categories. The first category is general CTR prediction models, which do not optimize for the cold user start problem: (1) FM [15] can capture high-order interactions information cross features. (2)

---

[1] https://grouplens.org/datasets/movielens/.
[2] https://tianchi.aliyun.com/dataset/dataDetail?dataId=56.

Wide&Deep [2] tries to combine the benefits of memorization and generalization for recommender systems. (3) DeepFM [6] is an improvement of FM, which learns both low-order and high-order interactions between fields. (4) DCN [25] is a DNN model which can be more efficient in learning certain bounded-degree feature interactions. (5) PNN [14] uses a product layer to capture interactive patterns between inter-field categories. Another major category is those CTR prediction models which specially optimized for cold-start recommendation. As mentioned above, these methods are roughly divided into two kinds: content-based and meta-learning-based. Since meta-learning-based methods are mostly model-agnostic, our CBVA can also be easily nested into a meta-learning framework for boosting. Hence we do not pay much attention to meta-learning-based methods. We choose the following three baseline models: (1) DropoutNet [23] is explicitly trained for cold start through the dropout method. (2) MWUF [33] proposes two modules to transform the cold ID embedding into warm and improve the model's anti-noise stability. (3) VELF [28] introduces VAE into cold-start CTR prediction in order to overcome the shortcoming that existing methods are all based on point estimate, which leads to a high risk of biased prediction.

5.3   Experiment Setup

All extra features provided by datasets are used. The hyper-parameter α which controls re-initialization is set to 0.2 and 0.1 for Movielens-1M and taobaoAD dataset respectively. All multi-layer perceptrons have a fixed layer number of 2. The embedding dimension and latent vector dimension are fixed to 16. Learning rate and mini-batch size are grid searched and set to 0.001 and 4096. All optimization is done by Adam optimizer over shuffled samples. For a fair comparison, we use the above setting for both our method and all other baseline methods. We report the mean results over 3 runs. Following [28,33], we choose AUC and F1 as metrics to measure model performance.

5.4   Performance Comparison

Table 1 gives a comprehensive comparison between CBVA and other baseline models. Models based on computational paradigms such as PNN and FM perform the worst because they neither overcome the shortcomings of biased point estimations nor sufficiently utilize content information. DeepFM and DCN use neural networks to better capture the interactions of features, so they perform slightly better than the former two. But they still cannot accurately extract user preferences from content information. DropoutNet and MWUF are specially optimized for cold start, and reasonably utilize content information to assist in modeling user preferences, but are still based on point estimations. VELF goes one step further and enhances the robustness of the model by introducing the VAE framework instead of point estimations. However, VELF does not integrate the two aspects of information well in the probability space. Our proposed CBVA uses VAE to jointly model historical interactions and content information, comprehensively considers user preferences from both sources, and

Table 1. Performance comparison on 'cold' users. 'Drop' is short for DropoutNet and 'W&D' is short for Wide& Deep. The best results of all are in bold, and the best results of previous works are underlined.

	Method	Cold-phase		Warm-a		Warm-b		Warm-c	
		AUC	F1	AUC	F1	AUC	F1	AUC	F1
Movielens-1M	FM	0.7124	0.7231	0.7140	0.7228	0.7168	0.7260	0.7204	0.7343
	W&D	0.6978	0.6809	0.6993	0.6879	0.7014	0.6866	0.7014	0.6731
	DeepFM	0.7142	**0.7397**	0.7160	0.7337	0.7191	0.7373	0.7209	_0.7450_
	DCN	0.7147	0.7324	0.7165	0.7252	0.7179	0.7280	0.7185	0.7393
	PNN	0.7083	0.7217	0.7106	0.7254	0.7135	0.7306	0.7150	0.7258
	Drop	0.7123	0.7365	0.7145	0.7353	0.7167	0.7404	0.7186	0.7433
	MWUF	0.7155	0.7395	0.7153	_0.7382_	0.7154	0.7390	0.7155	0.7404
	VELF	**0.7200**	0.7208	_0.7220_	0.7351	_0.7294_	_0.7433_	_0.7278_	0.7424
	CBVA	0.7076	0.7280	**0.7291**	**0.7418**	**0.7411**	**0.7520**	**0.7571**	**0.7621**
TaobaoAD	FM	0.5434	0.1322	0.5549	0.1360	0.5566	0.1362	0.5546	0.1361
	W&D	0.5396	0.1293	0.5541	0.1364	0.5543	0.1361	0.5556	0.1366
	DeepFM	0.5448	0.1328	0.5605	0.1383	0.5628	0.1407	0.5614	0.1411
	DCN	**0.5666**	**0.1421**	0.5687	_0.1405_	_0.5681_	_0.1411_	0.5692	_0.1413_
	PNN	0.5547	0.1366	0.5620	0.1393	0.5638	0.1392	0.5647	0.1406
	Drop	0.5469	0.1340	0.5605	0.1372	0.5626	0.1399	0.5634	0.1404
	MWUF	0.5412	0.1313	0.5581	0.1381	0.5390	0.1288	0.5406	0.1304
	VELF	0.5455	0.1318	0.5601	0.1374	0.5640	0.1397	_0.5696_	0.1402
	CBVA	0.5599	0.1377	**0.5665**	**0.1406**	**0.5687**	**0.1415**	**0.5728**	**0.1418**

effectively integrates their information in the latent space through an alignment strategy. Thus it effectively alleviates the problem of user cold start in CTR prediction and shows the best performance on two datasets. Our proposed CBVA outperforms all existing models in the vast majority of cases.

5.5 Ablation Study

We present an ablation study and evaluate several models based on DeepFM which is one of the lightest structures: (1) CBVA: The proposed model which contains all modules. (2) CBVA-V: CBVA without the variational framework. The encoders directly output the latent vector instead of the expectation and variance. (3) CBVA-A: CBVA without the alignment strategy. Shared decoders are no longer used, and user frequency is not inputted as a condition. (4) CBVA-S: CBVA without the self-attention balance module. Feature crosses at several orders will be added and sent into the sigmoid function directly. The mean AUC results over 3 runs are reported in Table 2. We can see that removing any of the proposed components directly leads to a decrease in the performance of CBVA, which proves the effectiveness of our variational joint modeling and the strategy of aligning in the latent space and finally using self-attention for better fusion.

Table 2. Variants comparison. 'ML-1M' is the abbreviation of Movielens-1M and 'TbAD' for TaobaoAD. The best results are in bold.

	Method	Cold-phase		Warm-a		Warm-b		Warm-c	
		AUC	F1	AUC	F1	AUC	F1	AUC	F1
ML-1M	CBVA	0.7076	0.7280	**0.7291**	**0.7418**	**0.7411**	**0.7520**	**0.7571**	**0.7621**
	CBVA-V	0.7060	**0.7338**	0.7197	0.7333	0.7206	0.7366	0.7261	0.7362
	CBVA-A	**0.7091**	0.7318	0.7221	0.7340	0.7285	0.7462	0.7331	0.7502
	CBVA-S	0.7088	0.7322	0.7251	0.7395	0.7357	0.7500	0.7535	0.7533
TbAD	CBVA	0.5599	0.1377	**0.5665**	**0.1406**	**0.5687**	**0.1415**	**0.5728**	**0.1418**
	CBVA-V	0.5577	0.1374	0.5652	0.1403	0.5665	0.1408	0.5700	0.1417
	CBVA-A	0.5567	0.1367	0.5644	0.1386	0.5686	0.1419	0.5702	0.1409
	CBVA-S	**0.5601**	**0.1382**	0.5647	**0.1406**	0.5683	0.1407	0.5714	0.1414

6 Conclusion

In this paper, we propose a novel content-based variational alignment model called CBVA for CTR prediction. As traditional is difficult to cope with user cold-start recommendations due to sparse historical data, we use VAE to model user preference into probabilistic distributions jointly from user historical interactions and content info and implement a variational alignment strategy in latent space. Moreover, to obtain a more reasonable fusion of feature crosses, we adopt a self-attention mechanism that integrates crossed features as a whole. Finally, we conduct extensive experiments on two real-world datasets and the results show that our proposed method is superior to several baseline models.

References

1. Chen, Y., de Rijke, M.: A collective variational autoencoder for top-n recommendation with side information. In: DLRS@RecSys, pp. 3–9 (2018)
2. Cheng, H.T., et al.: Wide and deep learning for recommender systems. In: DLRS@RecSys, pp. 7–10 (2016)
3. Dong, M., Yuan, F., Yao, L., Xu, X., Zhu, L.: MAMO: memory-augmented meta-optimization for cold-start recommendation. In: KDD, pp. 688–697 (2020)
4. Gaurav, P., Ambedkar, D.: Variational methods for conditional multimodal deep learning. In: IJCNN, pp. 308–315 (2017)
5. Gu, L.: Ad click-through rate prediction: a survey. In: DASFAA, pp. 140–153 (2021)
6. Guo, H., Tang, R., Ye, Y., Li, Z., He, X.: Deepfm: a factorization-machine based neural network for CTR prediction. In: IJCAI, pp. 1725–1731 (2017)
7. Hu, X., Xu, J., Wang, W., Li, Z., Liu, A.: A graph embedding based model for fine-grained POI recommendation. Neurocomputing **428**, 376–384 (2021)
8. Kingma, D.P., Welling, M.: Auto-Encoding Variational Bayes (2014)
9. Lee, H., Im, J., Jang, S., Cho, H., Chung, S.: Melu: meta-learned user preference estimator for cold-start recommendation. In: KDD, pp. 1073–1082 (2019)

10. Li, F., Chen, Z., Wang, P., Ren, Y., Zhang, D., Zhu, X.: Graph intention network for click-through rate prediction in sponsored search. In: SIGIR, pp. 961–964 (2019)
11. Li, Y., Xu, J., Zhao, P., Fang, J., Chen, W., Zhao, L.: Atlrec: an attentional adversarial transfer learning network for cross-domain recommendation. J. Comput. Sci. Technol. **35**(4), 794–808 (2020)
12. Liang, D., Krishnan, R.G., Hoffman, M.D., Jebara, T.: Variational autoencoders for collaborative filtering. In: WWW, pp. 689–698 (2018)
13. Pan, F., Li, S., Ao, X., Tang, P., He, Q.: Warm up cold-start advertisements: improving CTR predictions via learning to learn id embeddings. In: SIGIR, pp. 695–704 (2019)
14. Qu, Y., et al.: Product-based neural networks for user response prediction over multi-field categorical data. ACM Trans. Inf. Syst. **37**(1), 5:1–5:35 (2019)
15. Rendle, S.: Factorization machines. In: ICDM, pp. 995–1000 (2010)
16. Shen, F., Yan, S., Zeng, G.: Neural style transfer via meta networks. In: CVPR, pp. 8061–8069 (2018)
17. Sohn, K., Lee, H., Yan, X.: Learning structured output representation using deep conditional generative models. In: NIPS, pp. 3483–3491 (2015)
18. Song, J., Xu, J., Zhou, R., Chen, L., Li, J., Liu, C.: CBML: a cluster-based meta-learning model for session-based recommendation. In: CIKM, pp. 1713–1722 (2021)
19. Sun, H., Xu, J., Zheng, K., Zhao, P., Chao, P., Zhou, X.: MFNP: a meta-optimized model for few-shot next POI recommendation. In: IJCAI, pp. 3017–3023 (2021)
20. Sun, H., Xu, J., Zhou, R., Chen, W., Zhao, L., Liu, C.: HOPE: a hybrid deep neural model for out-of-town next POI recommendation. WWW **24**(5), 1749–1768 (2021)
21. Vahdat, A., Kautz, J.: NVAE: a deep hierarchical variational autoencoder. In: NIPS, pp. 19667–19679 (2020)
22. Vaswani, A., et al.: Attention is all you need. In: Advances in Neural Information Processing Systems 30 (2017)
23. Volkovs, M., Yu, G., Poutanen, T.: DropoutNet: addressing cold start in recommender systems. In: NIPS, pp. 4957–4966 (2017)
24. Wang, P., Jiang, Y., Xu, C., Xie, X.: Overview of content-based click-through rate prediction challenge for video recommendation. In: ACM Multimedia, pp. 2593–2596 (2019)
25. Wang, R., Fu, B., Fu, G., Wang, M.: Deep and cross network for ad click predictions. In: ADKDD@KDD, pp. 1–7 (2017)
26. Wang, X., Du, Y., Zhang, L., Li, X., Zhang, M., Dong, J.: Exploring content-based video relevance for video click-through rate prediction. In: ACM Multimedia, pp. 2602–2606 (2019)
27. Xie, R., Wang, R., Zhang, S., Yang, Z., Xia, F., Lin, L.: Real-time relevant recommendation suggestion. In: WSDM, pp. 112–120 (2021)
28. Xu, X., et al.: Alleviating cold-start problem in CTR prediction with a variational embedding learning framework. In: WWW, pp. 27–35 (2022)
29. Yu, B., Li, X., Fang, J., Tai, C., Cheng, W., Xu, J.: Memory-augmented meta-learning framework for session-based target behavior recommendation. WWW **26**(1), 233–251 (2023)
30. Zhao, J., Zhao, P., Zhao, L., Liu, Y., Sheng, V.S., Zhou, X.: Variational self-attention network for sequential recommendation. In: ICDE, pp. 1559–1570 (2021)
31. Zhou, R., Liu, C., Wan, J., Fan, Q., Ren, Y., Zhang, J., Xiong, N.: A hybrid neural network architecture to predict online advertising click-through rate behaviors in social networks. IEEE Trans. Netw. Sci. Eng. **8**(4), 3061–3072 (2021)

32. Zhu, Q., Bi, W., Liu, X., Ma, X., Li, X., Wu, D.: A batch normalized inference network keeps the KL vanishing away. In: ACL, pp. 2636–2649 (2020)
33. Zhu, Y., et al.: Learning to warm up cold item embeddings for cold-start recommendation with meta scaling and shifting networks. In: SIGIR, pp. 1167–1176 (2021)

Dual-Granularity Contrastive Learning for Session-Based Recommendation

Zihan Wang[1], Gang Wu[1,2(✉)], and Haotong Wang[1]

[1] School of Computer Science and Engineering, Northeastern University, Shenyang, China
{2101816,2171931}@stu.neu.edu.cn, wugang@mail.neu.edu.cn
[2] Key Laboratory of Intelligent Computing in Medical Image, Ministry of Education, Shenyang, China

Abstract. The data encountered by Session-based Recommendation System(SBRS) is typically highly sparse, which also serves as one of the bottlenecks limiting the accuracy of recommendations. So Contrastive Learning(CL) is applied in SBRS owing to its capability of improving embedding learning under the condition of sparse data. However, existing CL strategies are limited in their ability to enforce finer-grained (e.g., factor-level) comparisons and, as a result, are unable to capture subtle differences between instances. More than that, these strategies usually use item or segment dropout as a means of data augmentation which may result in sparser data and thus ineffective self-supervised signals. By addressing the two aforementioned limitations, we introduce a novel dual-granularity CL framework. Specifically, two extra augmentation views with different granularities are constructed and the embeddings learned by them are compared with those learned from original view to complete the CL tasks. At factor-level, we employ Disentangled Representation Learning to obtain finer-grained data, with which we can explore connections of items on latent factor independently and generate factor-level embeddings. At item-level, the star graph is deployed as the augmentation method. By setting an additional satellite node, non-adjacent nodes can establish additional connections through satellite nodes instead of reducing the connections of the original graph, so data sparsity can be avoided. Compare the learned embeddings of these two views with the learned embeddings of the original view to achieve CL at two granularities. Finally, the item-level and factor-level embeddings obtained are referenced to generate personalized recommendations for the user. The proposed model is validated through extensive experiments on two benchmark datasets, showcasing superior performance compared to existing methods.

Keywords: Session-based Recommendation · Contrastive Learning · Disentangled Representation Learning

© The Author(s), under exclusive license to Springer Nature Switzerland AG 2023
X. Yang et al. (Eds.): ADMA 2023, LNAI 14176, pp. 740–754, 2023.
https://doi.org/10.1007/978-3-031-46661-8_49

1   Introduction

Session-based recommendation (SBR) has gained significant attention recently and it provides recommendations solely based on information from an anonymous session.

The items that a user interacts with in a short period of time are arranged in chronological order to form a session. Session data bears some resemblance to text data in Natural Language Processing (NLP), and as a result, some classic NLP frameworks including Recurrent Neural Network(RNN) [1,2,5], Attention Mechanism [3,4,11,12] and Graph Neural Network(GNN) [6–10,14] have been adapted to SBR.

GNN-based models have been found to be more effective than other models because they can better model the complex relationships that exist between items. However, even with GNN-based models, SBR still faces the challenge of insufficient data to train accurate item embeddings, which can lead to suboptimal recommendation performance.

Contrastive Learning (CL) in Self-Supervised Learning [13,15,17] is widely regarded as a solution to the problem of data sparsity. The process of CL can be divided into three steps. Typically, CL models first create a new view through artificial data augmentation based on the original view. Then learn embeddings from both the original view and the augmentation view. Finally, the embeddings learned from the latter are partitioned into positive and negative samples relative to the embeddings learned from the former. By comparing the differences and similarities between them, models can adjust the embeddings during training following the principle that positive samples are close to each other and negative samples are mutually exclusive. As a result, the model can learn more accurate embeddings than with supervised training alone. Some researchers have applied CL to the SBR and have achieved promising results [16,18–20]. Nevertheless, we have observed that existing CL models often exhibit two flaws. First, existing CL methods are typically limited to coarse-grained comparisons at the item-level and/or session-level, often overlooking finer-grained relationships between instances. Second, these methods often rely on item or segment dropout as a form of data augmentation, which can exacerbate data sparsity and consequently, less effective self-supervised signals.

By addressing the two aforementioned limitations, we introduce a novel dual-granularity CL framework. Our approach typically involves incorporating two additional augmentation views, one at the item-level and the other at the factor-level, besides the original item-level view. The embeddings acquired from two augmentation views are compared with those acquired from the original view to finish CL tasks under two granularities.

At factor-level, to address the issue of fine-grained factor (e.g. brand and color) labels being often absent in SBR data, we propose the use of Disentangled Representation Learning(DRL) to obtain independent latent factor-level embeddings corresponding to items, which can replace the missing labels. So factor-level convolution channels that operate independently can be constructed and then can be compared with the factor-level embeddings converted from the embeddings learned in the original view to finish the factor-level CL. At item-level, the star graph has been planted as an augmentation of the original view. Star graph has the inclusion of an extra satellite node and non-adjacent nodes can communicate with each other via the satellite node, thereby enabling the acquisition of more information. Unlike traditional data augmentation techniques, it avoids further exacerbating data sparsity by promoting nodes to learn more information. Likewise, we compared the item-level embedding learned from the original view and star graph to finish item-level CL.

Moreover, we leverage the learned item-level and factor-level embeddings to model the user's overall and specific latent factor interests, respectively, in order to predict the user's next interaction at two granularities. In the end, we propose our model Dual-Granularity Contrastive Learning for Session-based Recommendation(DGCL-GNN) and summarize our main contributions as follows:

- We identify and address two challenges in existing contrastive learning methods for SBR, and propose a dual-granularity contrastive learning framework to improve the model's ability to learn embeddings.
- We innovatively introduce disentangled representation learning and star map augmentation to help us complete two granular comparative learning tasks.
- Eventually, we propose our model DGCL-GNN, and extensive experiments show that the proposed model has achieve statistically significant improvements on benchmark datasets.

This paper is organized as follows. 2 describes the related work for SBR and CL. 3 give formal definitions of the SBR. 4 presents the details of our model DGCL-GNN. 5 includes various experiments to demonstrate the effectiveness of our model. Finally, 6 presents our conclusions and suggestions for future studies.

2   Related Work

2.1   Session-Based Recommendation

Session-based recommendation(SBR) has gained considerable research interest in recent years. Some researchers have turned to neural network models inspired by NLP, which have shown promise in improving the effectiveness of SBR. Recurrent Neural Network(RNN) [21] was first noticed because it can capture sequential dependencies in a session, which is essential for SBR. However, RNN-based models [1,2,5] overlooks the global information that exists within the entire session. Some scholars have attempted to incorporate the attention mechanism [23] to capture global information and proposed some attention-based models [3,4,11,12]. Then Graph Neural Network(GNN) [22,24] model [6–10,14]emerged as a promising solution, showing a significant performance advantage. This is due to the GNN's strong ability to capture complex relationships between items, enabling enhanced representation learning capabilities of the model. Disen-GNN [25] has introduced DRL, which has become popular in other fields [26,27,30], into SBR for the first time. This creative approach refines the SBR problem to a fine-grained level, prompting a deeper analysis of the SBR problem. Although GNN has shown great advantages in SBR, it still faces a significant performance bottleneck, which is the sparsity of the SBR data.

2.2 Contrastive Learning

Contrastive learning(CL) is a type of unsupervised learning that aims to learn a representation space where similar samples are mapped to nearby points, while dissimilar samples are mapped to distant points. CL has been success-fully applied in various fields [29,31], such as computer vision and natural lan-guage processing, to improve the quality of learned embeddings and enhance the performance of downstream tasks.

In the field of recommendation systems like collaborative filtering recom-mendation and sequential recommendation, CL has also been widely used in recent years [28,32,33]. Some studies have also applied CL to the session-based recommendation scenario [16,18–20], where it has shown promising results in improving the performance of session-based recommendation models.

The CL task aims to address a pain point of SBR, as the data in this area is often sparse. Therefore, we firmly believe that improving CL is one of the key directions for enhancing SBR's performance in the future.

3 Preliminaries

In this section, we introduce the formal definitions of the session-based recom-mendation problem. Let $\mathcal{I} = \{v_1, ..., v_N\}$ denote the set of all items in a dataset and N is the number of items. Formally speaking, an anonymous session is rep-resented as $s = \{v_{(s,1)}, ..., v_{(s_n)}\}$, where n is the total length of the session. The task of the SBR models is to analyze the user's interests revealed by interactions in s and predict the next item $v_{(s,n+1)}$ that the user most likely interacts with. Finally, compare the simulation of user interests with the characteristics of each item to calculate the probability of its potential occurrence $p(v_i|s)$, which repre-sents the score of v_i in this recommendation. The Top-K items with the highest scores are selected and recommended to the user (Fig. 1).

4 Methodology

Next, we describe the framework of our proposed model DGCL-GNN. In general, we elaborate on three components of our model mainly including the basic rec-ommendation module, the factor-level CL module and the item-level CL module.

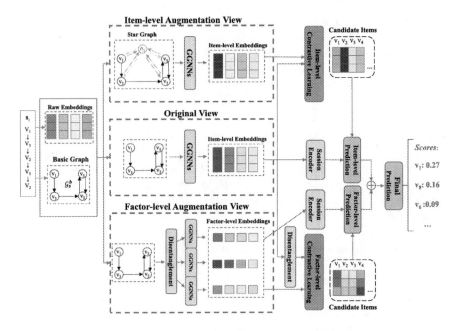

Fig. 1. Overview of DGCL-GNN

4.1   Initialization

Original Session Graph. As in the previous work, we use a directed session graph $\mathcal{G}_s^o =< \mathcal{V}_s^o, \mathcal{E}_s^o >$ to store the sequential relationship of each item in the session s. That is, when the next interaction of v_i is v_j, we set the i row j column of the adjacency matrix $\mathcal{A}_s^o$ to 1, which means that there is a pointing relationship.

Item-Level Embeddings. We first embed items in s into the same space to obtain the corresponding item-level embeddings $e^s = \{e_{(s,1)}, ..., e_{(s,n)}\}$.

Factor-Level Embeddings. We adopt DRL to acquire the representations of the independent latent factors underlying the observed data, which corresponds to the fine-grained, factor-level embeddings of the items.

The input of DRL is the item-level embeddings. Let $e_i \in \mathbb{R}^d$ represents a d dimension item-level embedding. Then DRL computes $\mathcal{K}$ factor-level embeddings from e_i by re-embedding it into corresponding spaces. The factor-level embedding $f_{(k,i)}$ on factor k is computed with Eq. 1.

$$f_i^k = \sigma(W_f^k e_i) + b_f^k (1 \leq k \leq \mathcal{K}) \tag{1}$$

Here, $W_t^f \in \mathbb{R}^{d \times d_f}$ and $b_t^f \in \mathbb{R}^{d_f}$ are the weight matrix and bias on factor k. And $d_f = \lfloor \frac{d}{k} \rfloor$ is the dimension of a factor-level embedding.

In order to avoid redundant information between factors, DRL uses the following loss function as the learning objective to generate independent factor-level embeddings.

$$\mathcal{L}_d = \sum_k^{\mathcal{K}} \sum_{t \neq k}^{\mathcal{K}} dCor(f^k, f^t) \tag{2}$$

$dCor$ is a formula to measure the correlation between variables in different spaces. For other details, please refer to [34].

4.2 Embedding Learning in Three Channels

We introduce convolutions in three channels in our model, which aim to learn embeddings for three different views.

Original View. The input to this view is the raw item-level embedding $e^s = e_{(s,1)}, ..., e_{(s,n)}$ and the raw session adjacency matrix $\mathcal{A}_s^o$. In Original view, we use the most commonly used Graph Gated Neural Network(GGNN) [35] method to update the item-level embedding.

$$c_{(s,i)}^l = Concat(W_{in}\mathcal{A}_{(s,in)}^o([x_{(s,1)}^l, ..., x_{(s,n)^l}]^\top) + b_{in},$$
$$W_{out}\mathcal{A}_{(s,out)}^o([x_{(s,1)}^l, ..., x_{(s,n)^l}]^\top) + b_{out}) \tag{3}$$

$$z_{(s,i)}^l = \sigma(W_z c_{(s,i)}^l + U_z x_{(s,i)}^{l-1}) \tag{4}$$

$$r_{(s,i)}^l = \sigma(W_r c_{(s,i)}^l + U_r x_{(s,i)}^{l-1}) \tag{5}$$

$$\tilde{x}_{(s,i)}^l = \tanh(W_h c_{(s,i)}^l + U_h(r_{(s,i)}^l \odot x_{(s,i)}^{l-1})) \tag{6}$$

$$x_{(s,i)}^l = (1 - z_{(s,i)}^l) \odot h_i^{l-1} + z_{(s,i)}^l \odot \tilde{x}_{(s,i)}^l \tag{7}$$

where $c_{(s,i)}^l$ is the information that $v_{(s,i)}$ can learn from its adjacent nodes in l-th layer, $A_{(s,in)}^o$ and $A_{(s,out)}^o$ are the in-degree and out-degree matrices corresponding to the adjacency matrix A_s^o. $x_{(s,i)^l}$ represents the embedding of node $v_{(s,i)}$ at layer l and $x_{(s,i)}^0 = e_{(s,i)}$. $z_{(s,i)}^l$ and $r_{(s,i)}^l$ are respectively the update gate and reset gate in the gating mechanism. σ and $\tanh$ are activation functions. Note that, $W_{in}, W_{out}, W_z, W_r, W_h, U_z, U_r, U_h \in \mathbb{R}^{d \times d}$ are all learnable weights and $b_{in}, b_{out} \in \mathbb{R}^d$ are biases.

Finally, we get the updated item-level embedding $\left\{ e_{(s,1)}^o, ..., e_{(s,n)}^o \right\}$.

Factor-Level Augmentation View. This view is subdivided into $\mathcal{K}$ independent sub-channels inside, which are respectively responsible for learning the embeddings corresponding to $\mathcal{K}$ hidden factors. According to 2, measure the distance similarity between factor-level embeddings of each item to generate loss $\mathcal{L}_d$, and finally we hope that each embedding is as independent as possible to reduce the redundant information between sub-channels

Since items may be similar in one latent factor but not in another, we transform the original session graph $\mathcal{G}_s^o$ to represent the relationship between items on different latent factors. We replace the weight in $\mathcal{A}_s^o$ with the corresponding similarity value to generete factor-level adjacency matrixes $\left\{\mathcal{A}_s^{(f,1)}, ..., \mathcal{A}_s^{(f,K)}\right\}$. $\mathcal{A}_s^{(f,k)}$ is the factor-level adjacency matrix on factor k.

$$a_{(s,ij)}^{(f,k)} = \frac{e_{(s,i)}^k e_{(s,j)}^k}{||e_{(s,i)}^k|| \cdot ||e_{(s,j)}^k||} \tag{8}$$

$a_{(s,ij)}^{(f,k)}$ corresponds to the value of row i and column j in $\mathcal{A}_s^{(f,k)}$.

The input of k-th sub-channel is , $\left\{e_{(s,1)}^k, ..., e_{(s,n)}^k\right\}$, the raw factor-level embeddings of items on hidden factor k and the factor-level adjacent matrix $\mathcal{A}_s^{(f,k)}$. Similar to the original channel, we perform convolution to learn the factor-level embedding based on the adjacency relationships stored in $\mathcal{A}_s^{(f,k)}$ to generate the final output of the sub-channel $\left\{e_{(s,1)}^{(f,k)}, ..., e_{(s,n)}^{(f,k)}\right\}$.

Item-Level Augmentation View. We decide to choose the star graph as the augmentation graph of the original graph at the item-level. In details, we set a satellite node O_s for session s and the embedding of it $e_{(s,O)}$ is set as the average pooling of the items in s.

$$e_{(s,O)} = \frac{\sum_1^n e_{(s,i)}}{n} \tag{9}$$

For the connections between the satellite node and other nodes, we adopt a completely random method that satellite node has an equal probability θ of pointing or being pointed to with other nodes. By implementing the above operations, we can initialize a star graph $\mathcal{G}_s^*$. Furthermore, we have the corresponding adjacency matrix $\mathcal{A}_s^*$. Then $e^s = \left\{e_{(s,1)}, ..., e_{(s,n)}\right\}$ and $\mathcal{A}_s^*$ are sent into the convolution channel and item-level embeddings are updated as $\left\{e_{(s,1)}^*, ..., e_{(s,n)}^*\right\}$.

4.3   Contrastive Learning

Once we have learned three embeddings from three views, we can leverage them to accomplish dual-granularity CL tasks. Combining two Contrastive Learning (CL) methods can enhance the embedding learning ability of our model.

Factor-Level Contrastive Learning. To begin, we re-embed the output of the original view $\left\{e_{(s,1)}^o, ..., e_{(s,n)}^o\right\}$ into the corresponding factor-level spaces to get factor-level embeddings $\left\{\left\{e_{(s,1)}^{(o,1)}, ..., e_{(s,n)}^{(o,1)}\right\}, ..., \left\{e_{(s,1)}^{(o,K)}, ..., e_{(s,n)}^{(o,K)}\right\}\right\}$. Then we use the factor-level embeddings learned by the factor-level augmentation view $\left\{\left\{e_{(s,1)}^{(f,1)}, ..., e_{(s,n)}^{(f,1)}\right\}, ..., \left\{e_{(s,1)}^{(f,K)}, ..., e_{(s,n)}^{(f,K)}\right\}\right\}$ to compare with the former.

To perform CL on $\mathcal{K}$ latent factors, we utilize K sub-channels. Likewise, we describe the operations on the k-th channel as an example. A standard binary cross-entropy (BCE) loss function has been chosen as our learning objective to measure the difference between the two.

$$\mathcal{L}_c^{(f,k)} = - \log \ \sigma(H(e_{(s,i)}^{(o,k)}, e_{(s,i)}^{(f,k)}) - \log \ \sigma(1 - H(e_{(s,i)}^{(o,k)}, e_{(s,j)}^{(o,k)})(i \neq j) \qquad (10)$$

The first half of the comparison involves positive examples, while the second half involves negative examples. and $H : \mathbb{R}^{d_f} \times \mathbb{R}^{d_f} \to \mathbb{R}^{d_f}$ is the discriminator function that takes two vectors as the input and then scores the agreement. between them.

Finally, we aggregate the losses generated by the K channels to obtain the total loss for factor-level CL.

$$\mathcal{L}_c^F = \sum_k^K L_c^{(f,k)} \qquad (11)$$

Item-Level Contrastive Learning. We utilize the two item-level embeddings learned from the Original view $\left\{ e_{(s,1)}^o, ..., e_{(s,n)}^o \right\}$ and the item-level augmenta-tion view $\left\{ e_{(s,1)}^*, ..., e_{(s,n)}^* \right\}$ to conduct item-level CL. Similarly, we adopt BCE loss as the learning object of item-level CL.

$$\mathcal{L}_c^I = - \log \ \sigma(H(e_{(s,i)}^o, e_{(s,i)}^*) - \log \ \sigma(1 - H(e_{(s,i)}^o, e_{(s,j)}^*))(i \neq j) \qquad (12)$$

The total loss of CL is set as follows:

$$\mathcal{L}_c = \alpha * \mathcal{L}_c^I + (1 - \alpha) * \mathcal{L}_c^F \qquad (13)$$

α is a hyperparameter controlling the ratio of item-level and factor-level CL losses.

4.4   Session Embedding

The item-level session embedding need to be generated to represent the user's overall preference for items. Inspired by Disen-GNN, we also generate factor-level session embeddings in addition, which represent users' interests for specific latent factors.

Item-Level Session Embedding. In order to mitigate the potential for mis-leading predictions, we opt to exclusively employ the embedding derived from the original view to compute the session embedding. Although the star graph's enhanced view contains valuable information for contrastive learning, there exists a risk of introducing inaccuracies. Therefore, to ensure more reliable predictions, we restrict our utilization to the item-level embeddings obtained from the original view during the session embedding calculation. The soft attention is employed as the session encoder.

$$\alpha_{(s,i)} = q^\top \sigma(W_s^1 e_{(s,i)}^o + W_s^2 e_{(s,n)}^o) \tag{14}$$

$$e_s^g = \sum_{i=1}^{n} \alpha_{(s,i)} e_{(s,i)}^o \tag{15}$$

$$e_s^o = \boldsymbol{W}_s^3 [e_s^l, e_s^g] \tag{16}$$

where $q \in \mathbb{R}^d$, $\boldsymbol{W}_s^1 \in \mathbb{R}^{d \times d}$, $\boldsymbol{W}_s^2 \in \mathbb{R}^{d \times d}$ and $\boldsymbol{W}_s^3 \in \mathbb{R}^{d \times 2d}$ are learnable parameters. e_s^l and e_s^g represent the session's local and global preferences for factor t respectively. And e_s^l is $e_{(s,n)}^o$, the last item's embedding, such setting can make the model pay more attention to the last clicked item, because usually the last item is the most related to the item that the user finally needs.

Factor-Level Session Embedding. Similarly, we use the same soft attention to compute the user's preference for $\mathcal{K}$ independent latent factors. The process will not be repeated, and finally we concat the user's preferences for $\mathcal{K}$ latent factors as $e_s^f = [e_s^1, ..., e_s^K]$.

4.5 Prediction and Optimization

As mentioned earlier, we score candidate items separately based on the item-level and factor-level interests, and then combine the results of them to make the recommendation. We use the inner-product as the scoring criterion. Note that, in order to calculate the factor-level scores, the item-level embeddings of the candidate items need to be converted into their corresponding factor-level embeddings, and then the inner-product is computed with the corresponding user's factor-level interests. The final score of a item is as follows:

$$\hat{y}_i = \frac{\hat{y_i^I} + \hat{y_i^F}}{2} \tag{17}$$

$\hat{y_i^I}$ and $\hat{y_i^F}$ represent the scoring results of item-level and factor-level respectively. $\hat{y}_i$ stores the final scores. For the prediction loss, we use the cross-entropy as the loss function, which has been extensively used in the recommendation system:

$$\mathcal{L}_p = -\sum_{i=1}^{N} y_i log(\hat{\boldsymbol{y}_i}) + (1 - y_i) log(1 - \hat{\boldsymbol{y}_i}) \tag{18}$$

So, the loss of the whole model consists of three parts: CL, the prediction and DRL. β_1 and β_2 are controlling their proportions.

$$\mathcal{L} = \mathcal{L}_p + \beta_1 \cdot \mathcal{L}_c + \beta_2 \cdot \mathcal{L}_d \tag{19}$$

Adam is adopted as the optimization algorithm to analyze the loss $\mathcal{L}$.

5 Experiments

In this section, we introduce the rich experimental content and outline some of the basic experimental settings.

5.1 Experiments Settings

Datasets. To verify the effectiveness of DGCL-GNN, we conducted experiments on two commonly used datasets in a session-based recommendation system, *Yoochoose* 1/64[1], and *Diginetica*[2]. The statistics of the two datasets are exhibited in Table 1.

Table 1. Statistical results of datasets

Statistics	Yoochoose1/64	Diginetica
#interactions	557,248	982,961
#training sess	369,859	719,470
#test sess	55,898	60,858
#items	16,766	43,097
#avg. length	6.16	5.12

Baselines. To demonstrate the comparative performance of DGCL-GNN, we choose several representative and/or state-of-the-art models. They can be categorized into three types: (1) **Non-GNN models**: NARM, GRU4Rec, and STAMP; (2) **Normal GNN models**: SR-GNN and Disen-GNN; (3) **GNN models with CL**: DHCN and COTREC.

- **GRU4REC** [1] adapts GRU from NLP to SBR. As an RNN-based model, it only cares about sequential relationships between items.
- **NARM** [3] combined attention mechanism with Gated Recurrent Unit(GRU) to consider both global relationships and sequential relationships to make recommendations.
- **STAMP** [4] emphasizes the impact of the short time and it designed a special attention mechanism with MLP.
- **SR-GNN** [10] introduces GNN to obtain item embeddings by information propagation.
- **Disen-GNN** [25] deployed DRL into SBR to learn the latent factor-level embeddings. Then use the GGNN in each factor channel to learn factor-level session embeddings.

---

[1] http://2015.recsyschallenge.com/challege.html.
[2] http://cikm2016.cs.iupui.edu/cikm-cup.

- **DHCN** [16] propose a new CL framework, which realizes data augmentation by learning inter-session information.
- **COTREC** [20] is an improved variant of DHCN, it integrated the idea of co-training into CL by adding divergence constraints to DHCN's CL module.

Evaluation Metrics. Following our baselines, we chose widely used ranking metrics P@K(Precise) and M@K(Mean Reciprocal Rank) to evaluate the recommendation results where K is 10 or 20.

Table 2. Comparing the prediction performance of DGCL-GNN with the baselines. The best results in them are highlighted in bold, and the second-best results are underlined.

Method	Yoochoose1/64				Diginetica			
	P@10	M@10	P@20	M@20	P@10	M@10	P@20	M@20
NARM	0.5920	0.2495	0.6811	0.2855	0.3544	0.1513	0.4970	0.1618
GRU4Rec	0.5011	0.1789	0.6063	0.2288	0.1789	0.0730	0.2939	0.0829
STAMP	0.6190	0.2583	0.6874	0.2967	0.3291	0.1378	0.4539	0.1429
SR-GNN	0.6197	0.2651	0.7055	0.3094	0.3669	0.1538	0.5059	0.1750
Disen-GNN	0.6236	0.2701	<u>0.7141</u>	0.3120	0.3981	0.1769	0.5341	0.1879
DHCN	0.6354	0.2635	0.7078	0.3029	0.3987	0.1753	0.5318	0.1844
COTREC	<u>0.6242</u>	<u>0.2711</u>	0.7113	<u>0.3121</u>	<u>0.4179</u>	<u>0.1812</u>	<u>0.5411</u>	<u>0.1902</u>
DGCL-GNN	**0.6509**	**0.2889**	**0.7469**	**0.3289**	**0.4305**	**0.1824**	**0.5501**	**0.1911**

5.2   Overall Performance

Table 2 shows the overall performance of DGCL-GNN compared to the baseline models. We take the average of 10 runs as the result. And we can make the following three observations:

(1) Compared with RNN-based and Attention-based models, GNN-based models obviously perform better. It exhibits the great capability of GNN in learning more accurate embeddings and modeling session data.
(2) Models trained through CL often exhibit superior performance and demonstrate consistent results across different datasets.
(3) DGCL-GNN outperforms all the baseline models in all datasets. Especially in Nowplaying, there is obvious performance improvement.

In summary, we can conclude that effective CL can lead to better model optimization, and the enhanced CL in our model has a positive impact on the model's overall performance (Fig. 2).

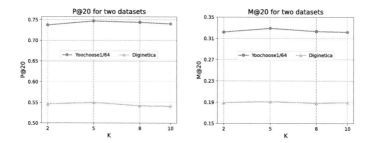

Fig. 2. Impact of the hyperparameter $\mathcal{K}$

5.3 Impact of Hyperparameters

The hyperparameter with the greatest impact on model performance is $\mathcal{K}$, which is the number of disentangled hidden factors. Having more hidden factors can lead to a finer granularity of factor-level CL, but it can also make it more challenging to train accurate factor-level embeddings. Thus, determining the appropriate value for the parameter $\mathcal{K}$ is a crucial consideration. We set it to 5 on the *Diginetica* and *Yoochoose1/64*. The following experimental results also prove the advantages of this setup.

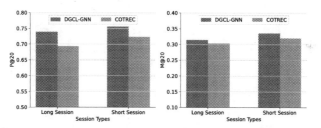

(a) P@20 and M@20 on Yoochoose1/64 of short and long sessions

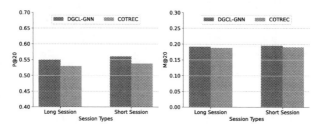

(b) P@20 and M@20 on Diginetica of short and long sessions

Fig. 3. Comparison on different lengths of sessions

5.4   Performance for Different Session Length

Next, as with most SBR models, we evaluated the performance of our model on both long and short sessions. We perform experiments on Yoochoose1/64 dataset and Diginetica dataset. We first split the test sets into *long* sessions and *short* sessions. Similar to [10], sessions with a length ≥ 5 are defined to be *long* sessions while the others are *short* ones. After that, we get two sets of long and short sessions and test the performance of MGCL-GNN and a state-of-the-art baseline, COTREC. The experimental results show that DGCL-GNN is superior to the other two models in both long and short sessions (Fig. 3).

5.5   Ablation Experiments

In order to verify the effectiveness of the main improvement in our model, the CL module, we have designed three variants of DGCL-GNN. The first one is DGCL-GNN-FCL, in which we remove the factor-level CL but maintain item-level CL. The second one is DGCL-GNN-STAR, we replace the star graph augmentation with the more conventional approach of deleting edges and points as a data augmentation method. The third one is DGCL-GNN-FP, We only use item-level user interest to predict the user's next interaction. The rest parts of the variants keep consistent with the original DGCL-GNN to ensure the fairness of the ablation experiments. The experimental results prove that any part of our CL module can improve the recommendation performance of DGCL-GNN (Table 3).

Table 3. Comparing the performance of DGCL-GNN with its three variants.

Model	Yoochoose		Diginetica	
	P@20	M@20	P@20	M@20
DGCL	**0.7469**	**0.3289**	**0.5501**	**0.1911**
DGCL-FCL	0.7391	0.3201	0.5432	0.1895
DGCL-STAR	0.7415	0.3227	0.5459	0.1902
DGCL-FP	0.7388	0.3211	0.5369	0.1889

6   Conclusion

Our DGCL-GNN model improves the exsiting CL strategy and the dual-granularity CL framework can effectively enhance the model's embedding learning capabilities and thus improve the recommendation accuracy of the model. Among them, the additional factor-level contrastive learning can enhance the effect of the original single-grained contrastive learning. The star graph augmentation method also ensures the validity of the self-supervised signal by avoiding destroying the original inter-session graph.

Acknowledgment. This work was supported by the National Key Research and Development Program of China (Grant No. 2019YFB1405302).

References

1. Hidasi, B., Karatzoglou, A., Baltrunas, L., Tikk, D.: Session-based recommenda-tions with recurrent neural networks. arXiv preprint arXiv:1511.06939 (2015)
2. Quadrana, M., Karatzoglou, A., Hidasi, B., Cremonesi, P.: Personalizing session-based recommendations with hierarchical recurrent neural networks. In: proceed-ings of the Eleventh ACM Conference on Recommender Systems, pp. 130–137 (2017)
3. Li, J., Ren, P., Chen, Z., Ren, Z., Lian, T., Ma, J.: Neural attentive session-based recommendation. In: Proceedings of the 2017 ACM on Conference on Information and Knowledge Management, pp. 1419–1428 (2017)
4. Liu, Q., Zeng, Y., Mokhosi, R., Zhang, H.: Stamp: short-term attention/memory priority model for session-based recommendation. In: Proceedings of the 24th ACM SIGKDD International Conference on Knowledge Discovery & Data Mining, pp. 1831–1839 (2018)
5. Tan, Y.K., Xu, X., Liu, Y.: Improved recurrent neural networks for session-based recommendations. In: Proceedings of the 1st Workshop on Deep Learning for Rec-ommender Systems, pp. 17–22 (2016)
6. Chen, T., Wong, R.C.W.: Handling information loss of graph neural networks for session-based recommendation. In: Proceedings of the 26th ACM SIGKDD International Conference on Knowledge Discovery & Data Mining, pp. 1172–1180 (2020)
7. Liu, L., Wang, L., Lian, T.: Case4sr: using category sequence graph to augment session-based recommendation. Knowl.-Based Syst. **212**, 106558 (2021)
8. Qiu, R., Li, J., Huang, Z., Yin, H.: Rethinking the item order in session-based recommendation with graph neural networks. In: Proceedings of the 28th ACM International Conference on Information and Knowledge Management, pp. 579–588 (2019)
9. Wang, Z., Wei, W., Cong, G., Li, X.L., Mao, X.L., Qiu, M.: Global context enhanced graph neural networks for session-based recommendation. In: Proceed-ings of the 43rd International ACM SIGIR Conference on Research and Develop-ment in Information Retrieval, pp. 169–178 (2020)
10. Wu, S., Tang, Y., Zhu, Y., Wang, L., Xie, X., Tan, T.: Session-based recommen-dation with graph neural networks. In: Proceedings of the AAAI Conference on Artificial Intelligence, vol. 33, pp. 346–353 (2019)
11. Luo, A., et al.: Collaborative self-attention network for session-based recommen-dation. In: IJCAI, pp. 2591–2597 (2020)
12. Xu, C., et al.: Graph contextualized self-attention network for session-based rec-ommendation. In: IJCAI, vol. 19, pp. 3940–3946 (2019)
13. Wen, Z., Li, Y.: Toward understanding the feature learning process of self-supervised contrastive learning. In: International Conference on Machine Learning, pp. 11112–11122. PMLR (2021)
14. Yu, F., Zhu, Y., Liu, Q., Wu, S., Wang, L., Tan, T.: Tagnn: target attentive graph neural networks for session-based recommendation. In: Proceedings of the 43rd International ACM SIGIR Conference on Research and Development in Informa-tion Retrieval, pp. 1921–1924 (2020)
15. Liu, X., et al.: Self-supervised learning: generative or contrastive. IEEE Trans. Knowl. Data Eng. **35**(1), 857–876 (2021)
16. Xia, X., Yin, H., Yu, J., Wang, Q., Cui, L., Zhang, X.: Self-supervised hypergraph convolutional networks for session-based recommendation. In: Proceedings of the AAAI Conference on Artificial Intelligence, vol. 35, pp. 4503–4511 (2021)

17. Xin, X., Karatzoglou, A., Arapakis, I., Jose, J.M.: Self-supervised reinforcement learning for recommender systems. In: Proceedings of the 43rd International ACM SIGIR Conference on Research and Development in Information Retrieval, pp. 931–940 (2020)

18. Pan, Z., Cai, F., Chen, W., Chen, C., Chen, H.: Collaborative graph learning for session-based recommendation. ACM Trans. Inf. Syst. (TOIS) **40**(4), 1–26 (2022)

19. Wang, L., Xu, X., Ouyang, K., Duan, H., Lu, Y., Zheng, H.T.: Self-supervised dual-channel attentive network for session-based social recommendation. In: 2022 IEEE 38th International Conference on Data Engineering (ICDE), pp. 2034–2045. IEEE (2022)

20. Xia, X., Yin, H., Yu, J., Shao, Y., Cui, L.: Self-supervised graph co-training for session-based recommendation. In: Proceedings of the 30th ACM International Conference on Information & Knowledge Management, pp. 2180–2190 (2021)

21. Medsker, L.R., Jain, L.: Recurrent neural networks. Des. Appl. **5**, 64–67 (2001)

22. Scarselli, F., Gori, M., Tsoi, A.C., Hagenbuchner, M., Monfardini, G.: The graph neural network model. IEEE Trans. Neural Netw. **20**(1), 61–80 (2008)

23. Vaswani, A., et al.: Attention is all you need. Adv. Neural Inf. Process. Syst. **30** (2017)

24. Zhou, J., Cui, G., Hu, S., Zhang, Z., Yang, C., Liu, Z., Wang, L., Li, C., Sun, M.: Graph neural networks: a review of methods and applications. AI open **1**, 57–81 (2020)

25. Li, A., Cheng, Z., Liu, F., Gao, Z., Guan, W., Peng, Y.: Disentangled graph neural networks for session-based recommendation. IEEE Trans. Knowl. Data Eng. (2022)

26. Chartsias, A., et al.: Disentangled representation learning in cardiac image analysis. Med. Image Anal. **58**, 101535 (2019)

27. John, V., Mou, L., Bahuleyan, H., Vechtomova, O.: Disentangled representation learning for non-parallel text style transfer. arXiv preprint arXiv:1808.04339 (2018)

28. Lin, Z., Tian, C., Hou, Y., Zhao, W.X.: Improving graph collaborative filtering with neighborhood-enriched contrastive learning. In: Proceedings of the ACM Web Conference 2022, pp. 2320–2329 (2022)

29. Dai, B., Lin, D.: Contrastive learning for image captioning. Adv. Neural Inf. Process. Syst. **30** (2017)

30. Tran, L., Yin, X., Liu, X.: Disentangled representation learning gan for pose-invariant face recognition. In: Proceedings of the IEEE Conference on Computer Vision and Pattern Recognition, pp. 1415–1424 (2017)

31. Khosla, P., et al.: Supervised contrastive learning. Adv. Neural. Inf. Process. Syst. **33**, 18661–18673 (2020)

32. Chen, Y., Liu, Z., Li, J., McAuley, J., Xiong, C.: Intent contrastive learning for sequential recommendation. In: Proceedings of the ACM Web Conference 2022, pp. 2172–2182 (2022)

33. Qiu, R., Huang, Z., Yin, H., Wang, Z.: Contrastive learning for representation degeneration problem in sequential recommendation. In: Proceedings of the Fifteenth ACM International Conference on Web Search and Data Mining, pp. 813–823 (2022)

34. Székely, G.J., Rizzo, M.L., Bakirov, N.K.: Measuring and testing dependence by correlation of distances (2007)

35. Li, Y., Tarlow, D., Brockschmidt, M., Zemel, R.: Gated graph sequence neural networks. arXiv preprint arXiv:1511.05493 (2015)

Efficient Graph Collaborative Filtering with Multi-layer Output-Enhanced Contrastive Learning

Keke Li⬤, Shaoqing Wang(✉)⬤, Shun Zheng⬤, Xia Wu⬤, Yao Zhang⬤,
and Fuzhen Sun⬤

School of Computer Science and Technology, Shandong University of Technology,
Zibo 255091, China
lyscoke@163.com, wsq0533@163.com, wuxia1143@sina.cn,
sunfuzhen@sdut.edu.cn

Abstract. Recently, Contrastive Learning (CL) is becoming a mainstream approach to reduce the influence of data sparsity in recommendation system. However, existing methods do not fully explore the relationship between the outputs of different Graph Neural Network (GNN) layers and fail to fully utilize the capacity of combining GNN and CL for better recommendation. Within this paper, we introduce a novel approach based on CL, called efficient Graph collaborative filtering with multi-layer output-enhanced Contrastive Learning (GmoCL). It maximizes the benefits derived from the information propagation property of GNN with multi-layer aggregation to obtain better node representations. Specifically, the construction of CL tasks involves considerations from both intra-layer and inter-layer perspectives. The goal of intra-layer CL task is to exploit the semantic similarities of different users (or items) on a certain GNN layer. The inter-layer CL task aims to make the outputs of different GNN layers of the same user (or item) more similar. Additionally, we propose the strategy of negative sampling in the inter-layer CL task to learn the better node representations. The efficacy of the suggested approach is validated through comprehensive experiments conducted on five publicly available datasets.

Keywords: Recommendation System · Collaborative Filtering ·
Contrastive Learning · Graph Neural Network

1 Introduction

As Web 2.0 gains widespread popularity, the issue of information overload is progressively intensifying. Recommendation systems, as an effective solution, can alleviate the problem of information overload. Collaborative Filtering (CF) can effectively recommend for users by learning user preference from various feedback, such as clicks, purchases, and adding to cart. Recently, powerful Graph Neural Networks (GNNs) have further enhanced CF by modeling interactive

© The Author(s), under exclusive license to Springer Nature Switzerland AG 2023
X. Yang et al. (Eds.): ADMA 2023, LNAI 14176, pp. 755–771, 2023.
https://doi.org/10.1007/978-3-031-46661-8_50

behaviors as graphs. GNNs can learn more effective node representations and make better recommendation performance for users, known as Graph Collaborative Filtering.

Despite the significant success of GNNs, two main problems remain. First, the interactive data of users is usually sparse or noisy, which will leads to inaccurate representations of learned users and items, as graph-based approaches may be more susceptible to data sparsity [24]. Second, existing GNN-based collaborative filtering methods are dependent on explicit interactions to learn node representations, while relationships of outputs of different GNN layers and user or item similarity are not used explicitly to enrich graph information. Contrastive learning methods have been adopted in recent studies to mitigate the scarcity of interaction data [22,24,26,29], however, they are still not fully exploited to mine potential various relationships among users (or items).

Apart from the evident interaction relations between users and items, there are various potential relations, e.g., structural neighbors and semantic neighbors, which are useful for recommendation tasks. NCL [14] takes these aspects into account by constructing contrastive pairs using rich neighbor relations. However, the potential of GNN in facilitating information propagation remains underutilized within NCL. Within the scope of this study, we design a more effective CL approach in order to fully utilize these potential relationships in graph collaborative filtering. Specifically, various potential relationships are utilized after aggregating multiple layers of GNN and further defined in two aspects: (1) intra-layer relationships pertain to the resemblance of output representations among distinct nodes within a given GNN layer, and (2) inter-layer relationships, which refer to the similarity of output representations of the same node at various GNN layers.

To harness the complete potential of output representations from diverse GNN layers, a model-agnostic framework based on contrastive learning for recommendation, called efficient Graph collaborative filtering with multi-layer output-enhanced Contrastive Learning (GmoCL), is proposed. Specifically, the proposed method constructs contrastive targets from two perspectives. From a macro point of view, there are potential relations between some nodes, which may not be explicitly connected on the graph. Inspired by NCL [14], we exploit to divide the similar nodes into same group by means of clustering algorithm. Each node and the cluster center which the node belongs to consist of positive pair, and we consider the other cluster center as the negative samples. In this view, we adopt the outputs of a particular GNN layer to perform cluster operations and construct contrastive targets. So, we denote it as intra-layer perspective and select the outputs of the second layer as representation of nodes in the experiments. From a micro point of view, the outputs of the kth GNN layer aggregate the information pertaining to the k-hop neighbors. Consequently, the outputs from distinct GNN layers of a given node are employed as positive pairs for contrastive learning. This is called inter-layer CL. The user-item interactive data can construct a bipartite graph. Considering the homogeneity between 2-hop neighbors, we divide the inter-layer CL into two kinds: inter-layer CL on odd layer and inter-layer CL on even layer. Furthermore, in inter-layer CL on even layers, we use a negative sampling strategy, with cluster centers as negative samples.

Within this paper, we fully utilize GNN to mine potential associations between users (or items) and combine these supplementary information and relationships to our CL framework. The outcomes of experiments conducted on five datasets demonstrate that the suggested method contributes to a discernible enhancement in recommendation performance. Within this paper, the contributions can be succinctly summarized as follows:

- We introduce a model-agnostic contrastive learning framework named GmoCL, which leverages the information propagation properties of GNNs and aggregates multiple layers of GNNs to improve graph collaborative filtering.
- We devise an intra-layer contrastive learning task and an inter-layer contrastive learning task, which effectively capture the similarities between outputs from distinct GNN layers, thereby enhancing representation learning. Furthermore, in the inter-layer CL task, we propose inter-layer CL on even-layer with negative sampling and inter-layer CL on odd-layer.
- We conduct experiments on five publicly available datasets, where the outcomes validate the rationality and efficacy of the proposed method. Subsequent ablation experiments demonstrate the individual contributions of each component to the enhancement in performance.

2 Preliminary and Definitions

2.1 Preliminary

Unlike traditional CF approaches, e.g., matrix decomposition [12,17] based approaches and autoencoder [13,18] based approaches, graph-based collaborative filtering constructs interactions within user-item interaction graphs and derives semantically valuable node representations from the structural information within the graph. Pioneering studies [1,6] extract structural information in the form of random walking on the graph, and later Graph Neural Networks (GNNs) are employed for collaborative filtering [9,18,23,27], where GNNs that introduce convolutional operations into the graph structure are called Graph Convolutional Networks (GCNs). The fundamental concept behind Graph Convolutional Networks (GCNs) is to acquire node representations by diffusing features throughout the graph. This is achieved through iterative graph convolutions, where features are progressively aggregated from neighboring nodes to form the representation of the focal node. For example, NGCF [22] and Light-GCN [9] use high-order relations on interactive graphs to improve the performance of recommendation. The effectiveness of NGCF is not significantly influenced by the inclusion of feature transformation and nonlinear activation, two operations inherited from GCN. Therefore, LightGCN with these two operations removed contains only the most essential part of GCN, i.e., neighborhood aggregation for collaborative filtering. This uncomplicated and linear model exhibits improved performance and is more straightforward to train. Therefore, Light-GCN is used as the base encoder Within this paper.

Applying a neighborhood aggregation scheme on the graph forms the core of the GCN-based collaborative filtering approach. This scheme involves updating the self-representation by aggregating the representations of neighboring nodes. It can be formulated as two phases.

$$z_u^{(l)} = f_{propagate}\left(\{z_v^{l-1} \mid v \in \{N_u \cup u\}\}\right),$$
$$z_u = f_{readout}\left(\left[z_u^{(0)}, z_u^{(1)}, ..., z_u^{(L)}\right]\right). \tag{1}$$

In the interactive graph G, where N_u represents the set of neighbors of user u, and with L denoting the number of GCN layers, $z_u^{(0)}$ is initialized vector. For user u, the propagation function $f_{propagate}$ aggregates its neighbors as well as its own $(l-1)$th layer representation to generate the lth layer representation, and there are also some works that aggregate only the neighbors' representations, such as LightGCN. Upon undergoing l iterations of propagation, the $z_u^{(l)}$ representation encapsulates the information derived from l-hop neighbors. The readout function $f_{readout}$ is used to receive the ultimate representation of user u. Similarly, the representation of item i can be received.

Predicting the probability of user u engaging with item i is the responsibility of the prediction layer. Here, z_u and z_i correspond to the ultimate representations of user u and item i, respectively. The prediction score is calculated illustrated as follows:

$$\hat{y}_{ui} = z_u^\top z_i, \tag{2}$$

To capture information directly from the interactions, the Bayesian Personalized Ranking (BPR) [17] loss in pairs, a ranking objective function for recommendations, is used. BPR loss forces unobserved interactions to have lower prediction scores than observed interactions. Outlined below is the objective function:

$$L_{main} = \sum_{(u,i,j)\in O} -log\sigma(\hat{y}_{ui} - \hat{y}_{uj}), \tag{3}$$

where $O = \{(u, i, j) \mid r_{u,i} = 1, r_{u,j} = 0\}$ is the training data in pairs, and $r_{u,j} = 0$ means that item j is not interacted by user u. L_{main} is used as the main supervised task for recommendation. By optimizing L_{main}, it possesses the capability to forecast interactions between users and items.

2.2   Problem Definition

In recommender systems, collaborative filtering aims to provide personalized recommendations to users by suggesting items that align with their potential interests, utilizing observed feedback as a foundation, e.g., click, adding to cart, purchase. To elaborate, considering the sets of users U and items I the observed feedback matrix is represented as $R \in \{0,1\}^{|U| \times |I|}$. When there exist interactions between the user u and the item i, then $r_{u,i} = 1$, otherwise 0. The recommender system can predict possible interactions based on the interaction matrix R. In addition, the GNN-based collaborative filtering method constructs the interaction matrix R as an interactive graph G, thus the objective of the problem involves mapping each node v within the set V into a lower-dimensional spatial representation. This mapping aims to facilitate the recommendation of items that might capture the user's interest.

2.3   Notations Definition

The notations used in the paper are shown in Table 1.

Table 1. The notations

Notation	Description
U, I	The set of users and items, respectively
V	$V = \{U \cup I\}$ the set of all users and items
E	$E = \{(u,i) \mid u \in U, i \in I, r_{u,i} = 1\}$ the set of relations
G	Interactive graph
R	$R \in \{0,1\}^{|U| \times |I|}$ represents the matrix of interactions
N_u	Set of neighbor nodes of user u
L	Layers of GNN
$z_u^{(l)}$	Representation of user u at layer l
$z_i^{(l)}$	Representation of item i at layer l
z_u	The ultimate representation of user u
z_i	The ultimate representation of item i
C	The set of cluster centers
K	The number of clusters
L_{main}	The main supervised loss
L_{intra}	The intra-layer contrastive loss
L_{inter}	The inter-layer contrastive loss

3   Methodology

The model architecture of GmoCL is illustrated in Fig. 1, where the annotation on the diagram takes the user as an example, and the item is similar. Our model has four important parts: 1) Multi-layer aggregation is to learn node representations by smoothing features on the graph, which performs graph convolution iteratively. 2) The intra-layer CL is to exploit the semantic similarity between nodes on a particular layer. 3) The inter-layer CL is to pull the outputs of different layer of same node together. 4) Multi-task learning is the joint training of BPR ranking loss and contrast loss.

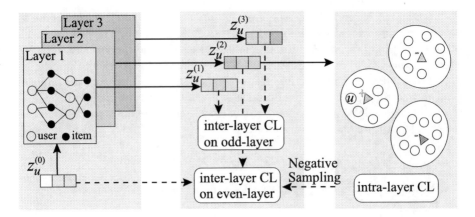

Fig. 1. The architecture of GmoCL. The annotation on the diagram takes the user as an example, and the item is similar.

3.1 Multi-layer Aggregation

The introduced GmoCL model exhibits model-agnostic properties, enabling its integration into numerous recommendation models based on graph neural networks. But for simplicity, we adopt LightGCN as the basic structure of the encoders. At each layer, we exploit LightGCN model to obtain embeddings of nodes. This whole process is repeated for $L(L = 3)$ times due to the over-smoothing problem. Then, corresponding to L layers, we obtain L output representations of each node.

3.2 The Intra-layer Contrastive Learning

The essence of collaborative filtering is to find similar nodes, so it is also essential to mine the relations between similar nodes that are unreachable (i.e., not directly or indirectly connected to each other) on the graph. Given output representations of a particular GNN layer, we construct a contrastive loss function to close the distance of nodes with similar features. It will facilitate nodes to learn a better representation. Specifically, a clustering algorithm is applied to the users and items output representations of a particular GNN layer to obtain the cluster centers, respectively. Our objective is to minimize the distance between the node and the cluster center of the cluster to which the node pertains. In the case of users, the aim of intra-layer contrastive learning is to minimize the subsequent functions:

$$L_{intra}^{U} = \sum_{u \in U} -log \frac{exp\left(cos\left(z_u^{(l)}, c\right)/\tau\right)}{\sum_{c_k \in C} exp\left(cos\left(z_u^{(l)}, c_k\right)/\tau\right)}, \quad (4)$$

where $z_u^{(l)}$ is the output representation of user u at lth layer, $cos(\cdot, \cdot)$ is the cosine similarity function, τ is the temperature hyper-parameter, and c_k is the cluster center obtained by applying the K-means algorithm on all user embeddings, while C represents the set of cluster centers, of which there are K in total. The contrasting learning objective for items is similar,

$$L_{intra}^{I} = \sum_{i \in I} -log \frac{exp\left(cos\left(z_i^{(l)}, c\right)/\tau\right)}{\sum_{c_k \in C} exp\left(cos\left(z_i^{(l)}, c_k\right)/\tau\right)}, \tag{5}$$

where $z_i^{(l)}$ is a representation of the output of item i at lth layer and c_k is the cluster center of item i. The ultimate objective of intra-layer contrastive learning is the summation of weights on both the user and item sides,

$$L_{intra} = L_{intra}^{U} + \alpha L_{intra}^{I}. \tag{6}$$

Here, α serves as the hyper-parameter for weight, maintaining a balance of the two losses.

By applying clustering algorithms, contrastive learning from the intra-layer perspective can alleviate data sparsity and effectively mine similar users or items, thus enabling the model to acquire an improved representation.

3.3 The Inter-layer Contrastive Learning

Viewing the interactive graph as a bipartite graph, the aggregation of information through the GNN-based model combines data from both homogeneous and heterogeneous nodes. We propose to exploit the similarity of the output representation of same node on odd or even layers through CL.

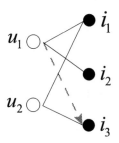

Fig. 2. An example of contrast on odd-layer.

Contrastive Learning on Odd-Layer. It is well known that multi-layer GCN operation on the graph aggregating information of high-hop neighbors can obtain more accurate node representations. In the recommendation scenario, There is a special bipartite graph of user-item interaction graph. And there are large amount of valuable information between the two-hop neighbors.

Illustrated in Fig. 2, user u_1 serves as a representative example. After one layer of GNN propagation, u_1 aggregates the information of the interacted items, such as i_1. After two layers of GNN propagation, u_1 aggregates the information of two-hop neighbors, such as u_2. After three layers of GNN propagation, u_1 aggregates the information of three-hop neighbors, such as i_3. Evident through the dashed line in Fig. 2, since u_1 and u_2 have interacted with the same item i_1, it is inferred that u_1 is more likely to interact with i_3, so the output representations of u_1 at 1st and 3rd layer are used as positive contrast pair, which naturally encodes the items that the user may interact with into the user interest. Specifically, the goal of inter-layer CL on odd-layer is to minimize the distance of outputs of the same node on two consecutive odd layers, such as 1st and 3rd layer. The contrastive loss of user and item are shown in Eq. 7 and Eq. 8.

$$L_{interO}^{U} = \sum_{u \in U} -log \frac{exp\left(cos\left(z_u^{(l)}, z_u^{(1)}\right)/\tau\right)}{\sum_{v \in U} exp\left(cos\left(z_u^{(l)}, z_v^{(1)}\right)/\tau\right)}, \tag{7}$$

$$L_{interO}^{I} = \sum_{i \in I} -log \frac{exp\left(cos\left(z_i^{(l)}, z_i^{(1)}\right)/\tau\right)}{\sum_{j \in I} exp\left(cos\left(z_i^{(l)}, z_j^{(1)}\right)/\tau\right)}, \tag{8}$$

where l is an odd number, and the value of l is 3 in this paper.

Combining the above two losses, contrastive learning objective of inter-layer on odd-layer is constructed as follows:

$$L_{interO} = L_{interO}^{U} + \alpha L_{interO}^{I}. \tag{9}$$

Here, α is the weight hyper-parameter utilized for harmonizing the two losses.

Contrastive Learning on Even-Layer with Negative Samples. The motivation of CL on even-layer is similar to that on odd-layer. However, in Sect. 3.2, we obtain the cluster centers based on the output representations of 2nd layer. So we adopt negative sampling strategy. To be specific, the negative samples are the centers of clusters where the target node is not belong to. This is different from the CL on odd-layer, which treats all other nodes in a batch as negative pairs and ignores the rest of the data in other batch, which suffers from data incompleteness. We constructed the contrastive learning objective for user and item as follows,

$$L_{interE}^{U} = \sum_{u \in U} -log \frac{exp\left(cos\left(z_u^{(l)}, z_u^{(0)}\right)/\tau\right)}{\sum_{c_m \in C} exp\left(cos\left(z_u^{(l)}, c_m\right)/\tau\right)}, \tag{10}$$

$$L_{interE}^{I} = \sum_{i \in I} -log \frac{exp\left(cos\left(z_i^{(l)}, z_i^{(0)}\right)/\tau\right)}{\sum_{c_n \in C} exp\left(cos\left(z_i^{(l)}, c_n\right)/\tau\right)}, \tag{11}$$

where c_m and c_n are cluster centers, l is an even number, and the value of l is 2 in this paper.

Merging the losses from both user and item sides, the contrastive learning objective function can be expressed as follows:

$$L_{interE} = L_{interE}^{U} + \alpha L_{interE}^{I}. \tag{12}$$

Here, α functions as the weight hyper-parameter, serving to balance the two categories of contrast losses.

Multi-layer aggregation constructs two contrastive loss objectives for nodes on even as well as odd layers of GNN, explicitly mining the information of homogeneous and heterogeneous nodes, which facilitates representation learning of nodes. Combining the even and odd layer contrastive learning objectives, the final multi-layer aggregation contrastive learning objective is:

$$L_{inter} = L_{interE} + \beta L_{interO}. \tag{13}$$

In this context, β represents the weight hyper-parameter that achieves a balance between the two types of contrast losses.

3.4 Multi-task Training

Given that the primary objective of collaborative filtering involves predicting potential interactions between users and items, the contrast loss serves as a supplementary component. To simultaneously train the InfoNCE contrast loss and the traditional BPR ranking loss, we utilize a multi-task learning strategy. The overall losses are as follows:

$$L = L_{main} + \lambda_1 L_{intra} + \lambda_2 L_{inter} + \lambda_3 \|\theta\|_2^2. \tag{14}$$

In this equation, λ_1, λ_2, and λ_3 serve as weight hyper-parameters, responsible for achieving a balance between the two proposed contrast losses and normalized terms. Meanwhile, θ represents the set of parameters within the GNN model.

4   Experiments

To validate the efficacy of our proposed model, an extensive array of experiments has been conducted, accompanied by thorough and detailed analyses.

4.1 Experimental Setup

In this section, we introduce the dataset used for the experiments, the baseline methods, the evaluation metrics, and some implementation details.

Datasets. To evaluate the performance of the proposed model, experiments are conducted using five public datasets: Yelp, MovieLens-1M (ML-1M) [7], Alibaba-iFashion [3], Gowalla [4], and Amazon-Books [16].

Baseline Methods. We conduct a comparative analysis of the proposed model in contrast to the subsequent baseline methods:

BPRMF [17] uses the matrix factorization (MF) framework to optimize BPR loss to acquire latent representations of users and items with potential.

NeuMF [10] uses multilayer perceptron instead of dot product in MF model to learn the matching function of users and items.

FISM [11] represents an item-based collaborative filtering model that combines historical interaction representations to capture user interests.

NGCF [22] utilizes a bipartite graph structure to integrate higher-order relationships between users and items, while also employing GNN to enhance the collaborative filtering-based model.

Multi-GCCF [19] facilitates the propagation of information among users (and items) with higher-order associations, extending beyond the user-item bipartite graph.

DGCF [23] disentangles representations of users and items, resulting in enhanced recommendation performance.

LightGCN [9] streamlines the GCN architecture for increased simplicity and compatibility within recommendation systems.

SGL [24] take advantage of contrastive learning to strengthen recommendation. In this paper, we adopt SGL-ED which is the best instance of SGL.

NCL [14] proposes the contrastive learning method involving both structural and semantic neighbors to enhance the recommendation performance. The method proposed in our paper enhances this model and maximizes the advantages of GNN to construct the contrastive learning loss function.

Table 2. Datasets

Datasets	#Users	#Items	#Interactions	Density
ML-1M	6,040	3,629	836,478	0.03816
Yelp	45,478	30,709	1,777,765	0.00127
Amazon-Books	58,145	58,052	2,517,437	0.00075
Gowalla	29,859	40,989	1,027,464	0.00084
Alibaba	300,000	81,614	1,607,813	0.00007

Evaluation Metrics. We employ two widely recognized metrics, NDCG@N and Recall@N, with N values set at 10, 20, and 50, respectively. These metrics are utilized to assess the top-N performance. Following [24] and [9], we employ a full ranking strategy, i.e., ranking all items with which the user has not engaged.

Implementation Details. Our model and all baseline methods are implemented using RecBole [28], a comprehensive open-source framework designed for the development and replication of recommendation algorithms. To ensure just comparisons, the Adam optimizer is employed for optimization across all methods. All parameters are initialized using the Xavier distribution, with a batch size of 4096 and an embedding size of 64. We use the early stop approach to prevent overfitting, the patience value is set to 40 epochs, and set NDCG@10 as the indicator (Table 2).

4.2 Overall Performance

The performance of the proposed method, along with other baseline models, is depicted in Table 3 across five datasets. From the results, several observations and conclusions can be drawn:

- Collaborative filtering models that encode representations of historical interaction behaviors as user interests, such as FISM, show better performance on all datasets, which demonstrates the effectiveness of collaborative filtering models. Of all the graph collaborative filtering baseline methods, LightGCN performs best in most datasets, indicating that the simple framework is more effective and robust. Additionally, the decoupled representation learning method DGCF performs less favorably than LightGCN, particularly when dealing with sparse datasets. This may be because, the dimensionality of the decoupled representation is too low to carry enough features.
- In terms of CL methods, SGL and NCL consistently exhibit superior performance over other supervised techniques. SGL obtains data augmented graphs for contrast by randomly perturbing user-item bipartite graphs, and although effective, this approach ignores other potential relationships (e.g., user similarity) in the recommendation system. NCL considers the importance of user (or item) similarity for representation learning from structural and semantic perspectives. While demonstrating effectiveness, it does not fully leverage the high-order information propagation properties inherent in GNNs.
- The superior performance of our model compared to all baseline models showcases the effectiveness of the contrastive learning approach through GNN multi-layer aggregation. Besides, the performance improvement of our method is more obvious on sparse datasets, such as Amazon-Books dataset and Alibaba dataset. This may be due to the fact that sparse datasets have too little interaction information, while our method explores the similarity between the outputs of discontinuous layers of nodes through contrastive learning, making the model to predict more accurate results.

4.3 Ablation Experiments

This subsection further analyzes the efficacy of the proposed model through ablation experiments. Figure 3 displays the results for Amazon-Books and ML-1M datasets, as constraints on space prevent the inclusion of additional data.

Table 3. Overall performance of different methods

Dataset	Metric	BPRMF	NeuMF	FISM	NGCF	MultiGCCF	DGCF	LightGCN	SGL	NCL	Ours	Improv.
ML-1M	Recall@10	0.1804	0.1657	0.1887	0.1846	0.1830	0.1881	0.1876	0.1888	0.2048	0.2106	+2.83%
	NDCG@10	0.2463	0.2295	0.2494	0.2528	0.2510	0.2520	0.2514	0.2526	0.2727	0.2771	+1.61%
	Recall@20	0.2714	0.2520	0.2798	0.2741	0.2759	0.2779	0.2796	0.2848	0.3032	0.3108	+2.51%
	NDCG@20	0.2569	0.2400	0.2607	0.2614	0.2617	0.2615	0.2620	0.2649	0.2842	0.2897	+1.94%
	Recall@50	0.4300	0.4122	0.4421	0.4341	0.4364	0.4424	0.4469	0.4487	0.4677	0.4785	+2.31%
	NDCG@50	0.3014	0.2851	0.3078	0.3055	0.3056	0.3078	0.3091	0.3111	0.3298	0.3365	+2.03%
Yelp	Recall@10	0.0643	0.0531	0.0714	0.0630	0.0646	0.0723	0.0730	0.0833	0.0912	0.0941	+3.18%
	NDCG@10	0.0458	0.0377	0.0510	0.0446	0.0450	0.0514	0.0520	0.0601	0.0679	0.0692	+1.91%
	Recall@20	0.1043	0.0885	0.1119	0.1026	0.1053	0.1135	0.1163	0.1288	0.1358	0.1411	+3.90%
	NDCG@20	0.0580	0.0486	0.0636	0.0567	0.0575	0.0641	0.0652	0.0739	0.0815	0.0837	+2.70%
	Recall@50	0.1862	0.1654	0.1963	0.1864	0.1882	0.1989	0.2016	0.2140	0.2171	0.228	+5.02%
	NDCG@50	0.0793	0.0685	0.0856	0.0784	0.0790	0.0862	0.0875	0.0964	0.103	0.1066	+3.50%
Amazon	Recall@10	0.0607	0.0507	0.0721	0.0625	0.0625	0.0737	0.0797	0.0898	0.094	0.0986	+4.89%
	NDCG@10	0.043	0.0351	0.0504	0.0433	0.0433	0.0521	0.0565	0.0645	0.0683	0.072	+5.42%
	Recall@20	0.0956	0.0823	0.1099	0.0991	0.0991	0.1128	0.1206	0.1331	0.138	0.1443	+4.57%
	NDCG@20	0.0537	0.0447	0.0622	0.0545	0.0545	0.064	0.0689	0.0777	0.0817	0.086	+5.26%
	Recall@50	0.1681	0.1447	0.1830	0.1688	0.1688	0.1908	0.2012	0.2267	0.2179	0.2267	+4.04%
	NDCG@50	0.0726	0.061	0.0815	0.0727	0.0727	0.0843	0.0899	0.0992	0.1028	0.1079	+4.96%
Gowalla	Recall@10	0.1158	0.1039	0.1081	0.1192	0.1108	0.1252	0.1362	0.1465	0.1502	0.1505	+0.20%
	NDCG@10	0.0833	0.0731	0.0755	0.0852	0.0791	0.0902	0.0876	0.1048	0.1082	0.1089	+0.65%
	Recall@20	0.1695	0.1535	0.1620	0.1755	0.1626	0.1829	0.1976	0.2084	0.2129	0.215	+0.99%
	NDCG@20	0.0988	0.0873	0.0913	0.1013	0.0940	0.1066	0.1152	0.1225	0.1263	0.1274	+0.87%
	Recall@50	0.2756	0.2510	0.2673	0.2811	0.2631	0.2877	0.3044	0.3197	0.3259	0.3274	+0.46%
	NDCG@50	0.1450	0.1110	0.1169	0.1270	0.1184	0.1322	0.1414	0.1497	0.1541	0.155	+0.58%
Alibaba	Recall@10	0.303	0.182	0.0357	0.0382	0.0401	0.0447	0.0457	0.0461	0.0484	0.0498	+2.89%
	NDCG@10	0.0161	0.0092	0.0190	0.0198	0.0207	0.0241	0.0246	0.0248	0.0264	0.0272	+3.03%
	Recall@20	0.0467	0.0302	0.0553	0.0615	0.0634	0.0677	0.0692	0.0692	0.0717	0.0748	+4.32%
	NDCG@20	0.0203	0.0123	0.0239	0.0257	0.0266	0.0299	0.0246	0.0307	0.0323	0.0335	+3.72%
	Recall@50	0.0799	0.0576	0.0943	0.1081	0.1107	0.1120	0.1144	0.1141	0.1155	0.1209	+4.68%
	NDCG@50	0.0269	0.0177	0.0317	0.0349	0.0360	0.0387	0.0396	0.0396	0.041	0.0427	+4.15%

With "w/o inter" and "w/o intra" denote the variables that remove inter-layer CL and intra-layer CL, respectively. As depicted in the figure, removing each aspect leads to a performance decrease, while both variants perform better than the LightGCN. Furthermore, these two relations mutually reinforce each other, contributing to performance improvement through distinct avenues.

4.4 Hyper-parameter Analysis

Within this specific section, we analyze the effects of hyper-parameter α and β. However, due to spatial constraints, we present the outcomes solely for the ML-1M and Amazon-Books datasets in Fig. 4 and 5 respectively.

Effect of Hyper-parameter α. The coefficient α is used for balancing the user side and item side on the intra-layer contrasts and inter-layer contrasts. To analyze the effect of it, we set its variation range between 0.1 and 2 and report the results in Fig. 4. This suggests that a proper α can be effective to increase the performance of our method. The best results are achieved on the ML-1M dataset with the value of 1.0 and on the Amazon-Books dataset with the value of 0.3, indicating that the homogeneity between the outputs of discontinuous layers is valuable for both users and items.

Effect of Hyper-parameter β. Here, we analyze the effects of hyper-parameter β for balancing odd- and even-layer contrasts through experiments. To analyze the effect of it, we set its variation range between 0.1 and 2 and report the results in Fig. 5. The results show that the value of β achieves the best results differs on different datasets. Specifically, the best results are achieved on the ML-1M dataset with the value of 1.0 and on the Amazon-Books dataset with the value of 0.5.

4.5   Distribution of Items Embedding

We show the effects of the proposed model on representation learning in Fig. 6, where our visualization is based on the SVD decomposition, which projects the embedding matrix into a two-dimensional space. From the figure we can see that the distribution of embeddings of low-frequency items is more balanced that they are located around the origin point in our proposed method compared to NCL, and we hypothesize that a more balanced embedding distribution better models different user preferences or item features.

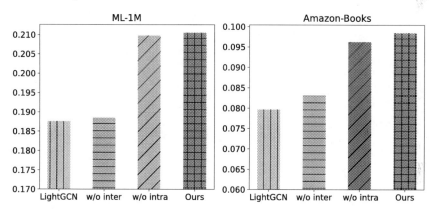

Fig. 3. Performance comparison without inter-layer CL and intra-layer CL on two datasets respectively (Recall@10).

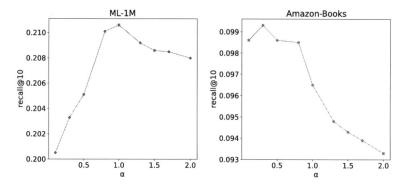

Fig. 4. Performance comparison for different α

5   Related Work

This section provides a concise overview of two pertinent studies: graph-based collaborative filtering and contrastive learning.

Graph-Based Collaborative Filtering. The CF-based approach has evolved by now to apply graph neural networks (GNN) to collaborative filtering [9,22,23,27]. For example, utilizing higher-order relations within interaction graphs, NGCF [22] and LightGCN [9] aim to enhance recommendation performance. Furthermore, [19] extends this idea by introducing the construction of multiple interaction graphs to attain more comprehensive association relationships between users and items. Although it is effective, they do not explicitly address the problem of data sparsity [24]. Lately, the integration of self-supervised learning into graph collaborative filtering has emerged as an approach to bolster the efficacy of recommendations. However, most graph-based approaches focus only on interaction history and ignore the potential neighbor relationships between users or items. Some recent self-supervised learning methods are proposed. NCL [14] proposes to consider users (or items) and their homo-

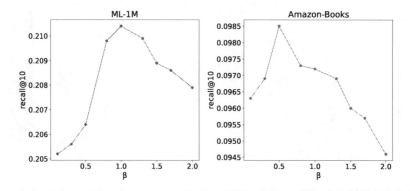

Fig. 5. Performance comparison for different β

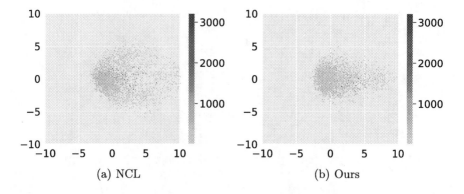

Fig. 6. Distribution of item embedding on ML-1M.

geneous neighbors as positive contrastive pairs and construct two self-supervised losses in the structure space and semantic space, respectively. However, the strong GNN is still not well utilized to exploit potential user (or item) relationships.

Contrastive Learning. Given the success of contrastive learning in Computer Vision (CV) [2], its application has extended across Natural Language Processing (NLP) [5], recommender systems [20], and graph data mining [15,25]. The primary objective of contrastive learning is to optimize the agreement between positive pairs while minimizing the agreement between negative pairs. DGI [21] regards graph-level representations and node-level representations of the same graph as positive pairs. CMRLG [8] realizes a similar goal by treating adjacency matrix and diffusion matrix as positive pairs. Recently, SGL [24] designs random data augmentation operations and constructed contrastive targets to improve the recommendation performance. More recently, NCL [14] considers the importance of user (or item) similarity for representation learning in structural and semantic aspects, and although effective, It falls short of fully capitalizing on the high-order information propagation characteristics inherent in GNNs. Within this paper, we introduce a contrastive learning framework that operates from both intra-layer and inter-layer perspectives. This framework is devised to comprehensively harness the output representations derived from various GNN layers.

6   Conclusion and Future Work

In order to take full advantage of the output representations of different GNN layers, a model-agnostic contrastive learning framework for recommendation, called GmoCL, is proposed. Specifically, contrastive goals are constructed in two respects. For intra-layer CL, we try to find the semantic similar contrastive pair by use of clustering algorithm. For inter-layer CL, we focus on the similar of output representations of two consecutive odd or even layers of the same node. Furthermore, we adopt negative sampling strategy for inter-layer CL on even-layer, which can make the model learn better representations. The effectiveness of the proposed method is underscored by an extensive array of experiments conducted on five publicly accessible datasets.

Going forward, we will place additional emphasis on the matter of positive and negative sampling. In addition, applying our contrast learning framework to different recommendation tasks is also part of our work, and we believe that the potential of GNN still deserves to be explored fully, and combining contrastive learning with heterogeneous graph neural networks will be more beneficial for representation learning and thus can be used better for downstream tasks.

Acknowledgements. This work was supported by Shandong Provincial Natural Science Foundation, China (ZR2020MF147, ZR2021MF017).

References

1. Baluja, S., et al.: Video suggestion and discovery for youtube: taking random walks through the view graph. In: Proceedings of the 17th International Conference on World Wide Web, pp. 895–904 (2008)
2. Chen, T., Kornblith, S., Norouzi, M., Hinton, G.: A simple framework for contrastive learning of visual representations. In: International Conference on Machine Learning, pp. 1597–1607. PMLR (2020)
3. Chen, W., et al.: POG: personalized outfit generation for fashion recommendation at Alibaba iFashion. In: Proceedings of the 25th ACM SIGKDD International Conference on Knowledge Discovery & Data Mining, pp. 2662–2670 (2019)
4. Cho, E., Myers, S.A., Leskovec, J.: Friendship and mobility: user movement in location-based social networks. In: Proceedings of the 17th ACM SIGKDD International Conference on Knowledge Discovery and Data Mining, pp. 1082–1090 (2011)
5. Giorgi, J., Nitski, O., Wang, B., Bader, G.: Declutr: deep contrastive learning for unsupervised textual representations. arXiv preprint arXiv:2006.03659 (2020)
6. Gori, M., Pucci, A., Roma, V., Siena, I.: Itemrank: a random-walk based scoring algorithm for recommender engines. In: IJCAI, vol. 7, pp. 2766–2771 (2007)
7. Harper, F.M., Konstan, J.A.: The movielens datasets: history and context. ACM Trans. Interact. Intell. Syst. (TIIS) 5(4), 1–19 (2015)
8. Hassani, K., Khasahmadi, A.H.: Contrastive multi-view representation learning on graphs. In: International Conference on Machine Learning, pp. 4116–4126. PMLR (2020)
9. He, X., Deng, K., Wang, X., Li, Y., Zhang, Y., Wang, M.: Lightgcn: simplifying and powering graph convolution network for recommendation. In: Proceedings of the 43rd International ACM SIGIR Conference on Research and Development in Information Retrieval, pp. 639–648 (2020)
10. He, X., Liao, L., Zhang, H., Nie, L., Hu, X., Chua, T.S.: Neural collaborative filtering. In: Proceedings of the 26th International Conference on World Wide Web, pp. 173–182 (2017)
11. Kabbur, S., Ning, X., Karypis, G.: FISM: factored item similarity models for top-n recommender systems. In: Proceedings of the 19th ACM SIGKDD International Conference on Knowledge Discovery and Data Mining, pp. 659–667 (2013)
12. Koren, Y., Bell, R., Volinsky, C.: Matrix factorization techniques for recommender systems. Computer 42(8), 30–37 (2009)
13. Liang, D., Krishnan, R.G., Hoffman, M.D., Jebara, T.: Variational autoencoders for collaborative filtering. In: Proceedings of the 2018 World Wide Web Conference, pp. 689–698 (2018)
14. Lin, Z., Tian, C., Hou, Y., Zhao, W.X.: Improving graph collaborative filtering with neighborhood-enriched contrastive learning. In: Proceedings of the ACM Web Conference 2022, pp. 2320–2329 (2022)
15. Liu, Y., et al.: Graph self-supervised learning: a survey. IEEE Trans. Knowl. Data Eng. (2022)
16. McAuley, J., Targett, C., Shi, Q., Van Den Hengel, A.: Image-based recommendations on styles and substitutes. In: Proceedings of the 38th International ACM SIGIR Conference on Research and Development in Information Retrieval, pp. 43–52 (2015)
17. Rendle, S., Freudenthaler, C., Gantner, Z., Schmidt-Thieme, L.: BPR: Bayesian personalized ranking from implicit feedback. arXiv preprint arXiv:1205.2618 (2012)

18. Strub, F., Mary, J., Philippe, P.: Collaborative filtering with stacked denoising autoencoders and sparse inputs. In: NIPS Workshop on Machine Learning for eCommerce (2015)
19. Sun, J., et al.: Multi-graph convolution collaborative filtering. In: 2019 IEEE International Conference on Data Mining (ICDM), pp. 1306–1311. IEEE (2019)
20. Tang, H., Zhao, G., Wu, Y., Qian, X.: Multisample-based contrastive loss for top-k recommendation. IEEE Trans. Multimedia (2021)
21. Veličković, P., Fedus, W., Hamilton, W.L., Liò, P., Bengio, Y., Hjelm, R.D.: Deep graph infomax. arXiv preprint arXiv:1809.10341 (2018)
22. Wang, X., He, X., Wang, M., Feng, F., Chua, T.S.: Neural graph collaborative filtering. In: Proceedings of the 42nd International ACM SIGIR Conference on Research and Development in Information Retrieval, pp. 165–174 (2019)
23. Wang, X., Jin, H., Zhang, A., He, X., Xu, T., Chua, T.S.: Disentangled graph collaborative filtering. In: Proceedings of the 43rd International ACM SIGIR Conference on Research and Development in Information Retrieval, pp. 1001–1010 (2020)
24. Wu, J., et al.: Self-supervised graph learning for recommendation. In: Proceedings of the 44th International ACM SIGIR Conference on Research and Development in Information Retrieval, pp. 726–735 (2021)
25. Wu, L., Lin, H., Tan, C., Gao, Z., Li, S.Z.: Self-supervised learning on graphs: contrastive, generative, or predictive. IEEE Trans. Knowl. Data Eng. (2021)
26. Wu, Y., et al.: Multi-view multi-behavior contrastive learning in recommendation. In: Bhattacharya, A., et al. (eds.) DASFAA 2022. LNCS, vol. 13246, pp. 166–182. Springer, Cham (2022). https://doi.org/10.1007/978-3-031-00126-0_11
27. Ying, R., He, R., Chen, K., Eksombatchai, P., Hamilton, W.L., Leskovec, J.: Graph convolutional neural networks for web-scale recommender systems. In: Proceedings of the 24th ACM SIGKDD International Conference on Knowledge Discovery & Data Mining, pp. 974–983 (2018)
28. Zhao, W.X., et al.: Recbole: towards a unified, comprehensive and efficient framework for recommendation algorithms. In: Proceedings of the 30th ACM International Conference on Information & Knowledge Management, pp. 4653–4664 (2021)
29. Zhu, Y., Xu, Y., Yu, F., Liu, Q., Wu, S., Wang, L.: Graph contrastive learning with adaptive augmentation. In: Proceedings of the Web Conference 2021, pp. 2069–2080 (2021)

Influence Maximization with Tag Revisited: Exploiting the Bi-submodularity of the Tag-Based Influence Function

Atharva Tekawade and Suman Banerjee[(✉)]

Department of Computer Science and Engineering, Indian Institute of Technology
Jammu, Jammu 181221, India
{2018uee0137,suman.banerjee}@iitjammu.ac.in

Abstract. Given a Social Network how to select a small number of influential users to maximize the influence in the network has been studied extensively in the past two decades and formally referred to as the Influence Maximization Problem. Among most of the existing studies, it has been implicitly assumed that there exists a single probability value that represents the influence probability between the users. However, in reality, the influence probability between any two users is dependent on the context (formally referred to as tag e.g.; a sportsman can influence his friends related to any news related to sports with high probability). In this paper, we bridge the gap by studying the TAG-BASED INFLUENCE MAXIMIZATION PROBLEM. In this problem, we are given with a social network where each edge is marked with one probability value for every tag and the goal here is to select k influential users and r influential tags to maximize the influence in the network. First, we define a tag-based influence function and show that this function is bi-submodular. We use the orthent-wise maximization procedure of bi-submodular function which gives a constant factor approximation guarantee. Subsequently, we propose a number of efficient pruning techniques that reduces the computational time significantly. We perform an extensive number of experiments with real-world datasets to show the effectiveness and efficiency of the proposed solution approaches.

Keywords: Social Network · Influence Maximization · Seed Set · Bi-Submodular Function

1 Introduction

Diffusion of information in a networked system has been studied extensively to answer several questions in different domains such as how infectious disease spreads in a human contact network [9], how malware, wormholes, etc. spread in computer networks [1], how innovation, concepts, ideas, etc. spread through

The work of Dr. Suman Banerjee is supported with the Seed Grant sponsored by the Indian Institute of Technology Jammu (Grant No.: SG100047).

social networks and many more [6]. The diffusion of information in social networks has got applications in different domains such as *viral marketing* [13], *computational advertisement* [10], *feed ranking*, etc. Hence, several diffusion models have been proposed in the literature. Among them, one of the popular diffusion models that have been studied extensively is the *Independent Cascade Model* (abbreviated as *IC Model*). According to this model, information is diffused in discrete time steps from a set of initially active nodes called as *seed nodes*. A node can be any one of the following two states: '*active*' (also called as '*influenced*') and '*inactive*' (also called as '*influenced*'). Every active node at time step t will get a single chance to activate its inactive neighbors with the success probability as the edge weight. The diffusion process ends when no more node activation is possible.

One important problem that has been studied in the context of information diffusion is the problem of *Influence Maximization*. Given a social network and a positive integer k the problem of influence maximization asks to choose a subset of k nodes whose initial activation leads to maximum influence in the network. This problem has got a significant applications in the domain of viral marketing. Consider a commercial house developed a new product and wants to prompt among people. They distribute a limited number of sample product (in free of cost or in discounted price) among a group of highly influential users with a hope that they will use the product and share information about it among their neighbors. Some of them will be influenced and buy the product. This cascading process will go on and at the end of diffusion process a significant number of people will ultimately buy this product and the E-Commerce house can earn revenue. Due to practical applications the problem of influence maximization has been studied extensively and several solution methodologies has been proposed in the literature.

One of the important drawback of the existing studies is that most of them considered that there exists a single influence value between any two users of the network. However, in practice the case is not exactly the same. The influence probability between any two user is always dependent on the context. As an example, a sportsman can influence related to any issue related to sports with high probability compared to any other contexts. These contexts are formalized as tags. In real-world situations between every pair of users there exists an influence probability value corresponding to every tag. Now it is easy to observe that the influence in the network will not only depends on the seed set we are choosing, but also the tags we are choosing. In this context the problem that arises is given a social network where each edges of the network is marked with an influence probability value corresponding to every tag and the aim is to choose a subset of k nodes and r tags such that the influence in the network gets maximized. Though this problem is quite natural in many realistic situations, however we observe that the number of studies which considers both selection of tags and seed nodes for influence maximization is very limited.

To the best of our knowledge, Ke et al. [7] were the first to study this problem and they proposed a sketch-based solution approach for this problem. Also, their experiments show that their proposed methodology can process significantly large datasets within reasonable computational time. Banerjee et al. [3,4] studied a similar problem where they considered the users and tags have non-uniform selection costs and a fixed amount of budget is given. The goal was to select a subset of the nodes as seed nodes and a subset of the tags within the budget to maximize the influence. They showed that in many keyword-based Social Network datasets the popularity of the tags varies a lot across different communities within the same network. Other than these two studies there are no studies that considers the same problem. In both these studies there is no mathematical analysis of the tag-based influence function has been done. One key observation of this study is that this analysis leads to efficient algorithms for optimization of this function. In particular, we make the following contributions in this paper:

- We study the problem of selecting influential users and tags simultaneously for which there exists limited studies in the literature.
- We do a mathematical analysis of the Tag-Based Influence Function and prove several theoretical results.
- We propose a Coordinate-wise Solution Approach and Community-based Solution Approach to solve the Tag-Based Influence Maximization Problem.
- We perform an extensive set of experiments with real-life datasets to show the effectiveness and efficiency of the proposed solution approaches.

Rest of the paper is organized as follows. Section 2 describes relevant preliminary concepts and defines our problem formally. Section 3 describes the proposed solution approaches with detailed analysis. Section 4 contains the experimental validation of the proposed solution approaches and finally, Sect. 5 concludes this study and gives future research directions.

2   Preliminaries and Problem Definition

In this section we describe some preliminary concepts and defines our problem formally. Initially we start by describing social networks.

2.1   Social Network

In this study we model a social network by a weighted and directed graph $G(V, E, P)$ where the vertex set $V(G) = \{u_1, u_2, \ldots, u_n\}$ represents the set of users of the network. The edge set $E(G)$ are the set of social ties among the users; i.e., there is an edge between the users u_i and u_j if there exists a social relation between u_i and u_j. We denote the number of vertices and edges of G by n and m, respectively. Consider there are a set of tags $\mathbb{T} = \{t_1, t_2, \ldots, t_k\}$ which are relevant to the users. For every edge $(u_i u_j) \in E(G)$ and for every tag $t \in \mathbb{T}$ there exists an influence probability denoted by $\mathcal{P}^t_{u_i \rightarrow u_j}$. This can be interpreted as as the influence probability of the edge (uv) when the tag t is used for the diffusion process. In the graph G, the edge weight function $\mathcal{P}$ maps each edge-tag pair to the corresponding influence probability; i.e.; $\mathcal{P} : E(G) \times \mathbb{T} \longrightarrow (0, 1]$. Now, we can observe that for all the edges their tag specific probability can be represented by a $m \times k$ matrix denoted by $\mathbb{P}$. (i, j)-th entry of the matrix $\mathbb{P}$; i.e.; $\mathbb{P}[i, j]$ contains the influence probability of the edge e_i for the tag t_j. Now, for a given subset of tags $T' \subseteq \mathbb{T}$, how to aggregate the influence probabilities for the tags in T' to obtain the influence probability of the edge. This depends on how we are aggregating the tags. In this study we are aggregating the tags considering they are independent to each other and this called as independent tag aggregation which is stated in Definition 1.

Definition 1 (Independent Tag Aggregation). *For an edge $(u_i u_j) \in E(G)$ and a subset of the tags $T' \subseteq \mathbb{T}$ the aggregated influence probability of this edge is denoted as $\mathcal{P}^{T'}_{u_i \rightarrow u_j}$ and defined using Equation No. 1.*

$$\mathcal{P}^{T'}_{u_i \rightarrow u_j} = 1 - \prod_{t \in T'} (1 - \mathcal{P}^{t}_{u_i \rightarrow u_j}) \tag{1}$$

Now, it is easy to observe that for all the edges in the worst case the independent tag aggregation can take $\mathcal{O}(k \cdot m)$ time.

2.2 Influence Diffusion in Social Networks

The diffusion process in a networked system has been studied extensively due to its applications in different domains including Epidemiology, Computer Networks, Social Networks, and many more. As this paper deals with social networks, here we discuss the diffusion of information in social networks. Due to several application domains such as viral marketing, computational advertisement, feed ranking, etc. there are extensive studies on the diffusion of information in social networks. Several models have been proposed and studied in the literature. Among them, two fundamental models that have been considered extensively are the *Independent Cascade Model (IC Model)* and *Linear Threshold Model (LT Model)*. In this study, we consider the diffusion in the underlying network is happening according to the rule of the IC Model which is stated in Definition 2.

Definition 2 (Independent Cascade Model). *The rules of the independent cascade model is as follows:*

- *Information is diffused in discrete time steps.*
- *A node can be either of the two states: 'inactive' (also refereed to as 'uninfluenced') and active (also referred to as 'influenced')*
- *An active node at time step t will get a single chance to activate its inactive neighbors at time step $(t + 1)$.*
- *A node can change its state from active to 'inactive', however not the vice versa.*

Let, T' be the set of tags that are used for the diffusion process. Also, these tags are aggregated as per Equation No. 6. In IC Model information is diffused in discrete time steps. It is assumed that initially (i.e.; at time step $t = 0$) a subset of the nodes $\mathcal{S} \subseteq V(G)$ are active and the diffusion process starts from the nodes in $\mathcal{S}$. We call these nodes as *Seed Nodes*. Every active node at time step t will get a single chance to activate its inactive neighbors with success probability as the aggregated edge probability. Now in the diffusion process, some of the nodes will be active. $I(\mathcal{S})$ denotes the set of nodes that are activated from the seed set $\mathcal{S}$. The number of nodes in $I(\mathcal{S})$ is called the *Influence of the Seed Set $\mathcal{S}$*. For any seed set $\mathcal{S}$, its influence is denoted as $\sigma(\mathcal{S})$ which is stated in Definition 3.

Definition 3 (Influence of a Seed Set). *Given a seed set $\mathcal{S}$, its influence is denoted by $I(\mathcal{S})$ and defined as the number of nodes that are activated at the end of diffusion process. Hence, $\sigma(\mathcal{S}) = |I(\mathcal{S})|$. Here, $\sigma()$ is the influence function that maps each subset of the nodes to its expected influence; i.e.; $\sigma : 2^{V(G)} \longrightarrow \mathbb{R}_0^+$ with $\sigma(\emptyset) = 0$.*

Now it is important to observe that in our problem we are dealing with both tags and influential users. Hence, we have to extend the influence function to the Tag-Based Influence Function which is described in the next subsection.

2.3   Tag-Based Social Influence

For any positive integer i, $[i]$ denotes the set $\{1, 2, \ldots, i\}$. As mentioned in Definition 3 given a seed set $\mathcal{S}$, the social influence function $\sigma(.)$ returns its influence. However, as in this study we are dealing with both users and tag we have to generalize the influence function that takes two arguments one is a subset of nodes and the other one is a subset of the tags. We denote the tag-based influence function as $\sigma^{T}(\mathcal{S}, T')$ and stated in Definition 4.

Definition 4 (Tag-Based Influence Function). *Given a subset of the nodes $\mathcal{S} \subseteq V(G)$ and a tag set $T' \subseteq \mathbb{T}$ the tag-based influence function returns the influence if the seed set $\mathcal{S}$ and the tag set T' is used. Hence, $\sigma^{T} : 2^{V(G)} \times 2^{\mathbb{T}} \longrightarrow \mathbb{R}_0^+$.*

It is an important point to observe that for any subset of nodes $\mathcal{S} \subseteq V(G)$ if no tag is selected the influence will be the cardinality of $|\mathcal{S}|$; i.e.; $\sigma^{T}(\mathcal{S}, \emptyset) = |\mathcal{S}|$. Now, based on the definition of tag-based influence function we define the problem of tag-based influence maximization which is stated in Definition 5.

Definition 5 (Tag-Based Influence Maximization Problem). *Given a social network $G(V, E, P)$, a set of Tags T, and two positive integers k and r the problem of* Tag-Based Influence Maximization Problem *asks to choose k seed nodes and r tags such that the tag-based influence function $\sigma^{T}(\mathcal{S}, T')$ is maximized. Mathematically, this problem can be presented using Equation No. 2.*

$$\sigma^{T}(\mathcal{S}^*, T'^*) = \underset{\substack{\mathcal{S} \subseteq V(G) \wedge |\mathcal{S}| \leq k \\ and \\ T' \subseteq \mathbb{T} \wedge |T'| \leq r}}{argmax} \sigma^{T}(\mathcal{S}, T') \tag{2}$$

Here, $\mathcal{S}^*$ and $T^{'*}$ denotes the optimal k size seed set and r size tag set, respectively.

It has been mentioned in [8] that the problem of influence maximization is NP-hard and hard to approximate beyond a constant factor under the both IC and LT Mdel of diffusion.

2.4   Set Function and Its Properties

Let $\mathcal{X} = \{x_1, x_2, \ldots, x_n\}$ be a set with n elements. A function is said to be a set function defined on the ground set $\mathcal{X}$ if f maps every subset of $\mathcal{X}$ to a real number. In this paper, we consider that range of f is the set of positive real numbers including 0; $f : 2^{\mathcal{X}} \longrightarrow \mathbb{R}_0^+$. We say that f is *non-negative* if for any $\mathcal{S} \subseteq \mathcal{X}$, $f(\mathcal{S}) \geq 0$, *monotone* if for all $\mathcal{S} \subseteq \mathcal{X}$ and for all $x \in \mathcal{X} \setminus \mathcal{S}$, $f(\mathcal{S} \cup \{x\}) \geq f(\mathcal{S})$; and submodular if for all $\mathcal{S}_1 \subseteq \mathcal{S}_2 \subseteq \mathcal{X}$ and for all $x \in \mathcal{X} \setminus \mathcal{S}_2$, $f(\mathcal{S}_1 \cup \{x\}) - f(\mathcal{S}_1) \geq f(\mathcal{S}_2 \cup \{x\}) - f(\mathcal{S}_2)$. We say that f is normalized if $f(\emptyset) = 0$. Now, f is said to be a bi-set function if no. of arguments of f are 2. For a bi-set function the ground set of the first and second argument may be same or different. A bi-set function is said to be normalized if for the both the arguments when it is $\emptyset$ then the functional value is 0; i.e.; $f(\emptyset, \emptyset) = 0$. It can be observed that the tag-based influence function $\sigma^T(\mathcal{S}^*, T^{'*})$ is a bi-set function and the ground set of the first argument is the set of nodes of G and the ground set of the second argument is the set of tags $\mathbb{T}$. Now we list down several properties of bi-set functions which be used to analyze the properties of the tag-based influence function [2,11]. Now, we state the notion of *Bi-Monotonicity* in Definition 7.

Definition 6 (Bi-Monotonicity). *A bi-set function f where the ground sets for the first and second arguments $\mathcal{X}$ and $\mathcal{Y}$, respectively is said to be bi-submodular if for all $(\mathcal{A}, \mathcal{B}) \in 2^{\mathcal{X}} \times 2^{\mathcal{Y}}$ and for all $x \in \mathcal{X} \setminus \mathcal{A}$ and $y \in \mathcal{Y} \setminus \mathcal{B}$, $f(\mathcal{A} \cup \{x\}, \mathcal{B}) \geq f(\mathcal{A}, \mathcal{B})$ and $f(\mathcal{A}, \mathcal{B} \cup \{y\}) \geq f(\mathcal{A}, \mathcal{B})$.*

Definition 7 (Bi-Submodularity). *A bi-set function f where the ground sets for the first and second arguments $\mathcal{X}$ and $\mathcal{Y}$, respectively is said to be bi-submodular if for all $(\mathcal{A}, \mathcal{B}) \in 2^{\mathcal{X}} \times 2^{\mathcal{Y}}$, $(\mathcal{A}', \mathcal{B}') \in 2^{\mathcal{X}} \times 2^{\mathcal{Y}}$ with $\mathcal{A} \subseteq \mathcal{A}'$ and $\mathcal{B} \subseteq \mathcal{B}'$, $x \notin \mathcal{A}'$ and $y \notin \mathcal{B}'$ if the following two conditions holds:*

$$f(\mathcal{A} \cup \{x\}, \mathcal{B}) - f(\mathcal{A}, \mathcal{B}) \geq f(\mathcal{A}' \cup \{x\}, \mathcal{B}') - f(\mathcal{A}', \mathcal{B}') \tag{3}$$

and

$$f(\mathcal{A}, \mathcal{B} \cup \{y\}) - f(\mathcal{A}, \mathcal{B}) \geq f(\mathcal{A}', \mathcal{B}' \cup \{y\}) - f(\mathcal{A}', \mathcal{B}') \tag{4}$$

3   Proposed Solution Approaches

In this section we describe the proposed solution approaches for our problem. Initially, we start by establishing several properties of the tag-based influence function which will be required for designing algorithms for optimizing the tag-based influence function.

3.1   Properties of the Tag-Based Influence Function

Lemma 1. *The tag-based influence function as defined using Equation No. 2 follows bi-monotonicity property.*

Proof. Consider $\mathcal{S} \subseteq V(G)$ and $u \in V(G) \setminus \mathcal{S}$. Also $T' \subseteq \mathbb{T}$ and $t \in \mathbb{T} \setminus T'$. So the tag-based influence function $\sigma^T(\mathcal{S}, T')$ and to show that $\sigma^T(.,.)$ is a bi-monotone function if both he following is true: (i) $\sigma^T(\mathcal{S} \cup \{u\}, T') \geq \sigma^T(\mathcal{S}, T')$, and (ii) $\sigma^T(\mathcal{S}, T' \cup \{t\}) \geq \sigma^T(\mathcal{S}, T')$. Case (i) can be observed very easily. Consider $\mathcal{S}' = \mathcal{S} \cup \{v\}$. As in both right and left hand side of Case (i) the tag set remains same, hence the aggregated influence probability will also remain the same. Now it is a fact that under the IC Model of diffusion the influence function is monotone; i.e.; $\sigma(\mathcal{S}') \geq \sigma(\mathcal{S})$. So the Case (i) is proved.

Now, to prove Case (ii), let us have the observation that for any two tag sets T' and T'' with $|T''| > |T'|$. Then for every edge $(uv) \in E(G)$, $\mathcal{P}_{u \to v}^{T''} \geq \mathcal{P}_{u \to v}^{T'}$. So, even if the seed set remains the same if more tags are used in the influence maximization then the influence when more tags are used influence will be more. This proves Case (ii) and as a whole the lemma statement.

Lemma 2. *The tag-based influence function as defined using Equation No. 2 follows bi-submodularity property.*

Proof. Consider $\mathcal{S}' \subseteq \mathcal{S}'' \subseteq V(G)$ and $u \in V(G) \setminus \mathcal{S}''$. Also, $T' \subseteq T'' \subseteq \mathbb{T}$ and $t \in \mathbb{T} \setminus T''$. Now, the tag-based influence function $\sigma^T(\mathcal{S}, T')$ will said to be a bi-submodular set function if both of the following are true:

- **Case I:** First we show that $\sigma^T(\mathcal{S}' \cup \{u\}, T') - \sigma^T(\mathcal{S}', T') \geq \sigma^T(\mathcal{S}'' \cup \{u\}, T') - \sigma^T(\mathcal{S}'', T')$. It is easy to observe that in both the left and right hand side of the inequalities, the tag set remains the same. That means the aggregated influence probability for all the edges remains the same. Hence, this case boils down to the simple influence function; i.e.; $\sigma(\mathcal{S}' \cup \{u\}) - \sigma(\mathcal{S}') \geq \sigma(\mathcal{S}'' \cup \{u\}) - \sigma(\mathcal{S}'')$ where $\mathcal{S}' \subseteq \mathcal{S}'' \subseteq V(G)$ and $u \in V(G) \setminus \mathcal{S}''$. It has been shown by Kempe et al. [7] that under the Independent Cascade Model the influence function is submodular. Hence, Case I is proved. In other words, the tag-based influence function $\sigma^T(.,.)$ is submodular with respect to the first orthent.
- **Case II.** Now, we want to show that $\sigma^T(\mathcal{S}', T' \cup \{t\}) - \sigma^T(\mathcal{S}', T') \geq \sigma^T(\mathcal{S}', T'' \cup \{t\}) - \sigma^T(\mathcal{S}', T'')$. In this case, we can observe that in both sides of the inequalities, the seed set remains the same only the tag set is changing. Now as mentioned previously, for any two tag sets T' and T'' such that $|T''| > |T'|$ for any edge $(u, v) \in E(G)$, $\mathcal{P}_{u \to v}^{T''} > \mathcal{P}_{u \to v}^{T'}$. Now consider the standard influence maximization problem in two different cases. In both cases, the topology and the structure of the graph is the same, however the edge probabilities are different. Let, $G^{T'}$ and $G^{T''}$ are the input social network with the aggregated edge probabilities for the tags T' and T'', respectively. Let, $\sigma_{G^{T'}}(\mathcal{S})$ and $\sigma_{G^{T''}}(\mathcal{S})$ denote the influence of the seed set $\mathcal{S}$ on the graphs $G^{T'}$ and $G^{T''}$, respectively. Also, it is easy to observe that $\sigma_{G^{T''}}(\mathcal{S}) > \sigma_{G^{T'}}(\mathcal{S})$.

This proves that the tag-based influence function $\sigma^{\mathcal{T}}(.,.)$ is bi-submodular.

The Bi-Submodularity property as mentioned in Lemma 2 has been exploited in the proposed solution methodology as described in the following sub-section.

3.2 Proposed Solution Approach

In this section we describe the proposed solution methodology which is based on the orthent-wise maximization of a bi-submodular function as described below.

Broad Idea of the Proposed Solution Approach. As mentioned in Sect. 2.3, the problem here is to maximize the bi-submoular set function $\sigma^{\mathcal{T}}(\mathcal{S}, T^{'})$ subject to the constraint $|\mathcal{S}| \leq k$ and $|T^{'}| \leq r$. Now, if we apply the coordinate wise maximization algorithm that works in the following way:

- **Step 1:** First, the tag set is initialized to an empty set and find out an optimal k-sized seed set that maximizes the tag-based influence function. Consider the obtained seed set is $\mathcal{S}^{*}$. So, we are solving the following optimization problem:

$$\mathcal{S}^{*} \longleftarrow \underset{S \subseteq V(G) \text{ and } |\mathcal{S}| \leq k}{argmax} \sigma^{\mathcal{T}}(\mathcal{S}, \emptyset) \tag{5}$$

- **Step 2:** Once the optimal seed set is found, we fix the seed set in the tag-based influence function with the optimal seed set, and we find out an optimal r-size tag set and let it be T^{*}. So, in this step, we are solving the following optimization problem:

$$T^{*} \longleftarrow \underset{T^{'} \subseteq \mathbb{T} \text{ and } |T^{'}| \leq r}{argmax} \sigma^{\mathcal{T}}(\mathcal{S}^{*}, T^{'}) \tag{6}$$

Intuitively, we can observe that after solving the optimization problems mentioned in Equations No. 5 and 6 we will obtain both the k-size seed set and r-size tag set. However, this is not possible if we apply both steps directly. The reason behind this is as follows. Consider the case of solving the optimization problem mentioned in Equation No. 5. When we assign the tag set to an empty set, the aggregated influence probability for all the edges of the network will be 0, which is equivalent to a graph with n nodes but no edges at all. Now, on such network we try to find an optimal k-sized seed set for the influence maximization problem using the incremental greedy approach based on marginal influence gain then any k-size subset of the vertex set can be returned which is not correct. So, in the proposed solution approach we tackle this problem and describe the proposed solution approach.

Description of the Proposed Solution Approach. We tackle the above mentioned problem in the following way. Initially, we choose the most popular tag and subsequently we select $(r-1)$ many tags incrementally after selecting the k-size seed set. So the working principle of the proposed solution approach is as follows. First we initialize the seed set and tag set to empty set. Then we select the most popular tag in the network and this can be done in the following way. Every user of the network is marked with the tags that they are associated with. Now, for every tag we count the frequency of every tag and choose the highest one. If there are ties that can be broken arbitrarily. This highest frequency tag is selected, and subsequently, we are left with to select k seed nodes and $(r-1)$-tags.

Algorithm 1: Co-Ordinate wise Maximization Algorithm for the Tag-Based Influence Maximization Problem

---

Data: The Social Network $G(V, .E, \mathcal{P})$, the tag set $\mathbb{T}$, Two positive integer k and r.

Result: $\mathcal{S} \subseteq V(G)$ with $|\mathcal{S}| = r$ and $T \subseteq \mathbb{T}$ with $|T| = r$ such that $\sigma^T(\mathcal{S}, T)$ is maximized

1   $\mathcal{S} \longleftarrow \emptyset; T \longleftarrow \emptyset;$

2   $t \longleftarrow$ The most popular tag; $T \longleftarrow T \cup \{t\};$

3   **for** $i = 1$ *to* k **do**

4     $u^* \longleftarrow \underset{u \in V(G) \backslash \mathcal{S}}{argmax} \ \sigma^T(\mathcal{S} \cup \{u\}, T) - \sigma^T(\mathcal{S}, T); \mathcal{S} \longleftarrow \mathcal{S} \cup \{u^*\};$

5   **end**

6   **for** $j = 1$ *to* $(r-1)$ **do**

7     $t^* \longleftarrow \underset{t' \in \mathbb{T} \backslash T}{argmax} \ \sigma^T(\mathcal{S}, T \cup \{t'\}) - \sigma^T(\mathcal{S}, T); T \longleftarrow T \cup \{t^*\};$

8   **end**

9   *return* $\mathcal{S}, T;$

Algorithm 1 describes the proposed solution approach in terms of pseudocode. In all the social network datasets where tags are associated every user of the network will be marked with the tags that they are associated with. From the frequency of tags the most frequent tag can be identified and taken into the tag set T. Consider the number of tags in the dataset is p. Now, finding the highest frequency tag can be obtained in $\mathcal{O}(p \cdot n)$ time. Line no. 6 computes the marginal gain and in the worst case in each iteration of the **for** loop of Line No. 5, the number of marginal gains computed is of $\mathcal{O}(n)$. Now, to compute the influence while computing the influence of a seed set under the independent cascade model of diffusion is of $\mathcal{O}(n \cdot (m+n))$. The **for** loop of Line No. 6 will execute for $\mathcal{O}(k)$ times. Hence, the time requirement to execute from Line No. 5 to 7 is of $\mathcal{O}(k \cdot n^2 \cdot (m+n))$. There is a little difference with the execution of the **for** loop of Line No. 8 because in each iteration the newly selected tag has to be aggregated for computing the marginal gain in the next iteration. As the number of tags is of $\mathcal{O}(p)$, then for all the edges to aggregate the influence probabilities will take $\mathcal{O}(p \cdot m)$ time. This additional time we have to bear in each iteration. Hence, the time requirement for executing from Line No. 8 to 10 will be $\mathcal{O}(r \cdot (pm + n^2 \cdot (m+n)))$. Hence, the total time requirement by Algorithm 1 is of $\mathcal{O}(p \cdot n + k \cdot n^2 \cdot (m+n) + r \cdot (pm + n^2 \cdot (m+n)))$. Now, the space requirement will be as follows will be of $\mathcal{O}(n)$ in the worst case to store $\mathcal{S}$, $\mathcal{O}(r)$ in the worst case to store T, $\mathcal{O}(m)$ to store the aggregated influence probabilities, also $\mathcal{O}(n)$ and $\mathcal{O}(r)$ to store the marginal influence gains while executing the **for** loop of

Line No. 5 and 8, respectively. Hence, the total space requirement will be of $\mathcal{O}(m + r)$. So, Theorem 1 holds.

Theorem 1. *Time and space requirement of Algorithm 1 will be of $\mathcal{O}(p \cdot n + k \cdot n^2 \cdot (m + n) + r \cdot (pm + n^2 \cdot (m + n)))$ and $\mathcal{O}(m + r)$, respectively.*

Community-Based Approach. In this section, we propose a community-based approach for solving the tag-based influence maximization problem. In this approach, the input social network is divided among the communities. The budget for both seed node and tags are divided among the communities based on the following criteria: "If the size of the community is large then it requires more seed nodes and tags to influence". Let, the network is divided into ℓ many communities and they are represented by $C = \{C_1, C_2, \ldots, C_\ell\}$. Let, for any community $C_i \in C$, $V(C_i)$ and $E(C_i)$ denote the set of vertices and edges of the community C_i. Also for any two communities $C_i, C_j \in C$, E_{C_i, C_j} denotes the set of edges between C_i and C_j; i.e.; $E_{C_i, C_j} = \{(uv) : u \in V(C_i) \text{ and } v \in V(C_j)\}$. So, $V(G) = \bigcup_{C_i \in C} V(C_i)$ and $E(G) = \bigcup_{\substack{C_i, C_j \in C \\ \text{and } C_i \neq C_j}} (E_{C_i, C_j} \cup E_{C_i})$. For any community $C_i \in C$, we denote the number of nodes and edges of this community are denoted by n_i and m_i, respectively. So, we have the following $n = \sum_{i \in [\ell]} n_i$. Also, for any pair of communities C_i and C_j, let m_{ij} denotes the number of edges between the community C_i and C_j. If the graph is undirected then m_{ij} and m_{ji}. Now, we divide the budget for nodes and tags among the communities as follows:

- **Budget Division for Seed Nodes** For any Community C_i, the number of maximum seed nodes that can be selected from this community can be given by $(\frac{k}{n} \cdot n_i)$.
- **Budget Division for Tags** For any Community C_i, the maximum number of tags that can be selected from this community can be given by $(\frac{r-1}{m} \cdot m_i)$.

It is important to observe that after the division of the budgets among the communities, the total budgets for all the communities do not exceed the allocated budget. Once the budget division is done the next step is to choose the seed nodes and tags using any algorithm. In this approach, we use the high-frequency tags in the community within the budget and we select the seed nodes based on the marginal influence gain. The proposed methodology has been described in terms of psudocode in Algorithm 2.

Now we describe the working principle of Algorithm 2. First, we initialize two sets S and T to store the seed nodes and tags, respectively. Next, we detect the communities of the network and for this purpose, we use the Louvian algorithm [12]. Here, *Community* is an array of size n where n is the number of nodes of G. Its i-th contains the community number to which the vertex v_i belongs to. So, Community[i] $= x$ means the vertex v_i belongs to the x-th community. Also, it is easy to observe that the maximum value among the numbers of this list gives the number of communities in which the network has been divided. Next,

---

Algorithm 2: Community-Based Approach for the Tag-Based Influence Maximization Problem

---

Data: The Social Network $G(V, .E, \mathcal{P})$, the tag set $\mathbb{T}$, Two positive integer k and r.

Result: $\mathcal{S} \subseteq V(G)$ with $|\mathcal{S}| = r$ and $T \subseteq \mathbb{T}$ with $|T| = r$ such that $\sigma^T(\mathcal{S}, T)$ is maximized

1 $\mathcal{S} \longleftarrow \emptyset; \ T \longleftarrow \emptyset;$
2 $Community = Community\_Detection(G); \ \ell \longleftarrow max(Community);$
3 $Seed\_Budget \longleftarrow array(\ell, 0); \ Tag\_Budget \longleftarrow array(\ell, 0);$
4 $Community\_Size \longleftarrow array(\ell, 0);$
5 **for** $i = 1$ **to** n **do**
6 $\quad$ **if** $Community[i] == x$ **then**
7 $\quad\quad |$ $Community\_Size[x] = Community\_Size[x] + 1;$
8 $\quad$ **end**
9 **end**
10 **for** $i = 1$ **to** ℓ **do**
11 $\quad$ $Seed\_Budget[i] \longleftarrow \frac{n_i}{n} \cdot k; \ Tag\_Budget[i] \longleftarrow \frac{m_i}{m} \cdot r;$
12 **end**
13 **for** $i = 1$ **to** ℓ **do**
14 $\quad$ $T_i \longleftarrow \emptyset;$
15 $\quad$ $Tag\_Popularity \longleftarrow$ Calculate the Tag Popularity of the i-th Community ;
16 $\quad$ $Sorted\_Tags \longleftarrow$ Sort the tags based on the $Tag\_Popularity$ value;
17 $\quad$ **for** $j = 1$ **to** $|Sorted\_Tags|$ **do**
18 $\quad\quad$ **if** $Sorted\_Tags[j] \notin T$ **and** $|T_i| < Tag\_Budget[i]$ **then**
19 $\quad\quad\quad |$ $T_i \longleftarrow T_i \cup \{Sorted\_Tags[j]\};$
20 $\quad\quad$ **end**
21 $\quad$ **end**
22 $\quad$ $T \longleftarrow T \cup T_i;$
23 **end**
24 **for** $i = 1$ **to** m **do**
25 $\quad$ $\mathcal{P}_i^T \longleftarrow$ Calculate the aggregated influence probability ;
26 **end**
27 **for** $i = 1$ **to** ℓ **do**
28 $\quad$ **for** $j = 1$ **to** $Seed\_Budget[i]$ **do**
29 $\quad\quad |$ $u^* \longleftarrow \underset{u \in V(G) \setminus \mathcal{S}}{argmax} \ \sigma^T(\mathcal{S} \cup \{u\}, T) - \sigma^T(\mathcal{S}, T); \ \mathcal{S} \longleftarrow \mathcal{S} \cup \{u^*\};$
30 $\quad$ **end**
31 **end**
32 $return \ \mathcal{S}, T;$

---

we initialize two arrays $Seed\_Budget$ and Tag_Budget of size ℓ, and the i-th entry of both store the budget for seed nodes and tags for the i-th community, respectively. As per the budget division policy described previously, the budget for both seed nodes and tags are divided among the communities from Line No. 6 to 8. The array $Community\_Size$ stores the size of each community; i.e.; x-th entry stores the number of nodes of the x-th community. Now, we subsequently proceed with the tag selection process in the following way. First, we initialize the set which will store the tag set selected from that community. Now, for each tag, we calculate the frequency of each tag, and subsequently, we sort the tags based on the tag popularity value. From this sorted tag list we scan over this list and while scanning we check whether the current tag has already been selected or not. If not and the allocated budget for that community still has not been exhausted then this tag is chosen. Once the tag selection of a community has been done then they are merged with the global tag set T. Once the tag selection is completed next the tag aggregation is done as mentioned in Equation No. 1. At last, from each of the communities we select the required number of tags based on the marginal influence gain. Next, we analyze the time and space requirement of Algorithm 2.

Initializing both S and T requires $\mathcal{O}(1)$ time. As mentioned in [5], the time requirement by Louvain Method for detecting communities of the network requires $\mathcal{O}(n \log n)$ time where n denotes the number of nodes of the network. Initializing both the arrays $Seed\_Budget$ and Tag_Budget require $\mathcal{O}(1)$ time. It is easy to observe that the number of times for loop of Line No. 8 will run for $\mathcal{O}(n)$ times. Also, the statements within this for loop will take $\mathcal{O}(1)$ time. Hence the time requirement for execution from Line No. 8 to 10 will take $\mathcal{O}(n)$ time. Also, the for loop of Line No. 11 will execute $\mathcal{O}(\ell)$ times. Within this loop, all the statements will take $\mathcal{O}(1)$ time. Hence, the time requirement for execution Line No. 11 to 13 is of $\mathcal{O}(\ell)$. Now, it is easy to observe that the time requirement for seed set and tag set selection from any community will depend on the number of nodes and edges that it contains, respectively. Consider the maximum number of nodes and edges of any community is denoted by n_{max} and m_{max}, respectively. So, $n_{max} = \max_{C_i \in C} n_i$ and $m_{max} = \max_{C_i \in C} m_i$. Now, it is easy to observe that if we analyze the time requirement for the tag selection process for the community containing m_{max} many edges and seed node selection process for the community containing n_{max} many nodes and multiply both the quantities with ℓ then we will get the time requirement for the tag set and seed set selection, respectively. First, let us consider the tag selection process. It is easy to observe that the size of T can be of $\mathcal{O}(t)$. Initializing the set T_i will take $\mathcal{O}(1)$ time. Now, as there are $\mathcal{O}(t)$ many tags, hence computing the frequency of each tag will require $\mathcal{O}(t \cdot n_{max})$ time. Now, to sort these tags based on the tag popularity value will take $\mathcal{O}(t \cdot \log t)$ time. Now, the for loop of Line No. 19 will take $\mathcal{O}(t)$ times. There are two conditions in the if statement. To check the first condition it will take $\mathcal{O}(t)$ time, however the second condition will take $\mathcal{O}(1)$ time. For the community having m_{max} many edges, the time requirement for the tag selection will take $\mathcal{O}(n_{max} \cdot t + t \cdot \log t + t^2)$ time which is reduced to $\mathcal{O}(n_{max} \cdot t + t^2)$. As mentioned previously, the time requirement for the tag selection for the whole network will take $\mathcal{O}(\ell \cdot (n_{max} \cdot t + t^2))$ time. Once the tag selection process is done, the nest step is to aggregate the tags to obtain the single influence probability and this will take $\mathcal{O}(m \cdot t)$ time. In the worst case the size of the seed set could be of $\mathcal{O}(n_{max})$. By extending the analysis of Algorithm 1, we can observe that the time requirement for seed set selection for the whole network will be $\mathcal{O}(\ell \cdot k_{max} \cdot n_{max}^2 (n_{max} + m_{max}))$. So, the total time requirement of Algorithm 2 will be $\mathcal{O}(n \log n + n + n + \ell + \ell \cdot (n_{max} \cdot t + t^2) + \ell \cdot k_{max} \cdot n_{max}^2(n_{max} + m_{max}))$. It is easy to observe that this quantity can be reduced to $\mathcal{O}(n \log n + \ell \cdot (n_{max} \cdot t + t^2) + \ell \cdot k_{max} \cdot n_{max}^2(n_{max} + m_{max}))$. Now, we can observe that ℓ can be of $\mathcal{O}(t)$ in the worst case, k_{max} and n_{max} can be of $\mathcal{O}(n)$, and m_{max} can be of $\mathcal{O}(m)$. Hence, in the worst case total time requirement will be of $\mathcal{O}(n \log n + \ell \cdot (n \cdot t + t^2) + \ell \cdot n \cdot n^2(n+m)) = \mathcal{O}(\ell \cdot (n \cdot t + t^2) + \ell \cdot n^3 \cdot (n+m))$.

Now, it is easy to observe that the extra space consumed by Algorithm 2 is to store the seed set and tag set which can be of $\mathcal{O}(n)$ and $\mathcal{O}(t)$, respectively. The arrays $Community$, $Seed\_Budget$, Tag_Budget, and $Community\_Size$ will take $\mathcal{O}(n)$, $\mathcal{O}(\ell)$, $\mathcal{O}(\ell)$, and $\mathcal{O}(\ell)$ space, respectively. Also, it is easy to observe that to store the arrays T_i, $Tag\_Popularity$, and $Sorted\_Tags$ will take $\mathcal{O}(n)$, $\mathcal{O}(t)$, and $\mathcal{O}(t)$ space, respectively. To store the aggregated influence probability for all the edges, we need $\mathcal{O}(m)$ space. Finally, we need to have $\mathcal{O}(n)$ space to store the marginal gain of the nodes. So the total space requirement by Algorithm 2 is of $\mathcal{O}(n + t + \ell + m)$. Hence, Theorem 2 holds.

Theorem 2. *The time and space requirement of Algorithm 2 will be of $\mathcal{O}(\ell \cdot (n \cdot t + t^2) + \ell \cdot n^3 \cdot (n + m))$ and $\mathcal{O}(n + t + \ell + m)$, respectively.*

4   Experimental Validation

In this section, we describe the experimental evaluation of the proposed solution approaches. Initially, we start by describing the datasets.

Datasets. In this study, we have used the following two datasets Last.fm, Delicious. Last.fm contains the social relations among the listeners of this online site. Delicious is a social book-marking web service for storing, sharing, and discovering web bookmarks. These datasets have been previously used by many researchers in the domain of social networks and recommender systems.

Experimental Set Up. In this study the following parameter values needs to be set up: Influence Probability, and the value of k and r. We have considered the following two probability setting, namely, count, and weighted cascade. In count probability setting, for every tag we compute its frequency for every user of the network. Consider two users u_i and u_j and one tag t_x. Their respective frequencies are $f_{u_i}(t_x)$ and $f_{u_j}(t_x)$, respectively. The influence probability for the edge $(u_i u_j)$ for the tag t_x under the count probability setting will be $\frac{|f_{u_i}(t_x) - f_{u_j}(t_x)| + 1}{f_{u_i}(t_x)}$. In the weighted cascade setting, this probability will be $\frac{1}{f_{u_j}(t_x)}$. In this study, we consider the following (k, r) value airs: $(5, 5)$, $(10, 10)$, $(15, 15)$, $(20, 20)$, $(25, 25)$, and $(30, 30)$.

Results and Discussions. Now, we describe the experimental results. Figure 1 shows the seed node-Tag Set budget pair Vs. Influence plots for Last.fm and Delicious dataset for two different probability settings, namely Count and Weighted Cascade. From this figure, we can observe that for most of the (k, r) pair values the seed and tag set selected by our proposed solution approaches lead to more influence compared to the baseline methods. As an example, for the Last.fm dataset with the weighted cascade setting, when the value of both k and r is set to 30, among the baseline methods, the Random method leads to the highest amount of influence and its value is 48.35. Between the proposed solution approaches, the seed and tag set selected by the community-based approach leads to the highest influence vale which is 57.29. Similar observations is made for the other probability settings as well. For the count probability setting, when the value of both k and r are set to 30, among the baseline methods, the seed and tag set selected by the high degree node-high frequency tag method leads to the influence value of 45.95. Between the two proposed methods, the seed node and tag set selected by the Coordinate-wise Method lead to the maximum value which is 65.65. Also, we observe that the computational time requirement is affordable. Hence, the proposed solution approaches lead to more amount of influence compared to baseline methods using reasonable computational time.

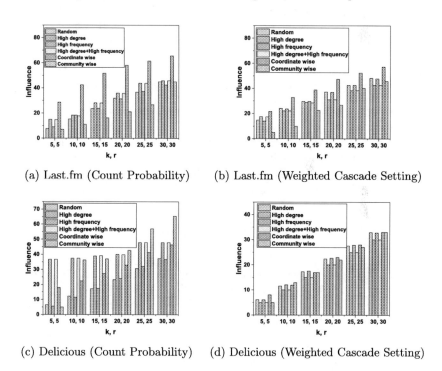

(a) Last.fm (Count Probability)   (b) Last.fm (Weighted Cascade Setting)

(c) Delicious (Count Probability)   (d) Delicious (Weighted Cascade Setting)

Fig. 1. (k, r) Pair Value Vs. Influence value for Last.fm and Delicious Dataset for Count and Weighted Cascade Probability Setting

5   Concluding Remarks

In this paper, we have studied the Tag-Based Influence Maximization Problem. First, we have shown that the tag-based influence function follows the bi-monotonicity and bi-submodularity properties. Subsequently, we have proposed two solution methodologies with a detailed analysis. Several experiments have been conducted with real-world datasets. Our future work in this study will remain concentrated on efficient pruning techniques.

References

1. Alghamdi, R., Bellaiche, M.: A cascaded federated deep learning based framework for detecting wormhole attacks in IoT networks. Comput. Secur. **125**, 103014 (2023)
2. Ando, K., Fujishige, S., Naitoh, T.: A characterization of bisubmodular functions. Discret. Math. **148**(1–3), 299–303 (1996)
3. Banerjee, S., Pal, B.: Budgeted influence and earned benefit maximization with tags in social networks. Soc. Netw. Anal. Min. **12**(1), 21 (2022)
4. Banerjee, S., Pal, B., Jenamani, M.: Budgeted influence maximization with tags in social networks. In: Huang, Z., Beek, W., Wang, H., Zhou, R., Zhang, Y. (eds.) WISE 2020. LNCS, vol. 12342, pp. 141–152. Springer, Cham (2020). https://doi.org/10.1007/978-3-030-62005-9_11
5. Blondel, V.D., Guillaume, J.L., Lambiotte, R., Lefebvre, E.: Fast unfolding of communities in large networks. J. Stat. Mech. Theory Exp. **2008**(10), P10008 (2008)
6. Guille, A., Hacid, H., Favre, C., Zighed, D.A.: Information diffusion in online social networks: a survey. ACM SIGMOD Rec. **42**(2), 17–28 (2013)
7. Ke, X., Khan, A., Cong, G.: Finding seeds and relevant tags jointly: for targeted influence maximization in social networks. In: Proceedings of the 2018 International Conference on Management of Data, pp. 1097–1111 (2018)
8. Kempe, D., Kleinberg, J., Tardos, É.: Maximizing the spread of influence through a social network. In: Proceedings of the ninth ACM SIGKDD International Conference on Knowledge Discovery and Data Mining, pp. 137–146 (2003)
9. Pastor-Satorras, R., Vespignani, A.: Epidemic spreading in scale-free networks. Phys. Rev. Lett. **86**(14), 3200 (2001)
10. Shukla, A., Gullapuram, S.S., Katti, H., Kankanhalli, M., Winkler, S., Subramanian, R.: Recognition of advertisement emotions with application to computational advertising. IEEE Trans. Affect. Comput. **13**(2), 781–792 (2020)
11. Singh, A., Guillory, A., Bilmes, J.: On bisubmodular maximization. In: Artificial Intelligence and Statistics, pp. 1055–1063. PMLR (2012)
12. Traag, V.A., Waltman, L., Van Eck, N.J.: From louvain to leiden: guaranteeing well-connected communities. Sci. Rep. **9**(1), 5233 (2019)
13. Zhang, Z., Shi, Y., Willson, J., Du, D.Z., Tong, G.: Viral marketing with positive influence. In: IEEE INFOCOM 2017-IEEE Conference on Computer Communications, pp. 1–8. IEEE (2017)

Multi-Interest Aware Graph Convolution Network for Social Recommendation

Zhengyi Guo, Yanmin Zhu$^{(\boxtimes)}$, Zhaobo Wang, and Mengyuan Jing

Department of Computer Science and Engineering, Shanghai Jiao Tong University,
Shanghai, China
yzhu@sjtu.edu.cn

Abstract. Social recommendation endeavors to harness users' social connections to enhance recommender systems. Graph Neural Network (GNN) has gained traction for its robust capacity in managing graph data. However, previous social recommendation works have failed to fully consider the crucial role of user distinct interests, which hinders their ability to accurately model complex user preferences and negatively affected the modeling of social influence. To tackle this challenge, we introduce a multi-interest approach to GNN-based social recommendation. Specifically, we firstly utilize a dynamic routing algorithm to cluster user interests from the items they have interacted with. Subsequently, we use a similarity-weighted GCN operation to capture user relationships within the social network. Finally, we use the aggregate the multiple interest representations for prediction. Our comprehensive experiments underscore the consistent superiority of our model compared with state-of-the-art competitors on real-world datasets.

Keywords: Recommender System · Social Recommendation · Graph Neural Networks

1 Introduction

With the swift rise of online services, recommender systems have emerged as a critical solution to the challenge of information overload and have become a vital pivot for providing personalized and high-quality recommendation to users. Moreover, with the surge of social networking applications, social-aware recommendation, which incorporates interpersonal connections to the recommendation process to enhance recommendation performance attracts increasing attention from both academic and industrial fields [6,11,17,32,33]. Social relations among users can enhance the modeling of user preferences by incorporating information from their friends, as well as offer additional perspectives for relevant items to users. The key functionality of social recommendation is to learn social influenced user preference, which typically includes distinct interests.

Existing methods for social recommendation try integrating social connections to the recommender system through various approaches. Conventional

X. Yang et al. (Eds.): ADMA 2023, LNAI 14176, pp. 787–801, 2023.
https://doi.org/10.1007/978-3-031-46661-8_52

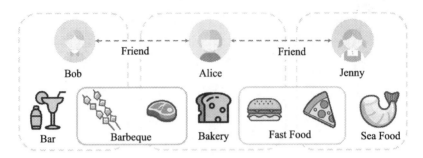

Fig. 1. An illustrative instance of our proposed framework. A social network user, Alice, has multiple interests including bakery, steak, and hamburger. Alice has 2 friends Bob and Jenny, who respectively share similar interests barbeque and fast food with Alice but respectively have unique interest, bar and sea food.

approaches [17–19,32] rooted in matrix factorization assume that users linked by social connections are likely to share similar preferences. Hence, these techniques employ social connections as a form of social regularization, constraining the process of user embedding learning. Other approaches [4,6] take into account the notion that interconnected users can exert mutual social influence, drawing from the principles of social influence theory [3,20]. These approaches include the opinions of friends regarding candidate items as part of modeling user preferences. Recently, Graph Neural Networks (GNNs) [7,13] achieve a desirable performance in the recommendation scenario. The fundamental concept is to propagate and aggregate the information from neighbor nodes, making it naturally suitable to depict social relation. Several recent studies have put forward applications of GNNs in social recommendation [5,29,30,34], attaining state-of-the-art performance.

Despite effectiveness, the crucial role of users' diverse interests in social recommendation has not been fully considered by the previous works. We propose that distinct interests of users need to be modeled separately at a fine-grained level, rather than compressed into a single preference representation as previous works. (1) First, a unified representation is insufficient to represent user multiple interests. As illustrated in Fig. 1, Alice has 3 distinct interests: bakery, barbeque, and fast food, which have little relationship with each other. The loss of key information can be resulted by representing these varied interests using a single representation. However, this problem are neglected by previous social recommendation related works. (2) Second, since social recommendation typically rely on preference propagation across social relation, integrally transmitting user all interests with a holistic representation can lead to unshared interests between friends wrongly spread. For instance, monolithically passing Bob's preference for bars and barbeque or Jenny's preference for seafood and fast food to Alice's user representation could lead to erroneous recommendation of bars and sea food. More importantly, the interests between friends in social networks generally share only partial interests. The data analysis conducted on a real world dataset indicates that only 28.18% of interests overlap within the friend pairs. We present the results in Fig. 2. Therefore, integrally propagating user preference would introduce a large amount of noise.

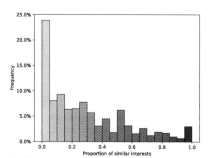 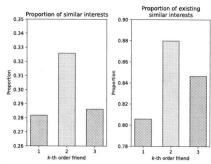

(a) Histogram of proportion of overlapping (b) Average proportion of overlapping in-interests for friend pairs in social network. terests for users with different levels of neighbor order.

Fig. 2. The data analysis on a real world dataset (Yelp) about interest overlap among friends in social network.

With the aforementioned reasons, we focus on augmenting the social recommender scheme with multiple user interest representations that can boost the modeling ability of distinct user interests and corresponding interest propagation. Hence, a novel Multi-Interest Aware Graph Convolution Network (MIAGCN) is proposed by us for social recommendation. Instead of only employ a unified preference representation and integrally propagate it, we propose to separate different user interest into multiple representations and propagate them separately across social relations. Specifically, the MIAGCN includes a multi-interest extractor layer, a GCN-based interest propagation layer and an interest aggregation and prediction layer. Firstly, the multi-interest extractor layer use a capsule network to identify different aspects of user preference and generate corresponding representations. Next, the interest propagation layer employ a similarity weighted GCN to model the impact of social relations on the extracted multiple user interests. Finally, the aggregation and prediction layer selects the most relevant interest representation of the target item to obtain the predicted score. Empirical results valid that our MIAGCN not only processes a strong ability to learn the multiple aspects of user implicit preference, but also achieves considerable effectiveness to model the propagation of interests among friends.

We summarize the main contributions of this paper as follows:

1. Our analysis in publicly available datasets (i.e. Fig. 2) demonstrates that friends in social network generally share only partial interests. Based on this finding, we propose to utilize multiple interest representations to better model the propagation process of interests at a fine-grained level.

2. We propose the Multi-Interest Aware Graph Convolution Network (MIA
   GCN), which leverages multiple interest representations and corresponding
   multiple nodes in graph to learn the diverse aspects of user preference and
   the influence of social relation.
3. Experiments on 3 benchmark datasets are conducted to demonstrate the effectiveness of the proposed model. The results consistently indicate that our
   model outperforms existing state-of-the-art models.

2   Related Works

In this section, we provide a concise overview of recent literature across the
following domains: (1) social recommendation and (2) multi-interest recommendation.

2.1   Social Recommendation

The objective of social recommendation is to utilize social links between users
to enhance the recommendation performance of recommender system. It works
by capturing social influence of social connections, which pertains to the notion
that the choices of a user can be influenced by the actions of their friends. As
an early attempt, SoRec [18] applies the matrix factorization technique to both
rating matrix and social relation matrix, incorporating social relation to user
preference representation. Due to each social relation corresponds to a different degree of preference similarity, a weighted social regularization method is
proposed by STE [17]. Further extensions [6,10,32] implicitly include information trust propagation mechanism to obtain user preference in the social network. Other research lines DeepSoR [4] and NSCR [26] significant advancements
through the utilization of deep learning-based techniques. However, these methods fail to effectively model social influence, for the lack of corresponding explicit
components.

Due to the remarkable effectiveness in managing graph-structured data,
GNN-based methods [8,21,22,28] become dominant in the field of social recommendation. GraphRec [5] pioneered the integration of GNNs into this endeavor,
treating user-user interactions as data within a social graph. DiffNet [30] and
its subsequent iteration, DiffNet++ [29], delve into capturing the influence of
users through iterative social diffusion processes. ESRF [34] introduces a deep
adversarial approach to tackle prevalent challenges within the realm of social
recommendation.

Overall, all existing methods compress user interest into a holistic representation, thus suffering from the large amount of noise produced by inaccurate
interest propagation. In contrast, our MIAGCN method explicitly decomposes
the user preference representation into distinct independent interest representations, empowering the proposed approach to individually model various facets
of user preferences and spread them precisely across social relation.

2.2   Multi-interest Recommendation

The restriction of using a single unified representation to express users' multi-faceted interests has been recognized by the community of recommender systems. To address this issue, plenty of work explore various approaches. As an example, MIND [15] introduces a capsule network coupled with a dynamic routing algorithm-based multi-interest extractor layer. This layer clusters historical behaviors and extracts a range of interests. ComiRec [1] adapts the dynamic routing algorithm for recommender systems, employing a self-attentive based approach that shows promise in real-world scenarios. Similarly, Octopus [16] constructs an elastic archive network to extract diverse user interests, facilitating recommendations of varied items to meet distinct needs. In contrast, SINE [24] integrates sparse-interest extraction and large-scale item clustering. This integration infers sparse concept sets for users from an extensive pool and generates corresponding multiple embeddings. However, the current multi-interest recommendation approaches do not take into account social relations, which limits them to fully realize the positive impact of finely-grained interest representations to accurately modeling of social influence.

3   Problem Definition

Within the realm of social recommendation, let $\mathcal{U} = \{u_1, u_2, \ldots, u_m\}$ represent the set of users, and $\mathcal{I} = \{i_1, i_2, \ldots, i_n\}$ symbolize the collection of items, where m and n denote the counts of users and items, respectively. The social graph $\mathcal{G} = (\mathcal{V}, \mathcal{E})$ is indicative of user-user connections. A link $(u_1, u_2) \in \mathcal{E}$ signifies a social connection between user u_1 and u_2. Consider the user-item interaction matrix denoted as $\mathbf{R} \in \mathbb{R}^{n \times m}$, where $\mathbf{R}_{ui} = 1$ signifies that user u has either made a purchase or provided a rating for item i, while a value of 0 indicates no such interaction.

Let $\mathcal{I}_u$ represent the collection of items that user u has interacted with, and $\mathcal{U}_u$ denote the group of users linked to u through social relations. With the interaction matrix $\mathbf{R}$ and social graph $\mathcal{G}$ at hand, the objective of social recommendation is to forecast the absent entries within $\mathbf{R}$.

4   Methodology

In this section, we firstly give an overview of our proposed model Multi-Interest Aware Graph Convolution Network. Then we proceed to introduce each of its components, which includes a multiple interest extract layer and a interest propagation layer. Finally, we will present its learning process.

4.1   Framework of MIAGCN

Our argument is that previous works have not fully considered the diverse range of user interests and have ignored the fact that interests among friends in social

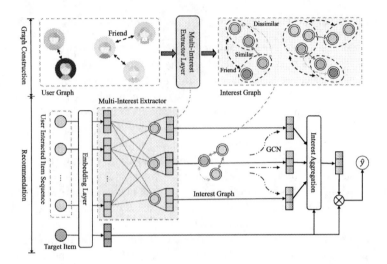

Fig. 3. The architecture of the proposed MIAGCN. It comprises 3 components: a Multi-interest Extractor Layer, designed to extract diverse interests from user-item interaction history; a Multi-interest Propagation Layer, formulated for capturing interest sharing among friends within the social network; and an Aggregation and Prediction Layer, responsible for generating prediction scores.

networks often exhibit only partial overlap. To overcome these limitations, we propose to separate user preference into multiple interest representations and propagate them separately across social relations.

With the aforementioned discussions, we propose Multi-Interest Aware Graph Convolution Network (MIAGCN). As shown in Fig. 3, our MIAGCN model includes (1) a multi-interest extractor layer, which clusters the user's historical behaviors into several interest groups to generate corresponding interest representation; (2) an interest propagation layer, which adaptively propagates user interests across social relations; (3) an interest aggregation and prediction layer, which chooses the most relevant interest representation for the candidate item to produce the prediction score.

4.2   Multi-interest Extractor Layer

For acquiring multiple interest representations, we employ a clustering procedure to categorize a user's historically interacted items into distinct clusters. Items sharing the same cluster collectively aim to represent a specific interest of a user's perference. Our approach employs the dynamic routing mechanism introduced in CapsNet [23], which iteratively computes high-level capsule based on the low-level capsules. Within our approach, item embeddings corresponding to the user interaction sequence are treated as the foundational low-level capsules, while the distinct user interest representations are considered as the resultant high-level capsules. Given an item embedding $\mathbf{e}_i$ representing the i-th capsule in the primary layer, we iteratively compute the j-th interest capsule $\mathbf{s}_j$ in the subsequent layer based on the primary capsules. During each dynamic routing iteration, we initiate by calculating the projection vector $\hat{\mathbf{e}}_{i,j}$ with

$$\hat{\mathbf{e}}_{i,j} = \mathbf{W}\mathbf{e}_i, \tag{1}$$

where $\mathbf{W}$ represents a trainable parameter matrix. Subsequently, the cumulative input to high-level interest capsule j encompasses a weighted summation of all projection vectors $\mathbf{e}_{i,j}$, denoted as

$$\hat{\mathbf{s}}_j = \sum_{i \in \mathcal{I}_u} c_{ij} \hat{\mathbf{e}}_{i,j}, \tag{2}$$

where c_{ij} corresponds to the coupling coefficient, a value ascertained through the iterative dynamic routing procedure. We employ softmax to get the coupling coefficients with initial logit b_{ij} and constrain its sum to 1 as

$$c_{ij} = \frac{\exp\left(b_{ij}\right)}{\sum_{k=1}^{K} \exp\left(b_{ik}\right)}, \tag{3}$$

where b_{ij} stands for the routing logit, signifying the logarithmic prior probability that associates primary item capsule i with output interest capsule j. The routing logit b_{ij} links previous-level capsule j and post-level capsule i commences with an initialization value of 0, and its values evolve through successive iterations via the following process:

$$b_{ij} = \mathbf{s}_j^\top \hat{\mathbf{e}}_{i,j}. \tag{4}$$

The squashing function [23] is proposed to reduce the modulus of vectors from $(0, \infty)$ to $(0, 1)$, which can protect the model from gradient explosion. After that, the high-level capsule j's value is obtained by

$$\mathbf{s}_j = \frac{\|\hat{\mathbf{s}}_j\| \cdot \hat{\mathbf{s}}_j}{1 + \|\hat{\mathbf{s}}_j\|^2}, \tag{5}$$

where $\mathbf{s}_j$ is the output of capsule j. The entire iterative routing procedure is listed in Algorithm 1. The collection of high-level interest capsules for u is assembled to a set $\mathcal{S}_u = \{\mathbf{s}_1, \ldots, \mathbf{s}_K\}$ for downstream interest propagation task, and K is the size of the set of interest capsules. In order to reduce the computational cost, we cut off gradient back-propagation except for the last iteration.

4.3   Interest Propagation Layer

After clustering the interest representations from item that user past interacted with through the multi-interest extractor layer, we apply an GNN-based method to aggregate user friends' interests for refining his/her interests representations under the influence of social relations.

Generally, GNN-based interest propagation operation can be abstracted as

$$\mathbf{h}_{u,i}^{l+1} = \text{Combine}\left(\mathbf{h}_{u,i}^l, \text{Aggregate}\left(\{\mathbf{h}_{v,j}^l : v \in \mathcal{U}_u, j \in \{1, 2, \cdots, K\}\}\right)\right), \tag{6}$$

Algorithm 1: Multi-Interest Extractor Layer.

Input: item capsules $\mathbf{e}_i$, size of the interest capsules set K, iteration times r
Output: the interest capsules set $\mathcal{S}_u$
initialize $b_{ij} = 0$ for each capsule (i, j) pair.
for $iter = 1, \cdots, r$ **do**

> for each capsule (i, j) pair: $c_{ij} = \frac{\exp(b_{ij})}{\sum_k \exp(b_{ik})}$.
>
> for each high-level capsule j: $\hat{\mathbf{s}}_j = \sum c_{ij} \mathbf{We}_i$ and $\mathbf{s}_j = \frac{\|\hat{\mathbf{s}}_j\| \cdot \hat{\mathbf{s}}_j}{1 + \|\hat{\mathbf{s}}_j\|^2}$.
>
> for each capsule (i, j) pair: $b_{ij} = \mathbf{s}_j^\top \mathbf{We}_i$.

return $\{\mathbf{s}_j, j = 1, \ldots, K\}$

---

where $\mathbf{h}_{u,i}^l$ is the l-th layer hidden state of user u's interest j, K is the number of interests, and user v is a neighbor of user u in social graph. Additionally, the initial layer hidden state embedding of user u's interest i is $\mathbf{s}_j$, which is extracted by the multi-interest extractor layer.

As recommended by certain recent investigations [2,9,25], intricate operations could potentially impose a heavy load on collaborative filtering. Conversely, a straightforward propagation approach might yield superior performance., we adopt a simple yet effective mean aggregation function in GCN propagation to implement our model,

$$h_{u,i}^{l+1} = \sum_{v \in \mathcal{U}_u, j \in \{1,2,\cdots,J\}} \frac{1}{|\mathcal{U}_u|} h_{v,j}^l. \tag{7}$$

However, users' directly connected social friends typically have a influence to their preference, particularly when they share similar interests. Therefore, the similarity between the different interests of friends should be considered in the aggregation step. Specifically, we introduce a similarity weight to select similar interests of social friends to characterize users' social interest exchange information,

$$h_{u,i}^{l+1} = \sum_{v \in \mathcal{U}_u, j \in \{1,2,\cdots,K\}} \frac{\text{Cosine Similarity}(h_{u,i}^k, h_{v,j}^k)}{|\mathcal{U}_u|} h_{v,j}^l, \tag{8}$$

with

$$\text{Cosine Similarity}(\alpha, \beta) = \frac{\alpha^T \beta}{|\alpha||\beta|}, \tag{9}$$

where $|\mathcal{U}_u|$ is the size of the neighbor set $\mathcal{U}_u$.

Following L layers of graph convolution operations, we proceed to amalgamate these representations from layers to construct the final user representations through

$$\mathbf{h}_{u,i} = \lambda_0 \mathbf{h}_{u,i}^0 \oplus \lambda_1 \mathbf{h}_{u,i}^1 \oplus \cdots \oplus \lambda_L \mathbf{h}_{u,i}^L, \tag{10}$$

where $\oplus$ is the element-wise addition operation. $\lambda_l \geq 0$ denotes the significance attributed to the l-th layer in composing the final representation. λ_l is set to model parameters and initialized to 1 for all layers.

4.4   Aggregation and Prediction Layer

After the interest propagation layer, we obtain K interest representations for each user based on his social relations. An argmax operator is employed to choose the most relevant interest representation with the target item i,

$$\mathbf{h}_u = \underset{\mathbf{h}_{u,j} \in \mathbf{h}_{u,1}, \cdots, \mathbf{h}_{u,K}}{argmax} \mathbf{h}_{u,j}^{\top} \mathbf{e}_i. \tag{11}$$

Besides users' interests, individual characteristics also need to be considered. To capture these individual traits, we use a user-specific embedding $\mathbf{e}_u$ to model them.

Finally, we get the prediction by performing inner product of user embedding $\mathbf{e}_u$ and relevant interest representation $\mathbf{h}_u$ with target item embedding $\mathbf{e}_i$,

$$\hat{y}_{u,i} = \mathbf{h}_{u,j}^{\top} \mathbf{e}_i + \mathbf{e}_u^{\top} \mathbf{e}_i, \tag{12}$$

where $\hat{y}_{u,i}$ denotes the ranking score utilized for the generation of recommendations in recommender systems.

4.5   Model Training

We employ BCE loss as our loss function, i.e.,

$$L\left(\theta_*\right) = -\sum y_{u,i} \ln \sigma\left(\hat{y}_{u,i}\right) + \left(1 - y_{u,i}\right) \ln \left(1 - \sigma\left(\hat{y}_{u,i}\right)\right) \tag{13}$$

where $\sigma(x)$ represents sigmoid function. Due to the presence of large unobserved values, we are limited to observing only positive feedback. To achieve balanced training, we randomly select 4 unobserved feedbacks as pseudo-negative samples for each positive feedback at every iteration.

5   Experiments

In this section, we execute experiments to assess the efficacy of the MIAGCN model. We begin by providing a overview of the datasets and the baseline methods employed. Subsequently, we compare the performance of our MIAGCN with these baselines. Finally, experiments are conducted to analyze the influence exerted by distinct components and hyper-parameters.

5.1   Experimental Settings

Dataset. We conduct experiments on 3 social recommendation datasets flickr, yelp and ciao. The data statistics are presented in Table 1.

Baseline Methods. For our baseline methods, we have selected several competitive baselines, which include MF [14], GrapgRec [5], NGCF [27], LR-GCCF [2], DiffNet [30]/DiffNet++ [29], and LightGCN [9].

Table 1. Data statistics of datasets

	Flickr	Yelp	Ciao
Users	8358	17237	7375
Items	82120	37378	105114
Ratings	327815	207945	282650
Rating density	0.048%	0.032%	0.036%
Social pairs	352952	259014	170410
Social density	0.505%	0.087%	0.313%

Parameter Settings. To ensure equitable comparison, we performed an exhaustive search to determine the optimal number of layers across all GNN-based approaches. We constrained the range to 2, 3, 4 while maintaining a consistent embedding size of 64 for all models. For optimization, we employed the Adam [12] optimizer with a fixed learning rate of 0.001 across all models.

5.2 Comparison with State-of-the-Art Methods

Table 2. Comparison between our method and baseline methods.

Dataset	Flickr		Yelp		Ciao	
Model	Recall	NDCG	Recall	NDCG	Recall	NDCG
MF	0.1815	0.1865	0.2031	0.1579	0.3352	0.3542
GraphRec	0.1676	0.1704	0.2170	0.1693	0.3327	0.3514
NGCF	0.1649	0.1705	0.2194	0.1723	0.3372	0.3559
LR-GCCF	0.2019	0.2057	0.2700	0.2106	0.3854	0.4069
DiffNet	0.2253	0.2290	0.2768	0.2159	0.3542	0.3820
DiffNet++	0.1650	0.1702	0.2131	0.1674	0.3381	0.3579
LightGCN	0.2009	0.2082	0.2878	0.2206	0.3875	0.4094
Ours	**0.2309**	**0.2387**	**0.2966**	**0.2319**	**0.4014**	**0.4222**

The performance comparison between our approach and the state-of-the-art competitors is presented in Table 2 through 2 evaluation metrics: HR@10 and NDCG@10. Our proposed MIAGCN consistently outperforms all baseline methods on all datasets, demonstrating the effectiveness of our proposed model. This proves that multiple user interest representations can not only capture multiple aspects of user interest, but also more accurately model the propagation of preferences between friends in social networks.

On the other hand, we noted that certain approaches incorporating intricate model architectures, such as GraphRec, NGCF, and DiffNet++, experience pronounced overfitting problems and exhibit subpar performance. This stems from

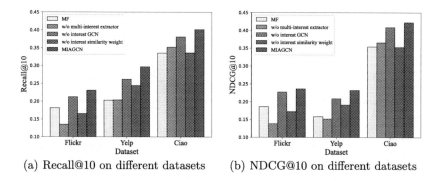

(a) Recall@10 on different datasets   (b) NDCG@10 on different datasets

Fig. 4. Performance comparison of MF, MIAGCN and its variants.

the assumption made by these models regarding the existence of raw user/item features, which might not be consistently available across various social recommendation datasets. Hence, we adopt the strategy of learning user/item representations from independent embeddings. In this scenario, models that eliminate extraneous intricate operations yield enhanced performance. This shows that our model has better robustness for we do only introduce merely additional parameters in Eq. 1) compared to traditional MF. All interest representations are inferred only by the embedding of items that users have interacted with, which reduces the over fitting risk of our model.

5.3   Ablation Study

To verify the effectiveness of the components on the recommendation accuracy of our model, we conduct an ablation study. We systematically removed each component from the original model and evaluated the resulting performance on the datasets. The results are shown in the Fig. 4.

Observing the results, it is evident that all components had positive impact on the model's overall accuracy. The multi-interest extractor layer has the most important impact on the model performance. This is because interest extraction is the cornerstone of the next algorithm, and its disappearance has brought confusion to the following interest propagation layer.

Surprisingly, compared with completely removing GCN, retaining GCN but remove interest similarity weight produce greater harm to the model. This shows that ignoring the correlation of different interests between friends and simply assuming that the interests between friends are related can extremely harm the recommendation performance.

Additionally, compared with conventional MF method, variant that only conduct GCN sometimes is less competitive on overall performance in partial cases. Controversially, introducing the proposed multi-interest extractor layer, our interests propagation layer have positive impact on recommendation performance. This indicates that coarse-grained preference propagation is not enough for modeling social influence as interests are partially shared by friends.

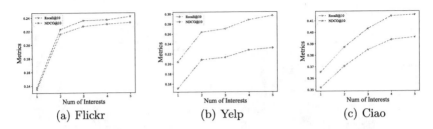

Fig. 5. Performance comparison of MIAGCN with various number of interests across different datasets.

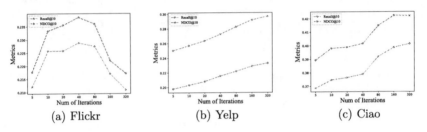

Fig. 6. Performance comparison of MIAGCN with various number of iterations of multi-interest extraction across different datasets.

5.4   Hyper-parameter Analysis

To assess the influence of various crucial hyperparameters on our model, we conduct hyperparameter experiments. Specifically, we check the impact of the number of interests and the iteration count of the multi-interest extractor on the MIAGCN's recommendation performance.

Initially, we investigate the influence of the size of the output interests capsules set. The outcomes are depicted in Fig. 5. It is obvious that greater interest representations tend to yield improved recommendation outcomes. This trend is attributed to that a higher number of interests can effectively enhance the modeling of diverse facets of user preference, thereby benefiting the proposed interest propagation phase.

We also perform experiments regarding the number of iterations in the multi-interest extractor, as illustrated in Fig. 6. The outcomes reveal that, across most datasets, the model's performance enhances with the increase of the number of iterations. This trend emerges is caused by the fact that a higher number of iterations tends to enhance the clustering of items users have engaged with into distinct user interests. As depicted in Fig. 7(b), with the increase in iteration count, distinct interest representations tend to diverge.

Nevertheless, in the case of the Flickr dataset, we note an initial increase followed by a decrease in recommendation accuracy as the number of interest iterations increases. To delve deeper into this issue, we conduct a thorough analysis of our model's behavior on this dataset. And the corresponding metric results are presented in Fig. 7. Initially, we isolate the remaining components of the model

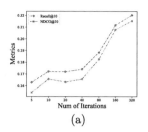 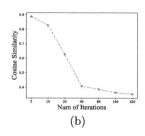 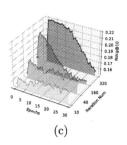

(a)   (b)   (c)

Fig. 7. Analysis of our MIAGCN on flickr dataset. (a) shows performance comparison of only multi-interest extractor with various number of iteration. (b) shows average cosine similarity between users different interests. (c) shows loss of different iteration number with different epochs.

and solely retain the multi-interest extractor layer for recommendation, aiming to assess the influence of iteration within the multi-interest extractor layer. The outcomes are depicted in Fig. 7(a). As observed, the recommendation performance progressively improves with an increase in the number of iterations. However, when examining the performance curves during training for varying iteration counts, as illustrated in Fig. 7(c), we discern that the metric curves for 160 and 320 iterations initially experience an ascent followed by a gradual decline. This behavior implies that the optimization algorithm encounters difficulties in achieving convergence at higher iteration counts.

The reason for this is that we cut off the gradients except for the last iteration, which causes the optimizer to ignore the previous iterations, ultimately resulting in worse training performance. However, not cutting off the gradients can lead to an increase in computational complexity, as well as problems such as gradient vanishing and explosion. On the other hand, in Fig. 7(b), we can find that as the number of iterations increases, differences between interest representations emerge, and multiple interest representations become effective. Therefore, a smaller number of iterations can also lead to the problem of multiple interest representations degenerating into one representation. In summary, the number of iterations of interest is a parameter that needs to be adjusted, and both too high and too low values can lead to problems.

6   Conclusion

In this paper, we propose the MIAGCN Network, which utilizes multiple interest representations and corresponding multiple nodes in the user-user graph to better model the diverse aspects of user preference and the impact of social influence in online social platforms. Experiments on 3 publicly available datasets prove that our method consistently surpasses state-of-the-art competitors.

Acknowledgements. This research is supported in part by National Science Foundation of China (No. 62072304, No. 62172277), Shanghai Municipal Science and Technology Commission (No. 21511104700, No. 19510760500, No. 19511120300), Shanghai East Talents Program and Zhejiang Aoxin Co. Ltd.

References

1. Cen, Y., Zhang, J., Zou, X., Zhou, C., Yang, H., Tang, J.: Controllable multi-interest framework for recommendation. In: KDD 2020, pp. 2942–2951. ACM (2020)
2. Chen, L., Wu, L., Hong, R., Zhang, K., Wang, M.: Revisiting graph based collaborative filtering: a linear residual graph convolutional network approach. In: AAAI 2020, pp. 27–34. AAAI Press (2020)
3. Cialdini, R.B., Goldstein, N.J.: Social influence: compliance and conformity. Annu. Rev. Psychol. **55**, 591–621 (2004)
4. Fan, W., Li, Q., Cheng, M.: Deep modeling of social relations for recommendation. In: AAAI 2018, pp. 8075–8076. AAAI Press (2018)
5. Fan, W., et al.: Graph neural networks for social recommendation. In: WWW 2019, pp. 417–426. ACM (2019)
6. Guo, G., Zhang, J., Yorke-Smith, N.: Trustsvd: collaborative filtering with both the explicit and implicit influence of user trust and of item ratings. In: AAAI 2015, pp. 123–129. AAAI Press (2015)
7. Hamilton, W.L., Ying, Z., Leskovec, J.: Inductive representation learning on large graphs. In: NIPS 2017, pp. 1024–1034 (2017)
8. Han, J., Tao, Q., Tang, Y., Xia, Y.: DH-HGCN: dual homogeneity hypergraph convolutional network for multiple social recommendations. In: SIGIR 2022, pp. 2190–2194. ACM (2022)
9. He, X., Deng, K., Wang, X., Li, Y., Zhang, Y., Wang, M.: Lightgcn: simplifying and powering graph convolution network for recommendation. In: SIGIR 2020, pp. 639–648. ACM (2020)
10. Jamali, M., Ester, M.: A matrix factorization technique with trust propagation for recommendation in social networks. In: RecSys 2010, pp. 135–142. ACM (2010)
11. Jiang, M., Cui, P., Wang, F., Zhu, W., Yang, S.: Scalable recommendation with social contextual information. TKDE **26**(11), 2789–2802 (2014)
12. Kingma, D.P., Ba, J.: Adam: a method for stochastic optimization. In: ICLR 2015 (2015)
13. Kipf, T.N., Welling, M.: Semi-supervised classification with graph convolutional networks. In: ICLR 2017. OpenReview.net (2017)
14. Koren, Y., Bell, R.M., Volinsky, C.: Matrix factorization techniques for recommender systems. Computer **42**(8), 30–37 (2009)
15. Li, C., et al.: Multi-interest network with dynamic routing for recommendation at tmall. In: CIKM 2019, pp. 2615–2623. ACM (2019)
16. Liu, Z., Lian, J., Yang, J., Lian, D., Xie, X.: Octopus: comprehensive and elastic user representation for the generation of recommendation candidates. In: SIGIR 2017, pp. 289–298. ACM (2020)
17. Ma, H., King, I., Lyu, M.R.: Learning to recommend with social trust ensemble. In: SIGIR 2009, pp. 203–210. ACM (2009)

18. Ma, H., Yang, H., Lyu, M.R., King, I.: Sorec: social recommendation using probabilistic matrix factorization. In: Proceedings of the 17th ACM Conference on Information and Knowledge Management, CIKM 2008, pp. 931–940. Association for Computing Machinery, New York (2008)

19. Ma, H., Zhou, D., Liu, C., Lyu, M.R., King, I.: Recommender systems with social regularization. In: WSDM 2011, pp. 287–296. ACM (2011)

20. McPherson, M., Smith-Lovin, L., Cook, J.M.: Birds of a feather: homophily in social networks. Ann. Rev. Sociol. **27**(1), 415–444 (2001)

21. Miao, H., Li, A., Yang, B.: Meta-path enhanced lightweight graph neural network for social recommendation. In: Bhattacharya, A., et al. (eds.) DASFAA 2022. LNCS, vol. 13246, pp. 134–149. Springer, Cham (2022). https://doi.org/10.1007/978-3-031-00126-0_9

22. Quan, Y., Ding, J., Gao, C., Yi, L., Jin, D., Li, Y.: Robust preference-guided denoising for graph based social recommendation. In: WWW 2023, pp. 1097–1108. ACM (2023)

23. Sabour, S., Frosst, N., Hinton, G.E.: Dynamic routing between capsules. In: NIPS 2017, pp. 3856–3866 (2017)

24. Tan, Q., Zhang, J., Yao, J., Liu, N., Zhou, J., Yang, H., Hu, X.: Sparse-interest network for sequential recommendation. In: WSDM 2021, pp. 598–606. ACM (2021)

25. Tao, Y., Li, Y., Zhang, S., Hou, Z., Wu, Z.: Revisiting graph based social recommendation: a distillation enhanced social graph network. In: WWW 2022, pp. 2830–2838. ACM (2022)

26. Wang, X., He, X., Nie, L., Chua, T.: Item silk road: recommending items from information domains to social users. In: SIGIR 2017, pp. 185–194. ACM (2017)

27. Wang, X., He, X., Wang, M., Feng, F., Chua, T.: Neural graph collaborative filtering. In: SIGIR 2019, pp. 165–174. ACM (2019)

28. Wu, J., Fan, W., Chen, J., Liu, S., Li, Q., Tang, K.: Disentangled contrastive learning for social recommendation. In: Proceedings of the 31st ACM International Conference on Information & Knowledge Management, CIKM 2022, pp. 4570–4574. Association for Computing Machinery (2022)

29. Wu, L., Li, J., Sun, P., Hong, R., Ge, Y., Wang, M.: Diffnet++: a neural influence and interest diffusion network for social recommendation. TKDE **34**(10), 4753–4766 (2022)

30. Wu, L., Sun, P., Fu, Y., Hong, R., Wang, X., Wang, M.: A neural influence diffusion model for social recommendation. In: SIGIR 2019, pp. 235–244. ACM (2019)

31. Wu, Q., et al.: Dual graph attention networks for deep latent representation of multifaceted social effects in recommender systems. In: WWW 2019, pp. 2091–2102. ACM (2019)

32. Yang, B., Lei, Y., Liu, D., Liu, J.: Social collaborative filtering by trust. In: IJCAI 2013, pp. 2747–2753. IJCAI/AAAI (2013)

33. Ying, R., He, R., Chen, K., Eksombatchai, P., Hamilton, W.L., Leskovec, J.: Graph convolutional neural networks for web-scale recommender systems. In: KDD 2018, pp. 974–983. ACM (2018)

34. Yu, J., Yin, H., Li, J., Gao, M., Huang, Z., Cui, L.: Enhancing social recommendation with adversarial graph convolutional networks. TKDE **34**(8), 3727–3739 (2022)

Enhancing Multimedia Recommendation Through Item-Item Semantic Denoising and Global Preference Awareness

Yanlong Zhang, Shangfei Zheng, Qian Zhou, Wei Chen$^{(\boxtimes)}$, and Lei Zhao

School of Computer Science and Technology, Soochow University, Suzhou, China
ylzhang828@stu.suda.edu.cn,
{sfzhengsuda,qzhou0,robertchen,zhaol}@suda.edu.cn

Abstract. Multimedia recommendation aims to predict whether users will interact with multimodal items. A few recent works that explicitly learn the semantic structure between items using multimodal features manifest impressive performance gains. This is mainly attributed to the capability of graph convolutional networks (GCNs) to learn superior item representations by propagating and aggregating information from high-order neighbors on the semantic structure. However, they still suffer from two major challenges: a) the noisy relations (edges) in the item-item semantic structure disrupt information propagation and generate low-quality item representations, which impairs the effectiveness and robustness of existing methods; b) the lack of an optimization objective that exploits informative samples and global preference information leads to suboptimal training of the model, which makes users and items indistinguishable in the embedding space. To overcome these challenges, we propose Enhancing *Multimedia Recommendation through Item-Item Semantic Denoising and Global Preference Awareness* (MMGPA). Specifically, the model contains the following two components: (1) a modal semantic representation network is carefully designed to learn the high-quality multimodal representation of items by modeling the denoised item-item semantic structure, and (2) a global preference-aware optimization objective prioritizes the most informative hard sample pairs while constraining the multiple preference distances to better separate the embedding space. Extensive experimental results demonstrate that the proposed method outperforms various state-of-the-art competitors on three public benchmark datasets.

Keywords: Multimedia recommendation · Semantic Denoising · Hard Sample Mining

1 Introduction

As the internet data continues to expand, users are increasingly faced with the task of sifting through vast amounts of information to find content that aligns with their interests. Recommender systems have become an essential tool in

X. Yang et al. (Eds.): ADMA 2023, LNAI 14176, pp. 802–817, 2023.
https://doi.org/10.1007/978-3-031-46661-8_53

information filtering, as they provide a means of personalized content for users. Nowadays, the trend of presenting information on the internet in various media formats (such as text and images) is becoming increasingly common. Multi-modal data reflects user preferences through multi-dimensional features. Therefore, multimedia recommendation methods that capture useful information from multimodal data have become increasingly popular.

In real-world scenarios, items are associated with rich content in multiple modalities, and similar items are more likely to arouse users' interest than dissimilar items [14]. Several recent studies [15,27,28] have explored different ways to mine the item-item semantic structure under multimodal content. With the capability of graph convolutional networks (GCNs) in propagating and aggregating information from high-order neighbors on the obtained semantic structure, better item representations can be learned to help recommendation models comprehensively discover candidate items. However, their recommendation performance is restricted due to the following two challenges.

Challenge I. The aforementioned methods fail to obtain satisfactory item representations. This is mainly because they incorporate noisy semantic relations [27,28] when modeling item representations. These noisy relations mainly originate from incomplete data during multimodal data acquisition, or from items that are modally similar but not actually similar (e.g., myopic glasses and presbyopic glasses). Moreover, these methods based on item-item semantic structures are more vulnerable to noisy semantic relations [26]. A major reason is that such noisy semantic relations interfere with the message-propagation of GCNs over semantic structures, which tends to amplify their impact on representation learning [2,18]. Therefore, it is crucial to enhance the robustness of such multimedia recommendation methods against noisy relations in the item-item semantic structure.

Challenge II. Optimization objectives of existing methods lead to suboptimal training of the model. This is because they cannot leverage informative samples and global preference information to generate more discriminative embedding spaces [5,6,12]. In general, the optimization objectives of existing multimedia recommendation models exploit Bayesian personalized ranking (BPR) loss with random negative sampling [15,24,27,28]. However, randomly sampled negative items are more likely to be redundant and uninformative for optimization [19,22], and BPR loss fails to account for the distance between other users and items in the embedding space beyond the sampled users and items [6]. This local preference exploration approach homogenizes the association between training samples. Therefore, the key to designing a reasonable optimization objective is to exploit both informative samples and global preference information.

To deal with the above challenges, we propose a novel model entitled MMGPA (*Enhancing* **M**ulti**m**edia Recommendation through Item-Item Semantic Denoising and **G**lobal **P**reference **A**wareness). The main difference between our model and existing multimedia recommendation models is that MMGPA not only obtains superior representations of users and items from denoised item-item semantic structures, but also better separates the embedding space by leveraging both informative samples and global preference information in the optimization objectives. Specifically, the model contains the following two components. (1) To solve the challenge I, a modal semantic representation network is proposed to obtain higher-quality multimodal representations of items utilizing denoised item-item modality semantic structures. Its semantic structure constructing module establishes the initial item-item semantic structure graphs by calculating the similarity of multi-modal items. The semantic relation denoising module of this network further prunes the potential noisy relations in the semantic structure with a certain probability and outputs more reliable item-item semantic structures. In addition, the semantic structure encoding module is designed to extract the final item representations, which has the ability to aggregate neighbor information by GCNs and fuse fine-grained multimodal representations of items based on an attention-aware multimodal fusion. (2) To solve the challenge II, a global preference-aware model optimization objective is designed to obtain a discriminative embedding space. We refer to those negative and positive items that are harder to distinguish from positive and negative items for user in the embedding space as hard samples, and introduce a hard sample mining strategy to select them as informative hard sample pairs. In addition, a multi-distance constrained loss that can correlate global preference differences other than sampled users and items is well designed to distinguish the similarities of learned user and item representations. To sum up, this study makes the following contributions.

- To the best of our knowledge, we are the first to point out the need for eliminating noisy item-item semantic relations as well as the necessity of jointly exploiting informative samples and global preference information when optimizing the training objective in multimedia recommendations.
- To resolve the above problems, we propose a novel framework MMGPA, which contains a modal semantic representation network that generates sufficient representations of items with less semantic relation noise, and a novel global preference-aware optimization objective that is designed to generate more discriminative embedding space by selecting the most informative samples while fully leveraging global preferences to optimize the model.
- We conduct extensive experiments on three real-world datasets. The results showcase better performance of MMGPA against state-of-the-art baselines.

2   Related Work

2.1   Multimedia Recommendation

Multimedia recommendation aims to overcome data sparsity and cold start problems by utilizing rich content information (such as images, text, etc.), and has been successfully applied in various scenarios such as e-commerce, short videos [27]. As an early representative of multimedia recommendation, VBPR [7] adds the visual features of items to collaborative embeddings to effectively complete recommendation tasks. Recently, GCNs [10] have been introduced into multimedia recommendation methods to better integrate multimodal features. MMGCN [24] constructs modality-specific graphs to capture user preferences for specific modalities using GCNs. GRCN [23] adaptively prunes noisy edges in the interaction graph structure and performs GCNs on the pruned graph. LATTICE [27] and MICRO [28] construct item similarity graphs to mine the latent semantic structure of items and employ GCNs to aggregate information from semantically similar neighbours of items. HCGCN [15] combines the user-item interaction graph with the item-item graph and performs GCNs on the co-clustering graph to capture different behaviour patterns of users. However, none of these works explicitly remove noisy semantic relations in the semantic structure of items, and these noisy relations can contaminate the final item representations and thus degrade the recommendation performance.

2.2   Optimization for Recommendation

BPR loss [17] with random negative sampling is usually used as the optimization objective of the model in multimedia recommendation methods [1,7,24,27,28], which encourages users to approach observed items (positive sample pairs) and move away from unobserved items (negative sample pairs). However, two points limit the effectiveness of BPR in recommender systems. The first point is that random negative sampling adopts many instances with uninformative or easy information [5]. This is because closer negative user-item pairs mean that it is harder to distinguish them from positive user-item pairs. The second point is that BPR loss cannot exploit other preference information beyond the sampled user-item pairs [20] to better seperate embedding space, which is benifitial for recommendation performance [22]. To address the limitations, SRNS [5] demonstrates the relationship between score variance and false positives and uses it to design a sampling strategy. MCL [6] eliminates unimportant samples and aggregates information from multiple users into their loss function. PMLAM [12] introduces explicit user-user/item-item similarity modelling in the objective function. Inspired by these methods, we design a novel global preference-aware optimization objective to mine and leverage informative samples and global preference information in the multimodal scenario.

3   Preliminaries

Let $u \in \mathcal{U}$ and $i \in \mathcal{I}$ represent a user and an item, respectively, where $\mathcal{U}$ and $\mathcal{I}$ denote the sets of all users and all items, respectively. We denote a user-item interaction matrix $Y \in \mathbb{R}^{|\mathcal{U}| \times |\mathcal{I}|}$ to represent the interaction relations, and the preference score of u over i is denoted as $y_{ui} = 1$ if user u has interacted with item i, otherwise $y_{ui} = 0$. In the initial step, let $e_u \in \mathbb{R}^d$ and $e_i \in \mathbb{R}^d$ denote the ID embeddings associated with user u and item i, respectively. Besides, we define the modal feature of item i as $e_i^m \in \mathbb{R}^{d^m}$, where d^m represents the dimension of e_i^m, $m \in \mathcal{M}$ is a modality, and $\mathcal{M}$ is the set of modalities. In this paper, we use visual and textual modalities, i.e., $\mathcal{M} \in \{v, t\}$.

Definition 1: Item-Item Semantic Structure. An item-item semantic struc-
ture is denoted as a graph $\mathcal{A}^m = (\mathcal{I}, \mathcal{R})$, reflecting modal similarities of items on
modality m, where $\mathcal{R}$ is the set of edges. Specifically, each edge $r_s^m (0 \leq s \leq |\mathcal{R}|)$
denotes the similarities between item i and item j on modality m.

Definition 2: Multimedia Recommendation. Given user $u \in \mathcal{U}$ and item
$i \in \mathcal{I}$, the goal of multimedia recommendation is to learn a model $\mathcal{F}$ that models
the representation of users and items using interaction matrix Y and features of
all modalities $\mathcal{M}$ of items such that $\mathcal{F}$ can output a preference score $\hat{y}_{ui}$ for user
u over item i, where a larger $\hat{y}_{ui}$ indicates that u is prefers more to i.

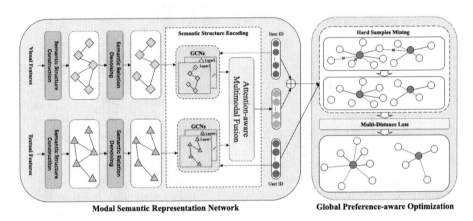

Fig. 1. The overall framework of our proposed MMGPA. In the hard sample mining
process, dark colored nodes represent users and light colored nodes represent items,
while the dashed line represents the distance of the negative sample sampled for the
user and the solid line's edge represents the distance of the positive sample for the
user.

4   The Proposed Method

In this section, we provide a detailed description of our model. As shown in Fig. 1,
our framework consists of two key components: a) a modal semantic representation network, which is designed to obtain superior multimodal representations of
items by constructing item-item semantic structure and pruning potential noisy
semantic relations, b) a global preference-aware optimization objective, which
is carefully designed to distinguish the learned representations by mining the
informative samples and correlating multiple preference differences.

4.1   Modal Semantic Representation Network

Semantic Structure Constructing Module. Based on the hypothesis
that similar items are more likely to interact with each other than dissimilar

items [14], we quantify the semantic relation between two items by their similarities and construct the raw item-item semantic structure. To avoid introducing unnecessary relation [4], we only select the top-k most similar items for each item based on their similarity scores. First, we construct a k-Nearest-Neighbor (kNN) raw modal semantic graph $A^m \in \mathbb{R}^{|\mathcal{I}| \times |\mathcal{I}|}$ for each modality m by calculating the cosine similarities between items:

$$A_{ij}^m = \begin{cases} a_{ij}^m & ,j \in top\text{-}k(a_{ij}^m) \\ 0 & ,otherwise \end{cases}, a_{ij}^m = \frac{(e_i^m)^\top e_j^m}{\|e_i^m\| \cdot \|e_j^m\|}, \tag{1}$$

where a_{ij}^m denotes the value of similarity between i and j, and A_{ij}^m corresponds to the values of the i-th row and the j-th column of the adjacency matrix A^m.

Then, to avoid the problems of gradient explosion or vanishing [10], we normalize the adjacency matrix as:

$$\bar{A}^m = (D^m)^{-\frac{1}{2}} A^m (D^m)^{-\frac{1}{2}}, \tag{2}$$

where $D^m \in \mathbb{R}^{|\mathcal{I}| \times |\mathcal{I}|}$ denotes the degree matrix of A^m, i.e., $D_{ii}^m = \sum_j A_{ij}^m$. Since the A^m constructed directly based on each raw modal features may be misleading for true semantic relations [27]. Thus, we use a linear layer to transform the raw modal features into high-level features and repeat Eqs. (1, 2) to dynamically learn high-level modal semantic graph $\widetilde{A}^m \in \mathbb{R}^{|\mathcal{I}| \times |\mathcal{I}|}$.

Finally, we employ learnable weights to adaptively assign different importance scores to fuse the two graphs, and obtain the adjacency matrix of initial semantic graph:

$$\hat{A}^m = w_0 \bar{A}^m + w_1 \widetilde{A}^m, \tag{3}$$

where $w_0, w_1 \in \mathbb{R}$ are importance scores of raw and high-level graphs, and let $w_0 + w_1 = 1$ to keep $\hat{A}^m$ normalized. In this way, we obtain the initial item-item semantic structure, i.e., the graph denoted by the adjacency matrix $\hat{A}^m$.

Semantic Relation Denoising Module. Recent studies [2,18] show that pruning potentially noisy relations in the graph structure can model more accurate representations. Inspired by these findings, we sparsify the graph by pruning redundant edges with a specific probability in the training stage to remove various noisy relations from the initial semantic structure. Specifically, we denote the adjacency matrix of the denoised semantic structure graph as G^m, which can be calculated as follows:

$$G_{i,j}^m = \phi(\gamma)\hat{A}_{i,j}^m, \tag{4}$$

where $\phi(\gamma)$ is a binary function that returns 0 with probability γ, and $G_{i,j}^m$ and $\hat{A}_{i,j}^m$ are the edge weights between node i and node j in the graphs corresponding to the adjacency matrices G^m and $\hat{A}^m$, respectively. Note that all the initial semantic structures are iteratively pruned in each training epoch, but the original $\hat{A}^m$ still be used in model inference.

Semantic Structure Encoding Module. In this module, we employ graph convolution to mine high-order item representations from the semantic structure and leverage attention-aware multimodal fusion with contrastive auxilary task to promote fine-grained multimodal fusion. First, we employ a simplified GCNs [9,25] for message propagation and aggregation on G^m, formulated as:

$$E^m_{(l)} = G^m E^m_{(l-1)}, \tag{5}$$

where $E^m_{(l)} \in \mathbb{R}^{|\mathcal{I}| \times d}$ represents the l-th layer item embedding matrix of modality m and is initialized by the item ID embedding matrix. After stacking L convolutional layers on G^m, we obtain the last layer modal embedding matrix $E^m_{(L)}$ of items, where the i-th row is the high-order modal embedding $\hat{e}^m_i$ of item i.

Then, we leverage an attention mechanism to fuse multi-modal embeddings:

$$\hat{e}_i = \sum_{m=1}^{|\mathcal{M}|} \alpha^m_i \hat{e}^m_i, \tag{6}$$

where $\hat{e}_i$ is the fused multimodal embedding of item i, and α^m_i is the attention score of modality m, which is formulated as:

$$\alpha^m_i = \frac{\exp(q^\top \tanh(Linear(\hat{e}^m_i)))}{\sum_{m=1}^{|\mathcal{M}|} \exp(q^\top \tanh(Linear(\hat{e}^m_i)))}, \tag{7}$$

where $q \in \mathbb{R}^d$ denotes the query vector, and $Linear(\cdot)$ represents a linear layer with parameter matrix $W \in \mathbb{R}^{d \times d}$ and bias vector $b \in \mathbb{R}^d$.

Following InfoNCE [3], a self-supervised contrastive learning is adopted to further force the fused item multimodal embeddings as well as learn shared important information from individual modality embeddings [15,28]. Formally, we define $(\hat{e}^m_i, \hat{e}_i)$ as the positive pair, and all other items in the same modality $(\hat{e}^m_i, \hat{e}^m_j)$ and fused modality $(\hat{e}^m_i, \hat{e}_j)$ as negative pairs:

$$\mathcal{L}_C = -\frac{1}{|\mathcal{I}|}\frac{1}{|\mathcal{M}|} \sum_{i \in \mathcal{I}} \sum_{m \in \mathcal{M}} \log \frac{e^{\rho(\hat{e}^m_i, \hat{e}_i)/\tau}}{e^{(\rho(\hat{e}^m_i, \hat{e}_i)/\tau} + \sum_{j \neq i}\left(e^{\rho(\hat{e}^m_i, \hat{e}_j)/\tau} + e^{\rho(\hat{e}^m_i, \hat{e}^m_j)/\tau}\right)}, \tag{8}$$

where $\tau \in \mathbb{R}$ is a temperature coefficient, and $\rho(\cdot, \cdot)$ denotes the agreement measure, which utilizes the negative Euclidean distance between two embeddings. By minimizing $\mathcal{L}_C$, the fused multimodal embedding can adaptively gathers information from all individual modality embeddings.

Prediction of Users' Preferences. To further enhance the item representation, we combine multimodal and collaborative representations of items as:

$$\tilde{x}_i = x_i + \frac{\hat{e}_i}{\|\hat{e}_i\|_2}. \tag{9}$$

To the end, we utilize the final embeddings of user u and item i to compute the preference score of u to i as follows:

$$\hat{y} = -\|x_u - \widetilde{x}_i\|_2^2. \tag{10}$$

where $x_u, x_i \in \mathbb{R}^d$ denotes the collaborative representation of user u and item i.

4.2   Global Preference-Aware Optimization

Hard Sample Mining Strategy. To fully utilize available preference information, we introduce a hardness mining strategy to select informative samples. Inspired by recent works in computer vision [19,22], we argue that a negative item closer to the user means that it is harder to distinguish that negative item from positive items that are also close to the user. Such pairs are referred as hard negative sample pairs, and vice versa for hard positive sample pairs. Such hard sample pairs are more informative and meaningful for learning discriminative embeddings [6,15]. Technically, we define hard positive sample as an item that is farther away from the user than at least one negative item, and a hard negative sample as an item that is closer to the user than at least one positive item. Formally, let $p \in \mathcal{P}_u$ and $n \in \mathcal{N}_u$ be any positive and negative item of user u, respectively, and $\mathcal{P}_u, \mathcal{N}_u$ denote sets of all positive items and all negative items of user u, respectively. A positive pair $\{u, p\}$ or a negative pair $\{u, k\}$ are retained during the mining procedure if they satisfy:

$$D_{up} > \min_{k \in \mathcal{N}_u} D_{uk} - \epsilon, \tag{11}$$

$$D_{uk} < \max_{p \in \mathcal{P}_u} D_{up} + \epsilon, \tag{12}$$

where D_{up}, D_{uk} denote the Euclidean distance between x_u and positive and negative items $\widetilde{x}_p$ and $\widetilde{x}_k$, respectively, and ϵ denotes a margin used to control the degree of separation. Since $\mathcal{N}_u$ is very large and severely weaken computational efficiency, we randomly sample a subset of $\mathcal{N}_u$ to replace it in practice. Afterwards, we obtain the sets of hard positive and hard negative samples for user u, denoted as $\mathcal{P}_u^s$ and $\mathcal{N}_u^s$, respectively.

Multi-distance Loss. To get the optimal model training, we design a Multi-Distance (MD) loss that allows the model to correlate differences of global preferences and better distinguish similarities between users and items. Let B denote a batch of user sets in the training process, and the MD loss is formulated as:

$$\begin{aligned}
\mathcal{L}_{MD} = {} & \frac{1}{\alpha} \log \left(1 + \frac{1}{|B|} \sum_{u \in B} \sum_{j \in \mathcal{P}_u^s} e^{\alpha(D_{up} + \lambda_p)} \right) \\
& + \frac{1}{\beta} \log \left(1 + \frac{1}{|B|} \sum_{u \in B} \sum_{k \in \mathcal{N}_u^s} e^{-\beta(D_{uk} + \lambda_n)} \right),
\end{aligned} \tag{13}$$

where λ_p and λ_n control the margin for positive and negative sample pairs, respectively, and α and β control the contribution of positive and negative sample pairs to the loss, respectively. The first term of the equation aims to shrink the distance of all hard positive sample pairs, while the second term aims to expand the distance of all hard negative sample pairs.

It has been mathematically demonstrated in computer vision with metric learning [19,22] that multiple similarities between different classes can be constrained by this type of loss to obtain a discriminative embedding space. In our model, MD loss is well designed to constrain multiple distance difference between users and items in the embedding space, i.e., global preferences. By doing so, the ability of the model to distinguish the distance differences between users and items is facilitated, enhancing the performance of multimedia recommendation.

Optimization. We combine MD loss with contrastive loss to obtain the final loss. The overall loss function can be formulated as follow:

$$\mathcal{L} = \mathcal{L}_{MD} + \mu\mathcal{L}_C + \eta\|\Theta\|^2 \tag{14}$$

where $\mu \in \mathbb{R}$ is a hyperparameter controlling the effect of the contrastive auxiliary task, Θ represents the input embeddings of users and items, which are also part of the model parameters, and η controls the L_2 regularization strength.

5 Experiments

5.1 Experiment Setting

Datasets. Our experiments are conducted on three real-world public Amazon datasets [13] which have been widely used in previous multimodal recommendation works [15,27,28]: a) Clothing, Shoes and Jewelry, b) Sports and Outdoors, and c)Baby, which we refer to as Clothing, Sports, and Baby in brief. For the visual modality, the datasets provide 4,096-dimensional visual features extracted by a pre-trained CNN [13]. For the textual modality, we extract 1,024-dimensional text features using a pre-trained Sentence-transformer [16] from concatenated item information [28]. The statistical analysis of these three datasets is shown in Table 1.

Table 1. Statistics of the datasets

Dataset	#Users	#Items	#Interactions	Density
Clothing	39,387	23,033	237,488	0.00026
Sports	35,598	18,357	256,308	0.00039
Baby	19,445	7,050	139,110	0.00101

Baselines. To evaluate the effectiveness of MMGPA, we compare it with ten competitors, including traditional recommendation methods that cannot use multimodal features and content-aware multimedia recommendation methods. For the former, we use four methods as baselines (i.e., BPRMF [17], NGCF [21], LightGCN [9], SGL [26], MCL [6]). For multimedia recommendation methods, we use six methods as baselines (i.e., VBPR [7], MMGCN [24], GRCN [23], LAT-TICE [27], MICRO [28], and HCGCN [15]), among which HCGCN is the SOTA multimedia recommendation method.

Table 2. Performance comparison of MMGPA with different baselines in terms of Recall@20(R@20), Precision@20(P@20), and NDCG@20(N@20). The best performance is highlighted in **bold** and the second is highlighted by underlines, respectively.

Model	Clothing			Sports			Baby		
	R@20	P@20	N@20	R@20	P@20	N@20	R@20	P@20	N@20
BPR	0.0191	0.0010	0.0088	0.0430	0.0023	0.0202	0.0440	0.0024	0.0200
NGCF	0.0387	0.0020	0.0168	0.0728	0.0038	0.0332	0.0591	0.0032	0.0261
LightGCN	0.0470	0.0024	0.0215	0.0803	0.0042	0.0377	0.0698	0.0037	0.0319
SGL	0.0598	0.0030	0.0268	0.0806	0.0043	0.0378	0.0745	0.0040	0.0328
MCL	0.0660	0.0034	0.0296	0.1030	0.0055	0.0478	0.0876	0.0046	0.0395
VBPR	0.0481	0.0024	0.0205	0.0582	0.0031	0.0265	0.0486	0.0026	0.0213
MMGCN	0.0501	0.0024	0.0221	0.0638	0.0034	0.0279	0.0640	0.0032	0.0284
GRCN	0.0631	0.0032	0.0276	0.0833	0.0044	0.0377	0.0754	0.0040	0.0336
LATTICE	0.0710	0.0036	0.0316	0.0915	0.0048	0.0424	0.0829	0.0044	0.0368
MICRO	0.0782	0.0040	0.0351	0.0988	0.0052	0.0457	0.0892	0.0047	0.0402
HCGCN	<u>0.0810</u>	<u>0.0041</u>	<u>0.0351</u>	<u>0.1032</u>	<u>0.0055</u>	<u>0.0478</u>	<u>0.0922</u>	<u>0.0048</u>	<u>0.0415</u>
MMGPA	**0.0873**	**0.0044**	**0.0387**	**0.1155**	**0.0061**	**0.0522**	**0.0994**	**0.0053**	**0.0439**
Imporv.	**7.78%**	**7.32%**	**10.26%**	**11.92%**	**10.91%**	**9.21%**	**7.81%**	**10.42%**	**5.78%**

Evaluation Protocols. We evaluate the performance of our model using three commonly used top-k recommendation metrics [9,21,24,27]: Recall@k, Precision@k, and NDCG@k [8]. For each evaluation metric, we set $k = 20$, and report the average value of the evaluation metric calculated by all users in the test set as the final evaluation metric. Additionally, we split all datasets into training, validation, and test sets in an 8:1:1 ratio, and we use the all-ranking protocol to compute the evaluation metrics for recommendation accuracy comparison.

Implementation Details. Following previous works [15,27,28], we fix the embedding dimension d of all methods to 64. The optimal hyperparameters for MMGPA are determined by cross-validation: The kNN sparsity parameter k is set to 10, the layer L is set to 1, and the regularization coefficient η is set to 1e-4. For contrastive loss parameters, the contribution coefficient μ was set to 0.03 and the temperature coefficient τ is set to 0.5. For multi-distance loss parameters,

$\alpha, \beta, \lambda_p, \lambda_n$ are set to $1.25, 5, 6.5, -0.5$, respectively. The pruning ratio γ is set to 0.3, and the number of negative sampling N is set to 10 as default. Moreover, we conduct experiments for all baselines and further report the best results of each baseline among different N.

5.2   Performance Comparison

The evaluation results are presented in Table 2. Specifically, we have the following detailed observations. (1) MMGPA significantly outperforms all baselines, demonstrating the effectiveness of our proposed method. Specifically, in Sports, Baby, and Clothing datasets, MMGPA improves over the best competitors (i.e., HCGCN) in terms of Recall@20 by 11.92%, 7.81%, and 7.78%, respectively. This is because MMGPA not only reduces the impact of noisy semantic relations in representation learning, but also gains an embedding space that is more conducive to distinguishing between user and item representation similarities. (2) Most of the recently proposed multimedia recommendation methods outperform MCL, which is the state-of-the-art non-multimedia recommendation method. One potential reason is that complementary semantic information about items from multimodal features can effectively improve recommendation performance [7,24]. (3) The overall performance improvement on Sports is higher than the other datasets. There are two potential reasons. On the one hand, compared to the other two datasets that rely more on visual modality to reveal item attributes [11], the Sports dataset is relatively less differentiated across modalities, which is more likely to result in noisy semantic relations. On the other hand, the Sports dataset has the largest number of interactions and the obtained embedding space is more likely to have poor separability [22]. Both of these issues can be effectively alleviated by MMGPA.

Table 3. Abalation Studies of MMGPA on each proposed component.

Model	Clothing			Sports			Baby		
	R@20	P@20	N@20	R@20	P@20	N@20	R@20	P@20	N@20
w/o. MD	0.0796	0.0040	0.0352	0.0989	0.0052	0.0457	0.0916	0.0048	0.0408
w/o. DN	0.0700	0.0035	0.0316	0.1045	0.0056	0.0485	0.0849	0.0045	0.0375
w/o. CL	0.0794	0.0040	0.0361	0.1087	0.0058	0.0491	0.0945	0.0050	0.0420
w. dot	0.0847	0.0043	0.0381	0.1055	0.0056	0.0492	0.0928	0.0049	0.0418
MMGPA	**0.0873**	**0.0044**	**0.0387**	**0.1155**	**0.0061**	**0.0522**	**0.0994**	**0.0053**	**0.0439**

5.3   Ablation Studies

To investigate the effectiveness of various key components in MMGPA, we compared MMGPA with different variants on three datasets: w/o. MD, a variant version uses the traditional BPR loss instead of the MD loss; 2) w/o. DN, which removes the semantic relation denoising module; 3) w/o. CL, which deactivates the contrastive loss; and 4) w. dot, a variant version measures all embedding agreement by dot product instead of the negative of Euclidean distance.

Table 3 shows the performance results of MMGPA and these variants, and we observe the following: (1) MMGPA outperforms all variants, indicating the effectiveness of all key modules proposed in MMGPA. (2) As reported in the results of w/o. MD, the proposed optimization objective is better than the traditional one, proving that sampling the most informative samples and considering global preference are both critical to multimedia recommendation. (3) The improvement between MMGPA and w/o. DN shows that pruning noisy semantic relations is an ingenious way to denoise the semantic structure of items, which is also a cornerstone for obtaining higher-quality item embeddings. (4) The report of w/o. CL shows the importance of multimodal fusion. We can capture item relations shared between modalities adaptively by this fusion. (5) The results of w. dot represent that using Euclidean distance as a measure of agreement between embeddings makes MMGPA perform better than using dot product similarity. This is because using Euclidean distance can separate the embedding space better [22]. Notably, even using dot product as agreement measurement, w. dot still outperforms all of the state-of-the-art baselines, indicating the intention of our method is effective and robust.

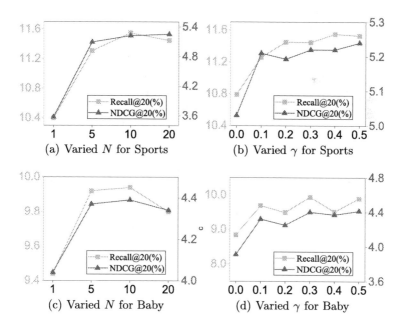

Fig. 2. Performance comparison of various hyperparameters N and γ on MMGPA.

5.4   Sensitivity Analysis

Since the semantic relation denoising module and the global preference-aware optimization with hard samples play critical roles in MMGPA, in this subsection, we mainly conduct sensitivity analysis on different hyperparameters related to them. Due to space limitations, we only show the experimental results of adjusting N and γ on the Sports and Baby datasets. The optimal selection of other hyperparameters is shown in the implementation details (Sect. 5.1), and their experimental results have shown that the model is not sensitive to their selection.

Impact of the Number of Negative Sampling N. The number of N is positively related to the amount of sample information, which means that more comprehensive global preference information can be considered into the optimization objective. We analyze the impact of different N adjusted in $\{1, 5, 10, 20\}$. As shown in Fig. 2, the performance of MMGPA gains significant improvement between $N = 1$ and $N = 5$, which indicates the rationality of the proposed global preference-aware optimization objective to mine hard samples and consider the global preference.

Impact of Pruning Ratio γ. The pruning ratio γ controls the proportion of randomly dropped edges in the semantic structure graphs before performing GCNs. This helps mitigate the negative influence of noisy semantic relations and improve the generalization ability of MMGPA. To analyze its impact, we adjusted γ in the range of $\{0, 0.1, 0.2, 0.3, 0.4, 0.5\}$. As shown in Fig. 2, all the performances for γ values between 0.1 and 0.5 are better than that of $\gamma = 0$, which verifies the necessity of pruning noisy semantic relations. Furthermore, the performance of MMGPA is not significantly affected by non-zero γ values, indicating that our method is not sensitive to γ.

Table 4. Performances comparison over different modalities.

Dataset	Modality	R@20	P@20	N@20
Clothing	Visual	0.0731	0.0037	0.0328
	Textual	0.0825	0.0042	0.0374
	Both	0.0873	0.0044	0.0387
Sports	Visual	0.1045	0.0055	0.0480
	Textual	0.1070	0.0059	0.0508
	Both	0.1155	0.0061	0.0522
Baby	Visual	0.0875	0.0046	0.0395
	Textual	0.0926	0.0049	0.0418
	Both	0.0994	0.0053	0.0439

5.5   Effects of Multi-modal Features

To investigate the effects of different modalities, we report the performance comparison using different modalities in Table 4. It can be observed that using multiple modalities performs better than using a single modality, which indicates using multiple modal features can learn more comprehensive information. In addition, the contribution of textual features is greater than that of visual features, this is because textual features contain information such as item titles, item categories, and item descriptions, which can provide more fine-grained information.

6   Conclusion

In this work, we propose a multimedia recommendation model, called MMGPA, that can mitigate the negative influence of noisy item-item semantic relations and consider more comprehensive global preferences during model optimization. We first construct the initial item-item semantic structure for each modality of items, and prune noisy semantic relations in an explicit way. Then, the item's denoised high-order information are injected into the item representation utilizing GCNs. Next, the higher-quality item multimodal representation is obtained by the designed attention-aware fusion approach. Finally, a hard sample mining strategy and a multi-distance loss are introduced for multimedia recommendation, which can mine the most informative samples and correlate multiple preference differences to better distinguish the learned representations. A large number of experiments demonstrate the recommendation advantages and effectiveness of MMGPA on three real-world public Amazon datasets.

Acknowledgments. This work is supported by the National Natural Science Foundation of China No. 62272332, the Major Program of the Natural Science Foundation of Jiangsu Higher Education Institutions of China No. 22KJA520006.

References

1. Chen, J., Zhang, H., He, X., Nie, L., Liu, W., Chua, T.: Attentive collaborative filtering: multimedia recommendation with item- and component-level attention. In: SIGIR, pp. 335–344. ACM (2017)
2. Chen, M., Wei, Z., Huang, Z., Ding, B., Li, Y.: Simple and deep graph convolutional networks. In: ICML. Proceedings of Machine Learning Research, vol. 119, pp. 1725–1735. PMLR (2020)
3. Chen, T., Kornblith, S., Norouzi, M., Hinton, G.E.: A simple framework for contrastive learning of visual representations. In: ICML. Proceedings of Machine Learning Research, vol. 119, pp. 1597–1607. PMLR (2020)
4. Chen, Y., Wu, L., Zaki, M.J.: Iterative deep graph learning for graph neural networks: better and robust node embeddings. In: NeurIPS (2020)
5. Ding, J., Quan, Y., Yao, Q., Li, Y., Jin, D.: Simplify and robustify negative sampling for implicit collaborative filtering. In: NeurIPS (2020)
6. Gao, Z., Cheng, Z., Pérez, F., Sun, J., Volkovs, M.: MCL: mixed-centric loss for collaborative filtering. In: WWW, pp. 2339–2347. ACM (2022)

7. He, R., McAuley, J.J.: VBPR: visual Bayesian personalized ranking from implicit feedback. In: AAAI, pp. 144–150. AAAI Press (2016)

8. He, X., Chen, T., Kan, M., Chen, X.: Trirank: review-aware explainable recommendation by modeling aspects. In: CIKM, pp. 1661–1670. ACM (2015)

9. He, X., Deng, K., Wang, X., Li, Y., Zhang, Y., Wang, M.: Lightgcn: simplifying and powering graph convolution network for recommendation. In: SIGIR, pp. 639–648. ACM (2020)

10. Kipf, T.N., Welling, M.: Semi-supervised classification with graph convolutional networks. In: ICLR (Poster). OpenReview.net (2017)

11. Liu, Q., Wu, S., Wang, L.: Deepstyle: learning user preferences for visual recommendation. In: SIGIR, pp. 841–844. ACM (2017)

12. Ma, C., Ma, L., Zhang, Y., Tang, R., Liu, X., Coates, M.: Probabilistic metric learning with adaptive margin for top-k recommendation. In: KDD, pp. 1036–1044. ACM (2020)

13. McAuley, J.J., Targett, C., Shi, Q., van den Hengel, A.: Image-based recommendations on styles and substitutes. In: SIGIR, pp. 43–52. ACM (2015)

14. McPherson, M., Smith-Lovin, L., Cook, J.M.: Birds of a feather: homophily in social networks. Ann. Rev. Sociol. **27**(1), 415–444 (2001)

15. Mu, Z., Zhuang, Y., Tan, J., Xiao, J., Tang, S.: Learning hybrid behavior patterns for multimedia recommendation. In: ACM Multimedia, pp. 376–384. ACM (2022)

16. Reimers, N., Gurevych, I.: Sentence-bert: sentence embeddings using siamese bert-networks. In: EMNLP/IJCNLP (1), pp. 3980–3990. Association for Computational Linguistics (2019)

17. Rendle, S., Freudenthaler, C., Gantner, Z., Schmidt-Thieme, L.: BPR: bayesian personalized ranking from implicit feedback. In: UAI, pp. 452–461. AUAI Press (2009)

18. Rong, Y., Huang, W., Xu, T., Huang, J.: Dropedge: towards deep graph convolutional networks on node classification. In: ICLR. OpenReview.net (2020)

19. Sohn, K.: Improved deep metric learning with multi-class n-pair loss objective. In: NIPS, pp. 1849–1857 (2016)

20. Song, H.O., Xiang, Y., Jegelka, S., Savarese, S.: Deep metric learning via lifted structured feature embedding. In: CVPR, pp. 4004–4012. IEEE Computer Society (2016)

21. Wang, X., He, X., Wang, M., Feng, F., Chua, T.: Neural graph collaborative filtering. In: SIGIR, pp. 165–174. ACM (2019)

22. Wang, X., Han, X., Huang, W., Dong, D., Scott, M.R.: Multi-similarity loss with general pair weighting for deep metric learning. In: CVPR, pp. 5022–5030. Computer Vision Foundation/IEEE (2019)

23. Wei, Y., Wang, X., Nie, L., He, X., Chua, T.: Graph-refined convolutional network for multimedia recommendation with implicit feedback. In: ACM Multimedia, pp. 3541–3549. ACM (2020)

24. Wei, Y., Wang, X., Nie, L., He, X., Hong, R., Chua, T.: MMGCN: multi-modal graph convolution network for personalized recommendation of micro-video. In: ACM Multimedia, pp. 1437–1445. ACM (2019)

25. Wu, F., Jr., A.H.S., Zhang, T., Fifty, C., Yu, T., Weinberger, K.Q.: Simplifying graph convolutional networks. In: ICML. Proceedings of Machine Learning Research, vol. 97, pp. 6861–6871. PMLR (2019)

26. Wu, J., Wang, X., Feng, F., He, X., Chen, L., Lian, J., Xie, X.: Self-supervised graph learning for recommendation. In: SIGIR, pp. 726–735. ACM (2021)

27. Zhang, J., Zhu, Y., Liu, Q., Wu, S., Wang, S., Wang, L.: Mining latent structures for multimedia recommendation. In: ACM Multimedia, pp. 3872–3880. ACM (2021)
28. Zhang, J., Zhu, Y., Liu, Q., Zhang, M., Wu, S., Wang, L.: Latent structure mining with contrastive modality fusion for multimedia recommendation. IEEE Trans. Knowl. Data Eng. (2022)

Resident-Based Store Recommendation Model for Community Commercial Planning

Kaiwen Wu, Yanhu Li, and Xiaofeng He[✉]

School of Computer Science and Technology, East China Normal University,
Shanghai, China
{51215901127,51215901017}@stu.ecnu.edu.cn, hexf@cs.ecnu.edu.cn

Abstract. The objective of community commercial planning is to identify appropriate stores to operate in a community shopping center, catering to the daily needs of residents and enhancing the appeal of the shopping center. However, obtaining data on the characteristics of all residents in the community is a major challenge, and practical methods for selecting suitable stores based on resident characteristics are unavailable. To address these issues, we propose a model that leverages mutual information maximization to learn representations of valuable residents in the shopping area and assess their value. Our key innovation is a value-ranking encoder-decoder that learns the characteristics of all residents in the community and recommends the most suitable store for each storefront. To balance the diversity and competition of businesses within the shopping center, we introduce a diversity loss function. Extensive experimental results show the effectiveness of our model.

Keywords: Community commercial planning · Encoder-decoder · Store recommendation

1 Introduction

Community shopping centers are becoming increasingly popular as people demand greater convenience than before. The primary goal of community commercial planning is to attract investment into vacant storefronts in a community shopping center and to identify suitable stores to operate there. Successful community commercial planning enhances the vitality of the shopping center and facilitates the lives of the residents in the community.

Although there have been several studies on shopping centers [5,32,33], most have focused on large shopping centers with high accessibility requirements, as their customers come from across the city. In contrast, community shopping centers primarily serve the residents of the community, who choose them for their daily dining and entertainment activities. Therefore, it is essential to consider the characteristics of the community's residents when planning a community shopping center. Investment targets must be accurate, as inaccurate targeting can affect the shopping center's profitability and reduce the convenience of residents' lives. For example, suppose an upscale restaurant opens in an ordinary residential area, and the price is significantly higher than the residents can afford. In this case, the residents would rather go further away to a regular restaurant than spend money at this upscale restaurant, which would quickly close down due to a lack of customers. When analyzing the characteristics of residents, it is crucial to consider their spending power, preferences, and needs. Focusing on high-value residents is more conducive to rational community commercial planning, as only a few residents frequently spend money in the shopping center. The profits of community shopping centers are mainly related to these residents. Therefore, when analyzing the characteristics of residents, it is necessary to distinguish their value. Storefront information, such as storefront size, rent, and floor level, should also be considered when selecting investment targets. Many stores have specific requirements for storefronts. For example, a gym generally requires a large storefront but only a little customer traffic. In contrast, a fast food restaurant needs a medium size storefront, but it requires to be in a high customer traffic location. So a large storefront on a high floor is unsuitable for a fast food restaurant. The distribution of stores in the shopping center also needs to be considered when attracting business. If the distribution of business forms in a community shopping center is too concentrated, the shopping center will be less attractive to residents, leading to unhealthy competition between businesses. For example, if there are too many food and beverage stores in a shopping mall, businesses will have to cut prices to compete for customers, which will reduce the store's profitability. Therefore, a reasonable distribution of business forms is more conducive to the sustainable development of community shopping centers.

To address these issues, we propose a method called Resident-based Store Recommendation (RSR) Model. We use the InfoNCE loss [20] to maximize mutual information and obtain representations of residents, stores, and storefronts. To capture the value that residents bring to a shopping center, we have developed a method for assessing the attention of each resident towards all stores. Using this method, we can calculate each resident's value to individual stores and assign weights to determine their overall value to the shopping center. To better understand the characteristics of residents in the community, we prioritize learning about residents based on their value, starting from the lowest and working our way up. Our decoder module uses the learned characteristics of both residents and storefronts to recommend the most suitable type of store. We also propose a diversity loss function to optimize shopping center distribution further and improve their attractiveness to residents. In summary, the contributions of this paper are as follows:

1) We propose a pre-training task based on mutual information maximization to enhance the representations of residents, stores, and storefronts.
2) We design a method to calculate the value of residents and storefronts to the shopping center.

3) We propose a novel model to derive the most suitable stores based on the representations of residents and storefronts.
4) We design a diversity loss function to optimize the distribution of businesses in a community shopping center.

2   Related Work

Shopping Centers. Numerous studies have been conducted on shopping centers, with some methods analyzing city traffic and population data to determine the optimal location for shopping centers to increase foot traffic and revenue [5,32,33]. Miao [18] applies the K-means algorithm [28] to classify stores based on three derived indicators and establish relationships between stores using the Apriori algorithm. This enables shopping center operators to arrange the layout of stores to balance customer traffic and maximize profits. Shim [26] utilizes data mining techniques such as decision trees, artificial neural networks, and logistic regression to develop a classification model that categorizes the value of customers to the shopping center. Previous research [3,10,12,13,24,27] has mainly focused on large shopping centers, whereas this paper explicitly examines community shopping centers. Community shopping centers vary significantly in size and range of services compared to large shopping centers, making it impractical to apply the previous methods directly to them.

Encoder-Decoder. The encoder-decoder architecture is widely used for various sequence-to-sequence tasks [1,4,6,15,16,30]. It comprises an encoder module that extracts information from input data and a decoder module that uses the refined information to generate the desired output. ABS [25] employs an attentive Convolutional Neural Network (CNN) encoder to access the sentence representations. Chopra [7] builds on this work by keeping the CNN encoder but replacing the decoder with recurrent neural networks (RNN). Their experiments show that the CNN encoder with the RNN decoder model outperforms ABS. Nallapati [19] further improves this model by replacing the CNN encoder with an RNN encoder, resulting in a full RNN sequence-to-sequence model. Zhou [31] proposes SEASS, a GRU-based encoder-decoder that uses the GRU encoder to read the input and the GRU decoder to produce the output summary. Building on these ideas, we present the Resident-based Store Recommendation Model, which leverages the encoder to learn resident characteristics and utilizes this information in the decoder to select appropriate stores.

3   Problem Formulation

We have three sets of entities: a resident set $\mathcal{P}$ with K residents, a store set $\mathcal{S}$ with M stores, and a storefront set $\mathcal{F}$ with N storefronts. To represent the relationship between residents and stores, we use a matrix $\mathbf{Q} \in \mathbb{R}^{K \times M}$, where $\mathbf{Q}_{ij} = 1$ if resident i spent money in store j, and $\mathbf{Q}_{ij} = 0$ otherwise. To represent the relationship between stores and storefronts, we use a matrix $\mathbf{U} \in \mathbb{R}^{N \times M}$, where $\mathbf{U}_{ij} = 1$ if store j is or was located in storefront i, and $\mathbf{U}_{ij} = 0$ otherwise.

At the encoder end, we input the features of all residents. At the decoder end, for each step, we provide a storefront representation $\mathbf{f}_i$ and then combine it with the features of residents to determine the most suitable store $\mathbf{y}_i$ for $\mathbf{f}_i$.

4   Method

In this section, we will present the proposed RSR model in detail. The general framework of the model is shown as Fig. 1.

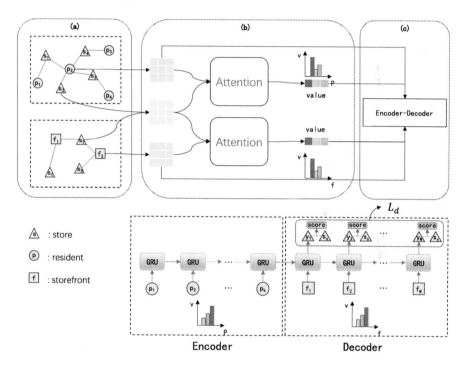

Fig. 1. The architecture of the RSR model. Part (a) is the MIM module. Part (b) is the value extractor. Part (c) is the encoder-decoder module, the details of which are shown in the bottom part of Fig. 1.

In RSR, the MIM module aims to obtain representations of residents, stores, and storefronts by maximizing mutual information. The value extractor uses the attention mechanism to access the value of both residents and storefronts. Then, the encoder-decoder module will learn the comprehensive features of the community based on the characteristics of residents and use this information to select the best store by considering the storefront characteristics. Meanwhile, the model will effectively control the degree of aggregation of business formats.

4.1   MIM Module

Mutual information is a measure of the dependence between two random variables. However, it can be challenging to calculate directly in practical settings. Previous work [22] established a connection between mutual information maximization and InfoNCE loss. For a community resident i, the set of all stores they have consumed is denoted as $S_i = \{j | Q_{ij} = 1\}$. Let $\mathbf{p}_i \in \mathbb{R}^d$ denote the embedding of community resident i and $\mathbf{s}_j^p \in \mathbb{R}^d$ denote the embedding of store j in the community shopping center. Using InfoNCE loss, we can maximize the mutual information of the resident and store representations. Specifically, we minimize the loss function as follows:

$$\mathcal{L}_{\mathrm{MIM}}(i) = -\sum_{j \in S_i} \left[\log \frac{f\left(\mathbf{p}_i, \mathbf{s}_j^p\right)}{\sum_{k \in \mathcal{S}} f\left(\mathbf{p}_i, \mathbf{s}_k^p\right)} \right]$$

where function f is implemented with dot product with activation:

$$f\left(\mathbf{p}_i, \mathbf{s}_j^p\right) = \sigma\left(\mathbf{p}_i \cdot \mathbf{s}_j^p\right)$$

where σ is the sigmoid function. The loss function can be easily extended from one resident to the entire resident set.

The embeddings of storefronts and stores can be obtained in the same way. Let $\mathbf{f}_i \in \mathbb{R}^d$ denote the embedding of storefront i and $\mathbf{s}_j^f \in \mathbb{R}^d$ denote the embedding of store j. $\mathbf{s}_j^p$ and $\mathbf{s}_j^f$ are the embeddings of store j obtained from residents and storefronts respectively. The relationship between store j and residents is contained in $\mathbf{s}_j^r$, the dependency between store j and storefronts is captured in $\mathbf{s}_j^f$, and the embedding of the store $\mathbf{s}_j$ is

$$\mathbf{s}_j = \mathbf{W}_s\left(\mathbf{s}_j^r \oplus \mathbf{s}_j^f\right) + \mathbf{b}_s$$

where $\oplus$ denotes the concatenation operator, $\mathbf{W}_s \in \mathbb{R}^{2d \times d}$ is a weight matrix and $\mathbf{b}_s \in \mathbb{R}^d$ is a bias term.

4.2   Value Extractor

The value that each community resident brings to a community shopping center varies, and we need to calculate the value of each resident to measure their contribution. The embeddings of residents contain their needs, while the embeddings of stores reflect the goods and services that stores can provide. When the similarity between $\mathbf{p}_i$ and $\mathbf{s}_j$ is higher, it indicates that the needs of resident i match more with the services provided by store j, and store j is more likely to attract attention from resident i. Likewise, the more value resident i has for shop j.

Given the store embedding $\mathbf{s}_j$, where j belongs to $\{1, ..., M\}$. Using the following formula, we can calculate the attention of resident i to store j as:

$$a_{ij}^p = \mathrm{Softmax}\left(\mathrm{sim}(\mathbf{p}_i, \mathbf{s}_j)\right)$$

$$= \frac{\exp(\mathrm{sim}(\mathbf{p}_i, \mathbf{s}_j))}{\sum_{j=1}^M \exp(\mathrm{sim}(\mathbf{p}_i, \mathbf{s}_j))}$$

where $\mathrm{sim}(\mathbf{p}_i, \mathbf{s}_j)$ is the similarity between resident i and store j, and it can be computed using the following formula:

$$\mathrm{sim}(\mathbf{p}_i, \mathbf{s}_j) = \frac{\mathbf{p}_i^\top \cdot \mathbf{s}_j}{\|\mathbf{p}_i\|_2 \|\mathbf{s}_j\|_2}$$

The value of a resident to a shopping center is the weighted sum of the value of the resident to each store within the entire community shopping center. So, the value v_i^p of resident i to the whole shopping center can be calculated by the following formula:

$$v_i^p = \sum_{j=1}^{M} a_{ij}^p \cdot \mathrm{sim}(\mathbf{p}_i, \mathbf{s}_j)$$

Similarly, the value v_i^f of storefront i to the shopping center can be obtained as:

$$v_i^f = \sum_{j=1}^{M} a_{ij}^f \cdot \mathrm{sim}(\mathbf{f}_i, \mathbf{s}_j)$$

$$a_{ij}^f = \mathrm{Softmax}(\mathrm{sim}(\mathbf{f}_i, \mathbf{s}_j))$$

Based on the operations above, we can calculate the value of each resident and storefront to the shopping center.

4.3   Encoder-Decoder Module

Encoder. There are many residents within a community shopping area, and these residents have different shopping needs and spending power. These characteristics are included in their embeddings. Denote the sequence of residents sorted by value in ascending order as $\mathcal{P} = \{\mathbf{p}_1, \ldots, \mathbf{p}_K\}$. We use an encoder based on GRU [9] to learn the features of the residents in the community as:

$$\mathbf{r}_t^p = \sigma\left(\mathbf{W}_r^p\left[\mathbf{h}_{t-1}^p, \mathbf{p}_t\right]\right)$$
$$\mathbf{z}_t^p = \sigma\left(\mathbf{W}_z^p\left[\mathbf{h}_{t-1}^p, \mathbf{p}_t\right]\right)$$
$$\tilde{\mathbf{h}}_t^p = \tanh\left(\mathbf{W}_{\tilde{h}}^p\left[\mathbf{r}_t^p \odot \mathbf{h}_{t-1}^p, \mathbf{p}_t\right]\right)$$
$$\mathbf{h}_t^p = \mathbf{z}_t^p \odot \mathbf{h}_{t-1}^p + (1 - \mathbf{z}_t^p) \odot \tilde{\mathbf{h}}_t^p$$

where σ is the sigmoid function, $\odot$ is the item-wise product, $\mathbf{W}_r^p$, $\mathbf{W}_z^p$ and $\mathbf{W}_{\tilde{h}}^p$ are the parameters of GRU network.

Decoder. The storefronts in the shopping center vary in size and are located on different floors, resulting in differences in their value. Denote the sequence of storefronts sorted by value in ascending order as $\mathcal{F} = \{\mathbf{f}_1, \ldots, \mathbf{f}_N\}$. The decoder takes a storefront embedding as input and generates the corresponding store recommendations for each input embedding based on the learned features of residents.

$$\mathbf{r}_t^f = \sigma \left(\mathbf{W}_r^f \left[\mathbf{h}_{t-1}^f, \mathbf{f}_t \right] \right)$$

$$\mathbf{z}_t^f = \sigma \left(\mathbf{W}_z^f \left[\mathbf{h}_{t-1}^f, \mathbf{f}_t \right] \right)$$

$$\tilde{\mathbf{h}}_t^f = \tanh \left(\mathbf{W}_{\tilde{h}}^f \left[\mathbf{r}_t^f \odot \mathbf{h}_{t-1}^f, \mathbf{f}_t \right] \right)$$

$$\mathbf{h}_t^f = \mathbf{z}_t^f \odot \mathbf{h}_{t-1}^f + \left(1 - \mathbf{z}_t^f \right) \odot \tilde{\mathbf{h}}_t^f$$

$$\mathbf{y}_t = \sigma \left(\mathbf{W}_o \cdot \mathbf{h}_t^f \right)$$

Here, $\mathbf{y}_t$ is the most suitable store embedding given by the model based on the characteristics of storefront t and residents.

4.4   Training

In our model, all store embeddings have been calculated in the MIM module, and we use dot production [21] to compute the store score. Denote the store recommended for a storefront as $\mathbf{y}_t$ and the actual store corresponding to the storefront as $\mathbf{s}_t$. The score $s(\mathbf{y}_t, \mathbf{s}_t)$ of store t opening in storefront t is computed as $s(\mathbf{y}_t, \mathbf{s}_t) = \mathbf{y}_t^\top \mathbf{s}_t$.

Inspired by Zhai [29], we adopt the negative sampling technique to train our model. For each store opened in a storefront (treated as a positive sample), we randomly select W stores from the shopping center that are not opened in this storefront as negative samples. Our model is designed to predict the scores of positive and negative stores, thereby transforming the store prediction problem into a pseudo $W+1$-way classification task. We minimize the sum of the negative log-likelihood of all positive samples during training as:

$$\mathcal{L}_{\text{ED}} = -\frac{1}{N} \sum_{t=1}^{N} \log \frac{\exp \left(s \left(\mathbf{y}_t, \mathbf{s}_t^p \right) \right)}{\exp \left(s \left(\mathbf{y}_t, \mathbf{s}_t^p \right) \right) + \sum_{j=1}^{W} \exp \left(s \left(\mathbf{y}_t, \mathbf{s}_{t,j}^n \right) \right)}$$

where N is the number of storefronts, $\mathbf{s}_t^p$ is the positive sample of the t-th storefront, and $\mathbf{s}_{t,j}^n$ is the j-th negative sample of the t-th storefront.

Suppose there is too much similarity between stores in the same shopping center. In that case, it can lead to fierce competition and reduce the diversity of formats in the shopping center, affecting its attractiveness to residents. Therefore, we introduce a diversity loss function to regulate the diversity of stores within a shopping center as follows:

$$\mathcal{L}_{\text{d}} = \frac{1}{N(N-1)/2} \sum_{i<j} \text{sim}(\mathbf{y}_i, \mathbf{y}_j)$$

In this way, a balance can be struck between diversity and healthy competition among the stores in the shopping center. Hence, the total loss function can be denoted as $\mathcal{L} = \lambda \mathcal{L}_{\text{ED}} + (1-\lambda)\mathcal{L}_{\text{d}}$, where λ is a hyperparameter.

5   Experiment

5.1   Dataset and Experimental Settings

As there is no readily available dataset for community commercial planning, we collaborate with businesses to obtain appropriate data and use it to train our model. The dataset includes information on the types of stores, the storefront where each store is located, and the spending history of customers in stores. The types of stores are divided into four main categories(i.e., restaurants, retail stores, services, and entertainment) and 17 sub-categories(e.g., fast food outlets, cafes, and gyms). The detailed statistical information is shown in Table 1. The time frame of the data spans from 2016 to 2022. We use 60% of the data for training, 20% for validation, and 20% for testing. In the experiment, we set the embedding dimension of stores, storefronts, and customers to 64, and λ is set to 0.7. The number of negative samples for each store is 4. The area under the receiver operating characteristics curve (AUC) and top-k accuracy are used as the evaluation metrics.

Table 1. Statistics of the dataset.

Stores	938	Expense Records	57325
Residents	12537	Avg. records per resident	4.57
Storefronts	353	Avg. records per store	61.11

5.2   Performance Comparison

As there has been no prior research investigating community commercial planning using deep learning methods, we evaluated the effectiveness of our model by comparing it to traditional machine learning methods, including:

1) **Decision Tree** [23] can be used to classify storefronts and residents based on various attributes of their features to arrive at the correct type of store.
2) **Random Forest** [2] is an integrated learning model that employs multiple decision trees to make predictions, effectively reducing overfitting and improving the accuracy of the model. When applied to the storefront and target user data, Random Forest can effectively mine the data for essential features, helping determine the appropriate store type.
3) **Logistic Regression** [17] can transform storefront and target user characteristics data into probability value, enabling the prediction of the most suitable store type.
4) **Support Vector Machine** [8] is a widely used classification model that can effectively mine critical features in the storefront and resident data to determine the appropriate type of store.

Table 2. The performance of different methods on community commercial planning. "DT", "RF", "LR" ', and "SVM" represent decision tree, random forest, logistic regression, and support vector machine. "Top-1", "Top-3", and "Top-5" represent top-1 accuracy, top-3 accuracy, and top-5 accuracy.

Method	Top-1	Top-3	Top-5	AUC
DT	45.23	54.22	58.49	63.07
RF	53.37	61.46	65.95	69.14
LR	46.39	55.91	60.54	63.86
SVM	52.53	60.81	64.93	68.72
RSR	**57.16**	**67.64**	**73.65**	**71.79**

The experimental results are shown in Table 2. As we can see, our proposed RSR model outperforms the other four models in all metrics, demonstrating its effectiveness for community commercial planning. Furthermore, the random forest outperforms the decision tree in all metrics. This can probably be attributed to the random forest's ability to construct multiple decision trees by randomly selecting features, which reduces the risk of overfitting to noisy data and improves the model's generalization and robustness. Similarly, the ability of SVM to maximize the distance between classifiers and different classes also helps it avoid overfitting, resulting in a performance comparable to that of random forests. Our model surpasses all traditional machine learning methods by learning and differentiating the value of each resident and using that information to determine the type of store. This contrasts traditional methods, which cannot select the most valuable residents. In addition, our method can effectively control the diversity of stores, which traditional methods cannot achieve.

5.3   Ablation Analysis

To evaluate the effectiveness of each module in RSR, we create several variants of the model and conducted a series of ablation experiments.

1) **RSR/MIM** removes the pre-training exercise process and utilizes an embedding layer to learn the representations automatically.
2) **RSR/VR** eliminates the notion of "value" associated with each resident and, instead, considers the value of each resident to the community shopping center as equal. To achieve this, we feed the representations of residents into the encoder in random order.
3) **RSR/R** feeds the resident representations into the encoder in reverse value order.
4) **RSR/CNN** replaces the GRU in the encoder-decoder with CNN [11].

The performance of the above variants of the model is shown in Table 3. From Table 3, we can see that (1) the original RSR performs best, which shows the effectiveness of each module in the model. (2) After removing the MIM modules

Table 3. Ablation study performance of RSR.

Method	Top-1	Top-3	AUC
RSR/MIM	55.37	62.29	68.48
RSR/VR	54.18	61.21	67.03
RSR/R	54.35	61.90	67.15
RSR/CNN	56.27	65.14	70.98
RSR	**57.16**	**67.64**	**71.79**

for residents, stores, and storefronts, the performance of the model decreases on all metrics, indicating that the MIM module enhances the representations of the model. (3) Removing the value of residents for the shopping center results in a significant decrease in the performance of the model, indicating that different residents have varying value for the community shopping center, and distinguishing residents by their value can improve the performance of the model. (4) Sorting residents in descending order of value reduces the performance of the model, indicating that sorting residents in ascending order of value can better capture the characteristics of all residents in the community and improve the performance of the model. (5) Comparing RSR/CNN with RSR shows that GRU learns resident features better.

5.4 Diversity Visualisation

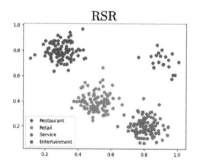 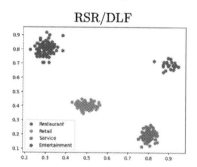

Fig. 2. Visualization of stores with and without diversity loss function (DLF).

To demonstrate the effectiveness of the diversity loss function, we utilized the t-SNE [14] algorithm to visualize the stores obtained by the model. When the number of stores is small, it can be difficult to observe their clustering degree.

To address this issue, we present the visualization results of the stores in three community shopping centers in a single figure. As shown in Fig. 2, the clustering degree is high when the diversity loss function is not utilized, indicating that numerous stores share a high degree of similarity. This intense competition among them can lead to a decrease in the diversity of the commercial district. However, when the diversity loss function is employed, the clustering degree is significantly reduced, demonstrating its effectiveness in alleviating the intense competition among stores and improving their diversity.

6   Conclusion

In this paper, we propose an encoder-decoder model for community commercial planning. Based on the principle of mutual information maximization, we utilize a pre-training approach to incorporate the relevance of stores to residents into their representations. Specifically, we distinguish the value of different residents to capture the characteristics of the entire community better. Meanwhile, we propose a diversity loss function to improve the diversity of stores in a community shopping center. Extensive experiments verify the effectiveness of our model. In the future, we will consider using more deep learning methods to explore the relationship between community residents and shopping centers and conduct community commercial planning based on this foundation.

References

1. Badrinarayanan, V., Kendall, A., Cipolla, R.: Segnet: a deep convolutional encoder-decoder architecture for image segmentation. IEEE Trans. Pattern Anal. Mach. Intell. **39**(12), 2481–2495 (2017)
2. Breiman, L.: Random forests. Mach. Learn. **45**, 5–32 (2001)
3. Chebat, J.C., Sirgy, M.J., Grzeskowiak, S.: How can shopping mall management best capture mall image? J. Bus. Res. **63**(7), 735–740 (2010)
4. Chen, L.C., Zhu, Y., Papandreou, G., Schroff, F., Adam, H.: Encoder-decoder with atrous separable convolution for semantic image segmentation, vol. 11211, pp. 833–851 (2018)
5. Cheng, E.W., Li, H., Yu, L.: The analytic network process (ANP) approach to location selection: a shopping mall illustration. Constr. Innov. **5**(2), 83–97 (2005)
6. Cho, K., et al.: Learning phrase representations using rnn encoder-decoder for statistical machine translation. In: Conference on Empirical Methods in Natural Language Processing, pp. 1724–1734 (2014). arXiv:1406.1078
7. Chopra, S., Auli, M., Rush, A.M.: Abstractive sentence summarization with attentive recurrent neural networks, pp. 93–98 (2016)
8. Cortes, C., Vapnik, V.: Support-vector networks. Mach. Learn. **20**, 273–297 (1995)
9. Dey, R., Salem, F.M.: Gate-variants of gated recurrent unit (gru) neural networks, pp. 1597–1600 (2017). arXiv:1701.05923
10. Kim, I., Christiansen, T., Feinberg, R.A., Choi, H.: Mall entertainment and shopping behaviors: a graphical modeling approach, vol. 32, pp. 487–492 (2005)
11. Krizhevsky, A., Sutskever, I., Hinton, G.E.: Imagenet classification with deep convolutional neural networks, vol. 60, pp. 84–90 (2017)

12. Laroche, M., Teng, L., Michon, R., Chebat, J.C.: Incorporating service quality into consumer mall shopping decision making: a comparison between English and French Canadian consumers. J. Serv. Mark. **19**(3), 157–163 (2005)

13. Lee, S., Min, C., Yoo, C., Song, J.: Understanding customer malling behavior in an urban shopping mall using smartphones. In: UbiComp (Adjunct Publication), pp. 901–910 (2013)

14. van der Maaten, L., Hinton, G.: Visualizing data using t-sne. J. Mach. Learn. Res. **9**, 2579–2605 (2008)

15. Malhotra, P., Ramakrishnan, A., Anand, G., Vig, L., Agarwal, P., Shroff, G.: Lstm-based encoder-decoder for multi-sensor anomaly detection. arxiv:1607.00148 (2016)

16. Mao, X.J., Shen, C., Yang, Y.B.: Image restoration using very deep convolutional encoder-decoder networks with symmetric skip connections. Adv. Neural Inf. Process. Syst. **29**, 2802–2810 (2016)

17. Menard, S.,: Applied Logistic Regression Analysis, vol. 45, p. 534 (1996)

18. Miao, Y.: A machine-learning based store layout strategy in shopping mall. In: MacIntyre, J., Zhao, J., Ma, X. (eds.) SPIOT 2020. AISC, vol. 1282, pp. 170–176. Springer, Cham (2021). https://doi.org/10.1007/978-3-030-62743-0_24

19. Nallapati, R., Zhou, B., dos Santos, C.N., Çaglar Gülçehre, Xiang, B.: Abstractive text summarization using sequence-to-sequence rnns and beyond, pp. 280–290 (2016)

20. Nguyen, H., Bougares, F., Tomashenko, N.A., Estève, Y., Besacier, L.: Investigating self-supervised pre-training for end-to-end speech translation, pp. 1466–1470 (2020)

21. Okura, S., Tagami, Y., Ono, S., Tajima, A.: Embedding-based news recommendation for millions of users. In: ACM Knowledge Discovery and Data Mining, pp. 1933–1942 (2017)

22. Oord, A.V.D., Li, Y., Vinyals, O.: Representation learning with contrastive predictive coding. arXiv preprint arXiv:1807.03748 (2018)

23. Quinlan, J.R.: C4.5: programs for machine learning (1993)

24. Rajagopal: Growing shopping malls and behaviour of urban shoppers. J. Retail Leisure Prop. **8**, 99–118 (2009)

25. Rush, A.M., Chopra, S., Weston, J.: A neural attention model for abstractive sentence summarization. In: Conference on Empirical Methods in Natural Language Processing, pp. 379–389 (2015). arXiv:1509.0068

26. Shim, B., Choi, K., Suh, Y.: CRM strategies for a small-sized online shopping mall based on association rules and sequential patterns. Expert Syst. Appl. **39**(9), 7736–7742 (2012)

27. Teller, C., Schnedlitz, P.: Drivers of agglomeration effects in retailing: the shopping mall tenant's perspective. J. Mark. Manag. **28**(9–10), 1043–1061 (2012)

28. Xie, J., Jiang, S., Xie, W., Gao, X.: An efficient global K-means clustering algorithm. J. Comput. **6**(2), 271–279 (2011)

29. Zhai, S., Chang, K.H., Zhang, R., Zhang, Z.: Deepintent: learning attentions for online advertising with recurrent neural networks. In: ACM Knowledge Discovery and Data Mining, pp. 1295–1304 (2016)

30. Zhang, J., Du, J., Dai, L.: A GRU-based encoder-decoder approach with attention for online handwritten mathematical expression recognition. In: 2017 14th IAPR International Conference on Document Analysis and Recognition (ICDAR), vol. 01, pp. 902–907 (2017)

31. Zhou, Q., Yang, N., Wei, F., Zhou, M.: Selective encoding for abstractive sentence summarization, pp. 1095–1104 (2017). arxiv:1704.07073

32. Zolfani, S.H., Aghdaie, M.H., Derakhti, A., Zavadskas, E.K., Varzandeh, M.H.M.: Decision making on business issues with foresight perspective; an application of new hybrid MCDM model in shopping mall locating. Expert Syst. Appl. **40**(17), 7111–7121 (2013)
33. Önüt, S., Efendigil, T., Kara, S.S.: A combined fuzzy MCDM approach for selecting shopping center site: an example from Istanbul, Turkey. Expert Syst. Appl. **37**(3), 1973–1980 (2010)

Author Index

A

Anwar, Adeem Ali 447

B

Bai, Xiangyu 553
Banerjee, Suman 772
Bose, Saugata 339
Bouquet, Fabrice 535

C

Cao, Ruyi 584
Chen, Guihai 3
Chen, Guisheng 494
Chen, Jiadong 3
Chen, Keyu 73
Chen, Lei 598
Chen, Lifei 48
Chen, Lina 200
Chen, Wei 802
Chen, Yakun 34
Chen, Ziqi 245
Cheng, Peng 598
Choy, S. T. Boris 661
Claus, Martin 677
Colonval, Jessy 535

D

Ding, Xuefeng 64
Do, Thanhcong 401
Dong, Xiao 293, 324
Du, Anan 417

F

Fan, Hao 352
Fang, Xiu 91
Feng, Jiamei 155
Fong, Simon 383

G

Gan, Mengjiao 293, 324
Gao, Hong 200
Gao, Hui 123
Gao, Junbin 661
Gao, Shiying 661
Gao, Xiangxiang 185
Gao, Xiaofeng 3
Ge, Qixiang 598
Gong, Yuyun 383
Guo, Hongjie 200
Guo, Na 569
Guo, Qiutong 277
Guo, Xi 598
Guo, Yafang 695
Guo, Zhengyi 787

H

Han, Fangyu 168
Han, Huaxu 431
Han, Yu 553
He, Wei 91
He, Xiaofeng 818
Hong, Song 155
Hou, Jingyu 524
Hu, Dasha 64
Hu, Xiaoyu 553
Huang, Liting 260

J

Ji, Xiayan 200
Jiang, Guifei 138
Jiang, Haowei 277
Jiang, Ying 709
Jiang, Yuming 64
Jin, Yuyuan 228
Jing, Mengyuan 787

K
Ke, Yan   584
Ke, Yuhua   245
Kong, Linghe   3

L
Landt-Hayen, Marco   677
Lei, Si   463
Li, Bo   479
Li, Jiahao   383
Li, Jiajia   569, 614
Li, Keke   755
Li, Li   107
Li, Qing   630
Li, Ruikun   661
Li, Xuliang   661
Li, Yang   509
Li, Yanhu   818
Li, Zhanshan   494
Li, Zhe   584
Li, Zihao   34
Lin, Hao   646
Liu, Guanfeng   91, 447
Liu, Guangya   293
Liu, Haibo   695
Liu, Haiyang   630
Liu, Li   339
Liu, Mengchi   155
Liu, Shuaipeng   309
Liu, Wei   277, 293, 324
Liu, Xueqing   509
Liu, Zehua   64
Long, Guodong   19
Lu, Jinhu   91
Luo, Wei   524

M
Meng, Rui   245
Mo, Tong   309
Mo, Wanghao   584

N
Nie, Tiezheng   228

O
Orgun, Mehmet   417

P
Pang, Shuchao   417

Q
Qiu, Shuang   324
Qiu, Tao   569

R
Rath, Willi   677
Ren, Jianyu   724
Ren, Yaru   553
Ruan, Sijie   431
Rui, Xiaobin   168

S
Sajjanhar, Atul   463
Shen, Derong   228
Shen, Fangyao   200
Song, Shihao   155
Song, Xin   695
Stefanzick, Julian   368
Su, Guoxin   339
Su, Xing   352
Sun, Chenchen   228
Sun, Dawei   463
Sun, Fuzhen   755
Sun, Guohao   91
Sun, Jinshan   598

T
Tang, Xingli   260
Tao, Wenjing   524
Tekawade, Atharva   772
Tran, Vanha   401

W
Wan, Guanglu   309
Wan, Ming   509
Wang, Chang-Dong   709
Wang, Haotong   740
Wang, Liang   695
Wang, Lizhen   401
Wang, Meng-xiang   293
Wang, Shaoqing   755
Wang, Shengrui   48
Wang, Shuliang   431
Wang, Shumei   168
Wang, Xianzhi   19, 34
Wang, Xiaoling   185
Wang, Xin   614
Wang, Xite   228
Wang, Zhaobo   787

Wang, Zhixiao   168
Wang, Zihan   740
Wei, Yinyi   309
Wölker, Yannick   677
Wu, Fan   3
Wu, Gang   740
Wu, Jia   352
Wu, Kaiwen   818
Wu, Renhui   168
Wu, Weiwen   479
Wu, Xia   755

X

Xiang, Mingrong   524
Xie, Yufan   383
Xing, Jiajie   213
Xu, Guandong   19
Xu, Ronghai   709
Xu, Yang   228
Xue, Shaojie   245

Y

Yan, Hailei   309
Yang, Chao   19, 34, 213
Yang, Chunming   479
Yang, Jian   91, 352
Yang, Yangfang   598
Yang, Yongyue   260
Yang, Yuping   138
Yang, Zhisheng   107
Yao, Junjie   646
Yao, Lina   19
Yao, Zhenfeng   73
Ye, Chunyang   260
Ye, Hengyu   3
Ye, Wei   309
Yen, Jerome   383
Yin, Jian   277, 293, 324
Younas, Irfan   447
Yu, Jianxing   277, 293, 324
Yuan, Hanning   431

Z

Zhang, Anzhen   614
Zhang, Detian   630
Zhang, Hui   479
Zhang, Jianfei   48
Zhang, Jinpeng   401
Zhang, Ruoqian   724
Zhang, Wendong   584
Zhang, Xianguo   213
Zhang, Xinyue   123
Zhang, Xuexin   614
Zhang, Xuyun   447
Zhang, Yanlong   802
Zhang, Yao   755
Zhang, Yuchen   352
Zhang, Yufeng   277
Zhang, Yuming   185
Zhang, Yuzhi   138
Zhang, Zhihong   73
Zhao, Changlin   569
Zhao, Ge   107
Zhao, Guoping   73
Zhao, Haijun   709
Zhao, Jiayu   168
Zhao, Jing   598
Zhao, Lei   802
Zhao, Qinghua   383
Zhao, Xin   368
Zhao, Xujian   479
Zhao, Ying   569
Zheng, Shangfei   802
Zheng, Shun   755
Zheng, Xiaochuan   352
Zhou, Hui   260
Zhou, Qian   802
Zhu, Chunjiang   630
Zhu, Rui   569, 614
Zhu, Wei   185
Zhu, Xun   598
Zhu, Yanmin   787
Zong, Chuanyu   614
Zu, Shuaishuai   107

Printed in the United States
by Baker & Taylor Publisher Services